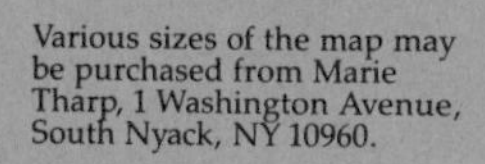
Various sizes of the map may be purchased from Marie Tharp, 1 Washington Avenue, South Nyack, NY 10960.

Mercator Projection 1:48,000,000 at the Equator.
Depth and Elevations in Meters.

# PRINCIPLES OF
# PHYSICAL GEOLOGY

PRINCIPLES OF

# PHYSICAL GEOLOGY

John E. Sanders

*Barnard College*

*Columbia University*

WILEY INTERNATIONAL EDITION

John Wiley & Sons

New York • Chichester • Brisbane • Toronto

**COVER DESIGN**
**Judith Fletcher Getman**

**COVER PAINTING**
*Bibémus Quarry*, c. 1895, by Paul Cézanne
Courtesy of Museum Folkwang, Essen



**LIBRARY OF CONGRESS CATALOGING IN PUBLICATION DATA:**

L.C. Card No. 81 501 25
ISBN 0-471-08424-7

Printed in the United States of America

10 9 8 7 6 5 4 3 2 1

# PREFACE

Geology has started off the 1980s with a bang, both literally and figuratively. Explosive eruptions from Mount St. Helens, Washington, and from other volcanoes; severe earthquakes; droughts; floods; disastrous slope failures; and new, exciting discoveries both from direct observations of the deep-sea floor and from images of the outer planets of the Solar System transmitted back to Earth are big news. Moreover, the end of the era of abundant and cheap energy from fossil fuels and the realization of the potentially great impacts that industrial activities have made on the natural environment have forced on every citizen the importance of knowing how the Earth works.

These new popular interests in the Earth have followed closely the profound and dramatic changes that the concepts of plate tectonics have introduced into the geological sciences during the past two decades. As a result, nearly every student who enrolls in a college course in physical geology has already heard much about the subject; many may even have studied earth science in their schools. Moreover, instructors have a wealth of new and lively material to present.

Although the new developments have prompted more and more students to major in geology, nevertheless, the fact remains that enrollment in physical geology includes large numbers of students who do not intend to become geologists, and who may not even take any other college course in science. I have written this book with the latter group particularly in mind. I feel very strongly that any student who studies physical geology with reasonable diligence will experience a new way of looking at the world. When this happens, students inevitably think about themselves with new perspectives.

Although most teachers of physical geology share the pleasure of watching their students' visions expand as students grasp this or that point about geology and begin to think about the Earth in a geologic context, geology teachers differ greatly in how best to organize their subjects. In this book I present a sequence that has proved to be effective for me in teaching more than 1000 Barnard and Columbia College students during the past decade.

The Introduction and the first three chapters attempt to sketch the broad outlines of geology, the concept of the geologic cycle, the Earth (both as a planet in the Solar System and as our home), plate tectonics, the pivotal significance of natural energy supplies and of natural resources, some important geologic subjects for modern industrial societies, and geologic time. I emphasize the idea, long ignored by nearly all astronomers, that *the axial point of the Solar System is not the center of the Sun,* but rather is the Solar System barycenter, and that the Sun itself orbits this barycenter with significant consequences for what happens on Earth. The orbit of the Sun around the Solar-System barycenter disposes of the old and oft-repeated bugaboo that the distribution of the angular momentum in the Solar System is peculiar and requires a remarkable and as-yet-not-formulated hypothesis to explain. It clarifies what should have been obvious all along—that half the angular momentum of the Solar System *must be in the Sun* and the other half distributed among the planets according to their masses, distances from the Solar System barycenter, and orbital speeds.

Several important themes are stressed here more than in other available texts. These include geologic time, the format for presenting rocks, and a pervasive emphasis on natural resources. In the following paragraphs I summarize these themes and include a synopsis of the general plan of the chapters.

I introduce the broader concept of geologic time with a discussion of the nineteenth-century "time dilemma" as set forth by F. C. Haber in 1959 and embellished here on the basis of my careful reading of the first edition of Charles Lyell's *Principles of Geology* and many books about Lyell's life and work. My purpose has been to try to convince even the most skeptical reader of the everlasting truth of John Playfair's remark that "no ingenuity can reconcile the natural history of the globe with the opinion of its recent origin." I present the use of radioactive methods in geologic dating in several places, starting with a discussion that does not require extensive knowledge of rocks and minerals. I begin radioactive dating methods with minerals. Later I show how minerals in rocks are used to date the rocks. The technical summary of radioactivity is in Appendix A ("Chemistry").

Each of the three rock chapters is organized in the same general way to include position in the rock cycle, processes of origin, occurrence, basis for recognition, method of geologic dating, and associated deposits of natural resources. The details required for identifying and naming rocks form Appendix C.

The subject of natural resources involves many aspects, such as impacts on society, geologic processes of origin, uses of individual products, and strategies for exploration. Some important societal topics and general definitions are located in Chapter 2. The geologic processes involved in forming natural resources are presented in appropriate chapters on rocks or processes. For example, hydrothermal processes are in Chapter 6 ("Igneous Rocks"); residual concentrations and pore spaces related to weathering, in Chapter 7 ("Weathering"); coal, petroleum, uranium, salt diapirs, and other sedimentary mineral deposits, in Chapter 8 ("Sedimentary Rocks"); contact-metamorphic deposits, in Chapter 9 ("Metamorphic Rocks"); and features related to faults, in Chapter 16 ("Geologic Structures and Mountains"). I bring together in Chapter 19 ("Exploration Geology") the how-to-find aspects using many examples of things to look for and ways of finding them.

I have placed groundwater much earlier than is customary, in Chapter 10. I did this so that it will be possible to discuss how interstitial water in the regolith influences the behavior of slopes (Chapter 11). The chapters on surface land-shaping processes follow what seems to have become a standardized pattern: rivers, glaciers, deserts and the wind, and oceans and shorelines.

My approach to plate tectonics begins early, but does not become too specific until near the end of the book after the chapters on structural geology and the interior of the Earth. Chapter 16 ("Geologic Structures and Mountains") includes descriptions of major structures, the concepts of strike and dip, unconformity and a related treatment of the timing of structural events, the effects of faulting on adjacent rocks, patterns of folds and faults, overthrusts, tectonic provinces, and kinds of mountains based on geologic structures.

Chapter 17 covers earthquakes and the Earth's interior, emphasizing the problems of earthquake prediction.

I approach the dynamics of the lithosphere (Chapter 18) first in the vertical sense, using natural loads, isostasy, subsidence, and epeirogeny; next with both vertical motions and lateral motions, based on the dynamic history of orogenic belts; and finally, on the large lateral movements, including continental drift and plate tectonics.

Chapter 20 summarizes the highlights of planetary geology. This chapter begins with the methods of studying planets and their satellites. It continues with results from many parts of the U.S. space program. I close with the significance of planetary geology for understanding the early stages of the Earth. Here I explain why James Hutton was unable to find "any vestige of a beginning," (or "marks of a beginning" in Playfair's words). Such Earthly "beginning marks" may have indeed existed formerly as impact craters, but erosion, sedimentation, and plate tectonics have erased all traces of them.

# ACKNOWLEDGMENTS

In compiling this book I have received considerable assistance and critical appraisal. This book would never have been started, much less completed without the stimulation, insights, and heroic labors in all phases of manuscript preparation, assembly of illustrations, and countless administrative tasks carried out by Mr. Robert Carola, Westport, Connecticut. Mr. Carola served as general critic and particularly as the advocate for the student. He wrote the first drafts of the chapter summaries and repeatedly kept me from straying off into some convoluted technical point. He also conceived the book's graphic design.

I thank many friends and numerous organizations for use of their photographs. Their names appear in the captions of each photograph.

I am extremely grateful to the following reviewers who made searching and useful criticisms on two drafts of the manuscript: J. Allen Cain (University of Rhode Island), R. L. Chase (University of British Columbia), Robert B. Furlong (Wayne State University), John T. Nicholas (University of Bridgeport), Susan Clark Slaymaker (California State University, Sacramento), and Edward J. Tarbuck (Illinois Central College). In addition, James T. Kirkland (University of Missouri at Kansas City), John Stolar (Cheyney State College), and Ronald R. West (Kansas State University) read a set of galleys. I carefully considered every critical reviewer's comment and tried to incorporate as many of them as I felt were consistent with my overall plan. Although several of these reviewers did not agree with my scheme of building toward plate tectonics late in the book, as contrasted with a fundamental organization built around plate tectonics first, last, and always, their comments resulted in my correcting many errors and in significantly improving my language.

In addition, I thank Mark R. Chartrand III (formerly of the Hayden Planetarium, New York City) for reading two early versions of Chapter 20, and Professor Charles P. Thornton (Pennsylvania State University) for reading the first drafts of Chapters 5 and 6.

Ms. Megan Newman (Barnard College) assisted in obtaining permissions for photographs and in assembling many of the prints. Mr. Philip E. Sanders (Dobbs Ferry, N.Y.) assisted in photographic processing and enlarging of my pictures. Mr. Thomas H. Sanders (Dobbs Ferry, N.Y.) assisted with many production errands. Mr. John Leahy (Columbia University) checked references and helped compile the manuscript of the glossary.

Donald H. Ritchie (Professor of Biology now *Emeritus*, Barnard College) extended the use of many departmental facilities in 1978–1979, enabling me to hide out from the rest of the College to do most of the work on the manuscript.

I want to express my appreciation to the many students who have enrolled in Physical Geology at Barnard College, Columbia College, Columbia School of Engineering, and Columbia School of General Studies, starting in the fall of 1968. Student responses have made my attempts to teach physical geology worthwhile; student questions have enabled me to find many new insights into the subjects we explored together.

Thanks also go to the staff of John Wiley and Sons, for numerous favors, kindnesses and forebearance during production: Bob Ballinger, Claire Egielski, Ed Starr, and Kathy Bendo. Donald H. Deneck, Wiley Geology Editor, has been a source of constant stimulation, good fellowship, much general assistance, and sharp judgments.

Finally, the world should realize that my wife, Barbara, managed to survive this project and to keep our household afloat while I concentrated on writing at the expense of sundry domestic tasks.

J. E. S.

# CONTENTS

# INTRODUCTION: THE STUDY OF GEOLOGY

MAJOR CONCEPTS AND CONCERNS OF GEOLOGY
OUR STRATEGY FOR PRESENTING PHYSICAL GEOLOGY
WHAT YOU CAN EXPECT FROM A COURSE IN PHYSICAL GEOLOGY
WHAT GEOLOGISTS DO

The geologic cycle in operation, high Andes along the border between Chile and Argentine, South America. In right foreground is the valley of the Rio Aconcagua. (Carl Frank, Photo Researchers, Inc.)

PHYSICAL GEOLOGY IS THE STUDY OF THE DYNAMICS OF THE PLANET Earth. The objectives of studying physical geology are to learn about the major interior and exterior features of the Earth and how these features respond to the sources of available energy, how human activities have affected the Earth and in turn are affected by natural processes, how geology must be applied to the searches for new supplies of natural resources, how geologic features record the passage of time, and how principles of geology learned with examples on the Earth can be applied to other planetary bodies.

Before we take up the detailed subjects covered in our 20 chapters, we want to review some of the major concepts and concerns of geology and then explain our strategy for presenting the subject matter, emphasizing what you should expect to gain from a study of physical geology.

## MAJOR CONCEPTS AND CONCERNS OF GEOLOGY

The most fundamental principle of geology is that the Earth is a dynamic body evolving through time. The Earth undergoes important reactions driven by various energy sources. The concept of the geologic cycle, developed in the late eighteenth and early nineteenth century, became the basis of one of geology's greatest triumphs, the proof of the enormous age of the Earth. Twentieth-century concerns include the great new series of insights into the behavior of the Earth's outer parts that come under the heading of plate tectonics; analyses of problems related to water, food, energy, mineral resources, and the condition of our environment; and the attempt to generalize to planetary scope knowledge of relationships between geologic materials and processes derived from studies carried out here on Earth.

### The Dynamic Earth Evolving Through Time

The idea that the Earth is dynamic implies that it is supplied with sources of energy which can drive chemical reactions and physical reactions. The Earth is subjected both to external supplies of energy and forces and to internal supplies of energy.

The Sun is our most important external supply of energy; it is responsible for much of what happens on Earth, including the continuation of life. (The details of the physics of solar energy are presented in Appendix D.) We are beginning to realize from space-age research that the Solar System behaves as a unit. Not only does the Sun send out energy that influences the surfaces of the planets in our Solar System, but the planets exert a force on the Sun that seems to affect the Sun's output of energy. That force is gravitational attraction. Because the planets follow elliptical orbits, their distances from the Sun change and thus the gravitational forces they exert and that are exerted on them are variable.

The interior of the Earth is the source of internal supplies of energy in the form of an outflow of heat; in fact, the Earth can be compared to a gigantic heat machine or heat engine. The interior consists of two parts; a dense, presumably metallic *core* and its thick, rocky surrounding layer, the *mantle*. Geologic theory suggests that at one time the entire core was liquid, but through time the inner part has become solid, whereas the outer part remains liquid. The slow movements of the liquid part of the core are thought to be the cause of the magnetic field that surrounds the Earth. Study of the Earth's magnetic field has become enormously important for geology, as we shall see in later chapters.

The effects of the incoming solar energy and the outflowing heat are many and varied. One of the first effects to be appreciated was what we now know as the geologic cycle.

1 **James Hutton** (1726–1797), the brilliant and observant Scotsman, who first stated the concept of the geologic cycle in 1785. (New York Public Library)

## The Geologic Cycle and the Geologic Record

The ideas that were put together into what became known as the *geologic cycle* represent one of the greatest intellectual achievements of all time. They were the work of one man, who surely ranks among those illustrious few individuals whose lives have changed the course of history. That man was James Hutton (Figure 1).

Hutton (1726–1797) was a gentleman farmer who set about attempting to apply the latest scientific methods in agriculture to his farm in Berwickshire, Scotland. He began to study rocks and the surface features of the Earth. About 1768 he left his farm, moved to Edinburgh, and devoted the rest of his life to what we would now call geologic research. Hutton proposed a whole new way of looking at the world, which he named a "theory of the Earth." His first public presentation took the form of two lectures to the Royal Society of Edinburgh, on 7 March and 4 April 1785, under the title of "Theory of the Earth, or an Investigation of the Laws Observable in the Composition, Dissolution and Restoration of Land upon the Globe." He expanded his ideas and published them in a two-volume book in 1795.

Hutton saw what was happening in the related activities we now know as the water cycle, or *hydrologic cycle* (Figure 2). The sun evaporates water from the oceans. Clouds form, drift over the lands, and drop their water as rain or snow. The rainfall seeps into the ground or flows over its surface and eventually returns to the sea. The moisture attacks the *bedrock,* the continuous solid rock that locally forms the surface of the Earth. As a result, the bedrock breaks down into masses of loose particles

2 **Hydrologic cycle** represented schematically by simplified and idealized profile through a segment of the land and the sea.

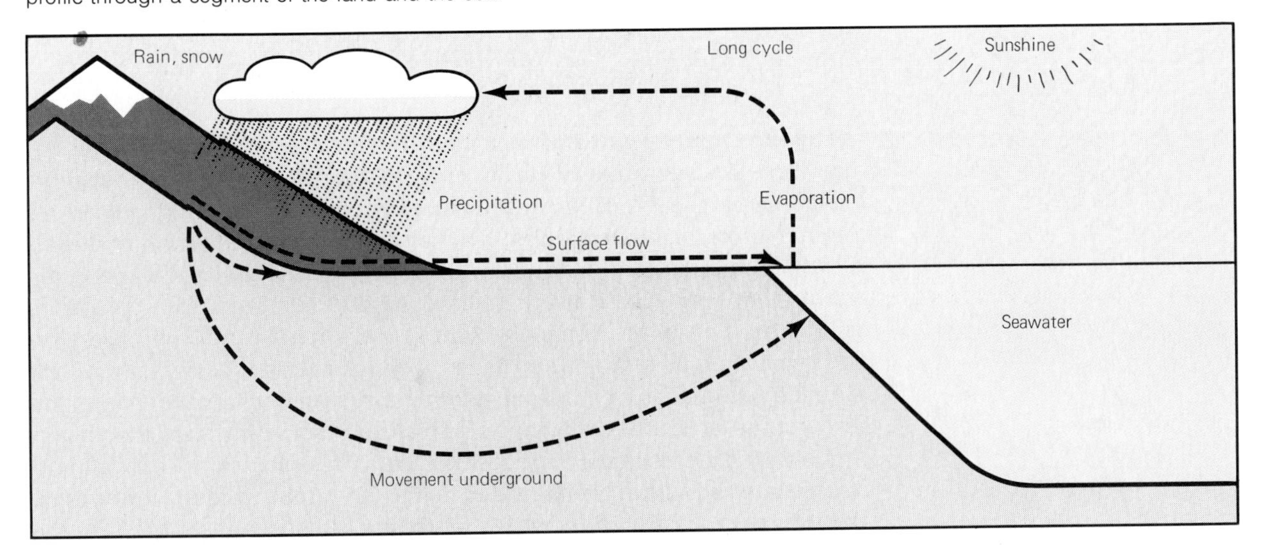

(a)

(b)

**3 Bedrock and regolith.** (a) Contrasting appearances of bedrock and regolith, view in Sixmile Canyon, central Utah. The horizontal layers of the bedrock terminate abruptly against the noncemented material forming the regolith, most of which here lacks layers. (J. E. Sanders) (b) Stratified bedrock in mountains, western wall of Val Lasties, Sella Group, Dolomites, eastern Alps, Italy. (Agi Castelli)

called *regolith* (Figure 3). When the regolith becomes transported, it is named *sediment.* When the flowing water returns to the sea, it carries with it a load of sediment particles and materials dissolved from the bedrock. The sediment is spread out in layers, or *strata.* The strata accumulate one after another, with the oldest at the bottom.

Hutton was not the only person to realize how the rainfall affects the regolith; any alert farmer must have noticed how the fields lost soil during a heavy rainfall. But Hutton was the first person to work out the full implications of what happened to the eroded material. He realized that the loss of material from today's lands was being counterbalanced by accumulation of sediment in the sea that would become the lands of the future.

What Hutton did was to point out that the geologic cycle operates an important subcycle that became known as the *rock cycle.* We will explore the operations of the rock cycle in Chapters 5 through 9, but a short summary of the three great groups of rocks and the rock cycle will be enough to show how both are related to the geologic record.

**The three great groups of rocks and the rock cycle** Geologists recognize three great groups of rocks: (1) igneous, (2) sedimentary, and (3) metamorphic. These differ in appearance, composition, and process of origin. *Igneous rocks* are formed by the solidification of molten material, or *magma;* molten material moves toward the Earth's surface and cools and hardens either beneath the surface or, if discharged from volcanoes as *lava,* on the surface. *Sedimentary rocks* are formed by the compression and cementing of sediment particles. The relationship between modern sediments and ancient strata of sedimentary bedrock (Figure 4) was the cornerstone of Hutton's "theory." He drew the significant conclusion that the continents being eroded today consist of layers of sedimentary materials which had been eroded from still-more-ancient continents. What is more, he suggested that the Earth's heat energy could cause the

(a)

(b)

**4 Layers in sediments and sedimentary rocks.** (a) Horizontal layers enclosing inclined layers, sedimentary rock south of Poughkeepsie, New York. (b) Horizontal layers enclosing inclined layers, sediments along north shore of Long Island, New York. (J. E. Sanders).

surface of the Earth to move up and down, thus elevating some areas (these would be eroded) and depressing others (where the sediments would collect). The third group, *metamorphic rocks,* consists of previously formed rocks which have undergone an extreme change in constitution caused by heat, pressure, or chemical action. Examples of the appearance typical of the three groups of rocks are shown in Figure 5.

**5 Contrasting appearances of typical examples of the three great groups of rocks.** (a) Igneous rock displays uniform interlocking of light-colored and darker-colored constituents. (b) Sedimentary rock displays a variety of constituents, each of which has been broken from older bedrock and has been rounded before being cemented into the body of rock shown here. (c) Contorted and deformed layers of light-colored and dark-colored constituents in metamorphic rock. (J. E. Sanders)

(a)

(b)

(c)

The concept of the rock cycle is that material from inside the Earth comes up to the surface and reacts with the atmosphere and water to become broken into pieces which form regolith or sediment. The sediment accumulates where parts of the Earth's surface are subsiding, becomes buried to greater and greater depths, and thus is returned to the interior. At depth the sediments may be compressed into sedimentary rocks, altered to metamorphic rocks, or melted into magma to form igneous rocks.

One of the great implications of the rock cycle is that rocks and the outer part of the Earth form a *geologic record,* a natural archive in which the history of the Earth is recorded. The geologic record consists of three parts: (1) the bedrock; (2) the regolith; and (3) the shape of the Earth's surface. We devote Chapters 10 through 15 to the various processes that shape the surface of the Earth. An important point about the geologic record is that it reflects the passage of time. And, perhaps more than anything else, geology is distinguished as a science by its concerns for time.

## Geologic Time

Geology did not progress much as a science until geologists were able to overcome the fierce, vigorous resistance to the idea that the history of the Earth is written in the rocks and that this history is unimaginably long. Geologists now consider that the age of the Earth is 4.6 billion years. The scale used to classify geologic time is given in Table 1. Later in the text we will give some details about how it was worked out. Readers should become familiar with the terms in Table 1 because they are used throughout the book. In Chapter 3 we discuss many aspects of geologic time.

## Plate Tectonics

When enough information had been assembled about the characteristics of the deep-sea floor and when worldwide investigations of earthquakes and volcanoes became possible, the science of geology entered a new era based on concepts of plate tectonics.

The subject of tectonics is the dynamics of rock bodies — movements that cause strata that were initially horizontal to be crumpled and broken. *Plate tectonics* is a concept which states that the great continental landmasses are constantly shifting, moving on huge slabs or plates. We refer to this exciting new theory throughout the book, and discuss it in detail in Chapter 18.

The concepts of plate tectonics deal with the distribution of geologic structures and the other major dynamic features of the Earth. Plate tectonics touches so many large-scale aspects of geology that some geologists have suggested that the name of the subject be changed from geology to plate tectonics.

The features formed by tectonic movements are *geologic structures.* We need to define a few kinds here for use in several chapters preceding Chapter 16, which is devoted to geologic structures and mountains. Geologists recognize two kinds of fractures: (1) joints and (2) faults. A *joint* is simply a crack or break — the blocks on either side of the break

| Eons | Eras | Periods | Epochs (of Cenozoic) | | Selected events in history of Earth | |
|---|---|---|---|---|---|---|
| PHANEROZOIC | Cenozoic | Quaternary | Holocene | | HOMO SAPIENS | |
| | | | Pleistocene | | Many mammals become extinct | Ice age<br>Widespread volcanic activity |
| | | | | 1.8 my | | |
| | | Tertiary | Pliocene | | Grazing mammals abundant<br>Grasses spread widely | Erosion of Grand Canyon begins |
| | | | | 5.2 my | | |
| | | | Miocene | | | |
| | | | | 24 my | | |
| | | | Oligocene | | Many early mammals become extinct | Lands elevated<br>Deformation in Alps |
| | | | | 38 my | | |
| | | | Eocene | | | |
| | | | | 54 my | | |
| | | | Paleocene | | | Climax of Rocky Mountain deformation |
| | | | | 65 my | | |
| | Mesozoic | Cretaceous | | | Dinosaurs become extinct | Coal swamps |
| | | | | 145 my | | |
| | | Jurassic | | | Modern plants; first primates appear | Atlantic Ocean (starts to open) |
| | | | | 190 my | | |
| | | Triassic | | | Time of dinosaurs<br>First birds and mammals | Palisades sheet intruded |
| | | | | 225 my | | |
| | Paleozoic | Permian | | | Many invertebrates become extinct | Ice age |
| | | | | 280 my | | |
| | | Carboniferous: Pennsylvanian | | | Mammal-like reptiles | Climax of Appalachian deformation |
| | | | | 325 my | | |
| | | Carboniferous: Mississippian | | | First reptiles | Coal swamps widespread |
| | | | | 345 my | | |
| | | Devonian | | | Oldest fossil footprint (amphibian) | Lands elevated |
| | | | | 395 my | | |
| | | Silurian | | | Oldest rooted land plants | Ice age |
| | | | | 430 my | | |
| | | Ordovician | | | First fish (oldest vertebrates) | Shallow seas cover continents |
| | | | | 500 my | | |
| | | Cambrian | | | First abundant fossils | |
| | | | | 570 my | | |
| CRYPTOZOIC | Precambrian | | | | Widespread ice age<br>Early ice age<br>Oldest one-celled plants | Abundant oxidized iron-bearing sediments deposited |
| | | | | 3500 my | | |
| AZOIC | | | | | Oldest rocks | |
| | | | | 4600 my | Origin of Earth | |

(a)

(b)

**6 Geologic structures (1): kinds of fractures in bedrock.** (a) Nearly vertical joints in massive granite, S wall of Cathedral Peak, just S of Tuolomne Meadows, Sierra Nevada, California. (Agi Castelli) (b) Steeply inclined fault, a fracture along which the opposite sides have been displaced in a direction parallel to the fracture. Exposure along King Hill Canal at Deer Gulch, about 13 kilometers SSE of King Hill, Elmore County, Idaho. (H. E. Malde, U. S. Geological Survey)

have not been shifted (Figure 6a). A *fault* is a fracture along which the broken blocks have been displaced in a direction parallel to the fracture (Figure 6b). Joints and faults imply that the rocks were brittle. A *fold* or a bend in rock strata, implies the opposite; it suggests that the rocks were ductile or pliable. Two contrasting kinds of folds are (1) archlike upfolds of strata, known as *anticlines,* and (2) troughlike downfolds of strata, named *synclines* (Figure 7).

(a)

(b)

**7 Geologic structures (2): Folds in bedrock.** (a) Anticline, an archlike feature in which the deformed layers are inclined away from the crest. (b) Syncline, a troughlike feature in which the deformed layers are inclined toward the lowest part of the trough.

### Resources and Environment

The human race is absolutely dependent on the Earth for the necessities of life. These necessities, and many of the materials that provide for enjoyments beyond the necessities, involve geology in countless ways. One of the great dilemmas faced by modern civilization is how to sustain the high standard of living enjoyed by many nations and at the same time not to pollute the Earth to the point where it becomes uninhabitable.

One of the grim realities of the modern era is that the materials derived from the Earth are being consumed at a faster rate than they are being discovered. The whole subject of the future of resources is complicated. But whatever the complications, one thing is clear: if new supplies are not found and soon, enormous changes in peoples' life styles are going to happen.

Geology deals with these necessities of life. It involves a study and understanding of the movements of water. Although food itself is not strictly geologic, understanding the soil on which food is grown is part of geology. Mineral resources and energy supplies are distributed according to natural principles being investigated and applied by geologists. Finally, the dynamic workings of the natural environment are studied by geologists.

Throughout this book, we emphasize the importance of mineral resources and energy supplies. We have placed material related to these topics in each of the chapters where we discuss some geologic process that concentrates minerals or energy supplies. In Chapter 19, we present a summary of exploration geology, the search for new resources. The future of civilization depends on the ability of geologists to find additional supplies of natural resources.

### The Planetary Scope of Geology

Through space-age research, we have discovered a great deal about the Moon and several of the planets of our Solar System. The principles of geology, learned here on Earth, have been vital parts of the scientific effort to understand the Solar System, because many of the fundamental lessons of geology apply as well to other planets. Conversely, the planetary aspects of geology are proving to be significant in learning more about the Earth. Chapter 20 summarizes some of the highlights of planetary geology.

## OUR STRATEGY FOR PRESENTING PHYSICAL GEOLOGY

In presenting the material of physical geology, we have adopted the strategy of beginning with some far-ranging ideas, then shifting down to individual minerals in order to lead up to large features again. The first three chapters deal with the basic ideas of geology: Chapter 1 with the larger features of the Earth; Chapter 2 with many of the ways in which geology and society interact; and Chapter 3, with geologic concepts of time.

Chapters 4 through 9 start small and work upward in scale. Chapter 4 deals with minerals, mineral reactions, and the application of radioactivity to finding the age of mineral specimens. Ultimately, we develop this idea to the point of using ages of minerals to work out ages of rocks and thus of understanding how geologists have calibrated the geologic time scale shown in Table 1. Chapters 5 through 9 cover different types of rocks. The discussion of rocks assumes that the reader is familiar with the details of classifying and naming rocks contained in Appendix C. Such familiarity is best based on the study of specimens both in the laboratory and in the field. In discussing a particular kind of rock we take up appropriate parts of the rock cycle, important rock-forming processes and principles, identification, occurrence, methods for finding the geologic age, and kinds of mineral deposits associated with it. Igneous activity — volcanoes and igneous rocks and associated valuable mineral deposits — are covered in Chapters 5 and 6. Chapter 7 summarizes weathering, with emphasis on preparing material to become sediments, on the origin of soils, and on related deposits of natural resources. Chapter 8 discusses sedimentary rocks, the origin of coal and petroleum, and sedimentary ore deposits. Chapter 9 concludes the overall treatment of rocks with a general view of metamorphic rocks and metamorphism.

Chapters 11 through 15 deal with the processes operating on the Earth's surface. We start this group of subjects with Chapter 10 the hydrologic cycle and water in the ground. We have done this to show how the understanding of slopes is increased if one knows how the interstitial water behaves. All the chapters that deal with the surface of the Earth involve both *erosion* and *deposition,* geology's inseparable "twins."

Chapters 16, 17, and 18 are devoted to tectonic subjects — structures, mountains, earthquakes, the Earth's interior, dynamics of the Earth's outer shell (the lithosphere), mountain building, and plate tectonics. We have tried to build toward an understanding of the concepts of plate tectonics rather than presenting them early in the book without any firm geologic background.

Chapter 19 covers exploration geology and the techniques that will be required in the search for future resources. Chapter 20 is a summary of space-age research about the Solar System.

## WHAT YOU CAN EXPECT FROM A COURSE IN PHYSICAL GEOLOGY

After you have completed a course in physical geology, you should have acquired a few important practical skills and should have learned to look at the world with new eyes — the eyes of a geologist. The skills include such things as being able to identify minerals and rocks, to read maps, to appreciate some of the principles governing the distribution of household water supplies and the behavior of slopes. You should know what the geologic cycle is, how its operation affects the environment and the geologic record, and how important climate is in its operation.

You should learn how the methods of science are applied to the Earth. This means that you will be exercising your powers of reasoning and trying to sharpen your skills in predicting how the world around you operates. This means that you will have a better appreciation of the problems and possibilities for predicting natural events, such as earthquakes, volcanic outbursts, and slope failures, and will also be able to relate to the problems of energy supplies and natural resources.

But, perhaps more important than anything else, the study of geology can expand your horizons almost without limit, as you finally begin to understand the vastness of geologic time. The question of time is so central to geology that we devote much of Chapter 3 to a discussion of how the nineteenth century's "great time dilemma" was finally resolved. After you have come to appreciate what we call a geologic point of view about the world, you will have acquired a marvelous new power and it should give you much joy in looking at the world. In fact, once you have learned some of the fundamentals of geology, you will be able to travel to any part of the world and not feel like a stranger. You will find something familiar in the landscape. Even though you may not know the people or their language, you should feel "at home" in your knowledge of the geologic setting of the place.

## WHAT GEOLOGISTS DO

In pursuing a subject matter that is so varied and so large in scope, geologists engage in many kinds of activities. Field work may be a major part of their study. Much field work includes mapping, which is the geologists' way of establishing large-scale quantitative data on "configurations." Mapping may concern the study and tracing of kinds of rocks. Or, the data mapped may consist of numbers determined by making measurements. Other field work includes collecting specimens, measuring the thickness and characteristics of layers of rock, or measuring the orientations of features that can be used to reconstruct the flow of ancient currents (Chapter 8) or that resulted from deformation (Chapter 16). Nowadays, with information from satellites available, much mapping can be done indoors using information that has been transmitted via radio.

In the petroleum industry, a subspecialty known as *subsurface geology* has been established by geologists who spend much of their time plotting the characteristics, thicknesses, and depths of layers that lie beneath the Earth's surface. Their information is derived from measurements taken in holes drilled in search of new supplies of energy or of materials. Much other subsurface information, usually from only shallow depths, is accumulated by engineering test borings. These are known as soil-test borings; they are made to test the bearing strength of the regolith, but also provide much geologic information. We summarize subsurface geology in Chapter 19.

The results of field work have to be compiled in the office and reports written about the results. This kind of work is centered around geologic libraries.

Much geologic work is done in the laboratory. Here, geologists attempt to establish the configurations of specimens by means of examination with microscopes, X-ray analysis, (Chapter 4), or bombardment with electrons or other particles. Some laboratory work concentrates on experimental duplication of the conditions under which natural reactions take place. Using high-pressure, high-temperature apparatus, for example, scientists have successfully created many substances, perhaps the most noteworthy being laboratory diamonds.

Geologic work of various kinds may be required in deciding how to use the land. This involves planning, engineering studies, and environmental-impact statements. Background information is required for making

**8 Diver-geologist examining marine organisms** on the shallow sea floor. (Robert F. Dill)

decisions about building roads, tunnels, bridges, dams, and energy-generating plants, for example.

Some geologists study natural processes, both in the field and in the laboratory. This may entail everything from learning to use SCUBA-diving equipment (Figure 8) to operating laboratory flumes to measure what effects are created in sand by the flow of water over it.

The main work of one group of geologists consists entirely of theorizing. These scientists devote their attentions to the study and analysis of various small-scale or large-scale patterns. Their purpose may be to compile special maps, to work out the budget of a particular element, or to try to understand how the Earth's magnetic field forms and varies. Although this list covers a wide range, it does not include everything geologists do.

# CHAPTER REVIEW

1. The Earth is a dynamic body that receives external energy from the Sun and is affected by the external forces of gravitational attraction, exerted by the Moon and the other planets, which vary periodically because the orbits are elliptical. The internal energy of the Earth is outflowing heat.
2. The Earth's metallic *core* contains a solid inner part and a liquid outer part. The slow movements of the liquid are thought to create the Earth's magnetic field. Surrounding the core is the *mantle,* a solid, rocky layer.
3. The concept of the *geologic cycle* was formulated by James Hutton in 1785. The geologic cycle refers to the interactions between the Earth's materials and its external and internal energy sources. In the *hydrologic cycle,* moisture evaporated from the sea by solar energy is precipitated, falls on the lands, and then flows over or under the ground back to the sea. The moisture on land reacts with the solid bedrock, breaking it down into regolith, and carrying sediment and dissolved substances back to the sea. Strata of sediment accumulated in the sea may become the continents of the future.
4. The three great groups of rocks — *igneous, sedimentary,* and *metamorphic* — are connected by *the rock cycle.* Igneous rocks form by the cooling of molten material. Most sedimentary rocks form when sediments are turned into stone. Metamorphic rocks typically form deep within the Earth at elevated temperatures and great pressures.
5. The continued operation of the geologic cycle creates a *geologic record,* which consists of the *bedrock,* the *regolith,* and *the shape of the Earth's surface.*
6. One of geology's important contributions to accumulated human knowledge is the concept of the vast length of geologic time and the organization of this time into a geologic time chart.
7. *Tectonics* deals with the dynamics of rock bodies. *Plate tectonics* is a series of concepts that relate dynamic activities to the shifting of huge plates formed by the outer parts of the solid Earth.
8. Important geologic structures formed by tectonic movements are *fractures* and *folds.* Two types of fractures are (1) *joints,* cracks along which the adjacent blocks have not been shifted and (2) *faults,* cracks along which the broken blocks have been displaced in a direction parallel to the fracture. Two contrasting kinds of folds are (1) *anticlines,* archlike upfolds of strata and (2) *synclines,* troughlike downfolds of strata.
9. Important necessities for human existence include water, food, supplies of natural resources, and a nontoxic environment — all of which are related to geology. A matter of great urgency is the need to discover additional supplies of natural resources, minerals and energy supplies.
10. The principles of geology learned on Earth have been found to be applicable to other planets. Also, space-age research about other planets has contributed to a better understanding of the Earth.
11. Geologists engage in both "field work" out of doors and in office work and laboratory work indoors. An important subspeciality is *subsurface geology,* which involves the study of materials beneath the Earth's surface as revealed in various kinds of borings and by other methods. Some geologists are involved in detailed laboratory analyses using various sophisticated equipment, in the studies for deciding how to use the land, in studying the effects of natural processes, and in theorizing.

## QUESTIONS

1. What is the Earth's chief external source of energy?
2. Name the chief external force operating on the Earth and explain why it varies periodically.
3. What are the two main interior parts of the Earth? Which of these two parts is thought to be responsible for the Earth's magnetic field and how?
4. Who formulated the concept of the *geologic cycle* and in what year?
5. What is the chief active agent in the geologic cycle? Explain how the cycle operates.
6. Name the three great groups of rocks and describe how each forms.
7. What are the three components which constitute the geologic record?
8. Does *plate tectonics* have anything to do with baseball or dinnerware? Explain your answer.
9. What is the difference between a *joint* and a *fault?* Between an *anticline* and a *syncline?*
10. Name the two categories of natural resources and explain why additional supplies of these resources are needed.

11. Explain the connections between geology and knowledge of the Moon and the other planets of the Solar System.
12. What are some of the kinds of work that geologists do?

**RECOMMENDED READING**

Eyles, V.A., 1972, James Hutton, p. 577–589 *in* Gillispie, C.C., *editor in chief*, Dictionary of Scientific Biography: New York, Charles Scribner's Sons, v. VI (Jean Hachette-Joseph Hyrtl), 619 p.

McPhee, John, 1980, Annals of the Former World. Basin and Range—I: The New Yorker Magazine, 20 October 1980, p. 58–60, 63–64, 66, 68, 70, 73–74, 76, 79–80, 82, 85–86, 88, 93–94, 96, 101–102, 104, 109–110, 112, 117–118, 120, 125–126, 128–132, 134–136.

Part II, The New Yorker Magazine, 27 October 1980, p. 57–60, 63–64, 67–68, 71–72, 75–76, 79–80, 83–84, 86, 91–92, 97–98, 103–104, 109–110, 115–116, 118–142, 144, 146, 148–155.

Wetherill, G. W., and Drake, C. L., 1980, The Earth and Planetary Sciences: *Science*, v. 209, Centennial issue, p. 96–104.

Wyllie, P. J., 1976, *The Way the Earth Works: An Introduction to the New Global Geology and its Revolutionary Development:* New York, John Wiley and Sons, 296 p.

THE EARTH AS A PLANET
MAJOR FEATURES OF THE EARTH
THE MAJOR GEOLOGIC FEATURES OF THE EARTH'S EXTERIOR
LITHOSPHERE PLATES AND PLATE TECTONICS

# CHAPTER 1 UNDERSTANDING THE EARTH

PLATE 1
Cracks formed by spreading on cooling crust of a lava flow, Hawaii. Movement on the narrow cracks nearly paralleling the hammer handle has been such that the block beyond the hammer moved toward the right, but the apparent offsetting of the wider cracks nearly at right angles to the narrow cracks seems to have been toward the left. This is a miniature version of what happens along a midoceanic ridge where new seafloor crust forms and spreads away from the crest of the ridge. (Dan Fornari, Deep-Sea Drilling Project, Lamont-Doherty Geological Observatory of Columbia University)

IN THIS CHAPTER YOU WILL BE TOLD THAT THE EARTH IS NOT A sphere, the continents are drifting across the oceans on huge plates, the North and South poles periodically change places, the world's longest and highest mountains are on the bottom of the oceans, and the "solid" Earth is partly liquid. Many, if not all, of these notions may seem strange to you. And yet, to the limits that current scientific knowledge can take us, all of these statements about the Earth are true. As our study of the Earth progresses through the text, we will present supporting arguments or evidence for these major concepts. For the time being, however, we ask you to place skepticism aside in favor of a simple wonder about *geology,* the study of the Earth.

## THE EARTH AS A PLANET

In order to understand certain large-scale processes and influences on the whole Earth, it is necessary to be familiar with a few aspects of astronomy. These necessarily include selected topics about the Solar System, the Sun, the orbits of the planets, meteorites, and some large-scale mechanical relationships about the Earth-Moon pair and about the Earth itself.

### The Solar System

The Solar System designates a luminous star, the Sun, and the various bodies, large and small, in orbit within the Sun's gravitational field. We shall deal here only with the nine planets (Figure 1.1a) and meteorites.

As a matter of mathematical simplicity, astronomers have long used the assumption that the center of the Sun is also the central point of all the planetary orbits (Figure 1.1b and c). The great mathematical genius Sir Isaac Newton (1642–1727) pointed out that the Sun moved with respect to the center of the mass of the Solar System (i.e., the *Solar-System barycenter*). He noted that the Sun never shifted away from this bary-

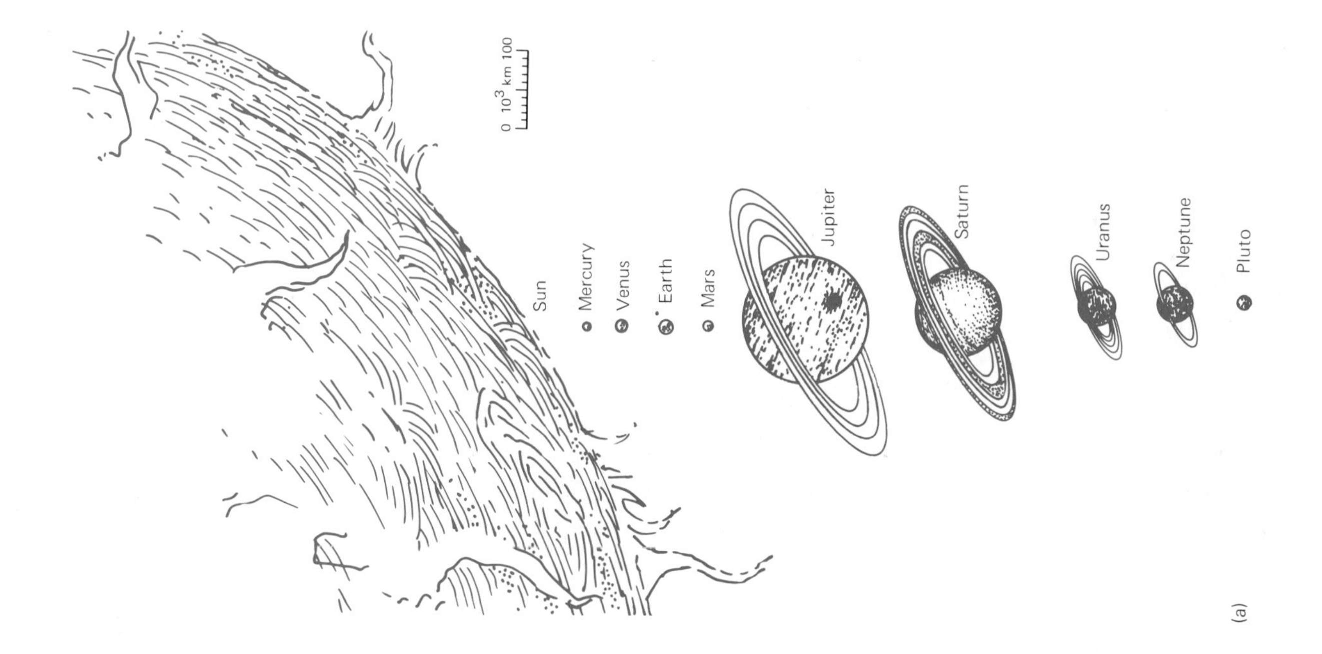


**1.1 Planets of the Solar System and their orbits.** (a) Planets arranged according to their positions outward from the Sun, with curvature of Sun and relative sizes of planets drawn to same scale. Distances of planets from Sun are schematic. Using the scale of diameters shown here, the scale-model Earth would have to be placed about 2 meters away from the scale-model Sun.

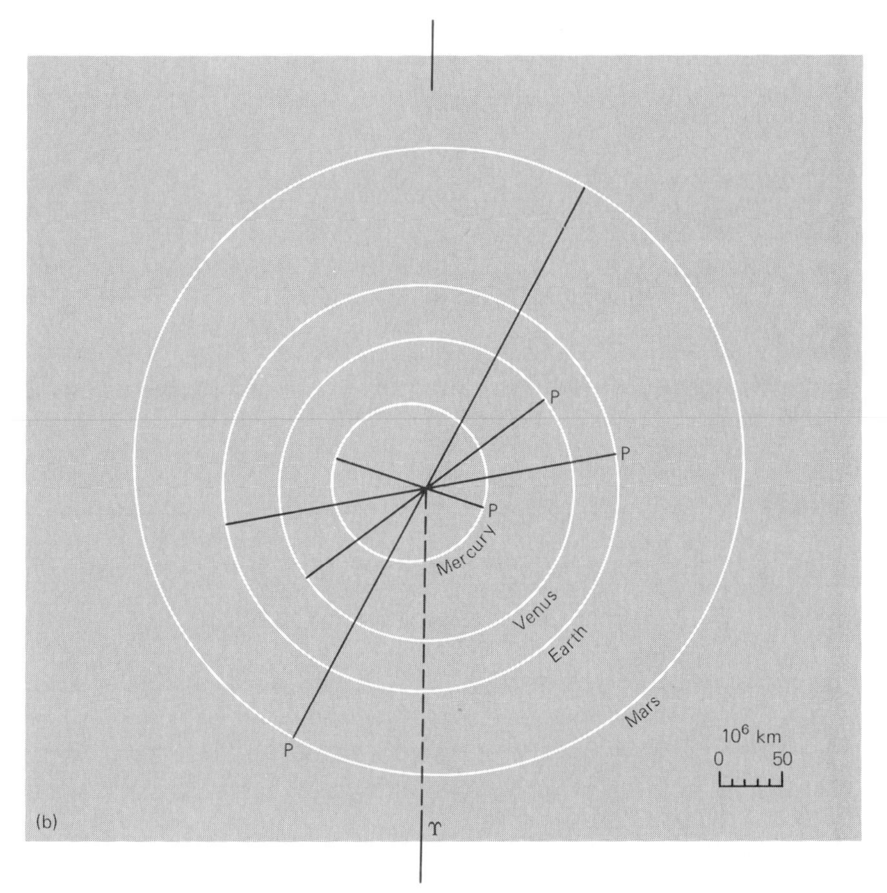


(b) Orbits of inner, or terrestrial planets viewed from above and drawn to scale and with major axes of 1978 orbital ellipses shown as blue lines with blue letter P's marking 1978 positions of closest approach to Sun (known as *perihelion*). Dotted line marks zero celestial longitude (first point in Ares), centered on Sun.

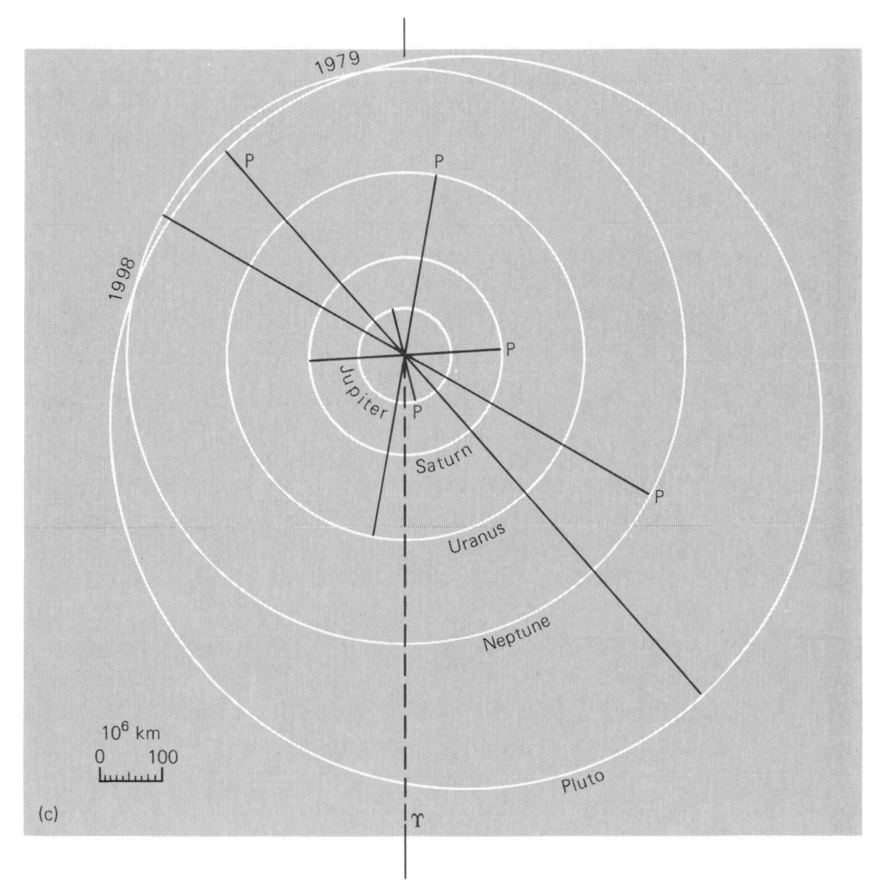


(c) Orbits of outer planets viewed from above and drawn to scale, using same color scheme as in Figure 1.1b. Starting in 1979 and continuing for a little more than 19 years, Pluto's elliptical orbit will bring it closer to the Sun than Neptune's orbital path. Because Neptune will not be in this part of its orbit, however, Pluto will not lose its rank as the planet farthest from the Sun.

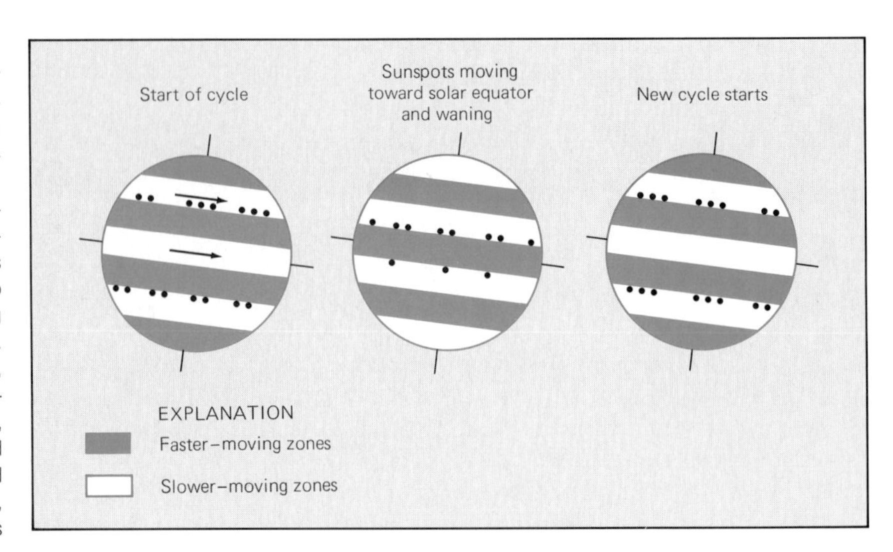


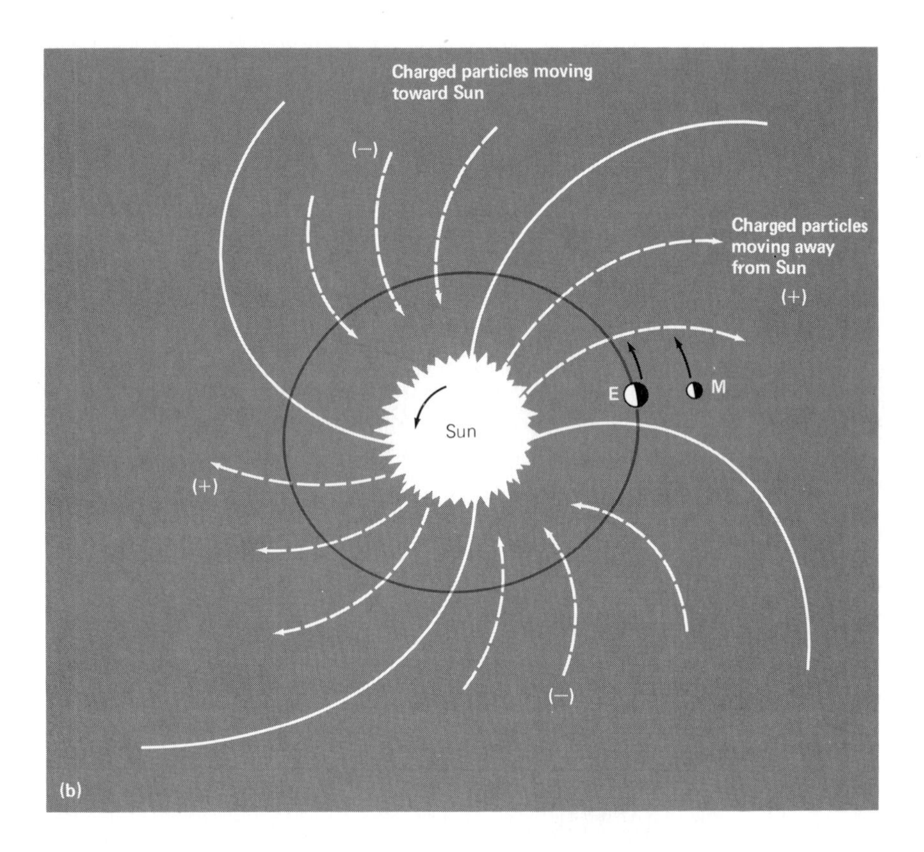


**1.2 Effects of rotation of the Sun seen in two views at right angles.** (a) View perpendicular to Plane of Ecliptic, showing zones of faster and slower rotation and relationship of these zones to sunspots and sunspot cycle. (b) View down from above Plane of Ecliptic showing radially curved zones within which the magnetic field surrounding the Sun causes charged particles of the solar wind to move outward from the Sun. In intervening zones, particles move toward the Sun. (Modified from J. A. Ratcliffe, 1970, Sun, earth and radio: an introduction to the ionosphere and magnetosphere: London, World University Library, Weidenfeld and Nielson, 256 p., Figure 7.4, p. 185; and Ralph Markson and Michael Muir, 1980, Solar Wind *(sic)* control of the Earth's electric field: Science, v. 208, p. 979–990, Figure 2, p. 981).

center by more than about one width of itself. As a result of ignoring the Sun's motion about the Solar-System barycenter, many astronomers have become convinced that the Solar System is peculiar because 85 percent of its angular momentum is supposedly possessed by the two largest planets, Jupiter and Saturn. The astronomers who reached such a conclusion assigned very little angular momentum to the Sun, only that associated with the Sun's rotation about its polar axis (Figure 1.2). When one realizes the significance of the Sun's movement around the

Solar-System barycenter, however, it is easy to show that the Sun possesses 50 percent of the angular momentum of the Solar System and that the orbiting planets collectively share the other 50 percent. Jupiter and Saturn's 85 percent refers to the planets' share, which means overall, Jupiter and Saturn possess 85 × 0.50, or 42.5 percent. We return to this topic in a following section.

**The Sun** The Sun is a luminous body within which nuclear reactions generate great heat and light. The Sun's mass is being consumed as it is converted into energy, which radiates outward in all directions. In the 4.7 billion years since the Sun has been creating such energy, the Sun's mass has diminished by an amount equal to about 109 times the mass of the Earth or about 1.5 times the masses of all the planets in the Solar System. This large quantity, however, is only 0.03 percent of the Sun's mass. Part of the energy is in the form of streams of charged subatomic particles, protons and electrons (see Appendix A for discussion of protons and electrons). These streams of charged particles have been designated as the *solar wind.* They are directed away from the Sun following curved paths, which resemble the paths of water sprayed out of a rotating garden sprinkler, that make four zones (Figure 1.2b). The intensity of the electromagnetic energy radiated by the Sun decreases outward, following an inverse-square law. This means that the intensity diminishes as the square of the distance out from the Sun (see Appendix D).

For our purposes, we can consider that the Sun is a sphere having an equatorial diameter of 1,392,000 km. The Sun's axis of rotation is tilted at an angle of about 7 degrees to a pole perpendicular to the plane of the Earth's orbit, known as the *Plane of the Ecliptic.* The equatorial parts of the Sun rotate faster than the polar parts (the equatorial rotation period is about 27 days; the polar period, about 34 days).

About 98 percent of the Sun consists of the elements hydrogen and helium; other significant constituents include oxygen, carbon, nitrogen, neon, and silicon. Because the main solar elements are gases (at least under conditions prevailing at the surface of the Earth), we presume that no clear boundary marks the surface of the Sun. In any case, what we can use for a boundary is known as the outer surface of the photosphere, which curiously enough, is the coolest part of the Sun; the estimated temperature is 4300°K (Kelvin temperature scale; graduated the same as the Celsius scale, but with 0°K corresponding to −273°C). Surrounding the photosphere is the chromosphere. Within the chromosphere, the temperature increases upward, reaching as high as 20,000°K.

The Sun's motion with respect to the Solar-System barycenter is controlled by the motions of the planets. Accordingly, we now turn to the planets and their orbits.

**Planets and planetary orbits** Orbiting within the Sun's gravitational field are nine planets, seven of which are themselves orbited by one or more satellite moons. Only the innermost two planets, Mercury and Venus, lack satellites. Important points about planetary orbits are dimensions and periods.

The great contribution of Johann Kepler (1571–1630) was the proof that the orbits of the planets are not circles, but ellipses. To specify the circumstances of an ellipse, one needs to know the length of the semimajor axis (*a* of Appendix Figure D.1), which astronomers designate the

**TABLE 1.1**
SELECTED DATA ABOUT PLANETS AND THEIR ORBITS

| Characteristic \ Planet | Mercury | | Venus | | Earth | | Mars | |
|---|---|---|---|---|---|---|---|---|
| Distance from Sun | A. U. | $10^6$ km | A. U. | $10^6$ km | A. U. | $10^6$ km | A. U. | $10^6$ km |
| Mean (=$a$) | 0.387099 | 57.909 | 0.723332 | 108.21 | 1.000000 | 149.598 | 1.523691 | 227.941 |
| Minimum (Perihelion) | 0.3075 | 46.001 | 0.7184 | 107.48 | 0.983 | 147.097 | 1.381 | 206.654 |
| Maximum (Aphelion) | 0.4667 | 69.817 | 0.7282 | 108.94 | 1.017 | 152.099 | 1.666 | 249.228 |
| Present eccentricity of orbit (=$e$) (Percent, rounded) | 0.205631<br>20.6 | | 0.006783<br>0.7 | | 0.016718<br>1.7 | | 0.093387<br>9.3 | |
| Distance from Earth Minimum Maximum | | 80.5<br>218.9 | | 40.9<br>259.1 | | —<br>— | | 56.3<br>399.1 |
| Inclination of orbit plane (measured from Plane of the Ecliptic) | 7°00′16″ | | 3°23′40″ | | — | | 1°50′29″ | |
| 1980 longitude of Perihelion Ascending node | 077°09′04″<br>048°05′58″ | | 131°17′45″<br>076°30′13″ | | 102°36′14″<br>— | | 335°41′56″<br>049°24′24″ | |
| Equatorial diameter (km) | 4844 | | 12,109 | | 12,756 | | 6782 | |
| Period of revolution (Earth days or Earth years) | 87.9693 | | 224.7009 | | 365.2564 | | 686.9796 | |

Measured distances in A. U. (astronomic units, based on mean distance between Earth and Sun, taken as 149,598,000 km). Maximum and minimum distances have been calculated from mean distances and values of eccentricity shown. Ascending node is defined by a planet's crossing of the Plane of the Ecliptic from below (celestial South) to above (celestial North). *Source:* 1980 World Almanac and Book of Facts, p. 764).

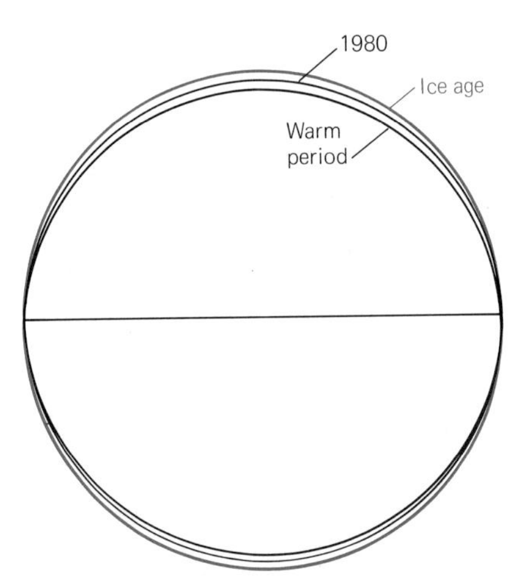


**1.3 Variations in orbital ellipse of Earth-Moon pair,** ellipticity much exaggerated. Although the contrast between the maximum and minimum ellipticity is not enough to show in small drawings, the actual difference in distance to the Sun is 124,120 kilometers.

*mean distance* of a planet and the value of the ellipticity, denoted $e$ (see Appendix D). At the present time, the mean distances range from (in rounded numbers) $58 \times 10^6$ km (Mercury) to $5960 \times 10^6$ km (Pluto) (Table 1.1).

Ellipticities range from the nearly circular orbits of Venus and Neptune ($e = 0.7$ percent) to Pluto's 25.5 percent. The present-day $e$ of the Earth's orbit is 1.67 percent, but through time, the value ranges between 0.5 and 4.1 percent, on a cycle of 102,000 years (Figure 1.3).

| Jupiter | | Saturn | | Uranus | | Neptune | | Pluto | |
|---|---|---|---|---|---|---|---|---|---|
| A. U. | $10^6$ km | A. U. | $10^6$ km | A. U. | $10^6$ km | A. U. | $10^6$ km | A. U. | $10^6$ km |
| 5.204377 | 778.56 | 9.577971 | 1432.85 | 19.26020 | 2881.3 | 30.09421 | 4502.0 | 39.82984 | 5958.46 |
| 4.956 | 741.34 | 9.047 | 1353.4 | 18.291 | 2736.3 | 29.894 | 4472.1 | 29.682 | 4440.34 |
| 5.453 | 815.78 | 10.109 | 1512.3 | 20.229 | 3026.3 | 30.294 | 4531.9 | 49.978 | 7476.59 |
| 0.0478063<br>4.8 | | 0.0554680<br>5.5 | | 0.0503139<br>5.0 | | 0.0066466<br>0.7 | | 0.2548743<br>25.5 | |
| | 592.2<br>965.6 | | 1199.0<br>1659.2 | | 2584.6<br>3143.0 | | 4292.1<br>4691.3 | | 4285.7<br>7473.8 |
| 1°18′20″ | | 2°29′11″ | | 0°46′13″ | | 1°46′23″ | | 17°08′14″ | |
| 014°35′05″<br>100°14′41″ | | 095°28′15″<br>113°30′41″ | | 172°55′20″<br>074°00′19″ | | 058°30′43″<br>131°31′22″ | | 223°00′51″<br>109°57′57″ | |
| 142,492 | | 120,057 | | 48,924 | | 50,212 | | 1287 | |
| 4332.1248<br>(11.86 yr) | | 10,825.863<br>(29.639 yr) | | 30,676.15<br>(83.985 yr) | | 59,911.13<br>(164.025 yr) | | 90,824.2<br>(248.659 yr) | |

**Periodic effects of orbiting planets; the solar cycle** Viewed from above their orbits, the planets all travel counterclockwise. Because of their differing periods of revolution (see Table 1.1), the distribution of the planets ranges from all but Neptune more or less in a single line (a configuration such as that of the "stellium" of early 1962, which is attained every 2224 years) to a scattering within all points of the compass. Because their orbits are ellipses, the planets experience periodic variations in their distances from the Sun as well as from one another. Two important Sun-related factors that are controlled by inverse-square laws and hence are sensitive to changes in distances, are (1) force of gravity exerted on a planet by the Sun, and (2) intensity of solar radiation received. Solar gravity affects tides, spin rate (which, in turn, is thought to control the magnetic fields of some planets), and may be a factor in triggering volcanic activity and earthquakes. Variations in the intensity of solar radiation received affect temperature contrasts between seasons and the overall climate, as for example, an alternation between ice ages and warm periods (Chapter 13).

The positions of the two largest planets, Jupiter, and Saturn, exert the dominant control on the motion of the Sun around the Solar-System barycenter. Viewed in terms of a three-body problem, the center of the Sun follows an elliptical orbit around the Solar-System barycenter, always remaining in tandem with but opposite from Jupiter, whose orbital period is 11.86 years. However, the major axis of the Sun's orbital ellipse is being swung around under the control of Saturn, hence shifts 360 degrees every 29.64 years, which is Saturn's orbital period (Figure 1.4). Because Jupiter and Saturn shift 90 degrees with respect to each other

**1.4 Orbital ellipse of Sun around Solar-System barycenter,** based on three-body approximation using Sun, Jupiter, and Saturn. (a) Jupiter and Saturn in line on same side of Sun; Sun at maximum range from Solar-System barycenter, which lies outside body of Sun. (b) Jupiter and Saturn in line but on opposite sides of Sun, and with major axis of Sun's orbital ellipse swung around by Saturn. Sun at minimum separation from Solar-System barycenter (now inside body of Sun).

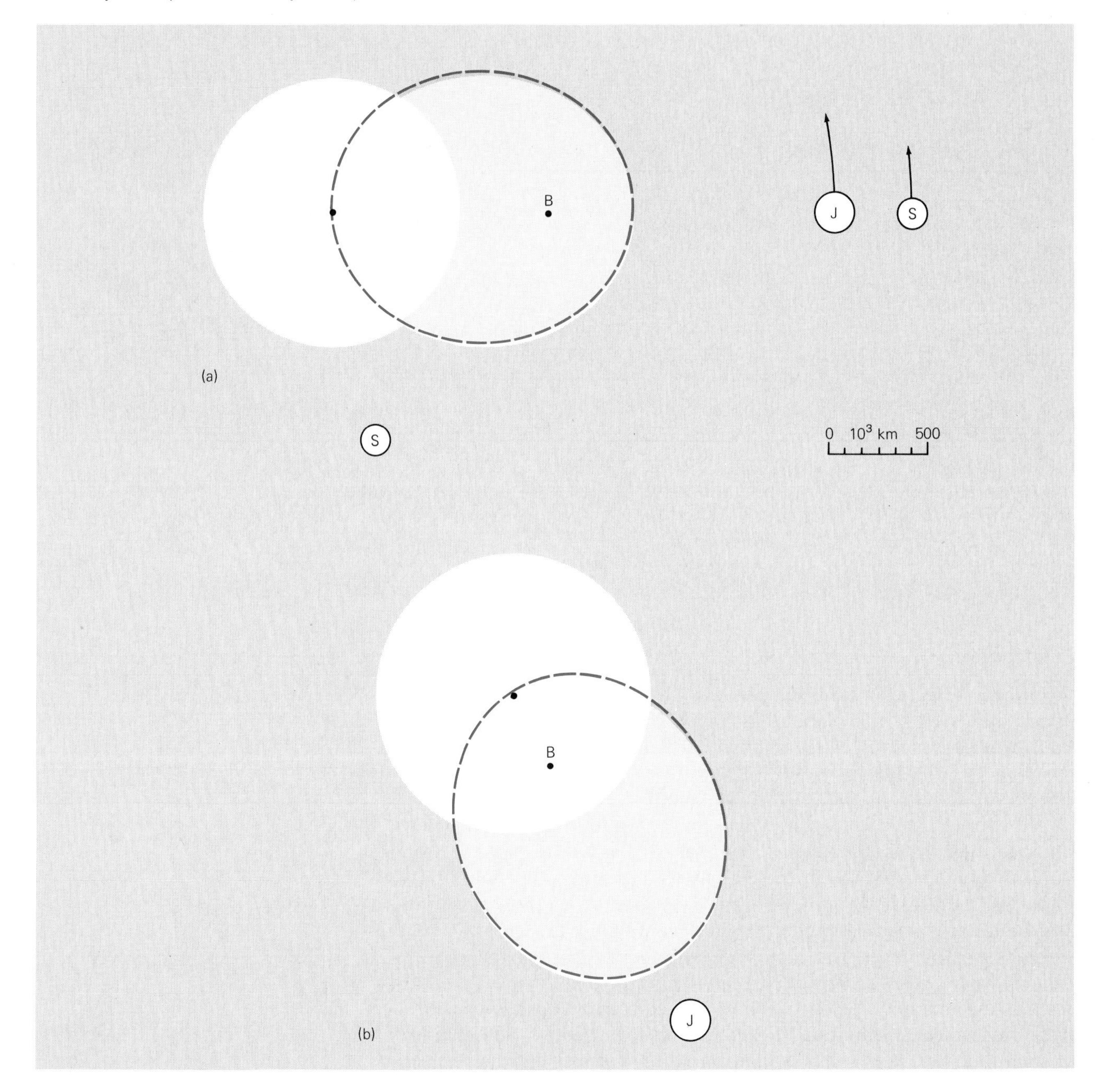

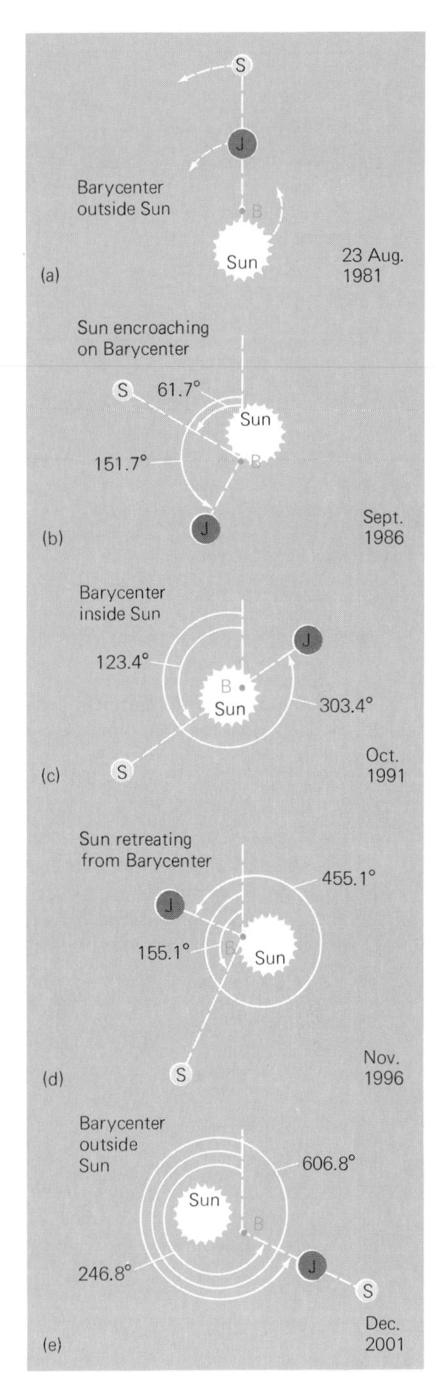


**1.5 Movement of Sun, Jupiter, and Saturn around Solar-System barycenter.** (a) Barycenter (blue dot marked B) 454,000 kilometers outside Sun; peak of sunspot cycle. (b) Sun approaches barycenter, which is at surface of Sun; low numbers of sunspots. (c) Sun at closest approach to barycenter, now 326,000 kilometers inside Sun; sunspot maximum but with polarities opposite those of 1981. (d) Sun retreating from barycenter, which is at surface of Sun (as in 1986); low numbers of sunspots. (e) Center of Sun at maximum distance from barycenter; peak of sunspot cycle (as in 1981).

about every 5 years, the Sun alternately crowds close upon the Solar-System barycenter and retreats to its maximum distance from this barycenter (Figure 1.5). During each of its orbits around the Solar-System barycenter, the Sun shifts position by about one width of itself. The other planets affect the motion of the Sun, but for our purposes, we shall ignore these effects.

A remarkable dynamic property of the Sun is its great loops of solar material governed by intense local magnetic fields, which when viewed from Earth appear as pairs of dark areas on the surface of the Sun, known as *sunspots*. The numbers of visible sunspots vary in a cycle whose period approximates 11 years; the magnetic polarities of the groups of spots alternate, thus making positive and negative cycles of 11 years, for a full cycle of about 22 years. The controlling mechanism of sunspots is not known, but the timing of the cycles is almost certainly dictated by the motion of the Sun with respect to the Solar-System barycenter. The times of maximum numbers of sunspots (Figure 1.6a) coincide with times when the Sun retreats to its greatest distances from the Solar-System barycenter (barycenter entirely outside the body of the Sun, as in 1980 and 1981; see Figure 1.5a and also 1.5e) or when the Sun makes its closest approach to the Solar-System barycenter (barycenter within the body of the Sun, as in Figure 1.5c). The times of minimum numbers of sunspots coincide with times when the Solar-System barycenter lies at the surface of the Sun (as in Figure 1.5b and 1.5d).

Thus, the movements of the planets not only cause variations related to changing distances, but also seem to introduce periodicities into the behavior of the Sun. Clearly, then, in modern terms, we should describe the Solar System as being a regularly repeating system having many internal feedback loops. One persistently cited loop is between numbers of sunspots and the Earth's weather, as recorded in temperatures (Figure 1.6b) and amounts of rainfall.

**Meteorites** The solid natural objects that reach the Earth's surface after falling through the atmosphere from outer space are known as *meteorites*. Depending on their compositions and the circumstances of their discovery, meteorites are classified into many groups.

Based on compositions, two contrasting end members of meteorites are: (1) *metallic meteorites,* which consist of intergrowths of iron and nickel; and (b) *stony meteorites,* which consist for the most part of the same minerals that form igneous rocks on Earth. Some stony meteorites contain small circular bodies named *chondri;* meteorites of this kind are

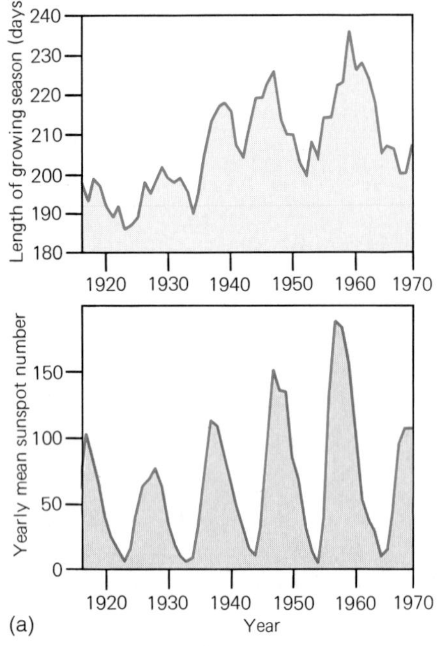


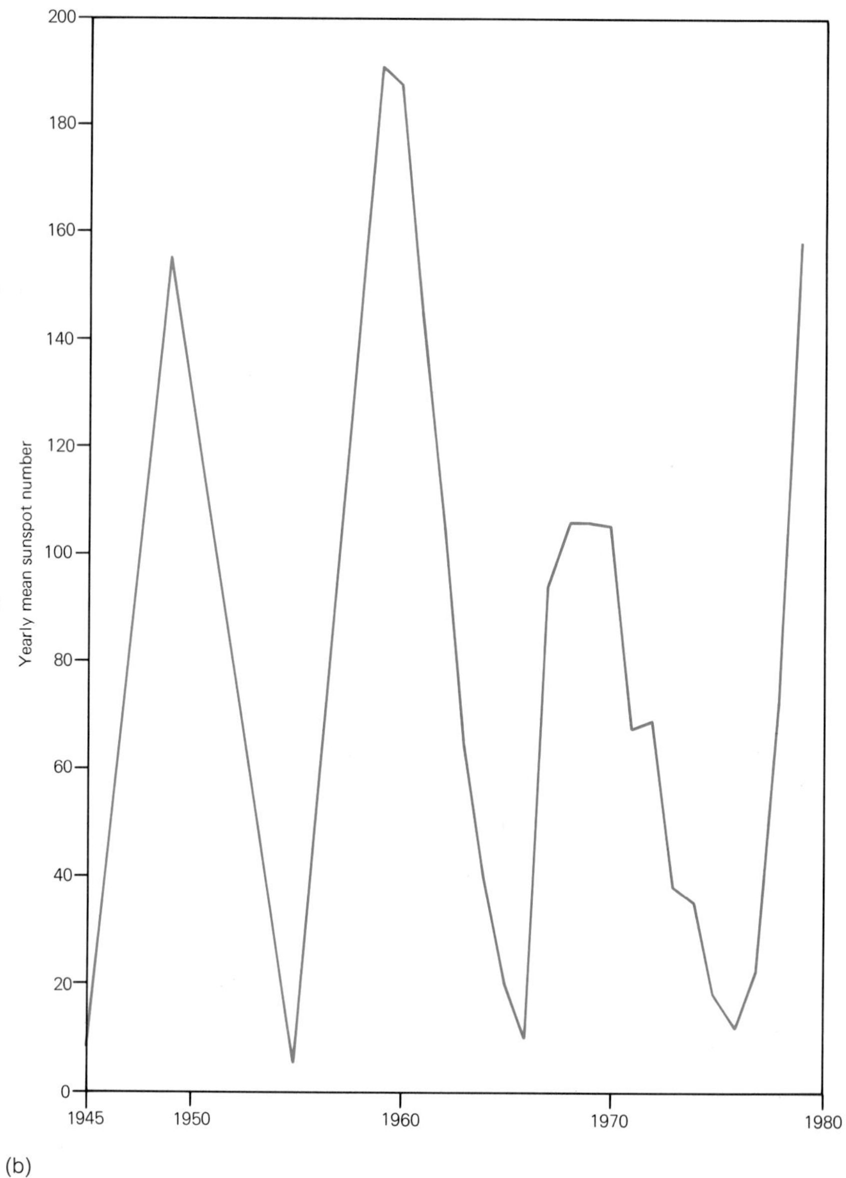


**1.6 Sunspots and temperatures on Earth.** (a) Temperatures as expressed by length of growing season in Scotland (Eskdalemuir, 55°N, 3°W), based on number of days when air temperatures 1.25 meters above the ground exceeded 5.6°C; 5-year running means (upper curve) compared with graph of sunspot numbers, yearly means (lower curve). (J. W. King, 1973, Solar radiation *(sic)* changes and the weather: Nature, v. 245, p. 443–446, Figure 1, p. 463). (b) Yearly mean sunspot numbers, 1945–1979. (*Source:* National Oceanic and Atmospheric Administration)

*chondrites.* The other stony meteorites lack chondri but contain instead angular particles; these are the *achondrites.* Other varieties of meteorites include *carbonaceous chondrites* (a variety of chondrites containing carbon compounds), *stony-iron meteorites* (mixtures of nickel-iron and rock-making minerals in rounded shapes), and *tektites* (glassy objects).

Based on the circumstances of their discovery with respect to their times of descent through the atmosphere, meteorites can be organized into *falls,* those actually seen falling to the ground, and *finds,* meteorites that were not actually seen falling to the ground but rather were discovered by someone at a later time and then reported to scientists. Among the falls, stony meteorites are by far the most abundant. By contrast, among the finds, metallic meteorites predominate.

Meteorites have been studied exhaustively to determine their chemical compositions in the utmost detail and to learn their ages. The calculated ages of meteorites are all about 4.7 billion years, a date considered to be close to the time of origin of the Solar System, hence also the time of origin of the Earth. Meteorites are older than the oldest rocks on Earth by about 0.9 billion years. This span of time between the presumed date of origin of the Earth (and the Solar System) and the oldest rocks on Earth is known as *pregeologic time* (see Table 1).

## The Earth-Moon Pair

The Earth and its comparatively large satellite, the Moon, orbit in two ways: (1) around their common center of mass, the *Earth-Moon barycenter,* and (2) around the Solar-System barycenter, hence around the Sun.

The Earth and the Moon orbit in a period of about 29.5 days (29 days, 12 hours, 44 minutes, and 2.8 seconds). The center of these orbits, the Earth-Moon barycenter, is a point that always lies within the Earth, but whose depth varies according to the changing distance between the Earth and the Moon. The Earth-Moon barycenter is situated in a depth zone 611 kilometers thick, which ranges from 1436.5 to 2047.5 kilometers beneath the Earth's surface (based on the maximum and minimum Earth-Moon distances shown in Table 1.2). When the Moon is at its mean distance from the Earth (384,400 kilometers), the depth to the Earth-Moon barycenter is 1707 kilometers.

The Earth-Moon barycenter traces an elliptical orbit about the Solar-System barycenter in 365.2564 days. This means that the center of the Earth must oscillate back and forth across the orbital path. The center of the Earth swings outside this ellipse at New Moon, is inside the ellipse at Full Moon, and lies on the ellipse at quarter phases (Figure 1.7).

The Earth's orbital plane (the Plane of the Ecliptic mentioned previously) is inclined 1°38′ to the mean plane of all the planetary orbits (Mean Solar-System Plane). The line of intersection between the Plane of the Ecliptic and the Mean Solar-System Plane pivots around 360° once every 362,880 years.

The plane of the Moon's orbit is inclined 5°09′ to the Plane of the Ecliptic. The line of intersection of these two planes is known as the Line of Nodes; it swings around 360° every 18.6 years. Full eclipses of the Sun and the Moon are possible at times when the Moon's orbit coincides with the Plane of the Ecliptic (Line of Nodes points toward the Sun). Only partial eclipses or no eclipses at all are possible when the Moon orbits above or below the Plane of the Ecliptic.

We discuss many other features about the Moon in Chapter 20. Here, we continue with the Earth.

The tilt of the Earth's axis (Figure 1.8) determines the trace across the Earth's surface of the *subsolar point,* the point on a planet where the Sun's rays strike the surface perpendicularly and thus where solar energy is most intense (Figure 1.9) as the planet revolves in its orbit, rotates on its axis, and the position of the axis changes.

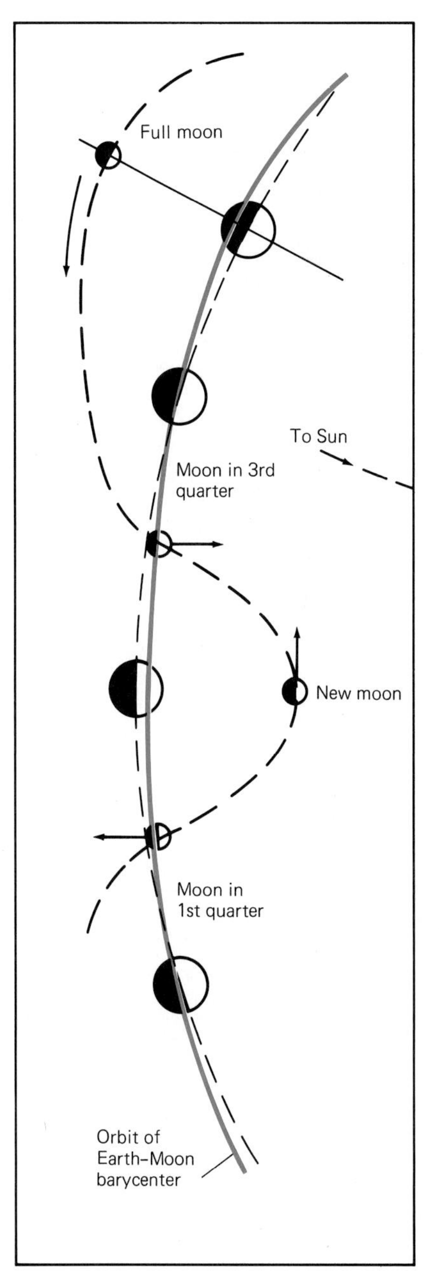


**1.7 Smooth elliptical orbit of Earth-Moon barycenter** compared with zig-zag path of center of Earth and center of Moon.

**TABLE 1.2**
SOME CHARACTERISTICS OF THE SUN, THE EARTH, AND THE MOON

| | Sun | Earth | Moon |
|---|---|---|---|
| Aspects of orbits | | | |
| Orbital distances (km) | (From Solar-System barycenter)[1] | (From Sun)[2] | (From Earth)[2] |
| Mean ($a$) | 759,888 | $149.598 \times 10^6$ | 384,402[3] |
| Minimum | 379,349 | $147.097 \times 10^6$ | 356,411 |
| Maximum | 1,149,427 | $152.099 \times 10^6$ | 406,699 |
| Ellipticity ($e$) | 0.5126 (Range not known) | 0.0167 (Range: 0.005 to 0.041) | 0.0549[4] (Range: 0.0430 to 0.0668) |
| Inclination of orbit plane with plane of the Ecliptic | ? | — | 5°09′ |
| Longitude (heliocentric) of major axis of orbital ellipse (28 Nov. 1978) | 158°14′55.8″ | 102°36′14″ | — |
| Orbital period (Earth years or days) | 11.1 years (mean) (Range: <10 years to 13.4 years) | 365.2564 days | 27.53 days |
| Rotation period (Earth days) | 27 (equator) 35 (polar areas) | 1 | 27.53 |
| Angle between equatorial plane and Plane of the Ecliptic | 7.2° | 23°10′00″ (mean) (Range 22°29′36″ to 23°50′30″; present value ~22°54′) | 1°32′ |
| Size, density, gravity | | | |
| Equatorial diameter (km) | 1,392,280 | 12,756 | 3460 |
| Volume relative to Earth | 1,303,730 | 1.00 | 0.0203 |
| Mass relative to Earth's mass | 332,958 | 1.00 | 0.0123 |
| Mean density relative to Earth | 0.26 | 1.00 | 0.61 |
| Mean density (grams/cm³) | 1.42 | 5.517 | 3.335 |
| Inward-acting gravitational acceleration relative to Earth | 27.9 | 1.00 | 0.17 |

Compiled from various sources, including F. L. Whipple, 1968, Earth, Moon, and Planets, 3rd ed.: Cambridge, Mass., Harvard University Press, 297 p., Appendix 3, p. 268; 1978 Ephemeris and Nautical Almanac, prepared by the U.S. Naval Observatory; 1980 World Almanac and Book of Facts, p. 764; Baker, David, 1978, The Larousse Guide to Astronomy: New York, Larousse and Company, Inc., 288 p., Table 2, p. 274–275; and original calculations of Sun's orbit).

[1] An approximation based on the solution of a three-body problem including the Sun, Jupiter, and Saturn in their present orbits.

[2] Earth-Sun and Moon-Earth distances are as shown in standard astronomical tables; both are based on the assumption that orbits can be measured from the center of the larger body. No adjustments have been made for oscillations of the center of the Earth around the Earth-Moon barycenter nor for the center of the Sun around the Solar-System barycenter.

[3] Based on ranges measured by radar.

[4] This is the mean value for $e$ cited in standard astronomical tables, but it does not match the maximum and minimum distances shown based on radar determinations. The calculated $e$ for the radar-measured distances shown here is 0.06541, nearly the maximum value. Using the mean distance shown here and the mean value of $e$ (0.0549), the computed maximum Earth-Moon distance is 405,503 km and the minimum, 363,296 km.

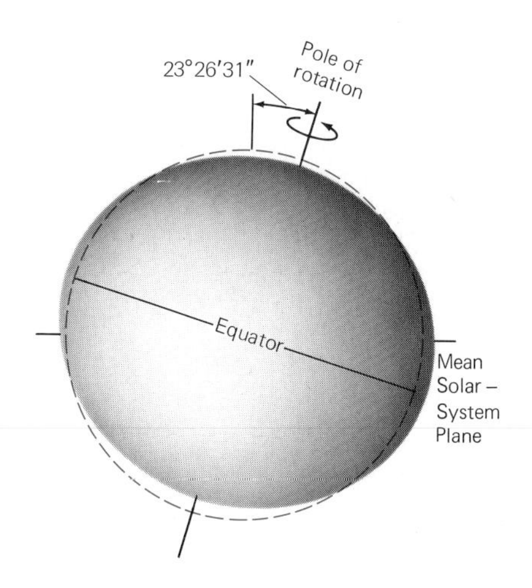


**1.8 Shape of the Earth as an oblate spheroid** schematic section through center of the Earth in a plane perpendicular to the Mean Solar-System Plane (mean plane of orbits of planets in the Solar System). Rapid rotation around Earth's axial pole causes the equatorial part of the Earth to extend outward beyond the outline of a sphere (indicated by the blue dashed circle in this section) and the polar parts to be drawn inward below the outline of a sphere. In 1979, the Earth's axis was tilted 23° 26′31″ from a line perpendicular to the Mean Solar-System Plane (named the Mean Solar Pole, but not labeled as such on the sketch). The amount of tilt changes by about half a second of arc per year.

On the scale of one revolution, we can consider that the direction of tilt of the Earth's axis remains the same (Figure 1.10). Therefore, during each orbit, the subsolar point migrates across the surface of the Earth through an angle of two times the angle of tilt (thus, in 1979, 46°53′02″). The range of migration of the subsolar point across the surface of a planet is what controls the seasons of the year. If other things are equal, then the greatest angle of tilt means greatest summer-winter contrasts and the least tilt means minimum summer-winter contrasts.

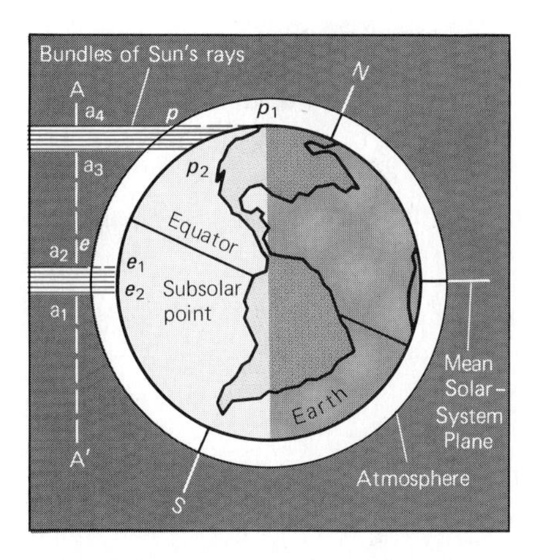


**1.9 Subsolar point migrates across the face of the Earth,** annually reaching its most northerly point (the Tropic of Cancer) during Northern-Hemisphere summer. Because of changes in the tilt of the Earth's axis, the latitude of the Tropic of Cancer varies between 20°53′36″ and 25°26′30″ N. The farther north the subsolar point migrates, the more solar energy is received in polar regions.

**1.10 Essentially constant orientation of Earth's axis on the time scale of one orbit,** schematic view of Sun, Earth, and Moon from above the Plane of the Ecliptic. Notice that the line marking the edge of the vertical plane containing the Earth's axis keeps its same orientation during the orbit and that this line is *not* parallel to the principal axes of the orbital ellipse of the Earth-Moon pair. When the Earth crosses the semimajor axis of the orbital ellipse, it is either at its closest approach to or its greatest distance from the Sun, depending upon which end of the ellipse is involved. Dates and seasons shown are for Northern Hemisphere in 1979. Although the seasons are determined by the times when the lines shown here pass through the center of the Sun, the *focus of the elliptical orbit of the Earth-Moon barycenter is not the center of the Sun,* but the Solar-System barycenter (not shown). The center of the Sun also orbits the Solar-System barycenter (see Figure 1.5).

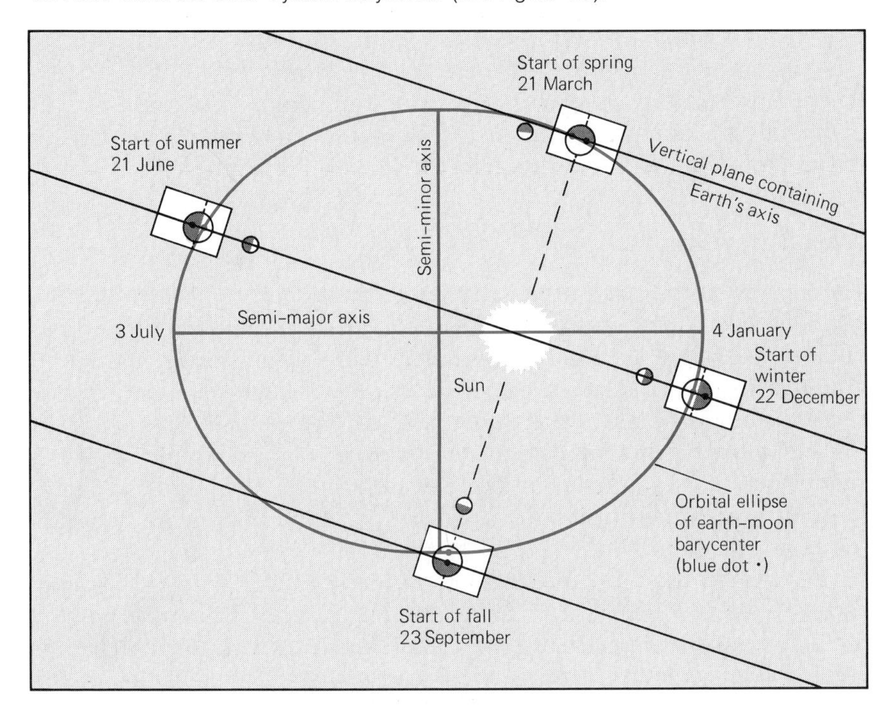

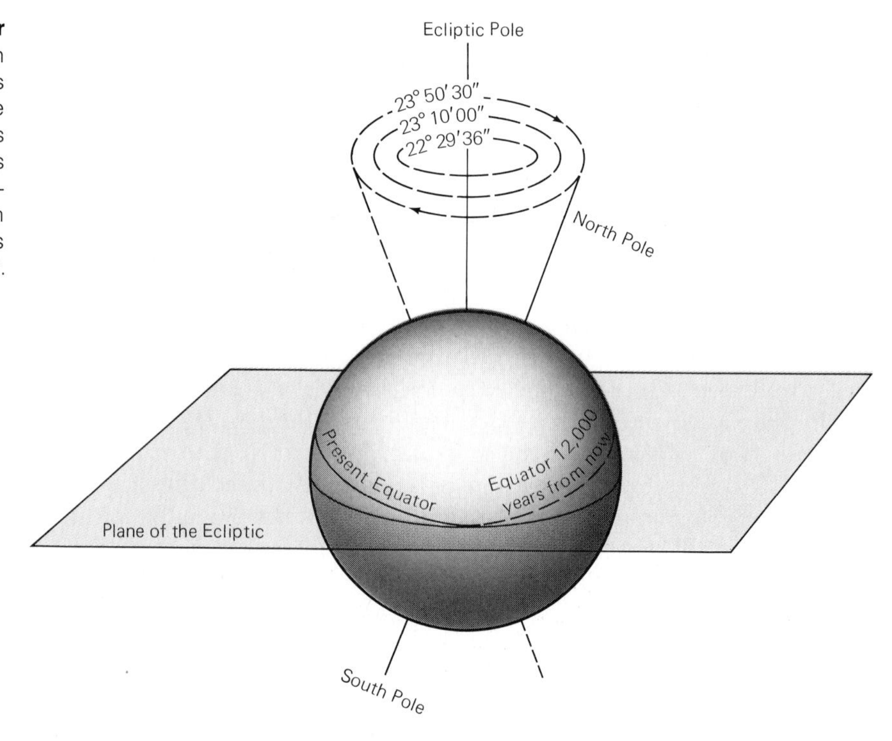


**1.11 Toplike gyration of Earth's polar axis.** When seen from above, the motion of the movement of the North Pole is clockwise, the opposite sense from the direction of the Earth's rotation and of its orbital revolution. As the pole gyrates, its angle from the Ecliptic Pole oscillates between the circles marking the maximum and minimum angles; the mean angle is 23°10′10″. Further explanation in text.

Despite its seeming constancy on the time scale of one year, in the long run, the direction of inclination of the Earth's axis does not remain constant. The Earth's axis describes a circle around the Ecliptic Pole, a motion named *precession,* which is comparable to the wobble of the axis of a spinning top. The time required for a full precession cycle is 25,920 years (Figure 1.11).

## MAJOR FEATURES OF THE EARTH

Under this heading we discuss the overall dimensions of the Earth, its major interior zones, the magnetosphere and ionosphere, the Earth's outward-flowing heat energy, and the Earth's atmosphere.

### Overall Dimensions

The figures on the size of the Earth have never been measured directly, but have been arrived at on the basis of precise measurements of lines oriented east-west and north-south that were painstakingly measured on different parts of the surface. From these we know that the Earth is not a sphere, but an oblate spheroid (see Figure 1.8). This shape is caused by the Earth's rotation around its own polar axis, which causes a slight bulge at the Equator and a minor flattening at the poles. Therefore, the equatorial diameter of the Earth is 12,756 kilometers, whereas the polar diameter is only 12,714 kilometers.

The mass of the Earth has been computed to be $5.976 \times 10^{27}$ grams, and its volume, $1.083 \times 10^{27}$ cubic centimeters. By dividing the mass by the volume, one can calculate the mean density, which for the Earth is 5.517 grams per cubic centimeter. Because the mean densities of the surface rocks are less than 5.517 (they average about 2.7 grams per cubic centimeter on continents and about 3.0 grams per cubic centimeter for ocean-floor rocks), we must infer that the density of the Earth's interior is greater than 5.517. This follows from the necessity of having a greater density at depth to counterbalance the lower-than-mean density of surface rocks in order to make the mean work out to be 5.517.

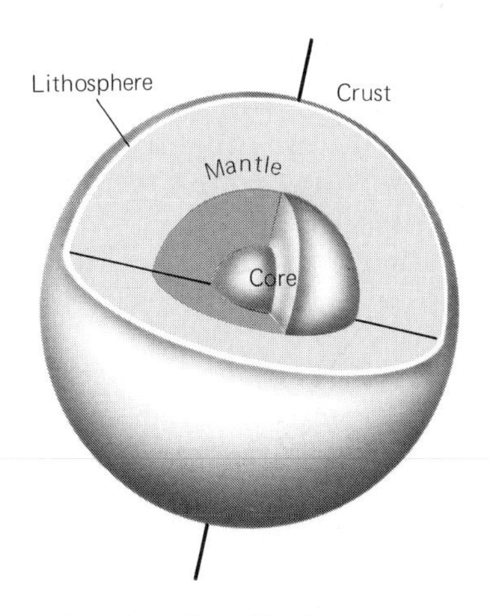


**1.12 Interior of the Earth, cutaway view of sphere.** Schematic, but with sizes drawn to correct proportion.

## Interior Zones

Based on behavior that we will examine in Chapters 17 and 18, geophysicists are convinced that the "solid" Earth consists of three concentric spherical parts (Figure 1.12). From the center outward, these are the core, the mantle, and the complex of lithosphere, crust, asthenosphere, and tectosphere.

**The Earth's core** The Earth's core (Figure 1.13) is thought to consist of two parts: (1) an inner, solid sphere with a diameter of about 2432 kilometers, and (2) an outer, liquid zone with a thickness of about 2259 kilometers. The entire core thus has a diameter of about 6950 kilometers. The movements of the liquid iron-oxygen outer core are thought to be the source of the Earth's magnetic field, and are probably also responsible for causing some movement in the mantle, the zone directly above the outer core.

**The Earth's mantle** Surrounding the Earth's core is a shell of rocky material called the *mantle.* The thickness of the mantle is about 2800 kilometers. If we were to compare the deep interior of the Earth to a furnace in which iron is prepared, the liquid part of the core would be the molten iron that is recovered from the bottom of the furnace. The mantle would be similar to the slag, a once-molten material that floats to the top and, after it has cooled, forms a solid crust.

The mantle is divided into two parts (see Figure 1.13). The upper mantle begins at a depth of about 100 kilometers from the Earth's surface; the lower mantle extends from a depth of about 700 kilometers to 2900 kilometers.

**1.13 Interior of the Earth, schematic pie-shaped slice** from the Earth's surface (top) to center of the Earth (below). Arrows and corresponding figures at left are diameters of various subdivisions. Figures at right show depths beneath surface of boundaries between zones indicated. The lithosphere includes all of the crust and the uppermost part of the upper mantle. Zones marked within lower mantle are depth ranges of the Earth-Moon barycenter.

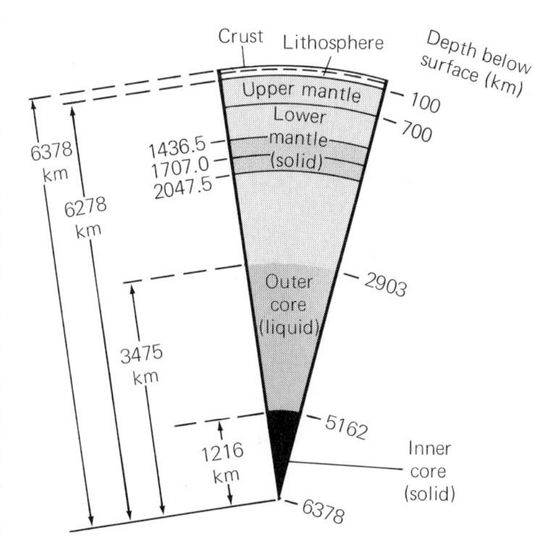


**The Earth's lithosphere, crust, asthenosphere, and tectosphere** The uppermost part of the mantle is the *lithosphere,* or "sphere of rock," which includes the rocky material of the outer part of the Earth. The outer part of the lithosphere is completely covered by the *crust,* a thin, rocky layer about 25 to 40 kilometers thick beneath the continents and about 5 to 10 kilometers thick beneath the deep oceans. The basic composition of the Earth's crust is shown in Table 1.3.

Until fairly recently, the contrasting kinds and thicknesses of the crust were thought to be the only information needed to explain the Earth's many large-scale geologic activities. In the mid-1960s, however, a totally different concept of the Earth's dynamics became almost universally accepted. According to this new concept, large-scale dynamics *do not take place at the crust-mantle boundary,* as was formerly supposed, but rather they operate at one or more deeper levels, and involve zones lying within the upper mantle. These zones are: (1) the *lithosphere,* a rock zone having some strength; (2) the *asthenosphere,* a zone within the upper mantle having negligible strength; and (3) the *tectosphere,* the zone

**TABLE 1.3**
COMPOSITION OF THE EARTH'S CRUST

| Element | | Proportion (Percent) | |
|---|---|---|---|
| Name | Chemical Symbol | By Weight | By Volume |
| Oxygen | O | 46.6 | 93.8 |
| Silicon | Si | 27.7 | 0.9 |
| Aluminum | Al | 8.1 | 0.5 |
| Iron | Fe | 5.0 | 0.4 |
| Calcium | Ca | 3.6 | 1.0 |
| Sodium | Na | 2.8 | 1.3 |
| Potassium | K | 2.6 | 1.8 |
| Magnesium | Mg | 2.1 | 0.3 |
| All other elements | | 1.5 | — |
| | | 100.0 | 100.0 |

*Source:* Brian Mason, 1966, Principles of Geochemistry, 3rd ed.: New York: John Wiley and Sons, Inc.

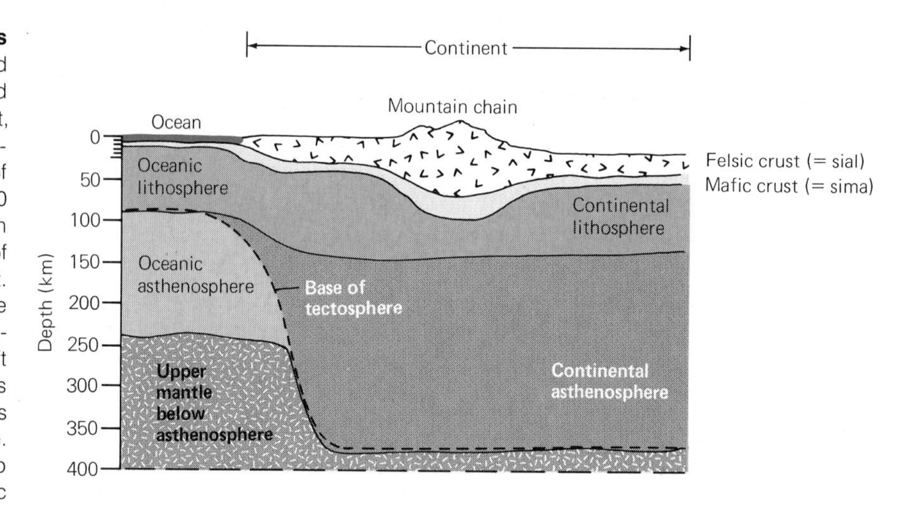


**1.14 Lithosphere and underlying parts of upper mantle,** schematic profile and section through edge of a continent and adjoining ocean basin. Continental crust, absent in oceans, overlies thick asthenosphere. At continental margin, the base of the tectosphere, at a depth of about 100 kilometers beneath the surface of an ocean basin, descends to a depth of nearly 400 kilometers beneath a continent. This change of depth of the base of the tectosphere implies that beneath deep-ocean basins, the lithosphere plates shift at the *top* of the asthenosphere, whereas beneath continents, any shifting takes place at the *base* of the asthenosphere. (After Thomas H. Jordan, 1979, The deep structure of the continents: Scientific American, v. 240, no. 1, p. 92–100; 103–107, Figures on p. 93 and 104)

**1.15 Magnetic lines of force around a magnet and around the Earth.** (a) Lines of force emphasized by iron filings (black) on a sheet of paper with magnet underneath. (Nat Messick) (b) In Earth's magnetic field, north-seeking pole points vertically downward at present North Pole (defined as positive) and vertically upward at present South Pole (defined as negative); lines of force are parallel to the Earth's surface at the Magnetic Equator.

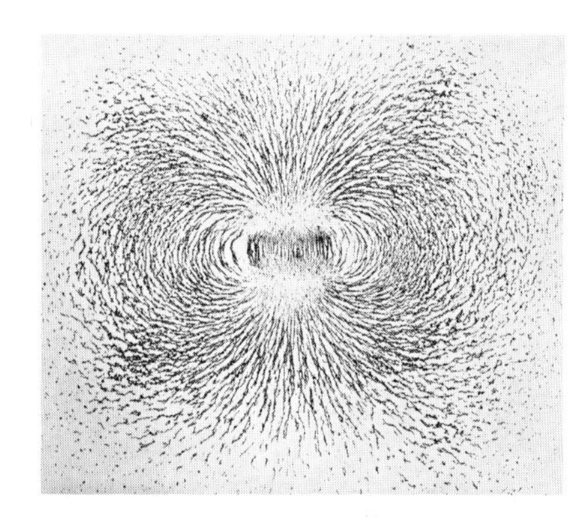

(a)

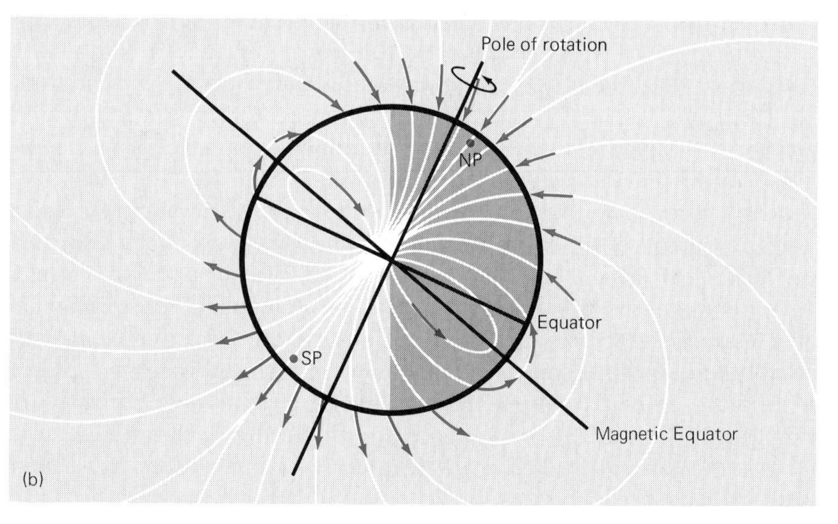

within which large-scale horizontal movements of the overlying materials can take place (Figure 1.14).

Beneath the oceans, the lithosphere extends down to a maximum depth of about 100 kilometers. Beneath continents, the thickness of the lithosphere is 110 to 200 kilometers. The underlying asthenosphere, about 200 kilometers thick, is a zone lacking long-term strength in which solid-state flow takes place easily. Large segments of the lithosphere beneath ocean basins evidently can move laterally at the level of the asthenosphere. The large, presumably movable segments of the lithosphere have been named *lithosphere plates,* which will be discussed in a following section of this chapter.

### The Earth's Magnetosphere and Ionosphere

Extending around the Earth and outward into space from it are two important "spheres" that are important for their magnetic, electrical, and high-energy reactions. These are the magnetosphere and the ionosphere. The *magnetosphere* is the zone of influence of the Earth's magnetic field and the *ionosphere* is a zone within which charged particles from cosmic radiation and from the solar wind react.

For centuries, scientists have struggled to understand the mysteries of *magnetism,* a force exerted through space by certain substances such as iron. Magnets have both positive and negative poles, and lines of magnetic force extend from pole to pole as a curved *magnetic field* (Figure 1.15a). The Earth itself behaves much like a giant bar magnet, generating lines of magnetic force that extend far out into space. Also like a bar magnet, the Earth has two magnetic poles, a positive pole toward the north and a negative pole toward the south (Figure 1.15b).

The magnetic behavior of the Earth is complex. For one thing, the magnetic poles are not aligned with the geographic North and South poles, which define the axis around which the Earth spins. The magnetic North Pole is located 11.4 degrees from the Earth's axis, or approximately the distance from New York to Minneapolis, Chicago to Tampa, or Denver to San Francisco. Also, the magnetic North Pole "wanders" slowly from place to place, and the strength of the Earth's magnetic field has dwindled about 6 percent in the last 150 years. Furthermore, geophysicists have recently discovered that periodically the polarity of the entire magnetic field reverses: the polarities of the North and South poles change. During the last 71 million years the poles have reversed 171 times. On the average, a reversal takes place about once each half million years. The last reversal took place 700,000 years ago, so we have reason to expect a polarity reversal in the near future.

The Earth's magnetic field is significant in many ways. It is used for navigation, it serves as a shield protecting organisms from harmful components of solar and cosmic radiation, and its fluctuating history is recorded in sediments and rocks. This last point is important for our overall study of plate tectonics. If a geologist finds an igneous rock 100 million years old, for instance, the rock can be used to test whether the Earth's magnetic field was "reversed" or "normal" when the rock formed. Whether the rock has moved from its original north-south orientation can also be determined, and in Chapter 18 we will see how important this knowledge can be in relation to continental drift and plate tectonics.

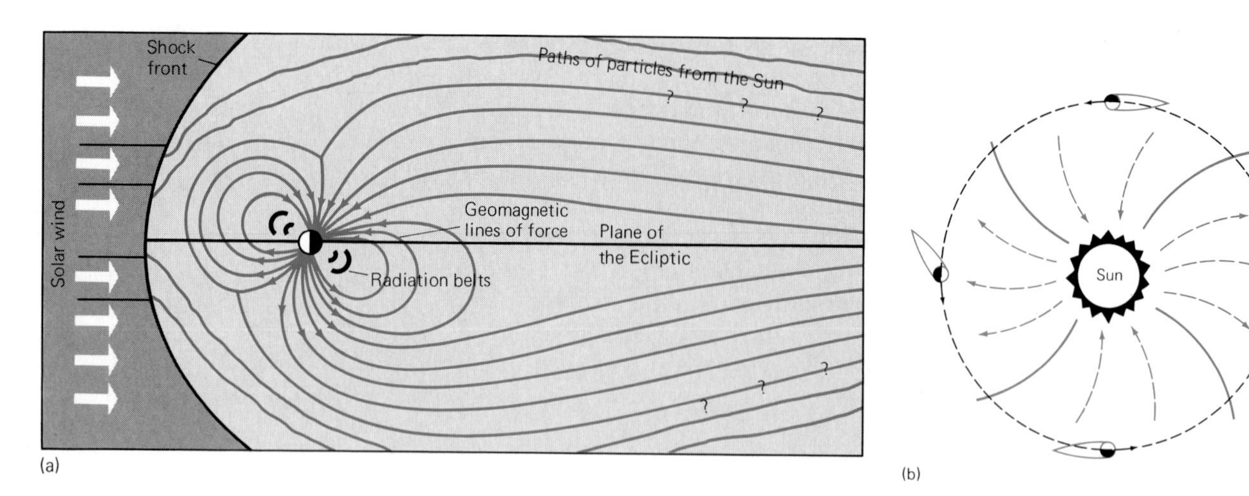


**1.16 Reaction between solar wind and Earth's magnetic field** in profile and in plan views. (a) Section perpendicular to Plane of the Ecliptic and normal to direction of Earth's orbit showing how solar wind (arriving from left) distorts the Earth's magnetosphere. (U.S. Navy) (b) Cometlike aspects of Earth's magnetosphere caused by solar wind plus Earth's orbital speed at left and right and by cosmic radiation and Earth's orbital speed at top and bottom where no solar wind streams away from the Sun. (Schematic)

The ionosphere shifts its position as the Earth rotates from daylight into darkness and also as variable blasts of the solar wind overwhelm the normal background of the particles known as cosmic radiation. The inflowing particles of the cosmic radiation and solar wind react with the magnetosphere; they deform its shape (Figure 1.16) and some charged particles are trapped within the magnetic field to form the Van Allen radiation belts. The electrical situation within the ionosphere may be related to the origin of certain clouds. If so, then variations in the electrical field surrounding the Earth might be able to affect the weather situation within the atmosphere.

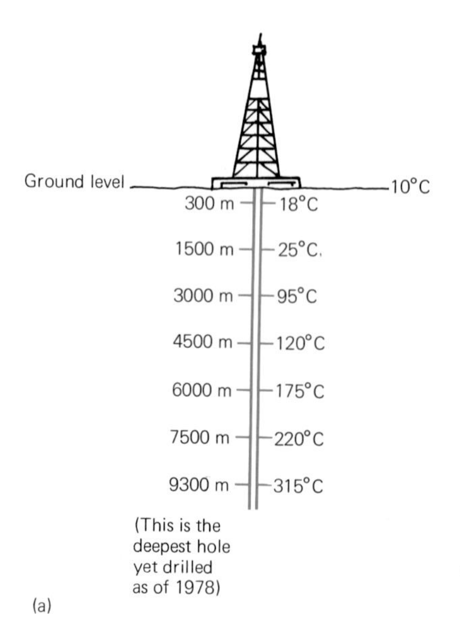


## The Earth's Outward-flowing Heat Energy

We can prove that heat flows upward from inside the Earth simply by looking at red-hot lava being discharged from a volcano or by taking the temperature underground in mine shafts and bore holes. The temperature increases downward, and this downward increase is known as the *geothermal gradient* (Figure 1.17). The average rate of downward increase is estimated to be about 1°C per 30 meters.

During the nineteenth century, geologists supposed that the only possible source of heat to explain the geothermal gradient was the original heat still flowing outward from the time when the Earth was presumed to have been a fiery-hot ball ripped loose from the Sun. All this changed in 1898, when the French physicist Henri Becquerel

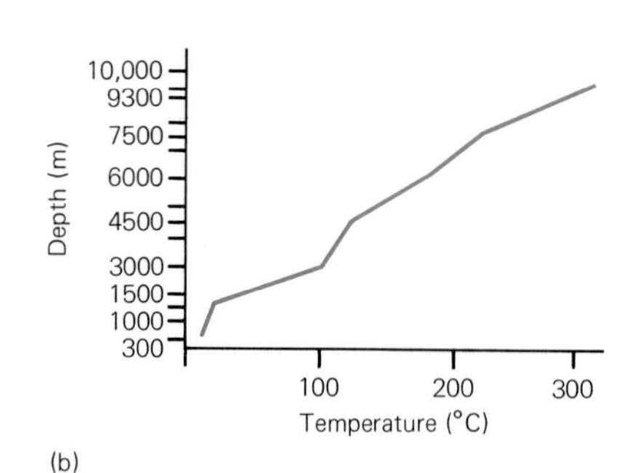


**1.17 Downward increase in temperature within the Earth** shown in two ways. (a) Temperatures encountered at various depths within a hole drilled to explore for natural gas. As of 1978, the world's deepest drillhole was in southwestern Oklahoma. (b) Temperatures from Oklahoma test boring expressed as a graph of depth plotted against temperature.

(1852–1908) discovered the process of radioactivity (see Appendix A) and also showed that radioactivity generates enormous amounts of heat. Shortly after Becquerel's discovery, scientists found that naturally occurring radioactive minerals are present in most rocks. Even though the amounts of natural radioactive materials in rocks are very small, the radioactive heat is very large. Calculations show that ordinary rocks give off about as much radioactive heat as the geothermal gradient implies is flowing upward.

Geophysicists are certain that radioactivity explains the flow of heat energy upward from the Earth's interior toward its surface, but they do not agree on exactly how radioactive heat is distributed.

## The Earth's Atmosphere

The term *atmosphere,* derived from the Greek *atmos,* "vapor," and the Latin *sphaera,* "sphere," refers to the gaseous envelope surrounding the Earth (or any other planet). Four-fifths of the Earth's atmosphere consists of nitrogen. Most of the remaining one-fifth consists of oxygen in the free state. In its combined form, oxygen is the most abundant element, forming 29.3 percent of the mass of the Earth. Of the various other trace gases in the atmosphere, carbon dioxide is of great importance because of its temperature-regulating capacity. Carbon dioxide in the atmosphere has an effect similar to the glass in a greenhouse. Incoming solar energy passes through carbon dioxide, but this gas blocks the passage of heat reradiated from the Earth (Figure 1.18). Hence, the

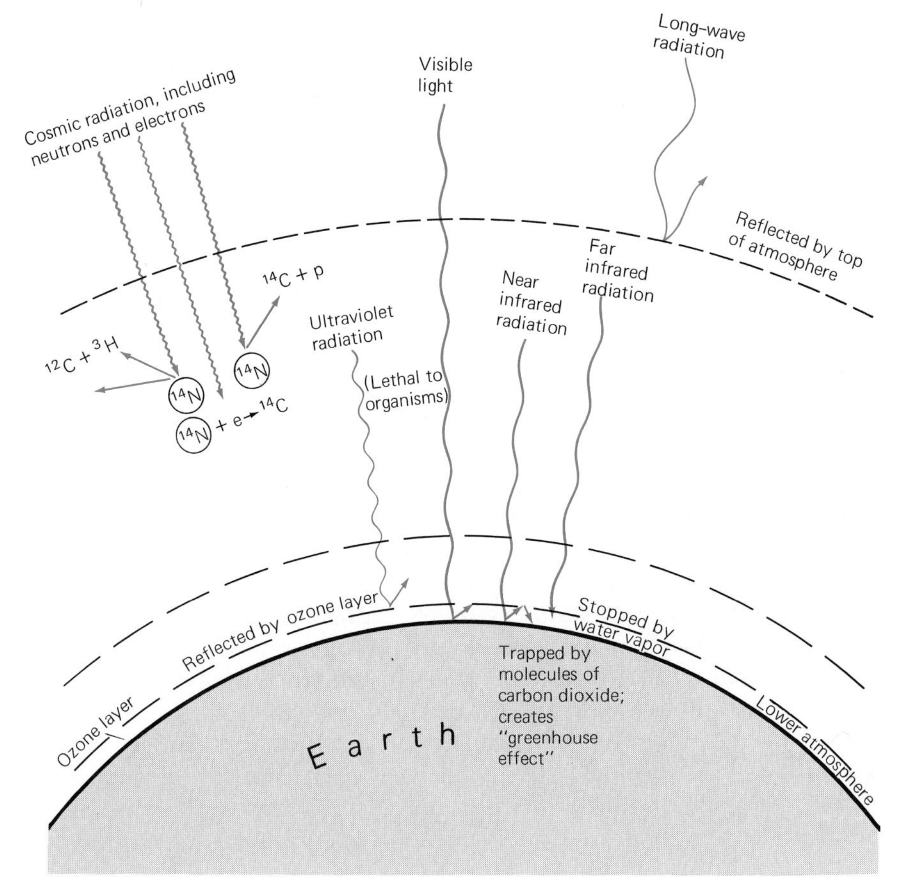


**1.18 Some reactions between solar radiation and the Earth's atmosphere** shown by schematic profile in vertical plane (relationship between curvature of Earth and thickness of atmosphere schematic). Wavelengths of energy waves increase from left to right (but their wavelengths shown here are schematic; further details on electromagnetic spectrum of energy waves are in Appendix D). Notations for products formed by reaction between cosmic radiation and atmospheric nitrogen explained in Appendix A. Radioactive carbon ($^{14}C$) formed in upper atmosphere is basis for calculating ages of many carbon-bearing materials (explained in Chapter 4). Some components of solar radiation convert atmospheric oxygen into ozone; the ozone, in turn, screens out the ultraviolet waves of the arriving solar radiation and keeps these lethal waves from destroying organisms living on Earth's surface.

heat-trapping effect of carbon dioxide has been named the "greenhouse effect." Because of the greenhouse effect, the proportion of carbon dioxide in the atmosphere is crucial to the habitability of the Earth. If the amount of carbon dioxide were to decrease, the Earth would be colder. If the amount of carbon dioxide were to increase greatly, the Earth's surface temperature would increase. Carbon dioxide is also essential because it is the basic material from which plants synthesize starch.

Incoming solar energy reacts with the upper part of the atmosphere to form *ozone,* a gas composed of molecules containing three oxygen atoms instead of the usual two in free oxygen. Ozone is important because it screens out long-wave ultraviolet rays from the Sun. Such rays are lethal to most organisms.

The Earth's atmosphere also contains variable amounts of water vapor. The movement of water from the hydrosphere into the atmosphere and back again is the *hydrologic cycle,* which we explain in detail in Chapter 10. James Hutton clearly realized that the water moving in the hydrologic cycle was one of the great factors in making the geologic cycle happen.

## THE MAJOR GEOLOGIC FEATURES OF THE EARTH'S EXTERIOR

The lithosphere is divided into two main groups of first-order relief features, continental masses and ocean basins. Scattered across parts of both are various volcanic edifices.

### Continental Masses

Continents, the major land areas of the world, constitute about 29.2 percent of the surface of the Earth. Continental masses are the major high-standing parts of the lithosphere, averaging about 800 meters in altitude.

Continents can best be analyzed by organizing them into large natural regions within which the surface morphology and the kinds of rocks form consistent associations. Such areas are known as *physiographic provinces.* Examples include mountains, plains, plateaus, and shields.

**Mountains** A *mountain* is considered to be any landmass that stands 400 meters or more higher than its surroundings.

The great mountain belts of the Earth have been named *cordilleras* (from the Spanish meaning "strings"). Individual parts of a cordillera based on morphology include mountain ranges, mountain systems, and mountain chains.

A *mountain range* is a large, complex, elongate ridge or a series of related ridges that form an essentially continuous and compact morphologic unit. Examples of mountain ranges include the Front Range of the Rockies in central Colorado and the Sierra Nevada in eastern California.

A *mountain system* refers to a group of mountain ranges whose forms, orientations, and geologic structures are closely similar, presumably because they are products of the same general causes that operated within a narrow span of geologic time. Two examples of mountain systems are the Rocky Mountain system, extending across western United States, northward into Canada, and southward into Mexico, and the Appalachian system, extending across eastern United States from Alabama northeastward into Newfoundland, Canada.

A morphologic term for mountain regions that is less exact than mountain range or mountain system is *mountain chain,* which refers to an elongate morphologic unit that includes many ranges or groups of ranges and implies no conditions about similarity of form, equivalence in age, or uniformity of conditions of origin.

**Plains** Features known as plains typically show small variations between the lowest and highest points, or *relief.* Commonly plains are underlain by sediments, and the surface of the plains area has grown upward by additions of new layers of sediment at the top. Plains characterize many coastal areas, large regions in the interiors of continents, and smaller tracts within mountain belts.

**Plateaus** A plateau is a high-standing area within which the rock strata are generally horizontal. Many plateaus have resulted from the cementation of the sediments underlying a former plains area and the subsequent elevation of the region. Some plateaus are underlain by layers of volcanic rocks.

**Shields** Areas referred to as *shields* or *continental shields* are generally low-lying regions whose rocks have formed by the joining together of numerous ancient mountain belts. The continental crust in shield areas is usually thick and rigid so that the shield behaves as a unit and its main movements are vertical–upward or downward.

Large areas of continents are underlain by truly ancient rocks, including the oldest-known rocks, dated at 3.8 billion years. A remarkable feature of continents is that they do not consist of any other kind of materials except those that have been through the geologic cycle. In other words, despite their ancient rocks, continents do not contain any materials left over from any earlier time when the Earth lacked an atmosphere similar to the present one; at such a time, the geologic cycle as we know it may not have been operating.

### Oceans and Ocean Basins

In considering the oceans and ocean basins, it is convenient to discuss the water first and then to describe the features of the basins containing the water. Accordingly, our headings in this section are ocean water and the hydrosphere, midoceanic ridges, and island arcs and trenches.

One of the great scientific achievements of modern time has been the compilation of detailed information about the characteristics of the bottom of the deep-sea basins, and one important discovery was that the sediment covering the ocean floor is not nearly as thick as had been supposed. The deep-sea floor has been found to consist of great bodies of sediment; of enormous world-encircling rocky ridges; of tremendous sets of parallel fracture zones; and of deep trenches. Scattered around in the deep-sea basins are numerous conical features built by volcanic activity; indeed, some are active volcanoes. Associated with the deep-sea trenches are rows of island arcs, including some chains of active volcanoes.

**Ocean water and the hydrosphere** Seawater covers about 71 percent of the Earth, but the aggregate area of the ocean basins totals only 60 percent of the Earth's surface. The mean water depth is four kilometers. Even though this depth is great by comparison with human standards, on a global scale it is trivial. Consider that if the Earth were scaled down to the size of a plastic beach ball, about a meter in diameter, and if this ball were dipped into water and taken out again, the thickness of the film of water adhering to the ball would represent to true scale the depth of the oceans.

The *hydrosphere* is the collective name of the water on and near the surface of the Earth in a liquid or solid (frozen) state. Besides the oceans, the predominant component, other liquid water forms lakes, rivers, and other surface waters, and occupies pore spaces beneath the surface. Immediately below the surface in most localities the underground water is fresh, but, deeper down, at an average of 300 meters or so, the water becomes salty. Solid water, in the form of ice, exists as glaciers, as icebergs, and as masses of pack ice in polar waters.

Of the many ways in which the hydrosphere is essential to the geologic cycle and to human beings, three stand out. First, water acts as transporter of regolith and as shaper of the Earth's surface. As we will discuss in Chapter 12 and elsewhere in the book, water falling as rain erodes mountains, carves river valleys, and transports sediment from one place to another. Without the counteracting mountain-building processes, the actions of water would quickly, in geologic terms, produce a nearly level surface. Second, the oceans act as a collection reservoir for nearly all the substances on Earth, including those dissolved from the rocks and regolith by moving water. Third, water is an essential ingredient for all forms of life.

**Mid-oceanic ridges** Coupled with the surprise of thin ocean sediment was the discovery of the mid-oceanic ridge system. In the nineteenth

**1.19 Great rocky mid-oceanic ridge (the Mid-Atlantic Ridge)** in center of Atlantic Ocean, profile having large vertical exaggeration (50X), as compiled from the strip charts recorded by a moving ship that was sending out sound waves every six seconds and recording and plotting the time required for the sound to reach the bottom, to be reflected off the bottom, and to return to the surface. This sound-down-and-echo-return time is a direct function of the water depth. (Bruce C. Heezen, Marie Tharp, and Maurice Ewing, 1959, The floors of the oceans. I. The North Atlantic: Geological Society of America, Special Paper 65, 122 p., 30pl., Plate 22, profile II)

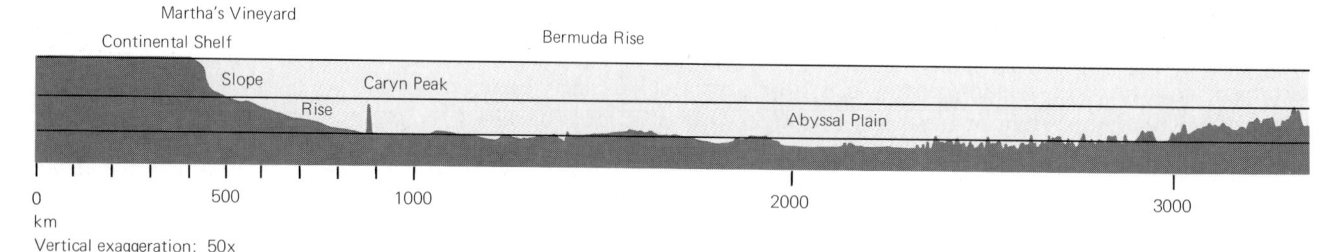

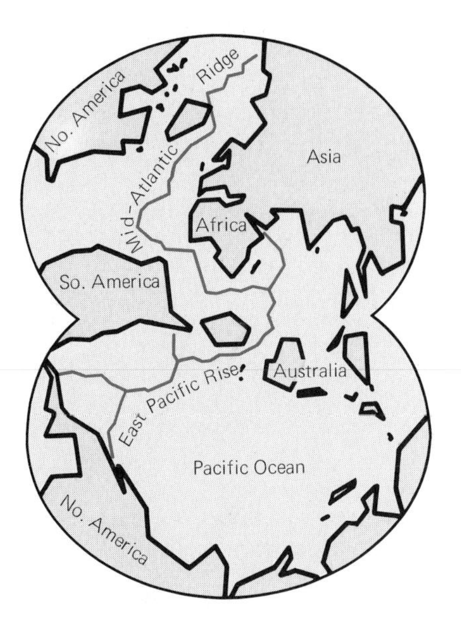


**1.20 Worldwide extent of continuous mid-oceanic ridge system** shown by special map projection centered on the South Pole and showing the Atlantic and Arctic Oceans in one hemisphere at top and Pacific Ocean in other hemisphere at bottom. Extensions of ridge system onto continents not shown. (After J. T. Wilson)

century, oceanographic ships had discovered unusual "mountain chains" in the mid-Atlantic Ocean, but the flurry of postwar bottom-sounding revealed these features to be far more extensive and regular than had been supposed (Figure 1.19). The mid-Atlantic Ridge proved to be part of an enormous, interconnected system that extends for some 60,000 kilometers throughout the oceans of the world (Figure 1.20). The mid-Atlantic Ridge actually protrudes above the water line (about 4 km from the floor of the ocean) near Iceland, the Azores, and in the South Atlantic, but in most places along its route, the rocky crest of the mid-Atlantic Ridge remains about 300 meters or more below the surface of the water.

Great linear systems of breaks in the Earth's crust are *fracture zones* (Figure 1.21). Conspicuous fractures with nearly east-west orientation are located in the eastern Pacific, tropical Atlantic, and central Indian oceans. Other fracture zones, diversely oriented, occur in other parts of the ocean basins and on land.

**Island arcs and trenches** Detailed mapping of the ocean floor revealed another puzzling feature: deep, regular trenches border certain continents and chains of islands (Figure 1.22). Why, oceanographers asked themselves, should these oceanic "deeps" be located at ocean margins? One would expect that material eroded from adjacent land areas would fill any marginal trenches with sediment. Although the trenches do contain sediment, more in some trenches than in others, it is clear that the trenches could not be as old as the continents and yet not be filled with sediment. Therefore, the trenches, and the ocean floor of which they are a part, must be much younger than the continents. We will examine other evidence for the relatively young age of the ocean floor in Chapter 18, where we discuss sea-floor spreading and the notion that the rocky crust beneath the sea floor is being generated by the rise of molten material at mid-oceanic ridges and the solidification of this material to make igneous rocks, and its lateral movement away from such ridges toward its presumed destruction beneath trenches.

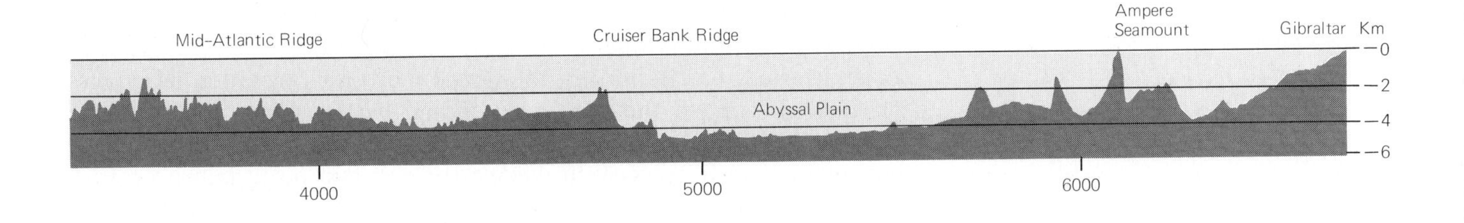

**1.21 Prominent linear fracture** cuts through rocks of northwestern Canada near Great Slave Lake. (Royal Canadian Air Force)

### Volcanic Edifices

When molten material from the Earth's interior comes out on the Earth's surface it forms various volcanic edifices ranging from cones to broad sheets. These may form on land or on the sea floor. Volcanic cones formed on continents rarely exceed a few tens of kilometers in basal diameter and up to 2 or 3 kilometers in height. Much larger cones have been built on the sea floor; their diameters range up to 400 kilometers and their heights to about 10 kilometers.

Both the cones and the great sheets of volcanic material are prone to collapse because of removal of material from below. The result is great circular or oval depressions, usually with steep marginal walls and flat floors.

The vast, rocky ridges found in ocean basins are products of long-persistent volcanic activity. These form irregular cones and sheets that are being pulled apart in the middle and building anew.

**1.22 The Puerto Rico Trench,** vertical profile from Puerto Rico on the S well out into the Atlantic Basin on the N, passes through the deepest part of the Atlantic Ocean. Instead of being within the middle of the basin as one might expect, the deepest part is near the island of Puerto Rico, as shown here. Vertical exaggeration is 100X. (Bruce C. Heezen, Marie Tharp, and Maurice Ewing, 1959, Reference cited in caption to Figure 1.19, Plate 24, Profile W32)

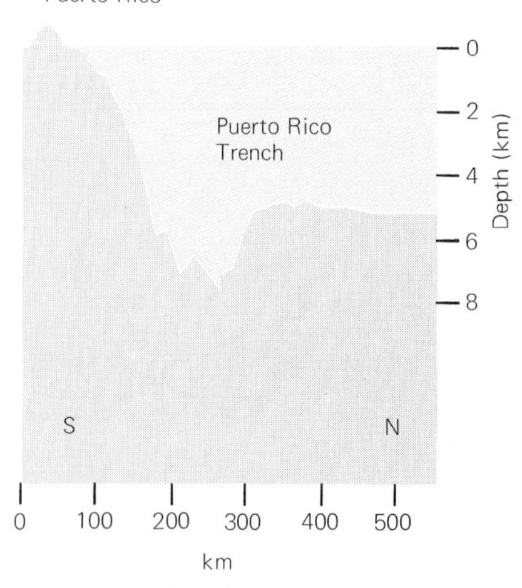


## LITHOSPHERE PLATES AND PLATE TECTONICS

Much of the evidence gathered in the study of magnetism, oceanography, and seismology has been united recently to formulate a new geological theory. The theory, called *plate tectonics,* touches nearly every aspect of geology and offers answers to many old and new mysteries about the Earth; it will be discussed throughout this book. The name, plate tectonics, has its root in the Greek word for carpenter, *tekton.* In the geologic context the root word should be translated as "builder"; "tectonic" may describe features that have been built as a result of plate movements, such as tectonic valleys or tectonic mountains. At this point, we shall briefly summarize how plates have been defined, mention how plate-tectonics ideas have been related to ideas about continental drift, and point out a few unanswered questions.

### Evidence For Mapping Plate Boundaries

The study of earthquakes, *seismology,* has expanded almost as rapidly as oceanography. With the aid of newly accurate space-age instruments and techniques, seismologists have been mapping the sites of earthquakes since the 1950s, noting them as dots on maps of the world (Figure 1.23). As they added more and more dots, an interesting pattern emerged: almost all earthquakes were occurring either along oceanic ridges or beneath oceanic trenches. As the pattern became clear in the mid-1960s, the idea of lithosphere plates was proposed. If earthquakes were distributed in such an obvious pattern, what underlying mechanism could explain the systematic seismic activity? The map shown in Figure 1.24 presents a reasonable possibility. Perhaps the areas of seismic activity defined the boundaries of separate sections, or moving plates, of the Earth's crust that are continuously straining against one another. (Compare Figures 1.13 and 1.14, and also compare both of these maps with Figure 5.4 on page 115, which shows the worldwide location of volcanoes.) Three kinds of plate boundaries have been recognized: (1) *divergent,* or spreading, (2) *convergent,* or coming together, and (3) *transcurrent,* or slipping past each other. Most geologists now think that energy released by sudden movements along the boundaries of these plates creates waves that we sometimes feel as earthquakes. However, not *all* earthquakes start at plate boundaries.

### Sea-Floor Spreading and Continental Drift

When the theory of plate tectonics first began to take form in the early 1960s, it was called *sea-floor spreading,* a term that describes the movement of the plates away from a spreading center. According to the theory, igneous rock arises continuously along the mid-oceanic ridges. The rock forms new sea floor, which spreads away from both sides of the ridge at rates of a centimeter or so a year. Eventually, hundreds or thousands of kilometers away and millions of years later, the sea floor is thought to plunge downward beneath the oceanic trenches (Figure 1.25). A single plate, then, is bounded by ridges or trenches. Where two or more plates adjoin, one plate may be forced under or past the other.

According to the theory, a plate may carry a load of continental material, much as a river in spring flood carries blocks of ice. The continental material, mostly granite, is lighter than the plate material, mostly basalt, so that the continents have remained "afloat" for hundreds of millions of years. The term *continental drift* is used to describe this alleged movement of large land areas.

### Unanswered Questions

The biggest remaining mystery in plate tectonics is what makes the plates move. What force or forces inside the Earth are so powerful and persistent that they can maintain plate movement? Many geologists

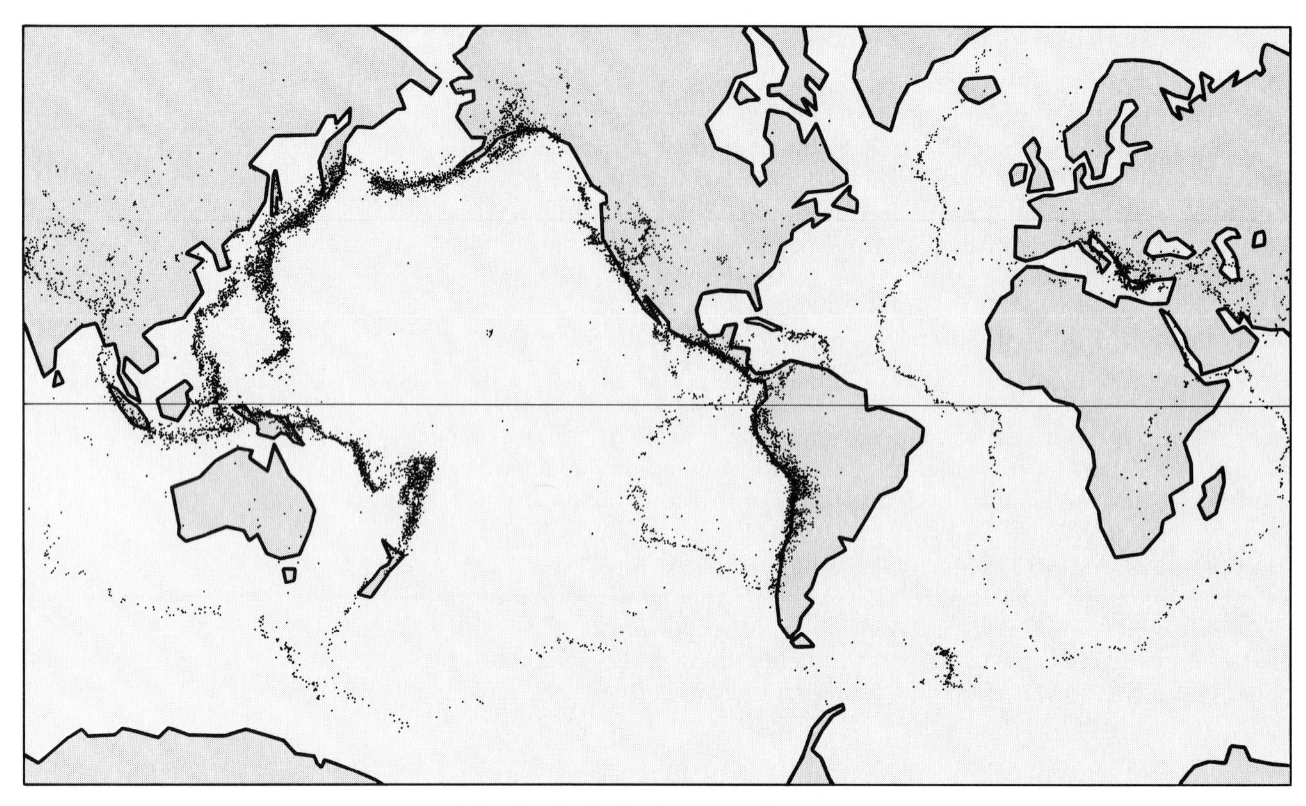

**1.23 World distribution of earthquakes** shown by a series of dots, one dot marking the point on the surface of the Earth directly above the place where the first trembling of each earthquake began. Such a point is defined as the earthquake epicenter (explained in Chapter 17). (U. S. Geological Survey)

**1.24 Lithosphere plates and plate boundaries,** with plate boundaries drawn to coincide with zones where earthquake epicenters are clustered closely together. Fifteen plates are shown here; more have been recognized by some geologists. (Warren B. Hamilton, 1979, Tectonics of the Indonesian region: U. S. Geological Survey, Professional Paper 1078, 345 p.; Figure 2, p. 8)

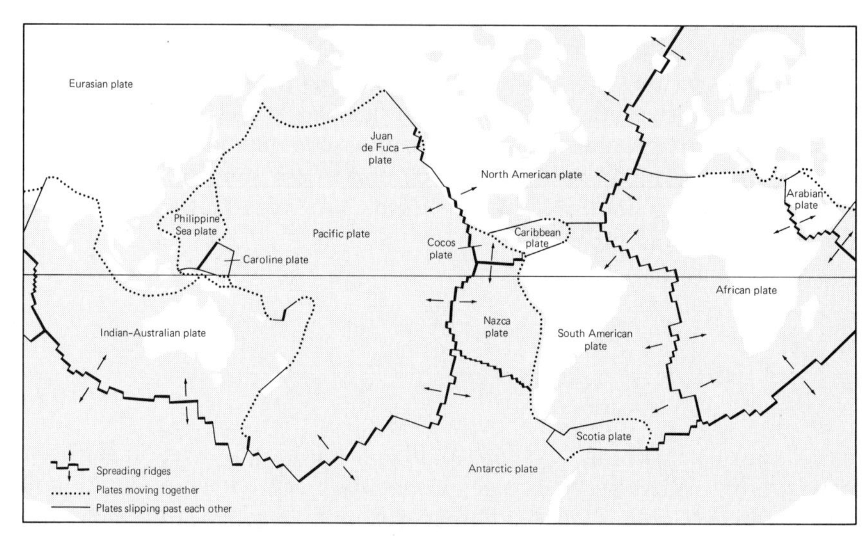

think that a layer of rock within the upper mantle is somewhat plastic and deformable. Heat from the Earth's interior, they say, can cause mantle rock to move, forcing it to rise toward the surface in certain regions, much as air rises when heated. The details of such motion are far from clear, however, and serious objections have been raised against the entire scheme. Only a few geologists believe that the Earth is expanding, pushing the continents outward. Many other suggestions abound and, as might be expected, the theory of plate tectonics has raised almost as many questions as it has answered. We will be discussing some of these questions throughout the book, particularly in Chapter 18.

## THE BIOSPHERE

The *biosphere* is defined as that part of the Earth in which life exists. Because most organisms require water and sunlight, it is limited to regions providing ample quantities of available water and energy from the Sun. The maximum range of the biosphere is about 20 kilometers thick with organisms variously distributed from the depths of the oceans to the highest mountain peaks (Figure 1.26).

Until recently, it may have seemed foolish to worry about harming the biosphere, with its vast forests, endless schools of fish, and, of course, human beings themselves. Yet, we see that forests, fish, and humans alike owe their existence to an exceedingly thin and fragile film of plant life spread unevenly over the Earth's surface. The energy source upon which this film of life depends is the Sun. Solar energy enters the biological cycle by way of green plants, which need the Sun's energy to manufacture starch through the process of photosynthesis. Animals, which are unable to manufacture food, depend upon consumption of plants or other animals for survival.

In *photosynthesis* the energy of sunlight powers the chemical reactions between carbon dioxide and water to produce oxygen and plant tissue, usually starch. Once again, nature provides a cycle of activity, this time utilizing the Sun's energy to help plants produce the food and energy

**1.25 Concept of sea-floor spreading** illustrated by schematic profile and section from Mid-Pacific Rise to continent of South America. Material rising from the asthenosphere at Mid-Pacific Rise creates new, comparatively thin lithosphere constituting an oceanic plate. Beneath the Peru-Chile Trench, off the west coast of South America, the ocean lithosphere encounters the much-thicker continental plate underlying South America. Boundary between oceanic lithosphere and continental lithosphere dips beneath continent at an angle of 45° from horizontal.

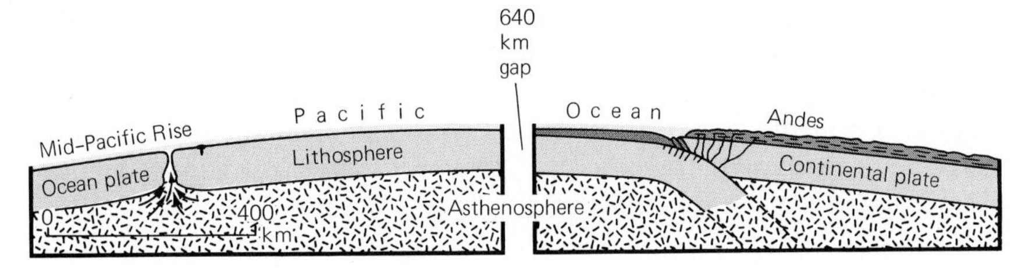

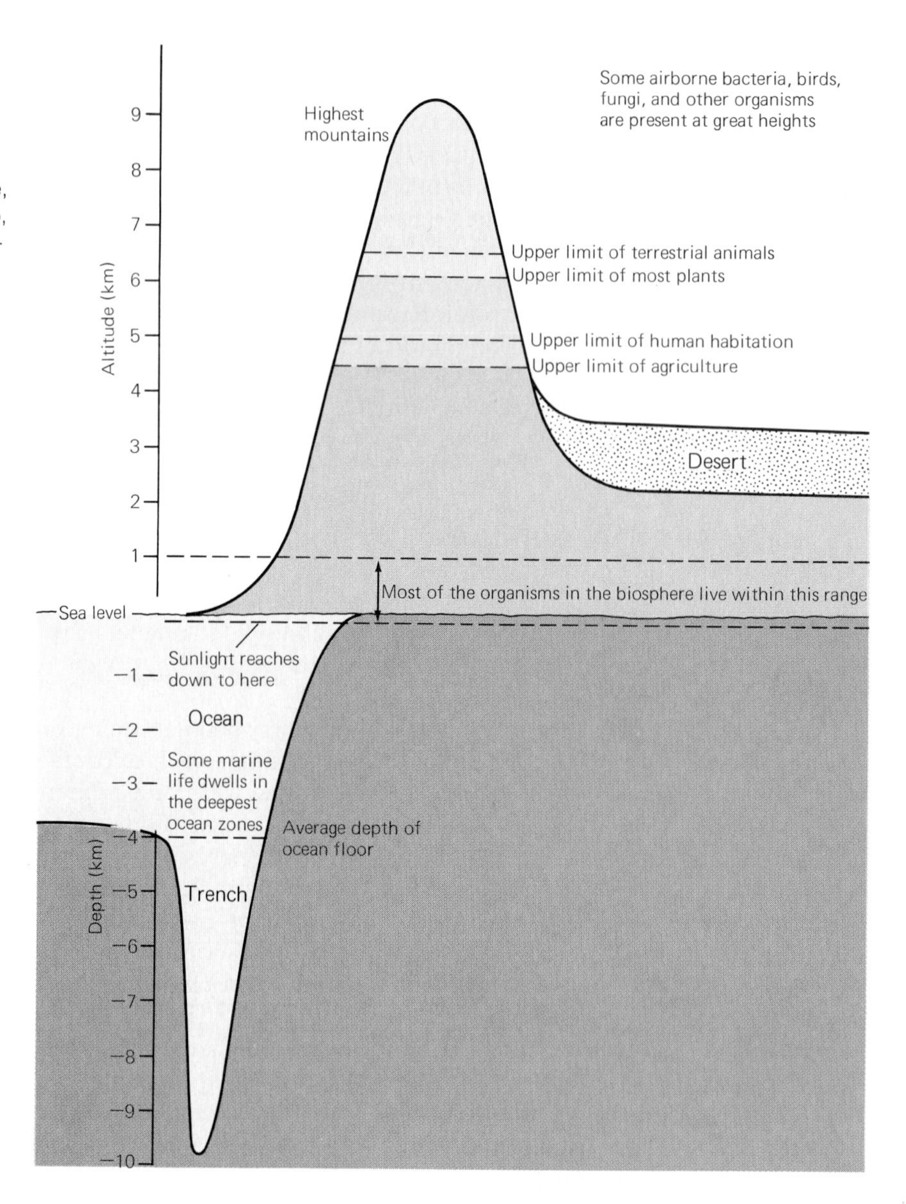


**1.26 The vertical span of the biosphere** shown by a schematic profile extending from the bottom of a deep-sea trench to the top of the highest mountains on Earth. (Suggested by a figure in G. Evelyn Hutchinson, 1970. The biosphere, p. 1–11 *in* The biosphere: San Francisco, W. H. Freeman and Co., A Scientific American Book, 134 p.)

they need to live. In the bargain, while they are transforming sunlight, plants "exhale" oxygen, which animals breathe; to pay our debt, we give off carbon dioxide as a waste product, the same carbon dioxide that plants need to start the cycle all over again. Of course, besides this simplified version of the exchange of gases, animals also benefit indirectly from the converted solar energy in the plants when they eat the plants.

It may appear peculiar to mention the biosphere in a discussion of physical geology. However, cycles of birth, death, and rebirth in the biosphere are sustained by energy from the Sun, and by grand-scale cycles of chemical elements. These cycles, which involve the lithosphere, hydrosphere, and atmosphere, are all affected by the activities of human beings — members of the biosphere. Human beings, in turn, depend upon these same cycles for their existence. Therefore, human activities, the biosphere, and the geologic cycle are delicately interlinked. We examine this linkage in Chapter 2.

# CHAPTER REVIEW

1. The Sun and the bodies held within its gravitational field, chiefly nine *planets* in elliptical orbits and various smaller satellites and other debris, some of which fall to Earth as *meteorites,* constitute the *Solar System.* The Sun lies near, but the center of the Sun does not coincide with, the center of this system.

2. Nuclear reactions inside the Sun convert the Sun's mass, consisting largely of hydrogen and helium, into energy, which radiates outward, its intensity decreasing with the square of distance from the Sun. During the 4.7 billion years since the Sun is thought to have been creating such energy, the Sun's mass has decreased by an amount equal to the mass of about 109 Earths, but which is only 0.03 percent of the Sun's mass.

3. Movement of the Sun with respect to the Solar-System barycenter is controlled chiefly by the two largest planets, Jupiter and Saturn. Jupiter's orbital period of about 12 years governs the speed of the center of the Sun around the solar orbital ellipse, whereas Saturn causes the long axis of this ellipse to spin around in a period of just under 30 years. The distance between the Sun and the Solar-System barycenter seems to regulate the *solar cycle* of sunspots and magnetic-field polarity reversals.

4. *Meteorites,* solid objects from outer space that strike the Earth's surface, are metallic (iron-nickel intergrowths), nonmetallic, or mixtures of metallic and nonmetallic materials. Nonmetallic meteorites predominate among *falls* (meteorites seen striking the Earth), whereas metallic varieties are most numerous among *finds* (meteorites discovered after their times of impact).

5. The Earth and its satellite the Moon orbit the Earth-Moon barycenter every 29.5 days. The Earth-Moon barycenter lies inside the Earth within a depth zone 611 kilometers thick, its mean depth value being 1707 kilometers. The orbital period of the Earth-Moon pair around the Solar-System barycenter is 365.2564 days.

6. Because of its rapid rotation on its polar axis, the Earth's shape is not exactly a sphere, but rather is an oblate spheroid with equatorial diameter of 12,756 kilometers and polar diameter of 12,714 kilometers. The mean density of the Earth is 5.517 grams per cubic centimeter. Because the densities of surface rocks are mostly less than 3.0 grams per cubic centimeter, the density of the Earth's deep interior must exceed 5.517 grams per cubic centimeter.

7. The tilt of the Earth's polar axis and the essentially constant orientation of the axis on the time scale of a few years cause the *subsolar point* (where the Sun's rays are perpendicular to the Earth's surface) to migrate through a range of latitudes equal to twice the tilt angle, and hence, are responsible for seasonal changes in temperature.

8. The Earth consists of a two-part *core,* a *mantle* of solid rock, and an outermost thin, rocky *crust.* The crust and the outer part of the mantle compose the *lithosphere,* which includes all the rocky material of the Earth's outer shell, extending from the surface to a depth of about 100 kilometers. The lithosphere is thought to consist of a mosaic of about a dozen giant "plates" that are in constant, slow motion over the underlying weak zone, the *asthenosphere,* moving along a zone named the *tectosphere,* which seems to lie at the top of the asthenosphere beneath oceans but at the base of the asthenosphere beneath continents.

9. Two important zones related to and surrounding the Earth are the *magnetosphere,* the zone of influence of the Earth's magnetic field, and the *ionosphere,* the zone within which charged particles from cosmic radiation and from the solar wind react. The polarities of the North and South magnetic poles have reversed numerous times in the past.

10. The *geothermal gradient,* the increase of temperature downward within the Earth, proves that heat is flowing upward. The source of this heat is thought to be radioactivity.

11. The Earth's *atmosphere* consists mostly of nitrogen and oxygen. Two important minor constituents are carbon dioxide and water vapor.

12. The lithosphere is divided into two groups of first-order relief features, *continental masses* and *ocean basins.* The water of the oceans covers about 71 percent of the Earth's surface; the average depth in the deep-ocean basins is 4 kilometers. Seawater forms the bulk of the *hydrosphere,* the name for all the water on and near the Earth's surface.

13. Important parts of ocean basins are *mid-oceanic ridges, deep-sea trenches,* and great *fracture zones.*

14. Conical volcanic edifices are found both on continents and in ocean basins; the largest volcanic cones are found in the ocean basins. Other volcanic features, such as continental plateaus or sea-floor sheets, are formed of tabular masses of igneous rock.

15. The theory of *plate tectonics* touches nearly every aspect of geology and offers answers to many geologic mysteries. The theory claims that the Earth's crust is constructed of moving plates. *Continental drift* describes the movement of large landmasses contained within these plates. *Sea-floor spreading* describes the movement of the plates away from a spreading center at a mid-oceanic ridge. One of the many unanswered questions about the theory is what makes the plates move.

16. The *biosphere* is defined as that part of the Earth in which life exists. Human activities, the biosphere, and the geologic cycle are delicately interlinked. Solar energy enters the biological cycle by way of green plants. Through the process of photosynthesis, they convert the energy into plant cells, the first level of the food chain through which the energy is distributed to all animal life.

## QUESTIONS

1. Is the center of the Sun the "pinwheel point" of the orbiting planets of the Solar System? Explain your answer.
2. What is the relationship between the Sun's mass and solar energy? What is the *solar wind?*
3. What is the *solar cycle?* What is thought to be the timing mechanism of this cycle?
4. Describe the two contrasting compositional groups of *meteorites.* What is thought to be the significance of the compositions and ages of meteorites?
5. Define the *Earth-Moon barycenter* and relate the location of this to the Earth and to the orbital path of the center of the Earth around the Solar-System barycenter.
6. Describe the shape of the Earth. What is the Earth's mean density? The density of surface rocks? The density of the Earth's deep interior?
7. At what level are the Earth's large-scale dynamic movements thought to occur? Why?
8. Describe the major features of the Earth's magnetic field and explain some of its variations.
9. What is the *geothermal gradient?* Explain what would happen if the Earth's geothermal gradient became zero.
10. What gas is responsible for the greenhouse effect? Explain how this effect operates.
11. What are the two main groups of relief features of the Earth?
12. True or false: The oldest rocks on Earth are found in the deepest parts of the oceans. Explain your answer.
13. What are the three kinds of plate boundaries?
14. How is the concept of sea-floor spreading related to continental drift?
15. Describe the relationship(s) between the *biosphere* and the geologic cycle.
16. What is the process by which plants use solar energy, carbon dioxide, and water?

## RECOMMENDED READING

Carrigan, C. R., and Gubbins, David, 1979, The Source of the Earth's Magnetic Field: *Scientific American,* v. 240, no. 2, p. 118–130.

Engel, A. E. J., 1969, Time and the Earth: *American Scientist,* v. 57, no. 4, p. 458–483.

Gamow, George, 1963, *A Planet Called Earth:* New York, The Viking Press, 257 p.

Jordan, T. H., 1979, The Deep Structure of the Continents: *Scientific American,* v. 240, no. 1, p. 92–107.

Markson, Ralph, and Muir, Michael, 1980, Solar Wind (*sic*) Control of the Earth's Electrical Field: *Science,* v. 208, p. 979–990.

Pollack, H. N., and Chapman, D. S., 1977, Flow of Heat from the Earth's Interior: *Scientific American,* v. 237, no. 2, p. 655–657.

Toon, O. B., and Pollack, J. B., 1980, Atmospheric Aerosols and Climate: *American Scientist,* v. 68, no. 3, p. 268–278.

# CHAPTER 2 GEOLOGY AND SOCIETY

NATURAL CATASTROPHES
NATURAL RESOURCES
THE ARITHMETIC OF EXPONENTIAL EXPANSION
PRESENT-DAY DILEMMAS INVOLVING NATURAL RESOURCES
A CHALLENGE TO US ALL

PLATE 2
Sequential views of Mount St. Helens, Washington, on Sunday, 18 May 1980, before and during the mighty eruption, as seen from a safe distance of 24 kilometers. Although the initial blast sent a dark debris-laden cloud upward, in a few moments the blast blew downward as well and a gigantic avalanche removed the north flank of the peak. (Copyright 1980 by Vern Hodgson and the Everett (Washington) Herald/Wide World)

| Date | Assumed density per square kilometer | Total population (millions) |
|---|---|---|
| 600,000 B.C. | .00425 | .125 |
| 200,000 B.C. | .012 | 1 |
| 20,000 B.C. | .04 | 3.34 |
| 8,000 B.C. | .04 | 5.32 |
| 4,000 B.C. | 1.0 | 86.5 |
| 1 A.D. | 1.0 | 133 |
| 1650 | 3.7 | 545 |
| 1750 | 4.9 | 728 |
| 1800 | 6.2 | 906 |
| 1900 | 11.0 | 1610 |
| 1950 | 16.4 | 2400 |
| 2000 | 46.0 | 6270 |

**2.1 The increase of the Earth's human population during the last twenty centuries.** The present cycle of growth coincides with the various steps in the fossil-fuel revolution, which began with the widespread burning of coal in the eighteenth century and surged forward starting at the beginning of the twentieth century with the discovery of large quantities of oil underlying the Gulf coastal plain.

AS A NATURAL SETTING FOR LIFE, THE WORLD SEEMS ALMOST BEYOND improvement. In most localities, air, water, food, and shelter are available to support a considerable population of humans (Figure 2.1).

People have always supposed that the operations of some parts of the geologic cycle would work in their favor. They have looked forward to finding fertile soils for growing food and trees. They have counted on rivers for washing away their dirt. And they have expected the wind to blow away the smoke from their fires.

Human history has been marked by ever-greater uses of materials derived from the Earth, materials that provide energy for doing things and the substances with which to do them. We have reached the point in our usage of Earth materials and sources of energy that modern industrial civilization cannot continue without supplies of resources from the Earth.

The power of nations has always been closely linked to the abundance of natural resources. Leading nations traditionally possess good supplies of metals and fuels, and the technological insight to put them to use. At the battle of Marathon in 490 B.C. the Persians, many of whom carried only leather shields and stone weapons, were soundly beaten by the Greeks, who had learned to make metal swords and shields. Throughout the nineteenth century the tiny island of Great Britain ruled much of the world largely because of her vast metal deposits; the rock and regolith of Britain held greater mineral wealth per acre than any other nation advanced enough to utilize it. Great Britain was the largest producer of iron, coal, lead, copper, and tin, and her scientists and builders knew how to fashion these riches into machines, mills, railroads, ships, and cities. Today, likewise, the two most powerful nations in the world — the United States and the USSR — are also richly endowed with natural resources.

Unfortunately, the Earth is not always peaceful and serene; various kinds of natural disasters have wiped out thousands of people in a few days, or even within a few minutes. Moreover, the extraction and consumption of natural materials have reached levels that have introduced significant environmental problems. Industrial impacts on the atmosphere, the hydrosphere, the lithosphere, and the biosphere pose serious threats for the future. What is more, doubts have been raised about the future supplies of resources now being used in such huge amounts.

In this chapter, we emphasize the importance of geology to modern society by reviewing some natural hazards and by examining the usages and problems associated with natural resources, notably water, natural energy supplies, and deposits of metals and nonmetals. We do this now with the intention of showing how important knowledge of geology is both for its connection to everyday problems facing us all and for its significance in dealing with the future.

## NATURAL CATASTROPHES

Many kinds of natural catastrophes can be cited. Drought wipes out water supplies and the vegetation withers away. As plants die, the food supply for animals vanishes. The animals are thus forced to migrate or they perish. Storms, with high winds, excessive rain, snow, sleet, or hail can create all kinds of problems for people. Rivers rise in flood,

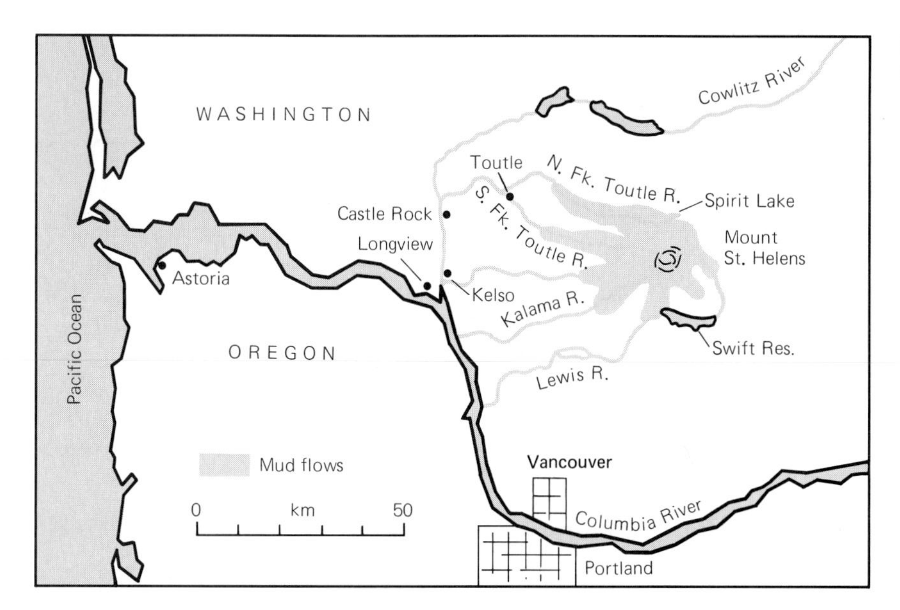


**2.2 Location map of Mount St. Helens and vicinity.**

people freeze to death in blizzards, water-soaked slopes crash down into lowlands. Earthquakes not only cause the otherwise-solid ground to move back and forth and up and down, but can also shake loose the materials underlying steep slopes. The fallen material can block valleys to form temporary lakes that can disappear suddenly in a catastrophic outburst flood when the earthen dam is breached. Earthquakes on the sea floor can generate great waves whose effects on some coasts can be devastating. Volcanic eruptions can spread molten liquid that destroys everything in its path. Volcanoes can also explode violently, discharging hot, lethal gases and mixtures of gases and solids that are as destructive as the hot liquids.

Although Hawaii and Alaska are widely known for their active volcanoes, the other 48 of the United States (conterminous United States of the usage of the U.S. Geological Survey) have been equally famed for their general lack of modern active volcanoes. This modern-day volcanic quiet was shattered by Mount St. Helens, in southwestern Washington state (Figure 2.2; for location see Figure 5.29), last active in 1857. The new cycle of activity commenced on 27 March 1980 and featured a mighty explosion at 0832 on Sunday 18 May 1980. The renewal of this volcano's activity began after a week-long series of moderate earthquakes. In fact, a sharp quake was recorded just 7 minutes before the volcano blew away one cubic kilometer from the northern part of its cone, lowering its peak by about 400 meters. The noise of the blast of hot gas and dust was heard 320 kilometers away to the north. Smaller explosions followed on 25 May and 12–13 June. On 15 June, a lava dome was forming.

As we present in Chapter 5, volcanic activity is varied and its destructive potential spans a wide range. The various damaging aspects of the Mount St. Helens explosion of 18 May 1980 are summarized in Table 2.1. So far in this renewed cycle of activity, Mount St. Helens has displayed for the first time ever since scientific study of volcanoes began the lethal effects of a *base surge* of hot gases and solid particles transported by the gases. Base surges began to be studied intensively in connection with

**TABLE 2.1**
SOME EFFECTS OF EXPLOSIVE ERUPTION OF MOUNT ST. HELENS VOLCANO, WASHINGTON, ON 18 MAY 1980

| Item | Estimated casualties or losses |
|---|---|
| Loss of life | 31 persons dead, 38 missing; victims included geologist David Johnston (1949–1980) and photographer Reid Blackburn (1953–1980), who was found in his car that had been covered to the windows by hot dust<br>5250 elk<br>6000 deer<br>200 bears<br>100 mountain goats<br>15 cougars<br>Thousands of smaller animals<br>70 million fish |
| Destruction of trees and timber | 44,000 acres of trees leveled; $10^6$ board feet of lumber washed away or destroyed; timber losses $500 million |
| Effects of mudflows | |
| (a) Dam built across the outlet of Spirit Lake; water level raised by 60 m | Posed threat of catastrophic flood if the new dam were to be breached suddenly |
| (b) Sediment by Toutle and Cowlitz rivers trapped 31 ships in harbors of Portland, Oregon, and Vancouver, Washington | A channel formerly 182 m wide and 12 m deep was reduced to 61 m wide and 4 m deep; restoration of channel by dredging will require 5 months and cost $15 million |
| Effects of airborne volcanic dust | |
| (a) 9600 km of highways in Washington State disabled | |
| (b) Airports closed; training base at Moses Lake, Wash., relocated | |
| (c) Crops and fruit trees covered by volcanic dust | Losses placed at $1 billion |
| Effects of heat | |
| (a) Melted ice; all glaciers gone from mountain peak | |
| (b) Water temperature in Toutle River raised from 10°C. to 32°C | Fish tried jumping out of the water to avoid exposure to hot water (see above) |

nuclear explosives detonated on Pacific Islands after World War II. The base surge in these explosions was formed of outward-moving mixture of water and gases that spread away from the base of the mushroom cloud blown upward by the blast.

The romantic attraction of having an active volcano nearby in the Pacific northwest soon gave way to the realization of the danger and difficulties associated with the explosive events. The clouds of volcanic dust darkened the sky at noon in Ritzville, Washington, 240 kilometers from the peak, and left a blanket of gray, powdery dust 10 centimeters thick. Internal-combustion engines ground to a halt as air filters were clogged. Aircraft jet engines were immobilized by the dust. Part of the dust cloud reached the stratosphere, more than 16 kilometers above the surface of the Earth. Dust from the lower-level clouds reached Montana; the high-altitude cloud circled the Earth and by 2 June 1980, had returned to western United States after crossing the Pacific Ocean. Only spotty fallout from the high-level dust cloud has been reported.

**2.3 Ruins of Galveston, Texas,** the aftermath of the hurricane of Saturday, 8 September 1900. (Rosenberg Library, Galveston; published in J. E. Weems, 1968, When the hurricane struck: American Heritage, v. 19, no. 6, p. 36–39, 74–77, photo on p. 36–37)

Mount St. Helens' behavior has proved to be difficult for geologists to predict. Its past history suggests that the 1980 activity will be only the beginning of a cycle that may last from 15 to 20 years and will undoubtedly include explosions whose violence will equal or exceed that of 18 May 1980. What is more, earthquakes at Mount Hood, Oregon, in July 1980 may mean other volcanoes in the Cascades are preparing to erupt, also explosively.

The most lives lost (6000 to 8000) in a natural disaster experienced within the United States was during the hurricane and associated flood that destroyed Galveston, then the fourth-largest city in Texas (population 27,000) on 8 September 1900 (Figure 2.3). During a 12-hour period, high winds (peak gusts estimated at 190 km/hr), torrential rainfall (an estimated 25 cm), and a wind-driven storm-surge flood poured raging water into the center of the city at levels about 4.5 meters higher than normal on top of which were 1.2-meter waves. Similar, but less-intense conditions have accompanied many other hurricanes that have ravaged other parts of the Gulf coast, the coast of Florida, and the east coast of the United States as far north as New England.

Coastal areas have become the focus of much study and concern as the demand to be at the water's edge has increased and more and more buildings have been constructed in exposed locations. The year 1980 has been designated as the Year of the Coast as a means of highlighting the conflicts and problems centered around the dilemma of how best to manage the nation's shorelines.

## NATURAL RESOURCES

In this section, we review people's dependence on water, energy, and materials from the Earth. Here, we emphasize problems of supplies and

**TABLE 2.2**
HOUSEHOLD USES OF WATER

| Daily Activity | Quantity of Water Required In liters | In gallons |
|---|---|---|
| Drinking water | 2 | 0.65 |
| Flush toilet | 11 | 3 |
| Dripping faucet (1 drop per second) | 15 | 4 |
| Wash dishes | 40 | 10 |
| Load of laundry | 75 to 115 | 20 to 30 |
| Take shower | 75 to 115 | 20 to 30 |

usage. In appropriate chapters on processes we take up the geologic aspects of the origin of these resources.

## Water

Water governs the activities of all living things. If a water supply is depleted, people and animals must find a new supply, move elsewhere, or perish. In this section, we include usage of water, sources of water, and water problems. In a following section we discuss the energy of river water.

**Usage of water** We can divide the use of water into three categories: household, industrial, and agricultural.

In planning for the water supply of residential areas in the United States, water engineers estimate that 400 to 600 liters (100 to 150 gallons) of water per person per day must be available. The most obvious use of fresh water is for drinking. Other household uses are shown in Table 2.2.

All industrial uses in the United States require 950 billion liters (950 million metric tons or 250 billion gallons) of water per day. By far the largest proportion of this, 90 to 95 percent, is used for cooling. Other

**2.4 Quantities of water required to produce various goods and services.** Notice that it takes about 5000 times as much water to grow a ton of cotton as it does to manufacture a ton of bricks. (Data from American Society for Testing and Materials, Committee D-19 on Water, 1969 Manual on water, 3rd ed.: Philadelphia, American Society for Testing and Materials, 355 p., Table 3, p. 320–322, converted to metric tons by J. E. Sanders)

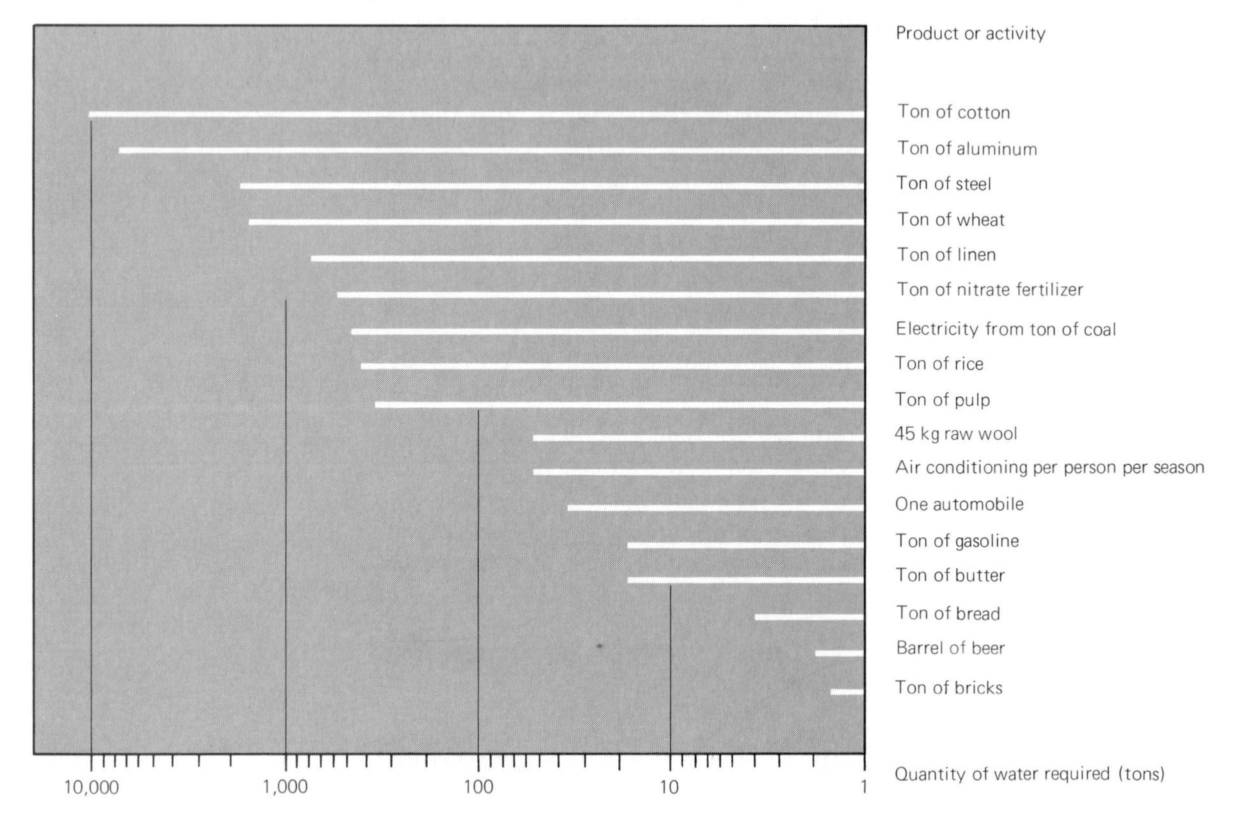

uses include cleaning, conveying, and the manufacture of beverages such as soft drinks and beer.

Even greater than industrial needs are demands for water by farmers in regions of the world where rainfall is inadequate for crops to grow. This water is said to be "consumed;" it is evaporated from the soil and transpired through plants into water vapor. Unfortunately, water losses during irrigation are highest where the water is needed most — in hot, arid climates. The quantities of water used for various industrial and agricultural purposes are shown in Figure 2.4.

**Sources of water** The two chief sources of water for human use are various bodies of water on the surface of the Earth and water underground.

The supply of surface water includes rivers, lakes, and artificial reservoirs. About four-fifths of the water used in the conterminous United States comes from surface sources.

The total volume of groundwater is estimated to be about 30 times as much as all the surface fresh water in lakes and streams and in the atmosphere (Figure 2.5). As supplies of surface water become over-used and polluted, supplies of clean groundwater are being used in increasing amounts. Groundwater is easily tapped with wells, is less exposed to pollution than surface water, is protected from evaporation, and requires no dams or surface reservoirs.

According to the U.S. Geological Survey, about one-fifth of the water used in the United States is groundwater. Irrigation takes 65 percent of all groundwater used. Some 91 percent of this is used in the extensive irrigation systems of the 17 western states where river flow is sparse. Industry uses about 22 percent of the groundwater, public water supplies about 10 percent, and rural needs other than irrigation about 3 percent.

**2.5 Volumes of fresh water** in various natural reservoirs within the hydrologic cycle shown by cubes of various sizes drawn to correct proportion.

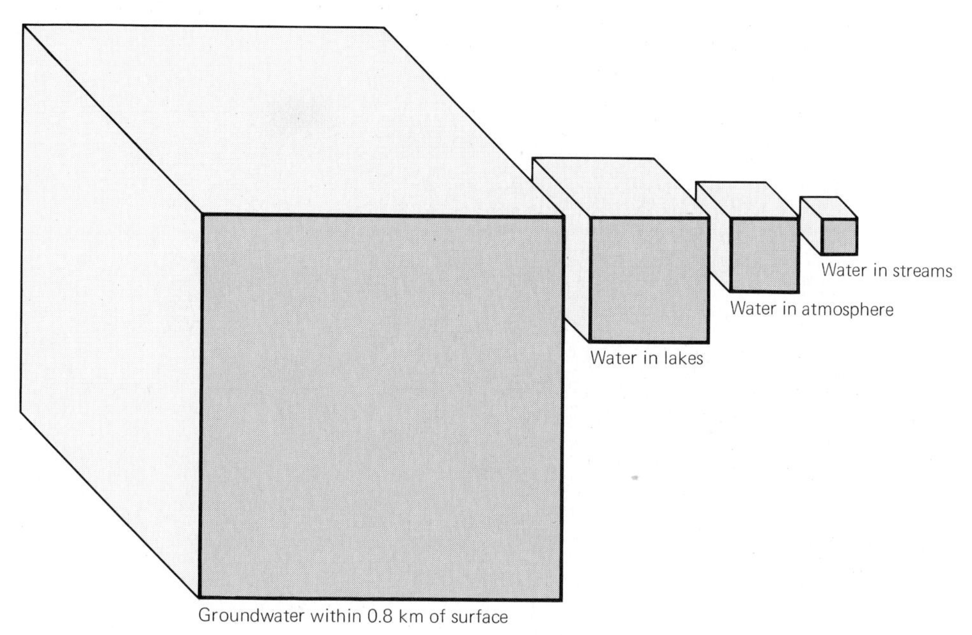

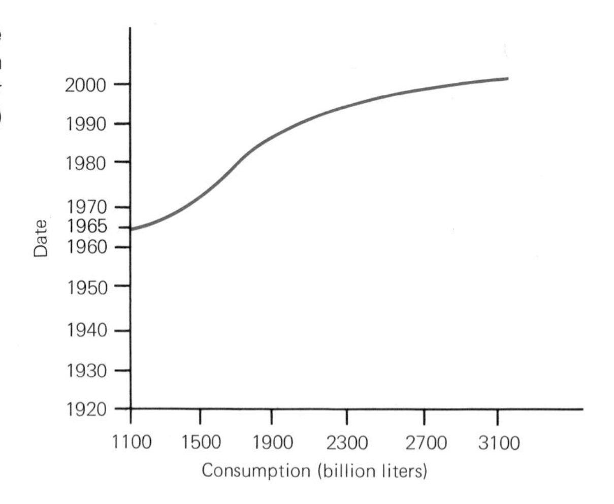


**2.6 Consumption of water in the United States,** based on usage through 1970 and on projection of future use for 1980 and 2000. (U. S. Geological Survey)

**Water problems** In contemplating the future, planners have to deal with many difficult problems of water supply. Four of these are increasing demand, uneven distribution, river floods, and pollution.

The rate of water use has increased sharply along with the rise of population, increasing industrialization, and expansion of irrigation. The recent history of increasing needs for water and a projection into the future are shown in Figure 2.6.

In the United States we are currently "consuming" one-quarter of the available water, even though much of this water has already been used more than once through recycling. Thus, our total usage is nowhere near the total water available. Nevertheless, water shortages exist locally because of the uneven distribution of the water. Acute shortages exist where the population density or agricultural expansion has exceeded the water supply.

River flooding is a natural process (Figure 2.7). The low, flat areas next to the stream channel that are occasionally inundated with water are

**2.7 Houses flooded on flood plain of the Meramec River,** a tributary of the Mississippi River, during high water of 5 April 1973. On 28 April 1973, the flood bulge in the Mississippi completely overpowered the Meramec and caused water in the lower Meramec to flow upstream. (Charles B. Belt)

called *flood plains.* They provide natural storing places for flood water and allowing the water to spread outward can be a great benefit to humans. Periodic flooding of the Nile was the basis of Egyptian agriculture for centuries.

However, people have consistently built buildings, roads, and railroads on flood plains, ignoring the geologists' dictum that "the flood plain belongs to the river." Many of the world's great rivers flood repeatedly; the floods have caused heavy loss of life and property. One of the greatest floods in history struck China in 1887. The Hwang Ho flooded 130,000 square kilometers and killed a million people. Destruction of crops caused a million more to starve.

Rivers and other water bodies have also been used as dumping grounds for everything from sewage to industrial and mining waste materials to human debris. Concern about water pollution, which began to be widely publicized during the 1960s, was first focused on the effects of household detergents (Figure 2.8), municipal sewage, and various industrial wastes. The unplanned bad effects on nontarget organisms of toxic pesticides, such as DDT, and the wide distribution of such chemicals throughout the hydrosphere, the atmosphere, the biosphere, and even the lithosphere were brought to light during the late 1960s. During the 1970s an even more complicated environmental problem arose when it was learned that certain industrial compounds, widely used because of their stability and electrical properties, were causing great difficulties in the biosphere. One group of such compounds, the polychlorinated biphenyls (PCBs), have been found worldwide. Pesticide residues have been discovered in the bodies of Antarctic penguins and high body burdens of PCBs have been measured in Arctic polar bears. The obvious conclusion is that certain materials, once released into the environment, may disperse and be transported throughout the biosphere.

**2.8 Detergent foam in the Guayre River,** came from household wastes in Caracas, Venezuela, 40 kilometers away from this spot. (United Nations, Food and Agricultural Organization, photo in 1971 by H. Null)

## Energy

The energy source used by primitive people was the Sun. Solar energy used by plants was recycled into the biosphere through decaying implements, animal hides, wood and bone implements, and excremental wastes from food. Energy requirements increased when fire came into use; the fuel was provided by chopping down trees.

The melting of metals from rocks greatly increased the need for energy. Copper was the first metal to be used on a large scale. Because of its low melting point, copper required little heat for refining. Iron, however, was much more difficult than copper to extract. Furnaces hot enough to do the job were not developed until about 1100 B.C. in the Middle East. In later times throughout Europe forests were hacked into firewood to fuel iron furnaces; many of the trees were never replaced.

Apart from trees and solar energy, natural energy supplies include water power, "fossil" fuels (coal and petroleum), uranium, and geothermal heat. (The term fossil fuel is applied to coal and petroleum, the collective designation of oil and natural gas, because these materials formed from the altered remains of ancient organisms, mainly plants. These organic remains escaped being recycled into the biosphere and, instead, accumulated with sedimentary strata.) Of these, solar energy, water power, and geothermal heat are *renewable,* that is, they are available in almost infinite amounts or are not permanently altered through use. The other sources, the fossil fuels and uranium, are *nonrenewable.* Once they have been used, they are gone forever.

**The energy of river water** The fundamental energy relationships in flowing water (a stream or river) begin with the mass of the water and its potential energy. The potential energy is represented by the quantity of water and the distance through which it will fall. As soon as the water begins to move, the energy becomes kinetic, and the kinetic energy is used up in the flow of the water, its geologic work, and heat losses because of friction. Heat losses are scarcely measurable, but they have been shown to exist.

**2.9 Bonneville Dam,** on the Columbia River, in Washington State, was built to enable the energy of the falling water to be converted into electricity. (Department of the Army, Corps of Engineers, North Pacific Division)

The kinetic energy of flowing water can be used to turn water wheels and such rotary motion has been put to various uses, from pumping water to turning the machines in factories to generating electricity. Most of the great dams in the United States, such as Hoover Dam, Bonneville Dam (Figure 2.9), and Grand Coulee Dam, were built for generating electricity. Nearly all of the suitable sites for large-scale electrical generators are already being used, and so, we cannot easily expand large-scale hydroelectric power facilities.

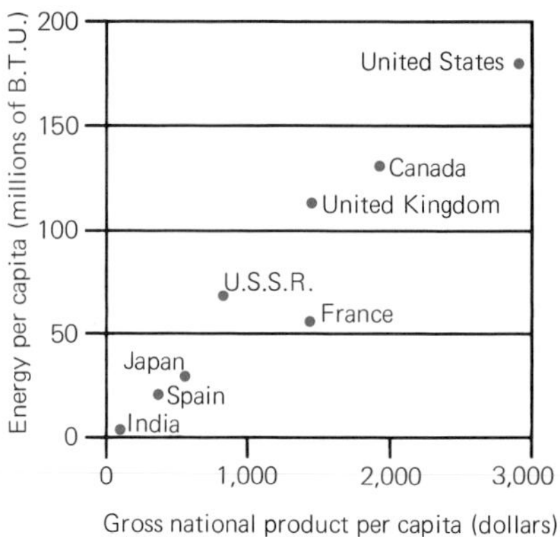


**2.10 Direct relationship between energy consumed and material standard of living** as expressed by individual shares of gross national product. (After S. Fred Singer, 1970) Human energy production as a process in the biosphere, p. 107–114 in The biosphere. A Scientific American Book: San Francisco W. H. Freeman and Company, 134 p., Figure on p. 109, ascribed to U.S. Office of Science and Technology, 1961)

**The fossil-fuel revolution** As England's supply of firewood diminished, industry, threatened with extinction, had to find another source of heat. The development of coal for this purpose in the late seventeenth and early eighteenth centuries in England launched the Industrial Revolution. In changing the course of human history, the linking of coal to iron production ranks second in importance only to agriculture.

The development of the steam engine rapidly overturned many aspects of life. Wooden sailing ships were quickly replaced by ships using iron as a structural material and coal as the energy source for propulsion. As the nineteenth century ended, steam combines appeared on farms and quickly took the place of horses and mules, only to be replaced themselves by internal-combustion machines powered by petroleum products.

The unique position of oil in modern industrial economic fabric scarcely needs any emphasis. A clear relationship has been demonstrated between standard of material prosperity and quantity of oil consumed (Figure 2.10). What is more, in the United States, our entire life style has been built around the automobile and the presumption that cheap gasoline for our cars would always be available.

Measured coal reserves in North America alone amount to some 1560 billion tons. The United States has between one-fifth and one-half of the world's coal deposits, concentrated mainly in 17 states (Figure 2.11). Even though underground mining methods can remove only about half of the coal present, accessible and usable coal is plentiful. The U.S. Geological Survey estimates total inferred reserves for the world at 8415

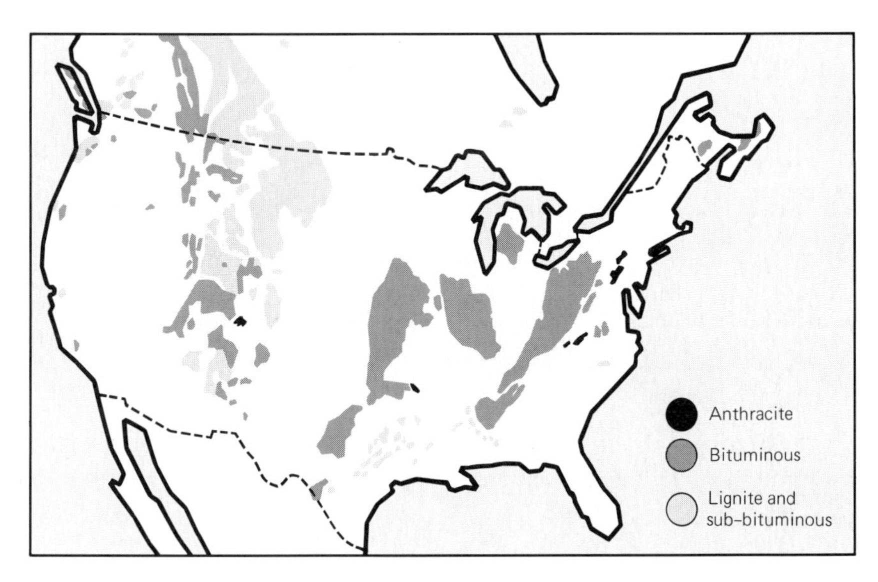


**2.11 Locations of principal coal and lignite deposits in United States and southern Canada.** (Compiled from maps published by United States Geological Survey and Geological Survey of Canada)

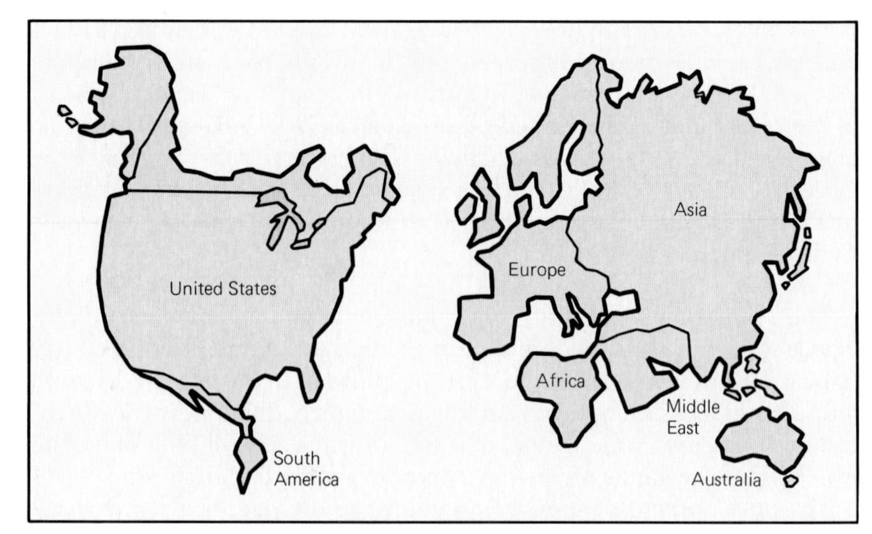


**2.12 Distribution of world coal resources** shown by distorted map in which the areas of the countries or regions have been drawn in proportion to the sizes of their coal resources. (Courtesy Exxon Corporation)

**2.13 World's yearly oil production,** organized by the four major producing regions, for selected years starting in 1947. In the two 13-year intervals between 1947 and 1960 and between 1960 and 1973, the total production more than doubled. It is not likely that another doubling will be achieved by 1986. (Data from publications of American Petroleum Institute)

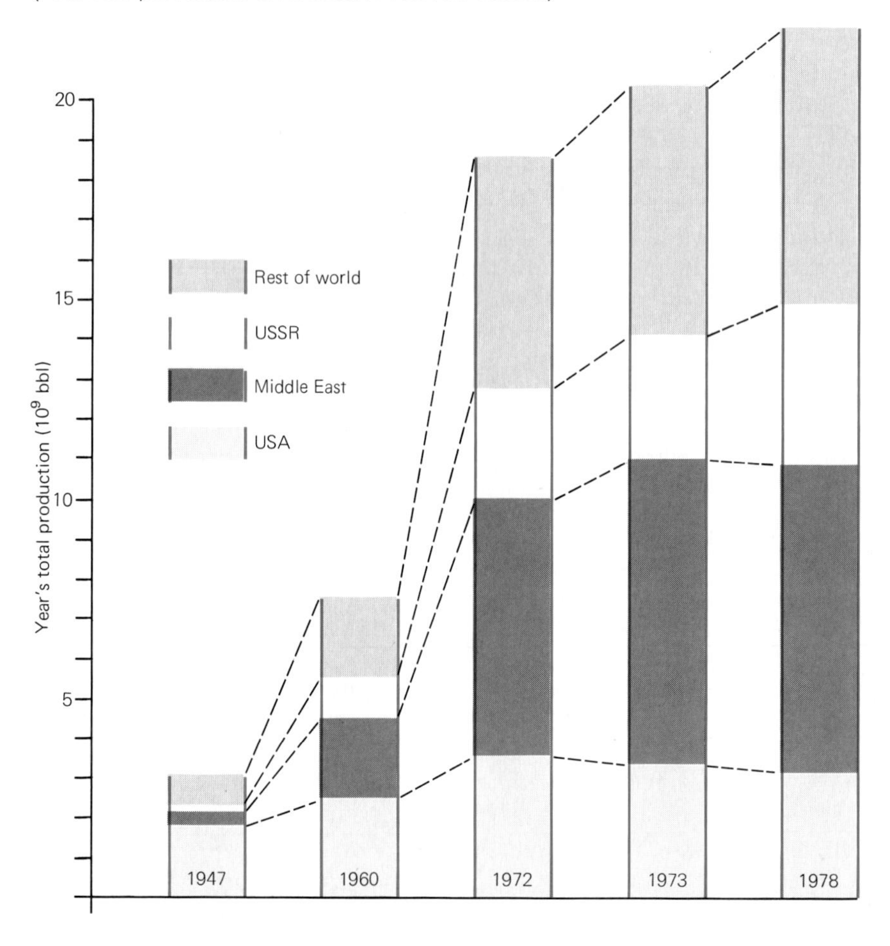

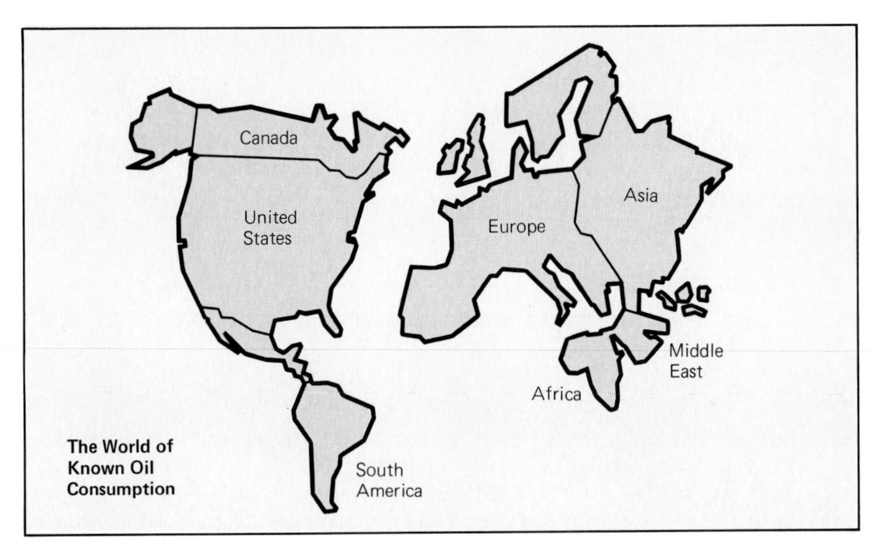


(a)

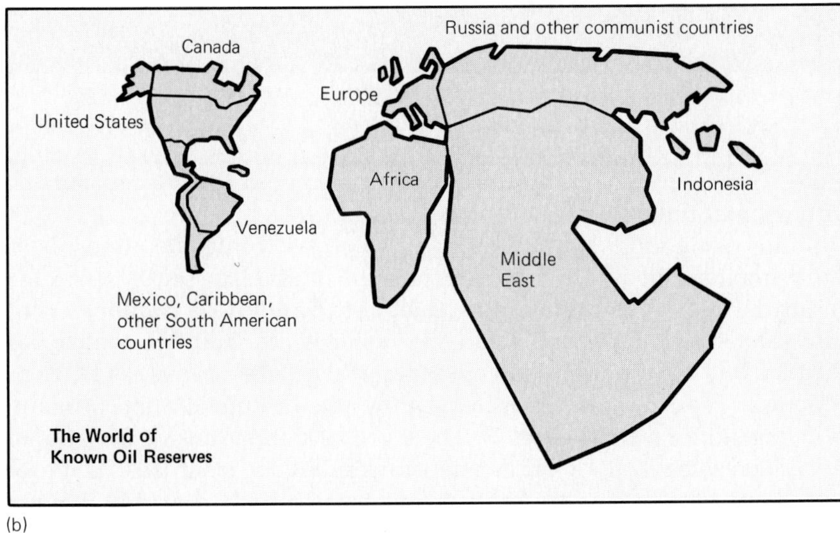


(b)

**2.14 World oil consumption compared with world oil reserves** using distorted maps with areas proportional to consumption and reserves. Consumption map (top) shows a large United States and a tiny Middle East. By contrast, reserves map (bottom) shows small United States and huge Middle East. (Courtesy Exxon Corporation)

**2.15 Sources of energy consumed in the United States 1850 to 2000.** (S. Fred Singer, 1970, reference cited in caption to Figure 2.10, p. 55; p. 111, based on estimates made by Hans H. Landsberg, Resources for the Future, Inc.)

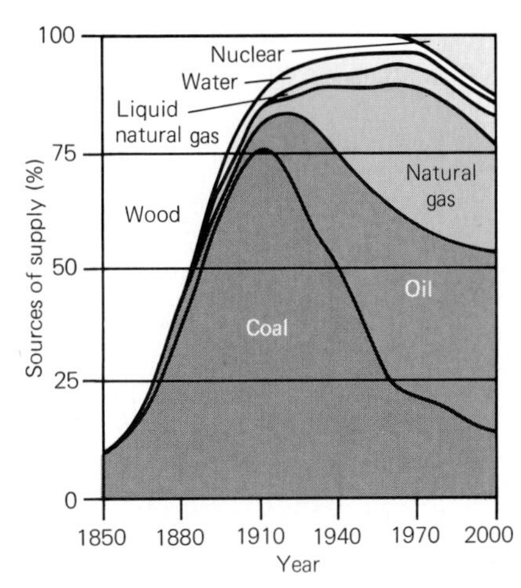


billion tons. Of this coal, 97 percent lies in Asia, North America, and Europe (Figure 2.12).

The United States, where the first oil well was drilled in 1859, was for a time the world's leading oil producer, but its reserves have dwindled. Currently, the Soviet Union is first, Saudi Arabia is second, and the United States is third. By 1970 the proportion of the world's oil production produced in the United States had dropped from its peak of 55 percent in 1944 to only 21 percent (Figure 2.13). Rate of consumption has passed rate of production, and the difference is made up by large imports from Venzuela and the Middle East, where oil is still abundant (Figure 2.14). Oil and natural gas now provide fully 78 percent of the energy used in the United States (Figure 2.15).

Although modern life-styles in the United States have been made possible largely because oil has been cheap and abundant, the combus-

**2.16 Polluted air;** plume of industrial smoke hovers thickly over town in Mosel Valley, near Cologne, West Germany. (United Nations photo, April 1963)

tion of both oil and coal yields smoke (Figure 2.16) and various exhaust gases. Of these, two, sulfur dioxide and carbon dioxide are causing environmental difficulties.

Sulfur compounds and others are also released into the atmosphere from smelters, in which raw materials are roasted to purify them for industrial uses. The oxidation product of sulfur dioxide combines with rainwater to form sulfuric acid. The result is acid rainfall, which has killed fish in otherwise-pristine mountain lakes located in areas far from industrial centers, and has increased the rate of deterioration of many building stones (see Chapter 7). The levels of carbon dioxide are known to be increasing in the world's atmosphere, but so far, at least, the predicted increase in world temperature from carbon dioxide's known "greenhouse effect" seems not to have been recorded.

Another combustion product, lead, which is toxic, has been released into the atmosphere and has spread throughout the world. Such environmental lead has come from the burning of tetraethyl lead, an antiknock compound widely used in gasoline, and from the smelting of lead-bearing minerals.

**Uranium** The fuel used in nuclear reactors is uranium. In 1975 nuclear reactors produced 8 percent of the electricity in the United States. The chief attraction of uranium as fuel is its tremendous efficiency. To fuel one 1000 megawatt power plant for a year requires about 2.3 million tons of coal, 10 million barrels of crude oil, or only 30 tons of uranium. In different terms, one pound of fissionable uranium produces as much energy as 6000 barrels of fuel oil. Thus, the use of uranium as fuel is highly desirable to most electrical-utility operators.

Coal resources are about 10 times the combined resources of oil, natural gas, and oil shale, and high-grade uranium resources are perhaps four times greater. But the reserves of uranium-235 are, like those of fossil fuels, nonrenewable. With the great expansion of uranium-fueled

**2.17 Natural steam** escaping from commercial geothermal plant at "The Geysers," Sonoma County, about 120 kilometers north of San Francisco, California. (U. S. Geological Survey)

reactors planned for the next two decades, most of the high-grade ore will be used up by the 1990s.

The accident at the Three Mile Island nuclear plant in Pennsylvania during March 1979 brought into sharp public focus the debate that had been raging over the safety situation in nuclear power plants. The theories of atomic physics on which the plants have been built are all very elegant, but the high-pressure pipes always seem to break or to corrode. Widespread opposition to the construction of nuclear power plants has been voiced, both for reasons of safety in and near the plants and for the serious environmental problems, not yet solved, over where to store the radioactive waste products so their radiation will not be hazardous.

**Geothermal energy** Geothermal heat, the heat being conducted outward from the interior of the Earth, escapes into space almost everywhere at a rate that is steady, but is so small that it is not noticeable. However, at volcanoes, hot springs, or perhaps one of the handful of power stations that generate electricity from geothermal heat, so much more heat escapes that its existence becomes spectacularly apparent.

Geothermal heat is a large and potentially useful source of energy (Figure 2.17). In many locations in the United States and elsewhere, hot magma chambers lie close beneath the surface. Engineers are experimenting with technologies to tap this huge heat source and bring its energy into the service of humanity. Even where no bodies of magma are present, deep, warm, salt water can be pumped to the surface for heating water and homes.

So far, only four locations in the world have produced meaningful amounts of geothermal energy: Italy, Iceland, New Zealand, and California. Other countries now attempting to develop steam resources include Mexico, Russia, Japan, Indonesia, and Chile (Figure 2.18).

Italy pioneered the use of geothermal energy at Larderello–Mt. Amiata in Tuscany. It has served as a kind of working laboratory for

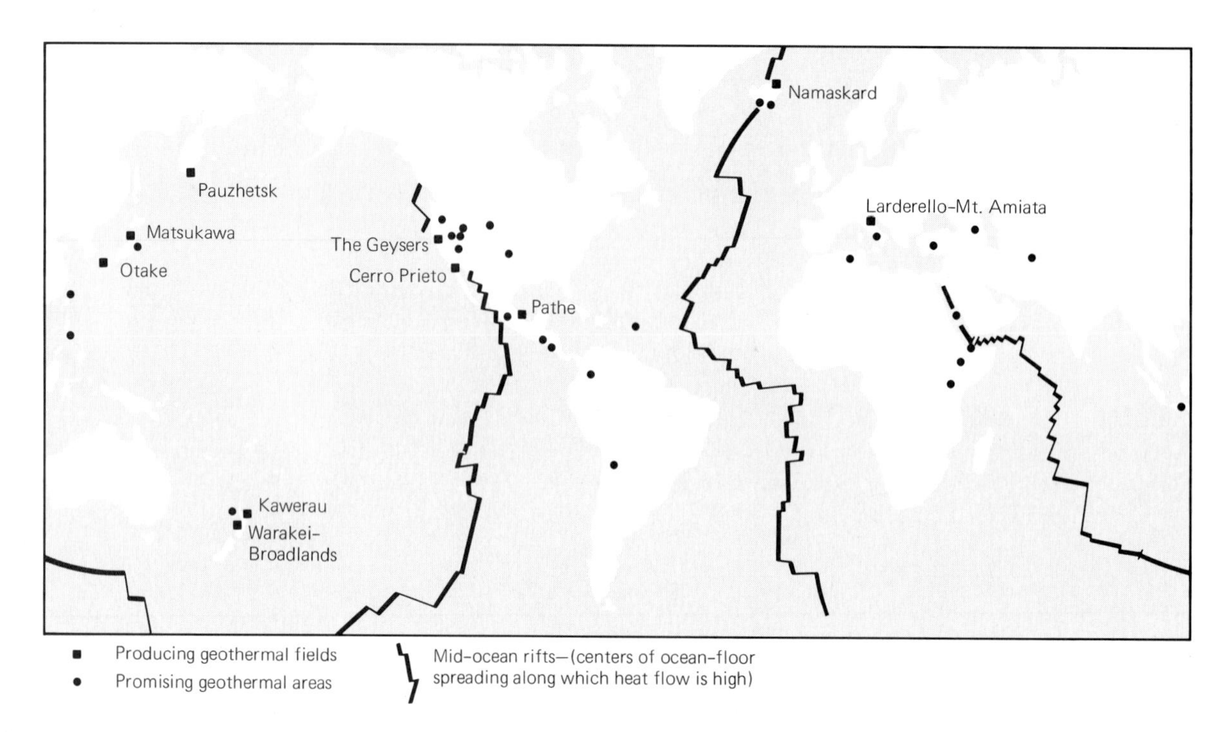


**2.18 Geothermal areas of the world** in relation to mid-oceanic ridges where heat flow is high and new sea-floor crust is forming. (U. S. Geological Survey)

others interested in the field. By 1923 Italian engineers had devised a system for removing most of the contaminating gases from the steam.

Iceland, built up entirely by volcanic processes, has abundant geothermal energy. Indeed, Iceland might better be called Fireland: it literally straddles the mid-Atlantic Ridge, where molten liquid mantle is thought to be rising continuously through the sea bed. The most abundant geysers and hot springs are in the vicinity of the capital of Reykjavik, where since 1925 natural hot water has been used to heat dwellings and greenhouses where tropical fruits are grown. Many Icelanders bake bread by burying the loaves in hot regolith. They also boil food by lowering containers into hot springs. Some 90 percent of the houses in the city are heated by water from hot springs 15 kilometers away. The first plant to convert geothermal steam to electricity was inaugurated at Hveragerdi in 1964.

The only region producing geothermal power in the United States is The Geysers, California, 150 kilometers north of San Francisco. Despite the name, the area has no geysers, but does seethe with 50 acres of fumaroles (vents that issue hot gases) and hot springs. In 1852 the first health resort was opened in The Geysers, and in 1921 the first wells were drilled for steam. The project was abandoned until 1955, when a 180-meter experimental well was sunk, and in 1960 the Pacific Gas and Electric Company opened the first commercial geothermal power station (12,500 kw) in this country.

### Ores and Economic Mineral Deposits

Under this heading belong all the metals and nonmetallic substances extracted from the Earth.

The field of *economic geology* is the subdivision of geology that deals with materials that can be extracted from the Earth for human use. Traditionally, the chief concern of economic geologists has been ore deposits.

**Ore deposits** An *ore deposit* is a natural deposit from which a metallic element can be extracted at a profit. Because the concept of profitability enters into this definition, the decision about whether a given deposit is an ore must take into account the concentration and form of the desired material, costs of removing it and concentrating it (if necessary), costs of transportation, and market value.

A deposit situated far from a suitable processing site may not be an ore even though its concentration is high. By contrast, new demands for a previously worthless metal may make an ore out of what was previously ordinary rock or new processing technologies may "create" ores. Price fluctuations and political situations also determine whether a particular deposit is an ore.

In a typical ore deposit, the metal to be extracted is contained in only a small fraction of the total bulk of the material and it must be separated from the other substances. The materials associated with an ore deposit that are of no immediate concern to the operation are known as *gangue.*

Many different metals are required by modern industrial societies. Heading the list are iron and the other elements mixed with (or alloyed with) iron to make steel. These so-called ferroalloy elements include manganese, molybdenum, chromium, tungsten, and zinc. Other useful metals include copper, lead, aluminum, and the precious metals, gold, platinum, and silver. The mere listing of these metals brings to mind the many products made from them that are integral parts of the modern world.

**Economic deposits other than ores** The term *ore* has been applied especially to deposits containing one or more metallic elements that must be separated from the gangue before the metal can be used. The term ore has not been applied to deposits such as limestone, coal, or halite, in which the entire amount of material extracted is processed for its contents. Such other non-ore deposits that can be extracted at a profit are referred to as *economic mineral deposits.* In these cases, the term "economic" carries the same significance as "ore;" it means that the selling of the material will bring in more money than must be spent to extract the material and ship the product to the point of sale.

Nonmetallic resources include a long list of materials ranging from asbestos to zeolites. Nonmetallic resources are vital in many industrial processes, in the making of steel, for fertilizers, in construction uses, as fillers and filters, in the making of bricks and pottery, as powders to mix with water to make dense drilling mud, as gemstones, for concrete, and for countless other applications.

The extraction and use of natural materials have raised significant problems for humans to face. Lead, cadmium, mercury, and asbestos are a few examples of natural materials that are hazardous to human health and that have been released into the environment in concentrations exceeding the levels of normal natural backgrounds. In addition, the extraction of natural materials has involved many changes in the lithosphere.

**2.19 Erosion of bare regolith where former plant cover has been removed.** Impact of falling raindrops and flow of water in rivulets downhill has scarred this area after a fire destroyed the trees. (U. S. Department of Agriculture, Soil Conservation Service photo by J. G. James)

Trees have cleared to make fields or to saw into lumber. The removal of the network of plant leaves, branches, tree trunks, and roots, by whatever method or for whatever purpose, lays bare the regolith to the impact of falling rain and erosion by water flowing downhill (Figure 2.19).

Roads and railroads have been built, including, where necessary, the cutting away of or tunneling through bedrock. Many such tunnels and cuts have opened to geologists' view rocks that might not otherwise be visible. The slopes at the sides of many roads do not always "stay put"; failure of these slopes creates many problems for the engineers (Figure 2.20; see also Chapter 11).

Mining, which has been practiced for thousands of years, involves everything from stone quarries (Figure 2.21) and gigantic open pits to underground rooms and tunnels. As the United States increases its output of coal from the western interior region, more and more strip mining will be done. Much strip mining has been done with little care for the environmental impacts. However, it is possible to do strip mining and then afterward to restore the land (Figure 2.22).

**2.20 Fallen rocks partially blocking eastbound lane of Interstate 70,** about 3 kilometers W of Bergen Park interchange, Jefferson County, Colorado, 8 May 1973. (W. R. Hansen, U. S. Geological Survey)

**2.21 Stone quarry where rock has been removed for use in construction,** view S from airplane. (Helipern photo, courtesy Victor Tomasso, Tilcon-Tomasso Company)

(a)

(b)

**2.22 Strip mining and land reclamation in southeastern Ohio** west of Caldwell, about half way between Zanesville and Marietta. (Ohio Power Company, American Electric Power System)

"Big Muskie," an electric dragline, and the world's largest excavating machine, strips away the overburden from a thick layer of coal.

Park and pond formed after land had been graded and trees planted in 1961. This photograph was taken in May 1972; the last strip mining took place nearby in 1959–1960.

## THE ARITHMETIC OF EXPONENTIAL EXPANSION

An exponential-type function that is familiar to many people is compound interest on a bank account, which amounts to a percent increase added to a base sum that expands each year as the percent increase is added on. One of the important aspects of any exponential-expansion function is the time required for the base number to double. This is determined by dividing the percent-expansion value (e.g., the interest rate on a bank account) into 70. Thus, 5 percent yields a doubling time of 14 years, 10 percent a doubling time of 7 years, and so forth.

### Exponential Expansion and Nonrenewable Resources

An important and somber aspect of exponential expansion is that it cannot continue forever, no matter how much one might wish for such expansion to continue. A complicated set of relationships evidently exists among rates of exponential expansion of natural resources, interest rates, and the levels of prices in a country's economy. A rapid exponential expansion of resources is inflationary. An example is the rapid local rise of prices that took place in California during the great gold rush of 1848 to 1850. When the exponential-expansion rates of natural resources and interest rates are equal, as they were in the United States for about a hundred years, from early in the nineteenth century until 1908, prices are stable (Figure 2.23). Inflation is evidently an effect of declining rates of resource expansion. Notice in Figure 2.23 that in 1908, when the exponential-expansion rates of the two basic measures of industrial activity, pig iron (for making steel) and thermal energy, dropped from about 7 percent to about 2 percent, and the interest rate did not drop correspondingly, but rather was steadily raised, a period of inflationary increases in prices began that still continues (see Figure 2.23).

## PRESENT-DAY DILEMMAS INVOLVING NATURAL RESOURCES

One of the urgent points for every educated citizen of a modern industrial society to understand is the situation regarding the future supplies of natural resources upon which that society has been based. We address this subject by first presenting some important definitions and concepts and then by examining some hard choices.

### Important Definitions and Concepts Relating to Natural Resources

In discussing natural resources, it is necessary to understand some distinctions about future conditions, which are subject to change. In particular, these conditions can change the circumstances about what can and cannot be done at a profit, and thus what are reserves and resources.

**Reserves and resources** *Reserves* are known, identified deposits from which some material can be extracted profitably with existing technology and under present economic conditions. *Resources* include reserves plus other mineral deposits that may become available in the future, known deposits that cannot be mined profitably with existing technol-

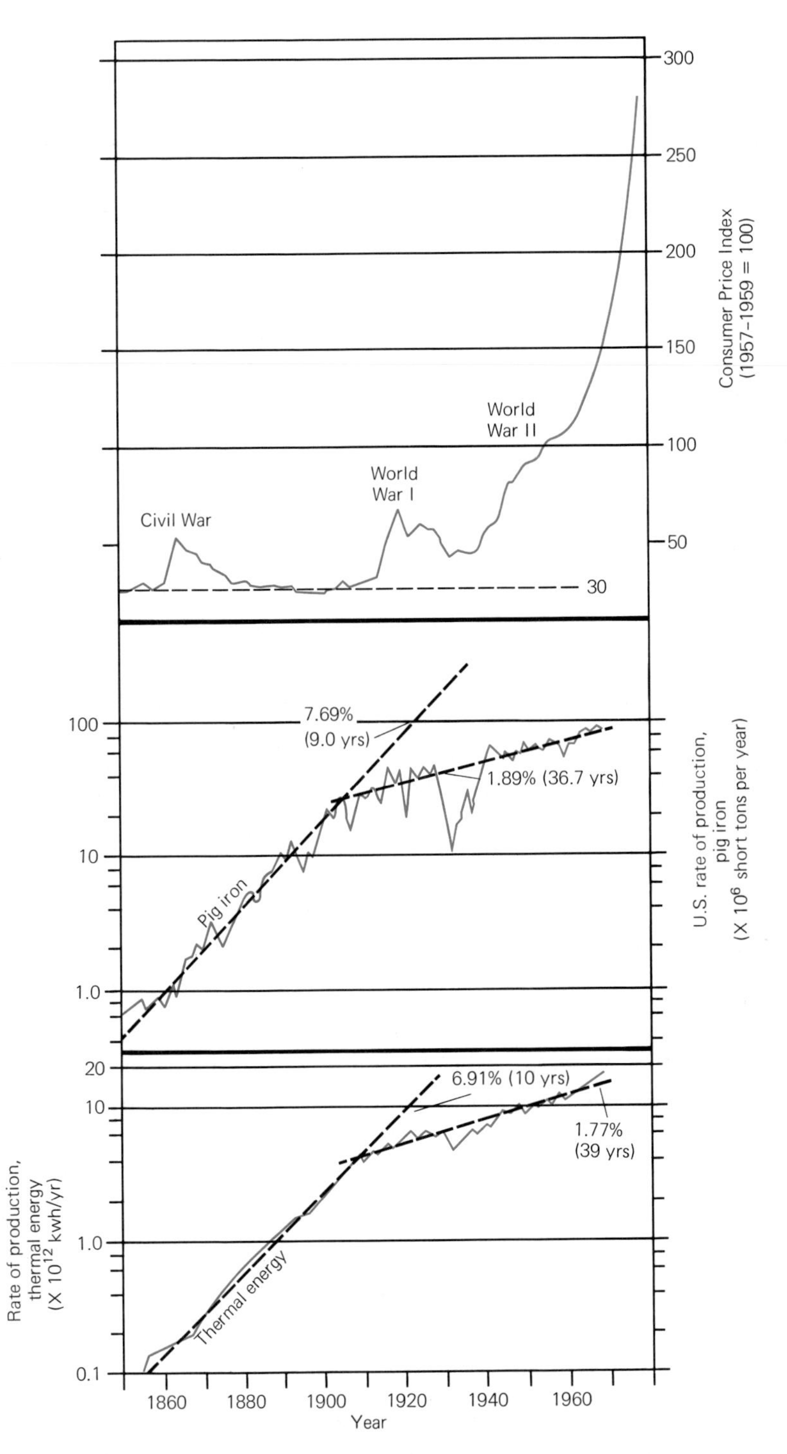


**2.23 Declining rates of increases in production of fundamental raw materials as a factor causing inflation in economic system based on compound interest.** Notice that a century of stable prices in the United States came to a close about 1910, when rates of increases of production of iron and thermal energy dropped below the interest rate. As production rates have continued to fall, the rate of inflation has continued to increase. (Lower two curves from M. K. Hubbert, 1973; graph of prices, to 1910 from U. S. Government Historical Almanac, 1910 to 1974, U. S. Bureau of Labor Statistics)

ogy or under present economic conditions, but that might become profitable with new technology or under different economic circumstances. A *potential resource* is a deposit, rich or lean, that may be inferred to exist, but that has not yet been discovered, or which is known to exist but cannot be extracted profitably under existing conditions. A good example of the second kind of potential resource is the large amounts of solid hydrocarbons in the so called "oil" shales of the western United States. The conversion of shale hydrocarbons to oil is so costly, in terms of dollars, energy, and water, that only pilot projects have been initiated so far.

## Reserves of Nonrenewable Resources and Use Curves

A great goal of human intelligence is to be able to foresee and, where necessary, to forestall future situations. As far as resources are concerned, this means gaining insights into the supply situation and adjusting use patterns accordingly.

A valiant attempt to do this for the draining of oil out of underground fields in the United States, which was generally not appreciated at the time, was made by M. King Hubbert (1903– ), an expert in the physics of the movement of fluids underground, who attempted to determine if any mathematical function could be applied to the history of use of nonrenewable resources. In 1956, Hubbert decided that the bell-shaped curve biologists had found applicable to the growth rates of individual mammals seemed to fit the use patterns of nonrenewable resources. The remarkable feature of the bell-shaped curve is that the area between the curve and the horizontal axis is equal to the total quantity of the particular resource (Figure 2.24). Therefore, if one knows a number for this total and has on record the production statistics for drawing the left-hand part of the curve, then one can make surprisingly accurate predictions about the trend of future production.

Because of the shifting conditions among reserves, resources, and potential resources, one might be skeptical of the value of Hubbert's approach. But, let us look at the record. Because of the lack of certainty not only about the shifting reserves situation but also of how much actual recoverable oil might be found, a spread of estimates exists among the experts who concentrate on these matters. Hubbert found a consensus lying between a low of 150 billion barrels and a high of 200 billion barrels. Therefore, he drew two curves, both of which implied that the production of oil in the United States would reach a peak in the early 1970s and would thereafter decline (Figure 2.25). If no new

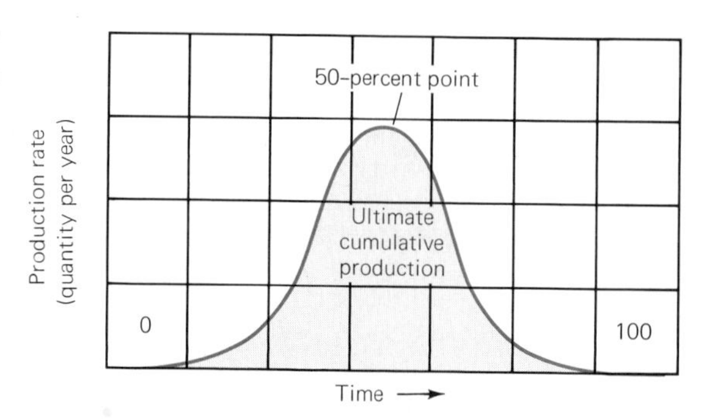


**2.24 Bell-shaped production-history graph of a nonrenewable resource that has been exhausted.** On such a graph, the number of squares enclosed between the curve and the time axis equals the total quantity of the resource recovered. (M. King Hubbert, 1957, Nuclear energy and the fossil fuels, p. 7–25 in Drilling and production practice, 1956: New York, American Petroleum Institute, 527 p., Figure 11, p. 12)

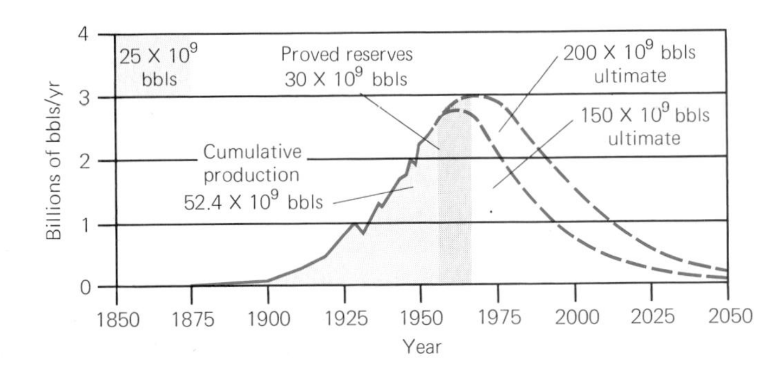


**2.25 Forecast, of crude-oil production in United States made in 1956,** based on bell-shaped curve and on two estimates of ultimate petroleum recovery, an upper value of 200 billion barrels and a lower value of 150 billion barrels. The decline in production predicted for the early 1970's actually happened during 1971–1972 (compare Figure 2.13). The production data seem to be defining a bell-shaped curve that is estimated to enclose an area equal to 180 billion barrels. Only major discoveries in areas now considered to lack petroleum can change the situation implied by this graph. (M. King Hubbert, 1957, Reference cited in caption of Figure 2.24, Figure 21, p. 17)

discoveries are made that drastically alter the concepts on which the above-cited estimates of reserves were made, then the use curve now being traced looks as if it will close around a figure of 180 billion barrels.

Notice that exponential expansion is possible only during the first 50 percent of a Hubbert-type curve. From the 50-percent-point onward, instead of more of a reserve each year, there is only less. In short, after 50 percent of a reserve has been extracted, it's downhill all the way to the point of exhaustion.

## Exponential Expansion and Fixed Rates of Use

The only way to "stretch out" a resource so that the supply lasts longer is to cut back on rate of use. In other words, this means that one shifts to a lower level of exponential expansion or even to a steady level in which the expansion rate is zero. As an example of what this means in the modern world, suppose that the Earth were nothing but a giant spherical steel tank full of oil and that for calculating the volume inside the tank one could ignore the wall thickness of the steel. At a fixed level of use of 22 billion barrels per year (the 1978 world consumption), such a volume of oil would last essentially forever — 344 billion years (or about 73 times the age of the Earth to date). However, at an exponentially increasing rate of 7 percent per year, the rate that has been sustained during the first 75 years of the twentieth century, the hypothetical Earth-size steel tank of oil would be pumped dry in 384 years.

## What About the Future?

If one admits the ultimate impossibility of continuous exponential expansion of natural resources, what can be done about it? Can we continue to live comfortably at a level comparable with that of the present? In order to do so, we must figure out ways to live with the same amounts of resources, not with more and more. In theory, we can recycle materials mined out of the Earth. For example, the photograph of an automobile junkyard shown in Figure 2.26 might be entitled an ore deposit of the future. Also, we could shift more and more to renewable energy sources, such as solar energy plant products (chiefly wood for fuel), and geothermal energy. Even while doing this, we need to increase our efforts at finding new sources. The main theme of this book is that geology has much to contribute toward the objective of finding new supplies of natural resources. If that proves not to be the case, then we

**2.26 Auto junkyard in Vermont,** where the discarded automobiles of the present day may be a source of metal in the future. (United Nations photo in 1972, by B. Brunzweig)

must start solving the dilemma of how to use less of all materials and at the same time not be ruined by inflation (see Figure 2.23).

Finally, we must know much more about human effect on the biosphere. In addition to the environmental problems previously mentioned, we must deal with such problems as the near extinction of the buffalo, the blue whale, and other animals and the alteration of global vegetation patterns by agriculture.

Other effects are not as easy to see, yet they may be even more far reaching. For example, all of us carry unnatural body burdens of human-made or human-liberated substances, such as PCBs, carbon monoxide, and lead. The meat we eat may contain minute amounts of hormones and antibiotics injected into livestock to make them grow faster. Fruits and vegetables bear residues of insecticides, herbicides, fungicides, and other chemicals that human bodies have never encountered before. Such chemicals are dispersed throughout the biosphere. They may threaten the existence of some species and favor that of others. By this means, we have been subtly changing the balance of insects and other members of the biosphere. We do not yet know whether such changes will be beneficial or detrimental to our own existence as a species.

## A CHALLENGE TO US ALL

The challenge before us today is what to do about our fast-dwindling resources and the environmental impacts that our previous practices have brought about. The problem is complicated by the fact that as the industrial nations of the world, the extravagant users of natural resources, begin to think about cutting their rates of consumption, the previously underdeveloped areas, who have not been such big consumers, now desire to enter the industrial age. In fact, some nations have argued that talk about conservation is nothing but a plot by the "haves" to keep the "have nots" dependent and noncompetitive.

Because of the space program, all people have had the opportunity to take a look at the world in the way geologists have been trying to visualize it. We not only have found out about other planets, but also have learned to consider the Earth from a much expanded point of view. As the Apollo 8 astronauts were returning to Earth on Christmas Day, 1968, after they had circled the Moon for the first time, the poet Archibald MacLeish recorded his impressions with these words:

> *For the first time in all of time men have seen the earth: seen it not as continents or oceans from the little distance of a hundred miles or two or three, but seen it from the depths of space — seen it whole and round and beautiful and small. . . . To see the earth as it truly is, small and blue and beautiful in that eternal silence where it floats, is to see ourselves as riders on the earth together, brothers on that bright loveliness in the eternal cold — brothers who know now they are truly brothers.*

Who can now doubt that the world is an entity, a system operating within the geologic cycle? Who has not been impressed with what a remarkable place the Earth really is? One of the most important lessons to be learned from study of geology is to understand this fragile thing called Earth and to know how it works. We must learn what can be done with it and what cannot be done to it.

# CHAPTER REVIEW

1. The Earth usually is a pleasant setting for life, but is subject to occasional natural catastrophes, such as drought, storms, earthquakes, slope failures, and volcanic outbursts.

2. After a quiet period lasting 123 years, the volcano Mount St. Helens, in Washington State, became active again, starting in March 1980. A mighty blast on 18 May 1980 released a lethal *base surge* of hot gases and solids that killed people, caused enormous damage to the north side of the mountain peak, and spread dust at low levels through several states and sent a high-level dust cloud into the upper-atmosphere wind belts so that it encircled the Earth with but sporadic fallout.

3. The greatest loss of life in a natural disaster in the United States occurred during the hurricane of 8 September 1900, which leveled most of Galveston, Texas, and killed 6000 to 8000 persons.

4. Human use of water, obtained from surface sources or from underground sources, can be divided into three categories: household, industrial, and agricultural. Four water problems are *increasing demand, uneven distribution, flooding,* and *pollution.*

6. The *fossil-fuel revolution* began in England when the rapidly diminishing supply of firewood used for energy threatened her iron industry.

7. Even though underground mining methods can remove only about half of a coal deposit, much accessible and usable coal is available.

8. By 1970 the proportion of the world's oil production pumped out of wells in the United States had dropped from its peak of 55 percent in 1944 to only 21 percent. In the United States, the rate of oil consumption has passed the rate of production; the difference is imported from foreign countries.

9. Combustion of fossil fuels has released sulfur dioxide, carbon dioxide, and lead into the atmosphere. Sulfur dioxide causes acid rainfall. If levels of carbon dioxide become high enough, they may cause a great warming of the Earth because of the "greenhouse effect" of carbon dioxide on incoming solar energy and on backscattered heat from the Earth's surface. Lead from gasoline and also from smelter exhausts, and which is toxic to humans, has polluted the entire biosphere.

10. The high efficiency of uranium has made it appear to some as an attractive alternative to oil as fuel for power plants used in the United States. However, reserves of uranium, like those of the fossil fuels, are finite and nonrenewable. Furthermore, problems of safety and of disposing of radioactive wastes are causing much public opposition to further development of nuclear power plants.

11. *Geothermal energy* from the Earth's interior is a source of power that is being exploited today in several localities around the world.

12. *Economic geology* is the scientific study of all aspects of the Earth's materials that can be extracted for human use. An *ore deposit* is a natural deposit from which a metallic element can be extracted at a profit; deposits other than ore deposits that can be extracted at a profit are *economic mineral deposits.*

13. *Reserves* are known, identified deposits from which some material can be extracted profitably with existing technology and under present economic conditions. *Resources* include reserves plus *potential reserves,* that is as yet undiscovered deposits that can be inferred to exist and known deposits that cannot be mined profitably with existing technology or under present economic conditions, but that might become profitable with new technology or under different economic circumstances.

14. *Resource-use curves* exert great influences on economic activities. During removal of the first 50 percent of a nonrenewable resource, the quantity produced can increase each year. However, after the 50-percent point has been reached, the quantity produced decreases each year.

15. Alternatives to present use patterns include development of renewable energy sources, such as solar energy, plant products, and geothermal energy, and increased recycling of materials.

## QUESTIONS

1. List some kinds of natural catastrophes. Discuss two examples of natural catastrophes that have struck the United States.

2. List the three chief uses of water in the United States. What four problems affect water usage?

3. Define and give examples of *nonrenewable* and *renewable* materials and energy supplies.
4. Define *fossil fuel.* What are the kinds of fossil fuels? What is meant by the *fossil-fuel revolution?*
5. Discuss the present supply situation of fossil fuels in the United States and in the world. How do the known amounts of fossil fuels affect the future energy situation in the United States and in the world?
6. What are some of the environmental hazards associated with the combustion of fossil fuels? Of using nuclear power stations?
7. Define *geothermal energy* and list three localities where such energy is being used today.
8. Define *ore deposit* and *economic mineral deposit.* List some examples of each and discuss some of the factors that affect the extraction of such deposits.
9. Give an example of an exponential-expansion function. How is the doubling time of such functions calculated? How are exponential-expansion functions related to natural resources?
10. Compare *reserves, resources,* and *potential resources.*
11. Discuss the impact on economic activities of the bell-shaped resource-use curves worked out by M. King Hubbert.

**RECOMMENDED READING**

Bartlett, A. A., 1980, Forgotten Fundamentals of the Energy Crisis: *Journal of Geological Education,* v. 28, no. 1, p. 4–35.

Bolt, B. A., 1977, *Geological Hazards; Earthquakes, Tsunamis (sic), Avalanches, Landslides, Floods* (2nd ed.): New York, Springer-Verlag, 330 p.

Heichel, G. H., 1976, Agricultural Production and Energy Resources: *American Scientist,* v. 64, no. 1, p. 64–72.

Keller, W. D., 1978, Drinking Water: A Geochemical Factor in Human Health: *Geological Society of America, Bulletin,* v. 89, no. 3, p. 334–336.

Kerr, R. A., 1980, Mount St. Helens: An Unpredictable Foe: *Science,* v. 208, p. 1446–1448.

Legget, R. F., 1973, *Cities and Geology:* New York, McGraw-Hill Book Co., Inc., 624 p. (Chapter 4, Hydrogeology of cities, p. 125–176.)

McCarthy, Joe, 1969, The '38 Hurricane: *American Heritage,* v. 20, no. 5, p. 10–15, 102–104.

McPhee, John, 1971, *Encounters with the Archdruid* (Narration about a conservationist and three of his natural enemies): New York, Farrar, Straus, and Giroux, 245 p. (paperback)

Pratt, W. P., and Brobst, D. A., 1974, Mineral Resources—Potentials and Problems: *U.S. Geological Survey, Circular 698,* 20 p.

# CHAPTER 3
# GEOLOGY AND GEOLOGIC TIME

PLATE 3
Polished slice of the transverse section of a tree trunk showing tree rings and radial cracks. Each year's growth of new wood consists of a thicker darker zone and a thinner light zone having the appearance of tiny dots. (U. S. Geological Survey)

THE UNIQUE ASPECT OF THE SCIENCE OF GEOLOGY IS ITS CONCERN for time. Geology deals with what has happened during past time, how what is happening today will leave a record, and how modern processes can be expected to create predictable effects in the future. The chief tangible records consulted by geologists are minerals, rocks, the regolith, and the Earth's surface features.

The domain of *historical geology* includes the methods by which geologists reconstruct what happened during the history of the Earth. However, the study of physical geology requires a student to acquire a sound perspective about geologic time and the ways in which processes that one can observe in the short term can leave their imprints in the record of the long term.

We begin our presentation about geologic time by reviewing the "time dilemma" that became a major focus of attention early in the nineteenth century. Bitter arguments raged when intellectuals in Western Europe finally became aware that the longevity of the Earth implied by geologists' observations was far larger than that based on scholarly analysis of written records and the Old Testament creation story.

Once geologists had established their case that the world's age is vastly greater than 6000 years, the question naturally arose as to just *how much* greater. Despite much ingenuity devoted to this subject during the second half of the nineteenth century, before radioactivity was discovered and applied to geology, about all geologists were able to do was to try to guess ages based on amounts of geologic work done (such as valleys eroded, sediments deposited, and so forth) using meager information about rates based on short-term observations in the modern world.

A method commonly employed to try to impress upon students the vast amounts of geologic time was to start from the precept that the only kinds of activities involved had been the ordinary, everyday, slow-acting, fair-weather processes that one can observe today. Given such imperceptibly slow-acting processes and impressive geologic results, the conclusion naturally followed that large amounts of time must have elapsed. What might have been the ultimate in this approach was a widely adopted view in the United States early in the twentieth century that we shall refer to as gradualism.

We illustrate examples of geologic work and relative time by citing two halves of the erosion-deposition partnership in the operation of the geologic cycle. We discuss the Grand Canyon as an example of erosion and the growth of the Mississippi River's delta in the Gulf of Mexico as an example of deposition. Next follows an account of an ancient flood in eastern Washington state and how geologists who had adopted gradualism refused at first to believe that anything so catastrophic as this mighty flood could have happened.

Finally, we examine how geologists can establish a relative chronology based on the analysis of rock units in a given area. Such analysis from many localities in Western Europe and North America constituted one of the main contributions of geologic work in the nineteenth century. Our example is based on the rock units that have been uncovered by the Colorado River's erosive activities in the Grand Canyon.

What we do not try to do in this chapter is to show how geologists have been able to calibrate their relative geologic time scale established from field studies and thus to present the basis for the geologic conclusion that the age of the Earth is 4.6 to 4.7 billion years. An understanding of how this calibration has been done is vital for all students of geology to acquire. However, before that understanding is possible, the student must be well versed in many topics of physical geology. These include minerals and their radioactive ages (Chapter 4), and the ways in which geologists can establish the ages of igneous rocks (Chapter 6), sedimentary rocks (Chapter 8), metamorphic rocks (Chapter 9), and episodes of deformation (Chapter 16). Accordingly, we think a detailed presentation of how the geologic time scale has been calibrated logically belongs in the domain of historical geology.

## JAMES HUTTON'S GEOLOGIC CYCLE POSES A TIME DILEMMA

In 1795, when James Hutton published his famous book *Theory of the Earth,* few people made any connection between rocks and the history of the world. Quite the contrary; the history of the world was thought to be fully and completely contained in the sacred writings of the Old Testament.

The Old Testament nowhere states how old the Earth is, but the days of creation are clearly described, and from the accounts of who lived when, biblical scholars attempted to work out a chronology. By making calculations based on the number of generations listed in the Bible (using what has been dubbed the "begat method"), as many as 200 different scholars had individually decided that the Earth had been created between 6000 and 3963 B.C. Archibishop James Usher (1581–1656) of Ireland decided on 4004 B.C. Usher's date became part of the marginal notations in the authorized edition of the King James Bible and came to be accepted as if it had been handed down by the prophets of old. The distinguished rabbinical scholar Dr. John Lightfoot (1602–1675) of Cambridge University even pronounced the very hour and day of creation: 9 A.M., October 23, 4004 B.C.

For many years nothing compelled anyone to doubt the age of the Earth derived by the begat method. However, as the early naturalists began to realize that study of the Earth's natural features formed the basis for unraveling the Earth's past history, a mighty dilemma began to appear. Evidence was uncovered for natural events not mentioned in the creation story. As the list of these well-documented natural events grew longer, the time dilemma became more acute. And to cap it all off, James Hutton was insisting that he could account for a very complex series of geologic events by processes that could still be seen happening in today's world. Hutton was careful to avoid stating just how old he thought the world might be. Just the same, it is clear that he was thinking about huge amounts of time when he wrote: "I see no mark of a beginning or of an end." He was referring to the geologic cycle, and everything that has been learned since Hutton's day proves that he was correct. In particular, what he meant was that he had not found any rocks that could not be explained by the operation of the "great geologic cycle." No one else has ever found such rocks on Earth either.

### Catastrophism and Day Stretching

As the time dilemma was beginning to make itself more and more apparent, various rationalizations were proposed by which the evidence being assembled by the natural philosophers could be fitted into the biblical scholars' chronology. In other words, the intellectuals tried to "have it both ways"; they wanted to bring together the natural philosophers' results and the clerical viewpoints without changing either one. Three possibilities were explored: (1) make all the inferred natural events happen suddenly, so that one could suppose that they all took place within the span of about 6000 years; (2) try to account for all the inferred natural events as results of the Old-Testament flood; (3) make a new interpretation of the six days of creation by substituting some other amounts of time for the "days."

The first two possibilities were argued by a group that became known as *Catastrophists*. Persons who tried to account for the geologic record using catastrophism were eventually forced to admit that too many and too vigorous catastrophes would be required to account for the geologic record within the brief time of 6000 years. Despite the stigma that eventually became attached to catastrophism, we now realize that the catastrophists were correct in drawing attention to vigorous natural activities.

The third possibility might be labeled "day stretching." Advocates of this rationalization argued that the creation days were allegorical and represented epochs of long duration rather than only 24 hours. The day stretchers might have convinced a few individuals, but they were opposed by the clergy. The church fathers would have nothing to do with any tampering, stretching, or nonliteral interpreting of the creation story.

## Cuvier's Compromise

**3.1 Baron Georges Chrétien Léopold Frédéric Dagobert Cuvier** (1769–1832), the distinguished French vertebrate paleontologist, whose "great compromise" that assigned geology to a dark, catastrophic time before the Old-Testament flood, was widely accepted and appeared to be the "perfect solution" to the time dilemma posed by geologists to theologians during the first third of the nineteenth century. (New York Public Library)

Into the midst of these vigorous arguments and difficult dilemmas came a brilliant Frenchman, Baron Georges Cuvier (1769–1832) (Figure 3.1). Cuvier proposed what seemed like the perfect way out of the difficulty. His scheme was widely adopted and became known as "Cuvier's compromise." Cuvier brought geology and the biblical scholars together by proposing that the history of the world included three periods: (1) POST-DILUVIAL (*after the Flood*), (2) DILUVIAL (*the Flood*), (3) ANTE-DILUVIAL (*before the Flood*).

According to Cuvier, the conditions described in the Old-Testament creation story were to be accepted exactly as written, including the 24-hour days. This eventful week began his Diluvial Period. Both the Diluvial and Post-Diluvial times were recorded by written documents; these documents described events that could be understood using scientific methods.

Geology, said Cuvier, referred to events that took place in the Ante-Diluvial Period. As a result, one was at complete liberty to suppose that almost anything might have happened during it without getting into any trouble. Best of all, no scholar had every tried to figure out how long Ante-Diluvial time was. This left the door open for assigning *any* amount of time to it. What a perfect situation! All one had to do to bring geology and the Old-Testament scholars together was to keep geology within the infinite bounds of Ante-Diluvial time.

Cuvier proposed to do exactly that. He was very specific; he sketched out the following conditions that could be used to define his Ante-Diluvial Period. First, it had been a time of eternal darkness. This followed from the implications of the words, "Let there be light." Second, no human beings lived then; they had not been created yet. Third, the world was inhabited by now-extinct monsters. Cuvier could prove this because he was one of the champion diggers of fossil bones; exhibits of his findings filled the halls of the great museums. Cuvier might have encountered a little difficulty with his third condition; after all, what he was arguing for was a world that existed in pre-Creation time. None of his followers seems to have been upset by this point, however. Finally, and perhaps most important to Cuvier, was the concept that everything about the Ante-Diluvial Period had been supernatural, catastrophic, and peculiar. Thus, by definition, there was no way anyone could understand anything about this dark time by trying to make comparisons with the modern world or by applying the principles of reason and of science. Shortly after Cuvier published his ideas, the term "geological time" was proposed to cover this peculiar Ante-Diluvial Period of darkness. The point of proposing the term was to contrast it with "historical time," as described by written records. (Only much later did geological time assume its present definition — very large amounts of time that have elapsed since the oldest rocks on Earth were formed.)

Cuvier was regarded as a great hero of his time by virtue of having solved the great time dilemma. His followers could rest easy with the vast amounts of time James Hutton was writing about by isolating geology within the Ante-Diluvial Period. However, they could not accept Hutton's fundamental concept, specifically, his argument that processes one could observe in the modern world formed a basis for reconstructing what had happened in the past. In Cuvier's scheme of things, there was no room for Hutton's great generalization about the geological cycle. Quite the contrary; geology was placed "out of bounds," so to speak; it went into the realm of superstition, darkness, and catastrophe.

### Lyell to the Rescue

If Cuvier's ideas had persisted, perhaps James Hutton would never be mentioned in geology textbooks, but Hutton's ideas found a champion in an Englishman named Charles Lyell (1797–1875) (Figure 3.2). Lyell devoted his life to proving that Cuvier was wrong and Hutton correct. He did this without attacking Cuvier directly; in fact, he scarcely mentioned Cuvier. Instead, Lyell meticulously established his case for what he termed "the efficacy of existing causes." His book, *Principles of Geology*, published in 1830, was a milestone in the progress of geology. It reestablished Hutton and laid the basis for all modern work in geology.

**3.2 Sir Charles Lyell** (1797–1875), English lawyer-turned-geologist, who without ever attacking Cuvier directly, completely refuted Cuvier's "great compromise" and restored James Hutton's ideas that had been laid aside by Cuvier. (New York Public Library)

Lyell was not involved in the ultimate resolution of the time dilemma. What Lyell did, however, was to make crystal clear to all his readers that the artificial wall Cuvier had built between the modern world and the geologic record was totally incorrect and never should have been admitted as a possible explanation.

Lyell clearly understood that the geologic record required long periods of time. About Hutton's viewpoint, Lyell wrote:

> *. . . The declaration was the more startling when coupled with the doctrine, that all past changes on the globe had been brought about by the slow agency of existing causes. The imagination was first fatigued and overpowered by endeavouring to conceive the immensity of time required*

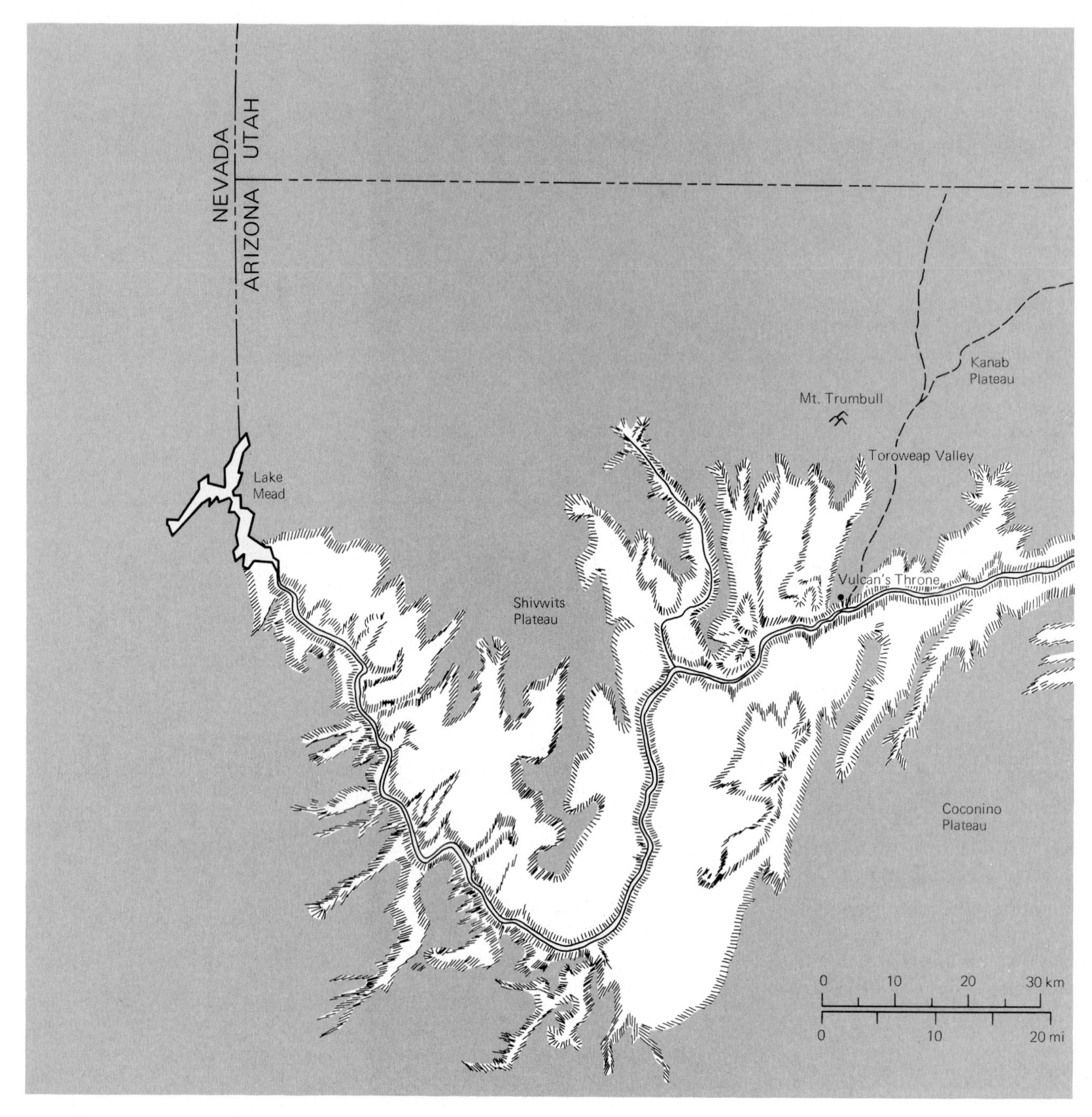


**3.3 The Grand Canyon of the Colorado River,** northwestern Arizona. Sketch map showing location of Grand Canyon and its chief tributaries.

*for the annihilation of whole continents by so insensible process. Yet when the thoughts had wandered through these interminable periods, no resting place was assigned in the remotest distance. The oldest rocks were represented to be of a derivative nature, the last of an antecedent series, and that perhaps one of the many pre-existing worlds. Such views of the immensity of past time, like those unfolded by the Newtonian philosophy in regard to space, were too vast to awaken ideas of sublimity unmixed with a painful sense of our incapacity to conceive a plan of such infinite extent.* (Charles Lyell, 1830, Principles of geology, v. 1, p. 63: London, John Murray.)

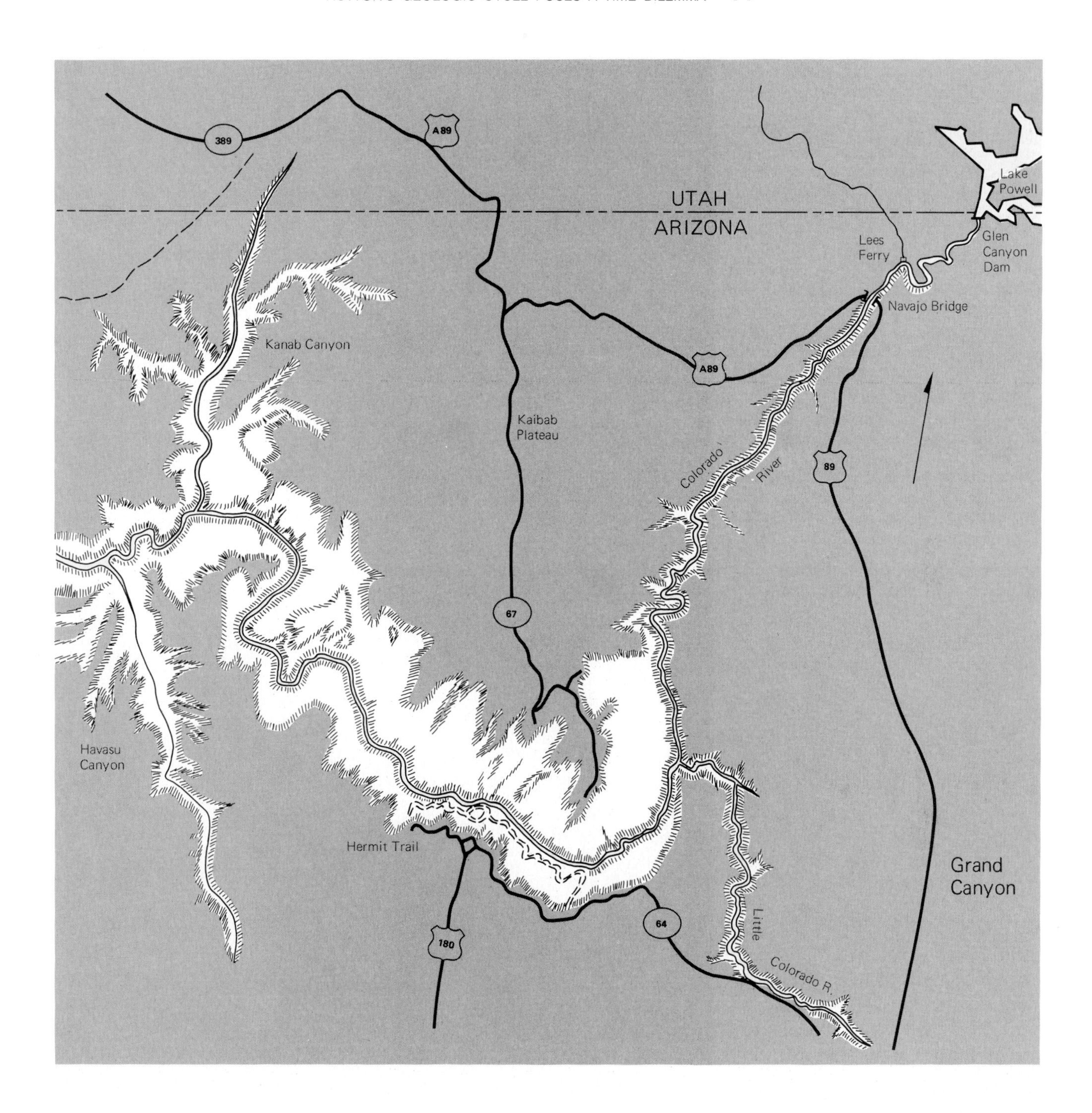


In the words of John Playfair:

> *"No ingenuity has been able to reconcile the natural history of the globe with the opinion of its recent origin. . ."* (1802, par. 125, p. 126).

Numerous individuals have participated in the calibration of the geologic time scale. The application of radioactivity in determining the ages of mineral specimens has brought about our present understanding of the duration of geologic time and of the age of the Earth. (We explain the details of radioactivity in Appendix A and the use of radioactive methods to obtain dates of minerals in Chapter 4.) However, as far as geologists are concerned, Sir Charles Lyell was responsible for establishing the modern science of geology.

## THE GEOLOGIC CYCLE THROUGH TIME (1): EROSION OF THE GRAND CANYON

The Grand Canyon of the Colorado River in Arizona is truly one of the great natural wonders of the world (Figure 3.3). Although the layers of rock that have been exposed by the erosion of the canyon date back to more than 500 million years ago, the actual erosion of the canyon has been shown to be a rather recent event. The uplift of the Colorado Plateau in the area of the Grand Canyon began about 20 million years ago.

The speed of eroding the canyon has recently been shown by study of the volcanic rocks that spilled into the canyon from the north rim. Great sheets of black igneous rock can be traced from the conical volcanic cones that mark the volcanic vents and over the cliffs into the Grand Canyon (Figure 3.4). Volcanic rocks of several ages are present. Lava of an older

**3.4 Ancient lava flows, now hardened into black igneous rock, and cone of loose volcanic debris** (Vulcan's Throne) along north wall of Grand Canyon at mouth of Tuweep Valley. The lava spilled into the Grand Canyon from the upper left, proving that the canyon had been eroded before the volcanic activity began. The volcanic rock blocked the river, but further erosion by the Colorado River has cut gorges through several such natural dams. (J. R. Balsley, U. S. Geological Survey)

episode of volcanic activity completely filled the inner gorge. Later, the Colorado River eroded a new inner gorge. In time, it, too, was partially filled by lava, now volcanic rock, from a younger episode of volcanic activity.

The material eroded from the Grand Canyon has been delivered to the delta of the Colorado River, which now empties into the Gulf of California.

## THE GEOLOGIC CYCLE THROUGH TIME (2): GROWTH OF THE MISSISSIPPI DELTA

Huge quantities of sediment eroded from the central interior of the United States have been delivered to the Gulf of Mexico by the Mississippi River. The detailed history of the Mississippi Delta (Figure 3.5) during the last 10,000 years has been reconstructed by means of thousands of borings, by careful analysis of the sediments, and by dating the

**3.5 The Mississippi River delta;** satellite images using different frequencies of a multispectral scanner (MSS). (a) Maximum contrast between land and water, MSS band 7, shows modern delta (lower right) and two sets of inactive deltas (at top, Nos. 8 and 9 of Figure 3.6e; and left, Nos. 7 and 10 of Figure 3.6e) now being eroded by the sea. (b) Suspended sediment (light gray areas in the water), emphasized by MSS band 5, is being spread along shore along both sides of modern delta. (NASA and U. S. Geological Survey, LANDSAT images on 16 January 1973, lower part of each pair, and on 9 October 1974, upper part of each pair)

widespread layers of peat using radioactive-carbon methods. The history is one of a rapid rise of sea level and then of rapid growth of delta lobes. Growth of a lobe stopped abruptly when the course of the river changed.

In the southern part of its broad valley, the Mississippi River has built two sinuous courses named *meander belts.* The modern river flows in the eastern meander belt. The western meander belt was made by a former course of the river, named the Teche (Figure 3.6).

## CATASTROPHISM REVISITED: THE GREAT SPOKANE FLOOD

During the nineteenth century, nearly all geologists came to regard catastrophism with contempt. This is not surprising, for geologists had to fight catastrophist thinking on many subjects. First of all, Cuvier had made geology synonymous with catastrophism and relegated it to the realm of darkness. Therefore, when Cuvier's ideas were challenged by Lyell, geologists were happy to be regarded as practitioners of science rather than as followers of the supernatural and nonrational. Other aspects of the battle between Catastrophists and Uniformitarians (those

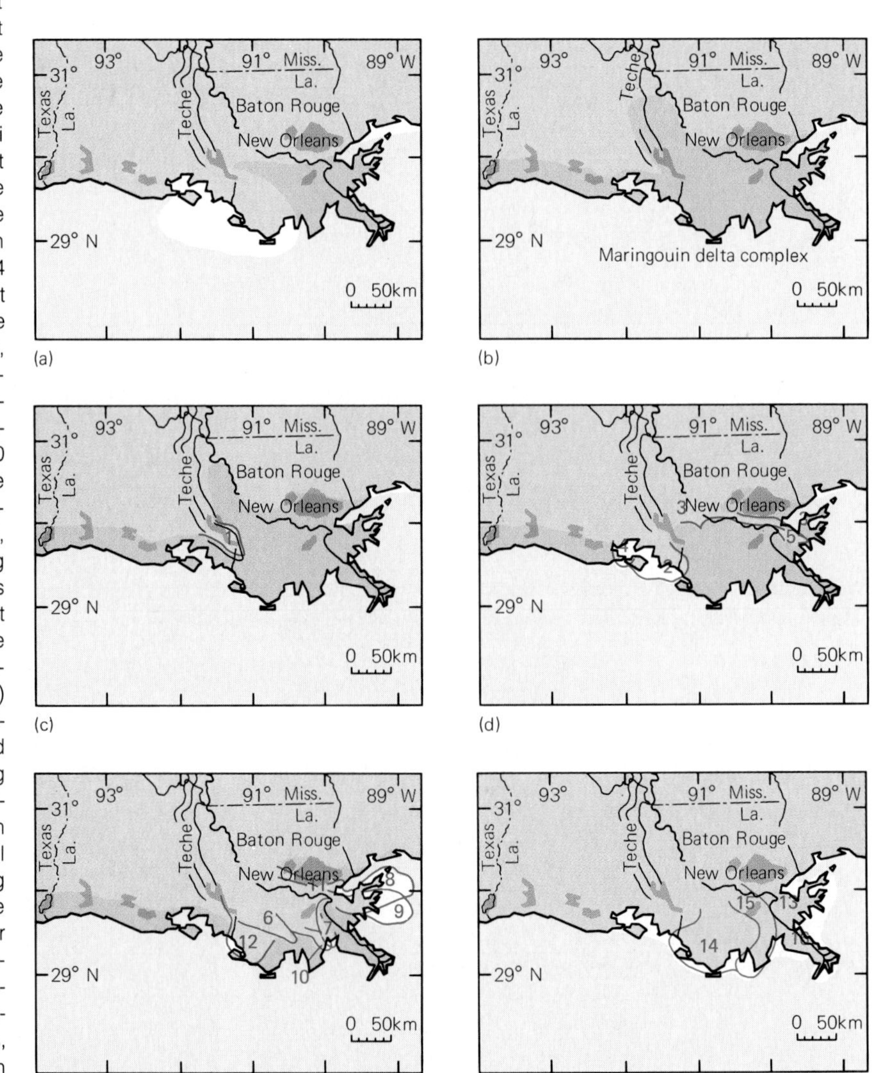


**3.6 Stages in the development of the Mississippi delta complex during the last 10,000 years.** (a) When the sea stood about 15 meters lower than its present level, the Mississippi flowed in the Teche meander belt and deposited the now-submerged Maringouin delta complex. (b) A rapid rise of sea level submerged the Maringouin delta and created an estuary that extended northward to at least Baton Rouge. (c) During the next 550 years, flow of water in the Teche meander belt filled the west half of the estuary and built a delta (No. 1) near the present coast. (d) The lower Mississippi drainage followed two paths for the next 1000 years, about half of the drainage from the Teche belt being captured by the present course of the river past Baton Rouge and New Orleans. Deltas 2 and 4 deposited by Teche drainage; the last Teche delta (No. 4) became inactive about 3900 years ago. Deltas 3 and 5, were built by water from the modern Mississippi channel. (e) After it had abandoned the Teche belt, the Mississippi divided near Donaldsonville, about 100 kilometers WNW of New Orleans. The eastern branch, flowing past New Orleans, built the St. Bernard deltas (Nos. 8, 9, and 11). The western branch, flowing southward, built the Lafourche deltas (Nos. 6, 10, and 12). Delta No. 7 was built by another branch of the river, the Lafourche channel, that flowed south from New Orleans (No. 7, 3400 to 2000 years ago). (f) About 800 years ago, the St. Bernard deltas were abandoned, and the river flowed southeastward from New Orleans, building the Plaquemine delta (No. 13) and, starting about 150 years ago, the modern delta (No. 16). The Lafourche channel built deltas 14 and 15 before becoming inactive. The north end of the Lafourche channel was dammed in 1904. (After D. E. Frazier, 1967, Recent deltaic deposits of the Mississippi River, their development and chronology: Gulf Coast Association of Geological Societies, Transactions, v. 17, p. 287–315; and G. M. Friedman and J. E. Sanders, 1978)

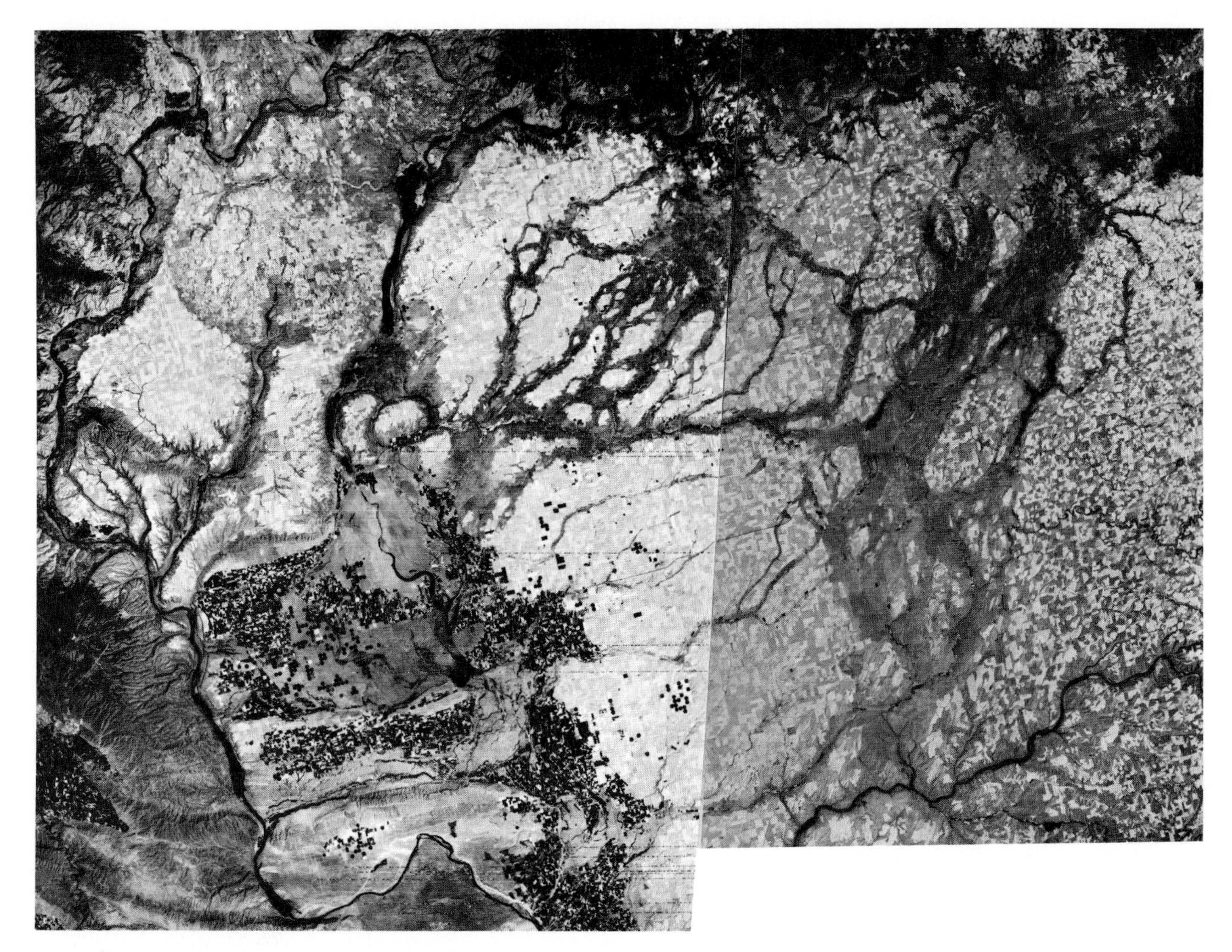

**3.7 Channeled scabland, Washington, an effect of the "Great Spokane Floods,"** viewed from satellite on 27 August 1973 (left frame) and on 8 August 1973 (right frame). Location, scale, and orientation on Figure 3.8. (National Aeronautics and Space Administration)

who followed the ideas of Hutton and Lyell) continued to appear and the Catastrophists were finally defeated.

One brand of geologic thinking that became popular in the United States as a method of emphasizing the great length of geologic time has been called *gradualism*. The gradualists tried to dramatize the effects of geologic time by arguing that time alone was necessary to accomplish great geologic work. The more time they had to work with, the more they were convinced that the pace of geologic processes was always slow and uneventful. In particular, gradualists emphasized that the everyday flow of water in a river could explain the erosion of a river valley. They were in principle opposed to notions that floods are important. After all, the whole series of arguments about the biblical flood and the use of this flood to "explain" the geologic record had been settled and the decision went against the flood advocates.

It is not hard to imagine what a difficult time a young geologist by the name of J. Harlan Bretz (1882–     ) must have had when he began in the 1920s to write papers about the hypothesis that big floods had been important events in Pleistocene history of the area southwest of Spokane, Washington, known locally as the channeled scabland.

After thirty years of being right for what seemed to many geologists to be no reasons or for wrong reasons, Bretz was at last able to be right for the right reasons. Aerial photographs and images from satellites (Figure 3.7) demonstrated that diagnostic features formed by tremendous flows

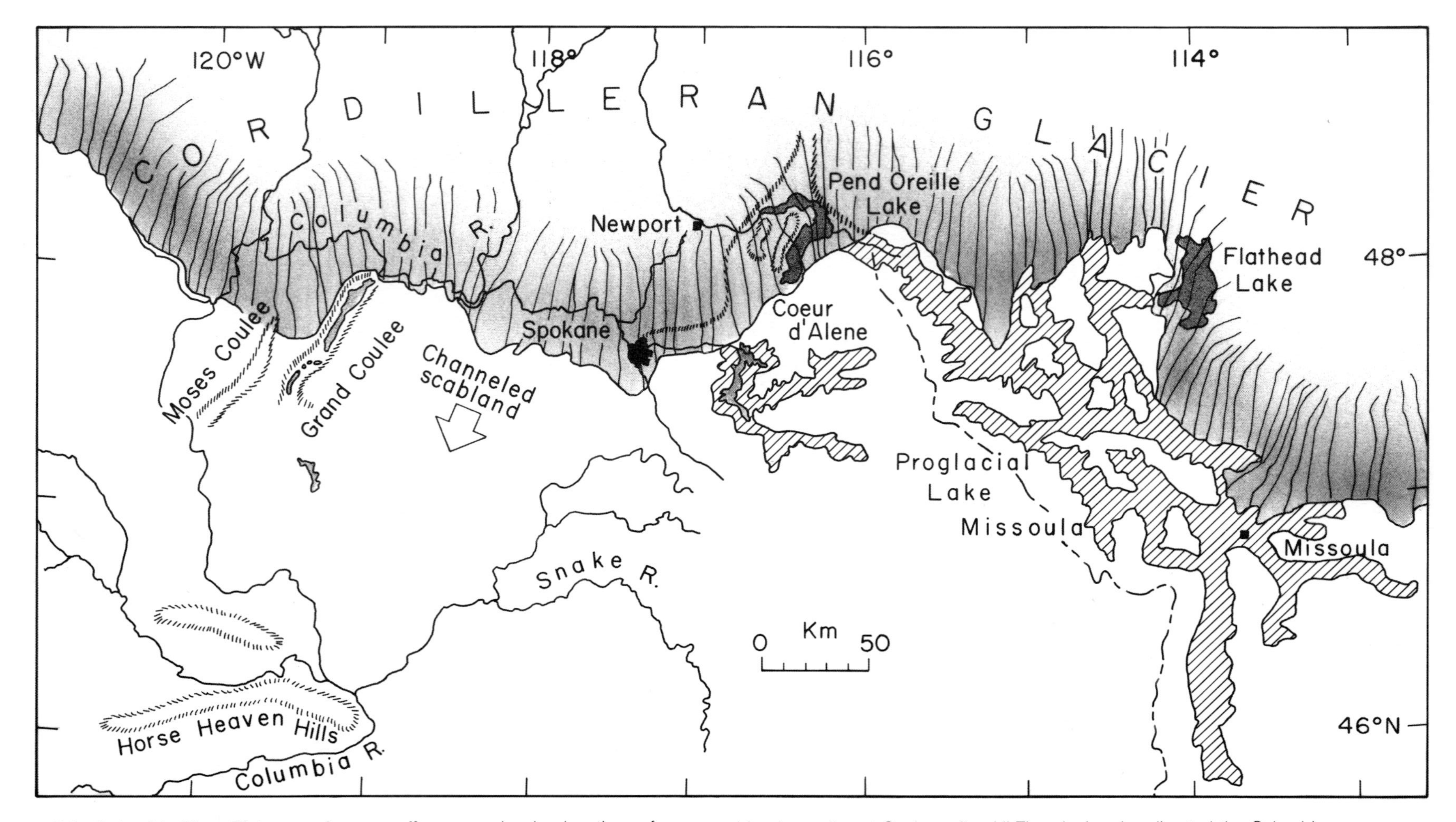


**3.8 Columbia River Plateau and surrounding area,** showing locations of channeled scabland, Cordilleran Ice Sheet (of Pleistocene age), and former proglacial Lake Missoula (also of Pleistocene age). When the Cordilleran Glacier blocked the rivers that flowed toward the northwest, Lake Missoula filled with water. When the edge of the Cordilleran Glacier retreated, even by a small amount, the water from Lake Missoula poured across the channeled scabland as a "great Spokane flood." The glacier also diverted the Columbia River, forcing it to erode new channels, two of which form Moses Coulee and Grand Coulee. (After G. M. Friedman and J. E. Sanders, 1968, Principles of sedimentology: New York-Santa Barbara-Chichester-Brisbane-Toronto, John Wiley and Sons, 792 p., Figure 2.9, p. 239, based on J. Harlen Bretz)

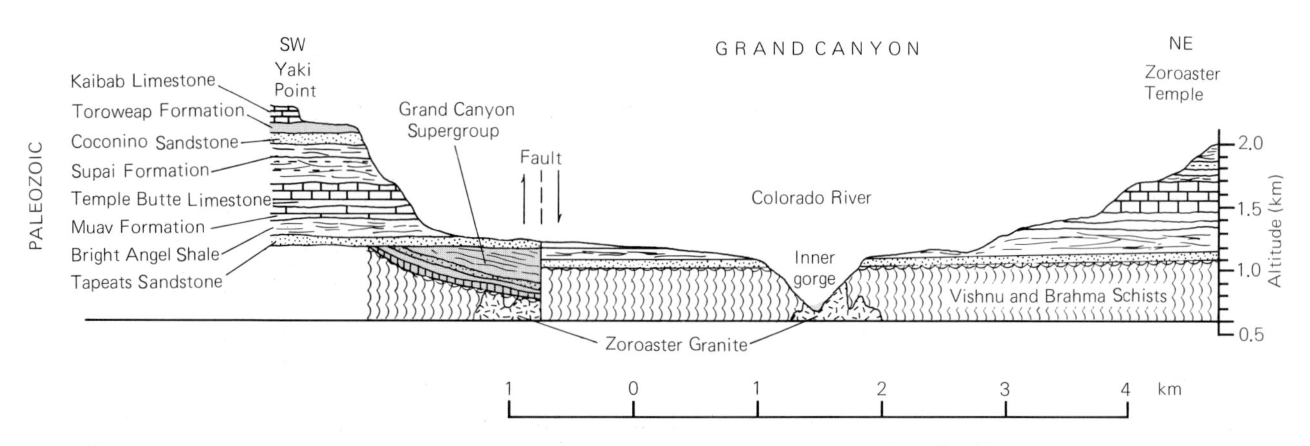


**3.9 Profile and geologic section across Grand Canyon of Colorado River,** northern Arizona, from Yaki Point to Zoroaster Temple. Topographic profile drawn from National Geographic Society's Grand Canyon topographic map, July 1978, on scale of 1/24,000. Further explanation in text.

of water are present. In addition, a geologically reasonable source for the water has been found. The water of ancient Lake Missoula was at times held in by the Cordilleran Ice Sheet. (Figure 3.8). When this glacier retreated, all the water flowed out of the lake and was discharged in great floods across the channeled scablands, as Bretz had written.

The Bretz case is only one of many examples of a shift in point of view. The importance of the geologic work of floods, storms, and other vigorous events has been neglected or denied in the past, but is now being accepted. The distorted viewpoint of the gradualists is now being corrected by a new look at natural "catastrophes."

## RELATIVE CHRONOLOGY OF THE ROCK UNITS EXPOSED IN THE GRAND CANYON

In the Grand Canyon, the Colorado River has uncovered four major rock units. The relationships among these units are so clearly exposed that we can illustrate how geologists analyze the relative ages of these units even before we take up the processes by which such rock units are formed. If we consider that the Grand Canyon itself forms a major feature, then our list includes five items. From youngest to oldest these are as follows. (1) the Grand Canyon itself and the Colorado River at the bottom, (2) the volcanic rocks that spilled into the Grand Canyon from the north side (see Figure 3.4), (3) the vast array of nearly horizontal rock strata, about 1 kilometer thick, that form the walls of the open part of the Canyon, from the Tapeats Sandstone at the base to the Kaibab Limestone at the top (Figure 3.9), (4) the tilted and gently bent strata of sedimentary rocks assigned to the Grand Canyon Supergroup that are truncated by the Tapeats Sandstone, and (5) a "basement complex" of metamorphic rocks (schists) and igneous rocks (granites), exposed in the V-shaped inner gorge of the Colorado River (see Figure 3.9), and that underlies either the Tapeats Sandstone (in most places) or the Grand Canyon Supergroup (in a few places). We describe these features in the order listed and then reconstruct their history, starting from the oldest event and proceeding in sequence toward the youngest event.

The Grand Canyon (Figure 3.10) is a composite feature consisting of a V-shaped (in profile) inner gorge that lies about in the middle of a much wider, steep-sided, and flat-bottomed valley (see profile of Figure 3.9). The broad valley has formed because the walls of what was doubtless a

**3.10 Grand Canyon, view south from an airplane at eastern end** where Colorado River leaves Glen Canyon (at left) and flows into the open upper canyon at Lees Ferry, Arizona (center). Paria Canyon and Vermilion Cliffs are in right foreground. (Myron J. Rand, 1978)

narrow, steep canyon, have retreated back from the river by distances of 1 to 3 kilometers.

Near the Grand Canyon, and in a few places located just at the canyon's rim, are volcanic cones and lobes of dark-colored igneous rock that have formed by the cooling of lava flows. Volcanic rocks of several episodes have been identified. An example of volcanic products of the most recent episode is Vulcan's Throne on the canyon's north wall (see Figure 3.4). Evidence from the inner gorge indicates that lava flows that spilled into the canyon dammed the Colorado River and that eventually, the river cut through the natural dam formed of igneous rock and continued to erode afterward (Figure 3.11). Several episodes of damming and dam breaching are indicated.

The rocks forming the upper walls of the Grand Canyon are sedimentary strata of Paleozoic age that begin at their base with the Cambrian Tapeats Sandstone (Figure 3.12) and end at the top with the Permian Kaibab Limestone, which forms the caprock of the canyon rim. These sedimentary strata can be followed along the river from one end of the canyon to the other and they can be matched exactly from the north wall to the south wall (see Figure 3.9). They illustrate three important points about sedimentary deposits that were first noticed in 1669 by Nicolaus Steno (1631–1686), a Danish naturalist who lived most of his life in Italy. These are as follows: (1) most strata are horizontal when deposited; (2) they are arranged originally with the oldest layers at the bottom and progressively younger layers toward the top; and (3) strata are usually

**3.11 Breached natural dam of volcanic rock** (black, in foreground) at Mile 245, Colorado River, Grand Canyon, Arizona. The volcanic rock plugged the lower part of the canyon; subsequently, the Colorado River cut a narrow gorge through the dam. (Agi Castelli, June 1979)

continuous throughout their areas of deposition; thus, the Paleozoic strata exposed in the north and south walls of the Grand Canyon formerly were continuous and have become separated because of erosion.

The Paleozoic strata were cited in early arguments over the question of whether any river could erode so much rock as has been removed from the Grand Canyon. Those who doubted that so much erosion could have taken place were invited to contemplate the significance of the sedimentary strata, which prove that some area, now inaccessible, must have been eroded to yield the sediment that forms the sedimentary rocks.

Two other aspects of the strata include subsidence and conditions of deposition. Because these strata have been derived from sediments of the kinds that are spread out in horizontal layers and not heaped up into great piles, the thickness of the layers implies that the sediments accumulated in a subsiding region. Moreover, the kinds of minerals and fossils within the strata contain clues about whether the strata were deposited in the sea or in a low-lying area near but outside of the sea. The limestones imply clear, warm seas; the shales have formed from compression of ancient muds, and the sandstones, from cementation of former sands.

**3.12 Paleozoic strata, Grand Canyon,** from Tapeats Sandstone at river level to Redwall Limestone (Mississippian) at crest of butte (compare Figure 3.9). (Agi Castelli, 1978)

Beneath the Tapeats Sandstone are found two contrasting kinds of rocks: (1) locally horizontal, but usually tilted and warped sedimentary strata and (2) contorted metamorphic rocks and associated bodies of granite. The strata belong to the Grand Canyon Supergroup; the contorted metamorphic rocks and granites to the "basement complex."

The scattered remnants of the Grand Canyon Supergroup record the deposition, deformation, and erosion of a group of strata older than the Tapeats Sandstone, yet younger than the "basement complex" (Figure 3.13). These relationships are proved by the situations at the boundaries or *contacts* (to use the geologic term) between adjacent bodies of rock. Both the Paleozoic strata and the Grand Canyon Supergroup overlie the "basement complex" along a former erosion surface. Moreover, the Tapeats Sandstone at the bottom of the Paleozoic strata, still horizontal, overlies the truncated and tilted strata of the Grand Canyon Supergroup. Accordingly, the indicated order of ages is (from oldest to youngest): (1) "basement complex," (2) Grand Canyon Supergroup, and (3) Tapeats Sandstone and overlying strata.

The implications of the Grand Canyon Supergroup include both originally horizontal positions and continuity of deposition, as with the Paleozoic strata. Similarly, the thickness of the Grand Canyon Supergroup implies an older episode of subsidence, and the kinds of strata within this supergroup suggest that they were deposited outside the sea, perhaps as an extensive valley filling. After the strata of this supergroup had been

**3.13 Grand Canyon Supergroup strata**/ (Precambrian) resting on Precambrian basement schists (foreground) and overlain by Tapeats Sandstone (Cambrian, forming cliff in skyline at center). (Agi Castelli, 1978)

**3.14 Precambrian basement rocks,** inner gorge of Grand Canyon at Mile 236. Light-colored granite at left has cut into dark-colored schist (at right). Later, the granite has been cut by narrow bodies of black igneous rock. (Agi Castelli, 1979)

deformed from their originally horizontal positions, they were eroded to a nearly flat surface. Still later, the region subsided to begin the episode of deposition of the Paleozoic strata.

"Basement complex" is a term geologists use to refer to whatever combinations of rocks, usually including various metamorphic varieties and possibly also granites, form the lowest and oldest units. In many localities, the rocks composing the basement complex are those that originated within ancient mountain belts. The rocks of the Grand Canyon basement complex are exposed only along the walls of the inner gorge. They include several kinds of schists and of granites (Figure 3.14). Because schist is a kind of metamorphic rock that is derived from a still-older sedimentary rock, the geologic history of the basement complex must begin with deposition of sediment. Later, the sedimentary strata were compressed and crowded together to form their present closely folded structures, and the material to make the granite was injected into them. Such changes in sedimentary strata and the formation of granites take place deep within the Earth, possibly at depths of 20 kilometers or so, where temperatures are high and pressures, enormous.

We can illustrate the successive stages in the development of the Grand Canyon by means of a series of block diagrams (Figure 3.15). The oldest event in the history is deposition of sediment on the bottom of a seaway that formed a trough (Figure 3.15a). After the sediments in the trough had accumulated to great thickness, they were deformed, changed, and invaded by granites; they formed an ancient chain of mountains named the Mazatzal Mountains, which trended NE–SW (Figure 3.15b). As the mountains were elevated, they were eroded;

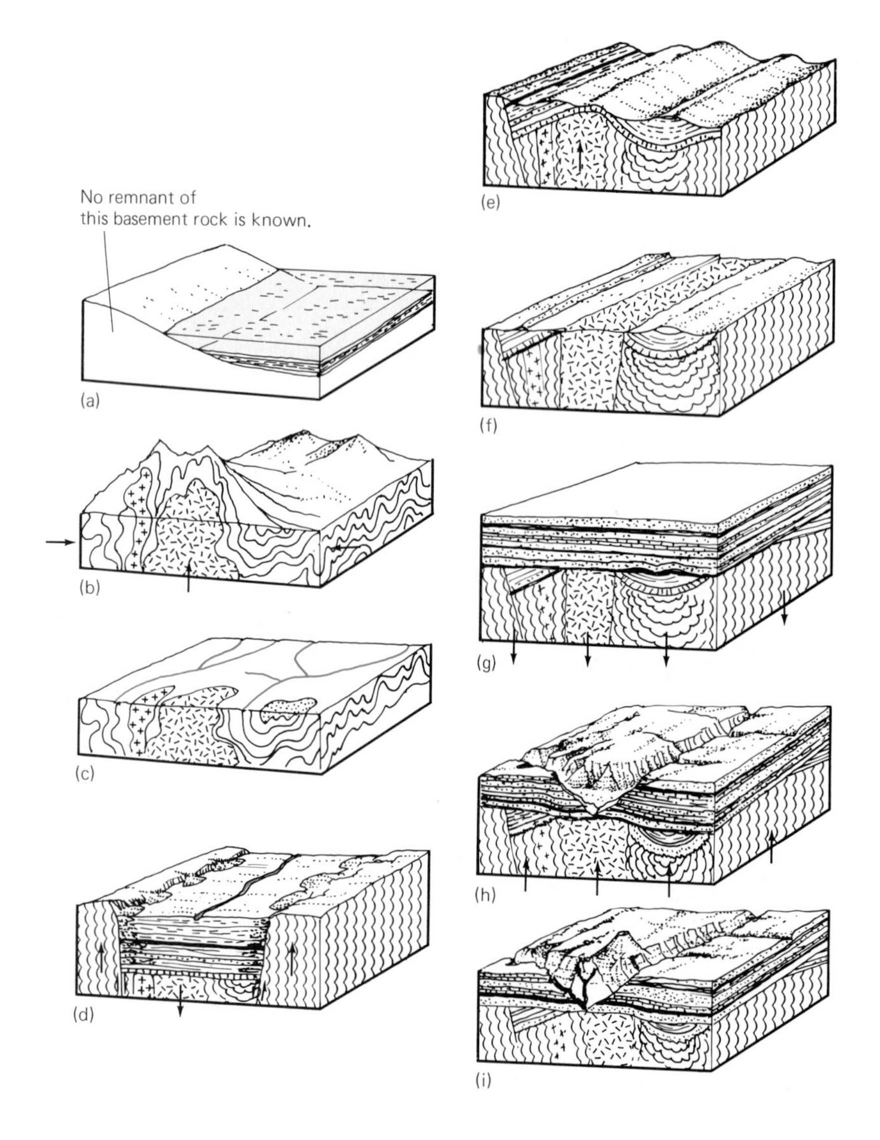


**3.15 Stages in development of rocks exposed in Grand Canyon,** Arizona, schematic blocks. (a) Trough in ancient seaway subsides, sediments destined to become schists at a later time, accumulate. (b) The Mazatzal Mountains, formed after deformation and metamorphism of the strata into schists and the emplacement of granites. (c) Erosion destroys Mazatzal Mountains and forms a nearly flat surface cut across the ancient basement rocks of the former mountains. (d) Faults form, blocks subside, and valley-fill strata of the Grand Canyon Supergroup are deposited. Black layer near top of fill represents a sheet of igneous rock that penetrated the strata. (e) Strata of Grand Canyon Supergroup are deformed; folds and faults are the result. (f) Erosion after deformation results in second low-relief erosion surface of regional extent. (g) Grand Canyon region after subsidence and deposition of Paleozoic strata about 1 kilometer thick. The region was also generally subsiding throughout much of the Mesozoic and early parts of the Cenozoic Eras, but strata of these ages are not exposed at the Grand Canyon. (h) Results of uplift and river erosion. (i) Volcanoes become active; some discharge lava into the Grand Canyon.

eventually elevation ceased and erosion reduced the peaks to ruins, forming a generally flat land surface (Figure 3.15c).

A new round of deposition began when the Grand Canyon region was pulled apart so that great faults formed and blocks between some faults subsided. In these ancient rift valleys the strata of the Grand Canyon Supergroup were deposited. After several kilometers of such strata had accumulated, they were invaded by sheets of molten matter that solidified to create dark-colored igneous rocks (Figure 3.15d). The episode of sediment accumulation evidently ended because new deformation and uplift commenced. The result was the bending and tilting of the strata of the Grand Canyon Supergroup and their general elevation and erosion (Figure 3.15e). Eventually, a second nearly flat erosion surface formed (Figure 3.15f). Parts of this surface were underlain by the schists and granites of the basement complex and parts, by the strata of the Grand Canyon Supergroup.

The Paleozoic cycle of deposition began when the new erosion surface started to subside on a regional basis. After regional subsidence of as much as 4 kilometers, lasting throughout most of the Paleozoic and Mesozoic Eras, a new body of flat-lying sedimentary strata had completely buried both the Grand Canyon Supergroup and the older basement complex (Figure 3.15g). Late in the Cenozoic Era, elevation of a broad area and establishment of a through-flowing river system set the stage for erosion of the Grand Canyon (Figure 3.15h). The last major event involved volcanic activity (Figure 3.15i).

From this review of the rock units exposed in the Grand Canyon one can see how geologists are able to organize rocks within a given area into units and to assemble these into a sequence within which relative ages can be demonstrated. The ultimate triumph of the science of geology took place when it became possible to calibrate the geologic time scale in terms of years. This great achievement is based on radioactivity; it includes knowledge of the radioactive ages of both minerals and rocks. We shall be discussing these topics in appropriate following chapters, starting with Chapter 4 devoted to minerals and their radioactive ages.

## CHAPTER REVIEW

1. The unique aspect of geology is its concern for time. Geology deals with what has happened in the past, how what is happening now will leave a record, and how modern processes can be expected to create predictable effects in the future.

2. Naturalists' recognition that study of the Earth's features might give clues to its history and Hutton's concept of the continuing processes of the geologic cycle challenged the prevailing ideas that the history of the world was fully and completely contained in the Old Testament and that the Earth was between 6000 and 8000 years old.

3. Attempts to resolve the time dilemma included *catastrophism,* the belief that the geologic record could be attributed to sudden, catastrophic events, such as the Flood, within a time period of about 6000 years; *day stretching,* a reinterpretation of the time meant by the six Biblical days of creation; and *Cuvier's compromise,* the concept that geology dealt with events that had taken place during a dark, supernatural time of unspecified duration, which had elapsed before the Creation as written in the Old Testament and which could not be understood by reason or science.

4. Publication in 1830 of Charles Lyell's *Principles of Geology* firmly reestablished Hutton's ideas and used these ideas to form the basis for all modern work in geology.

5. One of the remarkable contributions of geology has been the basis for calculating how old the Earth is and for reconstructing what has happened to the Earth during the passage of huge amounts of time.

6. Examples of geologic features formed by the long-continued operation of modern processes still at work are the erosion of the Grand Canyon and the deposition of the Mississippi delta.
7. The *gradualists,* a group of geologists who in the nineteenth century argued that the slow progress of geologic processes was sufficient explanation for all geologic work, represent a point of view as distorted as that of the catastrophists they opposed. With the accumulation in the twentieth century of evidence of past catastrophes, the importance of the geologic work of floods, storms, and other vigorous natural events is now being widely accepted.
8. Within the Grand Canyon are exposed four major rock units, from youngest to oldest: (a) volcanic rocks; (b) flat-lying Paleozoic sedimentary rocks; (c) tilted and bent strata of the Grand Canyon Supergroup; and (d) the basement complex of schists and granites.
9. The history implied by the geologic evidence in the Grand Canyon begins with deposition of sediments in a large marine trough. Great subsidence and deformation converted the sediments of the trough into schists; materials to make granites were injected. After great uplift and erosion, a flat surface was created. Faults created subsiding valley blocks on which sedimentary strata of the Grand Canyon Supergroup were deposited as valley-filling units up to 3 km thick. Tilting, folding, and uplift affected these strata, which were then eroded to a second flat surface. Regional subsidence and deposition prevailed during the Paleozoic and Mesozoic Eras; strata at least 4 km thick were deposited, many in an ancient sea. Late Cezozoic elevation raised some of these marine strata as much as 2 km above modern sea level. After the Colorado River became established, much of the Grand Canyon was eroded. During the Pleistocene Epoch, volcanoes discharged lava into the Grand Canyon, damming the Colorado River several times. Post-damming erosion has cut gorges in the volcanic rocks and continues today.

## QUESTIONS

1. Why did James Hutton's ideas create a "time dilemma"?
2. Discuss some of the ways that were proposed early in the nineteenth century for solving the time dilemma.
3. What was *Cuvier's compromise?* What was the impact on Hutton's ideas of the widespread acceptance of Cuvier's compromise?
4. Who was Sir Charles Lyell? What did Lyell do that was significant for modern geology?
5. List some examples of features that have been formed as the result of the operation of some geologic process through a period of time.
6. Explain the views of the group of geologists known as *gradualists.* How did gradualists react to the publication in the mid-1920s of Bretz's proposal of the occurrence of the "Great Spokane Flood"? What is the present status of Bretz's hypothesis?
7. List the four major rock units in the Grand Canyon area. Summarize the geologic history implied by these rock units and by the Grand Canyon itself.

## RECOMMENDED READING

Baker, V. R., 1978, The Spokane Flood Controversy and The Martian Outflow Channels: *Science,* v. 202, p. 1249–1256.

Bailey, Sir E. B., 1967, *James Hutton—The Founder of Modern Geology:* Amsterdam-London-New York, Elsevier Publishing Co., Ltd., 161 p.

Eicher, D. L., 1976, *Geologic Time,* 2nd. ed.: Englewood Cliffs, New Jersey, Prentice-Hall, Inc., 150 p.

Eiseley, L. C., 1959, Charles Lyell (1797–1875): *Scientific American,* v. 201, no. 2, p. 98–109. (offprint 846)

Haber, F. C., 1959, *The Age of the World, Moses to Darwin:* Baltimore, Maryland, The Johns Hopkins University Press, 303 p.

White, A. D., 1895, *A History of the Warfare of Science with Theology in Christendom:* New York, 1955 reprint, George Braziller, in 2 vols., v. 1, 415 p.; v. 2, 474 p.

Wilson, L. G., 1972, *Charles Lyell. The Years to 1841: The Revolution in Geology:* New Haven and London, Yale University Press, 553 p.

COMPOSITION, STRUCTURE, AND SOME PROPERTIES OF MINERALS
CHEMICAL GROUPS OF MINERALS
GROWTH OF MINERALS AND MINERALS AS INDICATORS OF ENVIRONMENT
RADIOACTIVITY: THE BASIS FOR DETERMINING AGES OF MINERALS

# CHAPTER 4 MINERALS AND THEIR RADIOACTIVE AGES

PLATE 4
Transparent slice of quartz showing hexagonal outline of crystal and dark, hairlike inclusions of rutile. Wide dark area in lower center marks a diagonal crack that cuts through part of the specimen. Outward growth is suggested by concentric hexagonal patterns starting with light area in center. (Barnard College specimen, photograph by J. E. Sanders)

MINERALS ARE THE FUNDAMENTAL BUILDING BLOCKS OF ROCKS and form various specialized deposits that are mined to recover useful materials. In a geologic context a *mineral* is defined as a naturally occurring solid having a definite chemical composition (either fixed or varying within a specific range) and an ordered ionic arrangement. Most minerals are formed by processes that do not take place within the soft parts of living organisms, but the hard parts, such as shells and bones, of some organisms, which are secreted by organisms are considered as minerals.

The key to minerals is their ordered ionic arrangement. Natural glass, which is composed mostly of silicon and oxygen, is not considered to be a mineral because its ions do not form a geometrically symmetrical network. By contrast, quartz, which like natural glass consists of silicon and oxygen, is a mineral because its ions do form a geometrically symmetrical network. Naturally occurring elements, such as the metals gold, silver, and copper, whose ions are packed in geometrically symmetrical networks, are considered minerals.

In the following section, we discuss the composition and structure of minerals as a basis for understanding the properties of minerals. The features and tests by which minerals can be identified are presented in Appendix B. Other major sections of this chapter include some of the chief chemical groups of minerals that are important in geology; important geologic information that can be obtained from the study of the shapes of minerals; analysis of minerals as products of natural environments; and measurements of radioactive isotopes contained in and radiation effects on mineral specimens, which enable the ages of the specimens to be calculated. The scientific study of minerals is *mineralogy*. *Crystal chemistry* refers to the study of the relationships between the properties of minerals and the ways the ions are organized. Readers not familiar with chemical concepts about the structure of matter, radioactivity, and radioactive reactions should read Appendix A before continuing with this chapter.

## COMPOSITION, STRUCTURE, AND SOME PROPERTIES OF MINERALS

Minerals can be understood if one first realizes that they are composed for the most part of the ions of only 12 different elements (see Appendix A) and that the ions are arranged in definite structural patterns that are controlled by the sizes and electrical charges of the ions. We first review the ions, examine the structural patterns they form, and then mention two important properties of minerals.

### Ions of Common Minerals

The ions of most minerals include those of the eight most abundant elements composing the Earth's crust (oxygen, silicon, aluminum, iron, calcium, sodium, potassium, and magnesium) plus those of carbon, hydrogen, chlorine, and sulfur. For the purposes of understanding minerals, we can consider that the ions form tiny spheres and that these spheres can be classified according to their diameters and electrical charges (Figure 4.1).

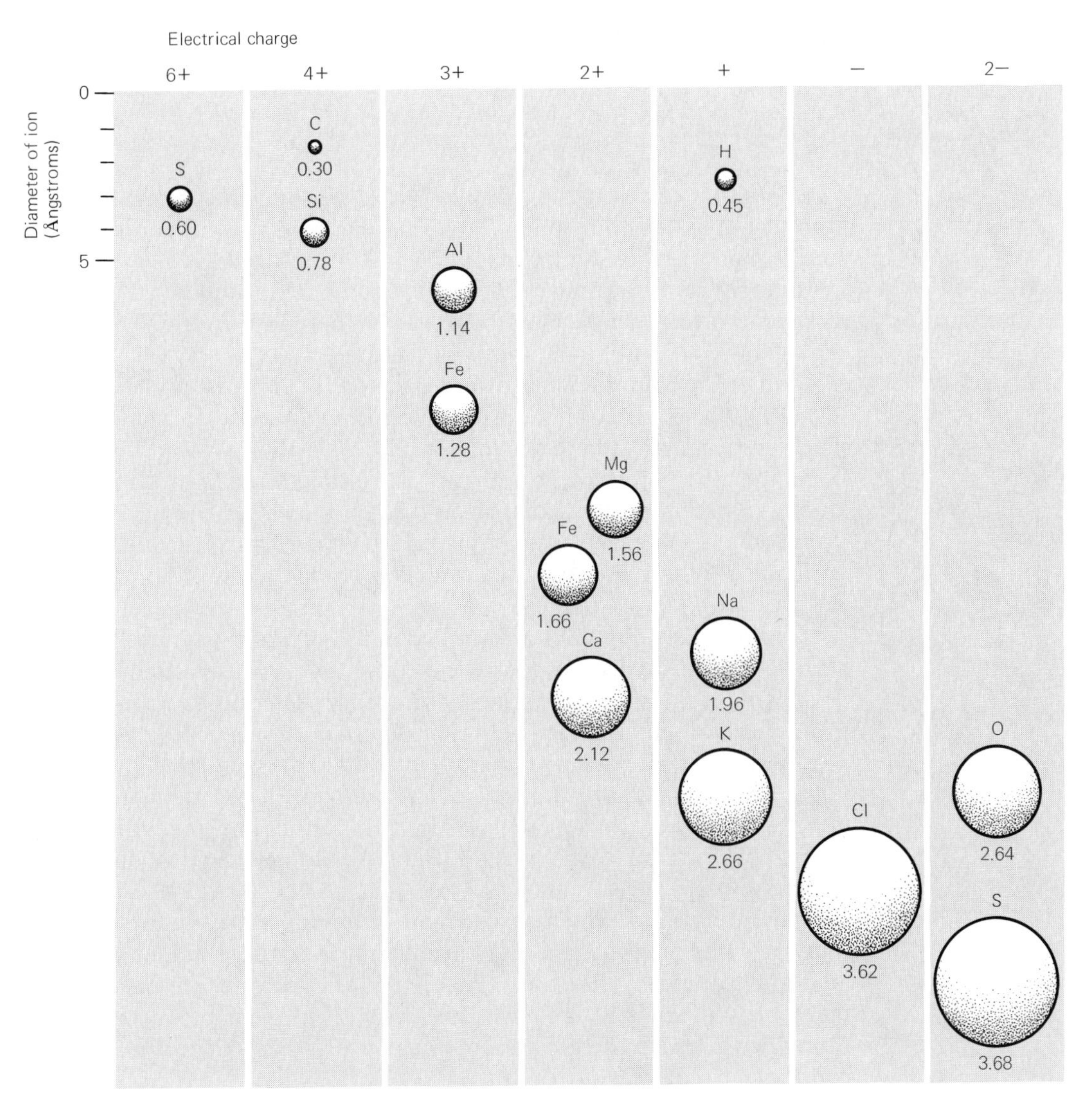


**4.1 Common ions of rock-forming minerals** arranged by their sizes and electrical charges.

## Structure of Minerals

All minerals consist of building blocks called *unit cells.* A unit cell is a regular arrangement of ions held together by electrical forces; it is the smallest unit displaying the properties of the mineral. Because, as we noted previously, the ions can be considered as tiny spheres, we can begin our discussion of the structure of minerals by examining some ways in which spheres can be packed together.

Within a single plane, spheres can be organized so that lines drawn through the centers of the spheres meet at right angles (Figure 4.2a). Additional planes of spheres can be laid on top of one another, so that lines drawn through the centers of the rows of spheres are mutually perpendicular (Figure 4.2b). This arrangement of spheres is named *cubic packing.* An example of a mineral having cubic packing is halite (Figure

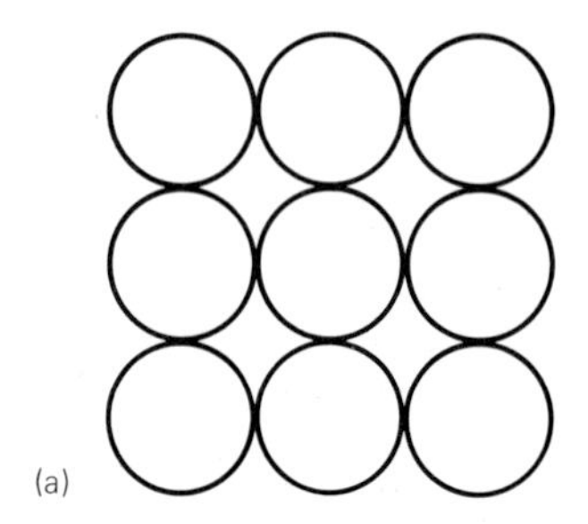


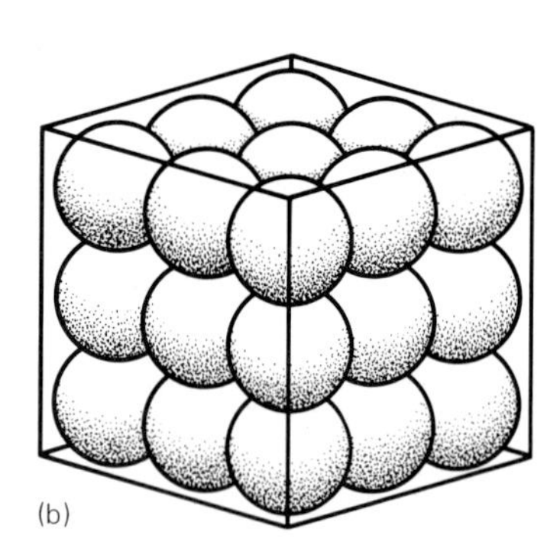


**4.2 Cubic packing.** (a) Spheres of one size in a single plane viewed from above. (b) Three-dimensional view.

4.3). The unit cell of halite (common table salt or sodium chloride), for example, contains four ions of chlorine and four ions of sodium. In a single grain of table salt are about $56 \times 10^{17}$ unit cells.

Another way to arrange spheres of the same size is known as *hexagonal packing*. In hexagonal packing the rows of spheres are offset, so that their centers attain their closest spacings (Figure 4.4).

Another important geometric arrangement of spheres is a *tetrahedron*. In a tetrahedron, three spheres forming a triangle are set in a single basal plane and a fourth sphere occupies a central position among these three but in the next-higher plane (Figure 4.5). Tetrahedra are very widespread among minerals. In particular, four oxygen ions arranged in a tetrahedron contain a central space that is just the right size to accommodate a silicon ion (Figure 4.6a), or, with a little stretching, an aluminum ion (Figure 4.6b).

The unit cells of minerals are stacked in a systematic way, like bricks in a wall, with virtually no spaces between them. The systematic ionic arrangement forms a *lattice* and the lattice determines most of the properties of a mineral. One of these properties is the beautiful geometric shapes of crystals. We define *crystals* as solid bodies bounded by natural plane surfaces that are the external expression of the regular interior lattice structure. Each crystal grows by additions of ions to its diagnostic unit cell; therefore each crystal's fully developed form is distinctive (Figure 4.7).

From careful study and measurements of angles, mineralogists realized that each time a crystal of a given mineral forms, the angles between adjacent planes are the same. This led to the formulation of a fundamental law of crystals; the *law of constancy of interfacial angles* states that the angles between corresponding faces on all crystals of one mineral are constant.

**4.3 Cubic packing of ions of two sizes in halite.** (a) Unit cell consisting of small sodium ions and large chlorine ions, enlarged view. (b) Crystal built of eight unit cells.

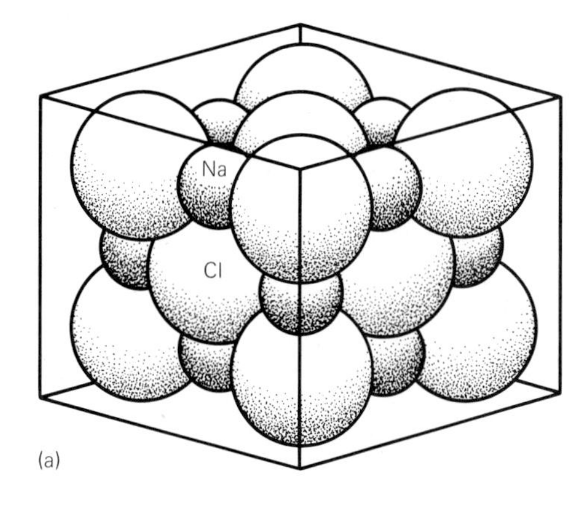


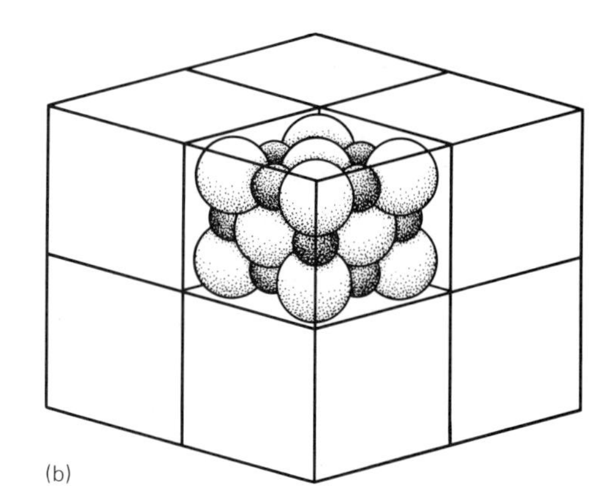

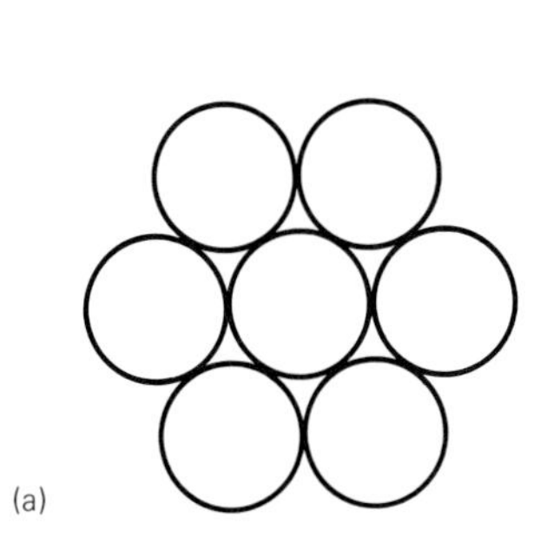


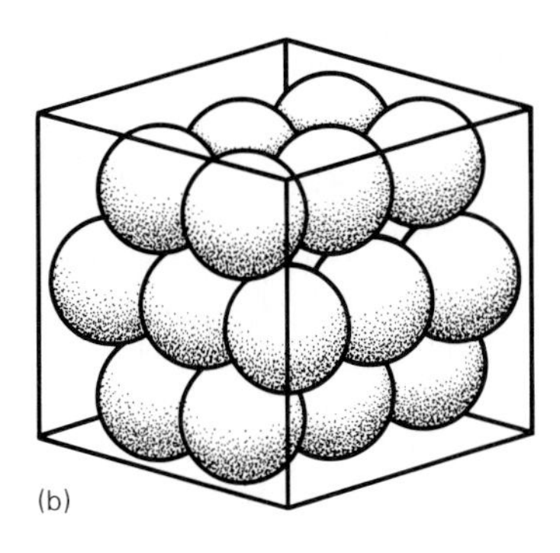


**4.4 Hexagonal packing.** (a) Spheres of one size in a single plane viewed from above. (b) Three-dimensional view.

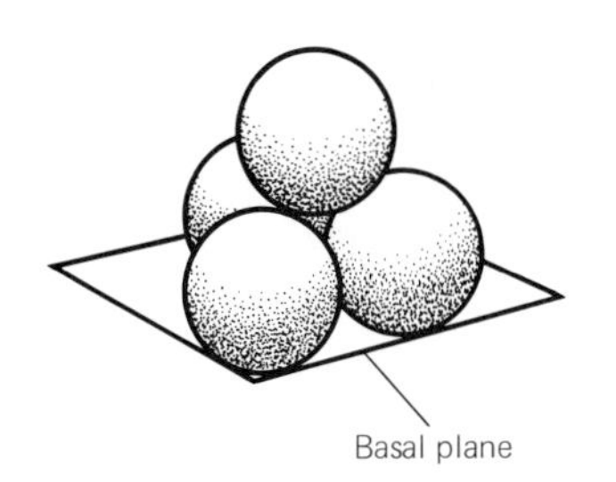


**4.5 Spheres packed as a tetrahedron.**

The reason why crystals formed in such regular ways was not understood until 1912, when an experiment suggested by Max von Laue (1879–1960) of the University of Munich revolutionized mineralogy. Von Laue passed X-rays through a crystal. When X-rays strike undeveloped photographic film, they affect the photographic emulsion much as light rays do. When a narrow beam of X-rays passes through a crystal, the ions within the crystal divert the X-rays and cause them to fan out so that they form a series of spots on a photographic film (Figure 4.8). The pattern of spots enables students of crystals (known as crystallographers) to calculate the sizes of the ions of the unit cells, the distances between ions, and the angles formed by lines drawn through rows of ions.

## Some Properties of Minerals

Two particular properties of minerals are discussed here because they are the bases for important industrial or scientific uses. These properties are electrical conductivity and piezoelectricity.

**Electrical conductivity** The *electrical conductivity* (or its inverse property, *resistivity*) refers to a mineral's response to an electric current. Metallic minerals, whose electrons can move freely through their crystal lattices, tend to be good conductors of electricity. By contrast, most nonmetallic minerals are poor conductors of electricity (or, put another way, they display high resistivities). The conductivity of silicate minerals increases as the temperature is raised. At a temperature of 920°C, the conductivity of silicates increases to 100 times the value at ordinary temperature. Minerals of a kind named semiconductors, especially those containing silicon or germanium, have lately become the basis of transistors and hence are the backbone of the modern electronics industry.

Electrical conductivity is a property used in exploration for minerals. The methods are discussed in Chapter 19.

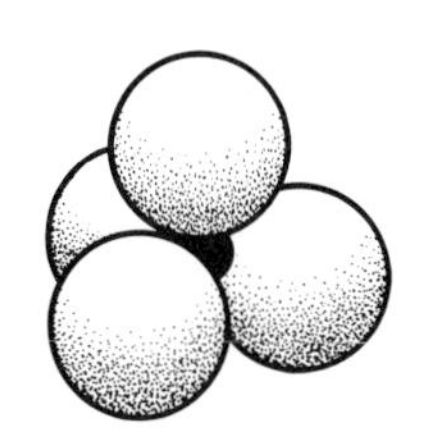


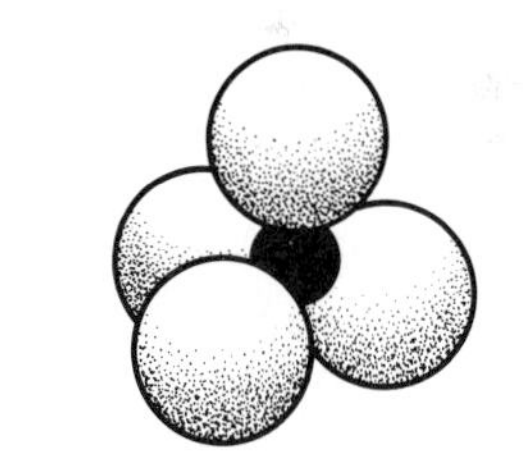


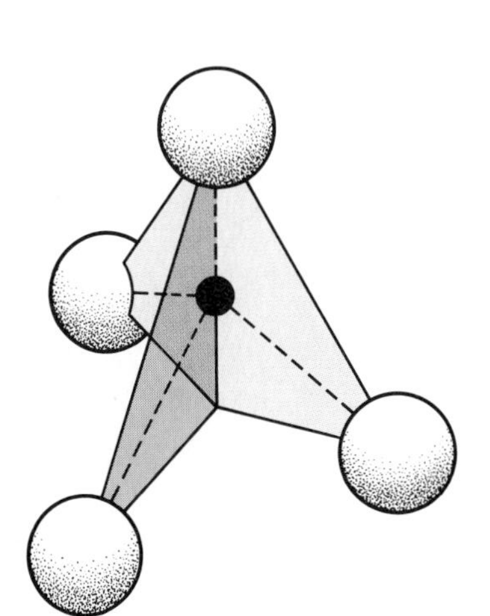


**4.6 Tetrahedron formed by clustering of four oxygen ions around a smaller ion.** (a) Oxygen ions enclose silicon ion and still touch one another. (b) Oxygen ions enclosing an aluminum ion do not touch one another. (c) "Exploded" view showing all ions of a silicon-oxygen tetrahedron.

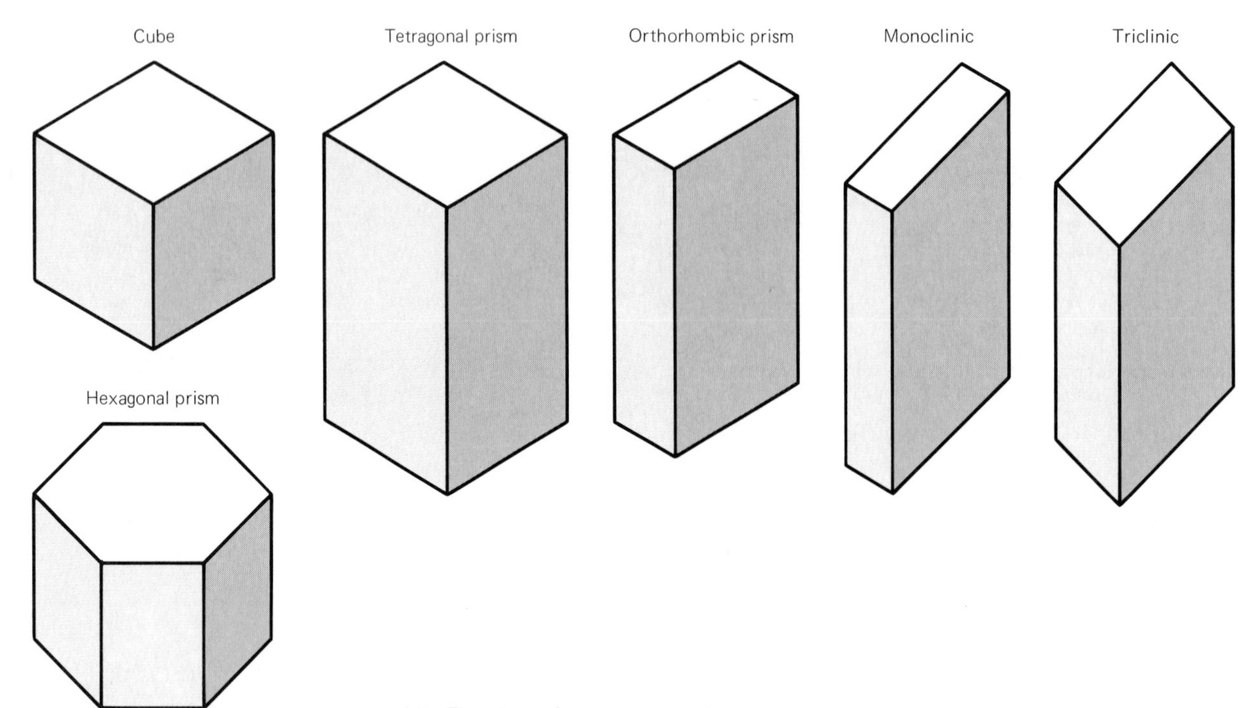


**4.7 The six major symmetry classes of crystals** illustrated by examples of solids that define them. In cubes, square faces meet at right angles. Tetragonal prism is defined by two square faces and four rectangular faces, all meeting at right angles. In an orthorhombic prism, all sides are rectangles and all meet at right angles. A monoclinic shape consists of three pairs of sides, including two pairs of rectangles and one of parallelograms (the largest sides). A triclinic shape consists of three pairs of parallelograms. A hexagonal prism is a six-sided figure of equal-sized squares or rectangles with plane ends that are parallel.

**Piezoelectricity** The property of *piezoelectricity* refers to the connection between mechanical movement or deformation of a crystal lattice and the charge of electricity within the lattice. (*Piezo* is derived from the Greek word for "squeeze." In the present context it can be translated as "pressure.")

When a piezoelectric mineral, such as quartz, is deformed, the lattice responds by storing an electrical charge. When the deforming forces have been taken away, the crystal discharges this electric current. The piezoelectric properties of quartz may be useful in predicting certain kinds of slope failures, as explained in Chapter 11, and earthquakes, as explained in Chapter 17.

## CHEMICAL GROUPS OF MINERALS

The major chemical groups of minerals include silicates, oxides, carbonates, sulfides, and sulfates. The minor groups include native elements, halides, and phosphates.

### Silicates

The minerals known as *silicates* consist predominantly of the elements silicon and oxygen arranged in tetrahedra (see Figure 4.6a) that are joined

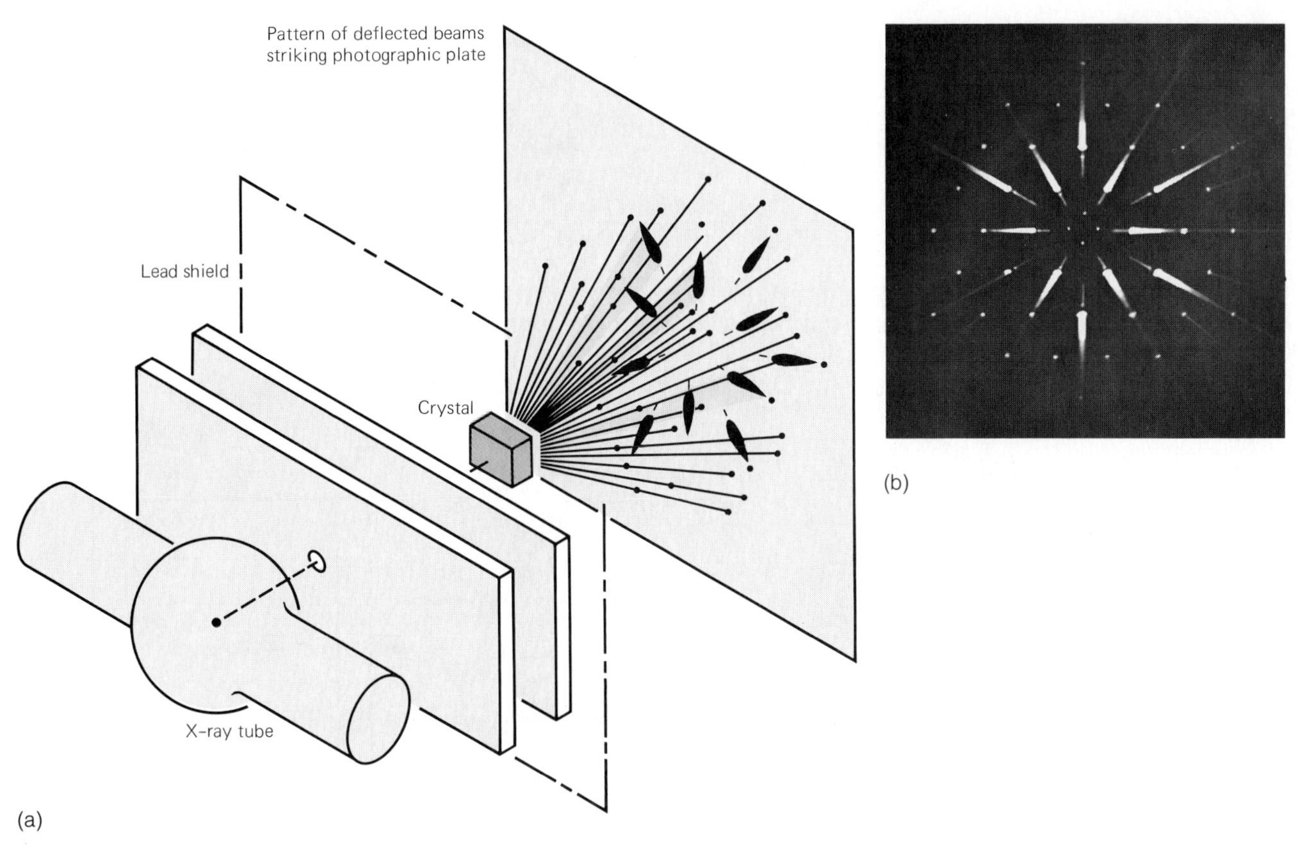


**4.8 Study of crystals using a narrow beam of X-rays.** (a) X-ray tube (left) emits rays in all directions but lead shield blocks all of those except for a narrow beam that is aimed at the crystal specimen. Ions in crystal lattice deflect the X-rays in a systematic, regular pattern, which can be recorded on photographic film. (b) Pattern of X-rays, deflected by ions of a crystal of ice, recorded on a photographic negative, later developed and printed. (I. Fankuchen, Polytechnic Institute of Brooklyn)

in various ways and to which are attached various ions, mostly of the metals magnesium, iron, calcium, sodium, and potassium (but including in some cases hydroxyl ion complexes and fluorine ions). The silicate minerals are so abundant in rocks that these minerals are usually designated as the rock-forming silicates. The structural patterns of silicate minerals are shown in Figure 4.9. The major groups of rock-forming silicates are discussed in detail in Appendix B.

### Oxides

The *oxide minerals* are defined as compounds in which positive ions, usually of a metal, are combined with negative oxygen ions that do not form tetrahedra. The specification about no tetrahedra is necessary in order to exclude quartz. If the definition were based merely on the chemical combination of oxygen with positive ions, then quartz would qualify; its formula is $SiO_2$. Before X-ray study disclosed the diagnostic tetrahedral arrangement among silicate minerals, quartz was classified not as a silicate but as an oxide mineral.

Oxides of interest to us are three varieties of iron oxide (magnetite,

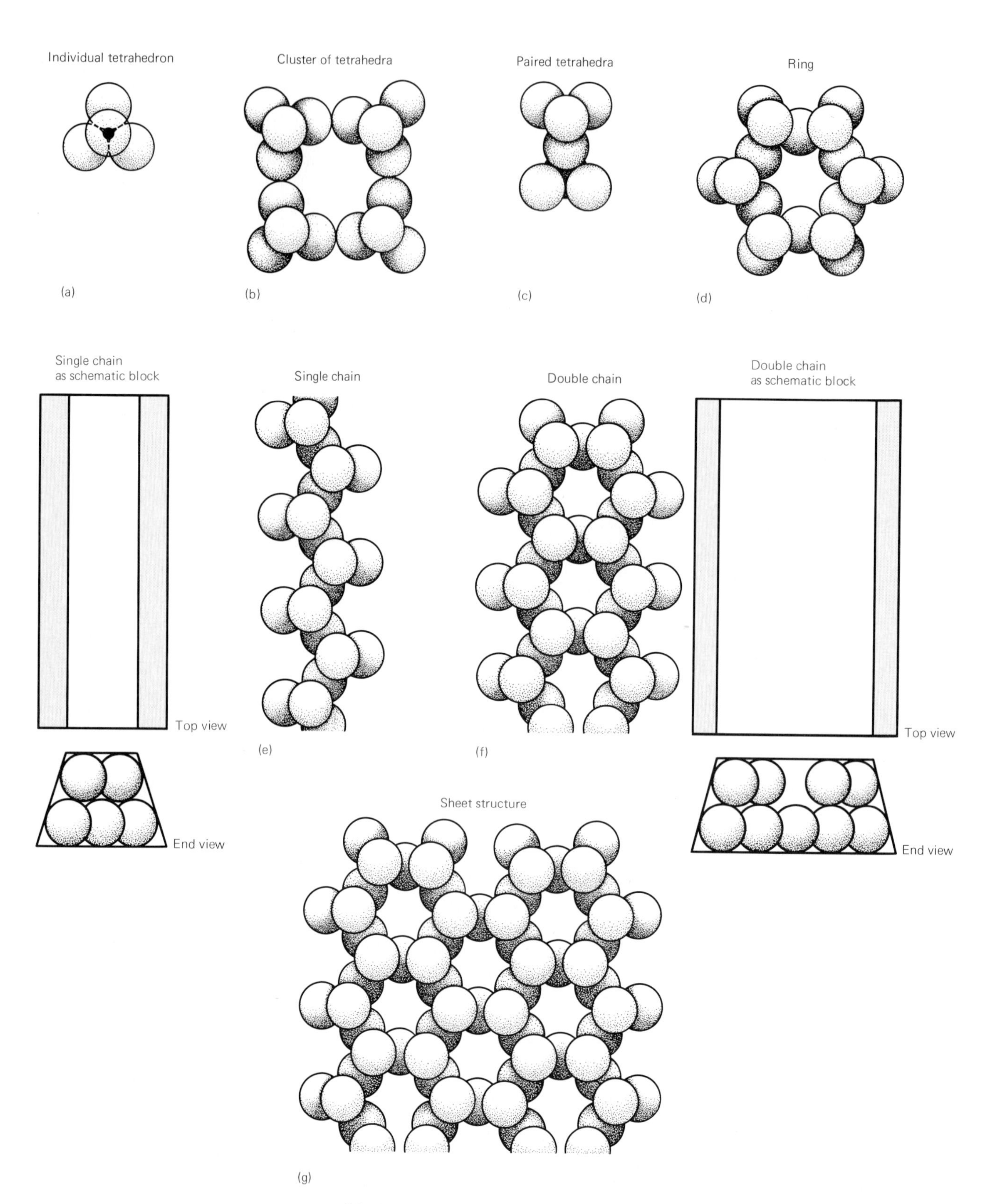


**4.9 Examples of arrangements of silicon-oxygen tetrahedra in the major groups of rock-forming silicate minerals,** viewed from above. Representations of single chains and double chains include schematic prisms in top views and end views.

hematite, and limonite), two aluminum oxides (bauxite and corundum), and the oxides of tin (cassiterite), manganese (pyrolusite), and uranium (uraninite). Iron-oxide minerals, particularly hematite, are mined worldwide as the raw materials for making iron and steel. Iron and steel left exposed to the atmosphere rust, which is another way of saying that they have become oxidized and that the rust-colored oxide mineral, hematite, or the yellowish brown oxide, limonite, or both have formed. (The root word of hematite, *hem,* is derived from the Greek word for "blood," signifying the reddish color of the streak of hematite.)

The other oxide minerals are also important sources of the element combined with the oxygen, namely, aluminum, tin, manganese, and uranium. Corundum, which ranks next to diamond on the Mohs scale of hardness, is used for abrasives and as a coating on stairways to form nonslip surfaces. Two gemstone varieties of corundum are deep red rubies and blue, transparent sapphires.

### Carbonates

The *carbonate minerals* are built of various combinations of other ions with carbonate ion complexes, which are clusters of three oxygen ions surrounding a carbon ion. Geologically significant carbonate minerals include two varieties of calcium carbonate, calcite and aragonite; dolomite (a double carbonate of equal numbers of $Mg^{2+}$ and $Ca^{2+}$ arranged in alternating planes); and the hydrated copper carbonates malachite and azurite.

**4.10 Some common sulfide minerals.** (All photographs by J. E. Sanders) (a) Cubes of pyrite (FeS) showing well-developed striae on cube sides. (b) Cubes of galena (PbS). (Frank Dunand)

(a)

(b)

### Sulfides

*Sulfide minerals* are formed by combination of sulfur ions with one or more metallic ions. Three widespread sulfide minerals are pyrite ($FeS_2$, Figure 4.10a), galena (PbS, Figure 4.10b), and sphalerite (ZnS). Pyrite is characterized by a yellowish metallic luster; this resemblance to gold has given it the nickname "fool's gold." Galena, identified by its gray metallic luster, great density, cubic crystals and cubic cleavage, black streak, and lack of magnetism, is an important source of lead. Sphalerite, the only sulfide mineral lacking metallic luster, can be identified by its yellowish brown to black color, resinous luster, numerous cleavages, and sulfurous odor of its fresh streak. Sphalerite is the chief mineral mined to recover zinc.

### Sulfates

Sulfate minerals are combinations of sulfate ion complexes $(SO_4)^{2-}$ with a 2+ ion, such as calcium ($Ca^{2+}$) or barium ($Ba^{2+}$). The best-known sulfate mineral is gypsum ($CaSO_4{\cdot}2H_2O$) which is readily identified by its hardness of 2 (softer than a fingernail), lack of effervescence with dilute hydrochloric acid, and lack of salty taste. The anhydrous form (without water) of calcium sulfate is anhydrite ($CaSO_4$).

Barite ($BaSO_4$) can be recognized by its unusually great density for a nonmetallic mineral (4.3 to 4.6), low hardness (2.5 to 3.5), two good cleavages and one imperfect cleavage, and lack of effervescence with dilute hydrochloric acid. Barite is ground to a powder and mixed with water or oil to make a dense drilling fluid (known as "mud") for use in petroleum-exploration borings. Because the density of the barite-fluid mixture is greater than that of the rock drilled out of the hole, gas blowouts are prevented.

### Native Elements

*Native elements* are minerals that consist of a single element that is not combined with any other kind of ion. Examples of minerals that are native elements include graphite and diamond (two varieties of carbon), sulfur, gold, silver, platinum, and copper. Native-element gold, native-element silver, and native-element copper are easy to recognize, easy to mine, and are highly valued. As a result, deposits of these three minerals began to be mined thousands of years ago.

### Halides

The *halide minerals* consist of positive ions of a metal, such as sodium, potassium, or calcium, joined together with a negative ion of the halogen group of elements, chlorine, fluorine, bromine, or iodine. We have already mentioned halite ($NaCl$) which gives the name to the halide mineral group. Halite is characterized by its salty taste, low density, extreme solubility in water, and cubic crystals and cleavage (see Figure 4.3).

A second common halide mineral is fluorite ($CaF_2$), which is a scale-of-hardness mineral (number 4 on the Mohs scale) and also a source of fluorine. Fluorite is characterized by cubic crystals; numerous cleavages with some not parallel to the cube sides (Appendix Figure B.14); high density for a nonmetallic mineral (3.18), and lack of effervescence with dilute hydrochloric acid.

### Phosphates

The *phosphate minerals* are important because they contain phosphorus, an element essential to living organisms. The only common phosphate mineral is apatite, a calcium phosphate also containing fluorine, chlorine, and hydroxyl ion complexes and having the chemical formula of $Ca_5(PO_4)_3(F, Cl, OH)$. Apatite is a scale-of-hardness mineral (number 5 on the Mohs scale) and is present as tiny crystals in many granitic rocks and in the bones and teeth of vertebrates. Apatite is used as fertilizer; some transparent varieties are gemstones. A thorium-bearing phosphate mineral is monazite.

## GROWTH OF MINERALS AND MINERALS AS INDICATORS OF ENVIRONMENT

Our discussion of minerals would be incomplete if we did not include the subject of how minerals grow and a summary of the patterns formed during the growth of minerals.

### Changes into the Solid State

In nature minerals form by changes into the solid state as a result of four principal processes: (1) solidification from a molten liquid, (2) precipita-

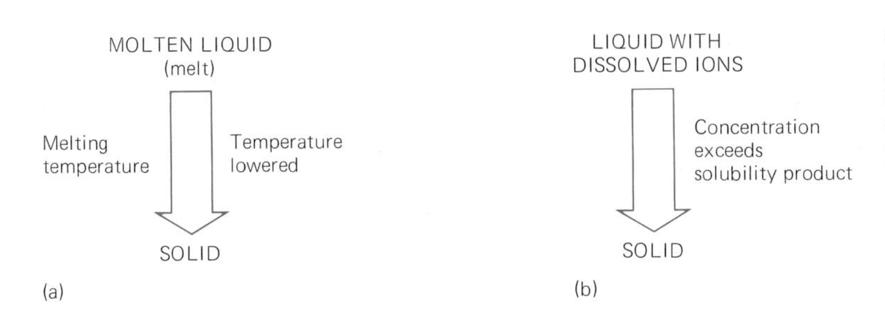


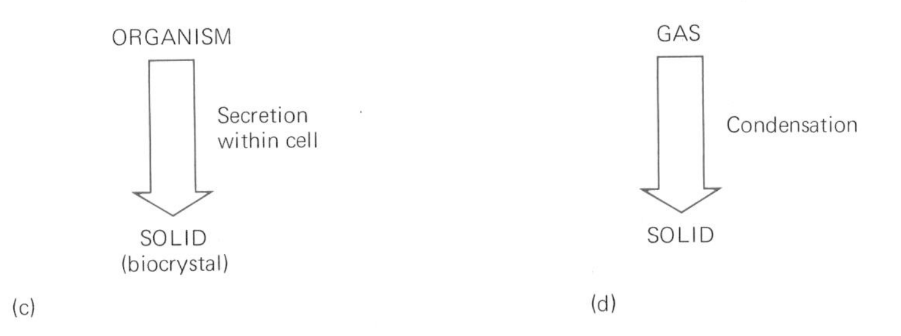


**4.11 Formation of crystals by various processes.** Arrows indicate process of crystallization; various starting materials are printed above the arrows, and various processes, alongside the arrows.

tion from a liquid containing dissolved ions, (3) secretion within the cells of an organism, and (4) condensation from a gas (Figure 4.11).

Solidification (or freezing) from a molten liquid takes place when the temperature decreases to a value less than the melting temperature of a given mineral. The melting temperature is affected by pressure and by the presence of water vapor or other gases. As pressure is increased, the melting temperature of most minerals is raised. An exception is ice, whose melting temperature is lowered by increased pressure. The presence of water vapor usually lowers the melting temperature of minerals.

As we will see in Chapter 6, complex reactions may take place between a growing crystal and the molten liquid.

Precipitation of a mineral from a liquid is related to many factors affecting the solubility of a given ion or group of ions. Some of these factors are: (1) salinity, or the concentration of the ions within the solution; (2) the temperature of the solution; (3) the pH, or the acidity or alkalinity of the solution determined by the concentration of $H^+$ ions; and (4) the dissolved gases in the solution, particularly the Eh, or oxidation-reduction potential, and the presence of hydrogen sulfide and carbon dioxide.

As water containing dissolved ions evaporates, the concentration of the dissolved ions progressively increases. Eventually, the concentration may exceed the solubility. The result is the precipitation of what are known as *evaporite minerals,* of which gypsum, anhydrite, and halite are typical examples.

Temperature may affect the point at which precipitation occurs. For example, some ions that stay in solution in hot waters are precipitated as the temperature drops. Mineral deposits formed from precipitation out of hot-water solutions are designated as *hydrothermal deposits.*

Calcite and the other carbonate minerals are especially sensitive to changes in pH. Carbonate minerals readily dissolve in acid solutions at

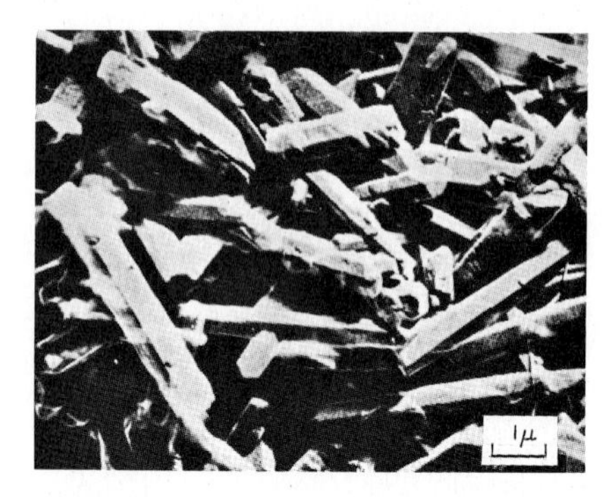


(a)

(b)

**4.12 Forms of calcium carbonate secreted by organisms,** scanning-electron micrographs. (a) Aragonite needles from stems of green alga *Penicillus*, Florida Bay. (K. W. Stockman, R. N. Ginsburg, and E. A. Shinn, 1967. The production of lime mud by algae: Journal of Sedimentary Petrology, v. 37, p. 633–648, Figure 2b, p. 635; courtesy R. N. Ginsburg). (b) Blocky calcite plates of skeletal plates of a coccolith, upper cretaceous of Israel. (A. Bein)

ordinary temperatures, but are precipitated in alkaline solutions.

Pyrite, a sulfide, responds to changes in Eh. When the oxygen in a solution becomes depleted and hydrogen sulfide gas forms, dissolved iron reacts with the hydrogen sulfide to form pyrite, which is then precipitated as a solid.

Secretions within the cells of an organism give rise to biocrystals (Appendix Figure B.8). Aragonite, calcite, and silica in the form of opal are the most common minerals secreted by organisms (Figure 4.12).

Under some conditions certain gases can condense to form solids without participation of the liquid state. An example is the reaction between hydrogen sulfide gas from a volcano and the oxygen of the Earth's atmosphere. The oxygen combines with the hydrogen to form water, and the sulfur is released to form crystals of native sulfur. Water vapor in the form of steam may contain dissolved ions that can crystallize as solids directly from the vapor phase. When minerals form directly from vapors they give off heat known as *latent heat of sublimation.*

## Free Growth of Minerals

The development of minerals involves two processes: (1) nucleation, or the formation of a single unit cell, and (2) enlargement, or growth. The free growth of minerals is responsible for the origin of beautiful crystals. We examine free growth of single crystals and growth of many crystals where the enlarging lattice or lattices are not being significantly agitated, and then compare these patterns with radial growth patterns, which form under various conditions, some involving vigorous agitation.

**Growth of single crystals** In the simplest possible case a mineral starts to grow by forming a single unit cell as a nucleus. With further addition of ions, from whatever source, the lattice enlarges by adding on ions in all directions. The result is a perfect crystal, such as the individual quartz crystals shown in Appendix Figure B.1. When only one nucleus is active, the single crystal will continue to enlarge as long as new ions are supplied and as long as no outside factors inhibit growth. The pattern established by the unit cell serves as a template to guide the positioning of the freshly arriving ions.

**Growth of many crystals; inclusions** Under most conditions when ions are regrouping themselves to form solids, many crystal nuclei may be active, all enlarging simultaneously. Fast-growing crystals may surround and engulf slow-growing crystals. The engulfed small crystals then stop growing. Such small crystals within larger crystals are named *inclusions.* A common example is zircon within biotite.

Circumstances under which perfect crystals can grow uniformly in all directions, either as single individual crystals or as groups of isolated individual crystals, are not particularly common. Two examples are silicate melts and soft muds.

**Concentric patterns and radial patterns** In contrast to the conditions under which one or more individual lattices enlarge to form crystals are the circumstances under which minerals form concentric patterns or radial growths. In such growths, many tiny individual lattices may be present, but rather than forming crystal faces by the outward enlargement from unit cells, they grow curved forms having concentric patterns or radial patterns or both. Commonly, such growth results in botryoidal specimens (Appendix Figure B.9).

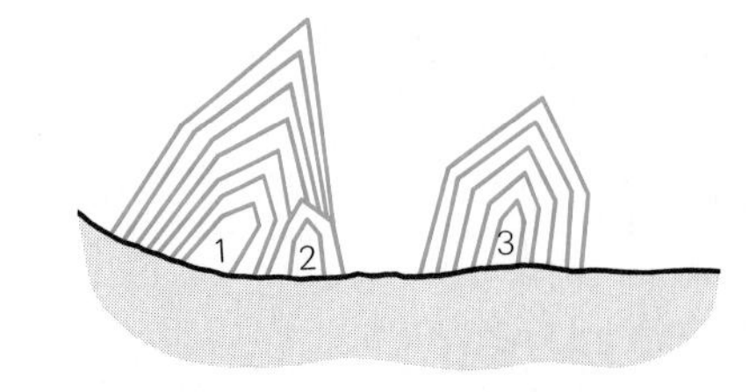


**4.13 Blocking of crystal no. 2 by growth of crystal no. 1.** By further growth, crystal no. 1 may eventually block crystal no. 3.

## Confined Growth: Compromise Boundaries and Mineral Aggregates

The growth of many crystal lattices eventually causes crystals to interfere with one another. One crystal may cut off another's progress and, after having blocked the other, may continue enlarging (Figure 4.13). If numerous growing lattices interfere with one another's growth, the result may be no crystal shapes at all, but rather a mineral aggregate in which the shapes of all the lattices are controlled by crowding, as in the minerals filling in the spaces between radial growths. The external surfaces of the lattices in this case are named *compromise boundaries* (Figure 4.14). Notice that despite the lack of crystal faces, each lattice possesses its diagnostic internal structure and thus displays the usual lattice-controlled properties, such as cleavage and hardness.

## Reactions within the Solid State

The mineral reactions we have been discussing are those that take place when crystals are newly growing in a space not previously occupied by a solid. This situation does not cover all the possibilities. Indeed, many important mineral reactions take place after one or more mineral lattices have already formed.

Some reactions within the solid state involve the reorganization of material already present, a process known as *recrystallization*. A material forming one pattern of crystals is reorganized so that it forms a different

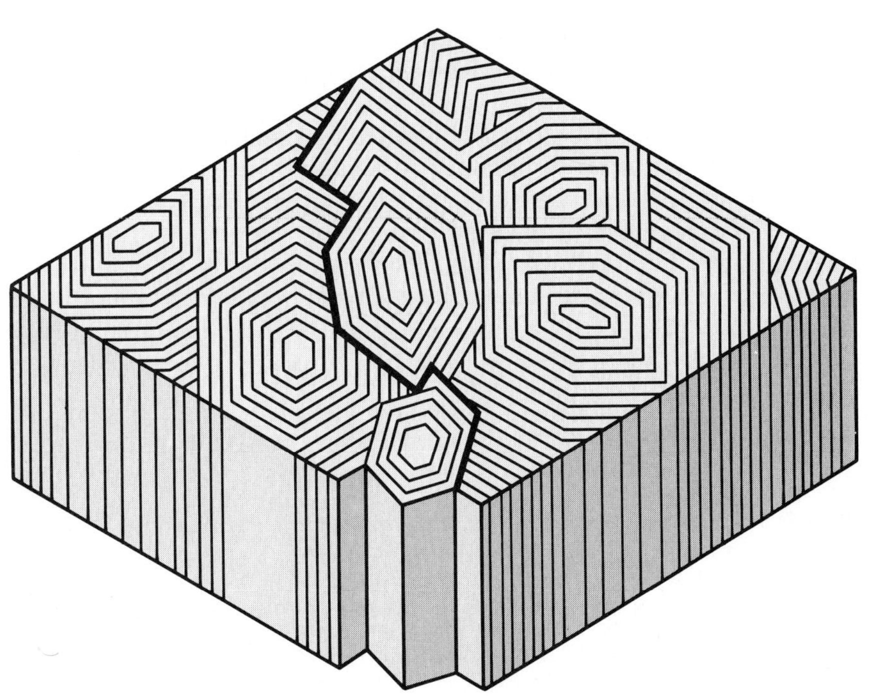

**4.14 Compromise boundaries** (one of which is shown as a thick line) result where enlarging crystals (eight shown here) have interfered with one another's growth. Thin lines show each enlarging crystal's crystal faces.

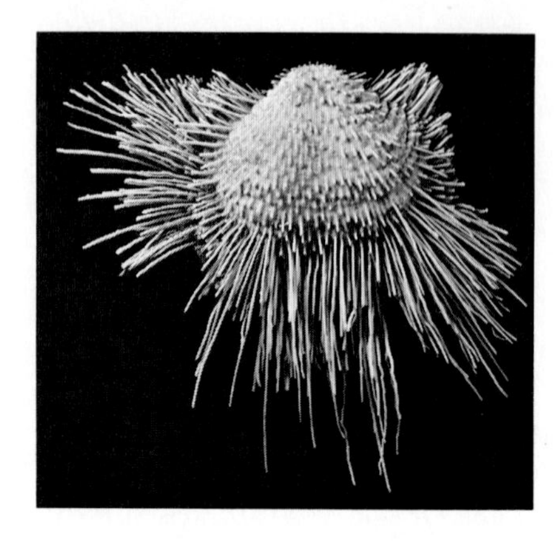

**4.15 Brachiopod shell, originally composed of calcite, now consists of microcrystalline quartz.** Where originally calcareous skeletal material in a carbonate rock has been replaced by silica (i.e., has been *selectively silicified*), exquisite specimens such as this one can be recovered by dissolving the enclosing matrix gently with dilute hydrochloric acid. (G. Arthur Cooper, Smithsonian Institution, U. S. National Museum)

pattern of the same mineral. Generally, the result of recrystallization is enlargement of the sizes of the minerals.

The *replacement* of one mineral by another is a process that has been repeated many times in numerous geologic settings. Calcite secreted by fossil organisms may be replaced by silica in the form of microcrystalline quartz (Figure 4.15), by pyrite, or by other minerals. Where the replacing mineral takes the form of the crystal it has replaced the result is a *pseudomorph*. Many pseudomorphs are formed when a crystal enclosed within some kind of supporting framework matrix is dissolved and the open space formed by its removal does not collapse. Later, a different mineral may fill in the crystal-shaped cavity. Evaporite minerals experience this circumstance commonly.

### Minerals as Indicators of Environment

Once we have determined, by laboratory experiments or by careful measurements in the field, the processes by which minerals are formed and the conditions necessary for those processes to occur, we can work backward from occurrence of a mineral to inferences about the environmental conditions at the time of growth.

Minerals formed by solidification can give us information about temperature and temperature change, pressure, and the presence of water vapor. Nearly any mineral that grew by precipitation from a solution containing dissolved ions can be used to make some specific inferences about environmental conditions based on our knowledge of the factors that affect solubility: salinity, temperature, pH, and Eh. Such minerals as aragonite, calcite, and silica result from the activities of certain organisms. Sulfur may yield some clues about bacterial action or about volcanic activity.

Mineral changes as well as free growth are related to conditions of pressure and of temperature. For example, the change from graphite to diamond requires high pressure and temperature (Figure 4.16).

## RADIOACTIVITY: THE BASIS FOR DETERMINING AGES OF MINERALS

One of the most outstanding scientific achievements has been the development of methods of calculating the ages of formation of minerals from measurements of certain isotopes that have accumulated within

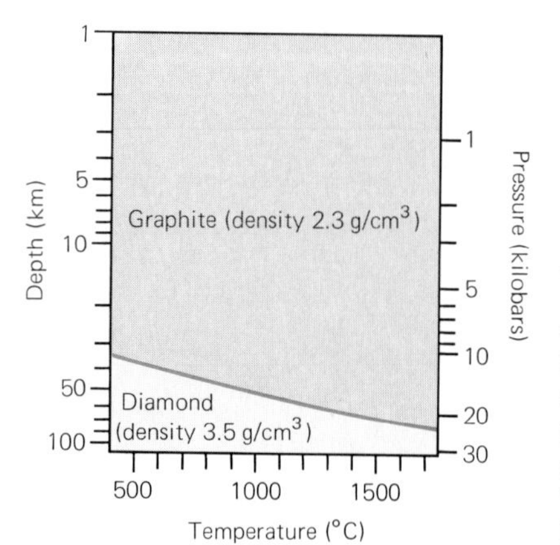


**4.16 Condition of pressure and temperature required to form diamonds,** based on laboratory experiments. We have shown the pressure on the vertical axis of the graph. Because pressure increases downward in the Earth, we have arbitrarily chosen the zero point on the vertical scale at the top, so that greater pressures are shown in the lower part of the graph. (Redrawn from W. S. Fyfe, 1964, Geochemistry of solids. An introduction: New York, McGraw-Hill Book Company, 199 p., Figure 11-1, p. 141)

mineral specimens. The physical principle upon which mineral dating is based is radioactivity (see Appendix A). Here we show how radioactivity can be used to date minerals. In Chapter 6 we relate mineral dating to igneous rocks, in Chapter 8 to sedimentary rocks, and in Chapter 9 to metamorphic rocks.

Radioactivity affects minerals in two contrasting ways. (1) Within crystal lattices, parent isotopes disintegrate into daughter isotopes, which may accumulate within the lattice as the amount of parent material is depleted. (2) The atomic particles generated by radioactive reactions, such as the large nuclei of daughter isotopes during nuclear fission, or alpha particles, leave various physical defects in the lattice. In both ways, evidence can be recovered from the minerals that makes possible calculations of an age that is related to the crystal lattice.

### Radioactive Series Used for Age Determinations

The reactions that lead from a radioactive parent isotope to a stable daughter isotope are known as a *radioactive-decay series.* The fundamental rule of radioactive decay is that the amount of time required for half the nuclei of the parent isotope to be converted into nuclei of the daughter isotope is a constant for each radioactive reaction. This amount of time is the same no matter what the temperature, the pressure, or other conditions may be; it is known as the *half life* of that particular radioactive transformation.

Another expression of the rate of a radioactive reaction is the *decay constant,* which expresses the proportion of the nuclei of the parent isotope that decay in a unit of time.

Because the number of nuclei of the parent isotope diminishes each time a parent decays into a daughter, it follows that the actual number of decay reactions will decrease with time. In other words, after one half-life time interval, an initial number of 100 nuclei of parent isotope will have been reduced to 50. After a second half-life time interval, these 50 will have become 25, and so on. During the passage of each half-life time interval, the number of nuclei of parent isotopes has diminished by one-half (Figure 4.17). Once the half life of a reaction is known, identifi-

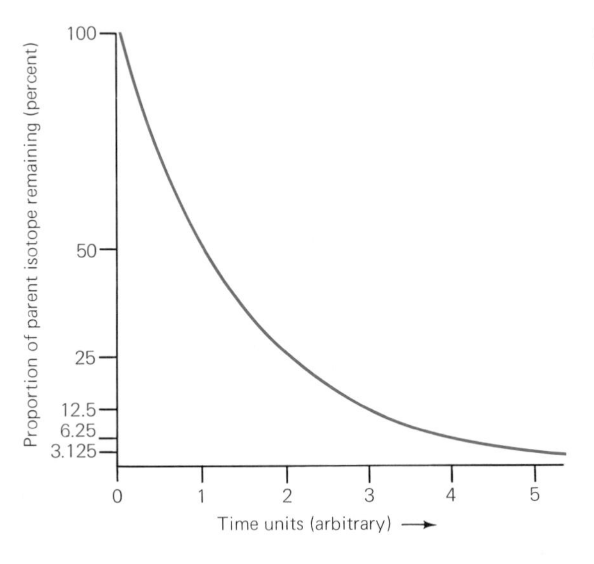


**4.17 Radioactive decay** shown by schematic graph of half-life time units.

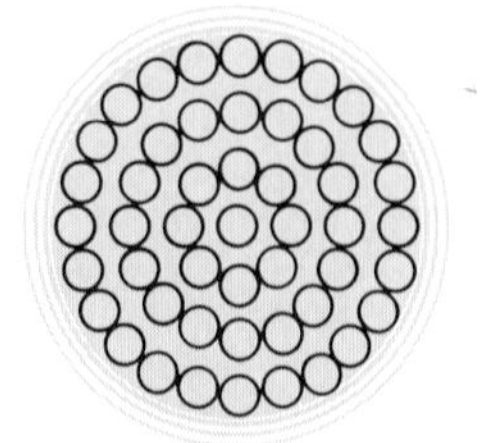

0 years—death of animal or plant

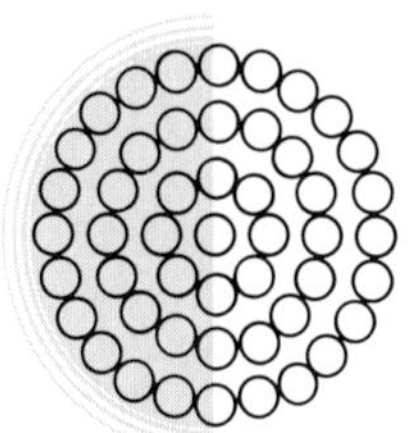

5,730 years—1/2 left

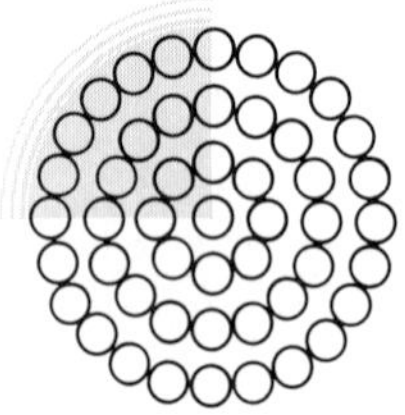

11,460 years—1/4 left

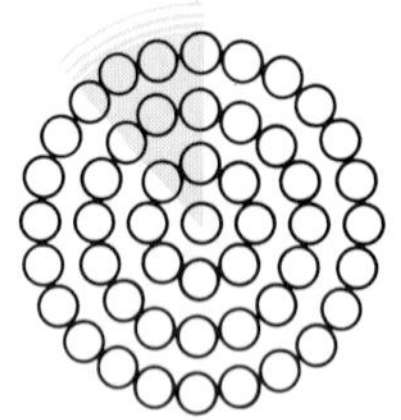

17,190 years—1/8 left

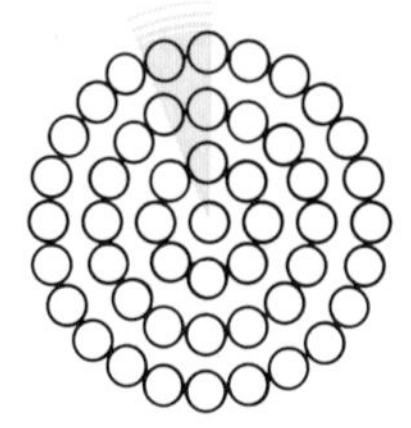

22,920 years—1/16 left

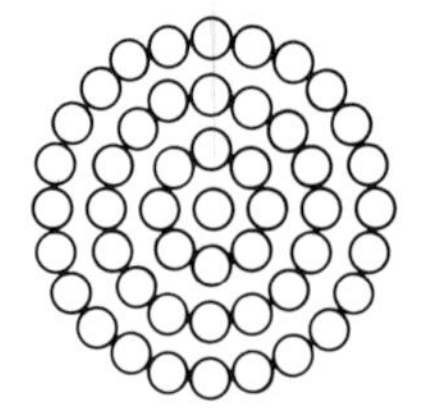

60,000 years—about 1/1,000 left

**4.18 Radioactive decay of radiocarbon** through four half-life time units (to 22,920 years) and extension to 60,000 years.

cation in a material of the isotopes involved in the radioactive series gives evidence of the age of that material. Appendix Box A.1 explains the mathematical theory of radioactive decay.

The seven radioactive series used to determine the ages of minerals and other geologic materials are shown in Appendix Table A.1. The chemical details about these series are discussed in Appendix A. Notice that three of the series proceed largely (but not entirely) by emission of alpha particles and that each of these series ends with a stable isotope of lead. Notice, also, that in each of these series, one of the intermediate daughter isotopes is radon, an inert gas. Whenever one of the isotopes involved in a radioactive decay series is a gas, the possibility exists that this gas will escape from the crystal lattice. The escape of gas from a lattice becomes easier as the temperature is raised. Therefore, in attempting to date a specimen, one must be aware of the possibility that the decay products may have escaped. The parent isotopes enter the crystal lattice when the lattice is first formed, but some or all of the daughter isotopes may escape if the lattice is heated. Accordingly, the calculated date of a specimen may be only the date of last heating and daughter-isotope escape and not the date of entry of radioactive parent isotopes into the lattice.

Dating by radiocarbon differs in some ways from dating by other radioactive isotopes. Unlike the isotopes of the six other radioactive series, radiocarbon is not confined to crystal lattices, but occurs as well in organic material. Another difference in use of radiocarbon is the laboratory method. One of the forms of energy that is released during radioactive reactions is gamma rays. Suitable instruments have been built for measuring gamma rays, and these are important tools in the use of radioactive carbon for dating specimens. In the six other radioactive series one measures the amounts of both parent and daughter isotopes. In the radiocarbon method one usually does not measure isotopes, but rather carefully counts the amount of gamma rays, which is a function of the amount of $^{14}C$ in the sample. As with other radioactive reactions the quantity of $^{14}C$ decreases with time (Figure 4.18). As a result, the amount of gamma rays one can count decreases progressively with older and older specimens.

The principle of using radiocarbon for dating is that the $^{14}C$ created in the upper atmosphere (see Figure 1.18) mixes uniformly throughout the atmosphere, hydrosphere, and biosphere. While an organism, either animal or plant, is alive, it is continually exchanging its content of carbon. This exchange maintains an equilibrium proportion of $^{14}C$ within the organism. Thus, even though the $^{14}C$ is decaying radioactively, its proportion in the organism is maintained by the exchange reactions. When the organism dies, however, the exchange ceases and radioactive disintegration causes the amount of $^{14}C$ present to decrease.

Because of its short half life (5730 $\pm$ 30 years) and low amount of $^{14}C$ in the equilibrium proportion within the biosphere, radiocarbon dating cannot be used on very ancient materials. The radiocarbon method has been checked against tree rings back to 8000 years. The method has been used on older specimens, but with less and less confidence as the practical limit of 50,000 years is approached.

**4.19 Fission-fragment tracks in coral skeleton** radiate outward from a place where uranium has been concentrated. Tracks have been made visible by etching and are shown here in a special replica of the original coral skeleton, collected at a depth of 10 meters off Elat in the Red Sea. (D. S. Miller, Rensselaer Polytechnic Institute)

## Ages Based on Effects of Radiation in Crystals

An indirect kind of dating of minerals is based on the effects of the radioactivity itself. The energetic bursting forth of particles from the disintegrating nuclei sends tiny missiles flying with great force outward into the surrounding crystal lattice. Radiation effects in crystals include fission-fragment tracks and radiohalos.

Fission-fragment tracks (Figure 4.19) are created in a crystal when a radioactive nucleus decays by fission into two daughter isotopes. Each track represents one decay event. By using a microscope to count the number of tracks within a given area and by comparing this count with the number of tracks made within an equal area when a specimen has been bombarded with a known quantity of radioactive material, one can calculate the amount of radioactive decay events that affected the specimen. Knowing this and the amount of radioactive material still present, one can calculate an age.

Radiohalos (Figure 4.20) result from the effects of alpha particles on various minerals. The halos are usually studied by using a high-powered polarizing microscope to examine paper-thin slices of biotite that contains inclusions of zircon. Uranium in the zircon propels alpha particles into the biotite. In such slices the halos appear as circles. Actually, the halos are concentric spheres, somewhat like the energy-level spheres of atoms (see Appendix Figure A.7). The spheres are formed because of the reactions between the alpha particles at the end of their journeys and the mineral. The ranges of alpha particles vary inversely with the half life of

**4.20 Radiohalo in biotite surrounds a zircon** that contains uranium. The darkness of this halo indicates that the age of the specimen is of Paleozoic age or older. In specimens of Cenozoic ages, halos are scarcely visible; they are faint in specimens of Mesozoic ages; and very pronounced in specimens that are of Paleozoic ages and older. (J. E. Sanders)

the disintegrating parent isotope. This important principle can be used to check on the idea that half lives have remained constant with time. One does this by measuring the diameters of the circles (or rings) of the radiohalos. Each circle indicates a particular parent isotope. Because the diameters of the halo circles have been found to be the same for specimens of all ages, geologists are confident that the half lives of the isotopes have not changed through time.

## CHAPTER REVIEW

1. Minerals are the fundamental building blocks of rocks; they form various specialized deposits that are mined to recover useful materials. In a geologic context a *mineral* is defined as a naturally occurring solid having a definite chemical composition and an ordered ionic arrangement.

2. The ions of most minerals include those of the eight most abundant elements composing the Earth's crust (oxygen, silicon, aluminum, iron, calcium, sodium, potassium, and magnesium) plus those of carbon, hydrogen, chlorine, and sulfur. For purposes of describing the structure of minerals, we can say that the ions form tiny spheres, and these spheres can be classified according to their diameters and electrical charges.

3. The structure of minerals is based on the *unit cell*. A unit cell is a regular arrangement of ions held together by electrical forces; it is the smallest unit displaying the properties of the mineral. Ions may be arranged in one of several ways: cubic packing, hexagonal packing, or packed to form tetrahedra.

4. *Crystals* are solid bodies bounded by natural plane surfaces that are the external expression of the regular interior lattice structure. Each crystal grows by additions of ions to its diagnostic unit cell, and the fully developed form of each crystal is distinctive.

5. A fundamental law of crystals, the *law of constancy of interfacial angles*, states that the angles between corresponding faces on all crystals of one mineral are constant.

6. Two properties of minerals with important scientific and industrial uses are *electrical conductivity* (or *resistivity*) and *piezoelectricity*. Conductivity, the ability to transmit an electrical current, is determined by the degree of freedom of movement of electrons within the crystal. Piezoelectricity is the storing of an electrical charge within the crystal in response to pressure (mechanical movement, deformation) and the release of that charge when the pressure is removed.

7. The major chemical groups of minerals include silicates, oxides, carbonates, sulfides, and sulfates. The minor groups include native elements, halides, and phosphates.

8. *Silicate* minerals consist mainly of the elements silicon and oxygen arranged in tetrahedra that are organized in various ways. Silicate minerals are usually designated as the *rock-forming silicates*.

9. *Oxide minerals* are compounds in which positive ions, usually of a metal, are combined with negative oxygen ions that do not form tetrahedra.

10. *Carbonate minerals* are built of various combinations of other ions with carbonate ion complexes, which are clusters of three oxygen ions surrounding a carbon ion.

11. *Sulfide minerals* are formed by combinations of sulfur ions with one or more metallic ions.

12. *Sulfate minerals* are combinations of sulfate ion complexes $(SO_4)^{2-}$ with a $2^+$ ion.

13. Minerals that consist of only a single element that is not combined with any other kind of ions are known as *native elements*. *Halide* minerals consist of positive ions of a metal joined together with a negative ion of the halogen group. *Phosphate* minerals contain phosphorus, an element essential to living organisms.

14. In nature, minerals form by changes into the solid state as a result of *solidification* from a molten liquid, *precipitation* from a liquid containing dissolved ions, *secretion* within the cells of an organism, and *condensation* from a gas.

15. The development of minerals involves two processes: nucleation and enlargement.

16. The external surfaces of growing lattices that mutually interfere with one another's growth are named *compromise boundaries*.

17. Reactions within the solid state may involve *recrystallization*, reorganization of the mineral already present, or *replacement* of one mineral by another.

18. Minerals can be indicators of environmental conditions such as temperature, pressure, salinity of solutions, pH, Eh, presence of water vapor, presence of types of life forms, and volcanic activity.

19. The phenomenon of *radioactivity* is defined as the spontaneous disintegration of the nuclei of certain unstable isotopes that converts these nuclei into the nuclei of other isotopes. The transformations that lead from a radioactive parent isotope to a stable daughter isotope are known as a *radioactive-decay series.*
20. It is possible to recognize three contrasting kinds of radioactive reactions: nuclear fission, emission of alpha particles, and the emission or capture of beta particles.
21. The fundamental rule of radioactive decay is that the amount of time required for half the nuclei of the parent isotope to be converted into nuclei of the daughter isotope is a constant for each radioactive reaction known as the *half life.* Ages of minerals can be calculated on the basis of measured amounts of radioactive isotopes whose half-life values are known.
22. Age determination may also be based on the effects on crystals of radioactive particles. Two such effects are *fission-fragment tracks* and *radiohalos.*

## QUESTIONS

1. Define *mineral.* What is a *unit cell?*
2. What is the single diagnostic feature of all minerals?
3. What is a *crystal?* Do minerals always appear as crystals? Explain.
4. State the *law of constancy of interfacial angles.*
5. Explain the effect of the ions of a mineral on a beam of X-rays.
6. What is *piezoelectricity?* Give an example of a mineral that displays piezoelectricity.
7. What is the fundamental structural unit of the silicate minerals? Explain how these units are arranged in the major groups of silicates.
8. Define *oxide mineral.* Explain why quartz, with the chemical formula $SiO_2$, is not considered to be an oxide mineral.
9. Discuss some of the factors that affect the solidification and growth of minerals.
10. Define *inclusion.*
11. Define *crystal face* and *compromise boundary.* Give an example of the conditions under which each of these may form.
12. State the fundamental rule of radioactive decay.
13. In a radioactive series, how are *parent isotopes* and *daughter isotopes* related and how does each change with time?
14. Define the *half life* of a radioactive isotope.
15. Explain how daughter isotopes that are gases may complicate radioactive dating.
16. Explain the similarities and differences between dating by radiocarbon and by other radioactive isotopes.
17. List and explain two side effects of radioactivity in crystals that can be used in age determinations.

## RECOMMENDED READING

Bragg, Sir Laurence, 1968, X-ray Crystallography: *Scientific American,* v. 219, no. 1, p. 58–74.

Broecker, W. S., 1965, Isotope Geochemistry and the Pleistocene Climatic Record, p. 737–753 *in* Wright, H. E., Jr., and Frey, D. G., *eds., The Quaternary of the United States:* Princeton, New Jersey, Princeton University Press, 922 p.

Grootes, P. M., 1978, Carbon-14 Time Scale Extended: Comparison of Chronologies: *Science,* v. 200, p. 11–15.

Inoué, Shinya, and Okazaki, Kayo, 1977, Biocrystals: *Scientific American,* v. 236, no. 4, p. 82–92.

Mason, Brian, 1966, *Principles of Geochemistry,* 3rd ed.: New York, John Wiley and Sons, 329 p.

Vanders, Iris, and Kerr, P. F., 1967, *Mineral Recognition:* New York, John Wiley and Sons, 316 p.

Zim, H. S., and Shaffer, P. R., 1957, *Rocks and Minerals. A Guide to Minerals, Gems, and Rocks:* New York, Golden Press, 160 p. (paperback)

# CHAPTER 5 IGNEOUS ACTIVITY AND VOLCANOES

HISTORICAL CONTROVERSY ABOUT IGNEOUS ROCKS: PLUTONISTS VS. NEPTUNISTS
THE RELATIONSHIP BETWEEN IGNEOUS ACTIVITY AND TECTONIC ACTIVITY
VOLCANOES
VOLCANIC ACTIVITY UNDER WATER AND UNDER GLACIERS
UNDERGOUND THERMAL ACTIVITY AND THE DISCHARGE OF VOLCANIC GASES
LEARNING TO LIVE WITH VOLCANOES
CAN VOLCANOES CHANGE THE EARTH'S CLIMATE?

PLATE 5
Aso Volcano, Japan, viewed vertically downward from an airplane. Small cone and crater at upper right mark an active vent that had discharged many lava flows, the youngest of which has flowed toward bottom center. (U. S. Geological Survey; photo taken on 6 November 1947)

IGNEOUS ACTIVITY IS THE GENERAL NAME FOR A TWO-FOLD process: first, the making of hot liquid material within the Earth and its rising within the lithosphere or onto the surface of the Earth; and second, the cooling of this liquid material, to form solid igneous rocks. Such rocks make up a vast bulk of the lithosphere. Accordingly, igneous activity is a process of fundamental importance in the geologic cycle. Igneous activity also includes the movements and effects of the molten rock before it solidifies. *Volcanic activity* is the surface expression of igneous activity: a *volcano* is the vent through which the molten rock material and associated gases pass upward to the Earth's surface.

The principles governing igneous activity apply not only to Earth, but also to other planets and to some of their satellite moons. This is a significant point, because on other bodies, igneous activity helped shape their surfaces.

We begin our discussion with some historical background of modern ideas about igneous rocks and some general remarks about the relationships of igneous activity to tectonic activity. We then proceed to volcanoes and volcanic activity. In Chapter 6 we take up igneous rocks.

## HISTORICAL CONTROVERSY ABOUT IGNEOUS ROCKS: PLUTONISTS VS. NEPTUNISTS

After many bitter arguments among geologists in the nineteenth century, it was finally concluded that many rocks have formed as a result of igneous activity, specifically the processes which take place when solid rock material becomes molten and then cools to form solid rock once again. The geologists involved in the arguments about igneous rocks all agreed that volcanoes erupt hot, molten material known as lava. The argument took place because one group of geologists insisted that no connection existed between modern volcanoes and ancient rocks. They believed that the only way any rock having a crystalline texture (explained in Appendix C) could form was by chemical precipitation of ions held in solution. This water-oriented group became known as the *Neptunists* (in reference to Neptune, god of the sea); they felt that the rocks being debated had formed by chemical precipitation at low temperature from a universal ocean that had formerly covered all the continents. The Neptunists even argued that the regular six-sided prisms of rock outlined by cracks in some rocks (Figure 5.1) were hexagonal crystals formed by precipitation out of water. The leader of the Neptunists was Abraham Gottlob Werner (1750–1817), a German mineralogist and mining geologist who lived in Saxony and lectured at the Freiberg Mining Academy. Werner's stimulating lectures, with emphasis on how knowledge of geology was useful in locating valuable ore deposits, attracted students from all over the world. Werner never traveled much; his traveling disciple was Leopold von Buch (1774–1853).

The other side of the argument was led by James Hutton. His chief ally, Sir James Hall (1761–1832), demonstrated that rocks having crystalline texture were formed when bits of granite were made molten in a forge and then were allowed to cool slowly. Hutton argued that what we would now name igneous rocks had formed because under one set of conditions the Earth's internal heat had caused some rocks to melt and

**5.1 Hexagonal prisms of igneous rock outlined by cooling joints.** Key at bottom gives scale. Devil's Post Pile National Monument, California. (Nelson King, August 1977)

then under other conditions the molten material had cooled and become solid. Because of their emphasis on what happened beneath the surface of the Earth, Hutton and his followers were called *Plutonists,* in reference to Pluto, god of the underworld.

The key locality in settling all these arguments proved to be the Auvergne district of central France (Figure 5.2a). In the Auvergne region are thousands of extinct volcanoes of Cenozoic ages, an older group of Miocene age, and a younger group of Late Quaternary age (radiocarbon dated at about 7500 years). Lava from these volcanoes had flowed down some valleys, in some cases completely filling them. The lava had cooled to form igneous rock. In one famous locality, the Montagne de la Serre (Figure 5.2b), a sheet of igneous rock from a lava flow of Miocene age

**5.2 Ancient volcanoes of central France.** (a) Index map. Arrow marks location of larger-scale map of Figure 5.2 b. C-F designates city of Clermont-Ferrand. (b) Volcanic features of two ages, Miocene and Late Quaternary, near Aydat, sketch map. Lava from volcanic cones flowed eastward from highland underlain by Precambrian granite, across fault at margin of lowland underlain by Pliocene sedimentary strata. Montagne de la Serre (center) is a ridge capped by a layer of igneous rock formed by the cooling of a lava flow of Miocene age that flowed down a now-vanished valley, as explained in Figure 5.2 c. (After Philippe Glangeaud, 1913, Les régions volcaniques du Puy-de-Dôme, II. La chaine des Puys: Service de la carte géologique de la France, Bulletin, v. 22, no. 135, 256 p., Figure 67, p. 209 and geologic map) (c) Montagne de la Serre with ancient valley reconstructed in a block diagram looking north. (After *Les Guides Verts Michelin,* Auvergne, 20th ed., 1970, p. 15.)

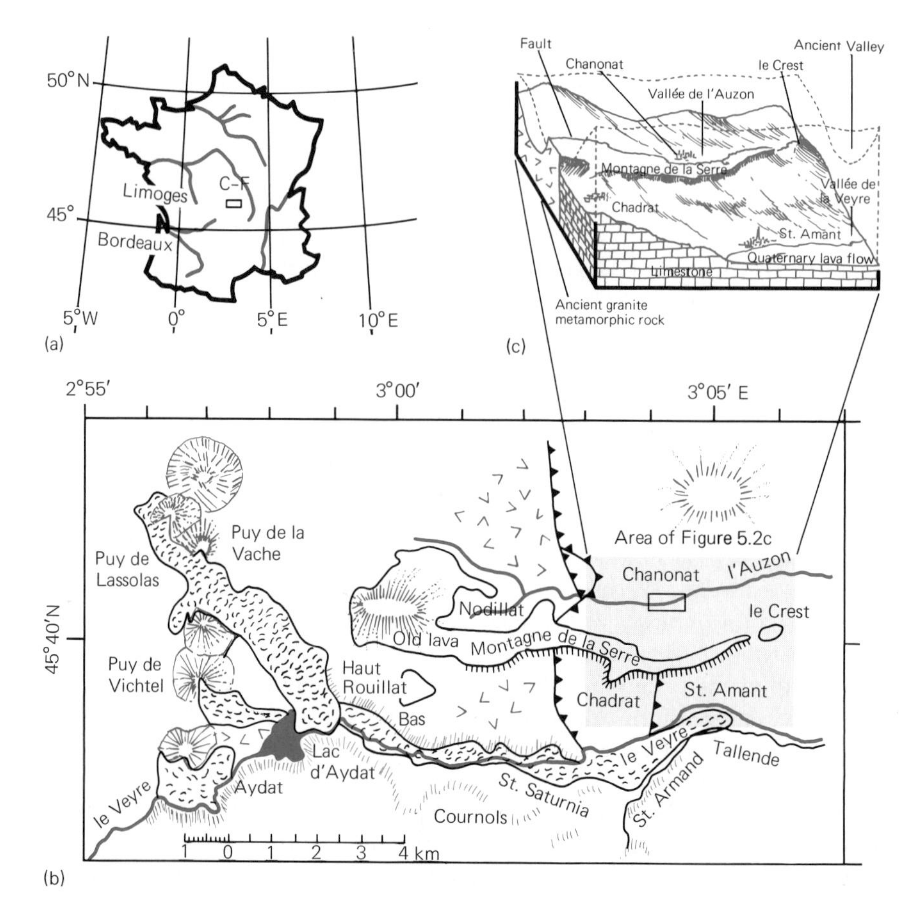

which had once been a valley filling is now the capping of a ridge. Because of great post-Miocene erosion, the landscape has been completely inverted. The former divides are the present-day valleys of the Auzon River on the north and the Veyre River on the south; the former valley is the present-day divide between them (Figure 5.2c).

But what is even more striking than this example of landscape inversion is the relationship between the tongues of igneous rock of Late Quaternary ages and the cones of loose cinders that built up around the former volcanic vents. Two distinct former lava flows are represented. An older one extends from its source down a tributary of the Veyre in a southeastward direction to the main valley of the river and then eastward down the main valley for another eight kilometers. This flow was covered by a younger one that extends from two cones, Puy de Lassolas and Puy de la Vache, southeastward down the tributary valley to the junction with the Veyre. Here the upper flow dammed the Veyre, forming Lac d'Aydat. South of the village of Chanonat one can see the Miocene igneous rock of Montagne de la Serre extending eastward to le Crest. Just south of Montagne de la Serre, Quaternary igneous rock partially fills the valley of the Veyre. Both rocks look much alike; both display all the characteristics that the Neptunists claimed had resulted from precipitation out of the "Universal Ocean." When Leopold von Buch was finally persuaded to visit this locality, Neptunism became a thing of the past. At le Crest and along the Montagne de la Serre, von Buch saw nothing to shake his faith in Neptunism. However, by the time he had finished walking upstream in the valley of the Veyre and saw the bodies of igneous rock connected to the obviously volcanic cones of Puy de la Vache and Puy de Lassolas, his opinion changed. He became a Plutonist on the spot, and the argument was settled.

## THE RELATIONSHIP BETWEEN IGNEOUS ACTIVITY AND TECTONIC ACTIVITY

According to the plate-tectonic view of the world, the Earth and all its important processes can be divided into two major categories: (1) the lithosphere plates themselves, thought to be comparatively inactive, and (2) the boundaries between lithosphere plates, characterized by great activities. Thus, the world's major belts of earthquakes and volcanoes coincide with plate boundaries.

Two contrasting belts are the mid-Atlantic, and the Mediterranean-Pacific border. Any process that causes part of the Earth's upper mantle to melt will create the kinds of fluid lavas that tend to erupt quietly and to form *basalt,* a dark-colored igneous rock. Basalt-forming volcanoes are located worldwide. Their chief patterns are (1) along mid-oceanic ridges, such as the mid-Atlantic Ridge, where plates are separating (Figure 5.3a); (2) along major fractures through the lithosphere named rift zones, located either on continents or on the sea floor; and (3) in scattered locations that are thought to indicate hot spots in the upper mantle, chimneylike plumes where much more heat than normal is flowing upward (Figure 5.3b).

By contrast, along boundaries where plates are coming together (Figure 5.3c), the volcanoes tend to erupt explosively and to discharge pasty lavas that form light-colored igneous rock. This contrasting kind of lava is

**5.3 Three common plate-tectonic settings of volcanoes,** rows along mid-oceanic ridges and island arcs, and scattered over hot spots, schematic profiles and sections. (a) Mid-oceanic ridge, where lithosphere plates are separating and molten material derived from the upper mantle rises upward and solidifies to form new sea-floor crustal rock. (b) Individual volcano above hot spot. (c) Island arc, where plates are thought to be coming together.

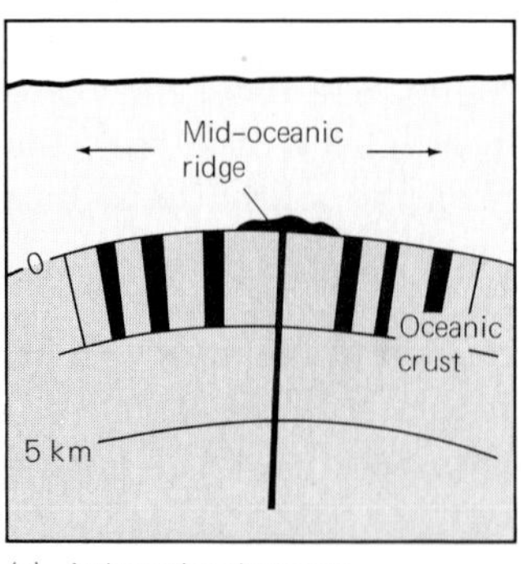

(a) At boundary between separating plates

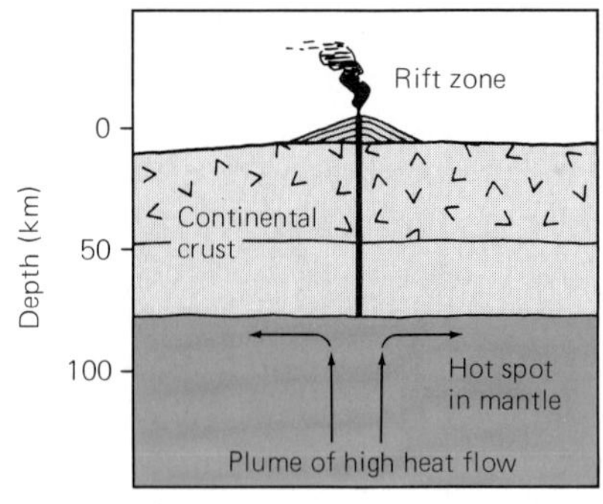

(b) Above mantle hot spot

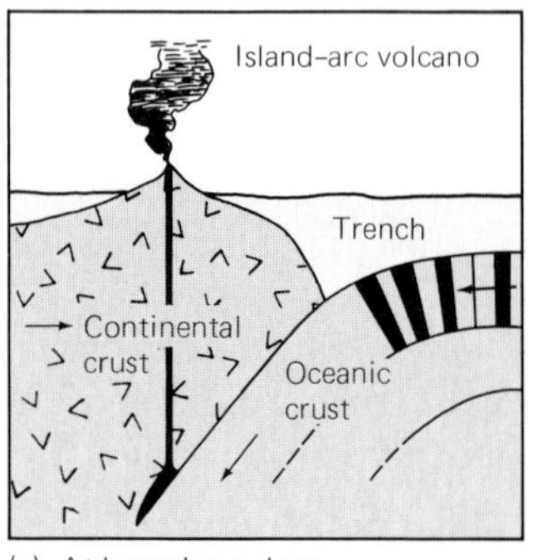

(c) At boundary where plates are coming together

thought to represent the melting of something other than upper-mantle material, for example, continental crust. Examples are found along the margins of the Pacific Ocean, along the northern and eastern margins of the Caribbean, and in the Mediterranean.

## VOLCANOES

A *volcano* is the vent through which molten rock material and associated gases pass upward from the Earth's interior onto the Earth's surface. Strictly speaking, the volcano is defined in terms of the existence of the vent, not the piling up of material; the volcano becomes active the moment gases start to come out of the vent. Some active vents never discharge anything but gases and thus do not affect the landscape in any significant way.

### Products of Volcanoes

Volcanoes have captured the popular fancy and have been incorporated into the folklore of many peoples. During the Middle Ages the "smoke and ashes" of volcanoes were thought to be caused by the fires of hell. Heklafeld in Iceland (Figure 5.4) was thought to be the gate to the underworld. As late as the eighteenth century some persons considered volcanoes to be the vents of great fires burning within the Earth. Only

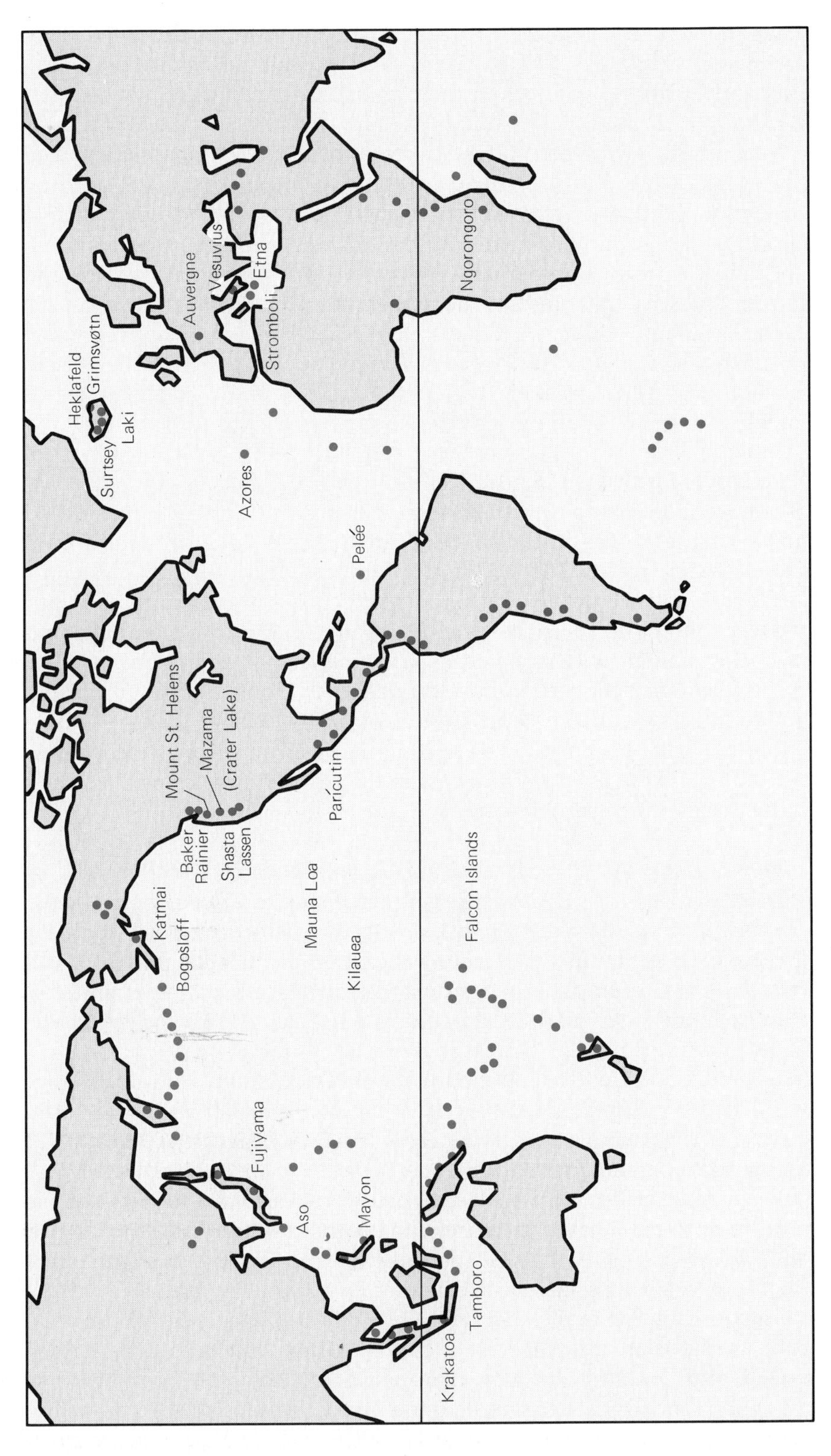


**5.4 Location map,** volcanoes mentioned in text.

gradually did early geologists discover that the "smoke" of volcanoes is not smoke at all and the "ashes" are not products of combustion. The complex and variable products of volcanoes include liquids, solids, and gases. Usually the gases and molten liquid are mixed together when they leave the vent and separate in the atmosphere. In some cases the gases are mixed with solids. The key process of volcanic activity is the separation and escape of gases from their associated liquids and solids.

**Volcanic gases** Virtually all the gases of the Earth's atmosphere and all the water of the Earth's oceans are thought to have been released by igneous activity. The release of these gases provides the driving force to expel the liquid and solid volcanic products. Although gases compose only 1 or 2 percent of the total weight of erupting magma, they expand into bubbles as the pressure decreases near the Earth's surface. The expanding bubbles behave in a way that resembles what happens when one takes the cap off a bottle of a carbonated beverage: gas formerly held in solution by the pressure within the capped bottle suddenly forms bubbles that rush to the top of the liquid and escape into the air.

Much volcanic gas consists of water vapor. The 1945 eruption of Paricutin in Mexico (see Figure 5.4) blew off 15,000 tons of steam per day. Steam from a volcano can come from two sources: (1) from water locked up within the lattices of some minerals in the Earth's mantle, and (2) from groundwater, water held within the underground pores of sediments and sedimentary rocks, which the rising magma vaporizes as it approaches the Earth's surface. The new water, that is; the water released from the mantle, is designated as *juvenile water;* it enters the geologic cycle by being discharged from volcanoes. Even if only a small proportion of the water vapor that escapes from a volcano is juvenile water, so many volcanoes have been active for so long that we can regard volcanoes as the ultimate sources of all the water in the hydrosphere as well as many of the gases in the atmosphere.

**Volcanic liquids** The lava ejected by volcanoes is a molten melt of silicates whose physical properties are controlled to a large degree by the chemical composition. Lava may be mafic or felsic in composition. *Mafic* (MAY-fic) lavas are those containing abundant magnesium and iron, and which, upon cooling, form ferromagnesian silicate minerals and calcium-rich plagioclase. Mafic lavas are hot (1000 to 1200°C) and tend to be liquid and free flowing. *Felsic* lavas are those containing small amounts of magnesium and iron, but abundant aluminum, sodium, and potassium, which solidify to form potassium feldspar, sodium-rich plagioclase, and quartz. Felsic lavas are somewhat cooler than mafic lavas (800 to 1000°C) and tend to be sluggish and viscous.

Volcanic activities vary with the viscosity (resistance to flow) of the lava. In fluid mafic lavas the gases can escape continuously. In sluggish felsic lavas the gases can escape only after they have built up great pressures and cause violent explosions.

Smooth-flowing mafic lava typically forms a pliable crust that can be bent and twisted into intricately curved forms, much as taffy can be pulled and folded. The Hawaiian word *pahoehoe* (pa-hoey-hoey) is applied to such twisted, ropy-looking lava (Figure 5.5a). A contrasting type of lava is *aa* (ah-ah), another Hawaiian word. Aa is more viscous than pahoehoe lava and has jagged, uneven surfaces. As the surface of aa lava cools rapidly and hardens, its brittle crust is broken continuously by the slow but constant movement of the still-fluid lava beaneath it. For reasons that are still not clear, aa is also characterized by spines that grow

(a)

(b)

(c)

**5.5 Kinds of lava based on physical appearances of resulting extrusive igneous rocks.** (a) Twisted, ropy-looking, curved surface of pahoehoe, Footprint Trail, Hawaiian Volcano National Park. (U. S. Geological Survey) (b) Rough-surfaced aa lava flow with pointed spines. Newberry Crater, Oregon. (R. L. Nichols) (c) Block lava, Yatsugatake Volcano, Japan. (H. Takeshita)

**5.6 Light-colored, porous pumice,** containing so many tiny vesicles that the rock will float on water. (J. E. Sanders)

upward from the rough blocks (Figure 5.5b). *Block lava,* rich in silica, resembles aa without the spines (Figure 5.5c). Some block lava is so viscous that it may move only a few meters a day, breaking itself into blocks as it flows over the ground.

Gas expanding within lava can create a froth that may solidify to a rock full of open spaces. These cavities are named *vesicles.* Vesicular felsic lava solidifies to light-colored *pumice* (Figure 5.6) that contains so many empty nonconnecting spaces that it may float on water. Gas expansion in mafic lava creates larger individual spaces that may or may not be connected to adjacent cavities. The resulting highly vesicular rock is a dark-colored jagged product named *scoria* (Figure 5.7a). If vesicles become filled with minerals, the resulting rock is *amygdaloidal* (Figure 5.7b).

**Volcanic solids** The jets of gas issuing from a volcano usually contain large quantities of solids, called *tephra,* including crystals of minerals that

**5.7 Vesicular basalt (scoria) and amygdaloidal basalt.** (a) Scoria, showing change in size of vesicles from about 1 millimeter in diameter in rapidly chilled crustal area to about 1 centimeter in diameter where cooling was slower. (J. E. Sanders) (b) Amygdaloidal rock, former scoria in which the vesicles became filled with chlorite and calcite. (J. E. Sanders)

(a) (b)

**5.8 Volcanic blocks** are usually smaller than this one at the base of Makaopuhi, Hawaii. (U. S. Geological Survey, Hawaiian Volcano Observatory)

had solidified within the magma before the eruption, pieces of rock through which the gas has passed, and bits of glass. The pieces of rock penetrated by the jets of gases may consist of whatever kind of rock underlies the volcanic landform as well as pieces of volcanic material that accumulated from earlier eruptions. The glass may have been ejected as a liquid but became solid rapidly in the atmosphere, or it may have been the crust of an older lava flow.

Various kinds of tephra are labeled according to shapes and sizes. Here we mention only the larger sizes (greater than 32 mm in diameter). Volcanic *blocks* are large, angular solids (Figure 5.8). Commonly associated with blocks are *bombs,* tephra that were ejected as clots of molten lava but solidified in the atmosphere. The friction between the air and the moving clots of lava and the rotation of the clots usually produces almond-shaped objects with twisted ends (Figure 5.9). Volcanic bombs are related to ropy pahoehoe lavas (compare Figures 5.9 and 5.5a); blocks

**5.9 Volcanic bombs,** showing typical spindle shapes with twisted ends. Cone N of Kalepeamoa, Mauna Kea, Hawaii. (C. K. Wentworth, U. S. Geological Survey, Volcano Observatory at Kilauea)

are akin to block lava or aa (compare Figures 5.8 and 5.5b and c).

Tephra from violently explosive volcanic eruptions can be spread through tens of thousands of square kilometers within a few days. As a result, the deposited layer of tephra is everywhere the same age. In some cases, its exact date of eruption can be determined, which gives the tephra layer great value in dating strata.

## Volcanic Landforms

The popular concept of a volcano is a conical mountain with a circular base and a pointed top. Indeed, many volcanoes have built features matching this image. However, depending upon the nature of the products and how they accumulate, volcanoes build a variety of landforms. These volcanic landforms include volcanic plains and volcanic plateaus, volcanic shields, various kinds of volcanic cones, craters, and calderas.

**Volcanic plains and volcanic plateaus** Flat sheets of extrusive igneous rocks are called *volcanic plains* or *volcanic plateaus.* These features are formed from extremely fluid mafic lavas or from mixtures of particles dispersed in volcanic gases that were highly mobile and flowed with great speed as if they were liquid. If their lavas issued from fissures, they formed on the Earth's surface through *fissure eruptions* (Figure 5.10).

The most voluminous fissure eruption in recorded history took place in 1783 at Laki, Iceland (see Figure 5.4), from a fissure 32 kilometers long. Altogether about 12 cubic kilometers of lava were discharged. The lava spread through an area of 558 square kilometers that extended 64 kilometers outward from the fracture on one side and nearly 48 kilometers on the other.

Volcanic plains and volcanic plateaus are usually regional in extent. A volcanic plain or volcanic plateau can cover more than 250,000 square kilometers. After thousands of individual eruptions, the thickness of the layers of extrusive rock may reach about a kilometer. Thus, the volume of extrusive igneous rock underlying a volcanic plain or a volcanic plateau can reach 400,000 cubic kilometers. The difference between a volcanic

**5.10 Fissure and fissure eruption.** (a) A fissure down which some lava flowed before solidifying. Mauna Loa, Hawaii. (M. R. Chartrand III) (b) Early stage in fissure eruption, with lava spraying along faults (fissures). (c) Later stage after lava has filled low block between two faults and has begun to spread laterally beyond the fissures.

(a)

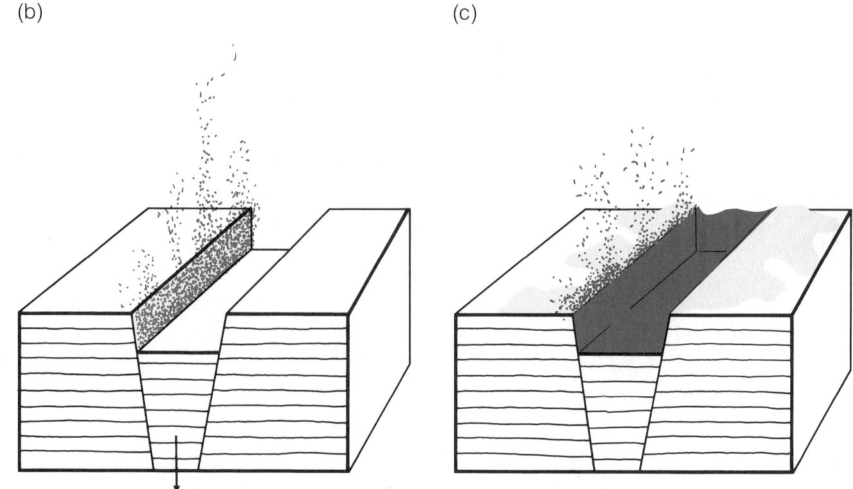

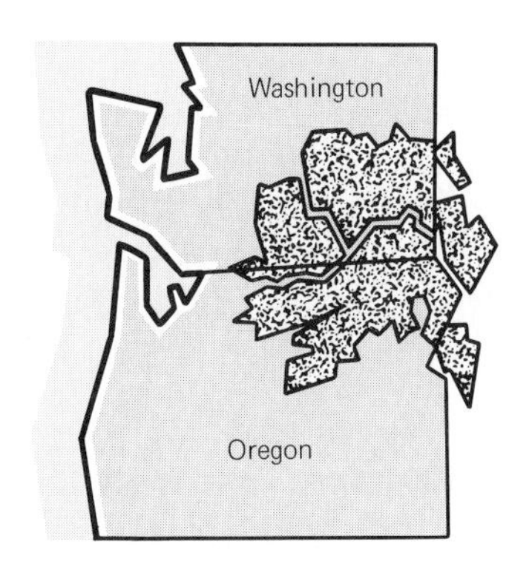


(a)

(b)

**5.11 Columbia basalt plateau,** Washington and Oregon. (a) Location map; textured areas are underlain by basalt that spread out in hundreds of successive fluid flows and eventually filled a former lowland. The maximum aggregate thickness of the basalt layers is about 1100 meters. (b) Sectional view of about 20 or more ancient lava flows, seen looking downstream along the Columbia River from Palouse Falls, Franklin-Whitman Counties, Washington. (F. O. Jones, U. S. Geological Survey, 1950)

plain and a volcanic plateau is the height of the surface compared to the surrounding countryside. A volcanic plain is a low-lying area in which the volcanic rocks usually are thin. A volcanic plateau is a high-standing area usually built by the accumulation of thick sheets of extrusive igneous rock.

During the Miocene Epoch in northwestern United States (Figure 5.11a) a large basin rimmed in by mountains was filled by the spreading out from various vents of hundreds of very fluid flows of mafic lava. These solidified to make sheets of basalt now exposed in section as a result of later uplift and deep erosion (Figure 5.11b). The areas underlain by these basalts are known as the Columbia River and Snake River plateaus.

Comparable piles of basalts are known from other continents, and we are only beginning to explore the largest volcanic plains of all — those found on the ocean floors. These thin layers of basalt are thought to be produced continuously along vast deep-sea volcanic ridges.

In many eruptions of felsic lavas the continuity of the liquid phase is disrupted by the violently escaping gases. As a result, hot blebs (blisters or bubbles) and clots of lava form a gas-and-particle mixture known as a *tephra flow.* In many ways, tephra flows behave as if they were very fluid, fast-moving lava flows. They can spread many kilometers from their vents. Tephra flows have been clocked at speeds of 30 to 100 kilometers per hour. The temperature within hot tephra flows may exceed 1000°C; the flows char trees and wooden buildings and may even melt glass and metals. Tephra flows tend to build up flat-lying layers (Figure 5.12) of particles that may be still so hot when deposited that they melt together as they settle, forming rock known as *welded tuff.*

**Volcanic shields** Repeated, quiet eruptions of highly fluid mafic lava from a circular vent or from a rift zone may create a broad, gently sloping conical mound of volcanic rock. Because of their resemblence to the shields of ancient warriors, these features have been named *volcanic*

**5.12 Flat-lying layers of volcanic rock** formed by extrusion of tephra flows. Bandelier Tuff (Pleistocene), near Los Alamos, New Mexico. (R. L. Smith and R. A. Bailey, U. S. Geological Survey)

*shields.* Although the term volcanic shields was first applied to examples in Iceland, the largest examples are the Hawaiian Islands and other mid-Pacific islands. In Hawaii, the volcano Mauna Loa was built up slowly from the sea floor during an interval of about a million years. This huge mass of basalt is 160 kilometers wide at the base and extends upward from the sea floor 5000 meters to sea level and an additional 4000 meters more above sea level. Its total relief, therefore, is more than 9000 meters. By comparison, the top of Mt. Everest, the Earth's highest point above sea level, stands at 8550 meters. Mauna Loa dwarfs Mt. Everest (Figure 5.13). So large is the Mauna Loa volcanic shield that even Kilauea and Hualalai, full-sized volcanic landforms by most scales, are mere flank craters on Mauna Loa (Figure 5.14).

**Volcanic cones** The most familiar of volcanic landforms are *volcanic cones.* Several varieties are recognized depending on the materials composing them. *Tephra cones* consist of fine-grained, usually uniformly sized tephra that were ejected from a circular volcanic vent, fell back as solids, and were piled up surrounding the vent (Figure 5.15).

The angle of the sloping sides of nearly every tephra cone in the world is about 30 degrees. Tephra cones rarely are higher than about 450 meters. This height is usually attained over a period of several months, but it may be reached in as short a time as a few weeks.

A tephra cone requires considerable reinforcing by lava before it is strong enough to support the weight of a column of lava high enough to flow out the top. Typically what happens is that the lava starts up the

**5.13 Mauna Loa Volcano,** Hawaii and Mt. Everest, viewed in profiles drawn to the same scale.

SW
Mauna Loa
Kilauea

**5.14 Summit of Mauna Loa,** Hawaii. Small circular depressions in foreground are craters; large circular depression in the background is a caldera. (U. S. Air Force)

**5.15 Tephra ("cinder") cone enlarging by addition of particles from eruption cloud,** Cerro Negro Volcano, west-central Nicaragua, November 1968. Small tephra cones such as this one are not strong enough to support a column of lava that could fill the central conduit all the way to the top and thus to allow the lava to spill from the apex of the cone. Instead, the lava has lifted the cone and has oozed out from beneath, forming the flat layers in the forground and at left rear. Continued discharge of lava sheets from beneath a tephra cone can build a broad, conical volcanic shield. A tephra cone atop a volcanic shield constitutes a composite cone (compare Figure 5.17). (U. S. Geological Survey)

central vent within the cone, but as it rises higher, the lateral fluid pressure at the base causes the lava to push sideways. The lava then may split out sideways as a flank eruption or may even lift the tephra cone and spread beneath the cone and beyond (see Figure 5.15, foreground).

Eruptions of small clots of liquid lava (*spatter*) may build small, steep-sided *spatter cones* (Figure 5.16). Rarely exceeding 30 meters in height, a spatter cone looks as though it were built by someone standing in a deep hole and hurling blobs of wet cement up onto the ground. The hot droplets of spatter tend to stick to one another and become welded together. This yields a structure that is more rigid than a tephra cone. Welded spatter may form walls that are nearly vertical.

*Complex volcanic cones* are composed of sloping layers of tephra and reinforcing layers of *extrusive igneous rock,* that is, igneous rock formed by the cooling of lava flows. In these cones, lava from the central vent pushes outward through the sides of the cones as flank eruptions.

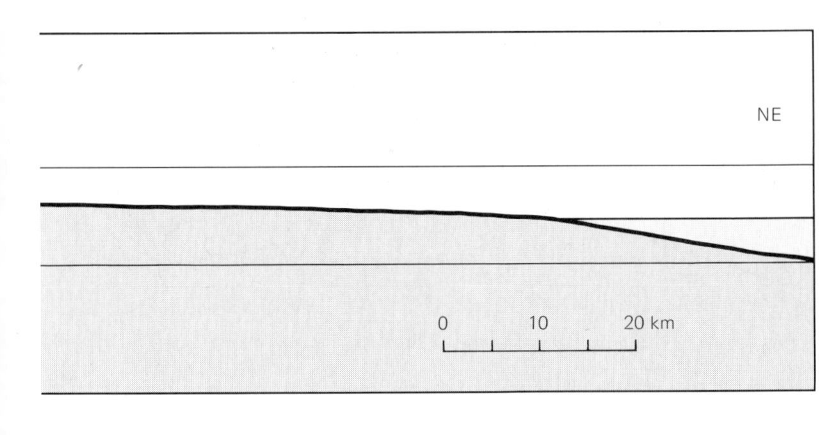



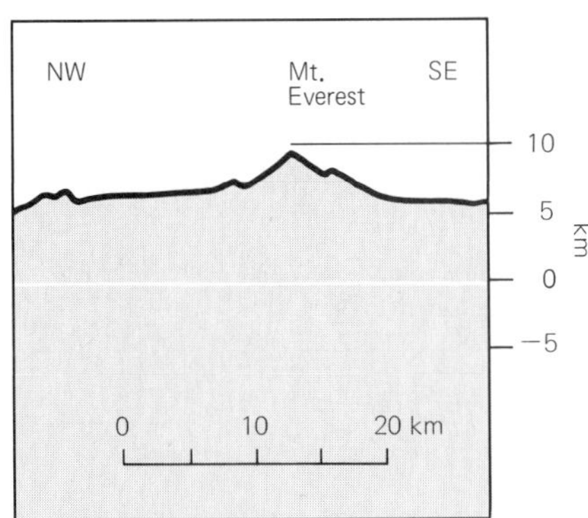

**5.16 Large spatter cones,** with nearly vertical sides, built at base of two active lava fountains at Surtsey, Iceland, August 1966. People at lower left indicate scale. (W. A. Keith)

Eventually, the cone's sides may be so strengthened by the sheets of igneous rock that it is possible for lava to rise to the top of the cone and to issue forth as summit eruptions (Figure 5.17).

*Composite volcanic cones* are combinations of complex cones built atop volcanic shields. The tephra cone in Figure 5.15 could be the beginning stage in the construction of a composite cone. If the tephra cone were eventually reinforced to become a large complex cone and if the layers of highly fluid lava issuing at the base accumulated to great thickness and formed a volcanic shield, the result would be a composite cone. Great conical peaks, such as Fuji in Japan, Vesuvius in Italy, Rainier in Washington, Mayon in the Philippines (Figure 5.18), and Shasta in California, all famed for their beauty and symmetry, are composite volcanic cones. The secret of such symmetry is eruption through a single central vent. If a second vent becomes activated or if flank eruptions should accumulate, secondary cones may form and destroy the symmetry.

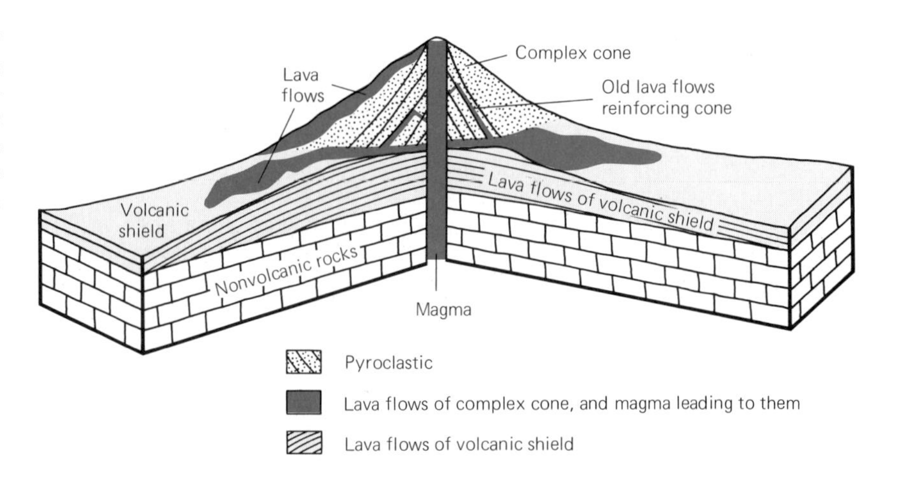


**5.17 Composite volcanic cone,** cutaway view, showing how sheets of igneous rock have made a complex cone by strengthening the sides of a tephra cone (at top) to the point where lava can spill from the apex of the cone. Sheets of lava at base are building a volcanic shield.

**5.18 Composite cone of Mt. Mayon,** Philippines, is considered by many to be the world's most symmetrical composite cone. Low slopes of volcanic shield at base contrast with steep slopes of complex cone at summit. (Peabody Association)

**Craters** A circular volcanic orifice having a diameter of less than 1.5 kilometers is a *crater* (see Figure 5.14). It is opened up by a concentrated jet of hot gases escaping upward through bedrock. Many craters are situated at the apexes of volcanic cones. Such craters may enlarge as a result of explosive eruptions or as a result of collapse following the withdrawal of lava from the cone.

**Calderas** A circular feature having a diameter greater than 1.5 kilometers and formed by collapse is a *caldera.* A caldera formed on an oceanic island may create an enclosed circular body of water that could be used as a harbor for ships (Figure 5.19). Calderas differ from craters not only in size, but also in their relationship to the life histories of volcanoes. Craters are active during most eruptive cycles. Calderas usually form after climactic explosions that terminate major eruptive cycles, as we shall see in the next section devoted to eruptive activities. The world's largest modern caldera is at Mt. Aso, Japan (Plate 5); it measures 16 by 22 kilometers. Another large caldera is Ngorongoro in Tanzania, Africa, which is 18 by 16 kilometers; its floor lies 600 meters lower than its rim. A caldera of Pleistocene age, measuring 68 by 45 kilometers, has been identified at Yellowstone National Park.

## Some Classic Volcanic Eruptions

Our ideas about the kinds of eruptive activities have been developed by assembling the case histories of particular volcanoes. These include Vesuvius, Krakatoa, Mt. Pelée, Mt. Mazama, the Hawaii volcanoes, and Stromboli.

**5.19 Caldera at top of volcanic island** has formed a potential harbor, Mang Island, northern Marianas Group. (American Museum of Natural History)

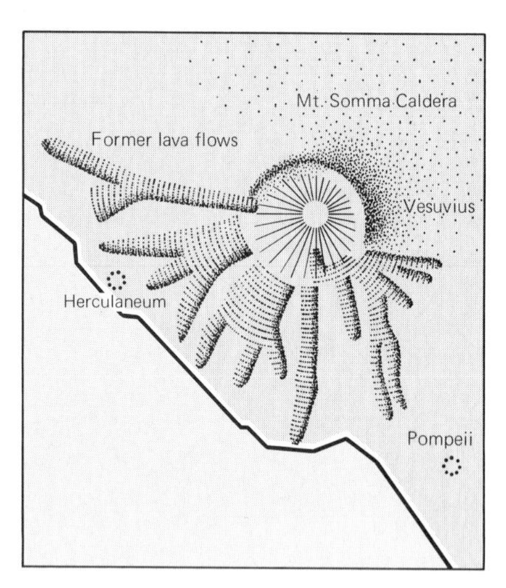


**5.20** Sketch map of Vesuvius and vicinity.

**Vesuvius** Ever since Roman times, the activity of Vesuvius has been observed and recorded. The modern Vesuvius is an average sized cone lying within the floor of the great caldera of a gigantic former volcano named Mt. Somma. The birth of Mt. Somma took place perhaps 100,000 years ago. After building a great cone, Mt. Somma became inactive. In the terminology of volcanology, a volcano is *inactive* if it has not erupted within historic time (approximately 10,000 years); *dormant* if it is known to have been active, but has been quiet during the past 50 years; *active* if it has erupted within the past 50 years; and *extinct* if all activity has forever ceased. The Romans knew that Mt. Somma was a volcanic peak, but because they thought its eruptive days were over, they built the cities of Herculaneum and Pompeii near it (Figure 5.20).

After a long period of no activity, Mt. Somma erupted in A.D. 79. The Roman historian Pliny the Younger wrote what is considered to be the first scientific description of a volcanic eruption. He had a strong personal reason for writing this description; his uncle, Pliny the Elder, died observing the activity.

Analysis of the thick layers of tephra forms the basis for reconstructing what happened. During a long and frightening series of violent explosions a layer of felsic tephra including much pumice, 2.5 meters thick, was deposited on Pompeii. This tephra buried the city and suffocated the residents as they sat or lay (Figure 5.21). The rising heat over the vent and the steam condensing from it created tremendous updrafts that triggered local thunderstorms. The torrential rainfall saturated the tephra on the western flank of the cone. This created volcanic mudflows that slid loose and buried the nearby city of Herculaneum. After the magma reservoir had been depleted, the entire top of Mt. Somma collapsed, leaving a giant caldera whose northeastern wall still stands, but whose southwestern half is open (Figure 5.22).

In the language of eruptive activities, a *Plinian eruption* now designates volcanic eruptions consisting of the explosive ejection of large amounts of lava (on the order of several cubic kilometers), which causes the top of the cone to collapse and thus to form a caldera. A Plinian eruption may or may not be accompanied by tephra flows.

**Krakatoa** The most famous Plinian eruption of recent times took place in 1883 from Krakatoa (Figure 5.23), which designates three small volcanic islands located in the Sunda Strait between Java and Sumatra. This eruption was so violent that the word *Krakatoan* has come to be used to describe particularly explosive eruptions. On 20 May 1883, after 200 years of being dormant, Krakatoa began a series of moderate eruptions to weak eruptions. Following three months of such activity, a climactic outburst came suddenly on 26 August at 1300 (1:00 P.M.). Explosions, heard throughout Java more than 150 kilometers away, continued all afternoon and all night, bringing a fallout of tephra that made it almost impossible to see, even with lamps. At 1002 the next morning a blast shot tephra 80

**5.21 Former residents of Pompeii,** Italy, victims of the explosive eruption of Monte Somma in 79 A.D., plaster casts of corpses. The shapes of these bodies were preserved as molds in the tephra that engulfed them. After the soft parts had oxidized and the bones had been removed from the cavities, plaster casts were made of the cavities and these specimens are the result. (SCALA)

(a)

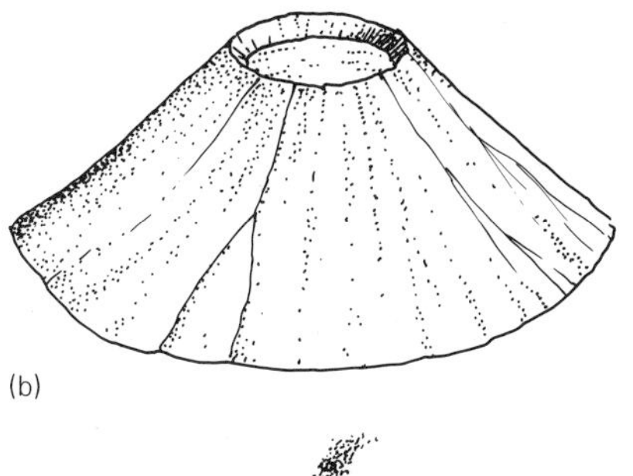

(b)

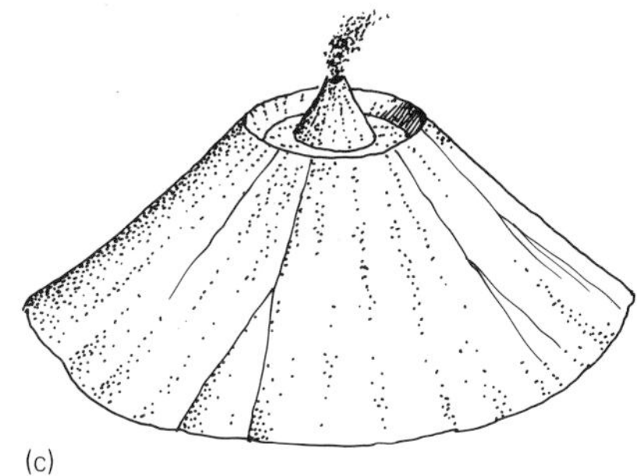

(c)

**5.22** Three episodes of caldera collapse and two of cone growth in history of Monte Somma and Vesuvius, Italy. (a) View from above of Vesuvius and remaining north rim of Monte Somma caldera. (C. B. Jacobson, U. S. Bureau of Reclamation.) (b) Monte Somma after last prehistoric Plinian eruption left a large flat-floored caldera at summit of cone. (c) A new tephra cone grows from floor of caldera. (d) Continued growth of new cone restores height and symmetry of peak. Dashed line marks rim of former caldera. (e) Plinian explosion in eighth century B.C. forms small flat-floored caldera open on the S side. This is the Monte Somma as known to the Romans. (f) At climax of catastrophic eruption of A.D. 79, a one-sided and broad, flat-floored caldera, open to the SW, is what remained of Monte Somma. (g) Post-A.D. 79 growth of Vesuvius, a new cone filling the open side of the Monte Somma caldera. Black areas are tongues of extrusive igneous rock from former lava flows; blue dots indicate former rim of Monte Somma caldera. (h) Profile and geologic section through Vesuvius and surroundings. Depth of magma chamber is based on inferred depth to top of Triassic carbonate rock. (After Alfred Rittmann, 1962, Volcanoes and their activity; translated from German by E. A. Vincent: New York-London Wiley-Interscience, 305 p.; Figure 67, p. 128)

(d)

(e)

(f)

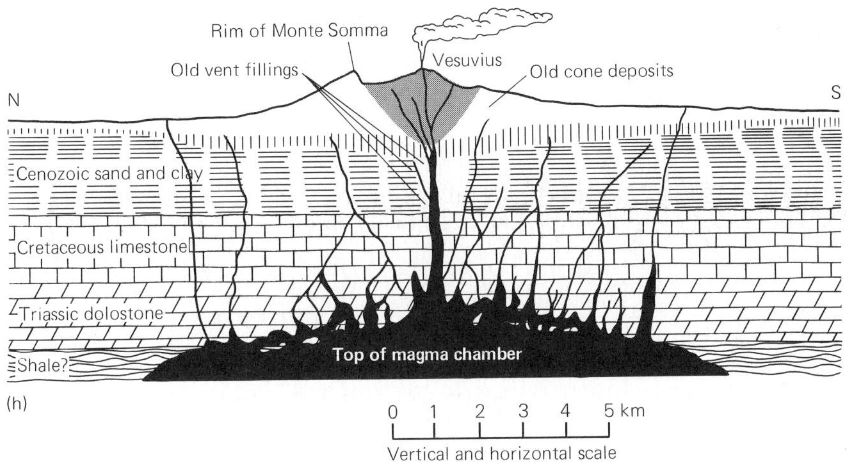


(h)

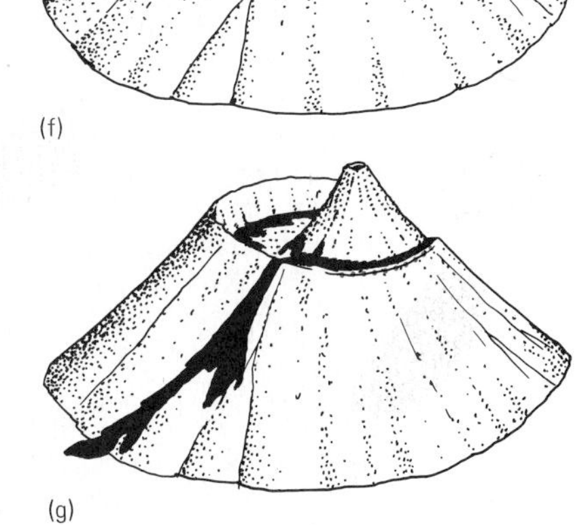

(g)

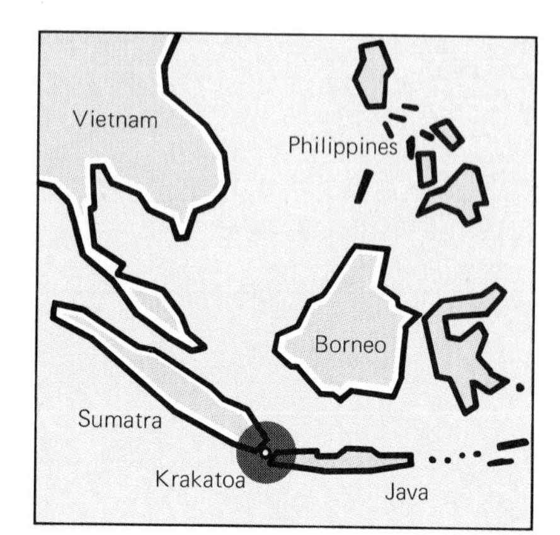


**5.23** Location map of Krakatoa.

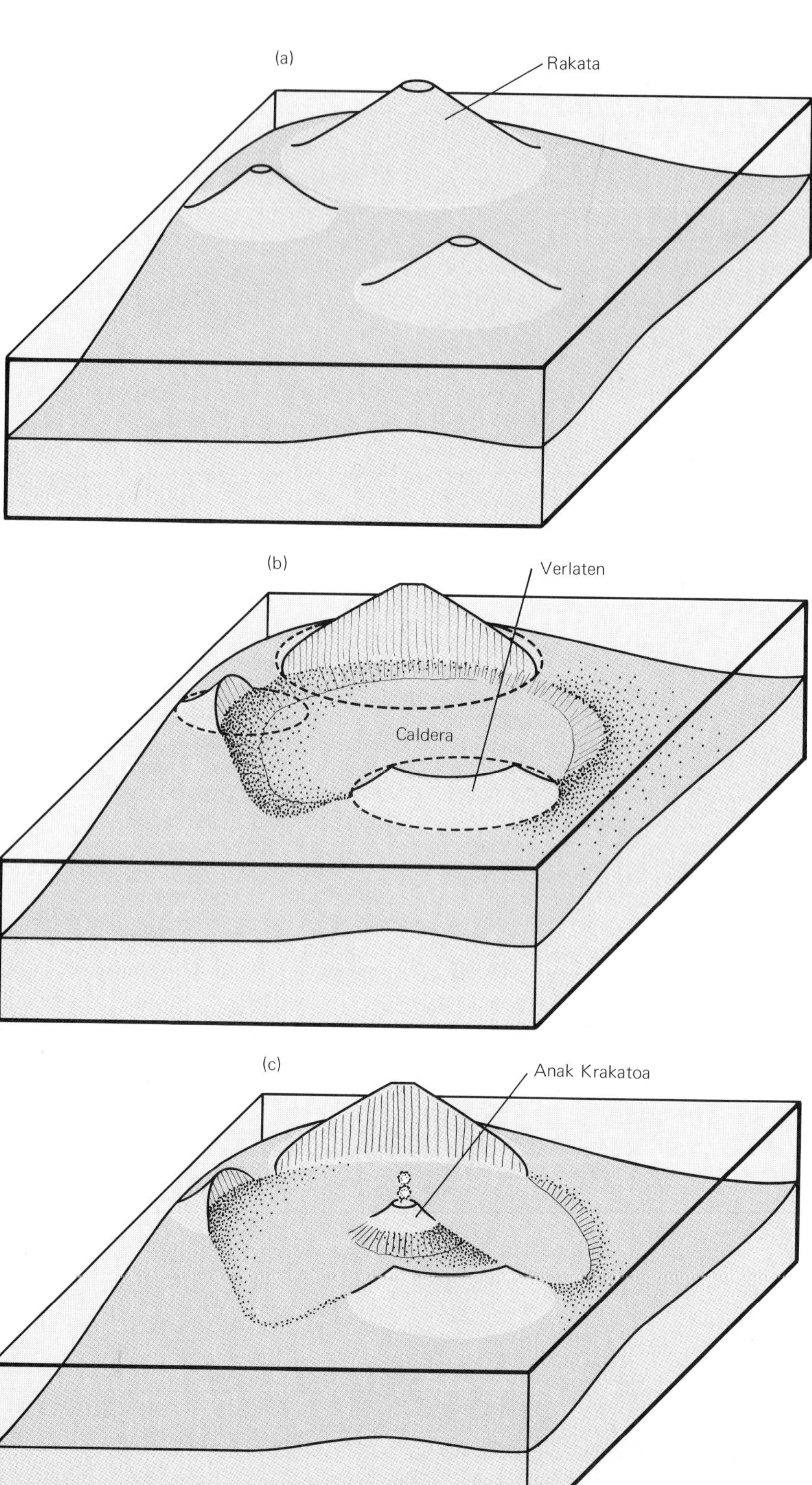


**5.24 Stages in history of Krakatoa,** Indonesia, schematic block diagrams. (a) Three small conical islands, restored to condition prior to 1883 explosion. (b) Violent eruption of 27 August 1883 and collapse formed a caldera and removed parts of all three island cones. (c) Later growth of a small cone in the caldera has formed a new island in the center of the three island remnants.

kilometers into the air and spread them through an area of 800,000 square kilometers. Residents of Rodriquez Island, nearly 5000 kilometers distant in the Indian Ocean, reported hearing sounds like heavy gunfire. The waning activity continued until February 1884. When it was all over, nearly 18 cubic kilometers of tephra had been spread through 4 million square kilometers.

When the top of Krakatoa collapsed and the caldera floor sank, a portion of the sea floor was displaced downward. When the bottom of a body of water is suddenly shifted, waves are created on the surface of the water. In the ocean such waves can reach gigantic size; they are named *tsunami,* from the Japanese meaning "harbor wave." (See Chapters 15 and 17 for further discussion of tsunami.) Some of the waves generated by Krakatoa's collapse and formation of a large caldera reached heights of more than 30 meters; water from them traveled as much as 150 kilometers inland. Altogether, the tsunami killed more than 36,000 persons. As for the island itself, more than two-thirds of it disappeared. All that was left was a tiny rock, three small islands, and a caldera 275 meters deep (Figure 5.24).

**5.25 Location map, Martinique,** the Caribbean island on which Mt. Pelée Volcano is situated.

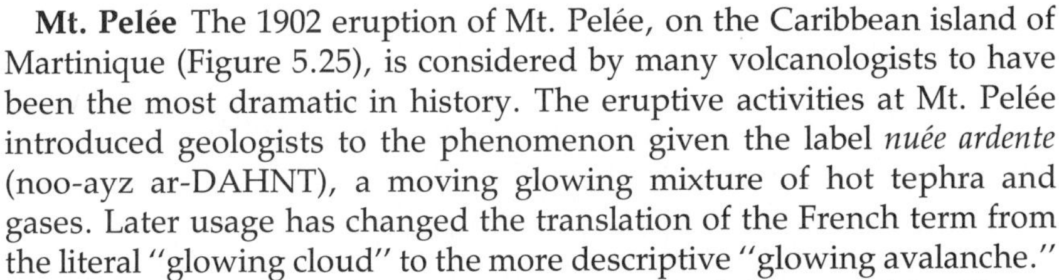
**Mt. Pelée** The 1902 eruption of Mt. Pelée, on the Caribbean island of Martinique (Figure 5.25), is considered by many volcanologists to have been the most dramatic in history. The eruptive activities at Mt. Pelée introduced geologists to the phenomenon given the label *nuée ardente* (noo-ayz ar-DAHNT), a moving glowing mixture of hot tephra and gases. Later usage has changed the translation of the French term from the literal "glowing cloud" to the more descriptive "glowing avalanche."

The summit of Mt. Pelée is situated north of the city of St. Pierre. Heading near the summit and leading generally southwestward is the valley of the Blanche River. Many glowing avalanches originating from the crest of Mt. Pelée traveled down the valley of the Blanche and ended in the sea (Figure 5.26).

**5.26 Nuée ardente,** headed directly toward the sea (at left) from summit of Mt. Pelée (mostly hidden from view behind cloud), Martinique. City of St. Pierre, safely out of harm's way on this particular occasion, is at lower right. See Figure 5.28 for the aftereffects at St. Pierre of a nuée ardente that did not flow directly into the sea. (Painting by Charles R. Knight; courtesy of the American Museum of Natural History)

The first signs of the fateful round of activity were observed by a high-school teacher on 2 April. On 23 April falls of tephra and sulfur fumes began to spread and to grow more intense daily. By the end of April falling tephra began to choke the streets in St. Pierre, 10 kilometers from the summit. Some businesses closed. Horses and birds dropped dead in the city and residents started to flee. However, the French governor wanted the people to remain so they could vote in elections scheduled for 10 May. He appointed a commission, which promptly reported that St. Pierre was in no danger from the volcano. This move failed to halt the flight of the populace and on 6 May the governor stationed troops around the city to keep people at home.

The city was bulging with several thousand panic-stricken refugees from the countryside when, at 0750 on 8 May, four loud explosions rocked Mt. Pelée. One of them shot a glowing avalanche sideways through a V-shaped notch in the summit crater. This avalanche started down the usual track toward the sea. However, the tephra flow was so

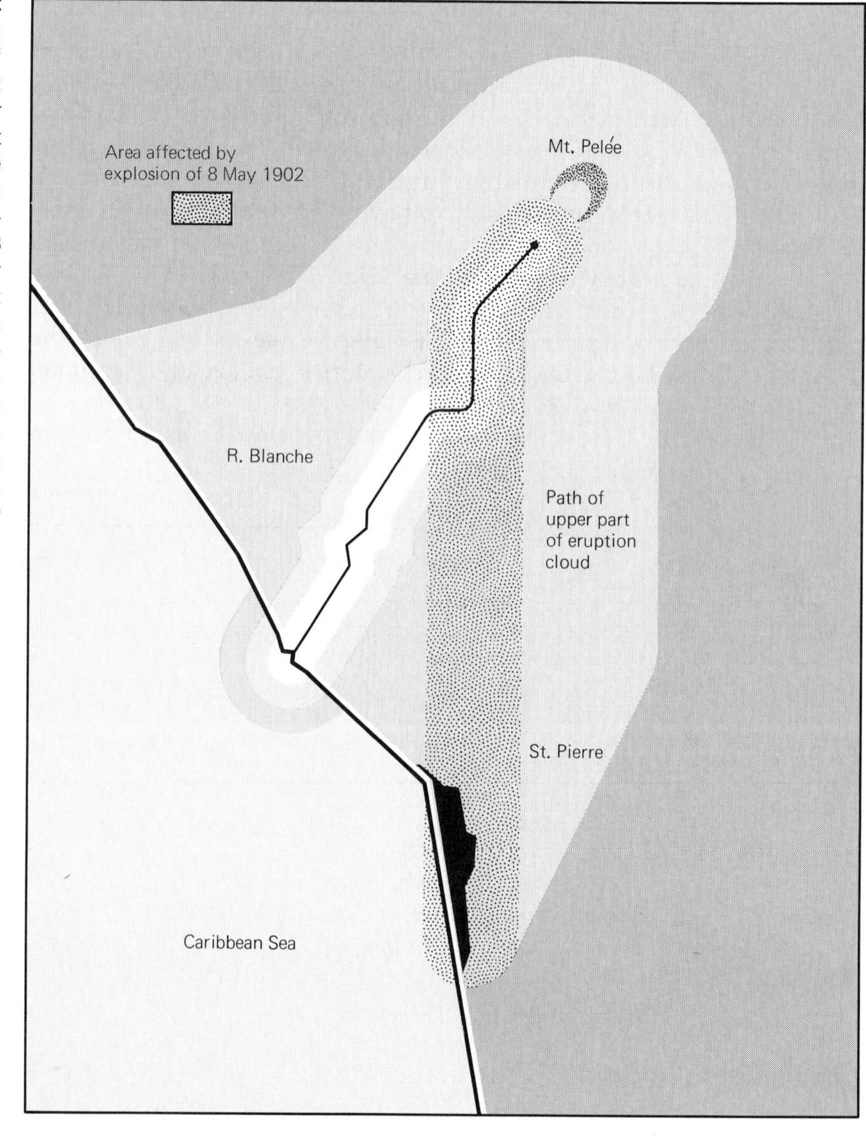


**5.27 Northwestern Martinique, West Indies,** schematic sketch map showing area devastated by explosive eruption of Mt. Pelée on 8 May 1902. The great eruption cloud started down River Blanche on its usual track to the sea, but halfway down, evidently did not make the sharp bend in the valley, but overflowed the valley and headed straight south toward St. Pierre. Because the eruption of 8 May was so large, the stretch of the river valley that trends N–S served as a rocket launcher pointed right at St. Pierre. The residents were caught unaware; they supposed that all glowing avalanches would flow harmlessly into the sea as in the past. (After Gordon A. MacDonald, 1972, *Volcanoes:* Englewood Cliffs, New Jersey, Prentice-Hall, Inc., 510 p., Figure 8–1, p. 144)

**5.28 Total devastation of St. Pierre** shown in photograph made with camera facing due north toward Mt. Pelée Volcano (mostly obscured by clouds in upper center). The only walls left standing are aligned N–S, parallel to the direction of flow of the fatal glowing avalanche. Harbor is at left. Compare with Figure 5.26. (American Museum of Natural History)

huge that most of it did not make the sharp turn in the valley of the Blanche, but instead headed straight south directly toward St. Pierre (Figure 5.27). Two minutes later, the glowing avalanche hit the city, stopping the clock on the Hospital Militaire at 0752 and instantly killing an estimated 30,000 persons, including the governor. The speed of this hot tephra flow was later calculated to have been around 150 kilometers per hour.

Destruction to the city was almost total (Figure 5.28). Only two persons are known to have survived, one a prisoner in a dungeon. The rest appear to have died in seconds from inhaling the deadly hot gases. From the fact that it softened glass, but did not melt copper (melting point 1058°C), the temperature of the hot tephra flow can be estimated to have been 650 to 700°C. Of the city itself, which had been the most important commercial center on Martinique, no building remained standing. Iron girders were twisted like licorice, burning rum flowed in streams through the streets, walls of cement and stone were torn apart, and a three-ton statue of the Virgin Mary was carried 15 meters from its base. The entire city was ignited as if by a giant match.

The term *Peléan* has been used for an eruption in which all the magma is disrupted by the gases and is expelled in the form of a *nuée ardente*. In such an eruption, almost no lava flows as a continuous liquid.

**Crater Lake and Mt. Mazama** In southwestern Oregon, near the southern end of the line of volcanic cones known as the Cascades (Figure 5.29), stands a most unusual feature: a great conical mountain that holds within its summit a body of water, Crater Lake. Careful geologic study of the shape of the present-day peak and of the layers of tephra and extrusive igneous rocks that were discharged indicate that the volcano, named Mt. Mazama, the conical mountain ancestral to the present Crater Lake, was violently explosive. The shapes of the valleys indicate that glaciers were present on Mt. Mazama's upper slopes.

The tephra from the great explosion spread outward both as high-level clouds and as fiery-hot, ground-hugging tephra flows (glowing avalanches). Radiocarbon dates on remains of charred trees indicate that the

**5.29 Major peaks of Cascade Range,** location map. Circles mark composite volcanic cones that grew during Quaternary Period.

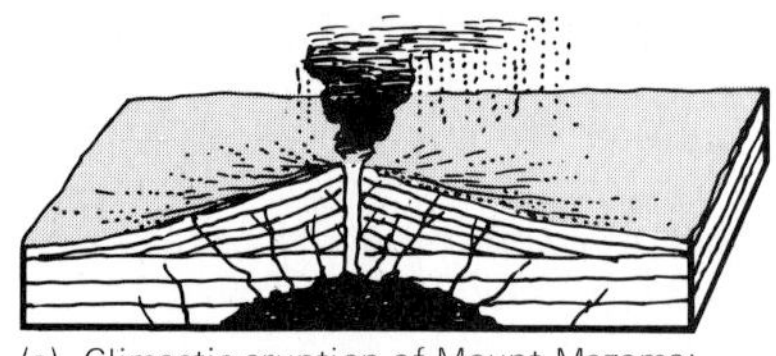

(a) Climactic eruption of Mount Mazama; magma chamber full

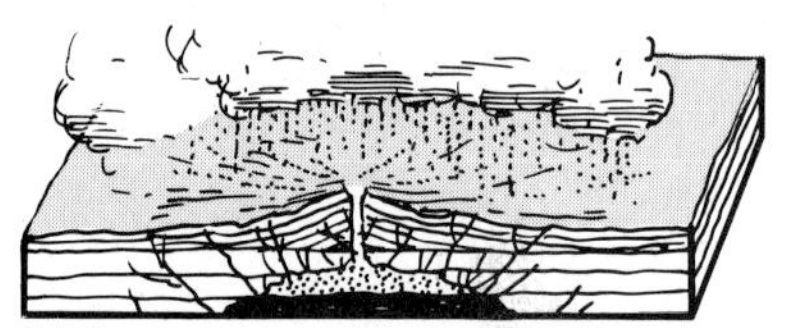

(b) Partly empty magma chamber before collapse

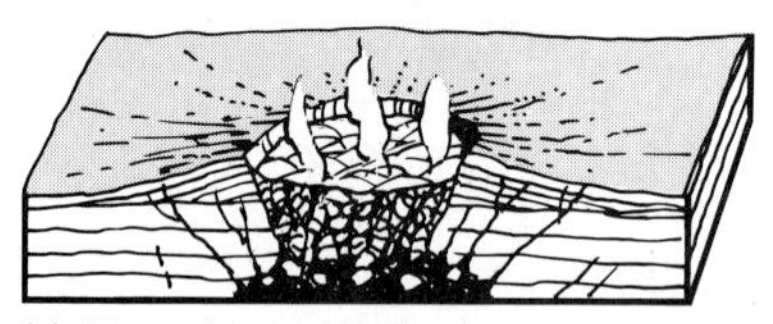

(c) Mount Mazama caldera

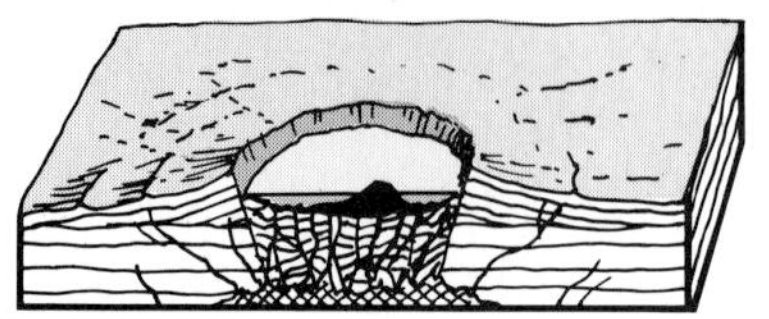

(d) Wizard Island cone in Crater Lake

**5.30 Stages in final explosive eruption and collapse of Mt. Mazama,** Oregon, and origin of Crater Lake. (a) Magma chamber fills; gas separates vigorously from liquid and begins to erupt from central vent, spraying tephra into the air. (b) Eruption intensifies; rapid and violent escape of gas and tephra partially drain magma chamber. (c) Weight of roof of partly empty magma chamber collapses. (d) Water collects in caldera. A second charge of magma, containing much less gas than the first, enters the magma chamber. Ejection of tephra forms a small tephra cone, Wizard Island, in Crater Lake. (Modified from Howel Williams, 1941, Crater Lake. The story of its origin: Berkeley and Los Angeles, University of California Press, 97 p., Figures on p. 84 and 85)

glowing avalanches descended about 6600 years ago. Tephra from Mt. Mazama have been found throughout an area of 900,000 square kilometers.

Comparisons have been made between the volume of the now-vanished summit of Mt. Mazama and the quantity and kind of tephra deposited. The volume of the missing top 2000 meters of the Mazama cone amounts to 70 cubic kilometers. Much of the missing volume of rock probably collapsed back into the magma chamber below at the climax of the violent eruptive cycle.

The caldera formed by this collapse now holds Crater Lake. At some time after the collapse, a new tephra cone was built; this is Wizard Island. This island compares to the Mazama caldera in the same way that modern Vesuvius compares to the caldera of Mt. Somma (see Figure 5.22). The chief geologic events leading to the formation of Crater Lake are shown in Figure 5.30.

**Hawaii** We have already mentioned the great bodies of mafic lavas that have cooled to form massive volcanic shields in Hawaii and elsewhere in the Pacific. The eruptions in Hawaii nearly always yield fluid lava in which the gases are liberated continuously and more or less without violent explosions. In recent eruptions from Hawaiian volcanoes it has not been unusual for jets of escaping gases to create fountains that may reach as high as 300 meters or more. Lava extruded from central vents or fissures has built some of the largest mountains on Earth.

**Stromboli** In the Aeolian Islands off the coast of Sicily (see Figure 5.4) stands Stromboli, whose long-continued discharges of red-hot lava have earned it the name of "lighthouse of the Mediterranean." Stromboli is unusual because it has been active almost continuously for many centuries. Its distinctive characteristics are the basis of the name *Strombolian eruptions* for those in which there are no discrete eruptive events between which the lava cools to form a solid crust.

## Distribution of Volcanoes

During the last 400 years, about 500 volcanoes are known to have erupted. Volcanoes are found in nearly all parts of the world, both on land (known as *subaerial* volcanoes because they erupt into the atmosphere) and on the sea floor. In the following sections we summarize the geographic distribution of subaerial volcanoes, describe volcanoes in the continental United States, and examine volcanic island arcs.

**Geographic distribution of subaerial volcanoes** The geographic distribution of subaerial volcanoes is related to boundaries of lithosphere plates. More than three-quarters of the world's subaerial volcanoes are distributed around the margins of the Pacific Plate (see Figure 1.24) in a pattern sometimes called the "Ring of Fire" (see Figure 5.4). An offshoot of this ring extends westward through Indonesia along the boundary of the Eurasian Plate and the Indian-Australian Plate. Fully 14 percent of the world's active volcanoes are located along this boundary.

Another series of volcanoes is scattered in the middle of the Atlantic Ocean along the spreading ridge between the Eurasian and North American plates and between the African and South American plates. An especially active cluster of this series is in Iceland. Many other volcanoes along the mid-Atlantic ridge have been active underwater.

The Mediterranean contains a group of active volcanoes, especially in Italy and the Aegean Islands south of mainland Greece. These lie along the boundary between the Eurasian and African plates.

Other active volcanoes are found within the African Plate along the Rift Valley in eastern Africa and bordering the South American Plate in Central America, along the Andes chain, and in Antarctica. Countless other volcanoes are scattered throughout the floor of the Pacific Ocean.

**Volcanoes in the continental United States** In the United States, as elsewhere, it is difficult to calculate the number of volcanoes that have been active during geologic time because most volcanic structures more than about 10 million years old have been eroded by wind and water. Some older volcanoes can be identified by the rocks composing their conduits or vents, but most ancient volcanoes have been eroded away.

Geologists estimate that during the past 60 million years several thousand eruptions probably have taken place in the Cascade Range, which extends from northern California to the Canadian border (see Figure 5.29). During the last few tens of thousands of years about 15 of these volcanoes have erupted. Of these, 7 have erupted since 1800, including Mt. Baker, Mt. Rainier, Mt. St. Helens, Mt. Hood, Mt. Shasta, Cinder Cone, and Lassen Peak. Mt. St. Helens has been more active and more violent during the past 2000 years than any other volcano in the continental United States. Geologists from the U.S. Geological Survey have compared Mt. St. Helens' activity to that of Vesuvius. Their 1976 prediction that Mt. St. Helens would erupt before the end of this century was dramatically fulfilled in early 1980 (see Plate 2).

Other volcanoes of Quaternary age have been found in Arizona near the Grand Canyon, in southern California, and in southern Idaho. Volcanic rocks of Cenozoic age are widespread in Nevada, western Utah, Yellowstone National Park, southwestern Colorado, and elsewhere in western United States.

**Volcanic island arcs** Many of the volcanoes that form the Pacific's Ring of Fire are associated with festoons of islands, called *island arcs*. Examples include the Aleutians, Kuriles, Ryukyus, Bonins, Marianas, New Guinea, and the Solomons-New Hebrides. Other island arcs are the Antilles, rimming the northern and eastern Caribbean; the group extending from the southern tip of South America at Cape Horn to the Palmer Peninsula, Antarctica, including the Falkland Islands, South Georgia, South

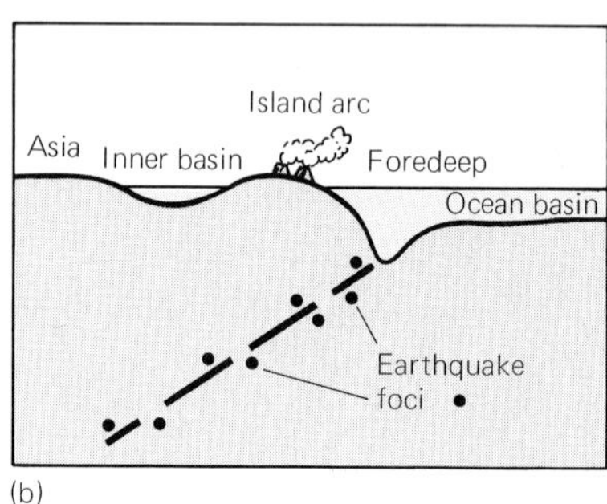


(b)

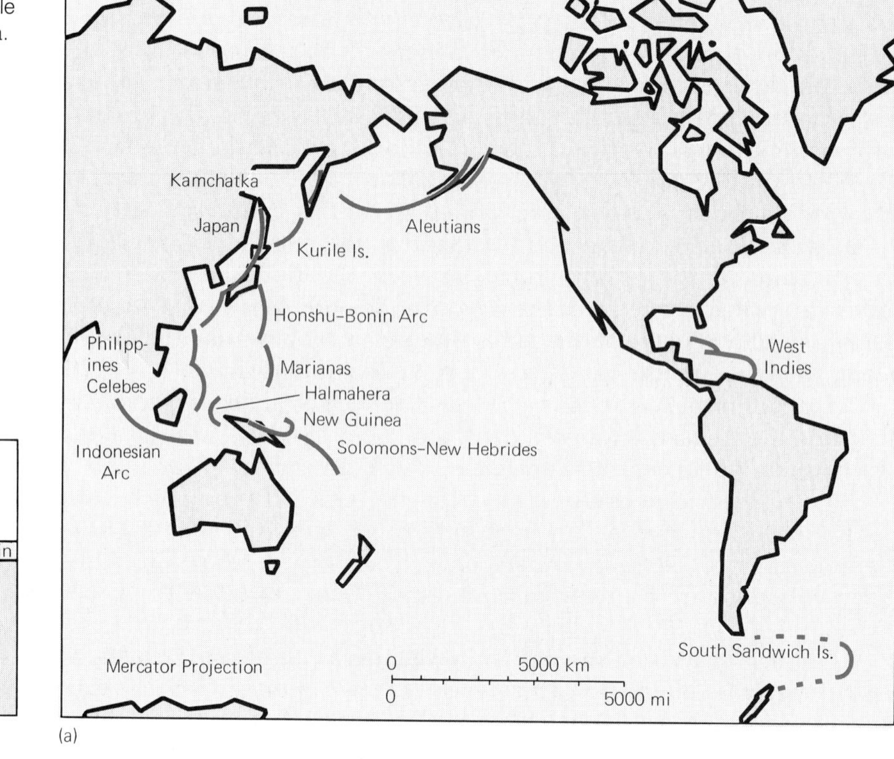


(a)

**5.31 Volcanic island arcs.** (a) Map of world's major arcs. (b) Schematic profile and section through arc lying E of Asia.

Orkney, and South Shetland Islands; and the Sunda Island group in Indonesia.

Island arcs typically are associated with an oceanic trench (Figure 5.31b). According to the concepts of plate tectonics, island-arc volcanoes are fed by magma that forms as great slabs of the lithosphere melt upon descending to depths of 100 kilometers or so. (Further discussion of these concepts is in Chapters 17 and 18.)

## VOLCANIC ACTIVITY UNDER WATER AND UNDER GLACIERS

Not all volcanoes come out in the subaerial realm; some are located under water and others beneath glaciers. Modern methods of investigating the deep-ocean basins have shown that the sea floor is dotted with tens of thousands of volcanic cones.

### Differences Between Subaerial and Underwater Eruptions

More than 80 percent of known volcanoes are subaerial, but this figure is misleading in reference to total volcanic activity. The estimated total of volcanic rocks that have accumulated on the floor of the Pacific Ocean is $2\frac{1}{2}$ times that erupted on all continents throughout geologic history. Also, geologists believe that gentle, but repeated eruptions along the mid-Atlantic Ridge are constantly creating new sea floor.

There are several important differences between subaerial and underwater volcanoes. Perhaps the most obvious is the difference in types of lava and, therefore, violence of eruption. The lava of subaerial volcanoes tends to be viscous and felsic, like the continental rock that melted to form the lava. Highly viscous, felsic lava results in explosive eruptions. Underwater lava is more fluid and mafic, allowing magmatic gases to escape freely before they build up enough pressure to explode. At depths of 600 meters or more the pressure of the water is great enough to prevent the gas from expanding, so that eruptions at such depths are never explosive, no matter what kind of lava is emerging.

The vertical build up involved in creating a submarine mountain is tremendous. The average depth of the world's oceans is about 4000 meters. Before an underwater mountain becomes visible, therefore, the volcano must build a mountain about 4000 meters high. The end result is the tallest mountains in the world, measured from sea floor to crater (see Figure 5.13).

### New Islands and Disappearing Islands

Very few submarine eruptions discharge enough material to build a mountain that reaches up to the surface of the ocean. Such a cone would probably be a volcanic shield, as in Hawaii (see Figure 5.13). Any island formed at the top of a submarine volcano leads a precarious life. Eruptions may change the shape, size, or location of such an island or even replace it altogether with a new one. In addition, the pounding waves may slice away the island, leaving only a submerged shoal to mark the site.

The best-known "new" island of recent years is Surtsey, born 14 November 1963, within sight of the coast of Iceland (Figure 5.32). Eruption began early on the morning of the 14th, and by the 16th an island had appeared. By the 19th it was 60 meters high and 600 meters long. Hydroexplosions (explosions caused by the contact of molten lava with water) continued for several months. Occasionally, bombs were thrown as high as 1000 meters. By April 1964 the cone had grown high enough so that water no longer reached the vents. The eruption became more similar to the Hawaiian type. Lava poured gently into the sea and armored the island against the waves. It is this armor that has made possible Surtsey's survival. A little more than a year later, on 7 May 1965, the eruption ended.

Two weeks after the Surtsey eruption ended, another tephra cone, Little Surtsey, was built about 600 meters to the northeast. By September it had built up to 70 meters above sea level and by 24 October it had washed away. Soon more eruptions began to the southwest of Surtsey. On 26 December these produced a visible island named Jølnir, or Christmas Island. Jølnir disappeared and reappeared five times. In the summer of 1966 it was washed away for good.

### Special Features of Lava Discharged Under Water

Lava that is erupted under water encounters a special environment characterized by great water pressure and cool temperature. On the deep-sea floor, the temperature of the water is nearly everywhere about 4°C. The water causes the cooling lava to form special features that are

**5.32 Stages in the growth of Surtsey,** a new volcanic island that appeared off the S coast of Iceland on 14 November 1963. (All photos by Sigurdur Thorarinsson) (a) Clouds of tephra conceal the island, which is 500 meters long and 37 meters high; 15 November 1963. (b) Clouds of steam and tephra billow from the enlarging cone, clearly visible in foreground. View from airplane, 22 November 1963. (c) Tephra cone is almost 700 meters in diameter. Lava appears for the first time and flows into sea through breached wall of cone; 30 November 1963. (d) Clouds, chiefly of steam, from reaction between

(a)

lava and sea water. Small, dark tephra cloud appears at center. Island is now 800 meters in diameter and 87 meters high; 16 December 1963. (e) Small steam cloud indicates declining activity. Island now occupies an area of 1.15 square kilometers; view N on 1 April 1964. (f) Lava entering sea forms protective terrace against wave attack; view W, 11 April 1964. (g) Whole island seen from the air at a distance. Protective terrace of igneous rock fully encircles island. Steam cloud indicates lava flowing into the sea. Area of island is 1.41 square kilometers; 18 June 1964.

**5.33 Pillowed volcanic rock,** sectional view showing three small complete pillows and parts of two large pillows. Two small circular pillows in center evidently had hardened before the third (upper) pillow appeared. Before the third pillow had hardened fully, it was surrounded by the two larger pillows, whose weight flattened the third pillow somewhat and pressed it down between the older two below, creating a rounded "bump" on its lower side. (Brad Hall, University of Maine)

preserved in the resulting extrusive igneous rocks. These features include pillow structure, broken-glass rocks, breccias, and lack of vesicles.

*Pillow structure* is the name given to underwater extrusive materials that are budded out into pillow- or sausage-shaped lumps having glassy but plastic exteriors. The sacklike blobs are named *pillows,* and they usually accumulate in piles. The bottom of each pillow is molded to the shape of the top of those below, but the top retains its original rounded shape (Figure 5.33).

The combined effect of rapid cooling and hydroexplosions creates great masses of glassy particles that may become cemented to form what we shall refer to as *broken-glass rocks.* Because the glass is not stable chemically, it may be altered into a waxy-looking, yellow-brown substance. If the particles formed are larger than a few centimeters, then only their margins become glassy. The breaking of fragments that later become cemented together again forms one kind of *volcanic breccia.* Lava discharged in water deeper than 800 meters generally lacks vesicles. Volcanic rock lacking vesicles can also form on land.

### Subglacial Eruptions

The effects of a volcanic eruption under a glacier can be catastrophic. Examples are known from Iceland, Antarctica, and the Andes in South America. The hot lava may melt enough ice to form a giant lake held in by ice. Eventually, the lake water may burst through its icy confines, sending water and icebergs cascading over the countryside. In 1934 and 1938 two eruptions of the Icelandic volcano Grimsvøtn (see Figure 5.4) melted enough water to form *jøkulhlaups,* the Icelandic term for lava-glacier floods. Water was discharged at the rate of about 50,000 cubic meters per second, which is five times the flow of the world's largest river, the Amazon. Jøkulhlaups are responsible for shaping much of the southern coast of Iceland.

**5.34 Old Faithful geyser,** Yellowstone Park, Wyoming, at peak of one of its regular hourly eruptions. Height of water column is approximately 20 meters. (U. S. National Park Service)

## UNDERGROUND THERMAL ACTIVITY AND THE DISCHARGE OF VOLCANIC GASES

In volcanic regions evidence of the Earth's interior heat often continues to surface for thousands of years after active discharge of solids and lava has ceased. Magma cools very slowly, giving off hot gases as it does. This gas is mostly steam, but it also contains carbon dioxide, hydrochloric acid, ammonia, and many other gases that rise from the mantle into the crust. In most places, the outer part of the crust contains pore spaces filled with water, much of which has seeped downward from the rainfall. Rising gases heat and react with this water, possibly changing some of it to steam. As a result of this movement of steam, fumaroles, hot springs, or geysers may reach the surface.

### Fumaroles

A volcano that is discharging gas nonexplosively is a *fumarole* (Italian for "smoking crack"). Typical gases from fumaroles are carbon dioxide and hydrogen sulfide. The former is a colorless, odorless lethal gas that is denser than ordinary air and thus tends to collect in low places.

### Hot Springs and Boiling Mud Pits

Waters circulating underground may come into contact with hot rock material and thus be heated to temperatures exceeding the average air temperature of the region. Where such water seeps out at the surface, it forms a *hot spring*. Where the hot water is mixed with mud, the result is a *boiling mud pit.*

### Geysers

A *geyser* is a special kind of hot spring that spouts water periodically (Figure 5.34). The governing principle of geyser action is that an increase in hydrostatic pressure raises the boiling temperature of water (Figure 5.35). A geyser can form only where a connected network of irregular underground openings exists in which the water can accumulate and from which it can escape quickly. Geologists have inferred that such a network consists of a central vent, more or less vertical, to which various side passageways are connected at different levels. Given the relationship between hydrostatic pressure and the boiling temperature and the necessary underground "plumbing" connections, a geyser is thought to work as shown in Figure 5.36.

**5.35 Basis of geyser operation** is the increase of boiling temperature with greater pressure, as shown in a graph in which greater pressures are shown in the lower part of the graph.

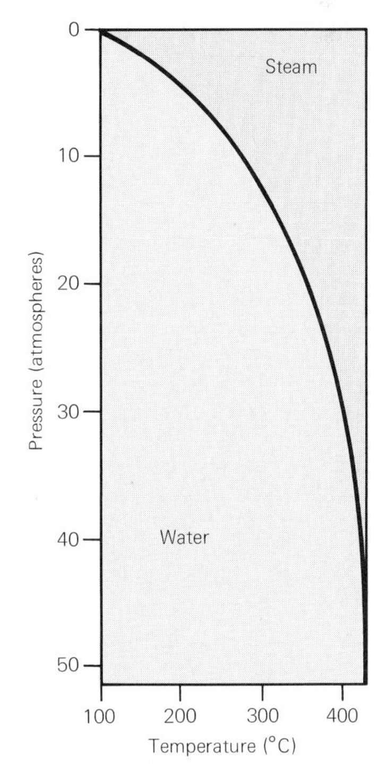


### Geographic Distribution of Fumaroles, Hot Springs, and Geysers

The chief places in the world where vents are discharging gases or hot water are Alaska, California, and Yellowstone National Park in the

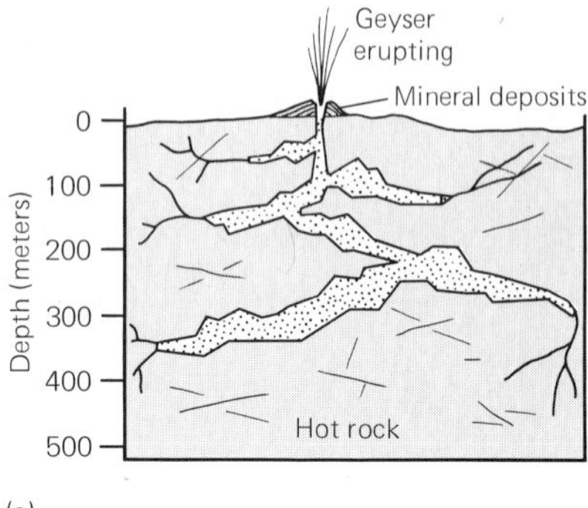


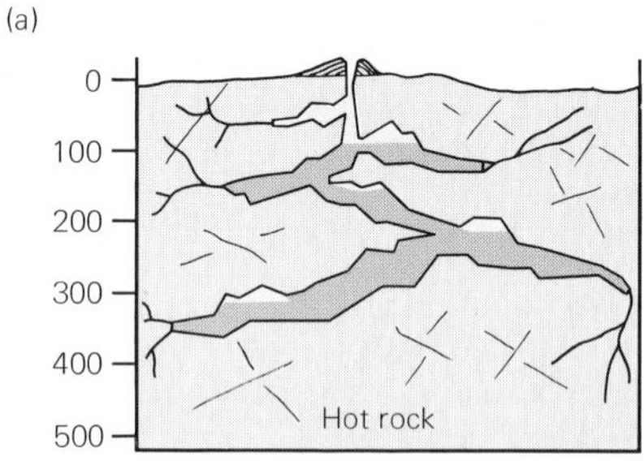


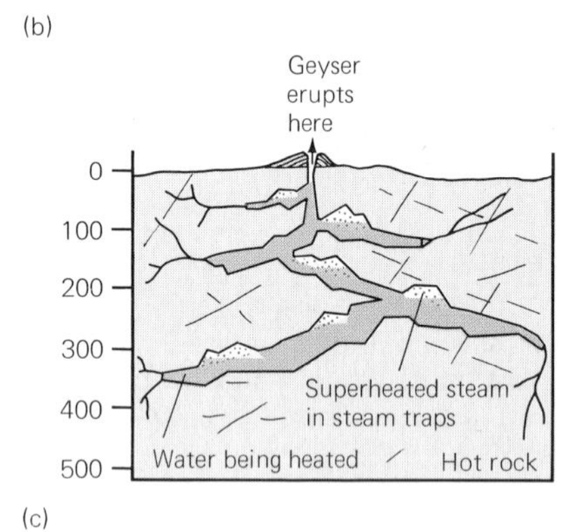


**5.36 Stages in cycle of eruption of a geyser,** schematic profile of Great Geysir, Iceland. (a) Geyser erupting. (b) After an eruption, water drains back into underground passageways. (c) As hot rocks heat the water, the water temperature is raised to the boiling point, which increases downward, as shown in Figure 5.35. Somewhere in the lateral passageways, one or more pockets comparable to inverted bowls are presumed to be present in which steam can accumulate and not escape upward as it forms. If the steam could escape as it forms, only a boiling spring would result. After steam has been trapped down below, its further increase will drive some of the water out of the lateral passageways and force some water to escape up the central conduit. The escape of some water lowers the level of the remaining water, thus decreasing the hydrostatic pressure. As a result, almost all at once, the whole mass of water down below is above its boiling temperature. Such water suddenly flashes into steam and expels all the water and steam vigorously out of the orifice as an eruption. (After T. F. W. Barth, 1950, Volcanic geology, hot springs and geysers of Iceland: Washington, D. C., Carnegie Institution of Washington, Publication 587, 174 p., Figure 36, p. 83)

United States; Iceland; New Zealand; Kamchatka Peninsula in the U.S.S.R.; and Italy. Yellowstone National Park stands out for the number and variety of its hot-water vents. Here, 3000 or so hot springs, boiling mud pits, and geysers are expressions of a slowly cooling mass of felsic magma just below the surface. Volcanic activity began here early in the Cenozoic Era, about 50 million years ago, and reached a climax during the Pleistocene Epoch, about 600,000 years ago when a large caldera formed.

## LEARNING TO LIVE WITH VOLCANOES

Each year, several dozen volcanoes on Earth erupt. Various amounts of damage are done by lava flows, glowing avalanches, showers of tephra, and noxious gases. Other destruction results from volcanic mudflows and tsunami. It is clear that an active volcano can devastate its immediate surroundings and cause damage transmitted by air or water hundreds or even thousands of kilometers away. Nevertheless, the aftereffects of many volcanic eruptions have been beneficial to people. Volcanic activity has built land to live on and fertile soils for agriculture. Small sprinklings of tephra provide farmers with "instant soil" containing many plant nutrients. If a new layer of tephra is more than a few centimeters thick, its initial impact is to blot out plants. However, new plants quickly colonize the tephra, so the setback is only temporary. The extrusive igneous rock formed from lava erupted in eastern Hawaii in 1840 is already densely covered with a tropical forest.

Partly because of these beneficial effects people have always chosen to live near volcanoes. To limit the hazards of that choice they have developed techniques for predicting eruptions and for diverting lava flows.

### Prediction of Forthcoming Eruptions

The ability to predict an eruption from a volcano would enable people to evacuate threatened areas in time to save their lives and portable possessions. Because the whole process of volcanic eruption is influenced by so many subsurface factors that cannot always be understood from the surface, the problem of making predictions is complicated. The best results so far have come from careful observations of the eruptive activities of individual volcanoes to learn about their peculiar characteristics and from geophysical measurements of things that are affected by the movement of magma underground and the increase in temperature caused by the approach of the magma to the surface.

**Seismic activity** The shifting of magma underground releases energy that travels upward to the surface as seismic waves. These can be detected by instruments known as *seismographs.*

Careful studies of seismic waves in Hawaii have shown that a typical eruption is preceded by two kinds of seismic wave trains, one from a "deep" source and the other from a "shallow" source. The deeper seismic waves come from depths of 50 to 60 kilometers below the ocean floor. These are thought to be related to the formation and collection of the magma within the upper mantle. The shallower seismic waves usually come from within about 8 kilometers of the surface. Based on study of seismic waves, R.H. Finch predicted in February 1942 that within a few months an eruption would take place along the flank of Mauna Loa at an altitude between 2500 and 3000 meters. The eruption started on 28 April at an altitude of 2800 meters.

**Changes in local magnetic field** When magnetic minerals, such as magnetite, are heated above their Curie temperatures (578°C for magnetite at 1 atmosphere pressure), their magnetism disappears. Thus, when a body of hot magma moves up into a volcanic cone, the intensity of the Earth's magnetic field in the vicinity decreases because the local amount of magnetic rock material has decreased. Even small changes in the intensity of the magnetic field can be measured by sensitive instruments called magnetometers that can be flown over a volcano in an airplane or used on the ground.

**Changes in temperature** As one might expect, the approach of magma to the Earth's surface increases the temperature. This change not only affects the magnetic field, as just mentioned, but also warms the ground and may raise the temperature of gases issuing from fumaroles. Not enough is yet known about the timing and significance of these temperature changes to enable them to be used for making accurate predictions of forthcoming eruptions. For example, in July 1975 steam began to escape from a newly activated vent along the southeast side of Mt. Baker, Washington (Figure 5.37). Heat from the vent melted the snow and ice

**5.37 Snow-capped peak of Mt. Baker,** Washington, showing steam cloud that became active early in 1975 from vent on southeast side. (Stephen Malone, University of Washington, courtesy Sigrid Schroeder)

that blankets the summit. The possibility of catastrophic avalanches from the melted snow and ice caused officials of the U.S. Park Service to close large areas of Mt. Baker North National Forest during the summer of 1975. Steam continued to escape, but, as of mid-1980, nothing drastic had happened. In 1980, the action shifted to Mt. St. Helens.

The increased temperature can be detected by heat-sensitive infrared equipment that can be mounted in airplanes or in Earth-orbiting satellites. Images recorded by such equipment show the pattern of the hot areas (see Figure 19.17).

**Changes in level of the land** Magma moving upward into a volcano causes the cone to swell, as if it were being inflated. Thus, just before an eruption, the ground surface is tilted slightly. This very slight tilting can be detected by extremely sensitive instruments called tiltmeters.

### Modification of Lava Flows

Assuming that a volcanic outburst does not bring immediate, total devastation as in St. Pierre in 1902, communities built on a volcano may be able to take steps to divert lava flows to save their homes. Perhaps the best-known early attempt at diverting a lava flow was made in 1669, when aa lava from Mt. Etna threatened Catania, Sicily. A group of men covered themselves with wet cowhides to shield themselves from the hot lava and used steel rods to punch a hole through the partially stiffened margin of the flow. This created an escape route for the lava, which flowed out the new opening instead of pushing ahead toward Catania at the frontal margin. This was successful — for awhile. Unfortunately, the diverted lava started flowing toward the town of Paterno. Infuriated residents of Paterno drove away the men from Catania and stopped the new flow of lava toward their town. As a result, the front of the lava flowed as before and destroyed part of Catania but none of Paterno.

The largest-scale attempt to modify a lava flow was made in Iceland in

**5.38 Houses on Heimaey, Iceland,** nearly engulfed by black tephra emitted during 1973 eruption of Eldfell Volcano. (Consulate General of Iceland, New York City)

1973, when a new volcano named Eldfell, next to Helgafell, erupted unexpectedly and lava threatened the fishing port of Vestmannaeyjar. Before any organized efforts could be made, lava burned many buildings and tephra blanketed many others (Figure 5.38). Thorbjorn Sigurgeirsson, an Icelandic volcanologist, directed efforts to save the important fishing harbor. Barriers were constructed and the lava was sprayed with seawater from high-pressure hoses. The idea was to cool the margins of the lava flow to make forward progress more difficult at given spots. The efforts were at least partially successful. A narrow, but usable harbor entrance was maintained around the new flows, and enough of the town was saved to rebuild and revitalize.

## CAN VOLCANOES CHANGE THE EARTH'S CLIMATE?

It has been suggested that tephra in the atmosphere have caused the climate to cool by blocking out solar radiation. The idea is an old one. It was proposed after explosive eruptions in 1783 at Laki (Iceland) and Asama (Japan). In 1784 Benjamin Franklin (1706–1790) suggested that the extremely cold winter of 1783–1784 had resulted from the volcanic dust in the atmosphere.

It is generally accepted that dust in the atmosphere can cause cooling. Geophysicists measuring the intensity of solar radiation have been able to monitor reductions caused by volcanic dust in the atmosphere. Measurements made at the Montpelier Observatory in southern France showed that during the three years following the arrival of the dust cloud from Krakatoa in 1883, ground-level solar radiation was 10 percent lower than usual. Measurements of ground-level solar radiation being made in Algeria were reduced by 20 percent when the dust cloud from the 1912 explosion of Katmai (Alaska) arrived.

The problem with attributing worldwide changes of climate to volcanic activity is complex. Geologists have argued that such volcanic activity would have deposited many more and thicker layers of tephra than have been reported. Without the worldwide tephra layers as evidence, geologists are naturally skeptical about accepting the idea that volcanic activity can influence climate on a global scale.

The possible effects of volcanic dust in the atmosphere are still being studied and debated. It has been suggested that changes in volcanic activity have contributed to climatic changes during the past several centuries, have had some influence on the last major ice age, and may be expected to produce major short- and long-term climatic changes in the future. (We discuss ice ages and volcanic activity in Chapter 13.) The answer may be known when the complex relationships between dust clouds in the stratosphere and distribution of tephra deposits have been determined. As mentioned in Chapter 2 and as discussed again in Chapter 14, small amounts of dust can be widely distributed in the stratosphere and thus might exert far-reaching influences on solar energy, yet not leave widespread continuous layers of tephra.

# CHAPTER REVIEW

1. Igneous activity is the general name for a two-fold process: (a) the making of hot liquid within the Earth and its rising within the lithosphere or onto the surface of the Earth; and (b) the cooling of this liquid material to form solid igneous rocks.

2. Volcanic activity is the surface expression of igneous activity. A *volcano* is the vent through which the molten rock material and associated gases pass upward to the Earth's surface.

3. The principles governing igneous activity apply not only to the Earth but also to other planets and to some of their satellite moons.

4. Early in the nineteenth century, what we now call igneous rocks became the center of a vigorous argument. One group, named *Neptunists* and led by Abraham Gottlob Werner and Leopold von Buch, believed that the only way to form crystalline-textured rocks was by precipitation from a saturated water solution. Thus, they argued that the rocks in question had been precipitated out of a universal ocean. Another group, the *Plutonists,* led by James Hutton and Sir James Hall, melted rocks and found that when the molten liquid cooled slowly, crystalline textures formed. The Plutonists argued that igneous rocks formed by the cooling of formerly molten material.

5. The Neptunists became Plutonists after Leopold von Buch visited the Auvergne district of central France, where volcanic rocks of Miocene age and Late Quaternary age are numerous. At Montagne de la Serre, a former valley-filling lava flow of Miocene age forms a sheet of igneous rock that caps a ridge between two modern valleys. In one of these valleys, lava flows only about 7500 years old formed fresh-looking sheets of igneous rock that can be traced to the bases of two obviously volcanic cones composed of loose cinders.

6. The concept of plate tectonics states that the Earth can be divided into two contrasting regions: (a) *lithosphere plates,* and (b) the *boundaries between lithosphere plates.* Many important geologic activities take place only at plate boundaries.

7. Basalt-forming volcanoes, discharging material melted from the upper mantle, are found (a) along mid-oceanic ridges where plates are separating, (b) along major fractures through the lithosphere named *rift zones,* located either on continents or on the sea floor, and (c) in scattered locations that are thought to indicate hot spots in the upper mantle, chimney-like plumes where much more heat than normal is flowing upward. Volcanoes that discharge pasty lava explosively characterize zones where plates come together.

8. The key process of volcanic activity is the separation and escape of gases from their associated liquids and solids.

9. Various gases are a characteristic volcanic product. The chief volcanic gas is water vapor, either (a) *juvenile water* derived from crystal lattices within the Earth's mantle or (b) *groundwater* held within the underground pores of sediments and sedimentary rocks.

10. Depending on their chemical compositions, volcanic liquids (*lava*) are designated as either *mafic* or *felsic.* Lava is classified according to form as *pahoehoe* (ropy), *aa* (spiny), or *block.*

11. The jets of gas issuing from a volcano usually contain large amounts of solids called *tephra.*

12. Volcanic landforms include *volcanic plains* and *volcanic plateaus, volcanic shields, volcanic cones* (including *tephra* or "cinder" *cones, spatter cones, complex cones,* and *composite cones*), *craters,* and *calderas.*

13. A volcano is *inactive* if it has not erupted within historic time; *dormant,* if it was once active but has been quiet during the past 50 years; *active,* if it has erupted during the past 50 years; and *extinct* if all activity has ceased forever.

14. Our ideas about the kinds of eruptive activities have been developed by assembling the case histories of such volcanoes as Vesuvius, Krakatoa, Mt. Pelée, Mt. Mazama, the Hawaiian volcanoes, and Stromboli.

15. A *Plinian eruption* refers to the explosive ejection of large amounts of lava (on the order of several cubic kilometers), with or without tephra flows, which causes the top of the cone to collapse and form a caldera.

16. A *Peléan eruption* refers to the disruption of all the lava by explosively expanding gases that form glowing avalanches or tephra flows (*nuées ardentes*), which cool to form welded tuffs. Almost no lava flows as a continuous liquid.

17. In *Hawaiian eruptions* the gases separate continuously from the fluid mafic lava and do not build up to violently explosive pressures.

18. In a *Strombolian eruption* activity continues for long periods. The lava in the vent remains red hot; there are no discrete eruptive events between which the lava cools to form a solid crust.

19. *Tsunami* are huge ocean waves created by abrupt displacement of the sea floor, as when the floor of a caldera collapses.

20. Volcanoes are found in nearly all parts of the world, both on land (*subaerial volcanoes*) and on the sea floor. Subaerial volcanoes are distributed around the edges of

the Pacific Ocean in a pattern sometimes called the Ring of Fire. Many volcanoes along the mid-Atlantic Ridge have been active under water, and many other active volcanoes are located in the Mediterranean, Indonesia, eastern Africa, central America, along the Andes chain in South America, and Antarctica. Most of the active volcanoes in the continental United States are located in the west.

21. Many of the volcanoes in the Pacific's Ring of Fire are associated with festoons of islands, called *island arcs.* According to the concepts of plate tectonics, island-arc volcanoes are fed by magma that forms where great slabs of the lithosphere melt as they descend into depths of about 100 kilometers.
22. Many volcanoes are located under water and beneath glaciers, and some submarine eruptions may form new (and usually temporary) islands. Subglacial eruptions may suddenly melt enough ice to cause vast floods.
23. Lava that is erupted under water may form special features including *pillow structure,* broken-glass rocks, breccias, and compact rock lacking vesicles.
24. A volcano that is discharging gas nonexplosively is a *fumarole.* Waters circulating underground may be heated by hot rock material to temperatures exceeding the average air temperature of the region; where such water seeps out at the surface, it forms a *hot spring.* A *geyser* is a special kind of hot spring that spouts water periodically.
25. In an attempt to predict volcanic activity, important geophysical measurements can be made of seismic activity, changes in the local magnetic field, changes in ground temperature, and changes in the level of the land.
26. The concept that dust in the atmosphere can cause cooling is generally accepted as being valid, but the possible relation between volcanic dust in the atmosphere worldwide changes in climate is still being studied and debated.

## QUESTIONS

1. Summarize the arguments between *Neptunists* and *Plutonists.* Which view finally became generally accepted? What locality provided the critical evidence for settling the Neptunist-Plutonist argument?
2. How are igneous rocks and lithosphere plates thought to be related?
3. Define *volcano.* What is the fundamental reaction related to all volcanic activity?
4. Do the "smoke" and "ashes" mean that volcanoes are burning in the same sense that wood burns in a fireplace? Explain.
5. Name the possible sources of volcanic steam.
6. Explain how the chemical composition of the lava, the physical characteristics of the lava, and volcanic activities are related.
7. Compare: *pumice* and *scoria; blocks* and *bombs;* and *aa* and *pahoehoe.*
8. Define *lava.* List three types of lava based on physical appearances.
9. Define *tephra.*
10. What is a *fissure eruption?* What volcanic landforms are usually related to fissure eruptions?
11. Compare *tephra* ("cinder") *cones, spatter cones, complex cones, composite cones,* and *volcanic shields.*
12. Define *crater.* Compare a crater to a *caldera.*
13. List the kinds of eruptive activities that take place in what is known as a *Hawaiian eruption, Plinian eruption, Peléan eruption, Strombolian eruption,* and in a *Krakatoan eruption.*
14. Define *nuée ardente.* With which kind(s) of eruption(s) are *nuées ardentes* associated?
15. What is the "Ring of Fire?"
16. Define *island arc* and name three modern examples.
17. What are *pillowed volcanic rocks?* How do pillows form?
18. Describe what may happen if a volcano erupts beneath a glacier.
19. Explain the principles that regulate the operation of a *geyser.*
20. Describe some of the methods that have been used for making predictions about where and when a volcano will erupt.
21. Name two beneficial aspects of volcanic activity.

## RECOMMENDED READING


Bullard, F. M., 1976, *Volcanoes of the Earth,* 2nd ed.: Austin, University of Texas Press, 563 p.

Crandell, D. R., 1969, The Geologic Story of Mount Rainier: *U.S. Geological Survey, Bulletin 1292,* 43 p.

Grove, Noel, 1977, Vestmannaeyjar: Up from the Ashes: *National Geographic Magazine,* v. 151, no. 5, p. 690–701.

Holmes, D. L., 1978, *Holmes Principles of Physical Geology,* 3rd ed.: New York, John Wiley and Sons, A Halsted Press

Book, 730 p. Chapter 5. Igneous Rocks: Volcanic and Plutonic Neptunists and Plutonists, p. 61–64.

Keefer, W. R., 1971, The Geologic Story of Yellowstone National Park: *U.S. Geological Survey, Bulletin 1347,* 92 p.

Macdonald, G. A., and Abbott, A. T., 1970, *Volcanoes in the Sea, The Geology of Hawaii:* Honolulu, University of Hawaii Press, 441 p.

Moore, J. G., 1975, Mechanism of Formation of Pillow Lava: *American Scientist,* v. 63, no. 3, p. 269–277.

Sillitoe, R. H., 1973, Environments of Formation of Volcanogenic Massive Sulfide Deposits: *Economic Geology,* v. 68, no. 8, p. 1321–1326.

Smith, R. L., 1960, Ash Flows: *Geological Society of America, Bulletin,* v. 71, no. 6, p. 795–841.

# CHAPTER 6 IGNEOUS ROCKS AND ASSOCIATED ORE DEPOSITS

PLATE 6
Giant columns of basalt formed by intersection of joints caused by shrinkage during cooling of Tertiary igneous rocks, north coast of Ireland.

THIS IS THE FIRST OF THREE CHAPTERS DEVOTED TO ROCKS. In each chapter we follow the same general format. We show how each kind of rock fits into the rock cycle; we discuss related rock-making processes; we summarize the occurrence of each kind of rock and present clues for recognizing it; we examine ways for obtaining geologic dates; finally, we review some of the related economic mineral deposits. Material on classification and description of individual kinds of rocks is presented in Appendix C. The discussion in these chapters is based on the assumption that the reader is familiar with the contents of Appendix C.

## IGNEOUS ACTIVITY IN THE ROCK CYCLE

In our previous discussions of James Hutton's concept of the geologic cycle, we included everything from the erosion of lands to melting of rocks under the Earth's surface. By isolating those parts of Hutton's great cycle that concern the making and taking apart of rocks, we have the *rock*

**6.1 The rock cycle,** represented by a schematic profile and geologic section through the outer part of the Earth. The materials at the Earth's surface, including sediments, volcanic ejecta, and older rocks of all kinds, are eroded and transported to subsiding areas, where they accumulate as bodies of sediment. After burial and lithification, the sediments become sedimentary rocks. At still-greater depths, sedimentary rocks can be heated and deformed into metamorphic rocks. At greater depths and higher temperatures, rock materials may melt to form felsic magma. In other locations, mafic mantle rock may melt. Whatever its origin, liquid magma deep within the Earth can be squeezed upward to zones where the general temperature is much less than that required for melting. The magma may stop within one of these higher levels or come all the way to the surface to form a volcano. Because of the mobility of magma, the presence of an igneous rock does not necessarily mean that all the materials at that level were hot enough to melt.

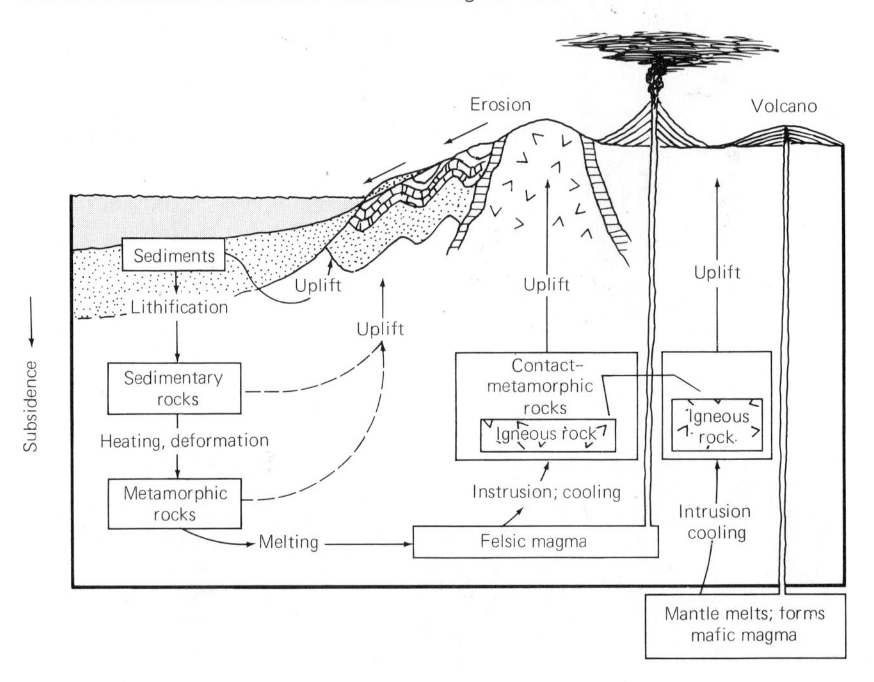

*cycle* (Figure 6.1). The rock cycle forms the basis for understanding the large-scale relationships among igneous rocks, sediments, sedimentary rocks, and metamorphic rocks; it is a major theme that we shall be pursuing in many subsequent chapters.

In tracing the history of an igneous rock, we usually cannot go farther back than the point where the material solidified from its parent molten material, whether this was magma within the Earth or lava at its surface. The reason the history cannot be extended farther is that the process of melting destroys the evidence of the kind of rock that was melted. In other words, it is possible to melt an igneous rock a second time and a sedimentary or metamorphic rock for the first time and not be able to tell which was which. If the chemical composition of all three started out more or less the same, then it would probably not be possible to figure out from the cooled product exactly what kind of substance had been melted to form the magma.

Igneous activity is a great geologic eraser; it cleans the record and forces us to start anew with the igneous rock itself. Igneous rocks can form from material that may have melted from a position deep within the Earth and is entering into the active rock cycle for the first time. Or, igneous rocks can form from sedimentary strata and associated volcanic materials that were spread out in layers at the Earth's surface but have reached great depths within the Earth and there have encountered temperatures and pressures which cause them to melt. Moreover, once a body of magma has formed, it may solidify more or less in the same place where the melting took place, it may move a little way or a long way, or it may even reach up to the Earth's surface to become a volcanic product.

## RELATIONSHIPS AMONG LAVA, MAGMA, HYDROTHERMAL SOLUTIONS, AND IGNEOUS ROCKS

The essence of igneous rocks is solidification — a change from the molten state to the solid state. The final product, the rock, reflects (1) the composition of the molten material and (2) the conditions under which this molten material cooled and became solid.

In Chapter 5 we discussed the behavior of volcanoes, which are expressions of the surfacing of lava and separation of formerly dissolved gases. Thus, in most lavas, crystals grow under dry conditions. Magma consists of molten material, chiefly silicates, and dissolved gases, of which water is usually predominant. What we now examine is the cooling and solidification of magma beneath the surface of the Earth, where the dissolved gases may or may not have been abundant and may or may not have escaped.

The study of igneous rocks in the field makes it strikingly apparent that the degree of freshness of all igneous rocks is not the same. Some rocks appear to be extremely fresh, hard, and durable. Others seem to have been "stewed in their own juices," so to speak, and have been extensively altered; their feldspars have become clay minerals of kinds formed at temperatures higher than those found at the Earth's surface. In addition, it is typical to find that notable bodies of sulfide minerals and perhaps other kinds of metal-bearing minerals have been deposited in or near the altered igneous rock.

The hot solutions that have altered the igneous rock and its surroundings and have deposited the metal-bearing minerals are known as *hydrothermal solutions.* The term hydrothermal simply means hot water. Some of these hydrothermal solutions arrive with the magma from deeper levels in the lithosphere, and others represent waters that the magma heats during its movement upward. In our analysis of igneous rocks, we begin with the situations where hydrothermal solutions were not important.

### Laboratory Experiments: Bowen's Reaction Principle

Experimental evidence about igneous rocks has been available since late in the eighteenth century, when Sir James Hall melted granite in a blacksmith's forge. More sophisticated experiments are now possible that re-create temperatures and pressures from well inside the Earth's core. Thimblefuls of ground-up rocks are placed in piston-type squeezers, heated to high temperatures, and then cooled. Small specimens of igneous rocks crystallize when the apparatus is cool, and these specimens can be studied.

Far-reaching results from a striking series of laboratory experiments were obtained by N. L. Bowen (1887–1956) early in the twentieth century. He discovered a fundamental principle of the behavior of silicate minerals that has been named in his honor.

Bowen experimented with two kinds of materials, the ingredients of plagioclase feldspars and of ferromagnesian minerals. He found that these minerals behaved quite differently. In Bowen's experiments, the dry minerals were heated until they melted and then were cooled for various lengths of time and at different rates. Experiments were ended by quenching, that is, by cooling abruptly to stop all reactions. Study of the results indicated that crystal lattices would form and then grow, as one would expect, but in addition, and unexpectedly, the growing crystals were found to have reacted with the melt. In the plagioclase experiments growing crystals and the melt exchanged ions. Sodium ions from the melt exchanged with calcium ions from the lattices. Plagioclase lattices remained intact throughout; the effect of the exchange was to increase the proportion of sodium in the lattices, not to change the type of lattice. This was named a *continuous reaction series* (the right-hand branch of the Y-shaped series of slant-topped blocks in Figure 6.2).

The behavior of the ferromagnesian minerals differed from that of the plagioclase. Growing olivine crystals reacted with the "magma" to exchange iron and magnesium, just as sodium and calcium exchanged in the plagioclase. However, as the temperature dropped, the olivine crystals surprisingly disappeared and pyroxene crystals began to grow. These pyroxene crystals exchanged ions with the liquid; presently the pyroxene vanished and in its place amphibole appeared. Therefore, as cooling took place, minerals not only exchanged ions with the melt, but crystals having one kind of lattice would disappear and crystals having another kind of lattice would take their places. Such a relationship was called a *discontinuous reaction series* (the left-hand branch of the Y in Figure 6.2).

In Bowen's experiments the minerals that crystallized first and at the highest temperatures from a mafic magma were calcium-rich plagioclase and olivine.

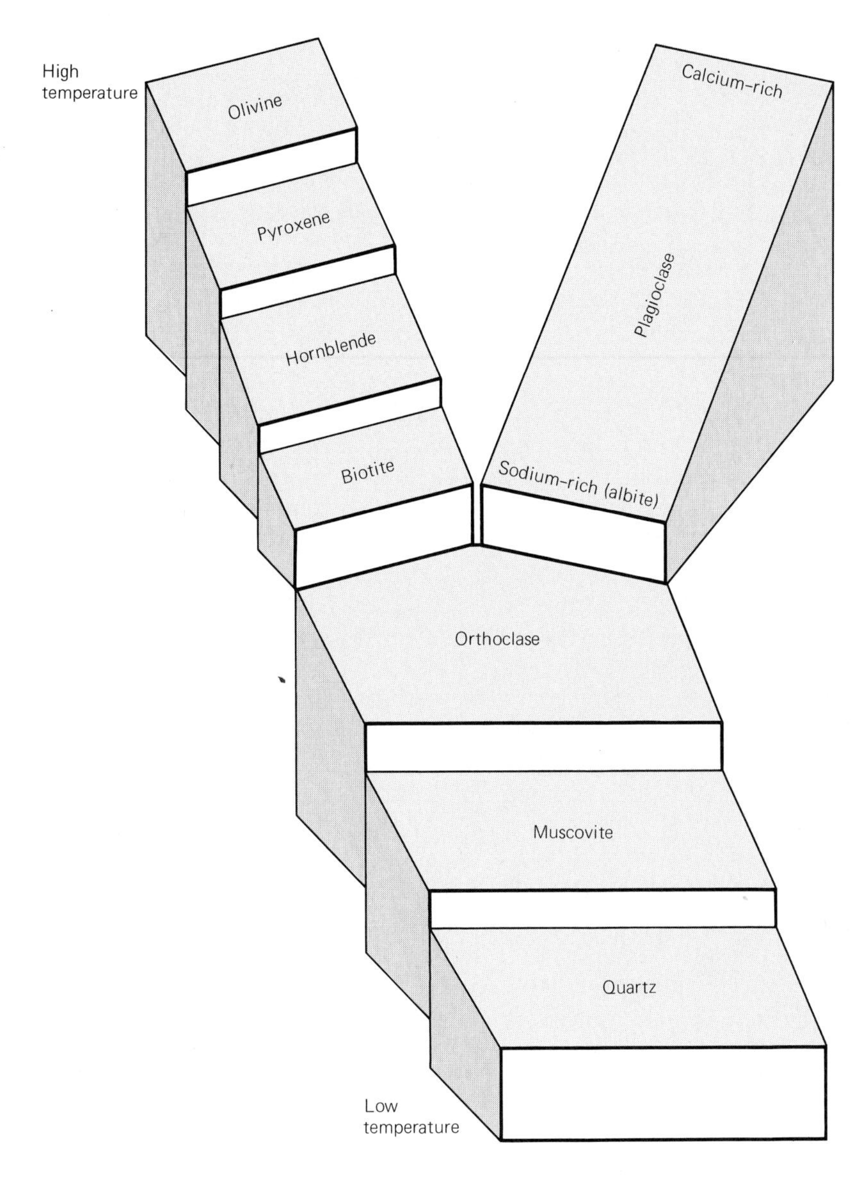


**6.2 Bowen's reaction** series shown by schematic blocks. *Continuous reactions,* in which ions of a given lattice simply exchange with ions in the melt, are shown by blocks having slanted tops. *Discontinuous reactions,* in which lattices of one mineral disappear whereas lattices of another mineral take their places, are shown by vertical faces along blocks whose bases have been offset.

(a)

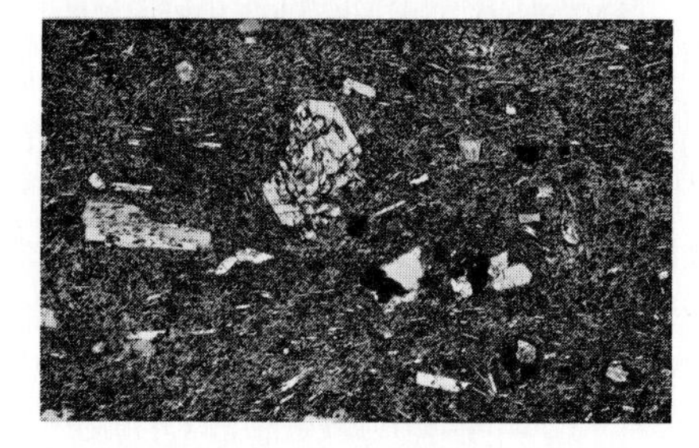

(b)

**6.3 Phenocrysts in porphyries,** views through polarizing microscope of paper-thin slices of rock. (a) Plagioclase phenocrysts (white, rectangular shapes) whose long axes are oriented randomly in a glassy groundmass (black). (b) Phenocrysts of plagioclase (small rectangles), pyroxene (largest shape left of center) and iron-titanium oxides (black) in a glassy groundmass (dark gray). The long axes of most plagioclase crystals are parallel. Summitville Andesite (Miocene), NE of Conejos Peak, Conejos County, Colorado. (P. W. Lipman, 1975, U. S. Geological Survey, Professional Paper 852; (a) is Fig. 38-C; (b) is Fig. 38-A)

## Growth of Early Crystals: Phenocrysts and Crystal Cumulates

Field study of igneous rocks indicates two contrasting situations as early crystals start to grow in a magma. One situation yields porphyritic rocks or porphyries (see Appendix C). Growing crystals remain suspended within the magma and continue to enlarge (Figure 6.3a). Evidently, the densities of the growing crystals do not differ significantly from the density of the magma, and the growing crystals do not sink. If the magma does not move, the arrangement of the long axes of elongate phenocrysts, such as plagioclase, is random. If the magma flows, however, then the long axes of phenocrysts become aligned parallel to the direction of flow (Figure 6.3b).

In the constrasting situation the densities of growing crystals exceed or are less than that of the magma. The denser early formed crystals sink through the magma and accumulate along or near the floor of the magma

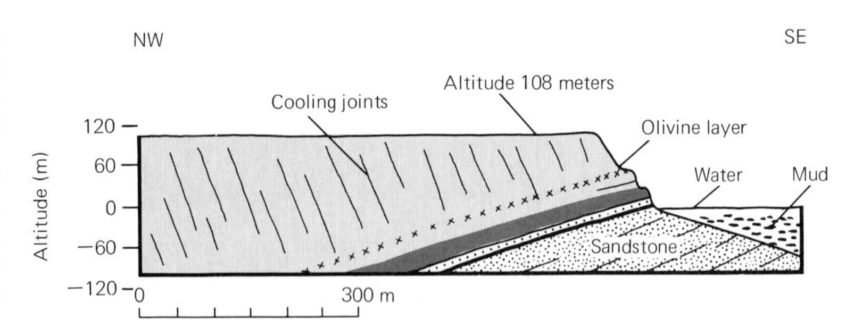


**6.4 Palisades sill and underlying strata,** profile and geologic section having same vertical and horizontal scales. Topographic information from U. S. Geological Survey, Central Park, New York, 7 1/2-minute quadrangle. (Geology modified by J. E. Sanders after C. P. Berkey, 1933, International Geological Congress, 16th, United States, Guidebook for trips in New York City and vicinity, Plate 8A)

chamber. The less-dense crystals float to the top. For example, the density of olivine is greater than that of mafic magma and, therefore, olivine tends to sink. In the sheet of igneous rock forming the Palisades, along the west bank of the Hudson River in New Jersey and New York, a layer of olivine crystals, 3 meters thick, exists in a zone lying about 15 meters above the bottom (Figure 6.4).

In other bodies of igneous rock, crystals of dense minerals such as olivine, pyroxene, chromite, and plagioclase have settled to form closely packed aggregates that make layers resembling sedimentary strata (Figure 6.5). Such igneous rocks, formed by accumulation of crystals, are named *crystal cumulates.* They differ from porphyries chiefly in their packing, layered arrangement, and, in some cases, absence of a groundmass.

## Order of Crystallization

The order of crystallization of minerals from a magma commonly follows the order determined in Bowen's experiments. In general, the early crystals display diagnostic crystal shapes. After the framework has formed, crystal shapes give way to compromise boundaries (see Figure 4.19). Early formed minerals that become engulfed by the rapidly growing lattices of the framework minerals appear as *inclusions* within the framework minerals.

**6.5 Intrusive body composed of layered ultramafic rock,** in which the settled crystals became arranged in definite layers comparable to sedimentary strata. Hall Cove Peridotite (Lower Cretaceous), Duke Island complex, southeastern Alaska. (T. N. Irvine, 1974, Petrology of the Duke Island Ultramafic Complex, Southeastern Alaska: Geological Society of America Memoir 138, 240 p., Frontispiece; used with permission of Geological Society of America)

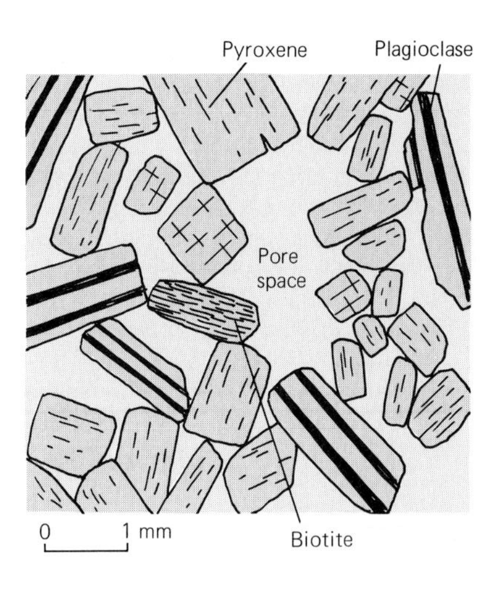


**6.6 Crystals in an igneous rock that have grown until they have begun to touch one another,** schematic sketch of view of thin slice using polarizing microscope. The space not occupied by crystals, marked pore space on the sketch, would usually be occupied by magma and would be filled by the last materials to solidify.

After the framework has been formed, perhaps 50 to 80 percent of the available space may be occupied (Figure 6.6). We refer to the minerals that fill up the remaining available space as interstitial space fillers. Typically quartz fills interstitial spaces; in some instances, quartz-feldspar intergrowths fill up the last spaces. Magnetite may form well-developed crystals appearing as inclusions within framework minerals or may fill interstitial spaces (Figure 6.7). Laboratory experiments suggest that water controls the crystallization of magnetite. Where water is not present, magnetite crystallizes early and appears as inclusions. Where water is abundant, the iron remains in solution, so that magnetite crystallizes last and thus appears as an interstitial space filler.

## OCCURRENCE OF IGNEOUS ROCKS

Igneous rocks are organized into two major divisions according to where they solidified: (1) *intrusive igneous rocks* solidified within the Earth and (2) *extrusive igneous rocks* solidified at the Earth's surface. Intrusive igneous rocks form bodies known as *plutons;* extrusive igneous rocks form layers.

### Plutons

A *pluton* (derived from Pluto, the Roman god of the underworld) is defined as any body of intrusive igneous rock. Other kinds of rock that surround a pluton are collectively named the *country rock.* Country rocks are usually layered, whereas plutons are commonly massive. The boundary between a pluton and the country rock forms one variety of *contact,* a term geologists use for boundaries between bodies of rock (Figure 6.8).

Typically, the intruding magma heats and alters the surroundings (Figure 6.9). An intruding magma may break off and surround pieces of the country rock. Broken-off pieces of country rock that become engulfed in the magma and thus eventually become surrounded by igneous rock are *xenoliths* (from the Greek, meaning "stranger rock".)

**6.7 Magnetite as an interstitial space filler,** view of thin slice of rock using polarizing microscope. The magnetite (black) coats plagioclase crystals (white) that had formed earlier. The final space-filling material is chlorite, which displays a radial fabric. Holyoke Formation, Lower Jurassic, North Branford, Connecticut. (J. E. Sanders)

**6.8 Geologic contact** between a body of igneous rock (right) and layers of country rock (left), sectional view.

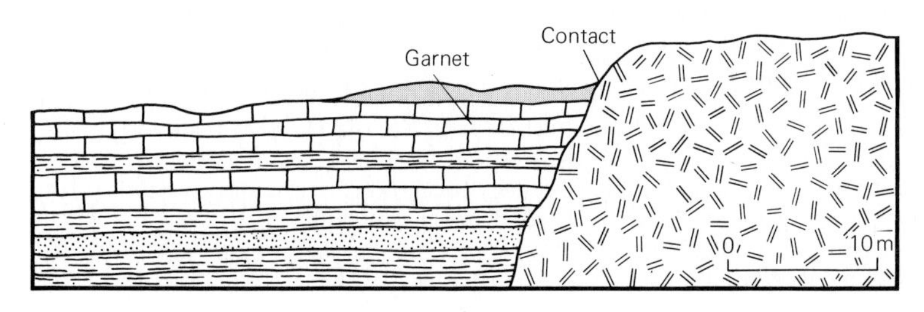

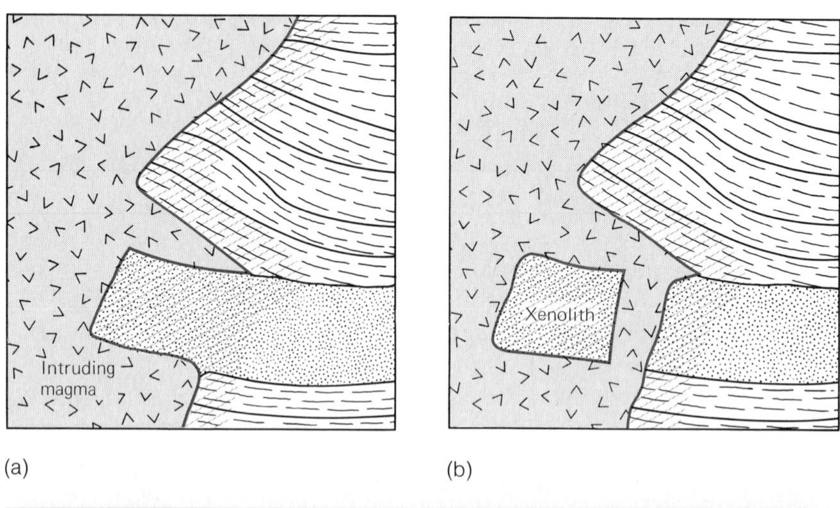


(a) (b)

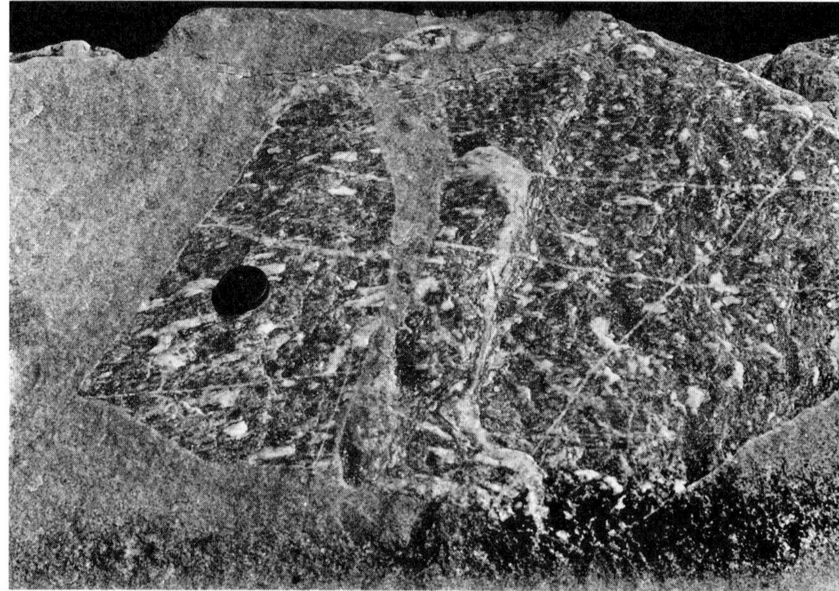

(c)

**6.9 Xenoliths.** (a) Intruding magma wedges itself under pressure into sedimentary strata, following along partings between strata. (b) Magma engulfs block of sedimentary rock and separates this block from the rest of the country rock. After the magma has cooled, the intruded igneous rock will completely surround a block of rock unlike itself (sedimentary, in this case). (c) Xenolith of coarse-grained metamorphic rock surrounded by light-colored, uniform, fine-grained massive granite. Diameter of lens cap is 50 millimeters (J. E. Sanders)

Plutons are classified according to: (1) the nature of their contacts and (2) their shapes. *Concordant* plutons have contacts that are parallel to the layers of the country rock. *Discordant* plutons have contacts that are not parallel to the layers of the country rock (Figure 6.10). The shapes of plutons range from tabular (parallel-sided slab) to lenticular (shaped like a double-convex lens) to cylindrical to irregular.

**Sills** Tabular concordant plutons are named *sills* (Figure 6.11). Sills vary greatly in size. The sills of Mt. Royal, in Montreal, Canada, measure only 5 to 10 centimeters thick and two meters in length. In Glacier National Park, Montana, visitors can study sills 30 meters thick that once extended at least 5500 square kilometers (Figure 6.12). Bare rock standing

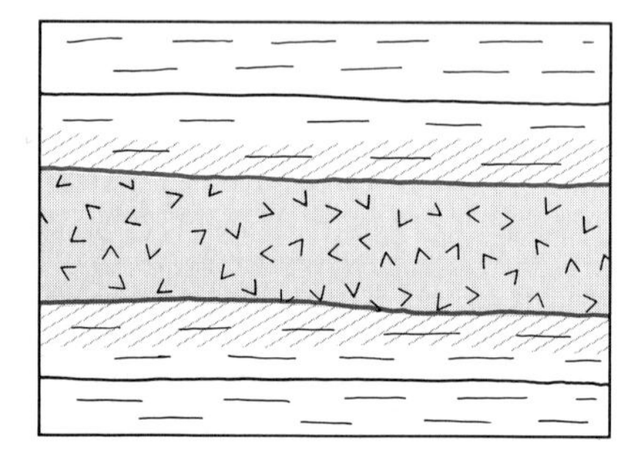

(a)

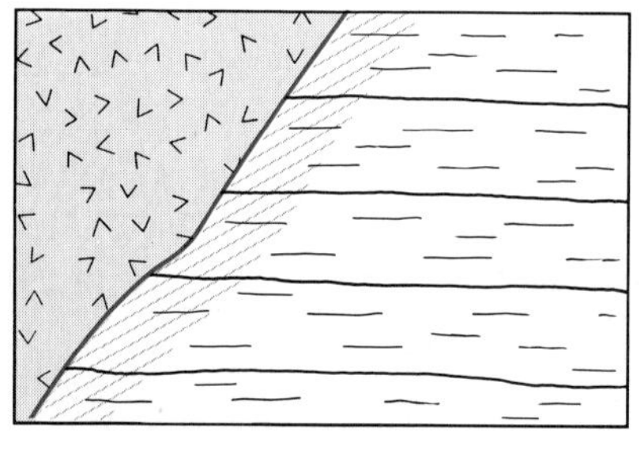

(b)

**6.10 Concordant contacts and discordant contacts.** (a) *Concordant contacts* are parallel to layers of country rock. (b) A *discordant contact* cuts across layers of country rock. Shaded areas indicate zones where hot magma has heated and altered the country rocks.

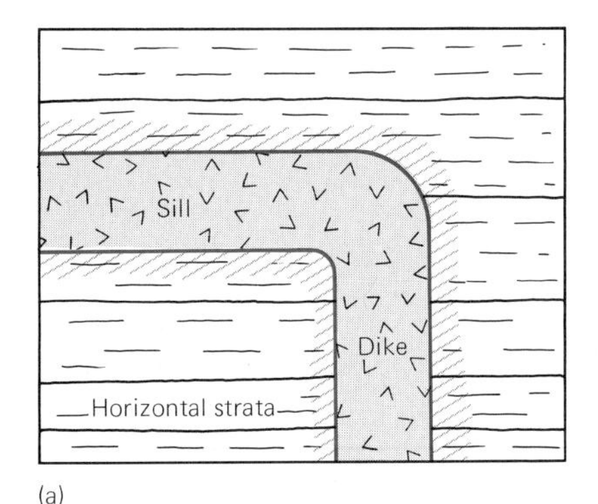

(a)

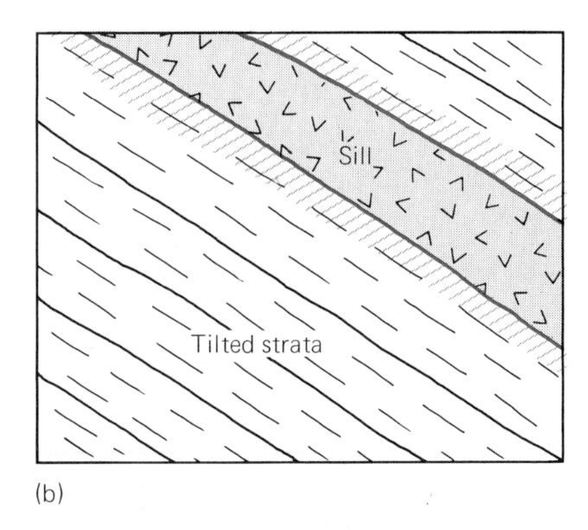

(b)

**6.11 Sills.** (a) Sill fed by a dike. (b) Sill intruded into strata that were horizontal originally but that have been tilted later. The essential characteristics of a sill, namely concordant contacts, remain whether the intruded strata are horizontal as in (a); inclined, as shown here; or are vertical.

in regular columns along the west bank of the lower Hudson River constitutes the impressive Palisades (see Figure 6.4). This striking example of the eroded edge of a tilted sill extends for 100 kilometers at thicknesses ranging from 100 to 300 meters. A sill in Ontario, Canada, is 1500 meters thick.

**Laccoliths** A *laccolith* is a concordant pluton with a flat floor and a convex-up top (Figure 6.13). In plan view laccoliths appear to be roughly circular or oval. They may range in diameter from several hundred meters to several kilometers, and in maximum thickness from a couple of hundred meters to almost two kilometers. Laccoliths were first identified in the Henry Mountains in the deserts of southeastern Utah.

**6.12 Sill visible high in the cliffs** underlain by ancient sedimentary rocks of Precambrian age. Mount Gould, Glacier National Park, Montana. (David W. Corson, from A. Devaney, New York)

**6.13 Laccolith.** (a) Newly intruded laccolith has arched the overlying strata, schematic profile and geologic section. (b) Laccolith core exposed after erosion of arched roof, schematic block diagram.

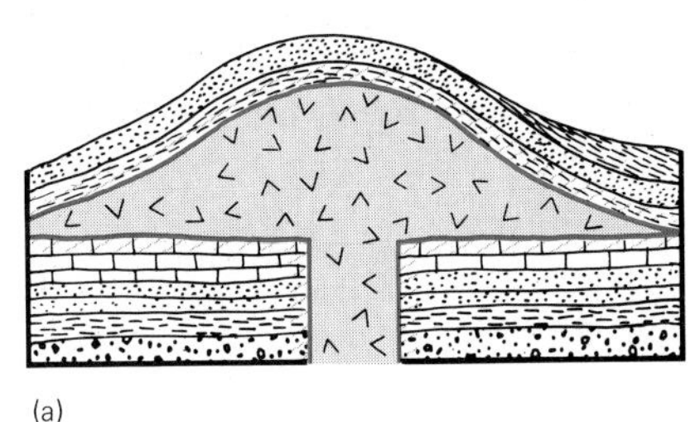
(a)

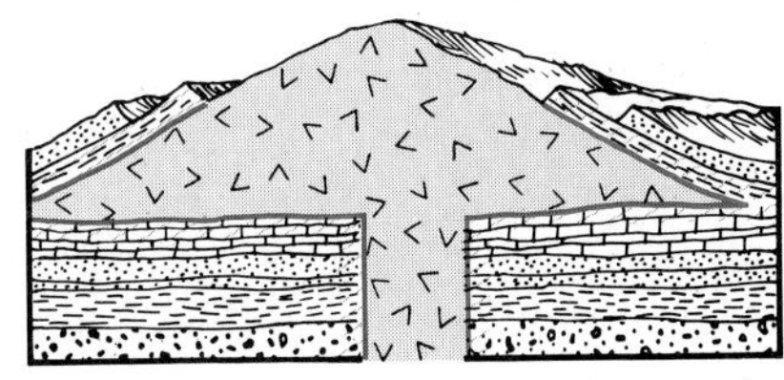
(b)

(a)

(b)

**6.14 Dikes of various sizes.** (a) Thin mafic dike cutting through light-colored granite, both seen in a hand specimen. (B. M. Shaub and J. E. Sanders) (b) Wall-like sheet of dark-colored rock in left foreground is a large dike several meters thick and hundreds of meters long. The steep-sided mass of rock forming the isolated, irregular hill is Ship Rock, New Mexico, a body of rock that solidified in the conduit of a former volcanic cone, now completely eroded away. (Courtesy Barnum Brown Collection, American Museum of Natural History)

**Dikes** A *dike* is a tabular discordant pluton. The margins of dikes typically are steeply inclined.

Dikes range in size from mere slivers, a centimeter thick and a meter or two long (Figure 6.14a) to huge bodies a kilometer thick and many kilometers long. Although one great dike in the north of England is more than 150 kilometers long, most dikes are much smaller. Where the country rock erodes more easily than the dike rock, the margins of the dike may form vertical natural walls (Figure 6.14b). Occasionally, the reverse happens. If the dike is less resistant than the country rock, the dike is eroded into the form of a natural ditch.

Some dikes extended upward to the Earth's surface and fed fissure eruptions (see Figure 5.10). Other dikes end upward in sills (Figure 6.15).

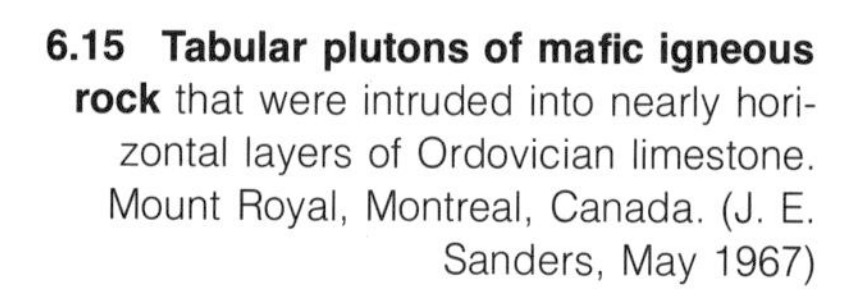

**6.15 Tabular plutons of mafic igneous rock** that were intruded into nearly horizontal layers of Ordovician limestone. Mount Royal, Montreal, Canada. (J. E. Sanders, May 1967)

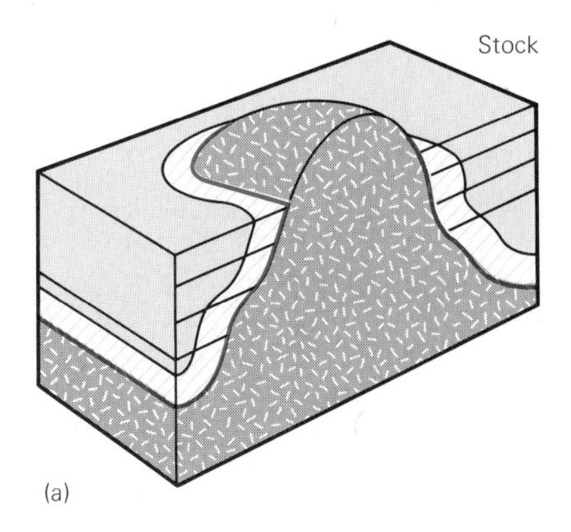


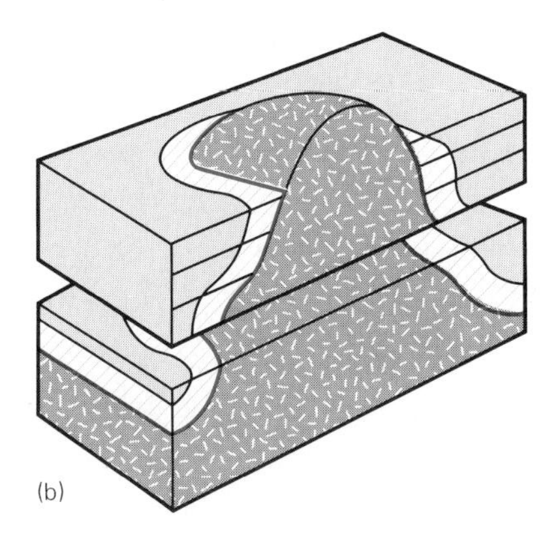


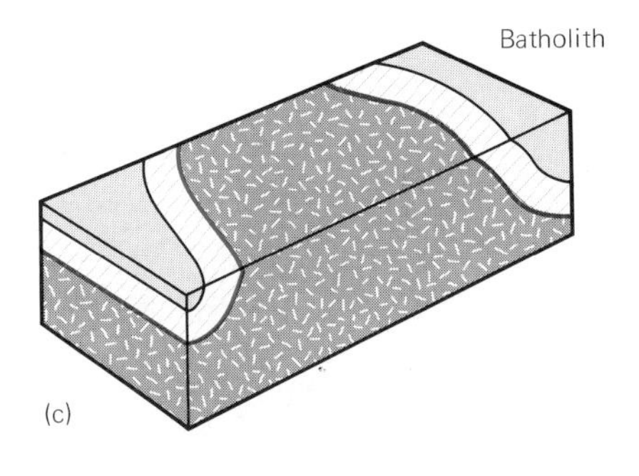


**6.16 Stock and batholith,** schematic block diagram. (a) Small exposed area (less than 100 square kilometers) qualifies the exposed part of this pluton as a stock. (b) Block sliced at a deeper level to indicate erosion after uplift. (c) Large exposed area of same pluton (more than 100 square kilometers) makes the designation of batholith appropriate.

**Volcanic necks and diatremes** Narrow, cylindrical discordant bodies of intrusive igneous rock that solidified in the throat of an ancient volcano are *volcanic necks.* After all vestiges of a former volcanic cone have been eroded, the protruding volcanic neck may be all that is left to mark the site. Usually, volcanic necks are associated with dikes that may radiate outward from the neck.

If the former vent contains a mass of breccia formed by explosive upward transport of a mixture of gases and solids, then the feature is named a *diatreme.* The famous diamond "pipes" of South Africa are examples of diatremes. Material from deep within the mantle, and containing the diamonds, was transported upward to higher levels in the lithosphere, where they are now found. In the United States, Ship Rock, New Mexico (see Figure 6.14) has been interpreted as a diatreme.

**Stocks** Massive discordant plutons having exposed areas of less than 100 square kilometers are *stocks.* Seen from above, the shapes of most stock are circular or oval (Figure 6.16a); the shapes of a few stocks are irregular.

One of the world's most carefully studied stocks is the Bingham–Last Chance stock, located at Bingham, Utah (Figure 6.17). The exact shape of the irregular stock is being revealed by the mining of the rock from what has become the world's largest open-pit excavation to recover copper, molybdenum, and other metals (Figure 6.18). The Bingham part of the stock is a *composite pluton,* that is, a pluton containing the products of more than one pulse of intrusion. The first magma cooled to form a quartz-poor, equigranular felsic rock. The second magma formed a quartz-rich porphyry. The Last Chance part of the stock contains only equigranular felsic rock.

In the Bingham part of the stock, the equigranular rock formed from the first charge of magma has been extensively altered by magmatic solutions that evidently followed shortly after the second charge of magma, which gave rise to the quartz-rich porphyry. Both the porphyry and the adjacent parts of the equigranular rock that were hydrothermally altered contain disseminated sulfide ore minerals. The uniform distribution of these ore minerals throughout the altered porphyry and equigranular felsic rock indicates that the interstitial spaces of both rocks were still open when the hydrothermal solutions appeared. These solutions first altered the feldspars by converting them into clay minerals. Afterward, the hot solutions deposited the sulfide minerals. The equigranular rock of the Last Chance part of the stock has not been hydrothermally altered and does not contain abundant sulfide minerals.

The Bingham–Last Chance stock intrudes sedimentary rocks of Late Pennsylvanian age that had been folded and broken by great overthrusts late in the Cretaceous Period, when large parts of the Rocky Mountain region were deformed. The minerals of the stock have been dated at $39.8 \pm 0.4$ million years (my), which places the date of intrusion in the

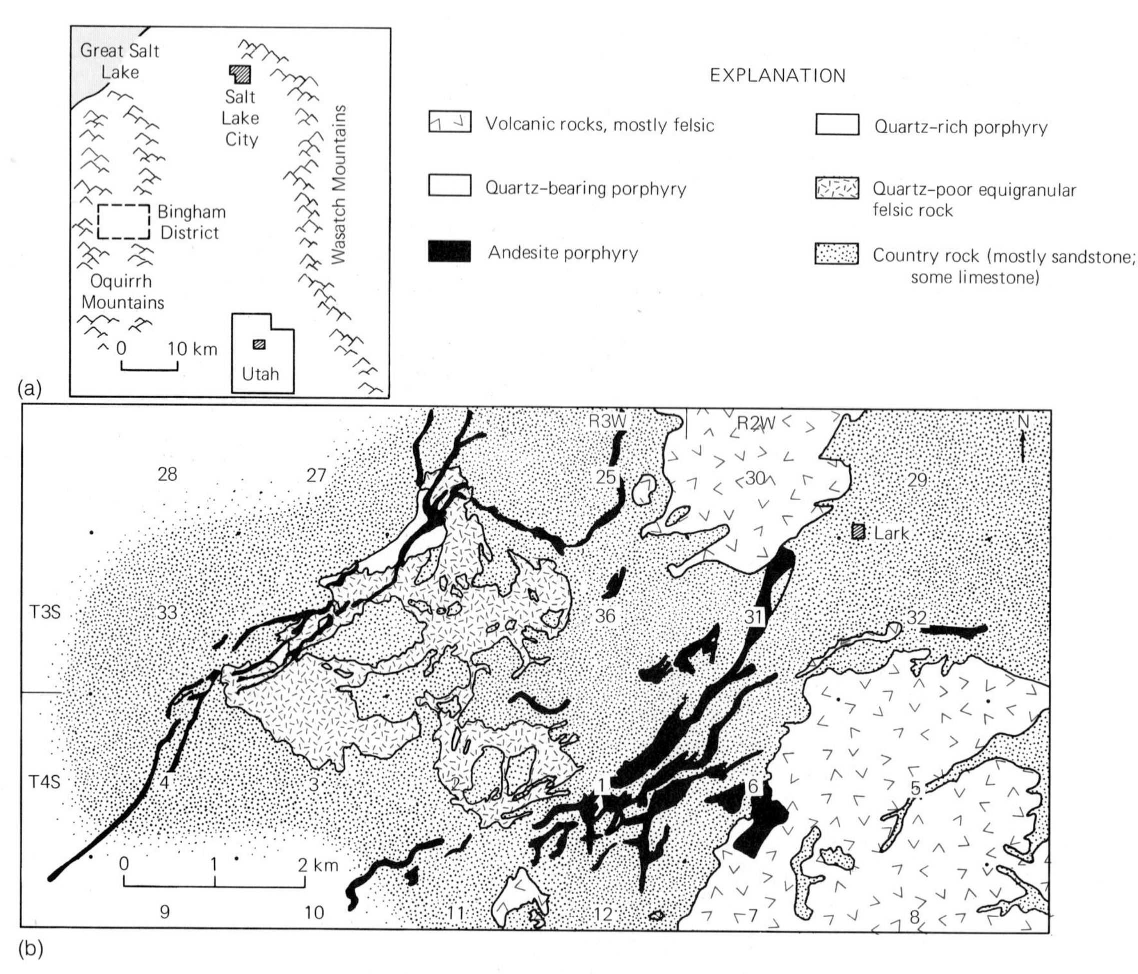


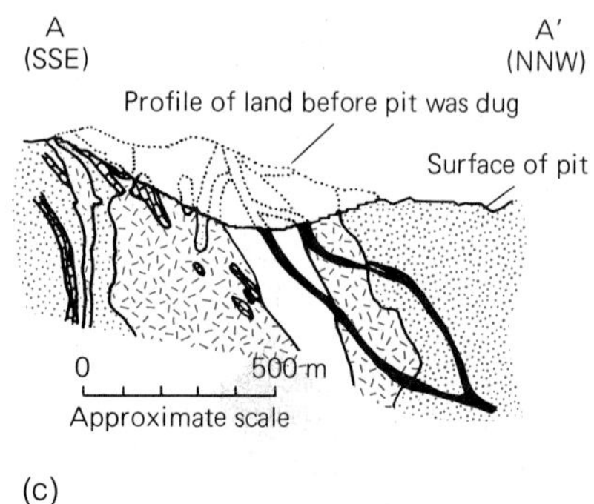


**6.17 Composite pluton of Bingham-Last Chance stock, Utah.** (a) Index map of Bingham district, about 30 kilometers SW of Salt Lake City. Small inset shows location in north-central Utah. (b) Principal rock bodies near the active mine complex. Dots mark corners of square-mile sections whose centers are numbered as parts of the township-range system of land surveys (explained in Appendix E). (Modified from F. W. Warnaars, W. H. Smith, R. E. Bray, George Lanier, and Muhammed Shafiqullah, 1978, Geochronology of Igneous Intrusions and Porphyry Copper Mineralization at Bingham, Utah: Economic Geology, v. 73, no. 7 (special issue), p. 1242–1249, (c) Profile and geologic section through mine pit (location shown on map of Figure 6.17b). Subsurface relationships are based on numerous borings. (After E. C. John, 1978, Mineral Zones in the Utah Copper Orebody: Economic Geology, v. 73, no. 7, (special issue), p. 1250–1259. Figure 6, p. 1256)

Oligocene Epoch (see Table 1, p. 7). Adjacent to the stock is a zone in which the country rocks have been extensively metamorphosed (see Chapter 9).

**Batholiths** *Batholiths* may be defined as massive discordant plutons occupying exposed areas of more than 100 square kilometers. Whereas laccoliths are usually emplaced between layers of rock near the surface, batholiths, like stocks, may extend downward indefinitely (see Figure 6.16).

Because the rocks composing batholiths originated at great depths,

**6.18 The world's largest excavation,** mine at Bingham, Utah. (Courtesy Kennecott Copper Company)

where cooling is extremely slow, they are coarse grained. Many of these huge plutons compose the central cores of mountain ranges. The Idaho batholith, extending through much of that state and part of Montana, occupies about 40,000 square kilometers. The Coast Range batholith of British Columbia and southeastern Alaska occupies an exposed area of about 250,000 square kilometers. Batholiths also form the Sierra Nevada in California and the granitic mountains of New England, Scotland, and many other areas. Batholiths composed of Precambrian rock (older than about 600 my) underlie about five million square kilometers of northeastern Canada and Greenland.

The massiveness of the batholiths raises the important question of how they made room for themselves. The "room problem" (or "space problem") needs to be considered in the analysis of any pluton. Many plutons made room for themselves by pushing apart the walls of a crack or a bedding-surface parting. Others, such as laccoliths, arched the overlying strata. Some others moved in by *stoping,* the spalling (or breaking off) of xenoliths and their disappearance down into the depths of the magma chamber. But the large size of some batholiths has convinced many geologists that the granite could not possibly have made room for itself by any of these mechanisms. They argue that the minerals forming the granite must have originated by the replacement of preexisting rocks during an episode of deep burial and general heating. Rock resulting

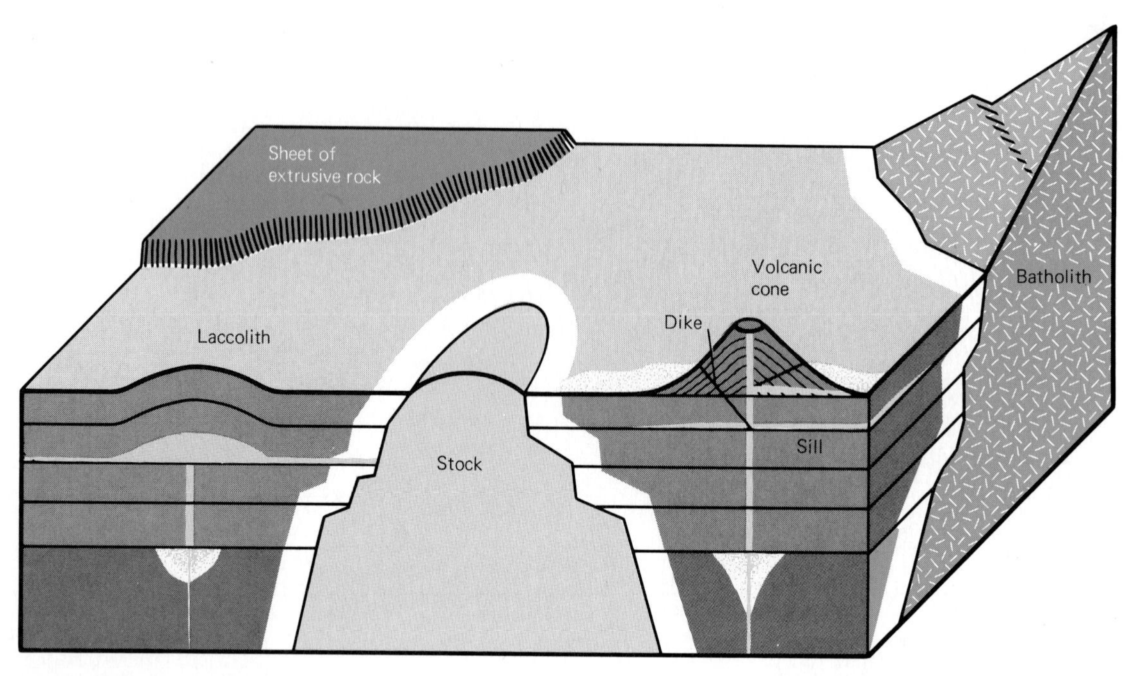


**6.19 Major kinds of plutons and various bodies of extrusive igneous rocks** shown on a schematic block diagram. White zones at margins of major plutons indicate thermal alteration of country rock. Thinner zones of altered rock exist at margins of smaller plutons, but are not shown here.

from this mechanism would be classified as metamorphic. Thus, granites could be formed by either of two processes, some being igneous, whereas others are metamorphic. We discuss the subject of the multiple origins of granite at greater length in Chapter 9.

The various kinds of plutons we have discussed are summarized schematically in Figure 6.19.

## Layers of Volcanic Rocks

As noted in Chapter 5, very fluid lavas or fluidized gas-lava-solid mixtures spread out on the Earth's surface to form layers which are one variety of strata (see Figures 5.11b and 5.12). In this section we examine the relationships of volcanic strata to other strata and summarize the paleomagnetic records found in ancient volcanic strata.

**Stratigraphic relationships of layers of volcanic rock** The term *stratigraphy* refers to the scientific study of strata. The term is usually associated with sedimentary strata, but includes strata of volcanic rocks as well. The fundamental rule of stratigraphy is that strata are spread out, one at a time, with the oldest at the bottom and successively younger ones on top.

Strata of volcanic rock are particularly useful for historical purposes where they were spread out over sedimentary strata and later, after cooling, were themselves buried by other sedimentary strata. For example, in the fillings of ancient inland basins within the Appalachians, in the Connecticut Valley in central Connecticut and Massachusetts, and in the Newark lowland of northern New Jersey and adjacent parts of southeastern New York, three thick units of uniform mafic extrusive igneous rock are interlayered with a complex group of sedimentary strata (Figure 6.20). This complex group of sedimentary rocks was not under-

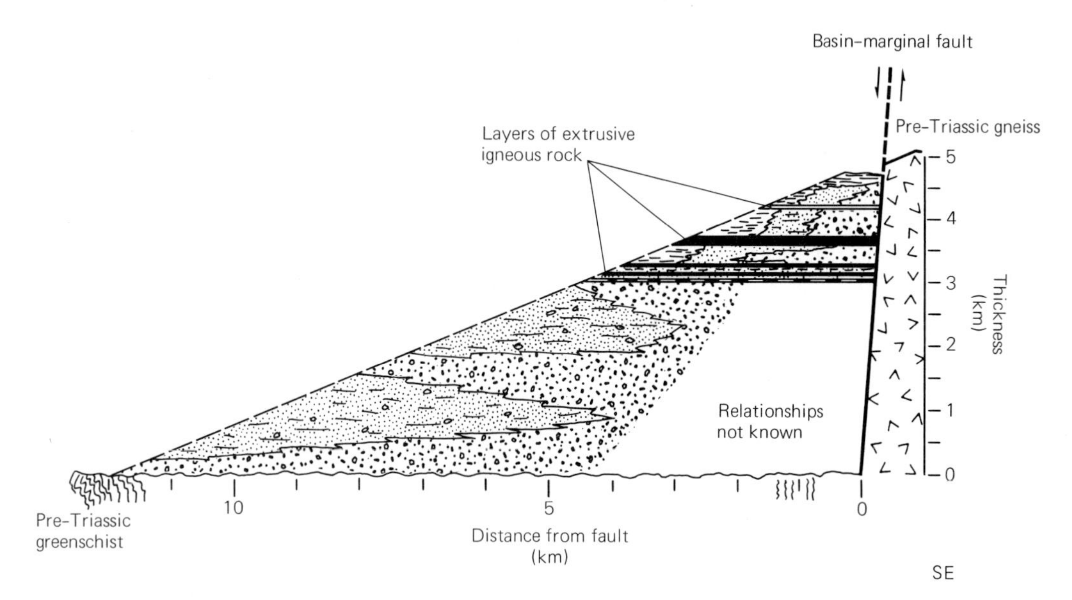


**6.20 Strata of volcanic rocks interbedded with sedimentary strata** filling the Connecticut Valley basin, south-central Connecticut, restored profile with strata shown in their former horizontal positions. (J. E. Sanders, 1968, Stratigraphy and primary sedimentary structures of fine-grained, well-bedded strata, inferred lake deposits, Upper Triassic, central and southern Connecticut, p. 265–305 *in* Klein, G. de V., *ed;* Late Paleozoic and Mesozoic continental sedimentation, northeastern North America. A symposium: Geological Society of America, Special Paper 106, 309 p., Figure 2, p. 271)

stood until the volcanic strata had been identified and their importance appreciated. The tilted edges of these resistant sheets of extrusive igneous rock form prominent ridges that extend from southern Connecticut north to the northern border of Massachusetts in the Connecticut Valley and that form the Watchung Mountains of northern New Jersey.

**Paleomagnetic records in extrusive igneous rocks** One of the outstanding scientific developments of the mid-twentieth century has been the discovery and analysis of the records of former magnetic fields of the Earth as contained within various rocks, but particularly within strata of extrusive mafic igneous rocks.

The principle of the method depends upon the relationship between the Curie temperature and magnetism. When lava arrives at the Earth's surface, its temperature (possibly 1100°C to 1200°C or so) lies well above the Curie point of magnetite (578°C). Therefore, the magnetite cannot be magnetic. However, as that lava cools and the temperature eventually drops below 578°C, the magnetite will become magnetized and will take on whatever magnetic field surrounds it. Usually, this will be the Earth's magnetic field. If a mafic lava were to cool in the Northern Hemisphere today, say in Iceland, its magnetite would take on a magnetic field similar to the Earth's at that latitude (see Figure 1.15b). A positive pole would be toward the north and a negative pole toward the south (Figure 6.21).

Study of the numerous layers of mafic igneous rocks exposed on Iceland revealed that the magnetite in the topmost layers is magnetized with fields the same as the Earth's present field (Figure 6.22). However, farther down the magnetite displays fields having polarities opposite to the Earth's present field. Such magnetite evidently cooled during a time when the polarity of the Earth's magnetic field was reversed, that is, the negative magnetic pole was located near the geographic North Pole and the positive magnetic pole near the South Pole. As studies of magnetite in ancient extrusive rocks progressed, it became apparent that the polarity of the Earth's magnetic field has reversed many times. As explained in Chapter 18, the combination of such magnetic data with

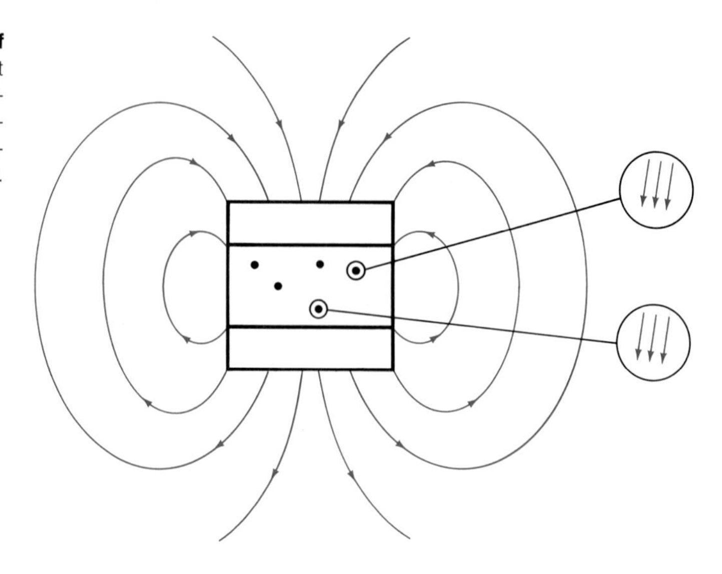

**6.21 Orientation of magnetic lines of force in magnetite of lava flow** that cooled in Northern Hemisphere surrounded by a magnetic field such as today's. Magnetic lines of force in background are for whole Earth.

calculations of ages of the minerals in the volcanic rocks based on radioactivity made possible the calibration of a magnetic-reversal time scale.

## RECOGNITION OF IGNEOUS ROCKS

Igneous rocks can be recognized by various diagnostic features.

### Small-Scale Characteristics of Igneous Rocks

As emphasized in Appendix C, the important small-scale features of igneous rocks are texture and mineral composition. The association of crystalline texture, either equigranular or porphyritic, with the small group of rock-making silicate minerals (olivine, pyroxene, amphibole, biotite, muscovite, feldspars, and quartz) is a strong hint that the rock is igneous. It is not an absolute basis for recognition, for many of these same minerals can be found in metamorphic rocks.

Other small-scale features include vesicles (see Figures 5.6 and 5.7a) and amygdales (see Figure 5.7b).

### Large-Scale Features of Igneous Rocks

Important large-scale diagnostic features of igneous rocks are the geometrically regular pattern of *joints* that form as a result of contraction during cooling (Figure 6.23) and the curved joints, known as *sheeting,* that appear as the covering country rock surrounding a massive pluton is

Thickness (m)
700
600
500
400
300
200
100
0
Pleistocene
Pliocene

EXPLANATION
Compact basalt
Globular basalt
Volcanic breccia
Till or tillite
Sand and gravel
Polarity same as today
Polarity opposite of today's

**6.22 Paleomagnetism in basalts of Late Cenozoic ages,** Iceland, columnar section. Small arrows at right of section indicate directions of north-seeking poles of magnets. In Iceland today, north-seeking poles of magnets point steeply downward. Shown here are Lower Pleistocene basalts with reversed polarities and Pliocene basalts having normal polarities in upper layers but reversed polarities below. (After H. Wensink, 1964, Paleomagnetic stratigraphy of younger basalts and intercalated Plio-Pleistocene tillites (*sic*) in Iceland: Geologische Rundschau, v. 54, no. 1, p. 364–384, Figure 6, p. 375)

**6.23 Regular pattern of cooling joints** in extrusive Banbury Basalt (Pliocene) about 6 kilometers NNE of Buhl, Twin Falls County, Idaho. Diameters of columns are 0.5 to 1 meter. (H. E. Malde, U. S. Geological Survey, 30 September 1960)

eroded away and the rock of the pluton expands upward (Figure 6.24).

Other large-scale features of igneous rocks include xenoliths (see Figure 6.9) and pillows (see Figure 5.33).

**6.24 Sheeting joints in massive pluton,** Liberty Cap, Yosemite National Park, California. (U. S. National Park Service)

### Distinguishing Intrusive Igneous Rocks from Extrusive Igneous Rocks

In many respects all igneous rocks display similar characteristics; thus, it is not always immediately obvious whether a given body of igneous rock is a pluton or part of an extrusive sheet. The general problem of distinguishing between intrusive and extrusive igneous rocks can be illustrated by comparing the relationships of a sill and a buried sheet of extrusive rock formed by the solidification of an ancient lava flow (Figure 6.25).

Both a sill and a buried sheet of extrusive rock may display columnar joints as a result of cooling (Figure 6.26). Both appear as a tabular concordant layer of igneous rock between layers of a sedimentary sequence (see Figure 6.12). In both, a zone of heated and altered rock lies below, and the zone near the basal contact of the igneous rock displays the textural effects of chilling. The finest sizes are at the contact, and sizes enlarge with distance away from the contact.

Several distinctive features of a sill, when present, may enable one to distinguish it from a former lava flow. A sill may include xenoliths of the surrounding country rock. In addition, the upper as well as the lower contact of a sill displays a chilled zone within the igneous rock and a zone of contact-altered material where the adjacent country rock was heated. The igneous rock forming a sill is almost never vesicular or amygdaloidal.

## GEOLOGIC DATING OF BODIES OF IGNEOUS ROCK

Bodies of igneous rock can be dated (1) by radiometric methods for determining ages of minerals composing the rock and (2) by determining the relative age of the igneous rock with respect to the ages of surrounding rock units.

**6.25 Differences between a sill and a buried sheet of extrusive igneous rock,** schematic sections. Further explanation in text.

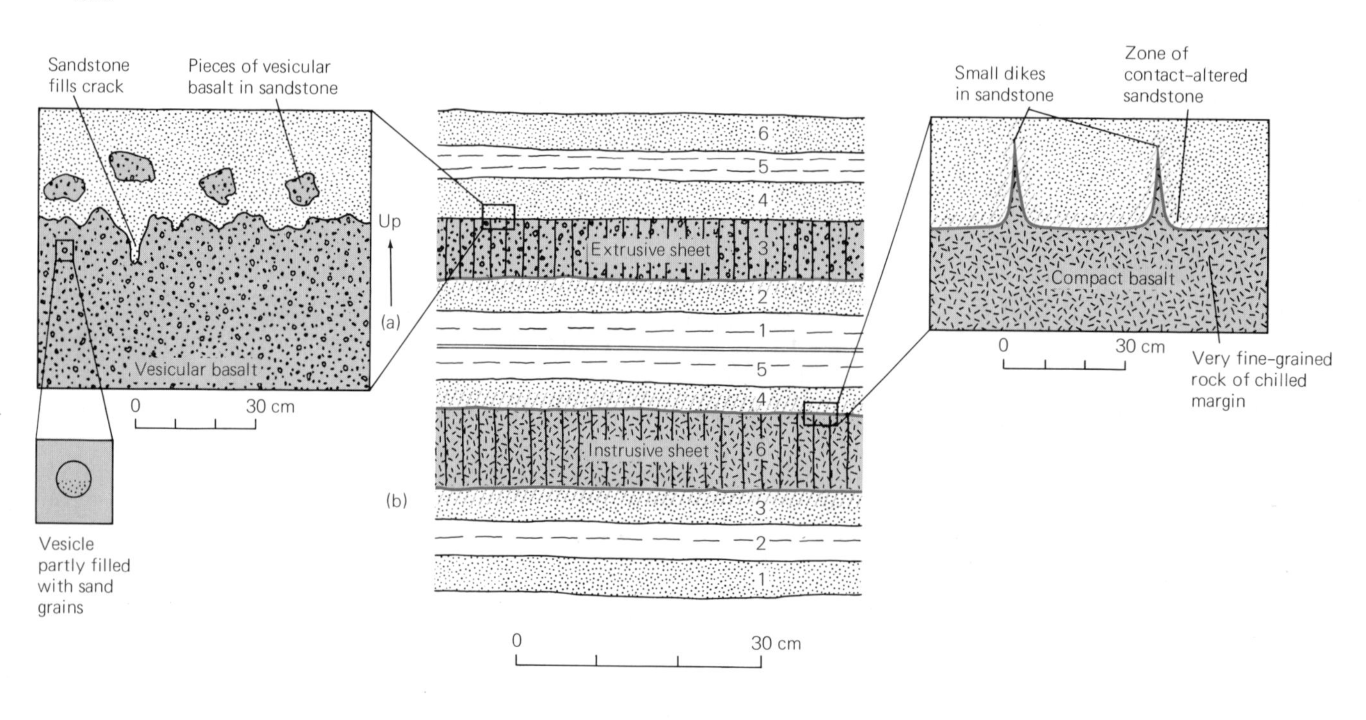

(a)

(b)

**6.26 Columnar joints in a sill and a buried sheet of extrusive igneous rock.** (a) Palisades, along W side of Hudson River N of New York City, displays columnar joints in a sill. (United Press International) (b) Columnar joints in ancient lava flow, Auvergne district, central France. (J. E. Sanders, May 1976)

### Radiometric Methods Applicable to Igneous Rocks

Nearly all the methods based on radioactivity for determining the ages of minerals can be applied to igneous rocks. The radiometric methods apply whether the body of igneous rock is a pluton or an extrusive sheet. As noted in Chapter 4, the radioactive isotopes decay spontaneously even in magma or lava. In magma or lava, however, the daughter isotopes that are gases escape. The time of lattice formation is the time when these decay products begin to accumulate. Provided the rock is not later heated to the point where the decay products escape, the lattices store the decay products.

In general, the radiocarbon method is not applicable to igneous rocks. An exception is charcoal formed by the effect of a lava flow or a *nuée ardente* on wood. A radiocarbon date on such charcoal can yield the date of the igneous rock formed by cooling of the lava or the *nuée ardente.*

### Relative Ages of Igneous Rock and Surrounding Rock Units

The relative geologic age of a pluton with respect to the surrounding rocks is determined by examining the contacts. A pluton is younger than the country rock that it cuts across and/or thermally alters. A pluton is older than an erosion surface that cuts it and is also older than sedimentary strata which bury its eroded surface (Figure 6.27). If the eroded upper surface of a pluton is not exposed, it may be possible to find pieces that were eroded from the pluton and incorporated within a layer of sand or gravel. In such cases the pluton is older than the sedimentary layer containing pieces eroded from it.

The layers in a sequence including one or more ancient lava flows or layers of welded tuff accumulated one at a time and in a regular order. Accordingly, the problem of finding the relative age of a sheet of extrusive igneous rock is easily solved: the sheet is younger than the underlying layer and older than the overlying layer (see Figure 6.25).

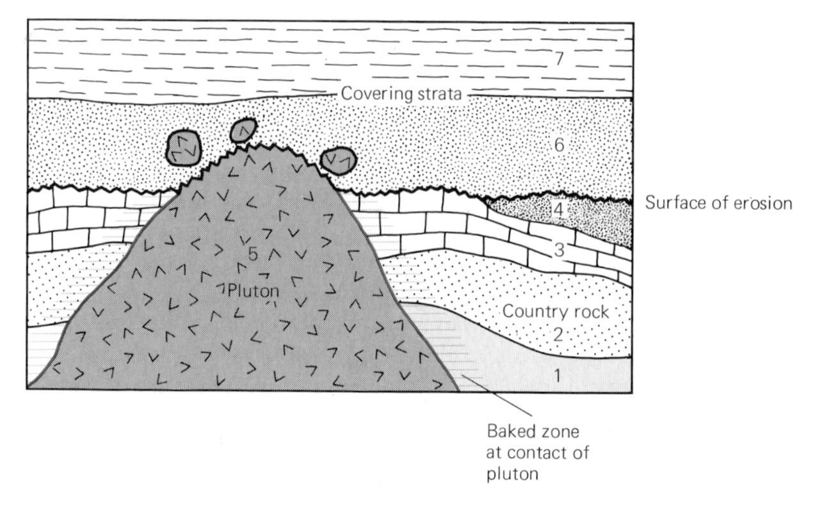


**6.27 Geologic age of pluton** is determined by finding youngest strata of country rock cut and altered by the pluton and oldest covering strata that bury the eroded pluton or contain particles eroded from the pluton. Schematic profile and section.

## THE ORIGIN OF MAGMA AND OF THE CHEMICAL VARIETIES OF IGNEOUS ROCKS

Two major problems about igneous rocks can be approached only indirectly. The answers to both lie hidden within the Earth. In attempting to analyze these topics, we have to rely on reasoning from principles of physics and chemistry, from geophysical measurements, geochemical analyses, laboratory experiments, and studies of volcanoes. We also apply geologic theories.

### The Origin of Magma

To understand the origin of magma we must determine what conditions would cause the solid material within the Earth to melt. The condition that seems most obvious is a rise in temperature. We know that in the laboratory we can melt rocks if we heat them to their melting points, and we also know that measurements taken within the Earth indicate that temperature increases downward (see Figure 1.17). Assuming that this principle of temperature increase applies even in deeper zones where temperatures have not been measured, one might suppose that an effective way to melt rock material would be to depress it to a great depth, where the temperature would exceed the material's melting point. Presumably, this process has happened many times.

Another factor to be considered in the change from solid to liquid is pressure. From laboratory experiments we know that as pressure increases, the melting point of a rock increases. As a body or rock is depressed to greater and greater depths, both the temperature and the pressure increase. The increase in pressure raises the melting temperature. If deep within the Earth pressure is locally reduced while the temperature and other conditions remain the same, material at that location which was near its melting point will suddenly be *above* its melting point. Thus, it will immediately begin to change from a solid into a liquid state. Geologists think such a process takes place in the mantle. We know that in most places, even under great pressure and high temperature, the Earth's mantle is solid. But we can speculate that if for some reason the pressure were to be reduced, the material would be above its melting point at the new pressure, and hence it would turn into liquid magma. This liquid magma would then tend to work its way upward toward the surface of the Earth.

Another physical phenomenon involving heat within the Earth is radioactivity. Even small amounts of radioactive isotopes generate huge quantities of heat. Therefore it is reasonable to suppose that wherever concentrations of radioactive isotopes exceed the normal low levels, the temperature will be higher than usual.

Another large-scale process that is probably affecting the Earth's mantle is upward transfer of heat from the liquid outer core. Theoretical analysis suggests that such upward movement of heat would not take place uniformly, but instead would give rise to chimneylike columns or plumes within which heat would rise more rapidly than in nearby areas and within which the temperature would be greater. Where such plumes intersect the base of the lithosphere they would create local "hot spots," within which rocks might be melted.

In many cases, it is not known why a given body of magma formed. The study of the origin of magma has been much advanced by modern laboratory experiments and theoretical analyses, but it remains one of several major geologic topics about which much is yet to be understood.

### Possible Causes of the Chemical Varieties of Igneous Rocks

One of the important theoretical aspects of igneous rocks centers on the question: "What causes the chemical variety of igneous rocks?" One of the most obvious expressions of this chemical variety is the great dominance of basalt among extrusive igneous rocks and of granite among intrusive igneous rocks. Why should this be so? Two contrasting answers have been proposed. One idea states that all igneous rocks are derived from a single kind of universal magma, but that the operation of some process or processes creates more than one kind of igneous rock. Any process by which one kind of magma gives rise to more than a single kind of igneous rock has been named *magmatic differentiation*.

The contrasting view is that each kind of igneous rock forms from a distinctive kind of magma and that the number of varieties of igneous rocks indicates how many kinds of magma there are. For example, simply based on their silica contents, igneous rocks (not including the crystal cumulates) can be divided into three groups (felsic, intermediate, and mafic). To some geologists, these three groups of igneous rocks imply that there are three kinds of magma.

**Kinds of lava** Perhaps the most direct evidence we will ever have about magma deep within the Earth comes from lava that we can see at the surface. The worldwide systematic pattern of the composition of lavas, however, suggests that three major kinds exist and that they are related to the kinds of materials that are melted. In addition, the kinds of materials that become melted are thought to be controlled by the behavior of lithosphere plates and the underlying upper part of the mantle.

**Plate-tectonics concepts** According to the concepts of plate tectonics, *mafic magmas* are considered to be the product of partial or complete melting of mantle material. Such magma, once formed, comes to the surface along deep-reaching fractures, such as are found along mid-oceanic ridges or along rift valleys, where the lithosphere is being stretched and pulled apart (Figure 6.28; see also Chapter 18). *Intermediate magmas* are thought to be the kind of melt that forms when a lithosphere plate descends into the mantle deep beneath a volcanic island arc. The evidence for this conclusion is the existence of the "Andesite Line" around the margins of the Pacific basin (Figure 6.29). *Felsic magmas* are inferred to be the result of melting of continental-type crust in places where tectonic forces cause such crust to be depressed to great depths.

**Processes of crystallization during solidification** According to the ideas of N. L. Bowen and other petrologists, the reactions between crystallizing minerals and still-molten magma could be a factor in causing more than one kind of igneous rock to be derived from a single magma. Bowen supposed that the formation and settling of crystals could be a mechanism by which one parent mafic magma could give rise to three kinds of

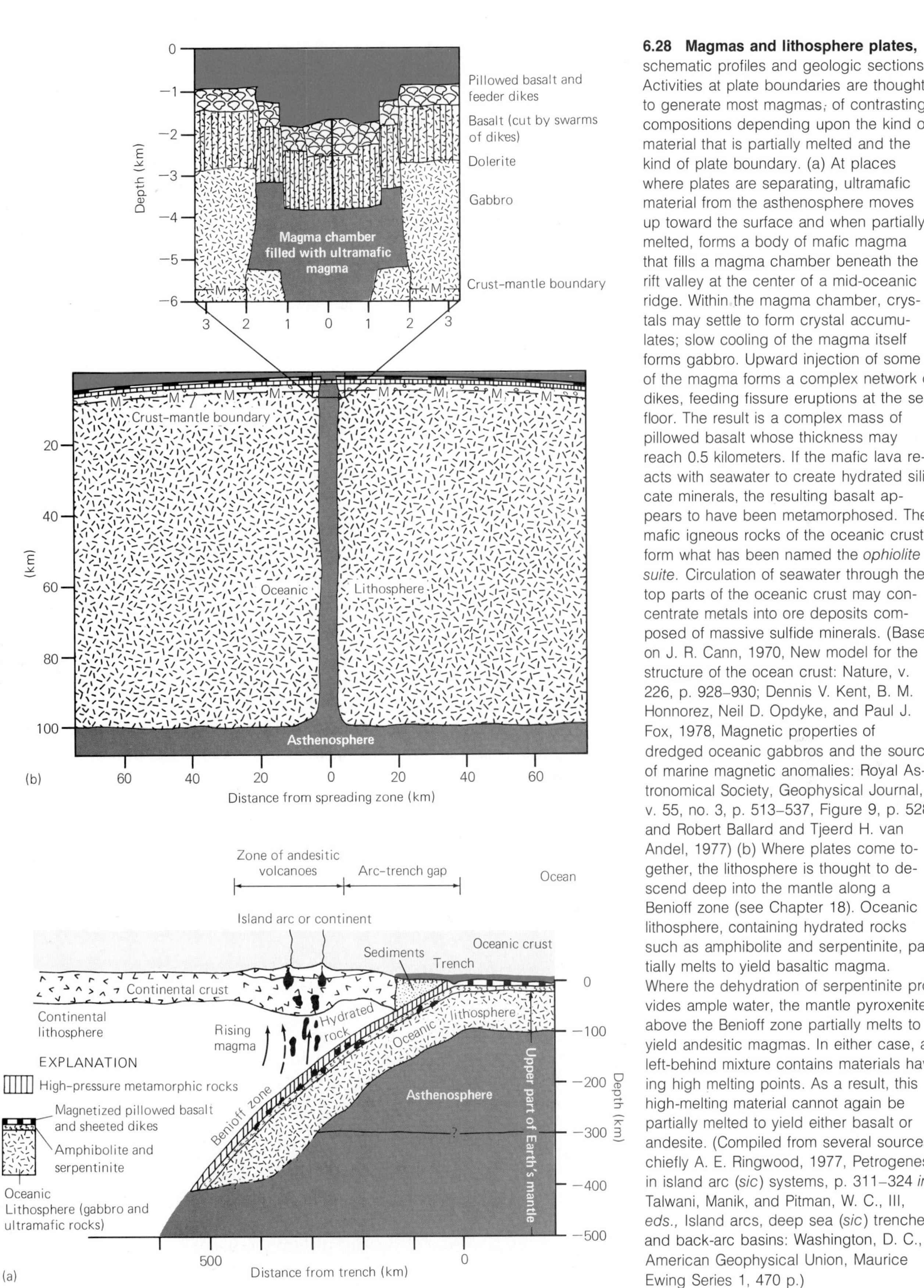


**6.28 Magmas and lithosphere plates,** schematic profiles and geologic sections. Activities at plate boundaries are thought to generate most magmas, of contrasting compositions depending upon the kind of material that is partially melted and the kind of plate boundary. (a) At places where plates are separating, ultramafic material from the asthenosphere moves up toward the surface and when partially melted, forms a body of mafic magma that fills a magma chamber beneath the rift valley at the center of a mid-oceanic ridge. Within the magma chamber, crystals may settle to form crystal accumulates; slow cooling of the magma itself forms gabbro. Upward injection of some of the magma forms a complex network of dikes, feeding fissure eruptions at the sea floor. The result is a complex mass of pillowed basalt whose thickness may reach 0.5 kilometers. If the mafic lava reacts with seawater to create hydrated silicate minerals, the resulting basalt appears to have been metamorphosed. The mafic igneous rocks of the oceanic crust form what has been named the *ophiolite suite.* Circulation of seawater through the top parts of the oceanic crust may concentrate metals into ore deposits composed of massive sulfide minerals. (Based on J. R. Cann, 1970, New model for the structure of the ocean crust: Nature, v. 226, p. 928–930; Dennis V. Kent, B. M. Honnorez, Neil D. Opdyke, and Paul J. Fox, 1978, Magnetic properties of dredged oceanic gabbros and the source of marine magnetic anomalies: Royal Astronomical Society, Geophysical Journal, v. 55, no. 3, p. 513–537, Figure 9, p. 528; and Robert Ballard and Tjeerd H. van Andel, 1977) (b) Where plates come together, the lithosphere is thought to descend deep into the mantle along a Benioff zone (see Chapter 18). Oceanic lithosphere, containing hydrated rocks such as amphibolite and serpentinite, partially melts to yield basaltic magma. Where the dehydration of serpentinite provides ample water, the mantle pyroxenite above the Benioff zone partially melts to yield andesitic magmas. In either case, a left-behind mixture contains materials having high melting points. As a result, this high-melting material cannot again be partially melted to yield either basalt or andesite. (Compiled from several sources, chiefly A. E. Ringwood, 1977, Petrogenesis in island arc (*sic*) systems, p. 311–324 *in* Talwani, Manik, and Pitman, W. C., III, *eds.,* Island arcs, deep sea (*sic*) trenches, and back-arc basins: Washington, D. C., American Geophysical Union, Maurice Ewing Series 1, 470 p.)

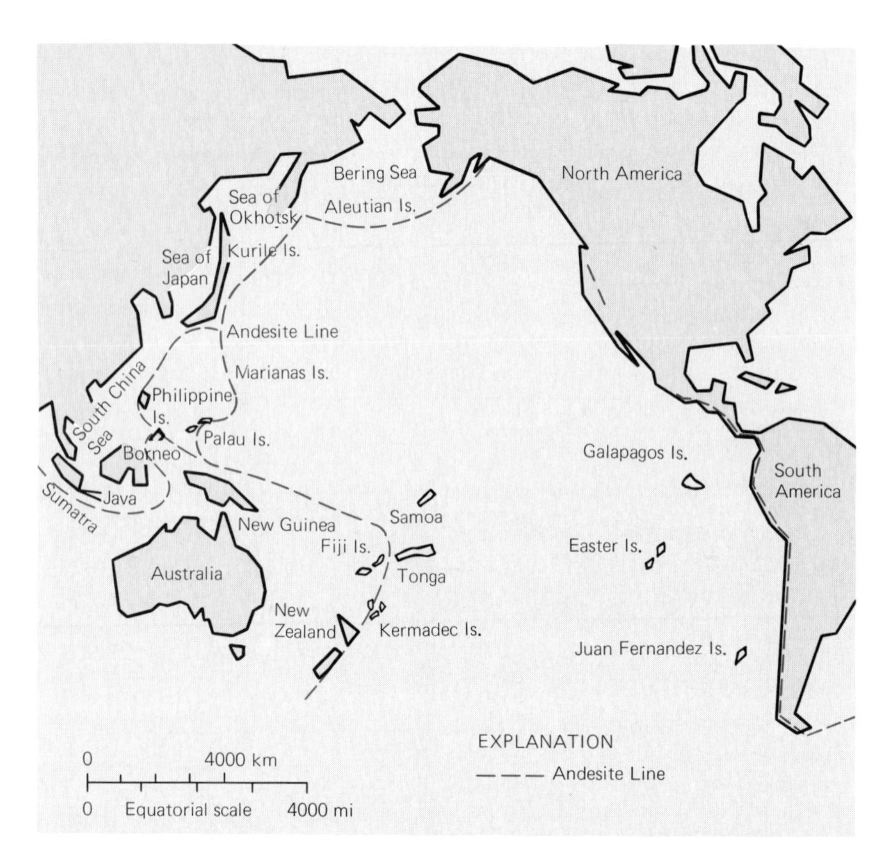


**6.29 The Andesite Line** in the marginal parts of the Pacific basin.

rocks: (1) ultramafic crystal-cumulate rocks formed by settling of early formed dense minerals; (2) mafic rocks, by the solidification of plagioclase and pyroxene; and (3) felsic rocks, by the solidification of the residual magma that remained after the ultramafic minerals had settled out and the mafic rocks had crystallized.

Bowen's concept may explain some associations of igneous rocks. However, the volumes of the various rocks that would form in this way are such that the felsic rocks would be a minor proportion of the total. The vast sizes of granitic batholiths, mentioned in a previous section of this chapter, has been cited as an argument against the origin of all granites by the process suggested by Bowen.

## ORE DEPOSITS AND ECONOMIC MINERAL DEPOSITS RELATED TO IGNEOUS ROCKS

Economic mineral deposits related to igneous rocks include ores that were concentrated in magmas, ores that were deposited by magmatic solutions, and nonmetallic mineral deposits of igneous origin.

### Ores Concentrated in Magmas

Ores that were concentrated by separation from a magma occur within or closely associated with various rock-making silicate minerals of mafic or ultramafic rocks. A worldwide systematic association exists between primary chromite, platinum, nickel, diamond, and some deposits of copper, magnetite, and ilmenite and ultramafic or mafic igneous rocks.

Examples include the diamond pipes of South Africa; the chromite ore of the Bushveld complex near Rustenburg and Lydenburg, South Africa; and the nickel deposit at Sudbury, Ontario, Canada. A comparable association exists between primary ores of tin and tungsten and granites, as in Cornwall, England (Figure 6.30).

### Ores Deposited by Magmatic Solutions

The ores assigned to this group are inferred to have formed from elements that were brought to the scene by a magma, but were not deposited among the rock-making silicate minerals at the time when these silicates crystallized. Instead, the metallic elements became concentrated within the hydrothermal solutions and were precipitated after the rock-making silicate minerals had solidified and possibly even had been altered to clay minerals by the hot solutions. Proof of derivation from magmatic hydrothermal solutions consists of position of the ore within or close to the plutonic rock. In some examples the hydrothermal minerals are zoned concentrically around the pluton.

Examples of hydrothermal deposits are the important group of so-called porphyry-copper ores, in which sulfides of copper and of other metals have been localized in an extensive network of pore spaces and tiny fractures that cut a porphyry. The large deposit at Bingham, Utah, is of this type (see Figures 6.19 and 6.20).

### Nonmetallic Mineral Deposits of Igneous Origin

Nonmetallic resources from igneous rocks that have not been hydrothermally altered include granite building stones and monumental materials, dolerite for crushed stone in ballast and concrete aggregate (see Figure 2.21), feldspar for use in ceramics, mica for use in electrical insulators, and pumice for use in lightweight aggregate.

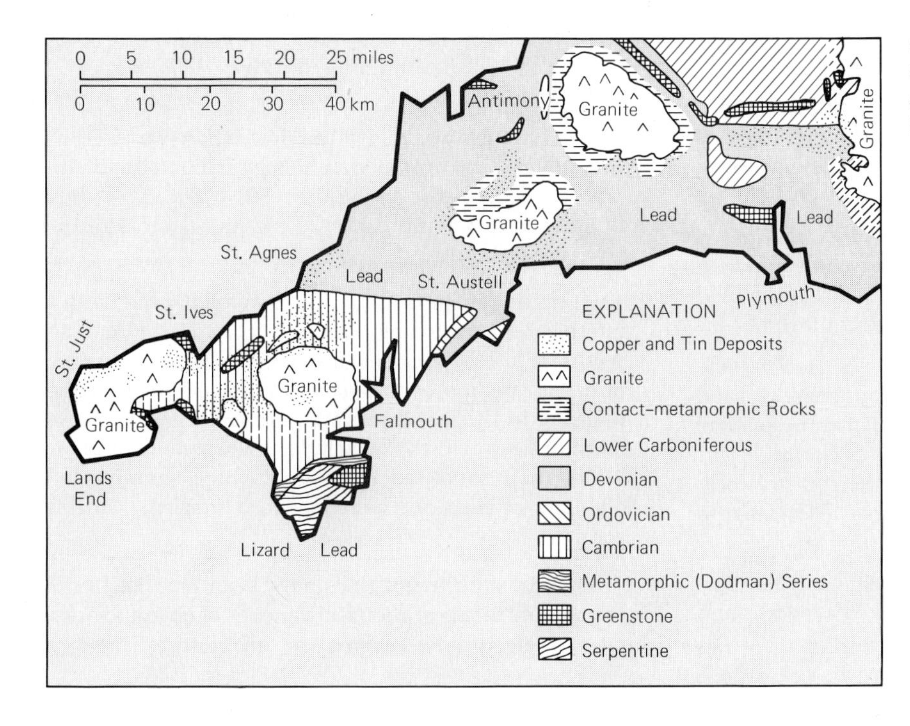


**6.30 Granite plutons and associated ores of tin, tungsten, and lead,** Cornwall, southwestern England. (After H. M. Geological Survey of Great Britain)

# CHAPTER REVIEW

1. Igneous rocks are important parts of the rock cycle for two reasons: (1) they represent products of melting of material that has been at the Earth's surface but later subsided to great depths and later became parts of mountain belts; and (2) they represent material from the mantle that forms parts of the deep-ocean floor and is also a widespread volcanic product.

2. Magma and lava may be accompanied by *hydrothermal solutions* that can escape or and thus not affect the silicate minerals or can soak through the rock and thus can alter the silicate minerals and possibly also deposit ore minerals.

3. *Bowen's reaction principle* describes the reaction between magma and crystals of different types of silicate minerals. In a *continuous reaction series,* growing crystals exchange ions with the melt, but remain intact. For example, plagioclase lattices remain while exchanging calcium ions for sodium ions from the melt. In a *discontinous reaction series,* growing crystals not only exchange ions but also crystals of one kind disappear and are supplanted by crystals of another kind. For example, ferromagnesian minerals appear in a series (olivine, pyroxene, hornblende, biotite) with each kind of lattice forming after a preceding lattice has disappeared.

4. Early formed crystals of a density different from the magma may settle to the bottom or float to the top of a body of magma. Early formed crystals of a density similar to the magma remain dispersed in the magma and become the *phenocrysts* of porphyries. Where many crystals collect by sinking or floating, they form *crystal cumulates.*

5. Growth of crystals from a magma forms a loose framework of individuals; much pore space remains. This pore space eventually disappears when quartz, various sulfides, or magnetite solidify. Early formed crystals enclosed in larger crystals are *inclusions.* Late-growing crystals that wrap around earlier-formed crystals are interstitial space fillers.

6. Intrusive igneous rocks occur in bodies named *plutons;* other kinds of rocks surrounding a pluton are named *country rock.* Extrusive igneous rocks form layers.

7. Plutons are classified according to: (1) *contacts,* the boundaries between a pluton and country rock, and (2) *shapes.* The contacts of *concordant plutons* are parallel to the layers of the country rock; those of *discordant plutons* are not parallel to the layers of the country rock. The shapes of plutons range from *tabular* to *lenticular* to *cylindrical* to *irregular.*

8. Concordant plutons include *sills* and *laccoliths.* Discordant plutons include *dikes, volcanic necks* and *diatremes, stocks,* and *batholiths.*

9. The Bingham-Last Chance stock, Utah, is a composite pluton that was intruded $39.8 \pm 0.4$ million years ago into deformed Pennsylvanian strata. Part of the stock is being mined in the world's largest open-pit excavation to recover copper, molybdenum, and other metals. The first rock in the pluton was a quartz-poor equigranular felsic rock; it was later intruded by a quartz-rich porphyry. After the porphyry had been emplaced, hydrothermal solutions altered the silicate minerals and deposited the ore minerals, which are chiefly sulfides.

10. Batholiths are so large that they raise the problem of how they made room for themselves. Mechanisms for making room in country rock include pushing the walls of cracks apart, arching the overlying strata, and stoping off numerous xenoliths that sink into the magma chamber. Some granitic rocks may have replaced preexisting rocks during extreme metamorphism.

11. The stratigraphic study of layers of volcanic rock is useful for historical purposes; volcanic layers form valuable reference layers for analyzing associated sedimentary strata. Layers of volcanic rock have been analyzed for their paleomagnetic records.

12. Igneous rocks can be recognized by various diagnostic features. Useful small-scale features include mineral composition, texture, vesicles, and amygdales. Large-scale features include geometrically regular patterns of cooling joints, sheeting joints, xenoliths, and pillows.

13. Sills can be distinguished from sheets of extrusive igneous rock (former lava flows) by study of the upper contacts. Magma from a sill heated the overlying sedimentary layers. By contrast, the top of an ancient lava flow had cooled by the time younger sedimentary strata were deposited on top of it.

14. Bodies of igneous rock can be dated in two ways: (1) by determining the ages of minerals by radiometric methods and (2) by determining the relative age of the body of igneous rock with respect to the country rock or the covering layers.

15. The origin of magma is a problem involving the Earth's internal heat, pressure, radioactivity, and the melting points of rocks. Magma may form where radioactive minerals are more abundant than usual, where the pressure in the mantle is locally reduced, where rocks subside deep within the Earth, or where the mantle is hotter than usual, as in plumes where concentrated amounts of heat are rising upward from the Earth's core.

16. Two contrasting hypotheses have been proposed concerning the origin of chemical varieties of igneous rocks: (a) a single type of magma can give rise to different

varieties of rocks through *magmatic differentiation;* (b) each variety of rock requires a different type of magma.

17. The three major kinds of lava (and thus presumably also of magma) are: (1) *mafic,* thought to be the result of melting of mantle material; (2) *intermediate,* considered to be the result of melting of descending lithosphere plates; and (3) *felsic,* inferred to be the product of melting of continental-type crust in zones deep within the Earth.
18. Ores concentrated in magmas include chromite, platinum, nickel, diamond, and some deposits of copper, magnetite, and ilmenite (concentrated in ultramafic rocks or mafic rocks); and tin and tungsten (in granites). Many ores related to igneous rocks were precipitated by hydrothermal solutions that altered the silicate minerals of the associated igneous rocks. The porphyry-copper ores at Bingham, Utah, and elsewhere are thought to be products of magma-related hydrothermal solutions. Nonaltered igneous rocks themselves are valuable for building stones, for concrete aggregates, and various other purposes.

## QUESTIONS

1. Explain how cooling history affects the appearance of igneous rocks.
2. What is the difference between a *crystal cumulate* and the *phenocrysts of a porphyry?*
3. Discuss the implications of the hydrothermal alteration of an igneous rock.
4. Define *Bowen's reaction principle.* Compare the reactions between growing plagioclase crystals and a melt and between growing crystals of ferromagnesian silicate minerals and a melt.
5. What is a *pluton?* Are plutons classified according to the kinds of igneous rocks forming them? Explain.
6. What are two kinds of tabular plutons? How do they differ?
7. Define *laccolith.* Illustrate your definition by drawing a schematic profile and section through a laccolith.
8. Compare *stocks* and *batholiths.*
9. What is a *composite intrusive?* Discuss the relationship between igneous rocks and ore deposits in the Binhgam–Last Chance Stock, Utah.
10. List some of the features that are diagnostic clues to igneous rocks.
11. Compare the geologic occurrence of intrusive igneous rocks and extrusive igneous rocks.
12. Explain in words and with idealized geologic sections the similarities and differences between a sill and a buried sheet of extrusive rock (ancient lava flow).
13. Discuss the methods for determining the geologic date of a pluton and of an ancient lava flow.
14. Define *magmatic differentiation.*
15. How many major kinds of lava have been recognized? What do these lavas imply about the variety of magmas?
16. List some of the ore minerals that are related to mafic igneous rocks and to felsic igneous rocks.

## RECOMMENDED READING


Bateman, A. M., 1950, *Economic Mineral Deposits,* 2nd ed.: New York-London, John Wiley and Sons, 916 p.

Bayly, Brian, 1968, *Introduction to Petrology:* Englewood Cliffs, New Jersey, Prentice-Hall, Inc., 371 p.

Cox, K. G., 1978, Kimberlite Pipes: *Scientific American,* v. 238, no. 4, p. 120–133.

Ernst, W. G., 1969, *Earth Materials:* Englewood Cliffs, New Jersey, Prentice-Hall, Inc., 149 p.

James, L. P., 1978, The Bingham Copper Deposits, Utah, as an Exploration Target: History and Pre-excavation Geology: *Economic Geology,* v. 73, no. 7 (special issue), p. 1218–1227.

Krauskopf, K. B., 1968, A Tale of Ten Plutons: *Geological Society of America, Bulletin,* v. 79, no. 1, p. 1–18.

Park, C. F., Jr., and MacDiarmid, R. A., 1975, *Ore Deposits,* 3rd ed.: San Francisco, W. H. Freeman and Co., 530 p.

Walton, M. S., 1960, Granite Problems: *Science,* v. 131, p. 635–645.

Watkins, N. D., 1972, Review of the Development of the Geomagnetic Polarity Time Scale and Discussion of Prospects for its Finer Definition: *Geological Society of America, Bulletin,* v. 83, no. 3, p. 551–574.

White, D. E., 1974, Diverse Origins of Hydrothermal Ore Fluids: *Economic Geology,* v. 69, no. 6, p. 954–973.

# CHAPTER 7
# WEATHERING, SOILS, AND CLIMATE ZONES



PLATE 7

Accelerated weathering of stone during twentieth century illustrated by two photos taken 61 years apart (1908 on left and 1969 on right). The statue, at Herten Castle, near Recklinghausen, Westphalia, Germany, was carved in 1702. Left photo shows effects of the first 206 years (1702 to 1908); right photo shows effects for next 61 years (1908 to 1969). (Dr. K. Schmidt-Thomsen, Landesverwaltungsdirekton, Der Landeskonservator von Westfalen-Lippe, Muenster, Germany)

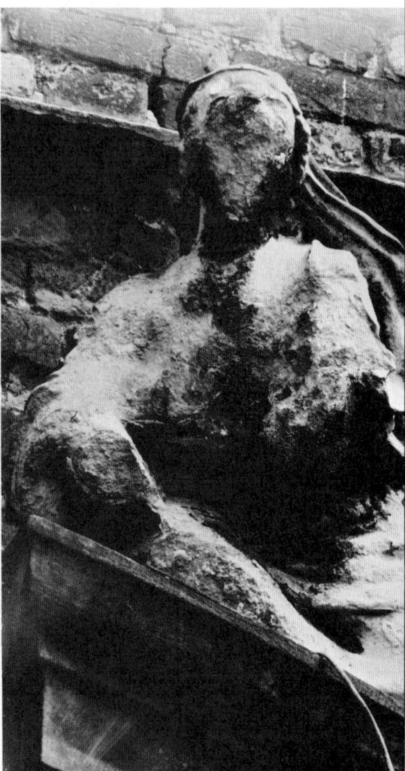

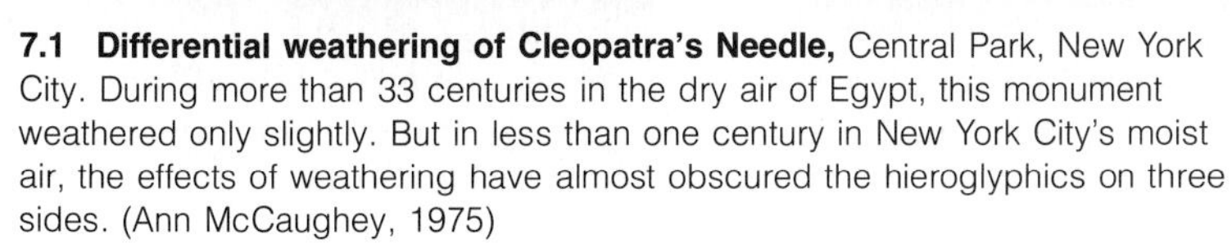

**7.1 Differential weathering of Cleopatra's Needle,** Central Park, New York City. During more than 33 centuries in the dry air of Egypt, this monument weathered only slightly. But in less than one century in New York City's moist air, the effects of weathering have almost obscured the hieroglyphics on three sides. (Ann McCaughey, 1975)

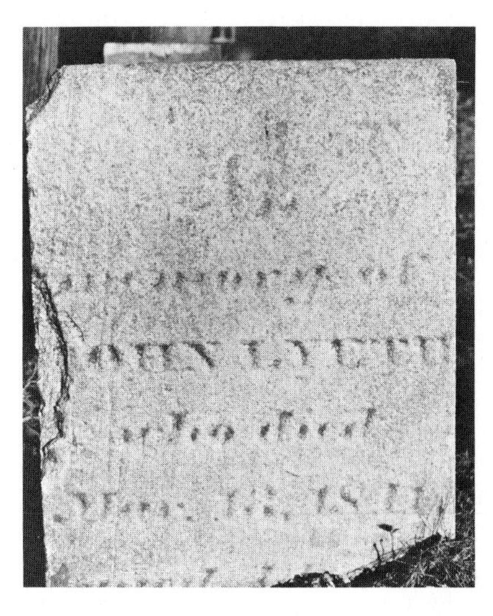

(a)

(b)

**7.2 Comparative weathering of different rocks** shown by preservation of tombstones. (a) Weathering in humid climate has blurred the inscriptions (now almost completely destroyed) on gravestones in Dobbs Ferry, New York. (J. E. Sanders) (b) Negligible weathering in dry climate of marble monument erected near Little Bighorn, Montana. (U. S. National Park Service)

ONE OF THE IMPORTANT LESSONS EMPHASIZED BY JAMES HUTTON concerns what happens to bedrock when it becomes exposed to the atmosphere. Everywhere he looked, Hutton saw the activities and effects of "a universal system of decay and degradation." In modern terms we would say that Hutton observed weathering and erosion. We use *weathering* as a general term that includes physical changes and chemical changes within rock material that take place as a result of its exposure to the subaerial environment. By contrast, *erosion* indicates changes in rocks at the Earth's surface as a result of the movement of water, wind, or ice. We discuss erosion in later chapters and concentrate on weathering in this chapter.

Weathering is important both in geologic theory and in many ways that affect our everyday lives. First of all, weathering offers visible proof that the geologic cycle is operating today and that the operations of this cycle have been influenced by the activities of industrialized societies. No matter where you are, you can see the effects of weathering. All you have to do is look carefully around you at statues that are crumbling (Plate 7), building stones or monuments that are decaying (Figure 7.1), and the carving on tombstones that is being blotted out (Figure 7.2).

Secondly, weathering illustrates how geologic materials and certain kinds of energy react to bring the materials into adjustment with conditions prevailing within what we refer to as the *environment of weathering*. This environment is at or close to the surface of the Earth, where materials are subjected to the effects of the atmosphere, of moisture, of temperature fluctuations, and of organisms.

Thirdly, weathering is a large factor in the conversion of bedrock into *regolith*. A regolith that has formed by the decay of the underlying bedrock is named a *residual regolith*. Weathering transforms the upper part of regolith into a condition where it becomes capable of supporting rooted plants; such regolith is known as *soil*. Without soil, there would be very few land plants and almost no land-dwelling animals. Thus, it is no exaggeration to state that without soil, we would not be here. Hence, because weathering affects soil, weathering affects us all.

In participating in the breakdown of bedrock, weathering creates the raw materials that become sediments and sedimentary rocks. Weathering loosens blocks of bedrock, breaks masses of bedrock down into their individual minerals, attacks and alters these individual minerals according to their tendencies to react, and takes ions into solution and carries them away.

Finally, extensive weathering creates various miscellaneous effects, such as complications for engineering projects, deposits that may contain materials valuable enough to be extracted for use in industry, and the creation of pore space in which valuable materials may accumulate.

## THE ENVIRONMENT OF WEATHERING

In this section we analyze the environment of weathering by summarizing climate, chemical composition of rainfall and effects of air pollution, pore spaces in bedrock and regolith, and the slope situation.

### Effects of Climate

We can define *climate* as the time-averaged weather in a given region. The environment of weathering depends intimately upon climate, particularly upon temperature, precipitation, and evaporation and on the way these factors influence the distribution and kinds of organisms. We can illustrate how these factors interact by summarizing four climate zones (Figure 7.3): (1) humid-tropical regions, (2) humid-temperate regions, (3) warm-arid regions, and (4) cold-arid regions.

**Humid-tropical regions** In general, the place on Earth where weathering is most intense is in the humid tropics. The bedrock and regolith are continuously moist. The air temperature rarely if ever drops below about 20°C. This means that snow and ice are never seen. The great rainfall (commonly more than 250 cm/yr) and high temperatures accelerate chemical reactions. For each rise of 10°C in the temperature, chemical activity doubles or triples. Biological activity increases as well. The cover of vegetation is dense and complete. A typical example is the tropical rainforest. Despite the lush vegetation growing overhead, the floor of a tropical rainforest lacks a carpet of decayed leaves. The tropical bacteria are so numerous and so active that little or no leaf litter is left to accumulate.

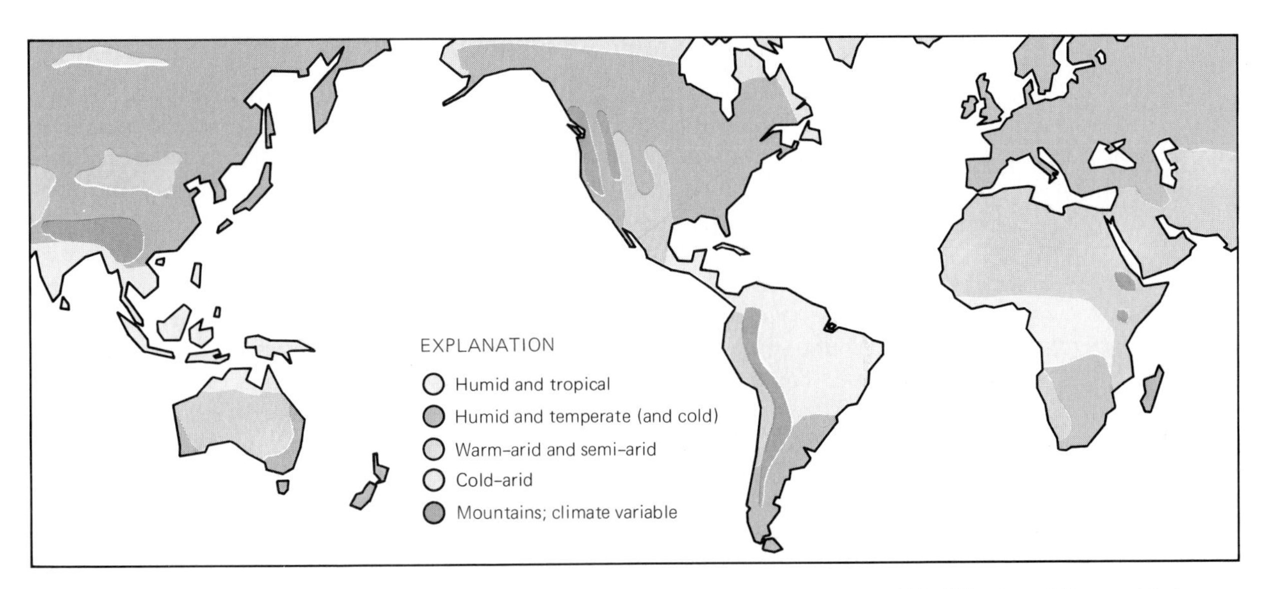


**7.3 Climates of the world** shown on generalized map.

**Humid-temperate regions** In humid-temperate regions, such as much of North America, rainfall usually is less than in the humid tropics, temperatures are more variable, and freezing during the winter creates ice and snow. The natural vegetation forms a complete cover, usually in the form of woodlands. In forests where the trees drop their leaves every fall, the forest floor becomes carpeted with a layer of decaying leaves known as leaf mold. The bacteria in these regions are active enough to cause the leaves to decay, but not so vigorous as to destroy them altogether as in the tropics.

**Warm-arid regions** In warm-arid regions temperatures are high and span a considerable range. Rainfall is scanty, and evaporation exceeds rainfall. As a result, water is drawn upward out of the regolith instead of trickling steadily downward through it, as in the humid regions. Without water, chemical processes of all kinds are inhibited and only a few specialized plants can survive. These are scattered and do not form a continuous cover. In desert regions, therefore, rocks may endure for many thousands of years without changing very much.

**Cold-arid regions** According to popular imagination, the polar lands of the Earth are swept by frequent blizzards and are buried beneath kilometers of ice and snow. In reality, however, many of the coldest places are also among the driest in the world. Their rain-equivalent precipitation is only a few centimeters per year (1 centimeter of rain is equivalent to 10 centimeters of snow). The little precipitation that falls remains frozen most of the time. In addition, the water in the top few hundred meters or so of deep regolith is frozen the year around, forming permafrost. Little liquid water is available for chemical reactions or for plant growth. Summer thawing of the top part of the permafrost layer and later refreezing of the thawed parts creates polygonal patterns in the regolith and pushes stones to the surface along the polygons.

### Chemical Composition of Rainfall; Effects of Air Pollution

Careful microchemical analyses have proved that rainwater is not as pure as distilled water, but contains many dissolved ions. From seawater it picks up ions of salt (sodium chloride or in the mineral form, halite). Gases dissolved in rainwater include nitrogen, oxygen, sulfur dioxide, carbon dioxide, and carbon monoxide. Rainwater also contains various organic compounds derived from seawater and from organic compounds released into the atmosphere by land plants. The substances it contains make rainwater a slightly acid solution, very active chemically, and a source of natural fertilizers and plant food. Rainwater contains a microflora consisting of bacteria and tiny spores and seeds of plants.

The corrosive effects of rainfall have been increased greatly as a result of industrial pollution. The most corrosive pollutant gas is sulfur dioxide, created by the burning of certain coals and fuel oils and by the smelting of sulfide minerals. Sulfur dioxide is oxidized by the oxygen in the atmosphere and the oxidation product reacts with atmospheric moisture to form sulfuric acid. In London, Venice, Rome, and many other cities statues and stone structures (see Plate 7 and Figure 7.1) have weathered much more rapidly since the beginning of the Industrial Revolution than they had previously.

### Pore Spaces in Bedrock and Regolith

The environment of weathering extends beyond the open surface of the Earth; it also includes the pore spaces of bedrock and regolith. In bedrock, pore spaces are located between individual minerals or along joints, faults, and other cracks or partings (Figure 7.4). In regolith, pore spaces exist among the framework particles (Chapter 10). Water tends to remain longer in pores than on surfaces exposed to sunlight.

### Slope Situation

*Slope situation* refers to the height of a slope, and its direction of inclination with respect to sunlight and rain-bearing winds. Elevated rocks are subject to more intense effects of rain, wind, and changes of temperature than are low-lying rocks. At high altitudes the climatic conditions include greater extremes. Furthermore, bits of weathered rock are more likely to fall away from elevated rocks, thus exposing additional surfaces of fresh bedrock to the effects of weathering. In regions of low relief weathered residues tend to remain in place and may accumulate to become tens of meters thick.

Rocks on the north slope of a mountain range weather at a different rate from those on the south slope. In the Northern Hemisphere the Sun's rays strike south-facing slopes far more directly than they do north-facing slopes. Solar energy evaporates more water from bedrock and from regolith underlying south-facing slopes. On north-facing slopes more ice is likely to form and to last longer during a thaw.

The prevailing direction of rain-bearing winds also affects weathering. Slopes receiving more rainfall tend to support more plant cover than other slopes and to erode faster regardless of plant cover.

(a)

(b)

**7.4 Space created in rocks by weathering along joints.** (a) Parallel joints, widely spaced, in massive limestone. The Burren, County Clare, Ireland. (Courtesy Irish Tourist Board) (b) About half of this massive sandstone has been removed by weathering along joints. Canyonlands, Utah. (American Airlines)

## PROCESSES OF WEATHERING

Weathering processes can be divided into those that involve physical activities and those that include chemical activities. Basically, the physical processes do the breaking and the chemical processes do the altering. However, as we shall see, in nature things are not as clearly separated as these two categories suggest. Part of this complexity results from the remarkable properties of water, which can be physical or chemical. Weathering processes tend to work together. Breaking rocks creates more surface area which becomes available for chemical attack. Chemical activities cause the physical release of minerals from some rocks. Growth of mineral crystals along cracks helps loosen rocks physically.

In order to include all these variables, we will discuss the processes of weathering under five main headings: (1) dual effects of water molecules, (2) physical weathering, (3) chemical weathering, (4) complex weathering, and (5) effects of organisms.

### Dual Effects of Water Molecules

In weathering, water is more important than any other single substance or process. It is abundant; it is the only compound that occurs naturally at the Earth's surface as a solid, a liquid, and a gas; and it is highly active chemically.

In water molecules two hydrogen ions, each having an "extra" electron, and one oxygen ion, needing two electrons to fill its outer shell, form what is termed a *hydrogen bond.* The two hydrogen ions line up on the same side of the oxygen ion (Figure 7.5) to create an uneven distri-

**7.5 Water molecule,** schematic sketch with ions drawn to correct size proportions.

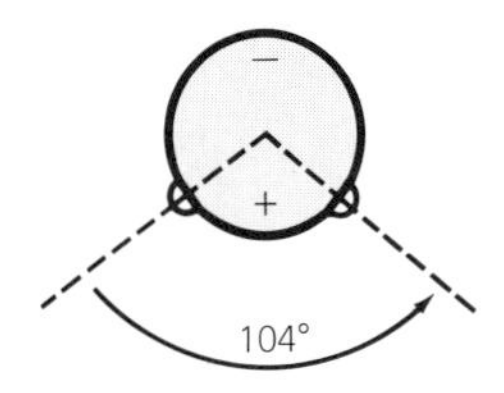

**7.6 Snow crystals,** despite great diversity, all display a six-sided (hexagonal) symmetry. Results of seeding experiment using nuclei of silver iodide and viewed through an electron microscope. (State University of New York at Albany, Institute for Atmospheric Sciences Research, courtesy Roger J. Cheng)

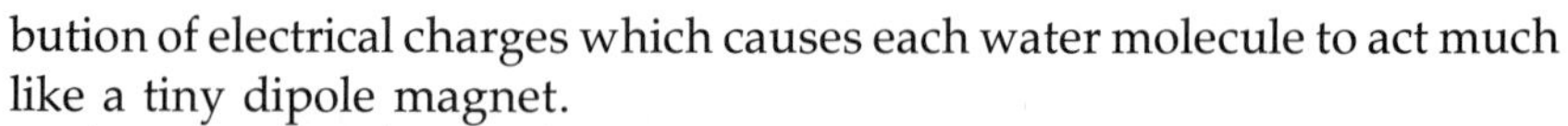

bution of electrical charges which causes each water molecule to act much like a tiny dipole magnet.

The major physical effect of water in weathering is expansion on freezing. As the temperature of water approaches 0°C, randomly arranged water molecules organize into characteristic lattices. The open-network structure of snow and ice (Figure 7.6) takes up more space than the random arrangement and thus the volume of water increases as it changes from liquid to solid. Freezing water expands by about 9 percent with such force that it can break apart solid masses of bedrock.

Water's major chemical effects in weathering are related to its uneven distribution of electrical charge. Water can detach and surround ions from mineral lattices. For example, water can remove $Na^+$ ions from halite lattices. The negative sides of the water molecules surround the $Na^+$ ions; the positive sides of other water molecules surround separated $Cl^-$ ions (Figure 7.7).

## Physical Weathering: Disintegration

*Physical weathering* or *mechanical weathering* (Syn. *disintegration*) refers to processes that break apart rocks without altering their chemical compositions. Although we discuss physical weathering as a separate process, we repeat that weathering processes work together. Also a physical change may result from either a physical process or a chemical process. For example, mechanical breakage can result from physical thermal effects or from crystallization of minerals. The chief effect of physical

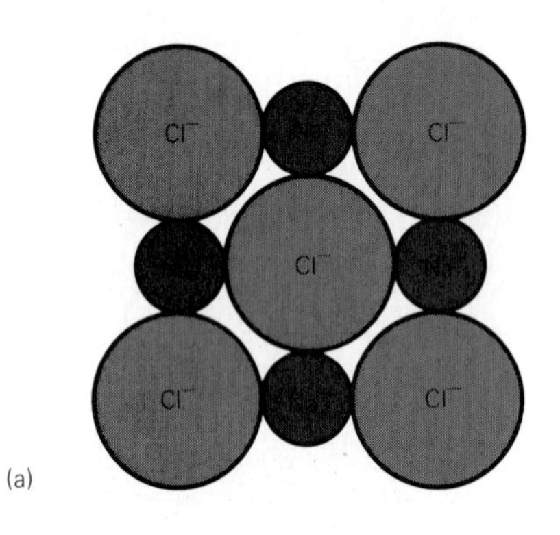


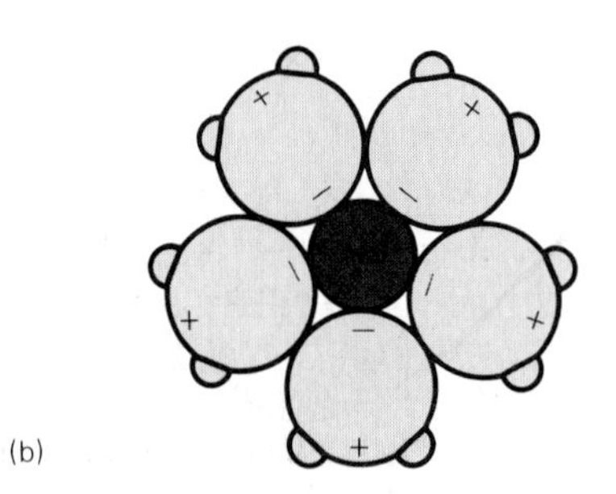


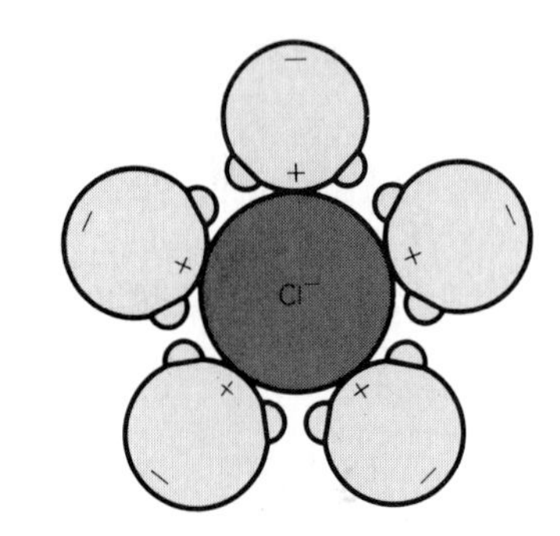


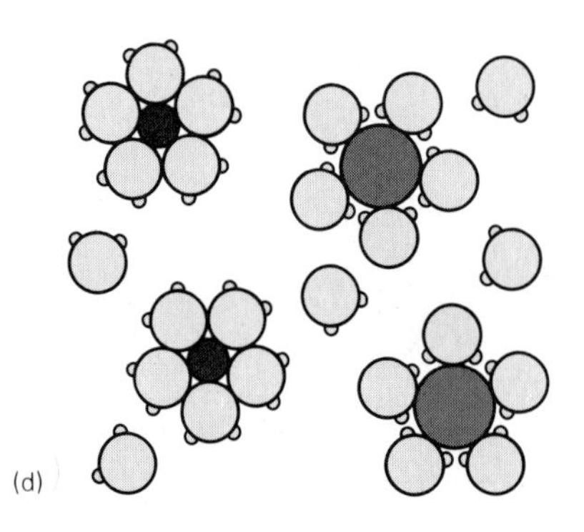


**7.7 Water dissolving sodium chloride.** (a) Water molecules surround segment of halite crystal. (b) Positive sodium ion surrounded by five water molecules attached by their negative sides. (c) Larger negative chlorine ion surrounded by five water molecules attached by their positive sides. (d) Several water clusters surrounding ions of sodium and chlorine.

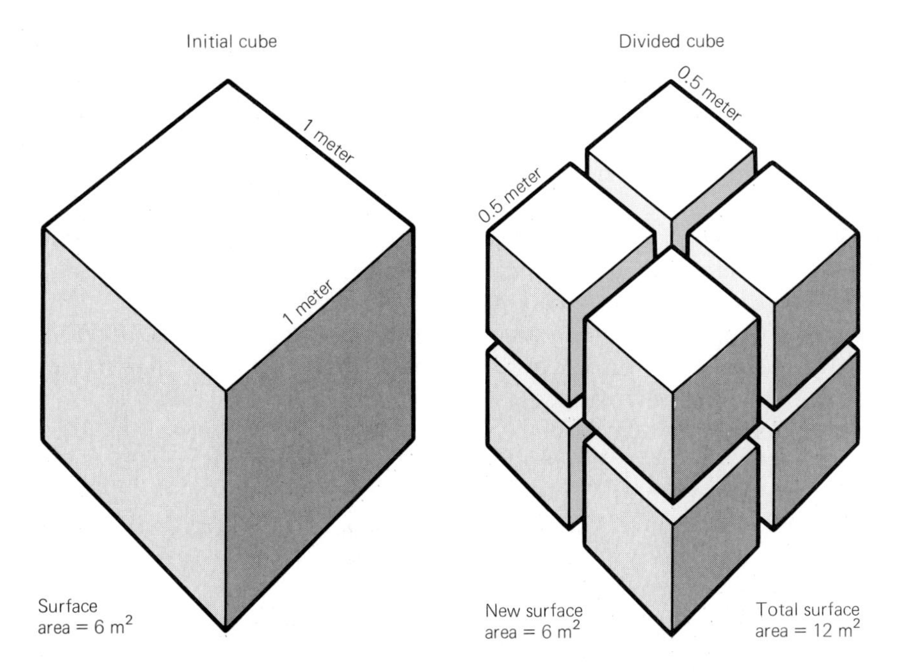


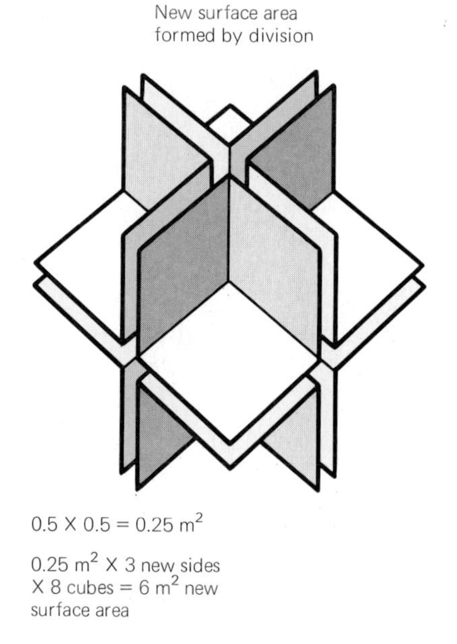


**7.8 Increase of surface area** results from dividing a cube in half along three mutually perpendicular planes. The lengths of the cube sides have been reduced by half but the surface area has been doubled.

weathering is to break rocks into smaller fragments and eventually into individual mineral particles. Each new break in a rock increases the surface area (Figure 7.8), thus creating more surfaces along which chemical activities can take place.

**Granular disintegration** The physical separation of the individual mineral particles of a rock from one another is a process named *granular disintegration.* A rock undergoing granular disintegration crumbles easily (Figure 7.9).

**Sheeting joints; exfoliation** Bedrock that is brought up to the Earth's surface is subjected to uneven pressures which cause the rock to fracture. The chief processes involved are pressure release from unloading and thermal effects.

Any rock formed deep inside the Earth is subject to uniform pressure which results from the weight of the overlying rocks (about 0.26 kg/cm$^2$ for each meter of depth). At a depth of 15 meters the pressure equals the air pressure inside a high-pressure bicycle tire. When relieved of the weight of its former overburden, bedrock exposed at the surface expands upward slightly. This upward expansion can be great enough to form *sheeting joints* parallel to the ground surface (see Figure 6.23). The massive granite "domes" of Yosemite National Park, California, are superb examples of landforms shaped chiefly by sheeting joints (Figure 7.10). The spalling off of bodies of bedrock along sheeting joints is *exfoliation.*

**Thermal effects: frost wedging and heat expansion** The effects of temperature on rocks are most pronounced where temperatures are extreme and fluctuate. The presence of moisture accentuates thermal effects.

Under cold conditions where air temperatures reach the freezing point of water (0°C), liquid water is transformed into solid ice and its volume

**7.9 Granular disintegration** is causing these massive granite blocks to crumble. Algeria. (United Nations, F. A. O. photo by P. Keen)

**7.10 Granite "dome" formed by weathering along sheeting joints.** Half Dome, viewed from Washburn Point, Yosemite National Park, California. (American Airlines)

expands by nine percent. The growth of ice crystals can produce (1) *frost heaving,* the lifting of overlying particles in regolith (to be discussed further in Chapter 11), and (2) *frost wedging,* the prying apart of solid bedrock. Frost wedging is most vigorous at high altitudes where alternate freezing and thawing results from temperature fluctuations from above freezing during the day to the freezing range at night.

The example of a campfire built on solid bedrock can demonstrate some thermal effects under warm conditions. Sheets of rock may break off when the rock is heated and when the fire is doused with water. The rock spalls off in sheets because the fire causes it to expand upward. More breaking takes place when cool water causes an abrupt temperature change in the heated rock.

In deserts, boulders have been found fractured as though by frost wedging. In addition, desert travelers have reported hearing rocks break with sharp, cracking sounds similar to gunshots. (It has been reported that when the French Foreign Legion patrolled the Sahara, legionnaires would often call a condition of battle alert upon hearing these sounds.)

In a laboratory experiment the dry heating and cooling of a small cube of granite was repeated for 89,400 cycles. The cube was heated for five minutes to 140°C and then cooled to 30°C. The routine was thought to be equivalent to 244 years of daily heating by the Sun and cooling at night. At the end of the three-year test no cracks were found in the specimen, even under a microscope. But when moisture was introduced, changes took place quickly. Presumably, the natural moisture in the desert can assist in the breaking of rocks. Where desert rocks remain absolutely dry, however, they change very slowly.

### Chemical Weathering: Decomposition

Chemical weathering, which changes the composition of rock materials, is known as *decomposition.* In order to understand how decomposition works we must consider chemical processes and geologic materials together, because a given chemical process exerts different effects on different materials. In this section, we describe some of the processes involved and in the next, give examples of how a few common rocks respond to chemical weathering.

**Oxidation** As used in discussing weathering, *oxidation* is the addition of oxygen to another element. The elements in rocks that are most easily oxidized are iron (found in ferromagnesian silicates and in sulfide minerals) and sulfur (found in sulfides). The rusting of a nail is a familiar example of oxidation. The iron of the nail combines with oxygen from the atmosphere to form a reddish iron oxide (the "rust"). The iron-oxide minerals that form during chemical weathering are colored red, yellow, orange, or brownish. If copper minerals are oxidized, green and blue minerals form.

As with other kinds of processes in chemical weathering, oxidation proceeds more rapidly if water is present. Oxidation in the desert is slow compared with oxidation where the rocks are exposed to water (see Figure 7.1).

Many rock surfaces underground are covered with a thin film of pyrite, an iron sulfide, which is easily oxidized. During oxidation of the pyrite, the iron forms limonite, and the released sulfur may collect as an element or may combine with water to create sulfuric acid, a powerful chemical agent. Other sulfide minerals are scattered through rocks or have been concentrated as deposits of sulfide minerals. These sulfides, too, are easily oxidized. Where copper minerals are present, oxidation may be responsible for creating ore deposits. Copper can be released through oxidation in the weathering zone and concentrated at greater depths.

Also present in some rock materials is organic matter. Natural organic matter can be classified into two large categories: (1) first-cycle organic matter and (2) polymerized organic matter. The first-cycle organic matter consists of the kinds of materials manufactured by organisms. Polymerized organic matter designates materials in which the carbon and hydrogen ions have formed long chains. Such chains are solids; the process of linking of carbon and hydrogen ions to form the chains is termed *polymerization*.

Most forms of first-cycle organic matter have such a great affinity for oxygen that they will unite with any oxygen that is available and may even use up the total supply. The end products of the oxidation of organic matter are carbon dioxide gas and water. When carbon dioxide is dissolved in water, it forms a weak acid (carbonic acid), which can attack minerals.

Most polymerized organic matter is exceptionally stable. One common variety is *kerogen,* a complex solid hydrocarbon found in many kinds of sedimentary rocks, but in greater-than-average amounts in so-called *oil shales.* Kerogen is able to survive weathering and thus to be recycled. Kerogen is thought to be the organic material from which petroleum is derived (Chapter 8).

**Dissolution** As we have described previously, water is capable of detaching and surrounding ions from mineral lattices; that is, it is capable of dissolving some solid rocks. Layers of rock salt (halite rock) are so easily dissolved that they are exposed only in the driest regions. Proof of water's ability to dissolve minerals has been provided by measurements of new-fallen snow in the Sierra Nevada, California. As soon as the snow melted and seeped into the mountain regolith, its chemical load increased seven and a half times. Almost immediately, the content of silica increased nearly a hundred times. This indicates that dissolution can begin almost as soon as water comes into contact with the minerals of the regolith and bedrock. If the water remains in the ground for several months, its chemical load may double again. The removal of minerals by their dissolution in water is a process named *leaching*.

**7.11 Clay-mineral crystals** (kaolinite) as seen using an electron microscope. (K. M. Towe, Smithsonian Institution)

**Hydrolysis** An important process in the weathering of silicate minerals is their combination with water, or *hydrolysis.* In this process water attacks the crystal lattices of silicates. Some of the $OH^-$ ions from the water become parts of the crystal lattices of the weathered silicates to form clay minerals (Figure 7.11). Clay minerals are relatively stable solids under all but the most humid tropical conditions. Therefore, these minerals are nearly universal ingredients of regolith; they are the chief secondary-alteration product of weathering. Feldspars weather so readily into clay minerals that only rarely do feldspars appear as survival products in the regolith. When they do, we can be certain that the mechanical breaking of the feldspars loose from the bedrock took place faster than chemical weathering could convert them into clay minerals. During the weathering of feldspars to clay minerals, sodium, calcium, and potassium ions are leached from the feldspar lattices and go into solution. In many cases, the potassium is taken up immediately by any plant rootlets; the ions of sodium and calcium are transported in solution to the sea.

**Carbonation** Carbon dioxide gas, which is present in the atmosphere and is a byproduct of the oxidation of organic matter, readily dissolves in water to form a weak acid, carbonic acid. The chemical statement for this reaction is:

$$\underset{\text{(carbon dioxide gas)}}{CO_2} + \underset{\text{(water)}}{H_2O} \rightleftharpoons \underset{\text{(carbonic acid)}}{H_2CO_3}$$

The reaction between carbonic acid and minerals is *carbonation*. Carbonate minerals, especially calcite, dissolve easily in carbonic acid. The chemical statement for the carbonation of calcite is:

$$\underset{\text{(calcite)}}{CaCO_3} + \underset{\text{(carbonic acid)}}{H_2CO_3} \rightarrow \underset{\text{(calcium ion)}}{Ca^{2+}} + \underset{\text{(bicarbonate ion complexes)}}{2(HCO_3)^-}$$

Because calcite is not only the chief ingredient of many carbonate rocks, but also a common cement in numerous sandstones, many kinds of rocks are greatly affected by carbonation.

**Exchange reactions** An *exchange reaction* is the exchange of positive ions between compounds resulting in the formation of new compounds. For example, sulfuric acid and calcite in the presence of water exchange ions to produce gypsum and carbonic acid:

$$\underset{\text{(water)}}{H_2O} + \underset{\text{(sulfuric acid)}}{H_2SO_4} + \underset{\text{(calcite)}}{CaCO_3} \rightarrow \underset{\text{(gypsum)}}{CaSO_4 \cdot H_2O} + \underset{\text{(carbonic acid)}}{H_2CO_3}$$

Because carbonic acid forms in this exchange reaction, further destruction of calcite by carbonation may take place. These two reactions may explain a strange phenomenon at the Lincoln Memorial in Washington, D.C. Visitors on rainy days have reported hearing a fizzing sound near the base of the monument. Presumably, they are listening to calcite effervescing in dilute acid. Sulfur dioxide from automobile exhaust fumes oxidizes in the atmosphere and reacts with rainwater to form sulfuric acid. The acid reacts with the calcite of the marble monument to form gypsum and carbonic acid, and the carbonic acid, in turn, destroys more calcite, releasing carbon dioxide gas. The gypsum is soluble in water, so it, too, can be dissolved easily. Unfortunately, this kind of reaction is happening with greater frequency around the world. Intensive programs are underway in an effort to preserve the building stones of precious structures before the stones are dissolved.

### Complex Weathering

Under this generalized heading we discuss the growth of crystals and the combined effects of chemical and physical weathering.

(a)

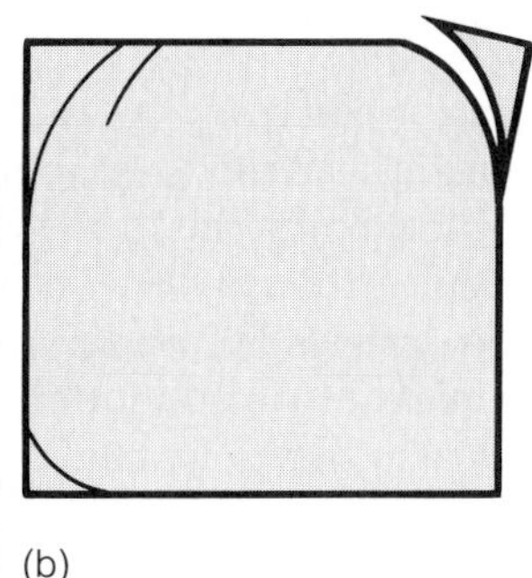
(b)

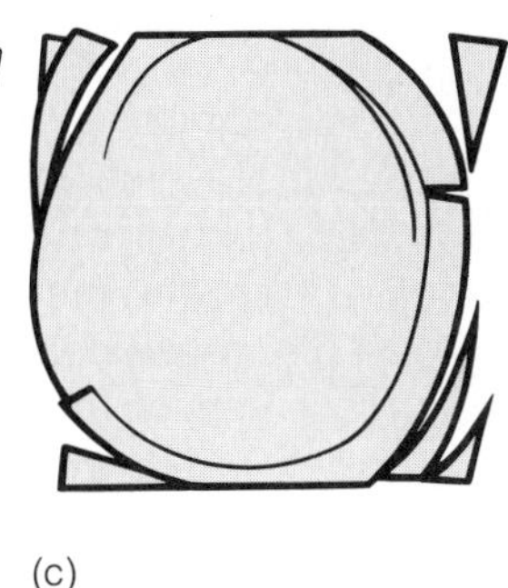
(c)

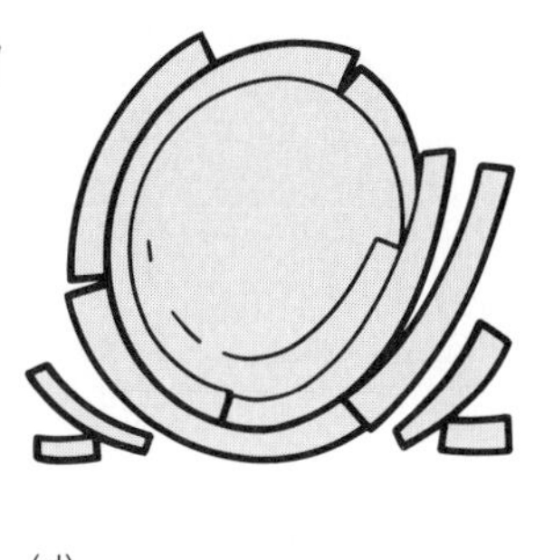
(d)

**7.12 Rounding of corners and edges of bedrock.**
(a) Edges between intersecting joints have become rounded, S wall of Cathedral Peak, Sierra Nevada, California. (Agi Castelli) (b through d) Cubes of rock become spheres by attack at corners and edges.
(b) Effects appear only at corners.
(c) Weathering has begun to affect sides as well as corners.
(d) Rock splits apart along concentrically curved shells, leaving a sphere within.

**Growth of crystals** As we have seen, the water circulating in the regolith contains many elements in solution. If this water evaporates, its dissolved materials crystallize as solids. These crystal lattices grow and exert large forces on their surroundings. Where minerals crystallize within the pore spaces of sediment, they cement the sediment into sedimentary rock (Chapter 8). Where the minerals crystallize along cracks, they pry rocks apart just as ice does.

The two common minerals whose crystal growth can wedge apart rocks are gypsum and halite. Gypsum, a hydrated calcium sulfate, is particularly troublesome where limestone or marble is exposed to sulfur oxides. As previously explained, sulfuric acid reacts with the calcite in the limestone and marble to form gypsum. As gypsum crystals enlarge in the cracks, they cause pieces of solid rock to flake off from buildings and monuments.

Halite, ordinary rock salt, crystallizes from the salty underground water in deserts. Intense evaporation pulls salt-bearing water of porous rocks upward to the surface where halite is precipitated. The growing halite crystals help make the rock crumble. In Egypt halite crystals have forced plaster away from the walls of buildings in Cairo and have severely weathered the temples at Luxor and Karnak. In London the Houses of Parliament have been damaged by crystallization of magnesium sulfate.

**Combined effects of chemical weathering and physical weathering** As feldspars are converted to clay minerals by hydrolysis, the mineral fabric of the rock swells. Such swelling may be strong enough to break the rock.

If a cube of rock is undergoing hydrolysis and breaking, the effects appear first at the corners, where the three surfaces join. The next place to show the effects is along the edges, where two planes meet. The last places affected are the sides, where only one plane is exposed. As the corners and edges are attacked and break more rapidly than the plane sides, the original cube becomes more and more spherical (Figure 7.12). The weathering then proceeds inward evenly from all directions. The end result is that a series of more-or-less regular sheets or shells peel off like the layers of an onion. These concentric partings have been named *spheroidal joints* (Figure 7.13).

**7.13 Spheroidal joints** in weathered mafic volcanic rock, Kanai, Hawaii. (American Museum of Natural History)

**7.14 Tree roots growing along joints** helped wedge apart the large block of rock that fell away in 1974, exposing the roots to view. In 1978, this tree was still alive. North side of Connecticut Turnpike, W of overpass for Todds Hill Road, Branford, Connecticut. Bedrock is a volcanic breccia of Late Triassic-Early Jurassic age. (Joanne Bourgeois, University of Washington, 1974)

### Effects of Organisms

The weathering created by organisms is both chemical and physical. Lichens and bacteria break down organic material and this liberates acids which attack silicate minerals. The growth of lichens on exposed bedrock is the first stage in the conversion of that bedrock into regolith. Other organic material affects the acidity of the waters and thus greatly influences the chemical environment of weathering.

Earthworms turn over vast amounts of soil and many animals burrow into the regolith. The turning over and burrowing bring rock particles to the surface where they can be weathered and admit gases and water into the regolith were these substances might not otherwise penetrate.

Plant roots follow fractures in rocks; they grow where they can find water (Figure 7.14). Growing roots widen the cracks, just as they cause considerable damage to sidewalks, walls, and foundations.

## WEATHERING OF SOME COMMON ROCKS

We can bring together our discussion of weathering by reviewing how a few common rocks respond to weathering. We include granite, mafic igneous rocks, limestone, and sandstone.

### Granite

The chief minerals of granite are feldspars and quartz. The minor (or accessory) minerals include biotite, zircon, ilmenite, and magnetite. In a humid climate, the feldspars weather by hydrolysis and become clay minerals. Much or all of their potassium, sodium, and calcium is removed as ions in solution. The quartz may not be much changed, but simply loosened. Thus, quartz tends to accumulate as a regolith of sand-size particles. Along with the quartz in this sandy regolith are the resistant minor minerals zircon, ilmenite, and magnetite. In the humid tropics, the magnetite may be oxidized to hematite. Any biotite present may be altered to chlorite in a cool climate, and to illite in warmer climates. In some tropical locations, chemical weathering dissolves quartz.

### Mafic Igneous Rocks

The chief minerals in mafic igneous rocks are plagioclase and pyroxene; the chief minor mineral is magnetite. Chemical weathering can destroy all of these minerals. Both plagioclase and pyroxene undergo hydrolysis and are leached. The result is clay minerals and ions in solution (chiefly calcium from the plagioclase and magnesium and calcium from the pyroxene). Oxidation changes the iron in the pyroxene and magnetite to hematite and limonite, which are not very soluble.

### Carbonate Rocks

The chief minerals of carbonate rocks (limestones and marbles) are calcite and dolomite. Other minerals include quartz, various other forms of silica ($SiO_2$), and pyrite. In moist regions, calcite readily undergoes carbonation, and the calcium is removed in solution. The pyrite is oxidized. The iron from the pyrite forms limonite and the sulfur may become sulfuric acid that can react with other calcite to create gypsum. The quartz and other silicate materials accumulate as a sandy residue.

In moist climates, the areas underlain by carbonate rocks are lowered rapidly and become valleys and lowlands. By contrast, in arid regions, carbonate rocks are among the rocks that are most resistant to weathering. In dry climates, carbonate rocks cap the highest mountain peaks.

### Sandstone

Because sandstone consists mostly of minerals that have already survived one or more cycles of weathering, not many more changes are likely to happen when sandstone is weathered. The chief mineral in sandstone is quartz. Others may include feldspar (which escaped destruction during a previous cycle of weathering) and various cements, such as calcite, iron oxides, pyrite, and even nondetrital quartz. The chief effect of chemical weathering on sandstone involves the cement. If the cement is calcite, the effects parallel those just described for carbonate rocks. In moist climates, the calcite is destroyed and the quartz particles form a sandy residue. In dry climates, the calcite persists and the rock is resistant. Cements composed of iron oxides and quartz are resistant nearly everywhere. Pyrite is oxidized, with results as explained previously.

## MISCELLANEOUS EFFECTS OF WEATHERING

Under this heading, we include the rates of weathering, the engineering aspects of residual regolith, and weathering reactions and deposits of economically valuable materials.

### Rates of Weathering

The *rates of weathering* depend upon the interplay among three factors: (1) the ease with which various minerals react; (2) the amount of surface area initially available; and (3) the intensity of the processes of weathering.

Comparative study allows us to determine how rock-making minerals react during weathering. Such studies indicate that minerals weather in the reverse order of their crystallization from the magma (Table 7.1).

Tephra deposited in warm, moist regions exemplify the combination of conditions that create the fastest weathering: susceptible minerals, large initial surface areas, and environment of maximum chemical activity. For example, in only 50 years the tephra that were ejected during the great explosion of Krakatoa in 1883 had lost about 5 percent of their initial silica, had formed clays, and had made a soil more than 2.5 centimeters thick.

The slow end of the weathering scale is found in dry regions. In polar areas having slight precipitation, many exposed rock surfaces remain nonweathered for thousands of years. Similarly, in desert regions such as the Nile Valley and the southwestern United States, inscriptions carved on rock faces thousands of years ago are still legible.

The rate of weathering can be different for different constituents of rock material. The weathering of a rock containing several minerals

**TABLE 7.1**
Order of weathering of minerals compared with order of crystallization of minerals in igneous rocks. (Source: S.S. Goldich, A study of rock weathering: Journal of Geology, v. 46, no. 1, p. 17–58.)

| | Order of crystallization in igneous rocks | | | Order of stability in weathering | | |
|---|---|---|---|---|---|---|
| Early ↓ | (High Temperature) | Olivine, | Ca-rich Plagioclase | Olivine, | Ca-rich Plagioclase | (Least Stable) ↓ |
| | | Pyroxene | ↓ | Pyroxene | ↓ | |
| | | Hornblende | | Hornblende | | |
| | | Biotite | Na-rich Plagioclase | Biotite | Na-rich Plagioclase | |
| | | Muscovite | | Muscovite | | |
| | | K-Feldspar | | K-Feldspar | | |
| Late | (Lower Temperature) | Quartz | | Quartz[1] | | (Most Stable) |

[1]Quartz is stable except under conditions of extreme tropical weathering, where it typically is dissolved.
Note: Many accessory minerals of igneous rocks that crystallize early, are very resistant to weathering. Examples include zircon, monazite, ilmenite, and magnetite. However, magnetite, as quartz, is destroyed under extreme conditions of tropical weathering.

commonly creates a pitted or otherwise irregular surface because some minerals are removed more rapidly than others. The uneven wearing away of rock materials is *differential weathering*.

## Engineering Aspects of Residual Regolith

Any major engineering project must take into account the effects of weathering on the bedrock of the site. Effects of weathering have been observed to depths of 100 meters in Nigeria, Uganda, and Czechoslovakia. Construction excavations and borings for a hydroelectric project in Australia revealed weathered granitic rock at depths of 170 meters, weathered schist at 180 meters, and weathered gneiss at 350 meters below the surface. Russian scientists have reported examples where weathering has reached a depth of more than 1200 meters. These thick zones of regolith and of "rotten rock" created by weathering have to be avoided in building dams and other heavy structures.

Another aspect of weathering that affects engineering projects is enlarged cracks and dissolved cavities in carbonate rocks (Chapter 10). When the Tennessee Valley Authority (TVA) was building dams in northeast Tennessee during the 1930s and 1940s, the excavations revealed that the carbonate rocks had been extensively dissolved during an episode of ancient weathering.

## Weathering Reactions and Deposits of Economically Valuable Materials

Weathering reactions involve two general kinds of processes that may be important in the origin of deposits having economic value: (1) the rearrangements of ions and thus the concentration of materials and (2) the formation of additional pore space, which may become the site where valuable materials accumulate.

**Ores concentrated by weathering** In the environment of weathering three kinds of processes can concentrate ore minerals: (1) residual concentration, (2) mechanical release of resistant minerals, and (3) secondary enrichment. The first two usually operate together.

*Residual concentration* is the removal in solution of soluble ions from rock material which leaves behind a material in which nonsoluble products are more concentrated. In the humid tropics, under conditions that cause extensive leaching not only of the ferromagnesian silicate minerals and feldspars but also even of the quartz, the residual regolith typically consists of iron oxides, bauxite, or kaolinite (a pure clay).

Scattered within a thick residual regolith thus formed may be resistant minerals that have been *mechanically released from the bedrock.* Such resistant minerals may include platinum, gold, monazite (a rare thorium-bearing phosphate mineral found in granites), cassiterite (an oxide of tin, also found associated with granites), and ilmenite (an iron oxide containing titanium). The downslope transport of these materials may concentrate them to the degree that they can be mined (Chapter 11). Still later, they may be concentrated in streams (Chapter 12) or on beaches (Chapter 15).

Most of the world's bauxite deposits that are mined to make aluminum have formed by residual concentration. Examples include the extensive bauxites on the island of Jamaica and those in the Deccan region of India. The Indian bauxite, up to 60 meters thick, has formed by the tropical weathering of dolerites that are about 60 million years old.

Many deposits of iron ores have been formed by the process of residual concentration. A famous example is the earthy hematite ores of the Mesabi district in Minnesota (Figure 7.15). Here, the initial material of Precambrian age, perhaps 2400 million years old, consists largely of silica and iron minerals, chiefly magnetite and some iron-bearing silicates, that form a hard, tough rock locally named taconite. Before the Upper Cambrian strata were deposited, about 530 million years ago, the taconite was subjected to a long period of intense tropical weathering. As a result, the magnetite and iron silicates were oxidized to hematite, and much of the silica was dissolved and removed. What was left consists largely of earthy iron oxide. Most of this hematite has now been mined. As the hematite was nearing depletion, a process was invented for crushing the taconite and for concentrating the magnetite. The process is named the taconite process; it yields spherical pellets that are shipped to the furnaces and can be smelted more efficiently than could the hematite.

*Secondary enrichment* is the downward migration of ions during weathering (Figure 7.16). It is best known for its effects on copper deposits. In many districts where copper is mined an initial deposit composed of copper sulfides, many of which contain iron, was weathered. When such sulfides are weathered, the iron forms limonite by oxidation and the sulfur is oxidized to form sulfuric acid. The sulfuric acid easily dissolves the copper from the weathered zone, takes it into solution, and transports the copper downward. In a lower zone, beneath the level where water is present in all the pores of the rock, the copper comes out of solution and is deposited. There, its concentration may reach high levels.

**Deposits left in pore spaces created by weathering** A typical result of the weathering of carbonate rocks is the enlargement of fractures and the

**7.15 Iron ores formed by residual concentration,** Mesabi Range, Minnesota. (a) Location map of Lake Superior and vicinity. (b) Intense weathering formed soft, earthy hematite in deep pockets along the exposed edges of the iron formations of Precambrian age. (Modified from Stephen Royce, 1942, Iron ranges of the Lake Superior district, p. 54–63 *in* W. H. Newhouse, *ed.*, Ore deposits as related to structural features: Princeton, New Jersey, Princeton University Press, 280 p., Figure 9, p. 61)

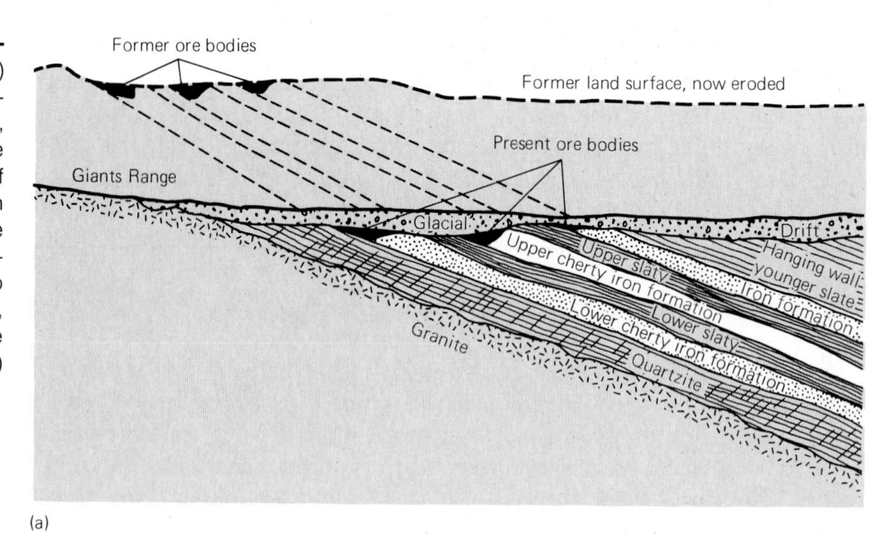


(a)

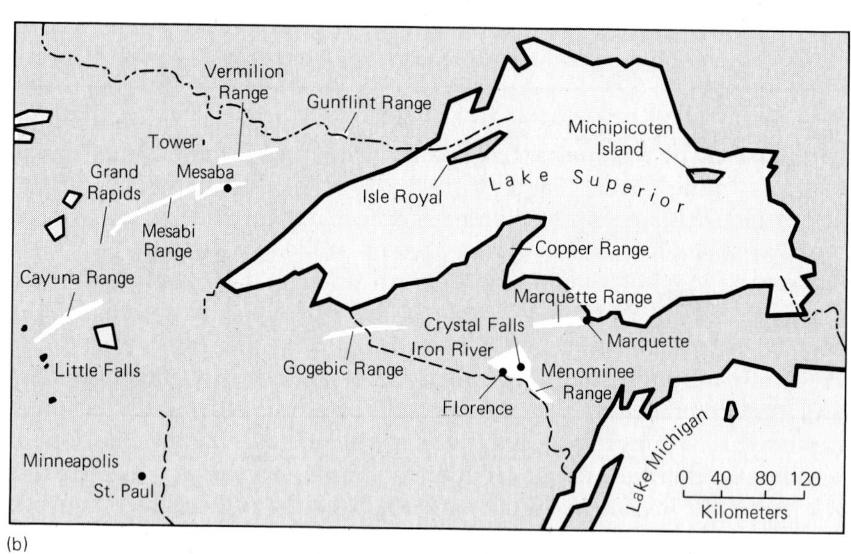


(b)

formation of various cavities in which valuable deposits can accumulate (we will refer to these features again in Chapter 10).

Many petroleum deposits in limestones occupy just such pore spaces formed during weathering. These are not the kinds of spaces in which petroleum is usually found; it is most commonly found filling the tiny spaces around sand grains. Nevertheless, many petroleum deposits in Texas, Kansas, Oklahoma, and elsewhere have filled open cracks and other large cavities that were weathered out of limestones. Afterward,

**7.16 Secondary enrichment of copper minerals** in a mineral-bearing vein cutting across granitic rock, schematic profile and section. Nonweathered vein (below) is overlain by enriched zone where downward-circulating water deposited minerals composed of metal ions leached from higher parts of the vein. At top is a *gossan*, a yellowish-brown to rust-colored, irregular mass of iron oxides formed by the weathering of sulfide minerals typical those containing both copper and iron. (After Alan M. Bateman, 1950, Economic mineral deposits, 2nd ed.: New York and London, John Wiley and Sons, 916 p., Figure 5.8-1, p. 245) the limestone was covered by clay or other material through which petroleum can not flow. Still later, when the petroleum migrated, it filled the spaces created by weathering but was trapped by the covering material. In all such deposits, the time of migration of the petroleum clearly came after the weathering, the formation of the pore spaces, and the deposition of the covering layer (Figure 7.17).

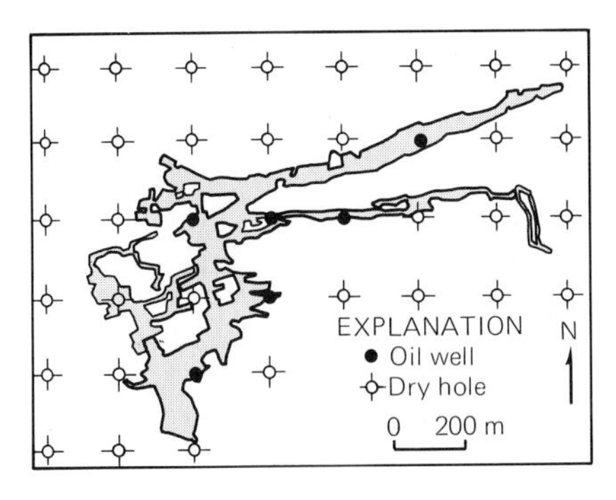


**7.17 Petroleum deposit in cavities created by weathering,** hypothetical example of Carlsbad Caverns, New Mexico. Of the 39 holes drilled on a standard oil-field grid pattern, only 6 would find the oil filling the cavern and 33 would be dry (assuming no other kinds of reservoirs were found). If Carlsbad Caverns were to be filled with oil, as suggested here, it would hold about 110 million barrels. (After H. P. Bybee, 1938, Possible nature of limestone reservoirs in the Permian basin: American Association of Petroleum Geologists, Bulletin, v. 22, no. 7, p. 915–918, Figure 1, p. 917)

## SOILS

As we have said before, weathering is a process that is essential to life as we know it on Earth. Without weathering, the surface of the Earth would be rocky, barren, and lifeless — much like that of the Moon, Mars, Mercury, or Venus (Chapter 20). Neither plant nor animal communities could exist if the Earth's bedrock were not converted into regolith and the regolith, in turn, converted into soil.

### Definition of Soil

*Soil* is defined as the upper part of the regolith that is capable of supporting the growth of rooted plants. Soil consists of solids, liquids, and gases, all of which must be in proper balance to support plant growth. Soil is a dynamic product; it is alive with complex chemical, physical, and biologic activities. We emphasize soil dynamics by discussing the factors involved in the origin of soils.

### Factors Involved in the Origin of Soil

Soils can best be understood by considering interactions among four factors: (1) parent material (kinds of regolith); (2) behavior of water; (3) organic material, such as bacteria, humus, and organic acids; and (4) time. These factors work together to form distinctive zones in the profile of a soil.

**Parent material: kinds of regolith** The parent material of soil is regolith. An important contribution of regolith is to provide mineral nutrients, shelter, and moisture for plant growth. We have discussed regolith repeatedly, but before we try to go any further in relating soil to regolith, we need to emphasize that regolith can be divided into two major categories: (1) *residual regolith,* which is formed as a result of thorough weathering of the underlying bedrock, and (2) *transported regolith,* which consists of sediment that has been transported from someplace else and may be altogether unlike the underlying bedrock. Whether similar to the underlying bedrock or not, a transported regolith is still ready to become a soil or was perhaps derived from another soil.

Important regolith-forming processes other than weathering include: (1) volcanic explosions, which create tephra, a kind of "instant regolith"; (2) downslope gravity transport of particles that break loose from bedrock (Chapter 11); and (3) glacial grinding of bedrock (Chapter 13).

That transported regolith has a headstart in forming a soil is evident from the sediment blown by the wind or deposited during floods. Such sediments may blanket nonweathered bedrock and enable plants to grow long before the bedrock can be weathered. Therefore, although we have mentioned the importance of the weathering of bedrock into regolith as fundamental to the geologic cycle, when we analyze the parent materials of soil, it is likely that many more soils important to agriculture have formed on transported regolith than on residual regolith.

**Behavior of water in regolith and soil** We have mentioned how important the distribution of moisture is in the environment of weathering. Another effect of moisture in the formation of soils is determined by the dominant direction of movement of water through the soil and the regolith. In the humid tropics, rainfall is so abundant and so frequent that the predominant movement of water is downward through the soil and the regolith. By contrast, in dry climates, where the rain falls infrequently and the air is dry between rains, the water moves downward through the soil and regolith only now and then; the water's predominant direction of movement is upward. Water is drawn upward and evaporates at the surface. Because water transports materials in the soil not only in solution but also by physical means, the direction of transport of the water is a major factor in determining the long-term development of a soil.

**Organic material: bacteria, humus, organic acids** The organic material of soil that affects its long-term development includes bacteria, humus, and organic acids. Bacterial action may clear the forest floor of detritus, as in the humid tropics, or such action may through partial decomposition lead to the development of *humus,* the dark-colored surface layer of soils in regions where trees shed leaves seasonally. Rainwater that has moved through the humus becomes charged with organic acids and thus becomes an especially effective leaching agent.

**Time** The reactions involved in the development of a soil do not happen all at once, but take place in a progressive sequence. Thus, time is a factor to be considered in discussing the relationship of soil to regolith. The time required for a soil to develop varies greatly and depends to a large degree on the kind of regolith. Only a few years are required for a soil to form from a layer of mafic tephra in tropical climates. If the regolith itself is residual and is being developed slowly by decomposition of the underlying bedrock, then the time required to form a mature soil may be thousands of years. By contrast, the silts deposited by the yearly floods of some rivers, such as the Nile, are "instant soils" that can be planted as soon as the flood waters dry off.

### Soil Zones and Soil Profiles

The conversion of the upper part of the regolith into soil creates layers. This happens whether the regolith is residual or transported. These layers are characterized by distinctive textures, colors, and compositions. The major layers are named *soil zones.* In the literature of soils, these zones are named *horizons.* We do not use that term, because in geology, horizon means a surface having zero thickness.

A mature soil is one that displays two distinctive zones, collectively named the solum. The two chief zones of the solum are designated by the capital letters A (for the upper) and B (for the lower). If subzones are recognized, they are numbered from the top downward with a separate set for each lettered zone ($A_1$, $A_2$; $B_1$, $B_2$; etc.). The solum overlies the parent regolith, or subsoil, which is designated zone C (Figure 7.18). The regolith of zone C may be residual or transported. If Zone C consists of residual regolith, then the lower boundary of this zone is extended downward to include the bedrock. Zone R designates the bedrock, whether or not it is related to the overlying regolith. Zone O refers to an upper organic zone, rich in decaying organic matter, that is present under conditions of abundant plant growth and not-too-vigorous bacterial activity.

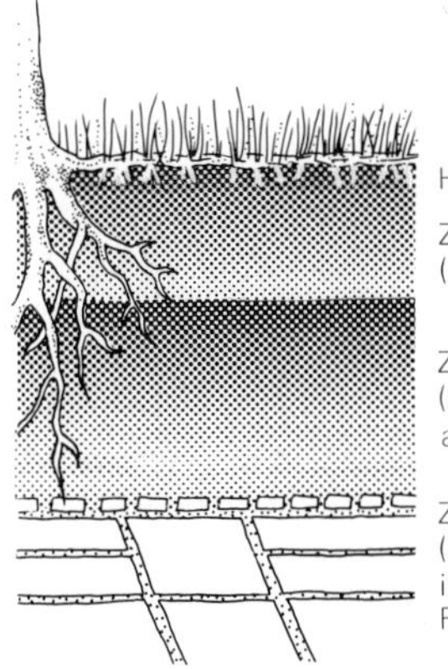


**7.18 Zonal soil formed on residual regolith** in a climate zone where trees can grow, schematic profile and section. Further explanation in text.

### Examples of Modern Soils

**Latosols** The reddish soils of the humid tropics are *latosols*. Because of intense bacterial activity, latosols typically lack humus. They contain abundant iron oxides and aluminum oxides and have been leached of nearly all other metallic ions and locally, even of quartz. Under conditions of extreme leaching, iron oxides and aluminum oxides may be so concentrated that they become valuable residual deposits of iron or aluminum. Regolith rich in aluminum oxide is *bauxite;* regolith rich in iron oxide is *laterite*. Lateritic regolith may become so hard that it can be used as a building material. The ancient temples of Angkor Wat, in Cambodia, have been constructed from carved laterite.

**Pedalfers** In humid-temperate regions, one encounters *pedalfer* soils. The name comes from *pedo,* Greek for "soil," and the chemical symbols for aluminum (Al) and iron (Fe), the two most characteristic components. (Soil scientists do not seem to be troubled by the fact that iron oxides and aluminum oxides are also the main ingredients of latosols; one would never guess this from the name.) As in the latosols, the predominant movement of water in pedalfer soils is downward. The downward movement (1) leaches away calcium carbonate and other easily soluble materials from zone A and (2) transports clay formed by decomposition of feldspars downward from zone A and deposits those clays in zone B. As a result, the A zone of pedalfers tends to be sandy and the B zone, clayey. Pedalfer soils predominate in the eastern half of the United States, where rainfall exceeds about 60 centimeters per year (Figure 7.19). Based on vegetation, pedalfer soils are divided into three varieties: (1) conifer soils, (2) leafy-tree soils, and (3) prairie (or grassland) soils.

**Pedocals** Where evaporation is intense, the balance of water movement in the soil shifts so that any soluble constituents that start downward can be drawn back upward again and, when the interstitial water has evaporated, can be precipitated at or near the surface of the soil. As a result, calcium, though soluble, is not removed from zone A, but tends to accumulate there. This calcium in the form of calcite is the basis for naming these soils *pedocals* from the Greek for "soil" and cal for calcium). In the conterminous United States pedocal soils are located in the western part, where annual rainfall is less than 60 centimeters (see Figure 7.19). As with pedalfers, we can use vegetation present to divide pedocals into three varieties: (1) grassland soils, (2) forest soils, and (3) desert soils (soils of sparse vegetation).

In soils where upward movement of evaporating interstitial water is intense, evaporite minerals are deposited within the soil and at its

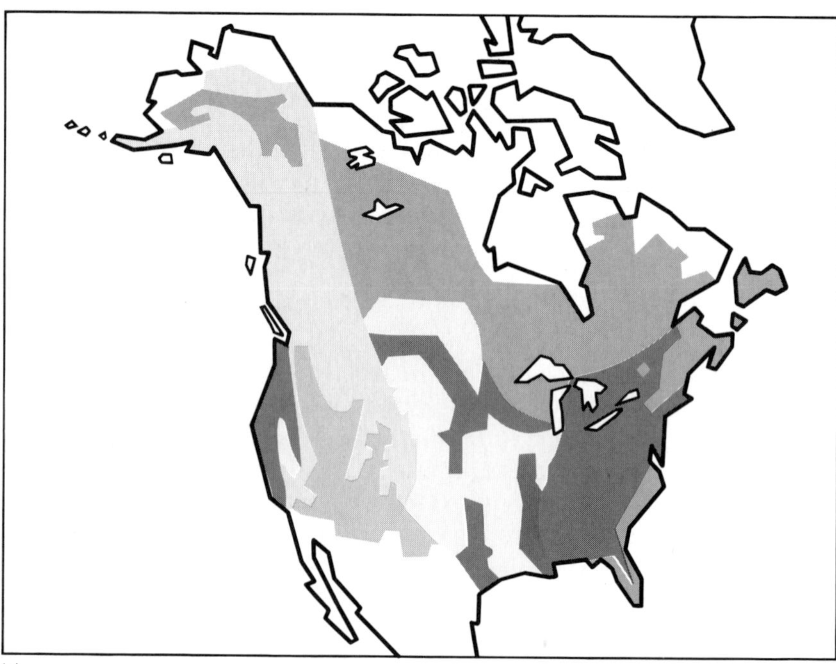

EXPLANATION

- Tundra
- Mountains; soils variable
- Pedalfers (conifers)
- Pedalfers (leafy trees)
- Alluvial soils
- Pedocals (grasslands)
- Desert soils
- Others (includes sand dunes, salt marshes, old lake beds)

(a)

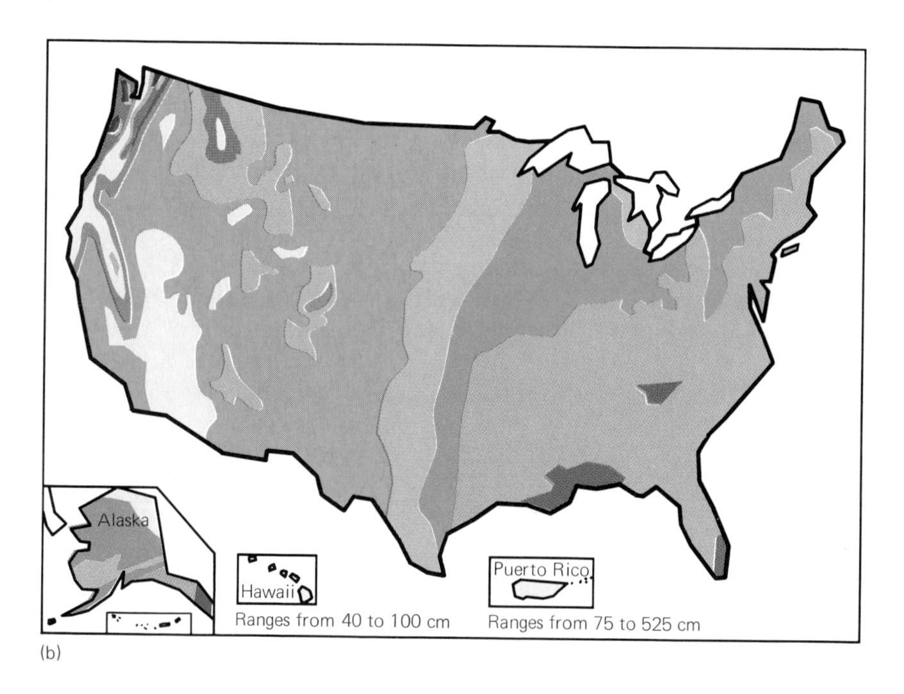


EXPLANATION

Average annual precipitation

- 0-15
- 25-50
- 50-75
- 75-100
- 100-125
- 150-250
- > 250

(b)

**7.19 Major soils and rainfall, United States.** (a) Map of major groups of soils. (Redrawn from U. S. Department of Agriculture map) (b) Average annual precipitation, expressed as centimeters of rainfall. (U. S. Water Resources Council, 1968, The nation's water resources: the first national appraisal: Washington, D. C., variously paged, Figure 3-2-1, p. 3-2-2)

surface. Nodules and crusts of calcite known as *caliche* are found in semiarid regions. In still-drier climates the entire surface may become cemented with various evaporite minerals, including silica, dolomite, gypsum, and halite.

**Other soils** A few other specialized soils should be added to our list of examples. *Bog soils* consist of partly decomposed plant matter that becomes peat. Such soils are common in low-lying sites of North America

**7.20 Ancient caliche,** southern Connecticut, south of Mount Carmel. (J. E. Sanders)

and Europe where plants have roofed over small ponds and lakes. *Tundra soils,* widespread in the Arctic, consist of sandy clay and raw humus. Many tundra soils are underlain by permanently frozen ground, or *permafrost.*

### Ancient Soils

Within the geologic record are found buried ancient soils, which have been named *paleosols.* These ancient soils prove that the area in which they are found was part of a land area which was subjected to subaerial weathering and may contain clues about the ancient climate. For example, numerous ancient caliche paleosols have been discovered in deep roadcuts recently excavated in the Triassic-Jurassic strata of southern Connecticut (Figure 7.20). They indicate a dry climate about 185 million years ago. The presence of ancient laterites in the Cretaceous strata of Ireland implies that about 100 million years ago Ireland enjoyed a humid-tropical climate. Paleosols in materials deposited by the advancing and retreating glaciers of the Quaternary ice ages (Chapter 13) prove that the deposits left by the glaciers were not continuously covered by ice, but rather at some times, were exposed and weathered. In many cases the characteristics of the paleosols suggest that while they were forming, the climate was warmer than the present climate in the area.

# CHAPTER REVIEW

1. *Weathering* is a general term that describes all the changes in rocks that take place as a result of their exposure to the atmosphere.
2. Weathering is important both in geologic theory and in many ways that affect our everyday lives. Weathering: (a) offers visible proof that the geologic cycle is operating today; (b) illustrates how geologic materials and certain kinds of energy react to bring the materials into adjustment with the *environment of weathering;* and (c) creates important products including regolith and soil.

3. Neither plant nor animal communities could exist if the Earth's bedrock were not converted into regolith and the regolith converted to soil. The upper part of the regolith that is capable of supporting the growth of rooted plants is *soil.*

4. *Climate* can be defined as the time-averaged weather in a given region. The environment of weathering depends intimately upon climate, particularly upon temperature, precipitation, and evaporation.

5. Rainwater contains many dissolved materials, including sodium chloride; dissolved gases such as nitrogen, oxygen, oxidation products of sulfur dioxide, carbon dioxide, and carbon monoxide; and various organic compounds. The corrosive effects of rainfall have been increased greatly as a result of industrial air pollutants.

6. The environment of weathering is located within the pore spaces of bedrock and regolith. In bedrock this means spaces between individual minerals or in joints, faults, and other cracks and partings in the bodies of rock.

7. The *slope situation* (height and direction of inclination with respect to sunlight and to rain-bearing winds) affects the intensity of weathering.

8. The processes of weathering include *dual effects of water molecules, physical weathering* or disintegration, *chemical weathering* or decomposition, and *complex weathering,* which consists of the growth of crystals, the combined effects of chemical weathering and physical weathering, and effects of organisms.

9. The *rate of weathering* depends upon the ease with which various minerals react, the amount of surface area initially available, and the intensity of the processes of weathering. The uneven wearing away of rock materials is *differential weathering.*

10. Weathering may create thick regolith, zones of "rotten rock," enlarged cracks, or cavities that must be avoided in the location of heavy engineering structures.

11. During weathering, valuable deposits of ore minerals can be formed by (1) residual concentration, (2) mechanical release of resistant minerals, and (3) secondary enrichment. Bauxite and iron ores have formed by residual concentration. Platinum, gold, monazite, cassiterite, and ilmenite are released mechanically during weathering. Many copper deposits have been concentrated by secondary enrichment.

12. Numerous oil deposits have accumulated in the pore spaces formed or enlarged during the weathering of limestones. The oil did not migrate into these pore spaces until after the limestone had been buried by a covering layer that could prevent the oil from escaping.

13. Soils consist of solids, liquids, and gases, all of which must be in proper balance to support plant growth. Soil is alive with complex chemical, physical, and biological activities. The factors of parent material, behavior of water, organic material, and time all work together to form distinctive zones in the profile of the soil.

14. Regolith can be divided into two major categories: *residual regolith,* formed by the deep weathering of the underlying rock, and *transported regolith,* sediment carried from someplace else and possibly unlike the underlying bedrock.

15. The conversion of the upper part of the regolith into soil creates layers, which are characterized by distinctive textures, colors, and compositions. The major layers are named *soil zones,* and are lettered from the top down A, B, and C according to the effects of migration of soluble materials during chemical weathering.

## QUESTIONS

1. Why is weathering important?
2. What is meant by the *environment of weathering?* Discuss the factors that determine the environment of weathering.
3. When "it rains," what kinds of materials are likely to accompany the water?
4. How does air pollution affect weathering? Can you find any examples where you live to illustrate your answer?
5. Explain ways in which water affects weathering. Does the structure of water molecules explain how water behaves in weathering?
6. What are the three chief processes of weathering?
7. Discuss the effects of oxidation on sulfide minerals.
8. Compare the effects of weathering on first-cycle organic matter and on kerogen.
9. Describe the exchange reaction between sulfuric acid and calcite.
10. What is the most important process in the weathering of silicate minerals? When a granite is weathered, what products are formed? What happens to these weathering products in the geologic cycle?
11. Compare the weathering of carbonate rocks in a humid climate with that in a dry climate. What landscape changes might take place after a long time in a region underlain by layers of limestone and of other kinds of rocks if the climate changed from moist to arid?
12. Describe some effects of weathering that have to be

considered in engineering construction projects.

13. Describe three effects of weathering on ore minerals. Compare *residual concentration* with *secondary enrichment.*
14. Does any connection exist between pore spaces created during weathering and accumulation of petroleum? Explain.
15. Define *soil.* How are soils related to *regolith?* How is regolith related to *bedrock?*
16. What is a *soil profile?* List the factors that affect the development of a soil profile.
17. How do plants, climates, and soils interact?
18. Define *pedalfer* and *pedocal.* Describe the conditions under which each forms. In which would one likely find *caliche?*
19. What is the significance of an ancient soil buried in the geologic record?

## RECOMMENDED READING

Berner, R. A., and Holdren, G. A., Jr., 1977, Mechanism of Feldspar Weathering: Some Observational Evidence: *Geology,* v. 5, no. 6, p. 369–372.

Bloom, A. L., 1978, *Geomorphology; a systematic analysis of Late Cenozoic land forms:* Englewood Cliffs, New Jersey, Prentice-Hall, Inc., 324 p.

Bybee, H. P., 1938, Possible Nature of Limestone Reservoirs in the Permian Basin: *American Association of Petroleum Geologists, Bulletin,* v. 22, no. 7, p. 915–918.

Comer, J. B., 1974, Genesis of Jamaican bauxite: *Economic Geology,* v. 69, no. 8, p. 1251–1264.

Gauri, K. L., 1978, The Preservation of Stone: *Scientific American,* v. 238, no. 6, p. 126–137.

Hubert, J. F., 1977, Paleosol Caliche in the New Haven Arkose, Connecticut: Record of Semiaridity in Late Triassic-Early Jurassic time: *Geology,* v. 5, no. 5, p. 302–304.

Hunt, C. B., 1972, *Geology of Soils: Their Evolution, Classification and Uses:* San Francisco, W. H. Freeman and Co., 344 p.

McNeil, M., 1964, Lateritic Soils: *Scientific American,* v. 211, no. 5, p. 96–102.

Simonson, R. W., 1954, Identification and Interpretation of Buried Soils: *American Journal of Science,* v. 252, no. 12, p. 705–732.

Winkler, E. M., and Singer, P. C., 1972, Crystallization Pressure of Salts in Stone and Concrete: *Geological Society of America, Bulletin,* v. 83, no. 11, p. 3509–3514.

# CHAPTER 8 SEDIMENTARY ROCKS, ORIGIN OF FOSSIL FUELS, AND SEDIMENTARY ORE DEPOSITS

SEDIMENTS AND SEDIMENTARY ROCKS IN THE ROCK CYCLE
RECOGNITION OF SEDIMENTARY ROCKS
OCCURRENCE OF SEDIMENTARY ROCKS
GEOLOGIC AGES OF SEDIMENTARY STRATA
BURIAL OF SEDIMENTS: LITHIFICATION AND THE ORIGIN OF FOSSIL FUELS
SEDIMENTARY ORES AND SEDIMENTARY ECONOMIC MINERAL DEPOSITS

PLATE 8
Flat-lying sedimentary strata of Mesozoic age, exposed after great erosion, Canyonlands National Monument, Utah.

SEDIMENTS AND SEDIMENTARY STRATA AFFECT OUR LIVES IN MANY ways. They provide most of our energy sources, including all of the fossil fuels and much of the uranium; they provide many building materials; they yield supplies of iron ore, fertilizer, and numerous other substances; they are the chief sources of water that comes from wells; and they contribute to soils that support plant life.

In this chapter we will focus on sediments and sedimentary rocks in the rock cycle, with emphasis on sediments as products of weathering. We explore the geologic occurrence of sedimentary rocks in strata, the concepts of sedimentary facies and sequences of strata, and diapirs. We review how to recognize sedimentary rocks (further material on classification and identification of sedimentary rocks is found in Appendix C), how to find the geologic ages of sedimentary strata, and the changes that sediments undergo when they are buried. We close with a section devoted to sedimentary ores and economic mineral deposits of sedimentary origin.

## SEDIMENTS AND SEDIMENTARY ROCKS IN THE ROCK CYCLE

Turn back to page 148 and look closely at Figure 6.1, the rock-cycle diagram. We have already discussed igneous rocks and some aspects of weathering, which helps convert bedrock into regolith and makes changes in the regolith. But, we have not yet explored the entire left side of the rock-cycle diagram, which shows sediments becoming sedimentary rocks and sedimentary rocks becoming metamorphic rocks. Sediment is broadly defined as regolith that has been transported at the surface of the Earth and deposited in low places as layers (strata).

James Hutton recognized that sedimentary rocks are formed by the "turning to stone" of sediments and that sediments, in turn, are formed by the breakdown of yet-older rocks. These two relationships are critical to the concept of both the whole geologic cycle and the rock cycle.

### Sedimentary Materials as Weathering Products both Physical and Chemical

Bedrock may become sediment without being weathered under special conditions such as an avalanche (Chapter 11) or grinding by a glacier (Chapter 13). Such nonweathered materials can be visualized as "escaped-weathering" products.

For the most part, however, weathering is the essential step that precedes the formation of sediments. Accordingly, we can subdivide sediments into two groups based on weathering products: *solid products* and *dissolved ions.*

The solid products of weathering include: (1) rock fragments; (2) resistant minerals, such as quartz, which survive weathering; and (3) secondary-alteration products, chiefly clay minerals, created during the decomposition of feldspars and ferromagnesian silicate minerals. The solid products travel and are deposited by mechanical means.

Ions travel in solution and become solids again by the activities of organisms or by crystallization under chemical controls.

**8.1 Rock fragments,** constituents of sediments that still display the characteristics of the parent rocks (in this example, older conglomerates and coarse-grained granitic rocks). Quaternary deposits, north shore of Long Island. (J. E. Sanders)

## Description of Sedimentary Materials

We can classify sedimentary materials into four major categories according to their origins: (1) detritus, (2) solids derived from ions carried in solution: (3) organic material; and (4) volcanic material.

**Detritus** Many of the particles derived from the breaking up of bedrock are transported and deposited physically as solids of varying sizes. All these particles are collectively designated *detritus.* A synonym for detritus is *clastic* sediment; clastic means broken, and the term is applied to any sediment whose particles have been broken. This breaking can be from the original bedrock or as a result of the processes of sedimentation.

Detritus is easy to recognize. It consists of one or both of two kinds of materials: (1) rock fragments, that is, particles which are large enough to retain recognizable minerals and textural characteristics of the original bedrock (Figure 8.1), and (2) particles that contain only one mineral; most of these are quartz, and include other rock-forming silicate minerals. Quartz is so abundant that it serves as an "index mineral" for recognizing detritus (Figure 8.2). All the kinds of sedimentary detritus have managed to escape being destroyed by chemical weathering.

A different kind of detritus, also solid and derived from bedrock, includes the secondary-alteration products that form during decomposition. Chief among these are the clay minerals that form by the hydrolysis of feldspars. Because feldspars are so abundant, this type of detritus is more prominent than all the other kinds (Figure 8.3; see also Figure 7.12). Other alteration products include chiefly iron oxides.

**Solids derived from ions carried in solution** The chief ions in solution are calcium, sodium, potassium, and magnesium. Because these ions are completely separated from their bedrock sources, they do not retain any clues about where they came from. Thus, ions that travel in solution to

**8.2 Quartz particles** appear clear and glassy, even in sand-size particles less than 1 millimeter in diameter; view through microscope. (B. M. Shaub)

**8.3 Clay minerals,** viewed using an electron microscope. (Technische en Fysische Dienst voor de Landbouw, Wageningen, The Netherlands)

the basin where they are deposited are not regarded as detritus. The ions may remain in solution or may form solids. The geologic information contained in these new solids relates only to how the ions came out of solution and got back into solid form once again. Usually nothing can be learned about where ions came from.

In the waters of the basin of deposition, the two chief processes for changing dissolved ions back into solids again are: (1) absorption by organisms and secretion as *skeletal material* and (2) crystallization when the water containing the ions evaporates.

Most skeletal material is calcareous (composed of calcium carbonate in the form of calcite or of aragonite), siliceous (composed of silica, $SiO_2$), or phosphatic (composed of calcium-phosphate compounds). Calcareous skeletal material includes shells (Figure 8.4a), bones, plates, spines, and other parts of organisms. All are secreted within the cells of organisms and serve as either external housing or as some kind of internal support.

Siliceous skeleton makers include single-celled marine and nonmarine plants, the *diatoms* (Figure 8.4b), and various single-celled marine organisms, the *Radiolaria,* and more complex marine organisms, the *sponges.* (Some sponges secrete calcareous materials; others secrete complex materials.) Scales from fish, the shells secreted by a few kinds of marine invertebrate organisms, and peculiar jawlike remains known as *conodonts* consist of phosphatic material. The teeth of many organisms are composed of tough, complex material that includes much calcium.

*Evaporites* are minerals that crystallize as the result of the evaporation of the water which contains them. Evaporites consist of well-formed crystals or of small spherical *oöids* (Figure 8.5).

**Organic material** Some sedimentary materials are the work of plants or animals. These are strictly organic in the sense that they consist chiefly of the elements carbon, hydrogen, and oxygen. The particles include the woody tissue of plants, such as tree trunks, branches, twigs,

(a)

(b)

**8.4 Skeletal materials.** (a) Clam shells washed up on a beach, south coast of Wadden Sea, The Netherlands. (J. E. Sanders) (b) Diatom remains, Pliocene, New Zealand. (Smithsonian Institution, U. S. National Museum)

**8.5 Sedimentary materials formed from waters in basin of deposition.** (a) Particles of calcium carbonate that enlarged by addition of concentric coatings. Sedimentary rock, viewed paper-thin slice through polarizing microscope. (b) Tiny halite cubes growing along a twig. (B. C. Schreiber) (J. E. Sanders)

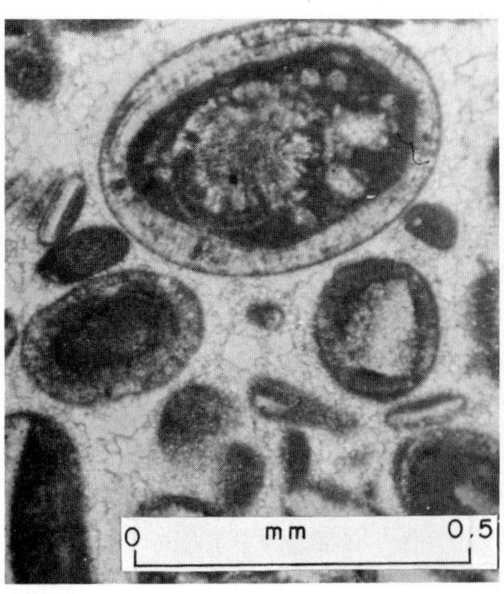


(a)

(b)

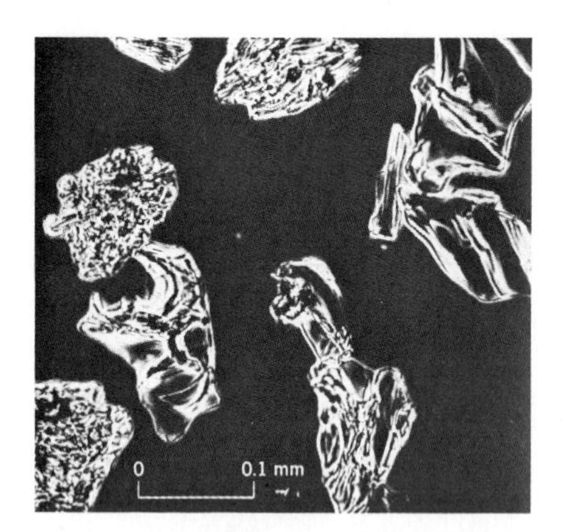


**8.6 Tephra,** illustrated by microscopic view of glass fragments exploded from Mount Mazama, Oregon, and deposited in sediments underlying Creston Bog, Washington. (R. B. Taylor, U. S. Geological Survey)

and leaves. Such plant material forms peat and coal. Other materials derived from plants include spores and pollen. These tiny objects are spread through the atmosphere and are so resistant to chemical alteration that they persist in sediments.

The fatty and waxy parts of woody land plants and the soft parts of aquatic plants, such as algae, and of animals of all kinds are a special category of sedimentary organic material. They are considered to be the ultimate raw material from which petroleum forms underground. In their first stages as sedimentary material, these soft organic products are fragile and are easily oxidized. After they have polymerized, some of them become stable solids (kerogen). We discuss sedimentary organic matter further in a later section of this chapter.

**Volcanic material** The materials ejected from volcanoes consist of tiny bits of glass, crystals, or volcanic rock fragments (Figure 8.6). Volcanic particles join the sedimentary materials in a basin of deposition either directly from the volcano (in the form of tephra) or by being eroded and transported at a later time.

## Sizes and Shapes of Sediments

Particles forming detritus are grouped and named on the basis of a standard size scale, using as diagnostic measurement the "diameter" of the particles. The main groups of sedimentary particles based on size are gravel, sand, silt, and clay (Figure 8.7).

The degree of uniformity of sizes of aggregates of sedimentary particles is known as *sorting*. The most important aspect of sorting is based on what happens to the fine particles, the silt and clay. These particles are readily transported by currents of water or wind. Hence, moving water or air nearly always separates these fine particles from the sand and larger sizes. Accordingly, we will consider that sediments are *well sorted* if the silt and clay have been separated from coarser material and *poorly sorted* if they have not (Figure 8.8).

Size is not the only basis of the sorting of sediment particles; density

**8.7 Sizes of sedimentary particles,** based on standard scale, with comparisons based on familiar objects (listed at top).

Volley ball
Tennis ball
Match head
Limit of visibility with naked eye
Particle diameter (mm)
256
64
2
1/16
1/256
1/4096
Gravel:
Boulders
Cobbles
Pebbles
Sand
Silt
Clay

(a)

(b)

**8.8 Contrasting sorting of sediments.** (J. E. Sanders) (a) Well-sorted sand. (b) Poorly sorted mixture of silt, sand, and pebbles. Diameter of camera lens cap is 53 millimeters.

(or *specific gravity*) is another factor. Dense minerals sink faster than less-dense minerals. Minerals denser than 2.8 grams per cubic centimeter (g/cm$^3$) are named heavy minerals. Such minerals may be concentrated into layers (Figure 8.9). Small-sized heavy minerals, such as magnetite (density of 5.2 g/cm$^3$), tend to be deposited along with larger but less-dense particles of quartz (density of 2.65 g/cm$^3$).

The shapes of sedimentary particles result from the initial shapes of the particles in their parent rocks and from the way the initial shapes are changed during transport. This change results from *abrasion,* the rounding and smoothing of particles as a result of collisions with other particles

**8.9 Layer of heavy minerals** (black), about 7 centimeters thick, concentrated at surface of Westhampton Beach, Long Island, New York, on 17 April 1977. (J. E. Sanders)

(a)

(b)

**8.10 Angular particles and rounded particles.** (a) Angular rock fragments, broken along joints and freshly fallen along gravel road in Yellowstone National Park. (U. S. National Park Service) (b) Rounded boulders in Tower Creek, viewed from bridge at Tower Falls, Yellowstone National Park. (U. S. National Park Service)

(Figure 8.10). The rate of abrasion depends both on the composition and on the size of the particles: Soft rocks, such as limestone (composed of calcite having a mineral hardness of 3), are smoothed quickly. Hard rocks (such as sandstone cemented by quartz having a mineral hardness of 7) become rounded slowly. Large particles become rounded after only a few kilometers of transport in a stream. By contrast, small particles may never become rounded. A single particle of silt, protected from collisions by the viscosity of the water, may move all the way from the Rocky Mountains to the Gulf of Mexico without changing shape.

## RECOGNITION OF SEDIMENTARY ROCKS

Sedimentary rocks can be recognized easily if they contain fossil skeletal debris. Other clues include strata (but other kinds of rocks contain layers so that layers alone do not prove that the rock is sedimentary); clastic texture; and abundance of certain minerals, such as evaporites and carbonates. Many sedimentary rocks display low mineralogic hardness. However, quartz-cemented sandstones and cherts are among the hardest rocks. Nevertheless, a positive effervescence in dilute acid and scratchability with a knife point are good clues to sedimentary rocks.

## OCCURRENCE OF SEDIMENTARY ROCKS

Sedimentary rocks usually occur as *strata*. However, in the case of density contrasts, where low-density material becomes buried by denser material, the low-density material may punch its way upward through the overlying strata as *diapirs*. In our review of the occurrence of sedimentary rocks, therefore, we include both strata and diapirs.

(a)

(b)

**8.11 Contrasting thicknesses of strata** (a) Thick stratum of sandstone extends from bushes along bank of Green River to top of cliff, Dinosaur National Monument, Utah. (Dick Frear, U. S. National Park Service) (b) Millimeter-thin laminae of calcite and gypsum, Castile Formation (Permian) west Texas. (J. E. Sanders, 1960)

## Sedimentary Strata

In presenting sedimentary strata, we emphasize features that are found both within sediments and sedimentary rocks. In fact, these common characteristics were important points involved in the realization that sedimentary rocks are derived from sediments.

Strata are named according to thickness. Strata thicker than one centimeter are usually referred to as *beds;* strata thinner than one centimeter are *laminae* (sing. *lamina*).

Strata exist because something changed. The change may involve color, size of particles, kind of cement, ways the particles are packed, or some other characteristic. In some cases a single stratum may become so thick that one's view of it is not large enough to detect the stratification (Figure 8.11). Typically, strata are distinctly set off from one another along parting surfaces that have been named *bedding-surface* (or bedding-plane) *partings.* The changes that create strata are related to the conditions of deposition. In later chapters, we will be examining some of the strata-making processes.

**Features within strata; sedimentary structures** Within a stratum the sizes of the particles may be uniform throughout or they may show various patterns. One possibility is that a systematic gradation in sizes of particles exists. The usual arrangement is larger particles at the bottom and progressively smaller particles higher up. A bed having such a gradient in the sizes of its particles is a *graded layer* (Figure 8.12). A graded layer forms when a current carrying particles of many sizes drops the

**8.12 Graded layer,** containing pebbles at base and fine sand at top. Pliocene of Ventura Basin, exposed along Santa Paula Creek, California. (J. E. Sanders)

**8.13 Horizontal layers in sand** of Pleistocene age deposited by Snake River when swollen with water draining Lake Bonneville, Utah, the ancestor of modern Great Salt Lake. Union Pacific Railway ballast pit, between King Hill and Glenns Ferry, Elmore County, Idaho. (H. E. Malde, U. S. Geological Survey, 2 September 1962)

**8.14 Bed forms in sand,** created by tidal current flowing from right to left, and exposed at low tide, Five Islands, Nova Scotia. Surveying rod is 12 feet (3.6 meters) long. (J. E. Sanders, July 1965)

largest ones first and, as the speed of the current decreases, deposits successively finer particles.

Many strata accumulate and build sequences that are parallel to the surface on which their sediments were spread out. In most cases this surface is almost perfectly horizontal (Figure 8.13). Some exceptions do occur. For example, the strata deposited on the sloping sides of a tephra cone are not horizontal.

One of the dramatic effects of a current flowing over sand is the creation of various regularly spaced features named *bed forms* (Figure 8.14); dunes are an example (see Chapter 14). The flowing current causes many kinds of bed forms to migrate in a downstream direction. As these bed forms move, they deposit sediment in layers that are inclined downstream.

Similarly, sand may be pushed along the bottom of a current until it encounters a local pocket of deeper water. In such a spot the current will form a growing sand embankment. Particles brought to the edge fall down the inclined front of the embankment, thus building it forward. As a result, successive layers are added at the angle of slope of the front of the embankment (Figure 8.15).

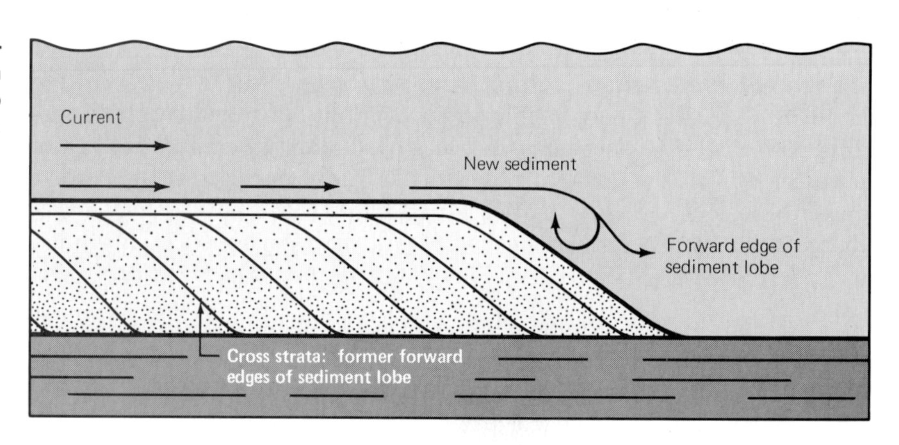


**8.15 Cross strata formed by water current transporting sand,** and forming an embankment that grows forward into deeper water. Schematic sectional view.

(a)

(b)

(c)

**8.16 Cross strata in modern sediments and ancient sedimentary rocks.** (a) Large-scale cross strata formed at front of dune that migrated toward left and built these layers at the angle of repose. Upper Medano Creek, Great Sand Dunes National Monument, Colorado. (Bob Hangen, U. S. National Park Service, 1961) (b) Cross strata of Navajo Sandstone (Jurassic) seen in section perpendicular to stratification, Zion National Park, Utah. (Mark R. Chartrand, III) (c) Small-scale cross strata in ancient carbonate rock (Lower Paleozoic), S of Poughkeepsie, New York. (J. E. Sanders, December 1977)

Eventually, the sloping layers of sediment from migrating bed forms or advancing embankments may become sandwiched between originally horizontal layers where a flat bottom built upward. The inclined layers are named *cross strata;* both recent sands and ancient sandstones and some carbonate rocks composed of sand-size sediment that was moved by currents display cross strata on various scales (Figure 8.16).

Cross strata form because the sediment particles are cohesionless, that is, they move readily as individuals without tending to stick together. Sands and coarser sediments are nearly always cohesionless; fine silts and clays are nearly always cohesive. In between are fine sands and

**8.17 Oversteepened and overturned cross strata,** deformed by drag of current and buried by cross strata having normal angles of inclination. Middle Ordovician, W side of Hudson River opposite Poughkeepsie, New York. (J. E. Sanders)

coarse silts that may be cohesionless at some times and cohesive at other times. Thus, sediment deposited as small-scale cross laminae, indicating a cohesionless condition, may be deformed, indicating cohesion. The deforming force may be the drag of the current. For example, as a result of current drag, the angle of the cross laminae, usually less than 30 degrees, can become oversteepened or even overturned. Such deformation of cross laminae proves two points: (1) the deposited sediments became cohesive and (2) they were dragged by the moving current or by sediment being transported. The deformed cross laminae may be buried by cross laminae having normal angles of inclination (Figure 8.17).

Some bodies of medium sand or coarse sand, perhaps as much as a meter thick, can be converted into massive sand flows (see Chapter 11). Such a sand flow can create a shearing drag on the substrate over which it passes. Where this substrate consists of cross-bedded sand, the cross beds may be folded (Figure 8.18).

Many evaporites become intensely deformed as the deposit accumulated. Such deformation is thought to be the work of changes in volume related to dissolution and to growth of crystals.

**8.18 Folded cross strata enclosed within layers that have not been folded,** Casper Sandstone (Pennsylvanian), SW of Laramie, Wyoming. (R. W. Fairbridge, Columbia University, July 1957)

Structural features formed in sediments during deposition are named *sedimentary structures.*

**Surface features of strata** The surfaces of modern sediments contain many features that are related to the movement of the water, passage of animals, or drying out of the sediment. These include such things as small-scale ripples (one of the many varieties of bed forms), footprints of various walking animals, the tracks and trails of crawling animals, and polygonal cracks made by the shrinkage of fine-grained sediment that has been dried out. Exactly comparable features can be found on the surfaces of strata of many sedimentary rocks (Figure 8.19).

## The Concept of Sedimentary Facies

The term *facies* comes from the Latin word for "aspect." As used in connection with sedimentary strata, facies refers to materials of distinctive appearance characteristic of particular conditions. Presumably, each time these conditions are repeated, products with the identical distinctive appearance would be deposited.

**8.19 Features of bedding surfaces of sediments and sedimentary rocks.** (a) Ripples created by waves in water only a few centimeters deep. (G. M. Friedman) (b) Ripple marks in ancient sandstone, Devonian, Catskills, New York (G. M. Friedman) (c) Cracks formed by the drying out of mud. (F. A. O., United Nations photo by H. Null) (d) Mudcracks on bedding surface of ancient rocks, Manlius Limestone (Devonian), Hudson Valley, New York State. (R. W. Fairbridge) (e) Bird tracks and snail trails in modern sediments. (G. M. Friedman) (f) Filling of dinosaur footprint on bottom of sandstone layer of Jurassic age that has been overturned, Dinosaur State Park, Rocky Hill, Connecticut. (J. E. Sanders)

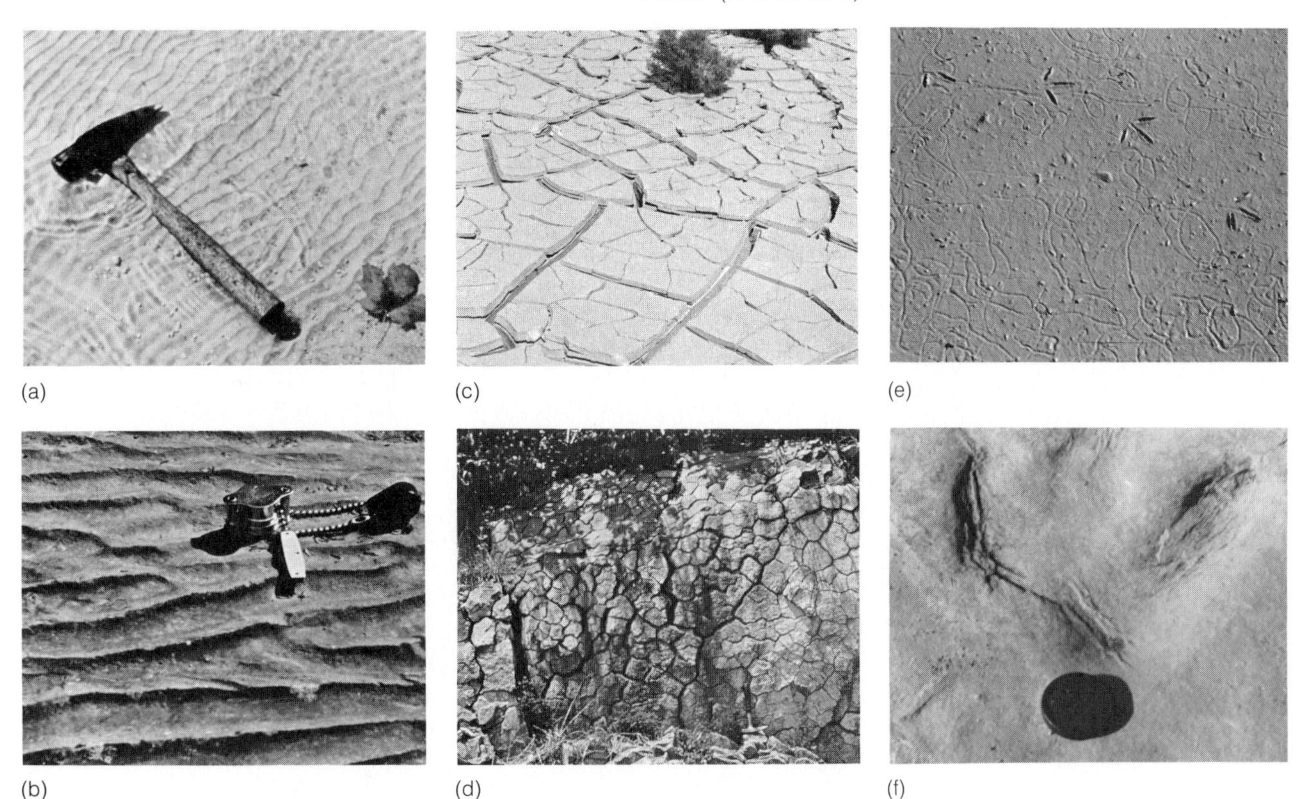

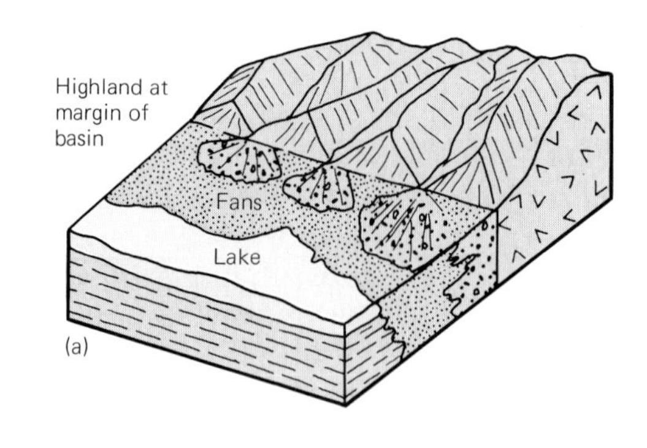


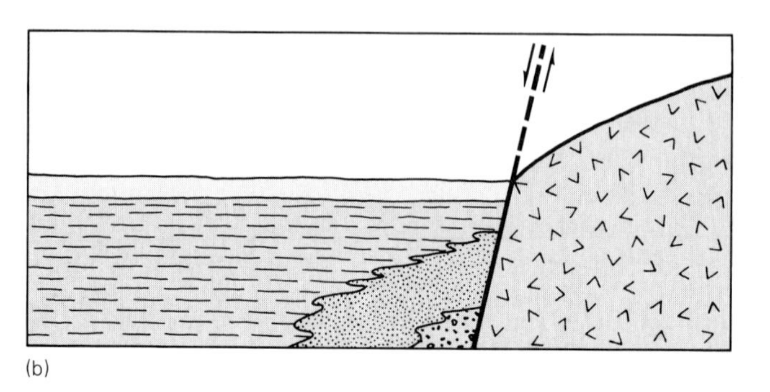


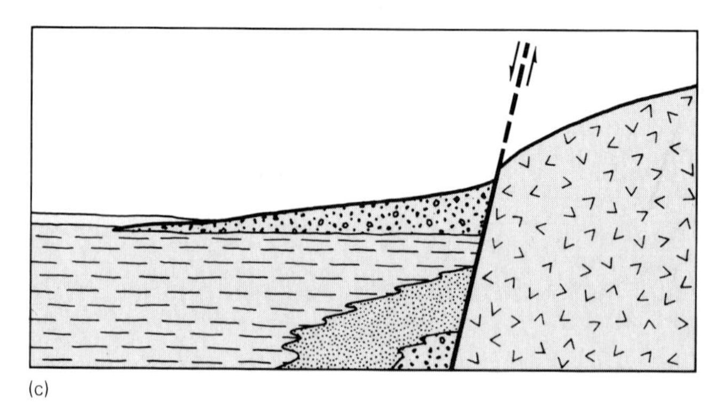


**8.20 Interbedding of contrasting kinds of sediments** at margins of a basin bordered by a highland. (a) General view of margin of basin, showing highland, fans, and lake; schematic block diagram. (b) Lake level rises; water submerges fans and deposits fine-grained strata at margin of basin. (c) New fan is deposited at margin of basin, pushing shores of lake away from highland. (b and c after J. E. Sanders, 1968, Stratigraphy and primary sedimentary structures of fine-grained, well-bedded strata, inferred lake deposits, Upper Triassic, central and southern Connecticut, p. 265–305 *in* Klein, G. de V., *ed;* Late Paleozoic and Mesozoic continental sedimentation, northeastern North America. A symposium: Geological Society of America, special paper 106, 309 p., Figure 7, p. 298)

Facies are identified by their distinctive appearances, and thus indicate the conditions, under which they were formed. At any given time different kinds of sediments are accumulating in various geographic locations. For example, in a sandy desert, dune sand accumulates; not far away may be a coast along which reefs are growing; and elsewhere, a river may be building a delta. If all this were happening during Late Pleistocene time, then one would say that the deltaic sediments, the reef rock, and the dune sand are facies of the Upper Pleistocene strata.

**Sequences of alternating kinds of strata** If two different sedimentary facies are being deposited in adjacent areas, the two kinds of sediments may become interbedded. For example, a fan of coarse and poorly sorted sediment may be present along the sides of a basin that is bordered by a steep highland. In the center of the basin, away from the highland, may be a lake in which fine-grained sediments, such as clays, are being deposited (Figure 8.20a). As the basin floor sinks, new sediment is added. However, the location of the shoreline of the lake, which is the boundary between the two kinds of sediments, may shift back and forth. When the lake level is high, the water may extend to the rocky cliffs at the margin of the basin. Lake deposits bury the fan sediments (Figure 8.20b). During dry times the fan may extend out to the center of the basin. Thus, fan deposits cover lake sediments (Figure 8.20c).

After many such changes, the succession of strata that is deposited consists of alternating layers of coarse fan sediments and fine lake deposits. The interlayering of these two kinds of sediments in a vertical profile through the deposits (as one might find in a bore hole, for example) is an expression of the *intertonguing* of the two facies, the fan facies and the lake facies. If a lava flow were to spread across the floor of the basin, part of the lava would cover fan deposits and other parts of it would bury lake sediments (see Figure 6.20). Such intertonguing of two facies is proof that the geologic ages of the two facies are the same.

## Diapirs

A *diapir* is a body of sedimentary material that has punched its way through the overlying strata (Figure 8.21). Diapirs typically consist either of halite in the form of salt plugs or of gas-charged mud or shale in the form of mud-lump islands or mud "volcanoes."

**Salt plugs** In the Gulf coastal region of the United States diapirs composed of halite are present beneath many of the circular features known as *salt domes* (Figure 8.22). The domes are caused by the salt, but the salt itself is not domed. Certain conditions of temperature and pressure must be present for salt to form diapirs. A temperature of 300°C is required before halite begins to flow plastically. The pressure of overlying material must raise the density of fine-grained sediments by compacting them and driving out the water. Both conditions are satisfied in a deeply subsiding basin.

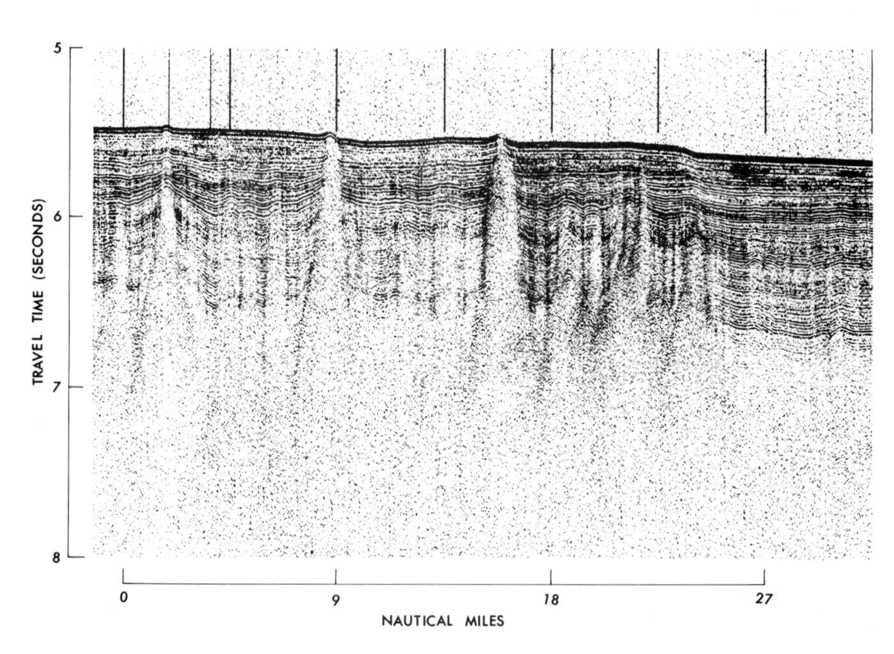


**8.21 Diapirs** shown on continuous seismic profile from Atlantic Ocean west of Senegal, Africa. The three largest diapirs have penetrated all the way to the top of the body of sedimentary strata. Numerous smaller diapirs have stopped before reaching to the top of the strata. (U. S. Navy, courtesy E. D. Schneider)

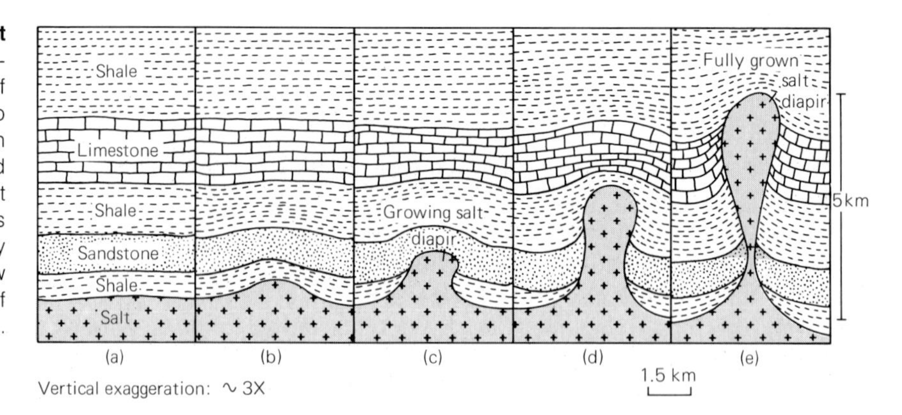


**8.22 Stages of growth in a typical salt diapir,** schematic sections. (a) Strata before diapir develops. (b) Initial elevation of top of salt layer. (c) Salt diapir begins to penetrate upward, driven by differences in density between low-density salt and greater-density overlying strata. (d) Salt has penetrated some layers and has domed the overlying limestone. (e) A fully grown salt diapir in which upward flow has reduced the diameter at the base of the diapiric mass.

At a depth of about 1.5 kilometers, the density of compacting fine-grained sediments has become 2.2 grams per cubic centimeter, which exceeds the density of halite (2.19 g/cm$^3$). At all greater depths, the density of the overlying sediment, which started out less than the density of halite, exceeds that of halite (Figure 8.23). As a result, less-dense halite underlies more-dense sediment. This creates a reversed density gradient, an unstable condition that gravity tends to adjust as soon as the flow temperature of halite is reached. Once the salt starts to flow, it moves upward as a great pillar or cylindrical body whose top may form a mushroomlike overhang. Some salt plugs have penetrated all the way to the Earth's surface. In dry regions, such as Iran, where rainwater is not present to dissolve the halite, the solid salt flows over the land surface, forming a feature known as a *salt glacier.*

**Mud-lump islands and "mud volcanoes"** Diapirs composed of gas-charged mud are common features in rapidly subsiding basins that are accumulating sediments. Examples include the mud-lump islands off the mouth of the Mississippi River and small "mud volcanoes" in many parts of the world. Presumably, the instability of the muds results from the weight of overlying sediments that were deposited rapidly. The change in density needed to make the mud mobile is thought to come about because methane gas is generated in the muds.

## GEOLOGIC AGES OF SEDIMENTARY STRATA

Analysis of the geologic record of sedimentary strata formed the basis for constructing the geologic time chart (see Table 1, p. 7). Most of the names shown on the chart have been in use for more than a hundred

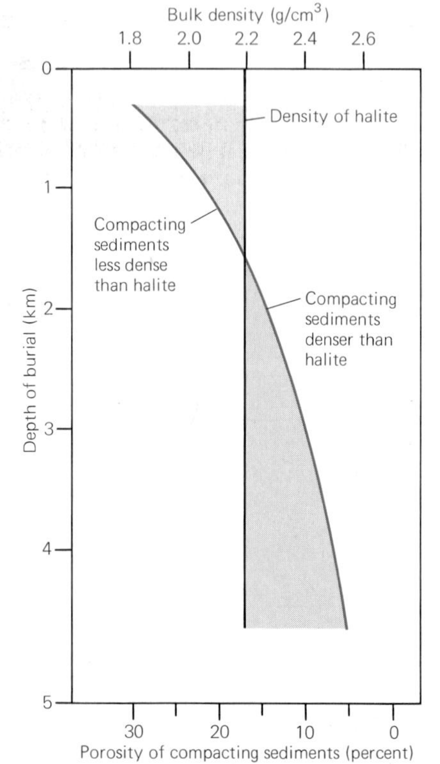


**8.23 Density of halite compared with densities of compacting fine-grained sediments.** At a depth of about 1.5 kilometers, the densities of the compacting sediments begin to exceed that of halite. This established one of the conditions that enable a salt diapir to grow. (Graph based on data from Cenozoic strata, Gulf of Mexico area, in W. C. Gussow, 1968. Salt diapirism: importance of temperature, and energy source of emplacement, p. 16–52 *in* Braunstein, Jules, and O'Brien. G. D., *eds.,* Diapirism and diapirs. A symposium: Tulsa, Oklahoma, American Association of Petroleum Geologists, Memoir 8, 444 p.)

years, and the general relationships that have been established are accepted worldwide. We first show how strata can be assembled into a relative geologic record, and then indicate how it has been possible to calibrate the stratigraphic record using radioactive dating.

### Relative Stratigraphic Record

The fundamental principle in assembling a relative geologic record is the *law of superposition of strata,* that is, strata are spread out, one at a time, with the oldest at the base and progressively younger strata higher up.

Another principle that is applied wherever possible is the succession of fossil organisms. This *law of faunal succession* states that the order of appearance of the various groups of fossil organisms in the geologic record is not haphazard, but follows a well-defined progression.

All exposed strata are intersected by the modern land surface. Moreover, because strata are spread out at the Earth's surface, it is not unusual for strata to cover a former part of the surface that had been eroded long ago. Along higher parts of former landscape, strata may end against the surface of erosion. Eventually, the younger strata may extend farther than the older ones and thus finally bury the old landscape.

The ages of some strata can be estimated by the kinds of debris they contain that may have been eroded from an uplifted block only during certain times in its history. In this case the place where the strata come into contact with the old landscape may not be visible, but the contents of the strata clearly indicate the connection with the area that was eroded.

### Calibration of Stratigraphic Record by Radiometric Dating

**Ages of minerals of sedimentary deposits** Sedimentary deposits usually contain minerals of at least two different ages: (1) the age(s) of the detrital minerals, fixed by the age(s) of the bedrock that was eroded, and (2) the age(s) of minerals that were precipitated as cement or grew within the sediments (known as *authigenic minerals*).

A radiometric date on the detrital minerals can be used only to determine the maximum age of the sedimentary rock. For example, if the rock is of Early Jurassic age (about 190 my old), it might contain detrital minerals eroded from Precambrian bedrock 1000 million years old and from bedrock of mid-Paleozoic age, say 365 million years old.

**Ages of interbedded volcanic rocks** Nearly any extrusive igneous rock contains minerals that can be used to calculate a radiometric age, usually by the K-Ar method. Radiometric ages of such volcanic rocks may give close approximation of the dates of associated sedimentary strata (see Figure 6.20). In some cases the ages of the volcanic rocks can be estimated by paleomagnetic methods based on the positions of the ancient magnetic pole or on polarity-reversal information (see Figure 6.22).

**Ages of strata in relation to radiometrically dated plutons** In the calibration of the geologic time scale, much use has been made of radiometric dates on plutons whose relationship to strata can be established. A search is made for the youngest strata cut by the pluton and for the oldest strata that cover the pluton or contain detritus eroded from it (see Figure 6.27).

## BURIAL OF SEDIMENTS: LITHIFICATION AND THE ORIGIN OF FOSSIL FUELS

A notable contrast exists between the kinds of sedimentary materials we can see at the surface of the Earth and the kinds of materials forming ancient sedimentary bedrock. We assume that at some point the modern sediments will become like the bedrock and at some previous time the materials forming the ancient sedimentary bedrock resembled modern sediments. In many cases sediments experience various changes as a result of being buried. These changes take place because of the higher temperatures, greater pressures, and chemically active fluids found beneath the surface. A general name for all changes that deposited sediments experience short of becoming metamorphosed is *diagenesis;* the changes are *diagenetic changes.*

As sedimentary strata accumulate in a subsiding basin, they become covered with more and more overburden. As a result, they are gradually shifted deeper inside the Earth, where with increasing depth the pressure increases, the temperature increases, and the interstitial water becomes saltier. Many of the sedimentary strata now exposed at the Earth's surface have been buried and subjected to the conditions within the subsurface environments.

At any one depth level the conditions of pressure, temperature, and salinity tend to remain constant. Conditions usually change only as the depth changes. A layer of sediment that has been spread out at the surface of a subsiding area sinks, and another layer is spread on top if it. With continuing sinking, covering, and spreading of new layers, the sediment progresses downward. As these materials descend, they pass through specific depth levels that are important environmental boundaries. The conditions at various levels cause sedimentary materials to react. Some of these reactions are important in converting sediments to sedimentary rocks, organic materials into coal or petroleum, and many sedimentary rocks to metamorphic rocks (see Chapter 9).

### Lithification

After sediment has been transported and deposited, one step remains in the production of rock, the conversion of loose sediment into a coherent aggregate. The overall term for the processes of conversion is *lithification* (from the Greek *lithos,* "stone," and the Latin *ficare,* "to make").

A key process in the production of a sedimentary rock is *cementation,* the crystallization of minerals from solution in the pore spaces of the framework particles. The *framework particles* are those of sand size or coarser. Fine particles associated with these coarser particles are called a

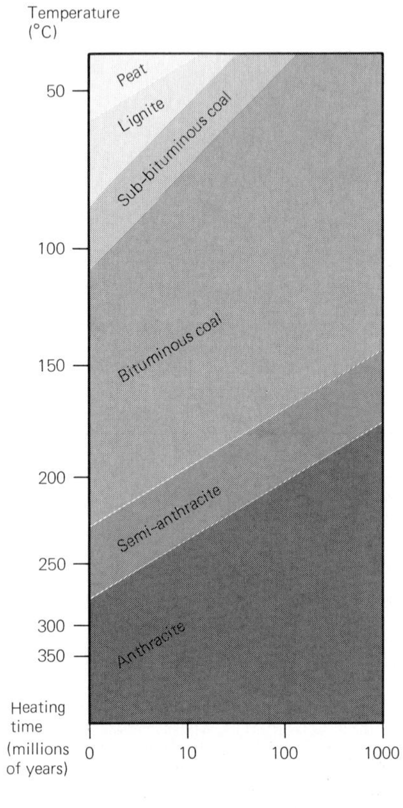


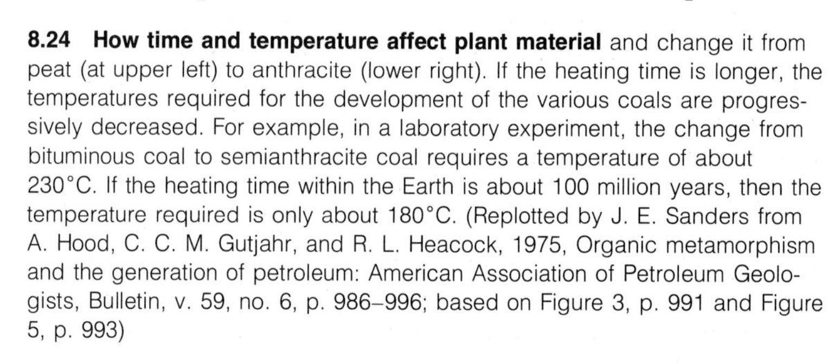

**8.24 How time and temperature affect plant material** and change it from peat (at upper left) to anthracite (lower right). If the heating time is longer, the temperatures required for the development of the various coals are progressively decreased. For example, in a laboratory experiment, the change from bituminous coal to semianthracite coal requires a temperature of about 230°C. If the heating time within the Earth is about 100 million years, then the temperature required is only about 180°C. (Replotted by J. E. Sanders from A. Hood, C. C. M. Gutjahr, and R. L. Heacock, 1975, Organic metamorphism and the generation of petroleum: American Association of Petroleum Geologists, Bulletin, v. 59, no. 6, p. 986–996; based on Figure 3, p. 991 and Figure 5, p. 993)

*matrix.* The minerals that form as cements may or may not be the same as those of the framework particles. The most common cements are calcium carbonate and silica. Others are iron sulfide (pyrite), iron carbonate (siderite), and iron oxide (hematite). These cements may eventually make up a large part of the rock itself. Hematite, in particular, may fill sandstone to such an extent that it is mined as iron ore. Pyrite cements of subsurface sandstones usually become limonite when oxidized at the Earth's surface.

As water circulates slowly through the remaining pore spaces of sediment being compacted, numerous chemical reactions may occur. Some minerals are easily dissolved, transported, and precipitated elsewhere to form new crystals. The particles of limestone may grow larger as calcium carbonate from muds or shells dissolves and recrystallizes.

### The Origin of Coal

Coal contains the carbonized remains of freshwater plants, including dead limbs, trunks, branches, leaves, and even roots in position of growth. Such plant material is thought to accumulate in a warm, densely vegetated swamp located in a basin that is slowly subsiding, perhaps on or near the delta of a large river. The bayous of south Louisiana are modern examples.

The raw plant material that falls into the water of the swamp is partially decomposed by bacteria. The result is *peat.* Further changes require subsidence and its accompanying compression by weight of overburden and progressively higher temperatures.

The effects of temperature, as related to depth of burial, and time on woody plant material are shown in Figure 8.24. As the temperature increases, volatile constituents are driven off, and the proportion of elemental carbon, also known as *fixed carbon,* becomes greater and greater. The greater the proportion of the fixed carbon to impurities, the higher the rank of the coal (Figure 8.25).

Lignite is the lowest grade of coal. It is a highly immature, brownish coal that contains 30 to 40 percent water and burns with a smokey, yellow flame and strong odor.

Next are various grades of bituminous coals, which are harder, blacker, and richer in carbon and burn better than peats or lignites. The highest grade is *anthracite,* a kind of coal usually considered to be a metamorphic rock (Chapter 9), but discussed here to complete the series. Most anthracite is found in the United States in eastern Pennsylvania. Anthracite is hard coal; it bears 80 to 90 percent fixed carbon and burns with a pale-blue flame without smoke or odor. Anthracite is formed by the heating of bituminous coals to a temperature of 170°C for 1000 million years, 210°C for 100 million years, or to 250°C for 10 million years. At higher temperatures anthracite may ultimately change to graphite, which is noncombustible.

The geologic occurrence of coal is a subject that has been well worked out by the studies of many geologists in all parts of the world. Coal is usually associated with patterned sequences of strata known as *cyclothems* (Figure 8.26). Typically the sequence is nonmarine strata below the coal, probably parts of an ancient delta, and mostly marine shales above, which are the sediments that accumulated after a shift in the distributary

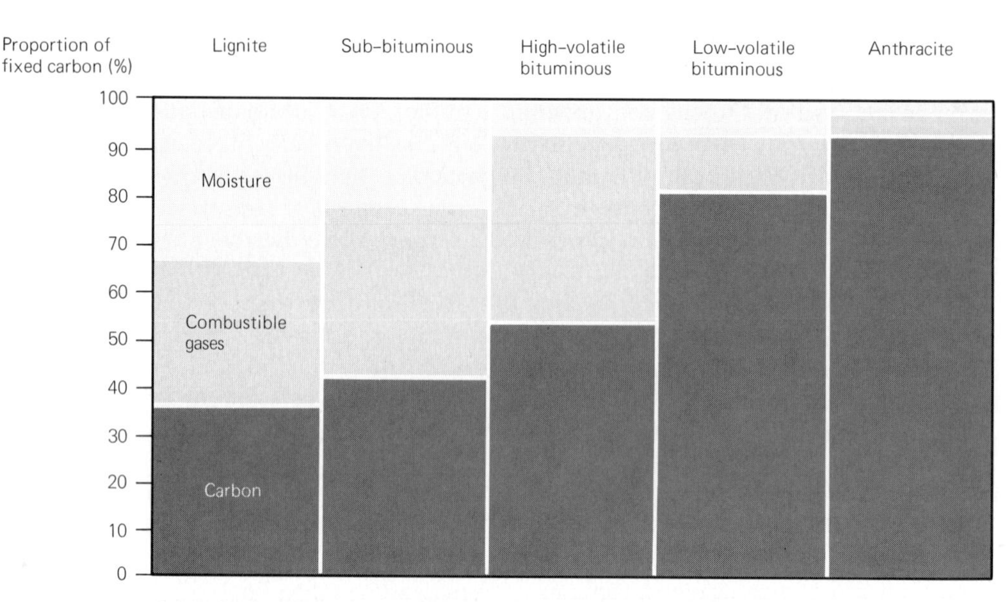


**8.25 Proportions of constituents in various grades of coal.** Fixed carbon refers to the amount of solid, elemental carbon as contrasted to the amount of carbon within combustible gases, such as methane. (U. S. Bureau of Mines)

pattern caused a delta lobe to become inactive and after it subsided further, became submerged by the sea. The thicknesses of coals range from only a few centimeters to 30 meters or more (Figure 8.27). Only coal beds more than a meter or so thick are mined. One of the great environmental problems related to the strip mining of coal is the pyrite in the black shales that usually overlie the coal. Oxidation of this pyrite after exposure to the atmosphere creates sulfuric acid (as explained in Chapter 7).

## Changes in Kerogen: The Origin of Petroleum

Modern research by organic geochemists has established several important principles about the origin of petroleum (oil and natural gas). First, the composition of petroleum has been revealed. The main components of petroleum are hydrocarbons, compounds of hydrogen and carbon. There are many kinds of hydrocarbons, and petroleum is a complicated mixture of them.

Second, the organic origin of most crude oils has been proved. The proof of organic origin is provided by the occurrence in oils, of small amounts of complex compounds called *porphyrins* (not related to and should not be confused with porphyries of igneous rocks). Porphyrins are formed only in the chlorophyll of green plants or in the hemoglobin

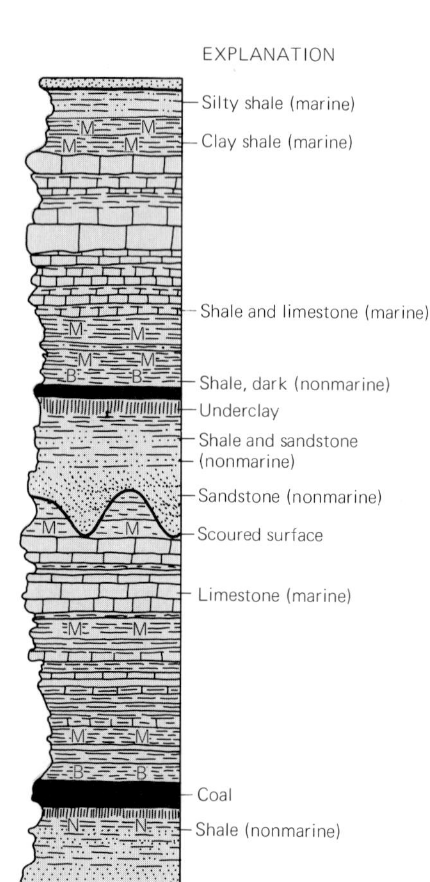


**8.26 Cyclothem,** a patterned sequence of strata associated with layers of coal, schematic columnar section based on observations in Pennsylvanian strata of Kansas. Comparable cyclothems have been found with coals of nearly all ages and in all parts of the world. (Raymond C. Moore, 1964, Paleoecological aspects of Kansas Pennsylvanian and Permian cyclothems, p. 287–380 *in* Merriam, D. F., *ed.*, Symposium on cyclic sedimentation: Kansas Geological Survey, Bulletin 169, v. 1, 380 p., Figure 154, p. 212)

**8.27 Coal** about 21 meters thick, being strip mined near Gilette, Wyoming, (National Coal Association)

of blood. In addition to showing the organic origin of crude oil, porphyrins place a definite upper limit on the temperatures that were involved in the origin of the oil. In the laboratory porphyrins break down at a temperature of 200°C. Therefore,we can be certain that oil containing porphyrins has never been heated above 200°C.

Third, organic geochemists have demonstrated that the immediate precursor of petroleum is not a fragile, easily oxidized, protoplasmic oozy liquid, as was formerly thought, but rather a group of stable, solid hydrocarbon materials classified as *kerogen.* The fine-grained rocks known as oil shales actually do not contain any oil. Instead, they contain kerogen, which is the world's most widespread form of organic carbon. Kerogen is 10,000 times as abundant as coal; it is found in sedimentary rocks of all ages and on all continents. It is most abundant in shales.

**Maturation of kerogen in shales** Because of the great abundance of kerogen in shales, organic geochemists have concentrated on studying the changes that kerogen in shales undergoes when the shales are buried and geothermally heated. The results differ depending on the composition of the kerogen. High-hydrogen kerogens, derived from algae, are the only kind that can generate crude oil. High-oxygen kerogens, derived from woody plant material, can generate only natural gas. (Natural gas can also be generated by the thermal cracking of the crude oil formed from high-hydrogen kerogens.)

The conversion of kerogen to petroleum is known as *maturation.* This conversion has been performed many times in the laboratory. It requires temperatures ranging from 350° to 500°C, much higher than the porphyrin limit of 200°C for natural petroleum. Presumably, in nature a long burial time reduces the temperature requirement. The progressive

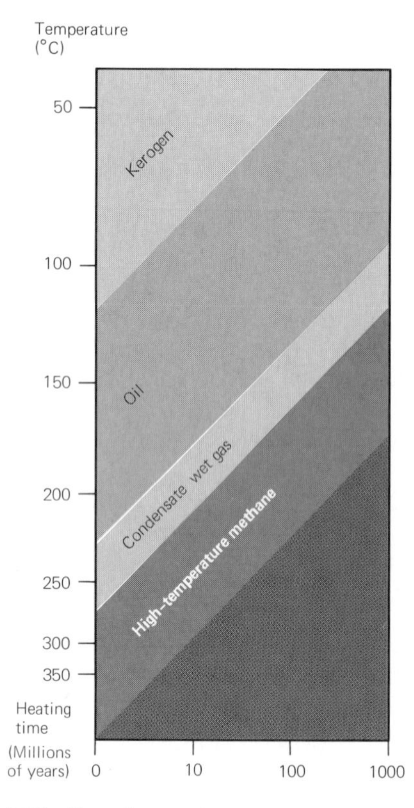


**8.28 How time and temperature change organic matter (kerogen) into oil and gas.** As in Figure 8.24, the effect of long heating times is to decrease the temperature required to create a given end product. (Replotted by J. E. Sanders from A. Hood, C. C. M. Gotjahr, and R. L. Heacock, 1975, Organic metamorphism and the generation of petroleum: American Association of Petroleum Geologists, Bulletin, v. 59, no. 6, p. 986–996; based on Figure 3, p. 991 and Figure 5, p. 993).

changes of high-hydrogen kerogen with temperature and time are shown in Figure 8.28.

As kerogen matures, its carbon content increases as in coals. This causes a color change. Fresh kerogen is light yellow. With heating the color changes to orange, orange-brown, dark brown, and finally to black, the color of kerogen that has generated petroleum. From careful study of these color changes, it is possible to infer the past temperature history of kerogen and to determine whether or not a kerogen has generated petroleum.

**Problem of the occurrence of petroleum in sand** Although all steps in the maturation of kerogen to petroleum can be duplicated in the laboratory and traced in shales, one more important problem remains to be solved before the origin of petroleum deposits can be understood. Oil is not usually found in shales; rather, it occurs as tiny droplets saturating the pore spaces of porous rocks. The shales are considered to be the *source beds* and the porous rocks, the *reservoir beds.* Perhaps the last final mystery of petroleum geology is how the organic matter, presumably starting in the shale source beds, manages to move into the porous reservoir beds.

One view states that the occurrence of oil in the reservoir beds can be explained by assuming an underground migration of liquid oil from the source beds. This process has been called *"primary migration"*. Numerous mechanisms have been proposed for explaining "primary migration." However, despite extensive research, the process has not been reproduced in the laboratory. A major difficulty with the concept of "primary migration" is that at the depth where geothermal heat can convert the kerogen in shales into crude oil (about 3 km), the weight of the overburden has reduced the pososity of the shales to about 10 percent (see Figure 8.23) and has destroyed even the small amount of permeability initially possessed by the clay.

Most petroleum geologists are convinced that the bulk of the world's oil forms in shales and, therefore, some process *must* exist to get the oil out of the shales, underground, and in the liquid form. The stability of kerogen introduces the possibility that this process may involve the large-scale recycling of kerogen, rather than the "primary migration" of oil.

Because petroleum is known to occur in sands, one possibility that needs to be examined is that the sands themselves can generate petroleum. This possibility was considered in the past, but the assumption that the precursor of petroleum was fragile, first-cycle organic matter led to the conclusion that sands do not contain enough organic matter. Now kerogen has been demonstrated to be the precursor and evidence that kerogen can be recycled from older shales has recently been found. If future research demonstrates that kerogen can be recycled on a large scale, then a whole new outlook may be possible about the origin of petroleum in sands.

The transfer of organic matter from shales to sands could take place: (1) in the solid state (as kerogen, a stable material); (2) at the Earth's surface; and (3) by a well-known process, erosion. Assuming that kerogen is stable, then one can suppose that the necessary transfer of organic matter from shales to sands need not take place during one cycle of sedimentation, would not involve the liquid state, and need not take place underground by a physically improbable process. Instead one could infer that during the first cycle of sedimentation, the organic matter in shales becomes kerogen. Later, the kerogen-bearing shale is uplifted and eroded. During erosion tiny solid particles of kerogen could be transferred from the shales and recycled. They might be deposited along with the quartz to form a sand.

After a second subsidence, this time involving the deep burial of the kerogen-bearing sand, the kerogen particles within the sand could become oil, just as they have been shown to do in shales. After they have become oil, the tiny droplets migrate up the dip of the sand. Left behind at depth would be the carbonized residues of the former kerogen. If this sequence of events based on recycled kerogen proves to be correct, then appropriate sands can serve as both source beds and reservoirs and the controversial primary migration need not be involved.

**Petroleum deposits** No matter where or how the oil originally forms, a deposit from which appreciable amounts of petroleum can be extracted requires that a large amount of petroleum be concentrated. If widely dispersed natural organic matter is the source, then some kind of large subsurface "drainage area" must supply and concentrate petroleum at the places where it is found.

Calculations based on the porosity and kerogen content of sand can yield an estimate of the subsurface volume of sand required to concentrate a specific quantity of petroleum. Taking production techniques and practices into consideration, petroleum geologists can then estimate the required area per produceable barrel. Further discussion of petroleum exploration is found in Chapter 19.

## SEDIMENTARY ORES AND SEDIMENTARY ECONOMIC MINERAL DEPOSITS

Sedimentary deposits include concentrations of valuable metallic minerals and of many other substances. We review these briefly under the headings of ores concentrated by mechanical processes, other sedimentary ore deposits, and sedimentary economic mineral deposits.

### Ores Concentrated by Mechanical Processes: Placers

As chemical weathering destroys the feldspars, the ferromagnesian silicate minerals, and the other easily changed minerals, the chemically resistant minerals are released from their positions within the bedrock. These resistant minerals include gold, diamonds, ilmenite, platinum, cassiterite, chromite, and monazite. The released minerals are shifted downslope by various gravity-transport processes and later may be concentrated mechanically in the bed of a stream or on a beach. The ore minerals in such deposits, known as *placers,* are usually found at the base of the sediment body, along the contact with the underlying bedrock. Diamonds along the coast of Africa, cassiterite in southeast Asia, ilmenite in New Jersey, and the now-mined-out gold in California are examples of placers.

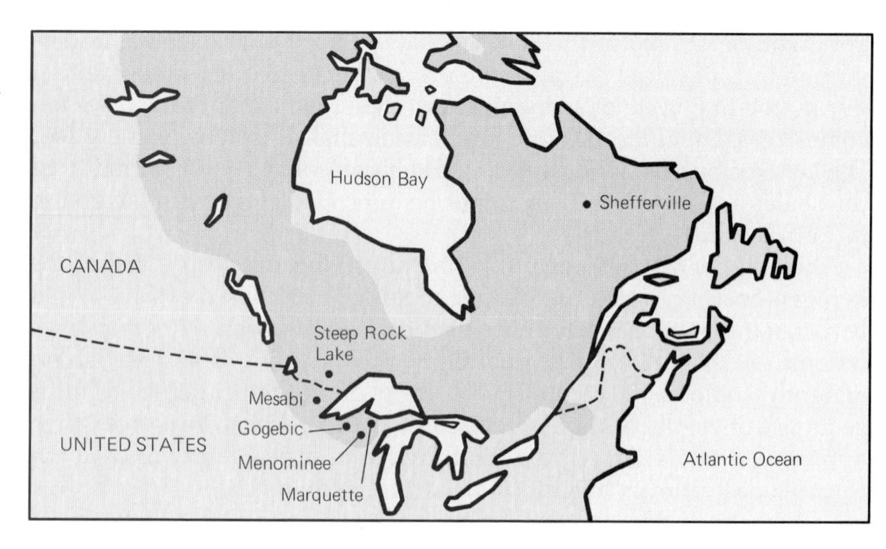


**8.29 Canadian shield,** north-central United States and southeastern Canada, showing locations of chief deposits of Precambrian sedimentary iron ores. (Geological Survey of Canada)

### Other Sedimentary Ore Deposits

The major metals included in the generalized category of sedimentary ore deposits are iron, copper, manganese, and uranium.

Many of the world's largest iron deposits are oxides or other iron minerals that were deposited as sedimentary strata. Many of these are of Precambrian age and are located in the Canadian Shield (Figure 8.29). In the Lake Superior region of Minnesota and Ontario, sedimentary iron deposits of Precambrian age were subsequently weathered. The originally hard, tough, cherty iron deposits were converted to soft, earthy hematite (see Figure 7.21). Thus, these hematite deposits are examples of sedimentary ores that were enriched by residual concentration.

Numerous copper deposits are mined from shales. The native copper deposits of the Keweenaw Peninsula, northern Michigan, are contained in a conglomerate. The copper fills the interstices, just as if it were a cement.

Manganese nodules are present on many parts of the modern deep-sea floor (Figure 8.30). The nodules have formed by deposition of the metals from seawater. Nodule-forming processes are complex and evidently include chemical activities, bacterial processes, and possibly also volcanic processes.

Most of the world's economic supplies of uranium minerals are contained in sedimentary rocks, including conglomerates, sandstones, shales, phosphatic deposits, and lignites. The geochemistry of uranium is complex, but its pattern has been established and can be used to explain the natural distribution of uranium. Where oxygen is present, uranium goes into solution. In waters lacking oxygen, uranium tends to be deposited. The deposition of uranium in waters lacking oxygen explains the association with black shales, lignites, phosphatic deposits, and some of the sandstones. In the Colorado Plateau of southwestern United States, uranium tends to be concentrated around fossil logs that were deposited at the bottoms of former stream channels. This association has been explained as an effect of the organic matter of the logs on the circulating underground waters. Decaying organic matter creates chemical conditions in which uranium tends to be precipitated. The locations of

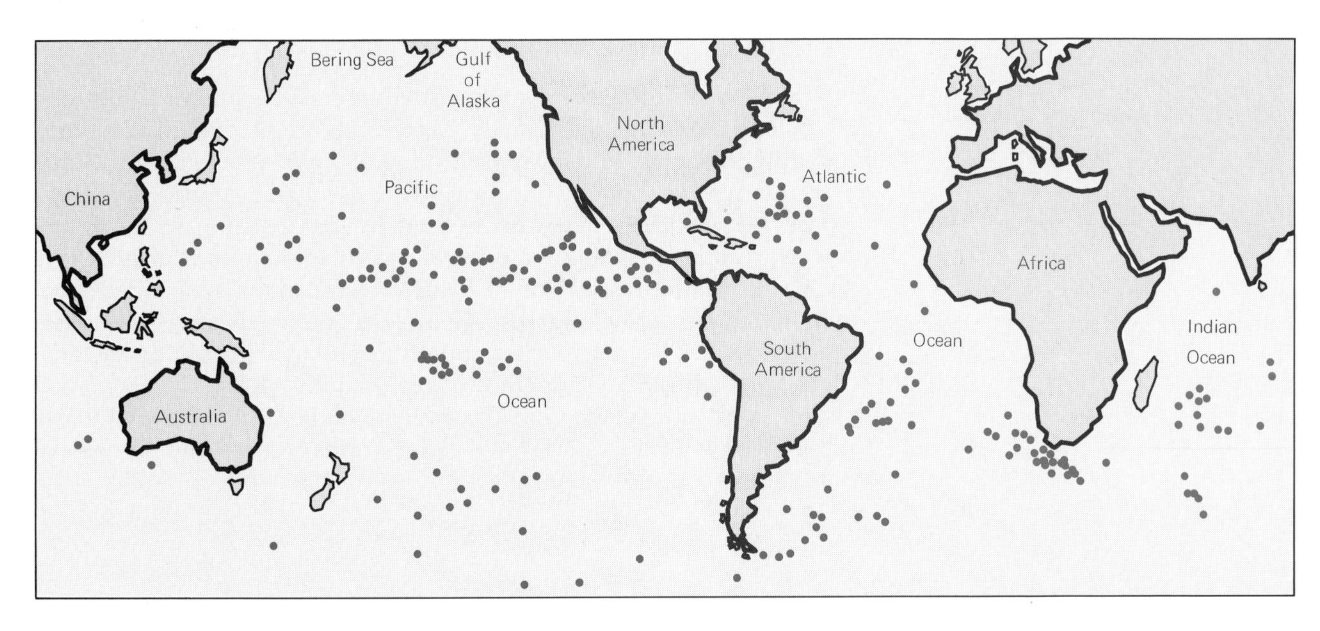


(a)

the chief sedimentary deposits of uranium in the United States are shown on Figure 8.31.

A large sedimentary uranium deposit in the southern part of the Canadian Shield is located in a conglomerate that has been mined in the Blind River district, Ontario. The uranium fills in what were formerly pore spaces among the rounded particles of quartz.

**8.31 Distribution of major deposits of uranium and thorium in the conterminous United States** showing proportions of uranium and estimated quantities of reserves. (M. K. Hubbert, 1957, Nuclear energy and the fossil fuels, p. 7–25 *in* Drilling and production practice, 1956: New York, American Petroleum Institute, 527 p., Figure 28, p. 23)

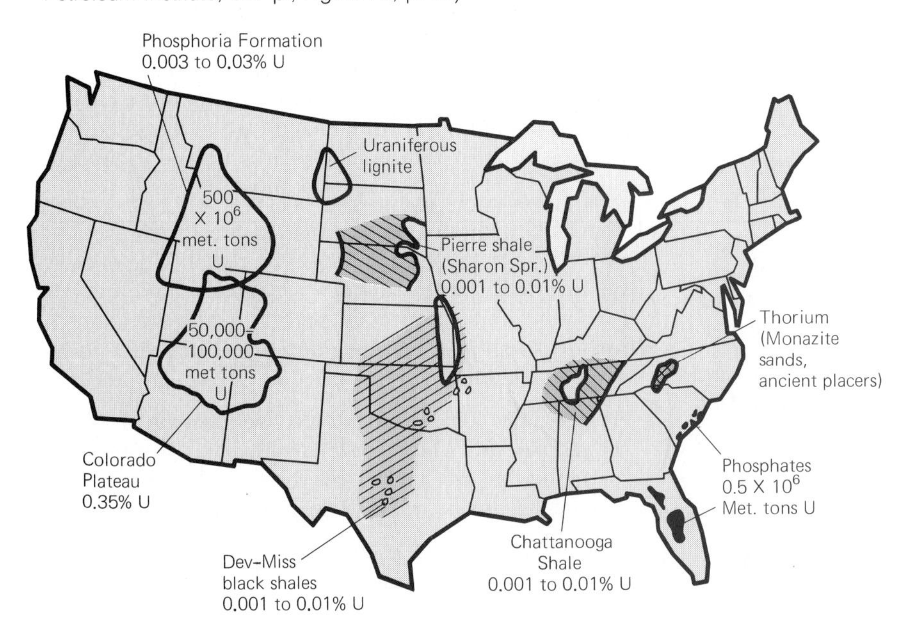

(b)

**8.30 Manganese nodules on the deep-sea floor.**
(a) Map showing locations of abundant manganese nodules.
(b) Baseball-size ferromanganese nodules photographed on the floor of the central Pacific Ocean, latitude 36°37′S, longitude 149°00′W, water depth, 5320 meters. (Lamont-Doherty Geological observatory of Columbia University)

### Sedimentary Economic Mineral Deposits

Mineral deposits in the sedimentary category include strata of limestone and dolostone; sand and gravel; concentrations of diatom skeletons (diatomite); zeolites; and evaporite minerals, gypsum and salt. Some sedimentary rocks, notably certain sandstones and limestones, can be cut or broken into pieces that can be used for construction stone. Many placers include nonmetallic materials such as diamonds and garnet. Clays may be transported strata or residual accumulations weathered from parent rocks containing feldspars. Sulfur is a nonmetallic resource that is found associated with evaporites; the sulfur probably formed by bacterial reduction of the sulfate minerals. Large quantities of sulfur are mined from the salt diapirs in the Gulf Coast region of the United States. Quartz sand is used to obtain silica for making ordinary glass and for special optical lenses.

## CHAPTER REVIEW

1. *Sediment* is broadly defined as regolith that has been transported at the surface of the Earth and deposited as strata in low places. Most of our energy sources come from the products of sediments and sedimentary rocks.

2. For the most part, weathering is the essential step that precedes the formation of sediments. Sediments can be divided into two groups based on weathering products: (1) *solid products,* such as rock fragments, resistant minerals, and secondary-alteration products; and (2) *dissolved ions.*

3. Sedimentary materials are classified according to their origins into four major categories: (1) *detritus,* (2) *solids derived from ions carried in solution,* (3) *organic material,* and (4) *volcanic material.*

4. The degree of uniformity of sizes of aggregates of sedimentary particles is known as *sorting.* Particles are *well sorted* if the silt and clay have been separated from coarser material and *poorly sorted* if they have not. The shapes of sedimentary particles result from the initial shapes of the particles in their parent rocks and from the way these initial shapes are changed by *abrasion* during transport.

5. Sedimentary rocks can be recognized if they contain skeletal debris, strata, clastic texture, and an abundance of certain materials such as evaporites and carbonates.

6. A characteristic feature of sedimentary rocks is that they form layers, or *strata.* The strata exist because something changed. The changes that create strata are related to the conditions of deposition.

7. A bed having a gradient in the sizes of its particles is a *graded layer.*

8. In connection with sedimentary strata, *facies* refers to distinctive-appearing materials that were deposited under particular conditions.

9. In some cases, low-density material punches its way upward through the overlying strata as *diapirs.*

10. The fundamental rule of strata is that they are spread out, one at a time, with the oldest at the base and progressively younger strata higher up.

11. The overall term for all processes of converting loose sediments into a rock is *lithification.* The chief process of lithification is *cementation,* the crystallization of minerals from solution in the pore spaces of the framework particles.

12. Coal derives from woody plant material transformed by decomposition, subsidence, compression, and high temperature. Decomposed plant material is *peat.* Varying levels of pressure and temperature produce types of coal graded by their fixed-carbon contents. The grades, from lowest to highest carbon content, are lignite, types of bituminous, and anthracite.

13. The main components of *petroleum* are hydrocarbons. The proof of the organic origin of crude oil is shown by the occurrence in oils of small amounts of *porphyrins.* The immediate precursor of petroleum is not an oozy liquid, but the group of stable, solid hydrocarbons classified as *kerogen.*

14. Sedimentary deposits include concentrations of valuable metallic and other minerals which may be ores concentrated by mechanical processes, other sedimentary ore deposits, or sedimentary economic mineral deposits.

## QUESTIONS

1. Why are sedimentary rocks more difficult to understand than igneous rocks?
2. Explain the relationship between weathering products and sediments.
3. List and briefly describe the four major kinds of sedimentary materials.
4. Describe *sorting*.
5. Define *strata*. How are strata named? How do strata form?
6. What is a *graded layer?* How do graded layers form?
7. Define *bed forms*. What are *cross strata?* Are cross strata related to bed forms? Explain.
8. Describe two ways in which currents create sedimentary structures in the sediments they deposit.
9. Define *sedimentary facies*. Discuss the significance of the interbedding of two sedimentary facies.
10. What is a *diapir?* Name two materials that form diapirs. Describe the relationship between diapirs and their surroundings.
11. List some of the characteristic features of sedimentary rocks.
12. State the *law of superposition of strata* and the *law of faunal succession*.
13. Explain the significance of radiometric dates of the minerals of a sedimentary deposit.
14. Define *lithification*. Describe three processes of lithification.
15. Trace the changes in plant material from *peat* to *anthracite*. What are the chief controls on these changes?
16. What are *porphyrins?* What is the significance of porphyrins to ideas about the origin of oil?
17. Summarize the maturation of kerogen into petroleum, including the color changes of the kerogen.
18. Define and discuss the pros and cons of the concept of "primary migration" of petroleum.
19. Comment on the following statement: "Oil occurs in sands; therefore, it originates in shales."
20. What is a *placer?* List some of the constituents of placers. Where are placers found?
21. List four metals that may occur as *sedimentary ore deposits* and briefly describe a geologic setting of each.

## RECOMMENDED READING


Blatt, Harvey, Middleton, G. V., and Murray, R. C., 1980, *Origin of Sedimentary Rocks*, 2nd ed.,: Englewood Cliffs, New Jersey, Prentice-Hall, Inc., 782 p.

Chowns, T. M., and Elkins, J. E., 1974, The origin of Quartz Geodes and Cauliflower Cherts Through the Silicification of Anhydrite Nodules: *Journal of Sedimentary Petrology*, v. 44, no. 3, p. 885–903.

Davidson, C. F., 1957, On the Occurrence of Uranium in Ancient Conglomerates: *Economic Geology*, v. 52, no. 6, p. 668–693.

Davidson, C. F., 1965, A Possible Mode of Origin of Strata-bound Copper Ores: *Economic Geology*, v. 60, no. 5, p. 942–959.

Epstein, Anita G., Epstein, J. B., and Harris, L. D., 1977, Conodont Color Alteration—An Index to Organic Metamorphism: *U.S. Geological Survey, Professional Paper 995*, 27 p.

Friedman, G. M., and Sanders, J. E., 1978, *Principles of Sedimentology*, New York, John Wiley & Sons, 792 p.

Harwood, R. J., 1977, Oil and Gas Generation by Laboratory Pyrolysis of Kerogen: *American Association of Petroleum Geologists, Bulletin*, v. 61, no. 12, p. 2082–2102.

Krinsley, D. H., and Smalley, I. J., 1972, Sand: *American Scientist*, v. 60, no. 3, p. 286–291.

LaPlante, R. E., 1974, Hydrocarbon Generation in Gulf Coast Tertiary Sediments: *American Association of Petroleum Geologists, Bulletin*, v. 58, no. 7, p. 1281–1289.

Sackett, W. M., Poag, C. W., and Eadie, B. J., 1974, Kerogen Recycling in the Ross Sea, Antarctica: *Science*, v. 185, p. 1045–1047.

# CHAPTER 9
# METAMORPHIC ROCKS AND METAMORPHISM



PLATE 9
Results of intense deformation during metamorphism, Precambrian rocks, Pelham Bay Park, Bronx, New York. (J. E. Sanders)

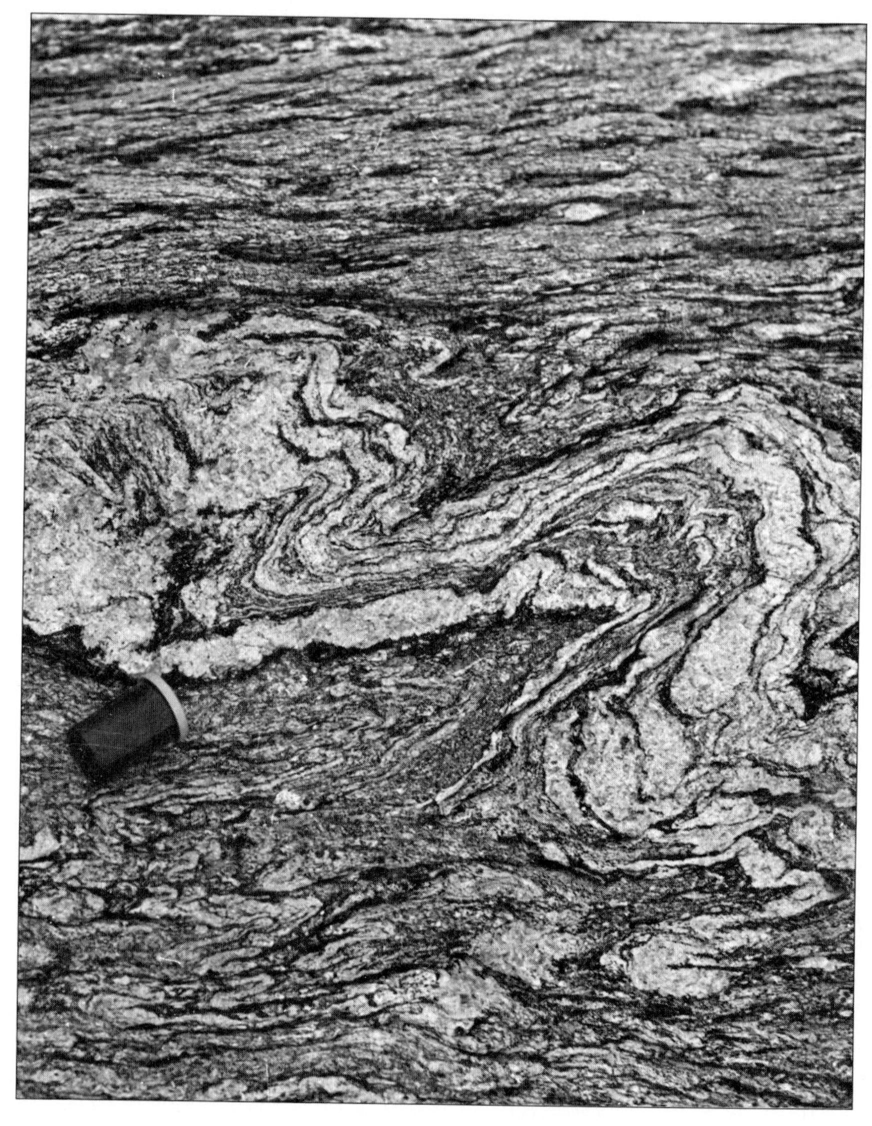

IN THIS CHAPTER WE EXAMINE SOME OF THE CONDITIONS WITHIN the Earth, usually at depths of several tens of kilometers, by which various rocks are so changed that they form a special group named metamorphic rocks. As with igneous rocks and sedimentary rocks, detailed description of individual metamorphic rocks and methods of classifying and naming them are contained in Appendix C. Our major topics of discussion are: the origin of modern concepts about metamorphic rocks, metamorphic rocks in the rock cycle, occurrence of metamorphic rocks, recognition of metamorphic rocks, the concept of metamorphic facies, geologic dating of metamorphic rocks, and economic products associated with metamorphic rocks.

## THE ORIGIN OF MODERN CONCEPTS ABOUT METAMORPHIC ROCKS

Metamorphic rocks display textures, minerals, and structural features that set these rocks apart from igneous rocks and sedimentary rocks. Originally, their distinctive features were thought to have formed as a result of special conditions that prevailed only on a primitive Earth, but that had long since ceased to exist. Accordingly, they were included among those rocks named "primitive rocks" and were considered to be among the oldest on Earth.

The basis for our modern ideas was established by Sir Charles Lyell in 1830. Lyell's observations in the Alps led him to conclude that certain of the rocks classified as primitive by Swiss geologists had formed by the alteration of ordinary shales which had been deposited as mud on the floor of an ancient sea during the Jurassic Period (approximately 180 million years ago). Fossils of Jurassic age proved to be the critical evidence. Such fossils were found both in ordinary shales and in mica-rich metamorphic rocks. Lyell was thus able to prove that the shales, an example of the preexisting rock that we refer to as a *parent rock,* had been transformed into the mica-rich metamorphic rocks and that this change had taken place after the Jurassic Period. In other words, the mica-rich metamorphic rocks were not the oldest rocks in the Alps, as had been supposed, but more like the opposite. The conversion of the parent-rock shale into the mica-rich metamorphic rock had taken place some time after the Jurassic Period.

Lyell proposed the term *metamorphic,* from the Greek meaning "changed form." Once geologists had accepted Lyell's concept, they began to study metamorphic rocks with a view toward understanding the conditions and situations under which various kinds of rocks had been transformed. They soon reached the conclusion that temperature was an important factor. Other controls on metamorphism include response of material to deformation, whether it is ductile and pliable or brittle and subject to breaking; high pressure, as in the transition of graphite to diamond; sudden elevation of both temperature and pressure that accompanies meteorite impact; and introduction of new materials and removal of old materials, commonly in hot solutions. The keys to metamorphic rocks, then, are knowledge of the composition of parent materials and how these materials react under different conditions.

Metamorphic rocks have become closely woven with concepts of plate tectonics. We mention a few of these connections here, but defer most of the details until Chapter 18.

## METAMORPHIC ROCKS IN THE ROCK CYCLE

Within the rock cycle, metamorphic rocks occupy an intermediate position between the kinds of materials formed at the surface, such as sediments and volcanic materials and their related sedimentary rocks and volcanic rocks, and the magmas that become igneous rocks or reach the surface as lavas. This does not mean that all sediments are destined to become metamorphic rocks. Some sedimentary materials, for example, have remained close to the Earth's surface for thousands of millions of years and thus have remained much as they were originally. What it does mean is that deep burial may subject sedimentary rocks to conditions of great heat, high pressure, and chemically active fluids, and as a result, the rocks react by forming new minerals and new textures. The changes occur without melting, for once the material has melted, it becomes igneous. Under certain conditions igneous rocks can be metamorphosed.

## OCCURRENCE OF METAMORPHIC ROCKS

Metamorphic rocks are distributed in two geologic and geographic patterns: local and regional extent. Rocks of local extent are found as fringes surrounding a well-identified local source of heat or of high pressure. The pattern of regional extent may take the form of belts, hundreds of kilometers wide and thousands of kilometers long, that are associated with the interiors of mountain chains or form major parts of continents. Within these belts, the rocks have been extensively deformed and recrystallized (Figure 9.1). Other metamorphic rocks of larger-than-local scale form bodies that may be tens or hundreds of kilometers long and up to a kilometer wide adjacent to a major fault. Both of these groups of metamorphic rocks of regional extent are thought to be related to the behavior of the plates of the Earth's lithosphere.

**9.1 Units of contrasting kinds of metamorphic rocks.** A distinct contact, presumably a former bedding-surface parting, separates dark-colored, massive rocks at left from light-colored, well-layered rocks at right. Belt of ancient metamorphic rocks of Precambrian (?) or Early Paleozoic ages, Pelham Bay Park, Bronx, New York. (J. E. Sanders)

## Metamorphic Rocks of Local Extent

Metamorphic rocks of local extent include the contact-metamorphic aureoles surrounding plutons and the zones of shock metamorphism adjacent to impact craters.

**Contact-metamorphic aureoles** In the areas immediately surrounding a pluton, the country rocks typically have been altered by the effects of the heat and possibly hot gases that were associated with the magma. Zones of altered rock surrounding a pluton are named *contact-metamorphic aureoles* (Figure 9.2). Contact-metamorphic aureoles are extremely useful for studying metamorphism; in many of them it is possible to trace all gradations from locations well away from the pluton where the country rock has not been altered to intensely metamorphosed rock adjacent to the pluton. The width of a contact aureole surrounding small dikes or sills may be only a few centimeters. By contrast, around a large batholith, the width may attain several thousand meters.

By bringing together the information about contact-metamorphic aureoles from many localities, geologists have been able to determine two things: (1) how one kind of pluton affected various kinds of country rock (Figure 9.3a) and (2) how a given kind of country rock has responded to different kinds of plutons (Figure 9.3b).

Where a large batholith has cut into country rock that includes many sedimentary varieties, it is possible to study how the temperature and fluids associated with that pluton affected different kinds of materials. The strategy of such studies is to determine the distribution of the various metamorphic products, starting at the contact with the pluton and progressing farther and farther away.

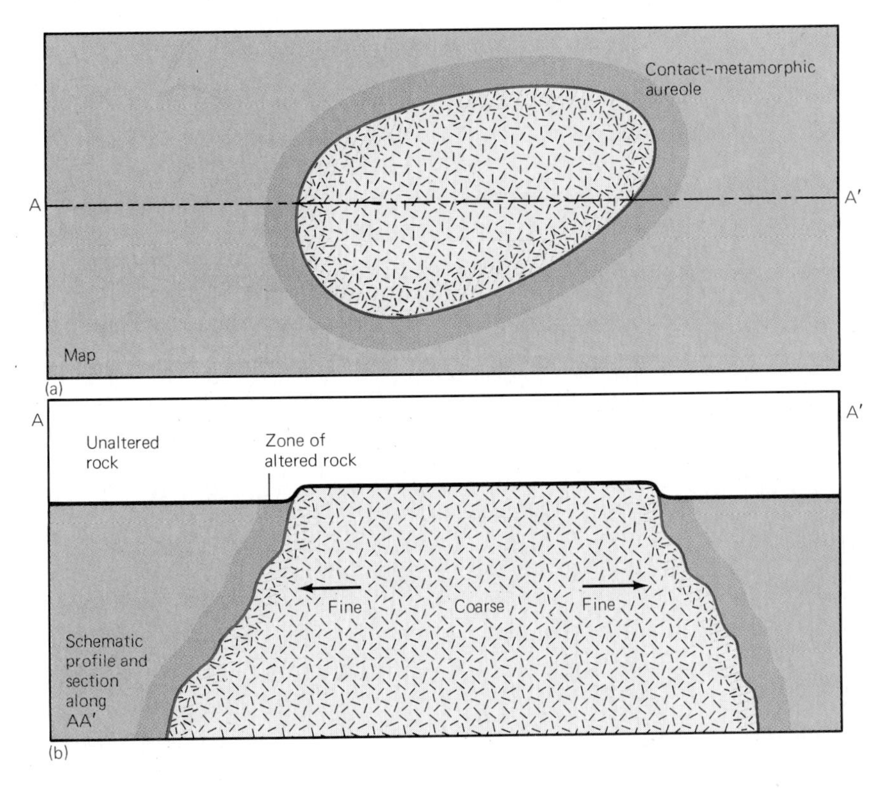


**9.2 Contact-metamorphic aureole surrounding a pluton,** schematic map view (above) and schematic profile and section along line AA′ on map, as the rocks would appear along a vertical slice at right angles to the Earth's surface, as in the wall of a canyon or steep roadcut.

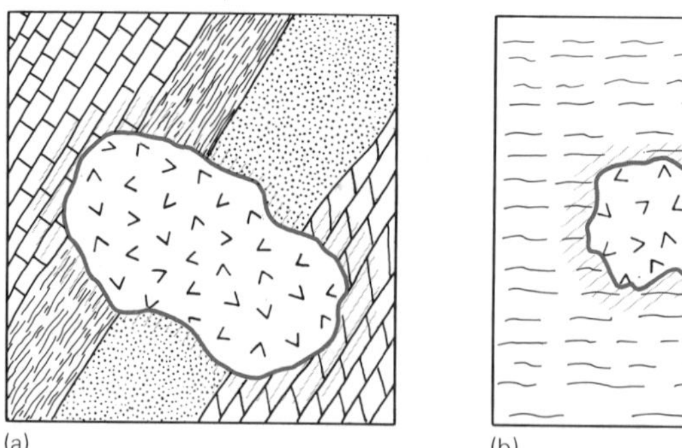

**9.3 Varied effects of plutons and country rocks,** schematic map views. (a) Where a pluton consisting of only one kind of igneous rock intrudes a variety of country rocks, such as limestone, shale, sandstone, and dolostone, shown here, one can study the effects of a single set of environmental conditions on different materials. (b) Where two or more plutons each consisting of a different kind of igneous rock cut into a single thick unit of country rock, one can study how two sets of environmental conditions affected a single material. (Based on relationships in Precambrian rocks of northern Minnesota)

The gabbro plutons on the isle of Skye, Scotland, cut into extrusive basalts (Figure 9.4). At the contact is a hornfels composed of pyroxene and plagioclase, the same minerals that form basalt. Next outward is a zone of amphibolite; then amphibolite containing epidote; chlorite schist; and finally, basalt that has not been altered.

Another useful strategy in studying contact metamorphism is to search for xenoliths (see Figure 6.9c). Within a xenolith, one may find meta-

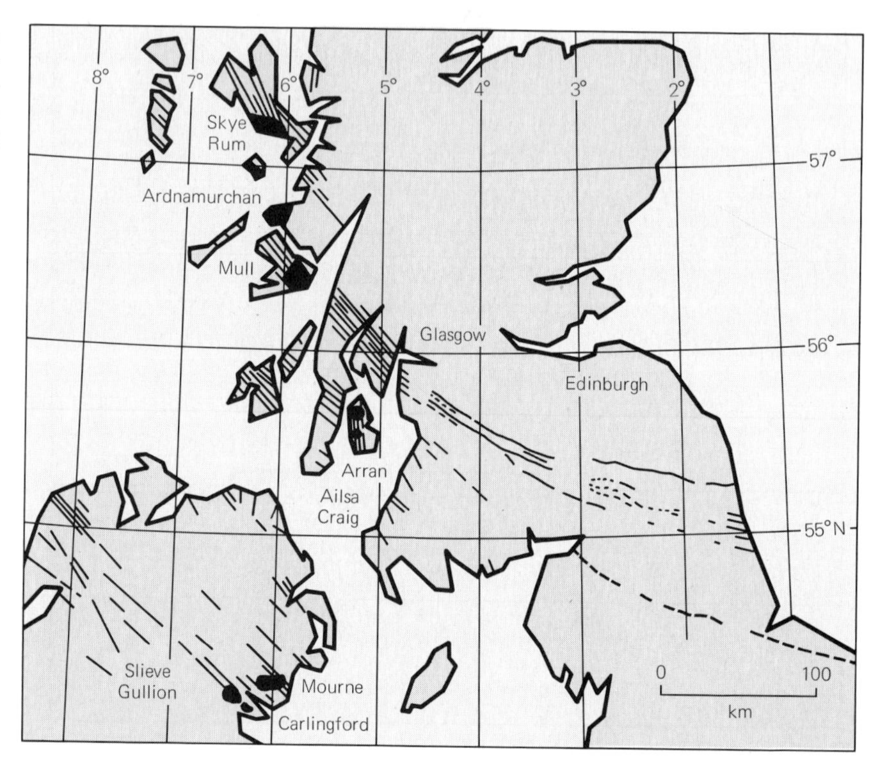


**9.4 Gabbro plutons** on Isle of Skye, Scotland, which intrude basalts and created high-temperature metamorphic effects on mafic materials. (H. M. Geological Survey of Great Britain)

**TABLE 9.1**
PARENT MATERIALS AND THEIR METAMORPHIC PRODUCTS IN CONTACT-METAMORPHIC AUREOLES

| Parent Material | Metamorphic Products in Contact-Metamorphic Aureoles |
|---|---|
| Pure quartz sandstone | Quartz recrystallizes; sandstone becomes **quartzite.** |
| Pure carbonate rocks | |
| Limestone containing calcite | Calcite recrystallizes; forms **calcite marble.** |
| Dolostone | Dolomite recrystallizes; forms **dolomite marble.** |
| Impure carbonate rocks | |
| Quartz + calcite | React to form **wollastonite** (a calcium silicate) or **diopside** (a calcium pyroxene) |
| Quartz + calcite + iron | React to form **garnet.** |
| Dolomite + quartz | React to form **magnesium-rich amphiboles** or **pyroxenes.** |
| Shales and slates | Baked to form **hornfels;** may contain porphyroblasts of biotite, andalusite, or other minerals. |
| Basalt | Reacts to form **chlorite schist, amphibolites** with or without epidote, or **gneisses containing pyroxene** |
| Granite | Little change of mineral composition |
| Coal | Rank increases; **anthracite** or **graphite** may form. |
| Kerogen | May form **petroleum** or **graphite.** |

morphic minerals that are not present in the contact-metamorphic aureole. Within a xenolith the temperature may have been higher than in any part of the contact zone. Thus, the only place where appropriate conditions existed for the formation of certain minerals may have been within the xenolith.

Table 9.1 lists some kinds of parent materials and how they react in contact zones. The changes that take place in contact zones begin with the driving out of gases, such as the carbon dioxide of calcite or dolomite or the methane, water vapor, and carbon dioxide of organic materials such as coal. In addition, a systematic progression outward from the contact involving ferromagnesian silicate minerals is from single-chain pyroxenes to double-chain amphiboles, and, finally, to sheet-structure mica minerals. Recall that this same order of formation characterized Bowen's reaction series as the temperature drops when a magma cools (see Figure 6.2).

**Zones of shock metamorphism surrounding impact craters** The great scientific debates during the 1960s over the origin of the Moon's circular features (see Figure 20.12 and Chapter 20) stimulated considerable new research to determine the sequence of events and conditions that are associated with the impact of a large meteorite. A sequence of typical events is portrayed in Figure 9.5. The shock of impact creates elevated

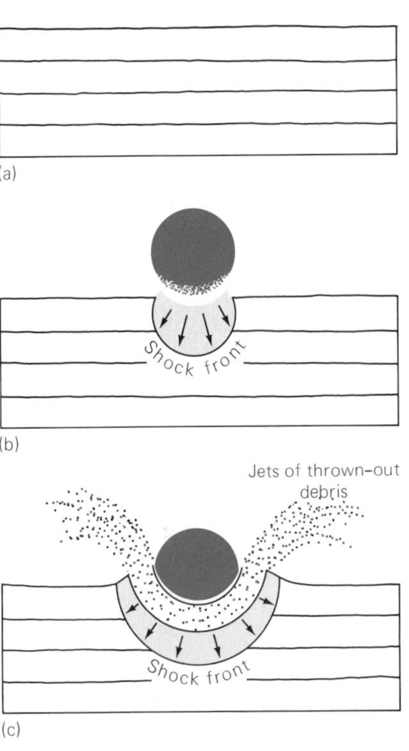


(d)

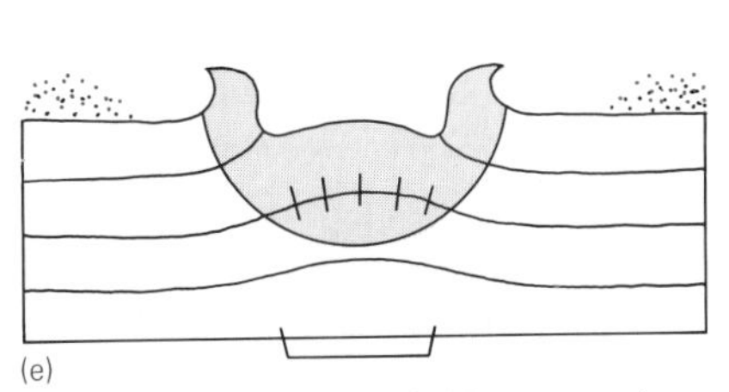


**9.5 Sequence of events during formation of an astrobleme.** (a) Meteorite about to strike Earth's surface. (b) Intitial impact and compression; meteorite starts to disintegrate; shock front advances radially downward (materials showing effects of shock metamorphism shown in gray). (c) Further disintegration of meteorite and advance of shock front; jets of debris begin to shoot out from the margins (compare Figure 10.3). Debris includes both meteorite fragments and pieces of shocked material struck by the meteorite. (d) Meteorite completely dissipated; debris begins to accumulate on ground outside the astrobleme and layers at margins are bent over upward and outward; faults and uplift begin in center of feature. (e) Final result; most meteorite debris outside the astrobleme. (After Donald E. Gault, William L. Quaide, and Verne R. Oberbeck, 1968, Impact cratering mechanics and structures, p. 87–99 *in* B. M. French and Nicholas M. Short, *eds.*, Shock metamorphism of natural materials: Baltimore, Maryland, Mono Book Corporation, 644 p., Figure 2, p. 88; Figure 11, p. 94, and Figures 19 and 20, p. 98; and H. B. Sawatzky, 1975, Astroblemes in Williston Basin: American Association of Petroleum Geologists, Bulletin, v. 59, no. 4, p. 694–710, Figure 2, p. 695)

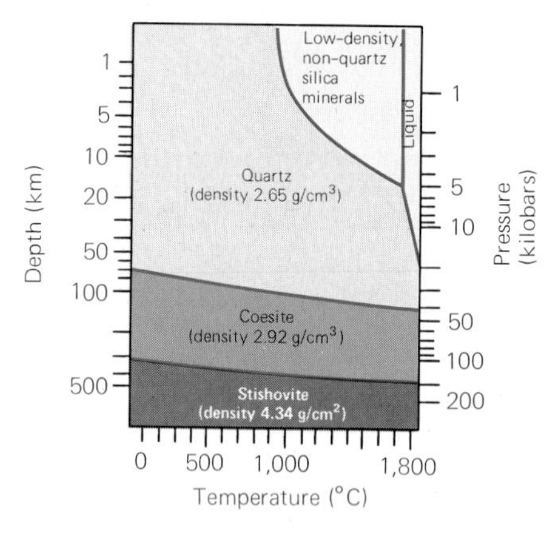


**9.6 Conditions required to form high-pressure varieties of $SiO_2$,** coesite and stishovite, shown on a graph of pressure *versus* temperature. (Replotted by J. E. Sanders from W. S. Fyfe, 1964, Geochemistry of solids. An introduction: New York, McGraw-Hill Book Company, 199 p.; Figure 11–4, p. 143)

temperatures and pressures that affect the rocks surrounding the point of impact. Breccias are formed and thrown out of the impact crater. High pressures may form the $SiO_2$ minerals coesite and stishovite (Figure 9.6).

The marks left on the Earth by the impact of a meteorite have been named *astroblemes.* Such features are not numerous on the Earth, but a few have been reported (Figure 9.7). The zones of shock metamorphism surrounding astroblemes are exceptions to the statement that metamorphic rocks form only deep beneath the Earth's surface.

## Metamorphic Rocks of Regional Extent

The largest volumes of the Earth's metamorphic rocks are distributed throughout two kinds of elongate zones of regional extent: (1) mobile belts that are related to mountain chains and (2) mylonite zones along major faults.

**Mobile belts** The great belts of metamorphic rocks associated with mobile belts are related to the origin of mountains and continents, subjects we analyze in Chapter 16 and 18. At this point, we merely mention a few examples to provide background for later sections of this chapter.

Within the United States belts of metamorphic rocks are exposed along the core of the Appalachians, from New England to Alabama; along the Sierra Nevada, in California; and in widely scattered circular or oval areas that have been greatly uplifted, such as the Adirondacks, the Ozarks, the Black Hills, the Wind River Mountains, and the Front Range (Figure 9.8). Although younger strata of sedimentary rocks blanket large areas between these zones of exposed metamorphic rocks, it is reasonable to infer that the "basement rocks" underlying these strata include many joined-together belts of metamorphic rocks. Many such belts have been found to compose the large region of Precambrian rocks known as the Canadian Shield. It is probably no exaggeration to state that the major belts of metamorphic rocks are characteristic parts of continents.

**Mylonite zones along major faults** Within some metamorphic rocks of mobile belts, major faults have not only displaced the rocks, but have brought about metamorphism. As a result, zones ranging in thickness or width from tens of meters to several kilometers have formed within which the rocks have been pulverized and so ground into pieces that they look like cherts or fine-grained volcanic rocks. This new kind of rock formed along the fault is named *mylonite* (see Appendix Figure C.5). Several well-developed zones of mylonite have been found within the belt of metamorphic rocks in eastern Connecticut.

**9.7 Astroblemes in North America** (dots on map). The large crescent near top center represents the world's largest astrobleme, whose diameter is about 450 kilometers. Arrow marks Barringer crater (Meteor Crater), Arizona, shown in Figure 20.25.

# CLUES FOR RECOGNITION OF METAMORPHIC ROCKS

Metamorphic rocks are recognized by their distinctive minerals, textures, fabrics, and structural features. Any rock that contains more than about 30 percent of mica and also displays large crystals of garnet almost surely is a metamorphic rock. Other clues include foliation, in the form of slaty cleavage and alternating layers of contrasting minerals, and a fabric

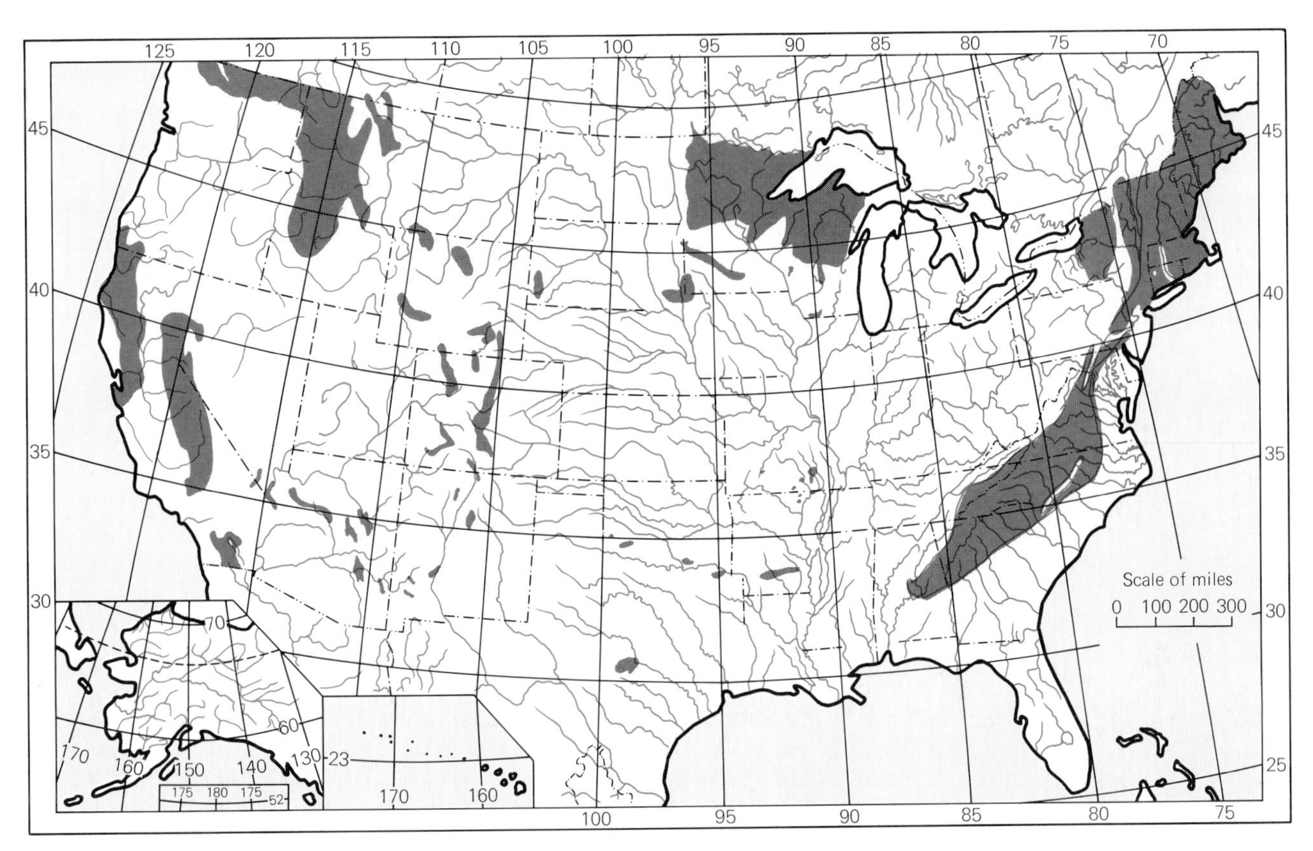


9.8 **Distribution of metamorphic rocks** in exposed mobile belts, conterminous United States. (U. S. Geological Survey).

displaying a pronounced preferred orientation of crystals. Extreme distortion of familiar shapes may have taken place, so that pillows (see Figure 5.32) or boulders (see Figures 8.10) appear as greatly elongated *blebs* or rods (Figure 9.9). If the rock consists largely of quartz, then the breaking test for distinguishing sandstone from quartzite should be applied. Sandstones break around the detrital particles, whereas quartzites break across their constituent quartz minerals (see Appendix Figures, C.22 and C.29).

9.9 **Distorted shapes** caused by the deformation that accompanies some kinds of metamorphism. (Both photographs by J. E. Sanders) (a) Elongated and flattened pillows in Precambrian rocks, southwestern Finland. (b) Distorted boulders, metamorphosed conglomerate of Carboniferous age, East Greenwich, Rhode Island.

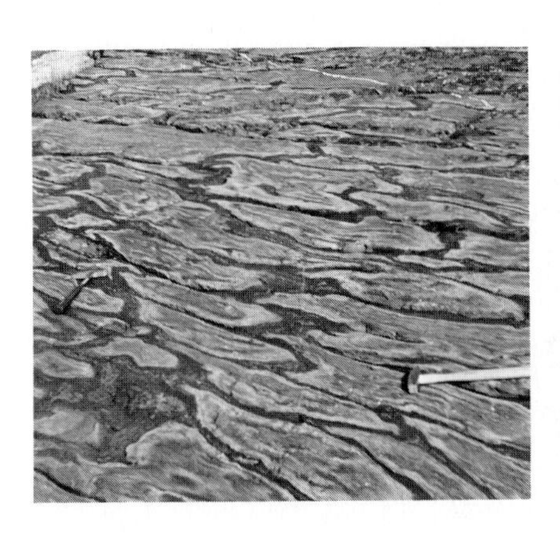

(a)

(b)

## THE CONCEPT OF METAMORPHIC FACIES

Taken literally, the term metamorphic facies should refer to the aspect of metamorphic rocks. However, as applied in the study of metamorphic rocks, the term means more than simply the aspect or appearance of the rocks. It refers to recurrent associations of minerals that have been found both in contact aureoles around plutons and in the major belts of metamorphic rocks.

### Metamorphic Facies in Contact-Metamorphic Zones

The basic principle of the concept of metamorphic facies is that if one begins with the same parent material, then one finds metamorphic mineral assemblages that are systematically related to metamorphic conditions. These conditions usually are temperature and pressure, but include as well the presence of various volatile materials, notably water vapor. We will illustrate the concept using only temperature and pressure.

If we make a plot of temperature versus pressure and show the pressure increasing downward from the top to the bottom of the graph, as it does in going deeper into the Earth, then we can summarize much of the information about the origin of metamorphic-mineral assemblages (Figure 9.10).

Along the top of the diagram, shown by the gray area, are conditions that are found in contact-metamorphic aureoles. In such aureoles, the chief variable is temperature; the pressure depends on the depth where the pluton intersects the country rocks. A position close to the pluton would be at the right side of the diagram. Notice the progression from pyroxene hornfels at the highest temperature to amphibole hornfels at lower temperatures. Not shown, but next on the left would be hornfels containing mica.

These changes are the basis for naming the distinctive mineral assemblages (or *facies*). The highest-temperature facies is the *pyroxene-hornfels facies.* In the middle range is the *amphibolite facies.* At the low-temperature end is chlorite schist, usually referred to as the *greenschist facies.*

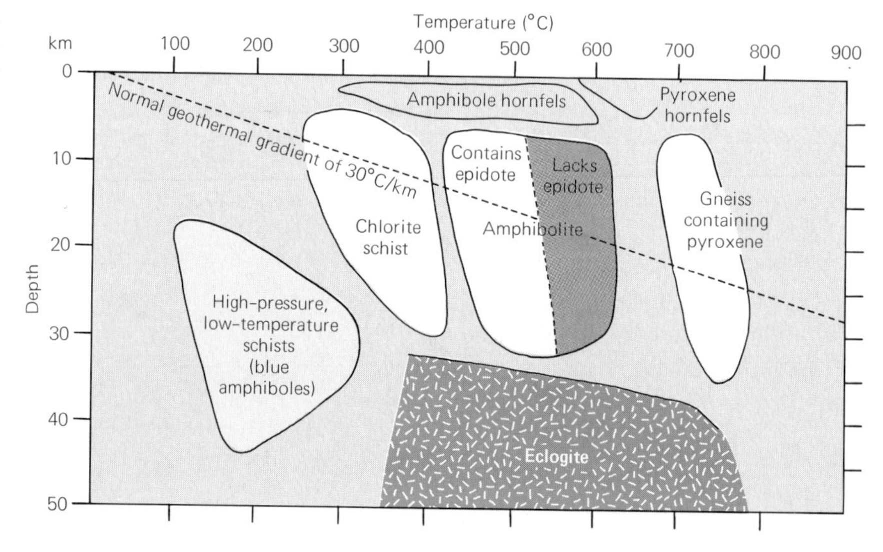


**9.10 Conditions of temperature and pressure related to contrasting assemblages of metamorphic minerals.** Eclogite is a high-pressure, high-temperature rock consisting of certain varieties of garnet and of pyroxene. Eclogite is thought to be present in the upper mantle; it is a kind of rock that does not fit easily into a simple classification of igneous rocks and metamorphic rocks.

## Metamorphic Facies in Mobile Belts: Isograds and Blue-Amphibole Schists

Results of regional studies of the metamorphic rocks of mobile belts have indicated that the same general associations of minerals which have been observed on a small scale in contact-metamorphic aureoles occupy large areas and are related to adjacent assemblages in the same kind of patterns seen in contact zones. The main difference is that in the mobile belts the volume of rock affected is enormous and the inferred temperature zones change more gradually from one to another than in contact zones. In mobile belts it has been found useful to subdivide the amphibolite facies into two categories, a lower-temperature *epidote-amphibolite facies* where epidote is present and *amphibolite facies* where epidote is not present. These facies appear in Figure 9.10 along the dashed line for the average geothermal gradient.

In mapping metamorphic rocks of mobile belts, a useful boundary can be drawn by plotting the limits of a given metamorphic mineral. On one side of such a boundary the rocks of appropriate composition contain the key mineral, whereas on the other side of this boundary the key mineral is not present. Such boundaries are named *isograds.* They are thought to represent surfaces of equal temperature at the time of metamorphism.

In many mobile belts a distinctive kind of schist containing blue amphiboles is present (represented by the blue area in Figure 9.10). These blue-amphibole schists are thought to be products of high pressure and comparatively low temperature, as might be experienced where a part of the lithosphere is depressed rapidly into the mantle. The plate-tectonics explanation of these schists is that they mark fossil subduction zones.

**Granitization** As mentioned briefly in Chapter 6, geologists have inferred that some granites are metamorphic rather than igneous. In the 1950s some geologists argued that not some, but nearly all granites had formed by extreme metamorphism. The name *granitization* was proposed for the processes of metamorphism that convert sedimentary rocks into granites.

A consensus was finally reached that recognized a feldspar-quartz-mica association which forms when appropriate materials encounter certain conditions of temperature and pressure. The feldspar-quartz-mica association, then, is analogous to one of the metamorphic facies, but it has never been "officially" named as such.

The conditions of granitization could be attained by *progressive metamorphism,* that is, by raising the temperature and the pressure on a body of sedimentary rocks. In Figure 9.11 progressive metamorphism is represented as movement from the upper left corner downward along the diagonal arrow. Granite-forming conditions are also attained by cooling a magma. This is equivalent to movement from the lower right corner upward along the arrow in Figure 9.11.

Viewed in this way, the theory of how a granite could form is generally agreed upon. However, the correct interpretation of any given granite may still be debated. Presumably, a granite formed by granitization would show some transitions between nonmetamorphosed sedimentary parent rock and the granite. Such granites might also display bedding features and a kind of combined granite-nongranite rock named

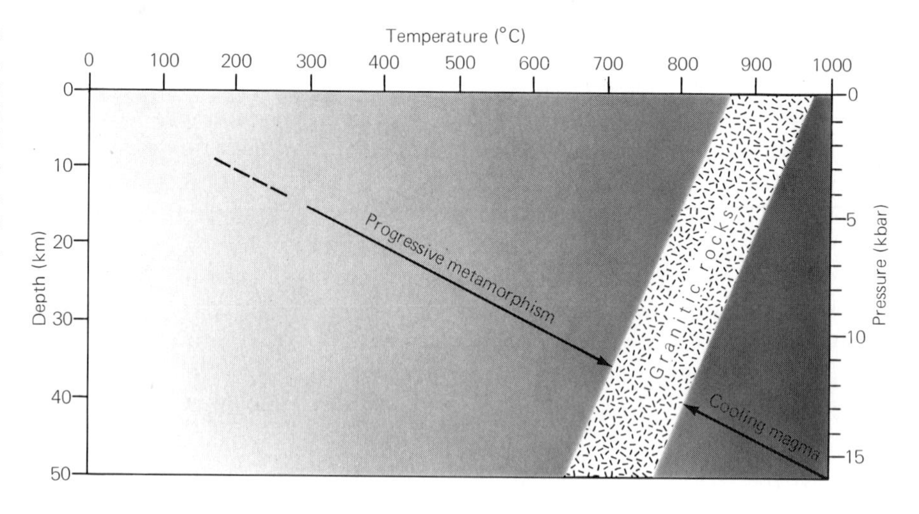


**9.11 Origin of granite** shown on a graph of temperature versus pressure, with pressure on vertical scale increasing downward, as in the Earth. The field marked granitic rocks can be entered from either side. During progressive metamorphism, entry is from the upper left. During the cooling of a magma, entry is from lower right.

*migmatite,* which means "mixed rock" (Figure 9.12). A typical migmatite shows the foliated appearance of a gneiss and, close by, the nonoriented, random textural pattern of granite. In migmatite thin stringers of schist or gneiss may fade imperceptibly into granite as if they had been dissolved. Wide zones of migmatite have been found around some granitic plutons.

## GEOLOGIC DATING OF METAMORPHIC ROCKS

Three geologically significant dates are associated with most metamorphic rocks: (1) the age of the parent materials, (2) the time of metamorphism, and (3) the age of exposure at the land surface. The dates may or may not be close to one another.

The geologic age of the parent rocks that have been metamorphosed, is usually determined by the geologic dating of sedimentary strata, using the methods discussed in Chapter 8.

The time of metamorphic reactions can be determined directly by using radioactive geochronology such as $^{40}K/^{40}Ar$ ratios. The date determined by this method is the time when the decay products stopped leaking out and began to accumulate in the crystal lattices. Usually this means the time of cooling. In some cases this time of cooling is the time of regional uplift, so that the lithosphere is elevated and the rocks brought up from deep, warm zones to shallower, cooler zones. In other cases, the change may result from some large-scale shifting of lithosphere plates so that the material that was hot is shifted into some cooler place, where the decay products would not escape from the crystal lattices.

**9.12 Migmatite formed during extreme metamorphism** under conditions that created granitic rocks. Ancient metamorphic rocks of Early Paleozoic (?) or Precambrian ages exposed in cuts along Interstate 95, Waterford, Connecticut. (J. E. Sanders)

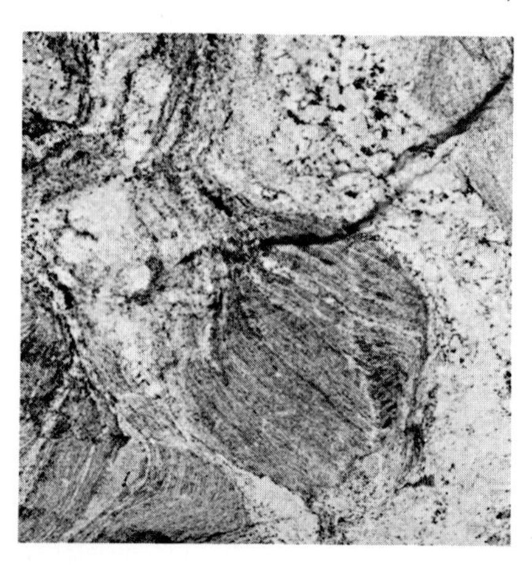

Multiple times of metamorphism may introduce complications. For example, in the New York City area, a rock unit named the Fordham Gneiss was metamorphosed during Precambrian time. Radiometric dates have been determined that give a result of approximately 1000 million years. No firm date is available on the age of the parent material of the Fordham Gneiss, but it was obviously older than the time of metamorphism. After being metamorphosed, the Fordham Gneiss was elevated and eroded. It was submerged by the sea in Early Cambrian time, and a body of quartz sand and pebbles, now the Lowerre Quartzite, was deposited. Next, a thick body of dolomitic carbonate materials accumu-

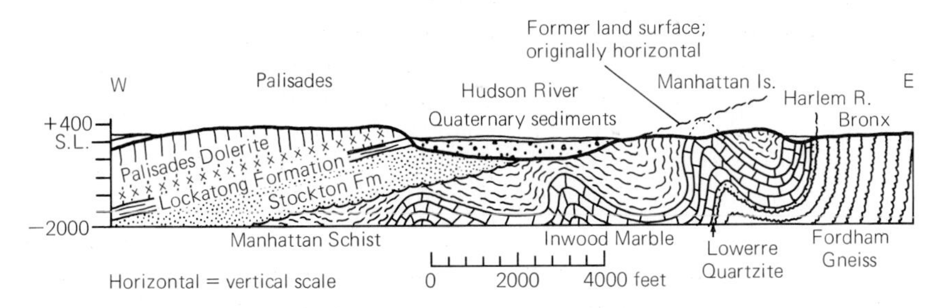


**9.13 Folded metamorphic rocks** of Paleozoic and Precambrian ages, New York City. Further explanation in text. (Redrawn by J. E. Sanders from C. P. Berkey, 1933, New York and vicinity: International Geological Congress, 16th, United States, Guidebook 9, New York Excursions, 151 p.; Plate 8 A, opp. p. 65)

lated, ending in Early Ordovician time. Then, the region subsided deeply, and a thick body of shale of Medial Ordovician age was deposited. Late in the Ordovician Period, another episode of metamorphism on a regional scale took place and the strata were folded (Figure 9.13). This event has been radiometrically dated at about 420 million years. Still later, at about 365 million years, which is in the middle of the Devonian Period, all the rocks were heated and thus metamorphosed once again. In most places in the vicinity of New York City, no matter what the rock unit, all radiometric ages come out to be about 365 million years. A radiometric date of 365 million years on a specimen of Fordham Gneiss, therefore, marks only the time of heating that has been the most recent. The so-called atomic clocks in the rocks were reset by the high temperatures to which the rocks were subjected in Medial Devonian time.

The upper limit on the relative geologic age of a body of metamorphic rock can be determined by finding the age of covering strata that have not been metamorphosed or by finding the age of covering strata that contain pebbles of the metamorphic rock. In either case, an important geologic indication is clear: the metamorphic rock had appeared at the Earth's surface after great uplift and was available either to be buried or to contribute sediment to accumulating strata. The age of the strata that cover the metamorphosed rocks of the New York City area (Figure 9.13) is Early Jurassic.

## ECONOMIC PRODUCTS ASSOCIATED WITH METAMORPHIC ROCKS

Both ore deposits and economic mineral deposits that are not ores are associated with metamorphic rocks.

### Ore Deposits of Metamorphic Origin

Two types of ores are associated with metamorphic rocks: ores in contact aureoles and ore deposits of various origin that have been metamorphosed.

In many contact aureoles, especially where a pluton cuts impure carbonate rocks, the products of metamorphism include ore minerals. We illustrate by describing the effects of two plutons, one mafic and the other felsic.

Our example of a mafic pluton is the body of dolerite (known also as *diabase*) that was emplaced along the northwestern and northern margin of the Newark-Gettysburg basin, near Cornwall, Pennsylvania (Figure

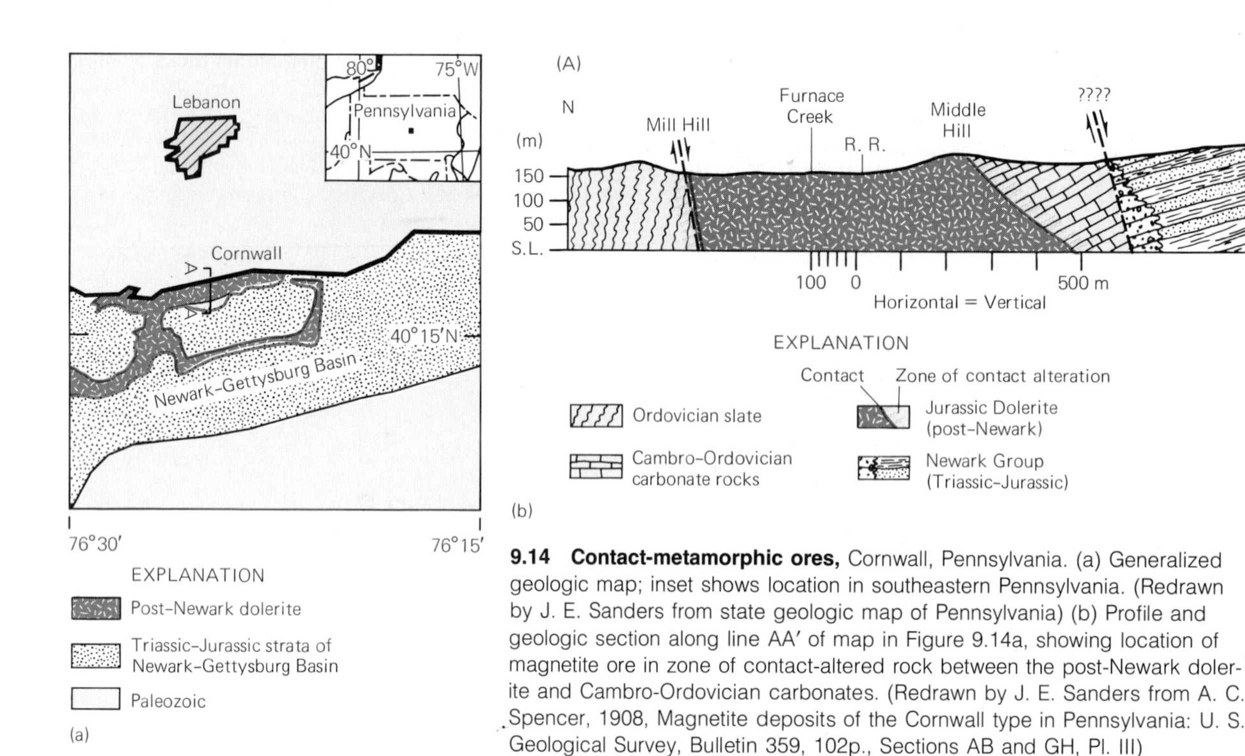


**9.14 Contact-metamorphic ores,** Cornwall, Pennsylvania. (a) Generalized geologic map; inset shows location in southeastern Pennsylvania. (Redrawn by J. E. Sanders from state geologic map of Pennsylvania) (b) Profile and geologic section along line AA′ of map in Figure 9.14a, showing location of magnetite ore in zone of contact-altered rock between the post-Newark dolerite and Cambro-Ordovician carbonates. (Redrawn by J. E. Sanders from A. C. Spencer, 1908, Magnetite deposits of the Cornwall type in Pennsylvania: U. S. Geological Survey, Bulletin 359, 102p., Sections AB and GH, Pl. III)

9.14). The mafic magma interacted with the Paleozoic carbonate rocks to form an ore deposit of magnetite.

Our example of a felsic pluton is the Bingham part of the Bingham–Last Chance Stock described in Chapter 6. Careful studies have been made of the effects of the felsic pluton on the carbonate rocks of Pennsylvanian age. The pattern of minerals formed in the contact zone is shown in Figure 9.15. As rich as these Bingham contact-metamorphic ore bodies are, they have contributed only about 23 percent of the total ore mined from the Bingham district.

In a few spectacular examples, ore deposits of one of the kinds previously discussed (in Chapters 6, 7, and 8) or even the numerous other kinds not mentioned, have been metamorphosed. During such metamorphism, the sulfide ore minerals do not react chemically with the silicates or carbonates, but may take on new textures as a result of the conditions of metamorphism. The unique assemblage of rare zinc minerals at Franklin, New Jersey, has been ascribed to metamorphism. Both the initial accumulation and the later metamorphism took place during Precambrian time. This age assignment is proved by the occurrence of the distinctive minerals as detritus in the overlying Lower Cambrian conglomerate.

## Nonmetallic Mineral Deposits of Metamorphic Origin

Nonmetallic resources associated with metamorphic rocks include slate, marble, graphite, kyanite, asbestos, talc, soapstone, and garnet. A few phyllites and schists break into flat slabs that can be used as sidewalks of patios.

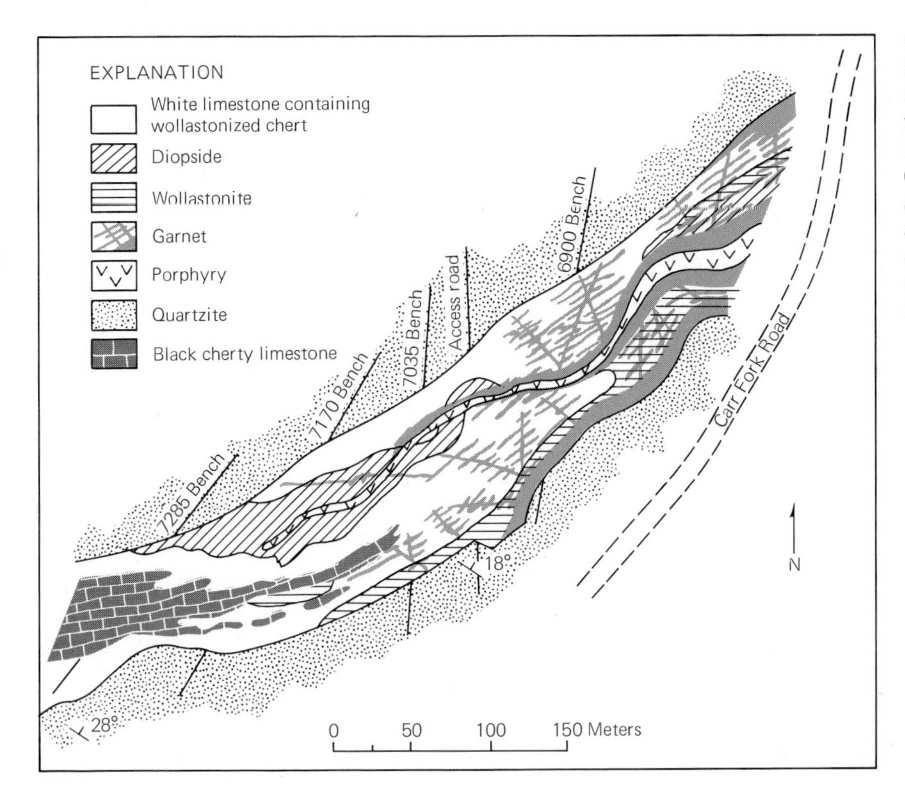


**9.15 Contact-metamorphic ore deposits** along NW margin of Bingham-Last Chance stock, Utah. (For further information about the geologic setting of the Bingham-Last Chance stock, see Figure 6.17.) (Modified from W. W. Atkinson, Jr., and M. T. Einaudi, 1978, Skarn formation and mineralization in the contact aureole at Carr Fork, Bingham, Utah, Economic Geology, v. 73, no. 7 (special issue), p. 1326–1365, Figure 10, p. 1347)

# CHAPTER REVIEW

1. Some controls on metamorphism are temperature, deformation, high pressure, the sudden elevation of both temperature and pressure that accompanies meteorite impact, and the introduction of new materials and removal of old materials.

2. Some metamorphic rocks are distributed in *local* fringes surrounding a well-identified local source of heat or of high pressure; others are of *regional* extent.

3. Zones of altered rock surrounding a pluton are named *contact-metamorphic aureoles.*

4. The material subjected to metamorphism is some kind of preexisting rock, called a *parent rock.* Parent rocks may be classified chemically into six major kinds: shales, sandstones and cherts, carbonate rocks, coal, mafic igneous rocks, and felsic igneous rocks.

5. The zones of shock metamorphism surrounding *astroblemes* are exceptions to the statement that metamorphic rocks form only deep beneath the Earth's surface.

6. The largest volumes of the Earth's metamorphic rocks are distributed throughout two kinds of elongate zones of regional extent: (1) *mobile belts* that are related to mountain chains and (2) *mylonite* zones along major faults.

7. Metamorphic rocks are recognized by their distinctive minerals, textures, fabrics, and structural features.

8. The basic principle of the concept of *metamorphic facies* is that if one begins with the same parent material, then one finds metamorphic mineral assemblages that are systematically related to metamorphic conditions.

9. Geologists have argued vigorously about the origin of granite. Those scientists who do not believe that granite was formed by *granitization* argue that at least some granites may result from the solidification of magma. One factor that illustrates how closely granites and metamorphic rocks are related is the existence of *migmatites,* or "mixed rocks."

10. Metamorphic rocks may be dated using techniques of radioactive geochronology that determine the time of cooling. Two other geologically significant dates are the age of the parent materials and the age of exposure at the land surface.

11. Ore deposits associated with metamorphic rocks may consist of ore minerals that are the products of metamorphism or may be preexisting ore deposits that have been metamorphosed. Economic mineral deposits that are not ores are also associated with metamorphic rocks.

## QUESTIONS

1. Explain the origin of the term *metamorphism*. Who introduced the concept of metamorphism into geology and when?
2. Describe the position of metamorphic rocks in the rock cycle.
3. Name two contrasting geologic settings in which metamorphic rocks are found.
4. What is a *contact-metamorphic aureole?* Discuss two kinds of information that can be obtained from studying these aureoles.
5. Describe the metamorphic products in contact-metamorphic aureoles of: (a) impure carbonate rocks, (b) shales, (c) basalt, (d) coal, and (e) kerogen.
6. Name two distinctive minerals that typically occur near impact craters.
7. Compare the kinds of metamorphic rocks in a mobile belt and along a major fault.
8. List some of the distinctive features of metamorphic rocks.
9. Define *metamorphic facies.* Compare the metamorphic facies of contact-metamorphic aureoles with those of mobile belts.
10. Define *isograd.* What are isograds thought to represent?
11. What is thought to be the geologic significance of *blue-amphibole schists?*
12. What is meant by *granitization?* Define *migmatite.*
13. Defend or attack the idea that granite should be added to the list of metamorphic facies.
14. What are the geologically significant dates of a metamorphic rock? Discuss radiometric dating of metamorphic rocks in terms of these significant dates.
15. List three kinds of economic products that may be associated with metamorphic rocks.

## RECOMMENDED READING

Ernst, W. G., 1969, *Earth Materials:* Englewood Cliffs, New Jersey, Prentice-Hall, Inc., 149 p.

James, H. L., 1955, Zones of Regional Metamorphism in the Precambrian of Northern Michigan: *Geological Society of America, Bulletin,* v. 66, no. 12, p. 1455–1488.

Hyndman, D. W., 1972, *Petrology of Igneous and Metamorphic Rocks:* New York, McGraw-Hill Book Co., 533 p.

Miyashiro, Akiho, 1972, Metamorphism and Related Magmatism in Plate Tectonics: *American Journal of Science,* v. 272, no. 7, p. 629–656.

Muffler, L. J. P., and White, D. E., 1969, Active Metamorphism of Upper Cenozoic Sediments in the Salton Sea Geothermal Field and Salton Trough, Southeastern California: *Geological Society of America, Bulletin,* v. 80, no. 2, p. 157–181.

Pecora, W. T., 1960, Coesite Craters and Space Geology: *Geotimes,* v. 5, no. 2, p. 16–19.

Vokes, F. M., 1969, A Review of the Metamorphism of Sulphide (*sic*) Deposits: *Earth-Science Reviews,* v. 5, no. 1, p. 99–143.

Walton, M. S., Jr., 1955, The Emplacement of "Granite": *American Journal of Science,* v. 253, no. 1, p. 1–18.

THE HYDROLOGIC CYCLE
GROUNDWATER
EXTRACTION OF GROUNDWATER
THE GEOLOGIC WORK OF GROUNDWATER

# CHAPTER 10 THE HYDROLOGIC CYCLE AND GROUNDWATER

PLATE 10
Cave deposits (speleothems), Temple of the Sun, Carlsbad Caverns, New Mexico. Ground water first dissolved away the rock to form the open space and subsequently has been filling in the space as calcium carbonate is precipitated out of the water dripping from above and flowing along the columns that have formed where stalactites hanging down from the ceiling have united with stalagmites growing upward from the floor. (U. S. National Park Service)

WITH THIS CHAPTER, WE BEGIN THE STUDY OF THE MANY PROCESSES that take place at the surface of the Earth, where the lithosphere, the atmosphere, the hydrosphere, and the biosphere all interact. A central theme will be water in all of its forms: vapor, liquid, and solid. In this chapter, we will review the hydrologic cycle and discuss *groundwater,* the water that is present below the surface of the ground within the pores of the regolith and the bedrock. In Chapter 11, we examine downslope of material by gravity transport, a subject closely connected to the pore-water situation. Chapter 12 summarizes the mechanics of flowing water in rivers and the geologic work of streams. Chapter 13 is devoted to solid water, as ice and glaciers, and to the vast cyclic changes in the hydrosphere that have taken place during the past few million years as the world's climate has oscillated back and forth between ice ages when the temperature was colder than now, and comparatively ice-free times, when the temperature was warmer than now. Chapter 14 discusses deserts, where water is scarce. Chapter 15 explores the oceans and shorelines.

## THE HYDROLOGIC CYCLE

As mentioned in Chapter 1, the hydrologic cycle (Figure 10.1) is the cyclic movement of all the Earth's water in all three states: (a) solid, as in polar ice caps, glaciers and snowfields; (b) liquid, including fresh water and salt water both on the surface of the Earth and beneath the surface; and (c) gas, as the water vapor in the atmosphere. The scientific study of water is *hydrology*.

The vast work done by water circulating within the hydrologic cycle was understood by James Hutton. In the words of Hutton's friend, John Playfair:

> *We have been long accustomed to admire that beautiful contrivance in nature, by which the water of the ocean, drawn up in vapour by the atmosphere, imparts, in its descent, fertility to the earth, and becomes the great cause of vegetation and of life; but now we find, that this vapour not only fertilizes, but creates the soil; prepares it from the solid rock, and, after employing it in the great operations of the surface, carries it*

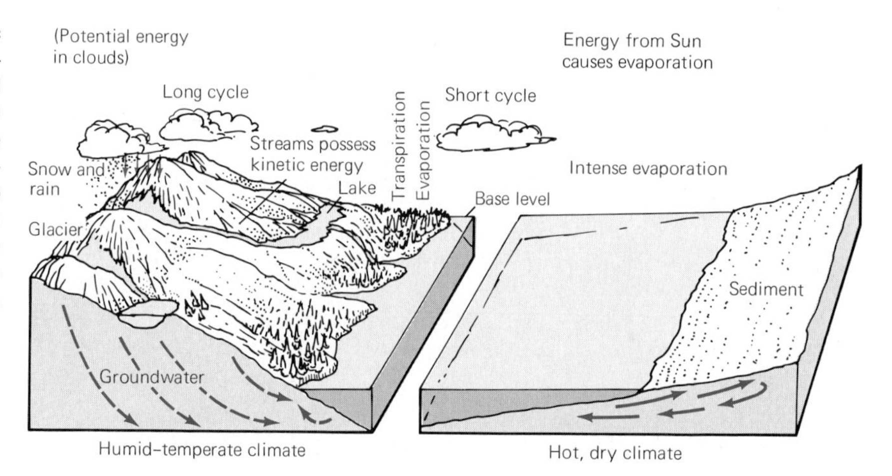


**10.1 The hydrologic cycle,** schematic block diagrams showing differences between circulation in hot, dry climate and humid-temperate climate. The *short cycle* refers to rain that falls directly into the ocean. The *long cycle* designates moisture that falls on the land and returns to the ocean via streams, evapotranspiration into the atmosphere from plants, via underground movement, or from melting icebergs.

*back into the regions where all its mineral characters are renewed. Thus, the circulation of moisture through the air, is a prime mover, not only in the annual succession of the seasons, but in the great geological cycle, by which the waste and reproduction of entire continents is circumscribed. (John Playfair, 1802, Par. 125, p. 128; facsimile reprint, University of Illinois Press, 1956.)*

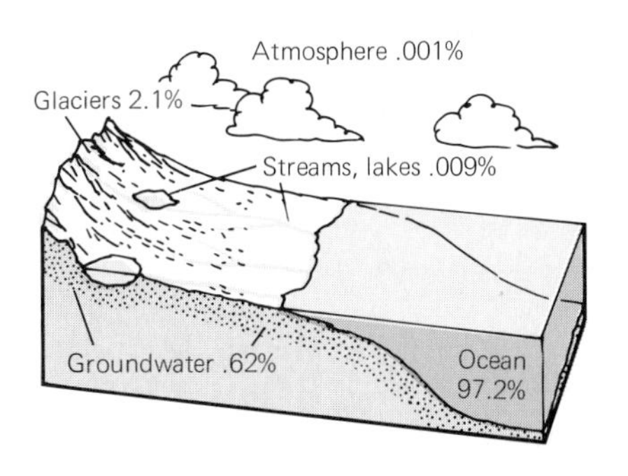


**10.2 Distribution of the world's water,** schematic block diagram. (Estimates from U. S. Geological Survey)

The total quantity of the Earth's water is about 1.36 billion cubic kilometers ($1360 \times 10^6$ km$^3$). Hydrologists estimate that this water is distributed as shown in Figure 10.2.

The water in the atmosphere may return to the Earth as precipitation almost immediately and near the point where it left the surface, or it may remain in the atmosphere for weeks and travel thousands of kilometers in upper-air currents. Water that evaporates from the surface of the sea may fall back directly into the sea. Water that falls on land returns to the sea by one of several paths. The adventures of water along these paths depend upon the climate and the kind of surface on which the water falls.

In our first close look at the hydrologic cycle, we have chosen the conditions within a humid-temperate climate, such as the climate that prevails within much of the conterminous United States. As we will see, important activities take place on the surface and several factors control the entry of water into the ground.

## Raindrop Spatter, Infiltration, and Runoff

Raindrops striking fine-grained regolith spatter silt upward with the splashed-up water surrounding the tiny impact craters (Figure 10.3). Such splashing is a powerful agent of erosion. A dramatic illustration of the effects of raindrop splash on silt can be provided by leaving a board on bare soil during a rainstorm. Afterward, the board's upper surface will be covered with a layer of silt.

In its natural, nondisturbed state, typical regolith is capable of absorbing most or all of a light rainfall or a moderate rainfall. By the process of *infiltration,* rainwater sinks into the Earth through natural passageways between particles of regolith or through cracks in bedrock (see Figures 6.23 and 7.4). Infiltration is governed by many factors, including the conditions within the pores, slope, vegetation, and human use of the land. Where the pores at the surface of the regolith are open and dry, maximum infiltration takes place. If the pores in the regolith are saturated with water, or if the water in the regolith is frozen, little infiltration takes place. Rainfall tends to flow down steep slopes before much of it can sink into the regolith. Vegetation can affect the amount and depth of infiltration. A dense growth of plants can prevent water from flowing rapidly across the surface, and thus can increase the amount that infiltrates. However, because the roots of rapidly growing plants soak up much or all of the water that infiltrates the regolith, vegetation can control the depth of infiltration. Water taken up by plants is stored or transpired back into the atmosphere via leaves. When plants are dormant, or where the vegetative cover is sparse, precipitation (except that which falls in frozen form) may infiltrate freely. If the regolith has been built on or paved over, infiltration cannot occur at all.

**10.3 Impact crater and radial lines of splashed-up material** from raindrop falling on wet soil. The tiny droplets travel upward and outward, carrying along silt particles. Compare Figure 9.5(c). (Official U. S. Navy Photograph)

Where rainwater accumulates faster than it can infiltrate, the excess flows downhill as a surface sheet or a surface film. Water flowing over the land is named *runoff.* Most runoff becomes organized into flows within definite channels, and then it is called *stream flow.*

After this brief review of how rainfall reacts with the surface of the Earth, we turn our attention to groundwater. We continue with runoff and stream flow in Chapter 12.

## GROUNDWATER

Our objectives in studying groundwater are: (1) to learn about the principles that govern its distribution and behavior; (2) to find out how such knowledge affects the daily use of groundwater; and (3) to examine the geologic effects of the water under the ground. We touched on some of these topics in connection with the effects of water in weathering (Chapter 7) and the effects of subsurface water in lithification of sediments (Chapter 8).

### Zones of Groundwater; The Water Table

Groundwater is stored in several zones (Figure 10.4). Immediately below the surface, and extending to depths as great as several hundred meters, is a zone in which the pore spaces are occupied partly by air and partly by water. This is defined as the *zone of aeration.* The groundwater within the zone of aeration is named *vadose water,* from the Latin word meaning "shallow." The top of the zone of aeration typically is a *zone of soil moisture.* The lower part of the zone of aeration consists of a zone of capillary water named the *capillary fringe.*

Underlying the zone of aeration is the *zone of saturation,* within which subsurface pores are completely filled with water under hydrostatic pressure. The upper boundary of the zone of saturation is the *water table.* The position of the water table is subject to considerable fluctuation. We examine such fluctuation in a following section after we have summarized the sources and storage of groundwater.

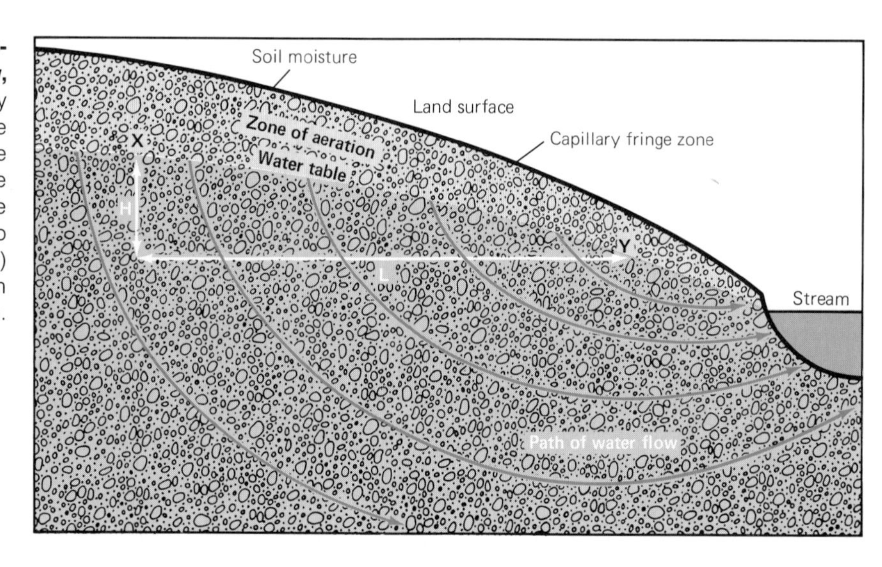


**10.4 Hydraulic gradient creates pressure that causes groundwater to flow,** schematic profile through hill underlain by porous, permeable sand and gravel. The hydraulic gradient equals the slope of the water table; this slope is calculated by the formula H divided by L, where H is the vertical difference in height between two points (such as X and Y on the sketch) and L is the horizontal distance between them.

### Sources of Groundwater

Groundwater can be divided into three categories: (1) fresh water, immediately beneath the surface and extending to variable depths; (2) salt water, below the level of the fresh water; and (3) juvenile water.

All shallow groundwater comes from *meteoric water,* which is defined as water that falls out of the atmosphere as rain or as snow. Part of this water infiltrates the regolith or the bedrock, where it is stored. The proportion of fresh groundwater to the other bodies of fresh water is shown in Figure 2.3. Meteoric groundwater generally is drinkable. Except near hot springs and very close to the surface, meteoric groundwater displays the remarkable property of constant temperature, usually varying less than a degree from the mean annual air temperature of an area.

The deeper, salty subsurface water is ancient seawater that has been buried with sediments and later has reacted so that it is very much saltier than ordinary seawater. In the oil fields, this water is named *formation water.* The name comes from the finding of this water in the same rock formations that contain oil or gas.

Juvenile water is new water that joins the hydrologic cycle for the first time. Juvenile water is thought to come from the mantle (see Chapter 5). Ultimately, all of the Earth's water should be classified as juvenile water; all of it is thought to have come from the mantle.

Most of the discussion of the groundwater in the remainder of this chapter centers on the shallow, meteoric groundwater. Where any other kind of groundwater is meant, we will so indicate.

### Storage of Groundwater

Groundwater is stored in three major kinds of underground openings or pores: (1) a framework of rigid openings around particles composing regolith and sedimentary rocks; (2) cracks, joints, and faults in massive bedrock (Figure 10.5); and (3) huge caverns, caves, and original lava tubes and other kinds of openings in volcanic rocks (Figure 10.6). Two general characteristics that affect water underground are porosity, the capacity of a material to store a fluid, and permeability, the capacity of a porous material to transmit a fluid. Most of the information on porosity and permeability has been obtained from study of tiny pores between the framework particles of regolith and of sedimentary rocks.

**Porosity** The *porosity* of a body of material is defined as the proportion of pore space to total volume. One can compute the porosity by dividing the volume of pores in a sample by the total volume of the sample (solids plus pores) and converting the result into a percentage. Stated as an equation,

$$\text{porosity (percent)} = \frac{\text{volume of pores}}{(\text{volume of pores} + \text{volume of solids})} \times 100$$

As an example, let us calculate the porosity of a cubic meter of sand within which it is possible to store 200 liters of water. Because 1 liter is equal to 0.01 cubic meter, 200 liters occupy 0.2 cubic meter. Therefore, the volume of the pore space is 0.2 cubic meter, and the porosity is 20 percent ($0.2/1 \times 100$).

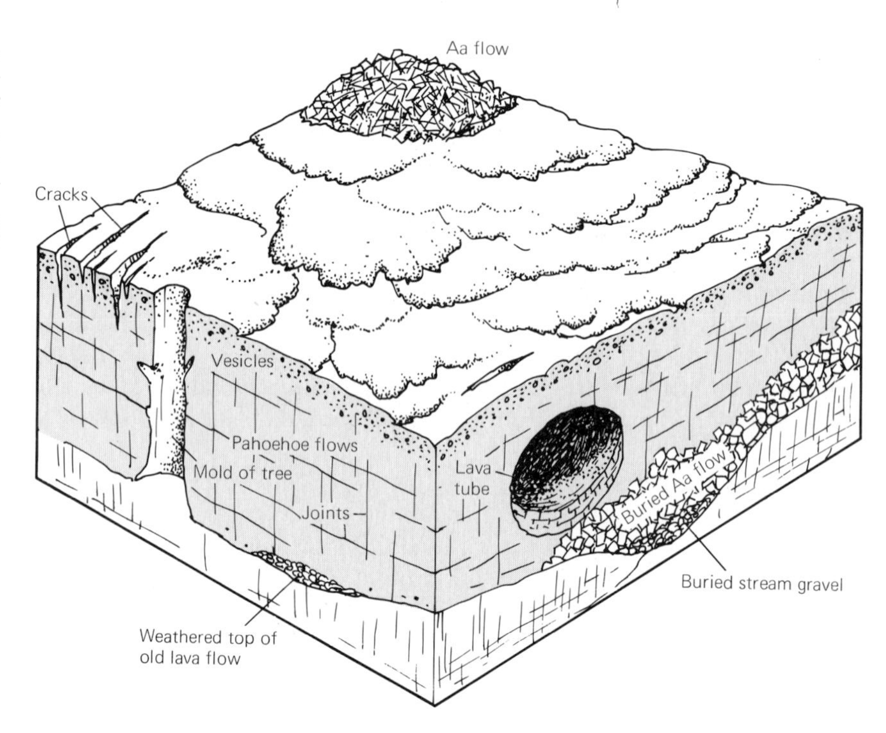


**10.5 Pore spaces in volcanic rocks,** succession of two sheets of extrusive igneous rock formed by cooling of former lava flows, schematic block diagram. (After S. N. Davis and R. J. M. deWiest, 1966, Hydrogeology: New York-London-Sydney, John Wiley and Sons, 463 p., Figure 9.8, p. 335)

**Permeability** The property known as *permeability,* a material's capacity to transmit fluid, is determined by the sizes and continuity of the openings. Highly permeable material is necessarily also highly porous, but the reverse is not always true. A highly porous material need not be highly permeable. This seeming paradox can be understood by considering the factors that affect porosity and permeability.

**Factors affecting porosity and permeability** Many factors work simultaneously to affect porosity and permeability. A few simple examples will show how some of these factors are interrelated. Important properties of sediments in this connection are sizes of particles, shapes of

**10.6 Large pore spaces in two lava tubes,** Surtshellir Cave, Iceland. (J. R. Reich, Jr., Speleo-Research Associates)

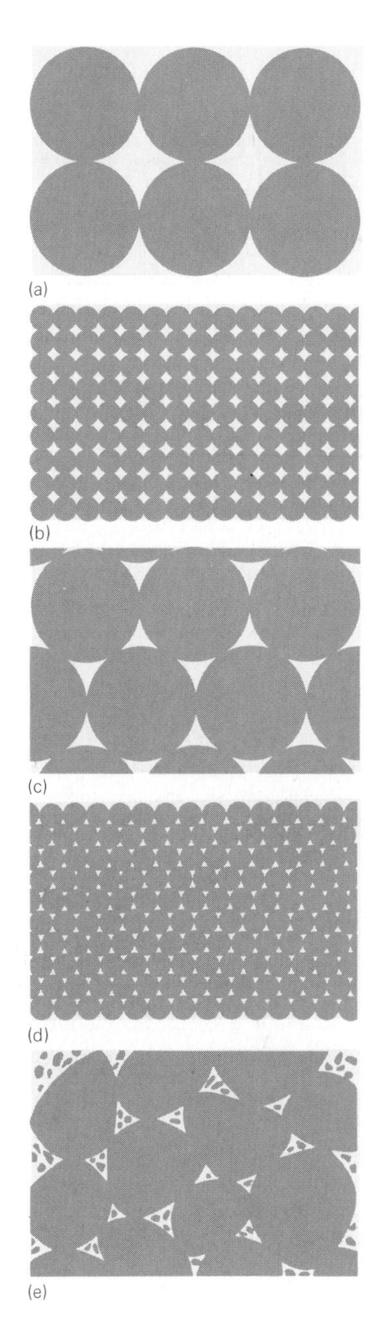


**10.7 Effects on porosity of various packing and sorting of sedimentary particles.** (a) and (b) Each sample with one-sized spheres in cubic packing. Porosity of (a) = (b) = 46.7 percent; permeability of (a) > (b). (c) and (d) Close-packed spheres. Porosity of (c) = (d) = only 26 percent; permeability of (c) > (d). (e) **Small particles filling spaces between larger particles in poorly sorted sample,** thus reducing both porosity and permeability over the values that would be present if the sample contained only the large-sized particles.

particles, packing, and sorting. Consider first various collections of spheres, each collection consisting of but a single size of spheres that are packed in the same way, cubic packing (see Figure 4.2). Because only spheres of one size are present, the sorting is perfect. The porosity of all such collections of spheres, whether large or small, is identical: 46.7 percent (Figure 10.7a, b). This is true because in each sample, the proportion of pores to total volume is constant. Although the porosities of all such collections of spheres are constant, the permeabilities are not the same. The permeability of the sample having the large spheres is greater because the openings between the larger spheres are bigger than the openings between the smaller spheres. More water can flow faster through the larger openings than through the smaller openings.

If we take the same collections of spheres and change the packing from cubic to hexagonal (see Figure 4.4), we find that the porosity has been reduced to 26 percent (Figure 10.7c, d).

If we reduce the sorting by adding other sizes of particles to our samples, the small particles fill the spaces between the large particles (Figure 10.7e). In this case, both porosity and permeability are reduced.

The factor of shapes of particles becomes significant when one compares the properties of sands, whose particles are usually spherical, with clays, in which the particles are flaky and irregular (see Figure 8.3). Because of the open packing of the clay flakes and the large spaces between flakes, the porosities of clays are high, 60 to 70 percent or so. Because of the small sizes of the pores in clays, however, the permeabilities of clays are low. Accordingly, clays may contain much water, but very little water will pass through. The permeability of a sand having a porosity of only about 25 percent is higher than that of clay having porosity of 70 percent.

Addition of cement reduces porosity and permeability. The cement is a solid that occupies former pore space.

## Mechanics of Groundwater Flow: Hydraulic Gradient

The movement of groundwater depends on the *hydraulic gradient,* which is a response to two factors: (1) permeability of the subsurface material and (2) the pressure gradient within the water. This relationship was discovered more than a century ago by a French hydraulic engineer named Henry Darcy (1803–1858). Known as *Darcy's Law,* it states that the flow rate through porous material is proportional to the pressure driving the water, and inversely proportional to the length of the flow path.

Groundwater flows at widely varying speeds. Where the hydraulic gradient is three meters per kilometer and permeability is low, groundwater may move only 0.00075 meters per day. Where the hydraulic gradient is 30 meters per kilometer and permeability high, groundwater may move more than four meters per day. Field tests indicate that the

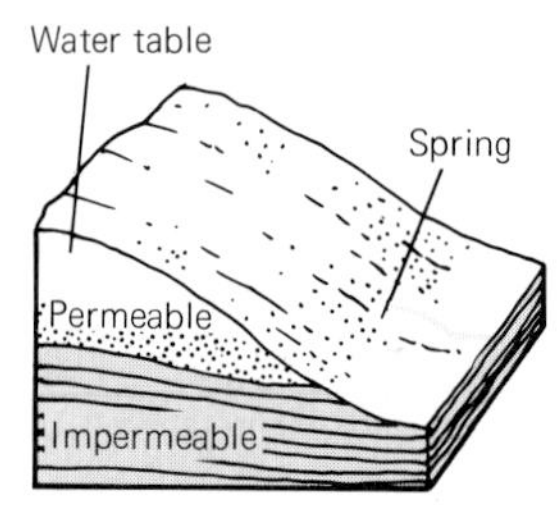

(a) At upper surface of an impermeable layer

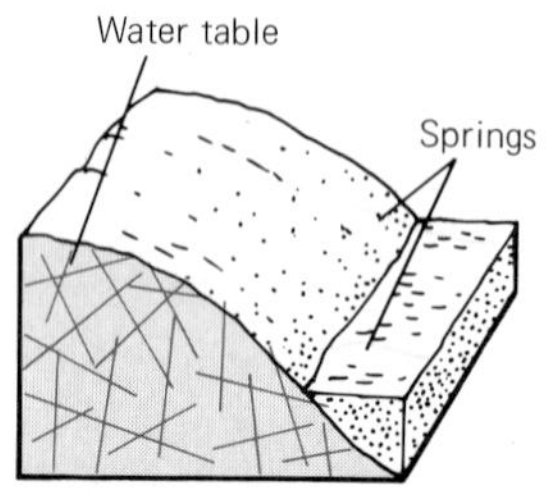

(b) In fractured massive rock such as granite

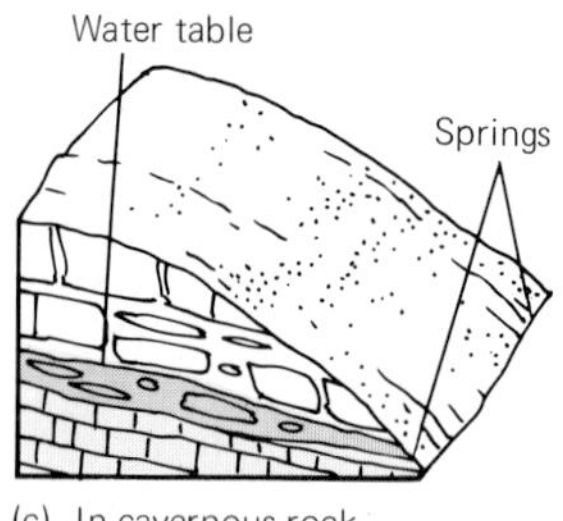

(c) In cavernous rock such as limestone

**10.8 Common geologic conditions responsible for springs.**
(a) Impermeable floor under permeable water-bearing layer.
(b) On hillside where water escapes from fractures.
(c) At base of hill where land surface intersects subsurface caverns.

normal range of groundwater flow is from 1.5 meters per year (with low pressure through fine sand) to 1.5 meters per day (with high pressure through coarse gravel), although speeds of 30 meters per day have been found under conditions of extremely high pressure and high permeability.

**Discharge of groundwater** Because of the hydraulic gradient, groundwater tends to flow from localities having high pressure to localities having lower pressure. Flow of groundwater to the surface or into a well tends to lower the water table; we speak of such flow as the *discharge of groundwater.* Discharge of groundwater takes place through seepage out of porous, permeable materials; through springs; or through wells.

Groundwater seeps out of the ground at numerous, widely spaced locations. Such seepage exerts a considerable influence on the stability of slopes, as shown in Chapter 11. Seepage also supplies water to swamps, lakes, and streams (Chapter 12).

A surface stream of flowing water that emerges from the ground is a *spring* (Figure 10.8). O. E. Meinzer (1876–1948), former head of the groundwater branch of the U.S. Geological Survey, classified *gravity springs* (springs flowing as a result of the hydraulic gradient) into eight categories according to their amounts of flow. He listed 65 "first-magnitude" springs in the United States, discharging more than 27 cubic meters per second. Several hundred second-magnitude springs yield 3 to 27 cubic meters per second, and thousands of third-magnitude springs discharge 1 to 3 cubic meters per second.

A *well* is a hole that is dug or drilled into the zone of saturation from which one can draw water. At one time farmers and homesteaders could retrieve all the water they needed from shallow cylindrical wells dug with pick and shovel and lined with bricks or stone. Nowadays, most wells are drilled by powerful machinery and equipped with electric pumps. Wells for agriculture and industry have been drilled to depths of 300 meters or more and their diameters may be 30 to 40 centimeters. Such wells may yield millions of liters per day. Household wells are smaller; they typically draw as little as about two liters per minute (120 liters/hr, 2880 liters/day).

Whenever water is removed from a well, the water table in the vicinity of the well is lowered. This is an effect known as *drawdown*. The extent of drawdown decreases at greater distances from the well, so that a *cone of depression* forms around the well (Figure 10.9). The cone of depression increases the hydraulic gradient; this causes groundwater near the well to flow faster toward the well. According to Darcy's law, at a certain point this increased-flow effect ceases. A cone of depression may extend as far as 16 kilometers from a big well. Overlapping cones of depression from many wells may lower the water table throughout a large area.

**10.9 Cone of depression in water table surrounding pumping well,** schematic profile. Further explanation in text.

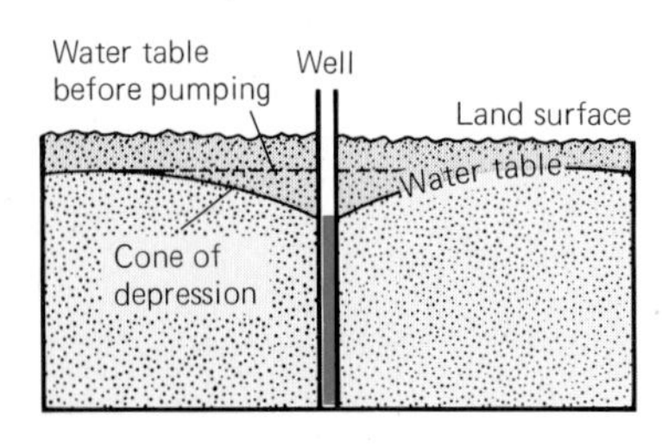


## EXTRACTION OF GROUNDWATER

In our consideration of the human uses of groundwater, we begin by analyzing aquifers and then review some of several conditions affecting use of groundwater.

### Aquifers

Saturated bodies of regolith or of bedrock that are porous and permeable and through which groundwater can flow readily are named *aquifers,* from the Latin meaning "water bearer." Two conditions must be satisfied: (1) saturation and (2) porosity and permeability. A porous, permeable gravel that is not saturated is not an aquifer, because no flowable water is present. A clay that is saturated with water is not an aquifer, either, because the water will not flow from the clay into a well.

The aquifers having the greatest flow capacities consist of clean, coarse gravel. Next are uniform coarse sands. In order of decreasing flow capacities are: some mixtures of sand and gravel, sediments of finer grain such as fine sand and silt, and finally residual regolith. Among consolidated rocks, sandstones form the most valuable aquifers. Next come limestones, which yield good water supplies from solution passages. Extrusive igneous rocks that possess fractures, connected vesicles, and other irregularities often yield as much water as limestones (see Figure 10.6).

Depending on their kinds of upper boundaries, aquifers may be classified as nonconfined or confined. In a *nonconfined aquifer,* the water table forms the upper surface of the zone of saturation (Figure 10.10a). Contour maps and profiles of a water table can be prepared from the elevations of water in wells drilled at different points to the surface of a nonconfined aquifer.

Groundwater that is "held in" by an overlying stratum having low permeability forms a *confined aquifer* (Figure 10.10b). Any well penetrating a confined aquifer is an *artesian well* (Figure 10.11). In an artesian well, the pressure on the confined aquifer drives the water upward, in some cases, to and above the Earth's surface.

The shapes, thicknesses, and extents of aquifers differ greatly. In Texas, the geologic formation known as the Carizzo Sand is an aquifer that ranges in thickness from about 30 to 60 meters and provides potable water to an area 30 to 80 kilometers wide and hundreds of kilometers long. By contrast, some of the permeable glacial deposits in the midwest form tiny aquifers only a few meters thick and less than a few square kilometers in extent.

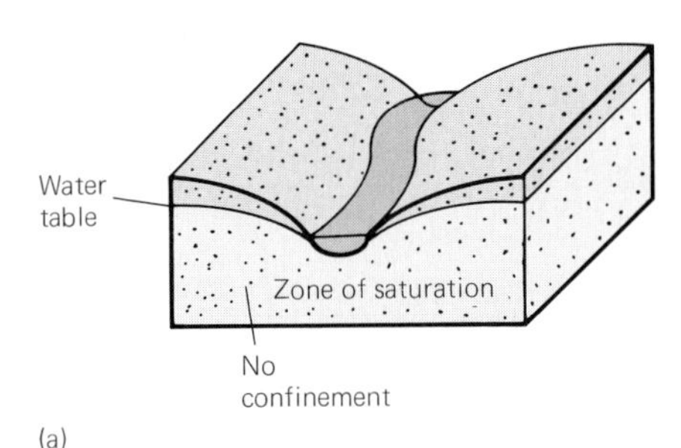


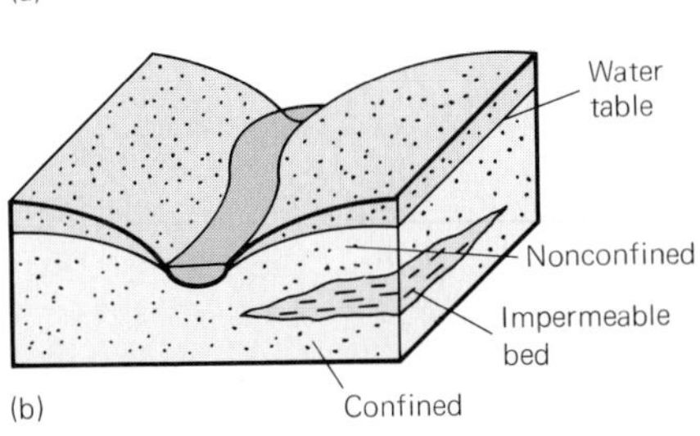


**10.10 Nonconfined and confined aquifers,** schematic block diagrams.
(a) In nonconfined aquifer, water table forms a subdued replica of the morphology of the land surface.
(b) In a confined aquifer, the top of the water table is the base of an impermeable layer through which water does not flow.

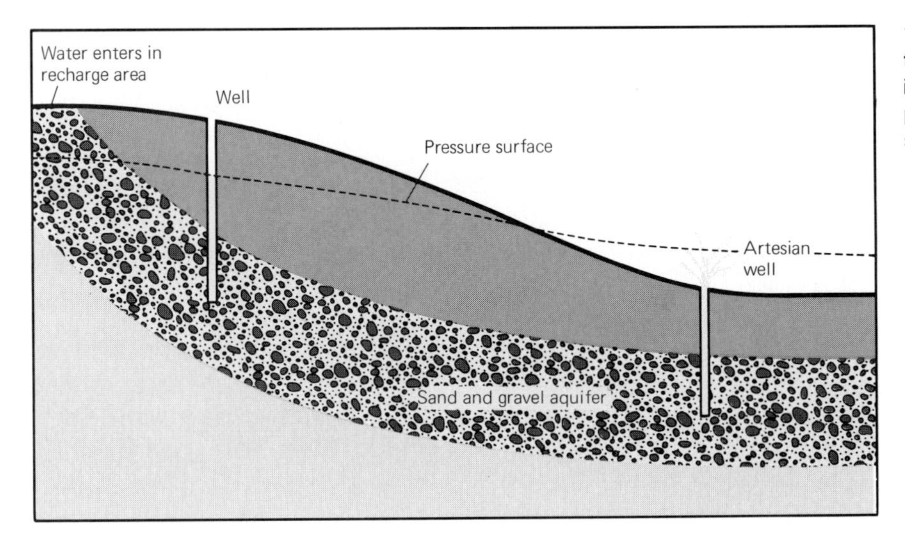


**10.11 Artesian well forms natural fountain** because level of ground at well head is at a lower elevation than that of the pressure surface within confined aquifer, schematic profile and section.

## Conditions Affecting Use of Groundwater

**Discharge and recharge of aquifers** The supply of groundwater in an aquifer may be actively involved in the operations of the hydrologic cycle or it may be a sealed deposit of water that is no longer participating in the cycle. The response of the aquifer to discharge differs drastically in the two situations.

Where an aquifer is actively participating in the modern hydrologic cycle, one can compare the water table to the balance in a bank account. The balance is a response to the deposit or withdrawal of funds. If deposits exceed withdrawals, the balance increases; if withdrawals exceed deposits, the balance shrinks. So it is with the water table of a nonconfined aquifer. The natural deposits to an aquifer consist of recharge by rainfall or other surface water connected to the underground pores. Withdrawals are natural discharge through seepage and springs and discharge water pumped from wells. The water table fluctuates up or down in response to the balance between recharge and discharge.

The water in an aquifer that is no longer participating in the hydrologic cycle is not being recharged. The situation can be compared to an inheritance, a lump sum to which no new funds are ever added. Spending reduces the sum until eventually the money is gone. The water can be drained only once; the supply situation resembles that in an oil well.

Prudent use dictates that the balance between pumping and recharge be maintained, but in some localities, the need for water is so great that aquifers are being pumped in excess of recharge. For example, since 1911, farmers and ranchers in the arid plains of Texas have pumped water from the Ogallala Formation, an aquifer 60 to 90 meters thick. More than 20 percent of the water has been extracted. At current rates of pumping the aquifer will eventually be drained dry.

**Land subsidence** In the San Joaquin Valley in California, the pumping of groundwater has created problems of land subsidence in an area extending nearly 3500 square kilometers. In some places, the land has subsided as much as 10 meters. Extracting oil or gas from the ground can also cause land subsidence (Figure 10.12). In Baytown, Texas, the land

**10.12 House flooded because of land subsidence.** Resulting from extraction of groundwater, Bay near Houston, Texas. (Dr. Charles W. Kreitler, University of Texas at Austin, Bureau of Economic Geology)

has been lowered about three meters since 1920 because oil has been removed.

Damage related to land subsidence can be extensive. Besides decreasing the efficiency of wells and aquifers, irrigation systems can be affected seriously, and flood possibilities may be increased along coastlines.

**Salt-water encroachment** In coastal aquifers, the presence of seawater introduces an additional consideration in the extraction of groundwater from wells. In such situations, the amount of usable groundwater is bounded both above and below by zones that can and do fluctuate. At the top is the water table; below is the interface between the fresh water and the salt water. The rise of the water table above sea level creates a head of hydrostatic pressure that tends to keep the salt water out of the aquifer (Figure 10.13). When the water table is lowered for whatever reason, such as pumping in excess of recharge, or natural discharge in excess of rainfall, the level of salt water rises. In effect, what is happening is that the amount of fresh groundwater is being "squeezed" from both top and bottom. The movement of salt water into an aquifer formerly containing fresh groundwater is known as *salt-water encroachment.* The importance of salt-water encroachment was learned the hard way on Long Island, New York, where the water comes from wells. After years of pumping groundwater from wells without regard for consequences, salt water encroached. After sensible regulation was introduced, rainfall renewed the aquifers and drove out the salt water.

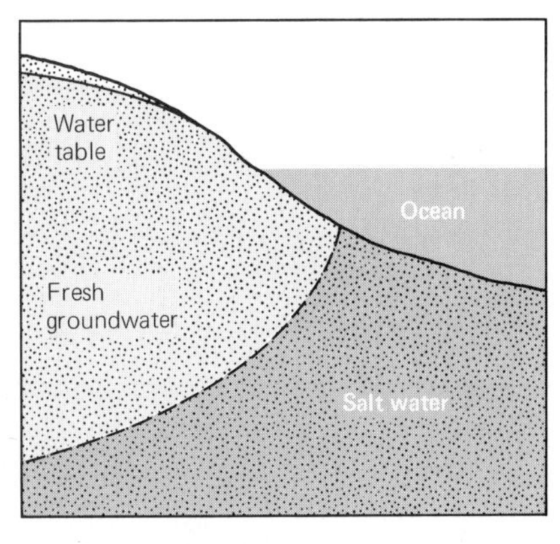


**10.13 Relationships between fresh groundwater and salt water in aquifer on oceanic island,** schematic profile and section. Height of water table above sea level determined position of underground interface between the two waters of contrasting densities.

**Pollution of groundwater** Groundwater supply must be guarded carefully against pollution. Just as an aquifer is recharged by seeping rainwater, it may be invaded by polluted seepage from the ground or from an updip location in a tilted stratum. The most common pollutant of household wells is sewage, which may seep from septic tanks, sewers, cesspools, barnyards, livestock areas, and polluted streams (Figure 10.14). Ordinarily, wells should be placed at least 15 to 30 meters away from such sources. Across greater distances, granular material such as sandstone acts as a natural purifier. By contrast, limestone is a poor purifier and may transmit polluted water great distances. The most

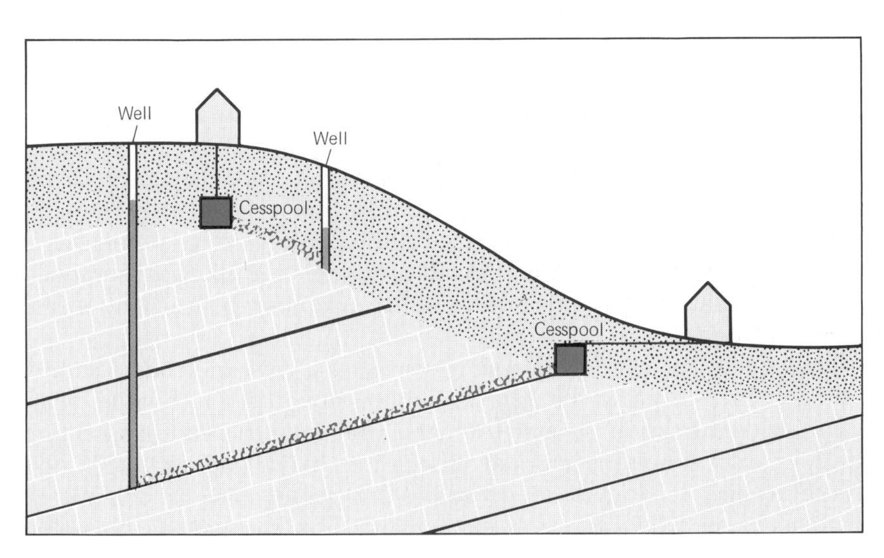


**10.14 Pollution of aquifer from cesspools,** schematic profile and section. Knowledge of the geologic relationships underground, obviously not applied in this case, can be used to locate wells where they will not be contaminated.

common disease communicated by polluted wells is typhoid fever, which is caused by a bacterium that can survive for a long time in groundwater.

The largest potential sources of polluted groundwater are urban wastes from dumps (or, as they tend to be called, "sanitary landfills"). Rainwater that seeps into these dumps becomes groundwater. In contact with solid household and other wastes, such groundwater becomes heated and acquires a chemical load by dissolving the garbage. The heated effluent, containing materials dissolved from the garbage, is *leachate.* In some places, leachate has locally contaminated the groundwater reservoirs near the landfills.

Some industrial wastes do not decompose with time and thus cannot be recycled. Some are toxic to organisms, or even radioactive. As unused land becomes more scarce, disposal of such "hard" wastes becomes difficult and costly. They can sometimes be injected deep into the ground in porous zones or other formations where they cannot be carried by groundwater into aquifers. The most acute waste-disposal problem facing the United States in the coming years is that of radioactive wastes from nuclear electricity-generating plants. Some of these wastes, such as plutonium, will remain "hot" for many thousands of years, so that any deep formations considered for disposal should be earthquake-free, tectonically stable, and dry — a difficult order. Because many energy experts consider that nuclear power is the only realistic long-term source of large amounts of electricity, the safe disposal of radioactive wastes is one of the most serious problems facing industrialized societies.

In arid regions, where other recharge water is not available, the treated water coming out of sewage-treatment plants is potentially valuable for artifically recharging aquifers.

**10.15 Water dowser conducting a "survey" using forked twig** (A), as seen in a sixteenth-century woodcut. (Georg Agricola, 1571, De Re Metallica, Basel)

**The search for new sources of groundwater** New groundwater supplies, hidden from sight, are seldom easy to find. However, as demand for water is rising quickly, the search for new sources is intensifying. Geohydrologists use a number of techniques in their search, including tools developed by petroleum and mineral prospectors. It is possible to estimate the amount of water in a formation by the behavior of electricity passing through it. Seismic waves may be speeded up or otherwise affected by water. Knowledge of the depositional history of an area, the types of rocks to be found, and other basic geologic information are crucial clues in locating aquifers. As detailed satellite photographs have become available in recent years, geologists have learned to use such clues as vegetation, drainage patterns, erosion, and color as indicators of subsurface features. The seven vast basins of groundwater beneath the Sahara were predicted in 1958 on the basis of data obtained from aerial photographs. Infrared instruments enable geologists to map minute temperature differences at the surface of the Earth. Desert aquifers have been found in Iran and Saudi Arabia during oil-drilling operations. Deep-sea drilling has revealed the presence of aquifers beneath the ocean floor off Florida.

The ancient practice of dowsing, or using a divining rod to locate water, persists even to the present day. Although the technique lacks scientific confirmation, "water witchers" diligently trudge to and fro in many countries, holding a forked stick in both hands (Figure 10.15) until the butt end is drawn downward — supposedly by groundwater. Whether or not there is any scientific merit in what they do, dowsers have at least one condition in their favor: in most parts of the world, almost any hole dug deep enough will yield water.

## THE GEOLOGIC WORK OF GROUNDWATER

So far, we have discussed the physical aspects of groundwater and its use by human beings. Now we will survey some of the geologic work groundwater does in the rock cycle. Groundwater dissolves minerals from bedrock and from regolith and transports the ions in solution, both to other formations in the Earth and to streams and the ocean. In moving great amounts of dissolved material from place to place, groundwater creates both distinctive morphologic features and many features in sedimentary strata.

### Morphologic Features Related to Groundwater

As we read in Chapter 7, most groundwater contains carbon dioxide in solution and thus is a weak acid capable of dissolving calcite, the chief mineral of limestone and other carbonate rocks. Such dissolution creates many features at and near the surface of the Earth. These include caves and caverns and their deposits, sinkholes, and karst landscapes.

**Caves and caverns and their deposits** Percolating groundwater charged with carbon dioxide can be involved in contrasting processes. (1) On the one hand, the groundwater may erode underground openings. (2) On the other hand, it may also fill in the openings previously made.

By dissolving away the calcite of carbonate rocks such as limestones and marbles along fractures and bedding planes, the water slowly widens these openings and builds an intricate drainage network of channels and chambers. When these chambers can be reached from the surface and are large enough for a human to enter, they are called *caves*. When the openings make a large system that interconnects underground, they are called *caverns*.

Most of the world's large caves and caverns form in gently dipping strata of limestone, just below the water table, where water may move as slowly as 10 meters per year. Because the water follows the principal joints in the limestone, which intersect nearly at right angles, the "floor plan" of a cave often resembles that of an ancient city (Figure 10.16).

Limestone caves being filled in are usually decorated with a variety of *speleothems*, elaborate whitish deposits in the shape of icicles, slabs, and mounds. Most of these deposits are calcium carbonate, precipitated when groundwater drips into the cave. Perhaps the most familiar forms are *stalactites*, which grow vertically downward from the ceiling like icicles, and *stalagmites*, which grow upward from the cave floor toward a drip

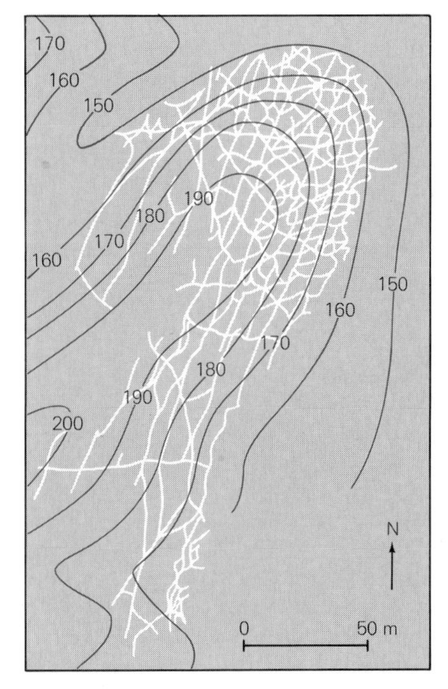


**10.16 Underground passageways formed by enlargement along intersecting joints,** map of Cameron Cave, Missouri. (Missouri Speleological Society, in J. N. Jennings, 1971, *Karst:* Cambridge, Massachusetts, and London: MIT Press, 252p., Figure 50(b), p. 154)

**10.17 Speleothems formed where previously enlarging cavern is being filled in by dripping from above and by flowing along columns** formed where enlarging stalactites and stalagmites join. Temple of the Sun, Carlsbad Caverns, New Mexico. (U. S. National Park Service)

source on the ceiling (Figure 10.17). Stalactites and stalagmites range from slender strawlike structures to the massive columns that are created when the two structures meet. Undulating "drapes" or "curtains" may grow when rivulets of groundwater trickle down an inclined roof or wall.

**Sinkholes** Under the persistent attack of groundwater, a limestone region becomes riddled with underground channels and caves that may drain the region more efficiently than surface streams. The landscape is pockmarked with *sinkholes* (Figure 10.18), formed by the collapse of cave roofs, which funnel more water below ground.

**Karst landscapes** The landscape in the Kars region in Yugoslavia shows the effects of dissolution of limestone by groundwater. This region is characterized by an almost complete lack of surface streams. This distinctive type of landscape has been named *karst morphology* (Figure 10.19). Karst morphology is present in the United States in Kentucky, northern Florida, and elsewhere in the south.

**10.18 Sinkhole** in which recent collapse has ruined a house in Bartow, Florida. (U. S. Geological Survey)

## Features in Sedimentary Strata Related to Groundwater

In addition to its work of creating distinctive landscapes, groundwater fashions many unusual features in sedimentary strata. These include: dissolution of evaporites and formation of solution-collapse breccias, deposits around hot springs and geysers, geodes, concretions, stylolites, and the replacement of one mineral by another.

**Solution-collapse breccias** Groundwater is capable of dissolving entire formations composed of readily soluble minerals, such as halite. As long as the halite is not in contact with water or is in contact with only salty formation water, it does not dissolve. But, if a deep layer of halite is uplifted and encounters fresh groundwater, the halite will soon be dissolved away completely. As the halite dissolves, the overlying mate-

**10.19 Features of karst morphology,** including dry valleys and lines of sinkholes formed by groundwater. Near Timaru, New Zealand, view from low-flying airplane. (S. N. Beatus, New Zealand Geological Survey)

**10.20 Solution-collapse breccia** consisting of particles of dolostone that were broken up when underlying evaporites were dissolved. Victoria Formation (Upper Devonian), Eureka quadrangle, Utah County, Utah. (H. T. Morris, U. S. Geological Survey, ca. 1952)

rial collapses and breaks into angular pieces. After such material has been cemented into a sedimentary rock again, it is known as *solution-collapse breccia* (Figure 10.20).

**Tufas and Geyserites** Water bubbling out of hot springs and geysers carries many dissolved salts, chiefly ions of sodium and potassium mixed with carbonates, silicates, chlorides, and sulfides. As the water emerges, it usually deposits calcareous products called *tufa* (TOO-fa) around the vent. The hotter the water, the greater the amount of calcareous tufa, because hot water can carry more calcareous material in solution than cooler water can. Siliceous hot-water deposits around geysers are termed *geyserites* (Figure 10.21). Around both hot springs and geysers, the hot

**10.21 Siliceous deposits around a hot spring.** Jupiter Terrace, hot springs, Yellowstone National Park, Wyoming. (U. S. National Park Service photo by C. H. Hanson)

water may construct a variety of mounds, terraces, or bowl-like structures.

**10.22 Small geode,** polished section through center of mineral-filled cavity. At least five contrasting generations of mineral growth are recorded as cavity was progressively reduced in size. (E. Dwornik, U. S. Geological Survey)

**Geodes** A *geode* is a small cavity that has been partly or wholly filled by the inward growth of minerals from the walls (Figure 10.22). The history of a geode resembles that of caves that are being filled in. First, the open space must be formed; underground water dissolves out an opening that does not collapse. Later, water seeping into the open space begins to deposit minerals. Many geodes are thought to have formed where evaporite minerals, such as anhydrite or gypsum, were dissolved.

**Concretions** A *concretion* is a feature formed by chemical precipitation of minerals around a nucleus such as a bone, a leaf, a pebble, or an entire fossil organism (Figure 10.23). Concretions may consist of calcite, quartz, gypsum, barite, calcium phosphate, or other minerals.

**Stylolites** Many carbonate rocks are irregularly seamed by *stylolites,* dark, zig-zag surfaces along which insoluble material has been concentrated (Figure 10.24). Stylolites form where the soluble carbonate minerals have been dissolved from places where pressure reduced the solubility of the minerals and transported away in solution to places where the pressure is less. The nonsoluble material accumulates where the carbonates were dissolved.

**Replacement deposits** Solution and deposition are often simultaneous; one material may be dissolving as another is deposited in its place. This exchange of one solid element for another is *replacement.* Fossils often show the effects of replacement; the calcite of a seashell may be replaced by silica (see Figure 4.12), by pyrite, or by other minerals.

A substance is said to be *petrified,* or turned to stone, if organic molecules have been replaced by inorganic minerals. Thus, if a log is buried in a bed of sand which later becomes saturated with groundwater, the wood may slowly be replaced by hydrated silica, or opal. Eventually, an entire tree trunk may be converted to a solid mass of silica and exposed millions of years later after the overlying sediment has washed away. Trees that were buried during the Mesozoic Era 160 million years

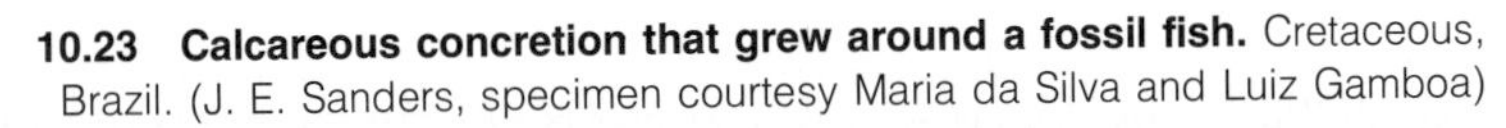

**10.23 Calcareous concretion that grew around a fossil fish.** Cretaceous, Brazil. (J. E. Sanders, specimen courtesy Maria da Silva and Luiz Gamboa)

**10.24 Stylolite** seam in limestone. (J. E. Sanders)

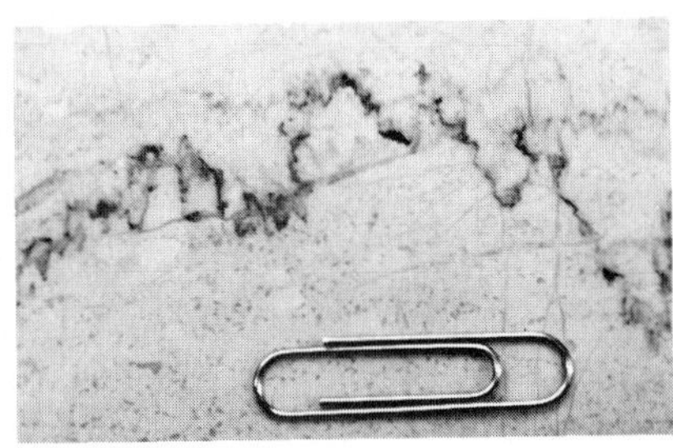

ago, converted to opal, and later exposed can be seen today in Arizona at the Petrified Forest National Park.

### Ore Deposits Related to Groundwater

The minerals of many kinds of ore deposits were precipitated out of hot waters within the temperature range of 50° to 500°C. Many different kinds of hot waters exist in subsurface environments, and perhaps all of them at some place or time have formed ore deposits. Subdivisions within this group are based on the kind of water (if that can be determined) and on the type of pore space or host material within which the ore minerals solidified. Examples of such ore-depositing solutions include juvenile water, hydrothermal solutions associated with plutons, and various kinds of groundwater.

The deposits of hot springs, geysers, and other hot waters may have sources that may not always be known with certainty. Also, some waters may have started out as fresh groundwater, but later became extremely saline by dissolving out layers of halite in the subsurface. If salty waters of this origin were to circulate deeply enough within the crust, they would be heated by the Earth's geothermal heat and might become mineralizing solutions. The minerals deposited by such water might be the same as those deposited by magma solutions, yet no magma was involved. For example, a solution that dissolves halite finds a ready source of metals in the small amounts of impurities associated with halite and also finds ready-made spaces in the collapse breccias within which to deposit minerals. The somewhat-puzzling deposits of lead and zinc minerals within the Paleozoic carbonate rocks of the upper Mississippi Valley (Wisconsin and Illinois), of the Tri-State district (Missouri, Kansas, and Oklahoma), and of northeast Tennessee may have formed by geothermally heated waters that dissolved evaporites. These deposits are generally classified as *hydrothermal,* and some geologists have attributed these hydrothermal solutions to plutons that have not yet been exposed.

In the next chapter we examine how groundwater affects the behavior of material forming slopes.

# CHAPTER REVIEW

1. The water that is present below the surface of the ground, within the pores of the regolith and bedrock, is *groundwater.*
2. The *hydrologic cycle* is the cyclic movement of all the Earth's water in all three states: solid, liquid, and gas. The scientific study of water is *hydrology.*
3. The seepage of rainwater into the ground is known as *infiltration.* Infiltration is governed by such factors as type of material, previous amount of water in the ground, slope, vegetation, and human use of the land.
4. Groundwater is distributed in two zones: the *zone of aeration,* containing an upper area of soil moisture and a lower area of capillary water, and the *zone of saturation.* The upper boundary of the zone of saturation is the *water table.*
5. Water flowing over the land is *runoff.* Most runoff becomes organized into flows within definite channels, whereupon it is named *stream flow.* The water flowing in streams is collectively called *surface runoff.*
6. Groundwater can be divided into three categories related to depth and source. The source of fresh water at shallow depths is *meteoric water.* The deeper source of salty, subsurface water is ancient buried ocean water or *formation water. Juvenile water* is thought to come to the surface from the mantle.
7. Groundwater is stored in three major kinds of under-

ground openings or *pores:* a framework of rigid openings around the particles of regolith and sedimentary rock; cracks, joints, and fractures in massive bedrock; and huge caverns, caves, and lava tubes.

8. The two general characteristics of openings that affect the flow of water underground are *porosity,* the proportion of pore space to total volume, and *permeability,* the size and degree of connections among the spaces. Factors that affect porosity and permeability are *sorting* and *packing*.
9. The movement of water depends upon the *hydraulic gradient,* a combination of the two factors of permeability of the material and the pressure gradient. This relationship is *Darcy's law:* the flow rate through porous material is proportional to the pressure driving the water, and inversely proportional to the length of the flow path.
10. Discharge of groundwater occurs through seepage, springs (surface streams of flowing water that emerge from the ground), and wells (holes dug or drilled into a water source).
11. An *aquifer* is a saturated body of porous and permeable regolith or bedrock through which groundwater can flow readily. Aquifers may be classified as *nonconfined* or *confined.*
12. Conditions and problems associated with the human use of groundwater include balancing discharge and recharge, land subsidence, salt-water encroachment, pollution from urban and industrial wastes, and the need to find new sources.
13. Groundwater dissolves minerals from bedrock and regolith and transports the ions in solution, both to other formations in the Earth and to streams and the ocean. In moving great amounts of dissolved material from place to place, groundwater creates both distinctive morphologic features and many features in sedimentary strata.
14. Morphologic features created by groundwater include caves, sinkholes, and karst landscapes. The dissolution of the calcite of carbonate rocks may widen underground openings and build an intricate drainage network of channels and chambers. A *cave* is a chamber that can be reached from the surface and is large enough for a human to enter. Groundwater may also fill in a cave by precipitating minerals, usually calcium carbonate. Two forms of *speleothems,* or cave deposits in elaborate shapes, are *stalactites* and *stalagmites. Sinkholes* are formed by the collapse of cave roofs.
15. A distinctive type of landscape that is characterized by an almost complete lack of surface streams is named *karst morphology.*
16. Features made by groundwater in sedimentary strata include solution-collapse breccias, tufa, geyserites, geodes, concretions, stylolites, and replacements. Solution-collapse breccia is sedimentary rock formed from broken pieces of overlying formations after underlying soluble material has been dissolved.
17. A calcareous product deposited by hot water around the vent of a spring is called *tufa;* a siliceous hot-water deposit around a geyser is called *geyserite.*
18. A *geode* is a small cavity that has been partly or wholly filled by the inward growth of minerals from the walls. A *concretion* is a feature formed by chemical precipitation of minerals around a nucleus such as a bone or an entire fossil organism.
19. A *stylolite,* is a dark, zig-zag surface on carbonate rock, formed by pressure solution, along which insoluble material has been concentrated by the removal of soluble carbonate minerals.
20. Solution and deposition are often simultaneous, and the exchange of one solid element for another is *replacement.* A substance is said to be *petrified* if organic molecules have been replaced by inorganic minerals.
21. Many different kinds of hot waters exist in subsurface environments, and perhaps all of them at some place or time have formed ore deposits.

## QUESTIONS

1. Define *groundwater.* Discuss the relationship of groundwater to the hydrologic cycle.
2. What is *infiltration?* What factors influence infiltration? Discuss the relationships among infiltration, *runoff,* and *stream flow.*
3. Name the two zones of groundwater. What is the *water table?*
4. Compare *meteoric water, formation water,* and *juvenile water.*
5. Define *porosity* and *permeability* and discuss their relationship to the kinds of openings present under the surface of the land.
6. State *Darcy's law.* Show how the water table and Darcy's law affect the flow of groundwater.
7. What is meant by *discharge of groundwater?* Name three ways in which groundwater is discharged.
8. Define *aquifer* and discuss the two chief kinds of aquifers.
9. Explain what is meant by *drawdown* and a *cone of depression.*
10. Define *saltwater encroachment* and give an example of a locality where such encroachment has taken place.

11. Discuss the problem of pollution of groundwater.
12. Discuss some of the methods being used in the search for new sources of groundwater.
13. How are *caves* formed?
14. Define *speleothems, stalactites,* and *stalagmites.*
15. Define *karst morphology.* List two localities in the United States where one can see karst morphology.
16. What is a *solution-collapse breccia?* Where would one expect to find such breccias?
17. Define *geode, concretion,* and *stylolite.*

## RECOMMENDED READING

Davis, S. N., and deWiest, R. J. M., 1966, *Hydrogeology:* New York, John Wiley and Sons, Inc., 463 p.

Howard, A. D., 1963, The Development of Karst Features: *National Speleological Society, Bulletin,* v. 25, no. 1, p. 45–65.

Idso, S. B., Jackson, R. D., and Reginato, R. J., 1975, Detection of Soil Moisture by Remote Surveillance: *American Scientist,* v. 63, no. 5, p. 549–557.

Jennings, J. N., 1971, *Karst:* Cambridge, Massachusetts, and London: The Massachusetts Institute of Technology Press, 252 p.

Kendall, A. C., and Broughton, P. L., 1978, Origin of Fabrics in Speleothems Composed of Columnar Calcite Crystals: *Journal of Sedimentary Petrology,* v. 48, no. 2, p. 519–538.

Leopold, L. B., 1974, *Water: A Primer:* San Francisco, W. H. Freeman and Co., 172 p.

Middleton, G. V., 1961, Evaporite Solution Breccias from the Mississippian of Southwest Montana: *Journal of Sedimentary Petrology,* v. 31, no. 1, p. 189–195.

Moore, G. W., and Moore, N. J., 1964, *Speleology: The Study of Caves:* Boston, D. C. Heath and Co., 120 p.

Pantin, H. M., 1958, Rate of Formation of a Diagenetic Calcareous Concretion: *Journal of Sedimentary Petrology,* v. 28, no. 3, p. 366–371.

Waltham, A. C., 1975, *Caves:* New York, Crown Publishers, Inc., 240 p.

PHYSICAL PRINCIPLES AFFECTING SLOPES
GRAVITY-TRANSPORT PROCESSES
BEHAVIOR OF SLOPES AND USE OF THE LAND

# CHAPTER 11 GRAVITY AND DOWNSLOPE MOVEMENT

PLATE 11
Results of slope failure about 100 meters wide and 200 meters long, near Oakland, California on 9 December 1950. Cracks (dark lines) and spoon-shaped upper limit of the failed zone indicate movement by slumping, but rounded lower end indicates that slumping changed downslope into a debris flow. (Bill Young, from *San Francisco Chronicle*)

THE SURFACE OF THE EARTH CONSISTS OF LOW-LYING, NEARLY horizontal plains, typically underlain by sediment, and inclined areas, or *slopes,* which may be underlain by bedrock or regolith. Slopes include the sides of mountains, of volcanoes, of isolated hills, and of valleys. The material underlying slopes is actively moving; as the slopes move, the landscape changes. Gravity is what makes slopes change. In many instances, gravity is powerfully assisted by groundwater. The study of the landforms and the changes they undergo is *geomorphology,* from the Greek *geo,* meaning "earth," and *morphe,* meaning "form."

Almost everyone has observed some of the processes that change the Earth. Shepherds have watched "landslides" (a popular term for any slope failure), fishermen have seen waves gnaw at coastlines, and farmers have struggled to prevent erosion of topsoil by rainwater. These changes are the familiar, visible parts of the geologic cycle.

In coastal California the hills are young, geologically speaking. They are so new that even though they consist of soft shale bedrock, their slopes are steep and their relief is great. Hillsides that face the Pacific Ocean offer spectacular scenic views and are considered prime sites for residential development. However, nearly every heavy rain creates a series of local disasters. The newspapers are filled with reports of houses that have slid down unstable slopes (Figure 11.1).

The results of downslope gravity transport are perhaps less spectacular in other areas, but they are no less real. Retaining walls tip over, boulders crash down on houses, and pieces of lawn or garden slip away. Slope failures affect highways, sewer lines, and other public property. Knowledge of the potential hazards resulting from the effects of gravity is of great importance in any construction project in terms of both safety and expenditure.

In this chapter, we will summarize the physical principles affecting slopes, study the kinds of materials forming slopes and the ways these materials move downslope, and finally, discuss some ways in which the behavior of slopes affects the use of the land.

**11.1 House in danger of falling into Pacific Ocean** as gravity pulls the underlying material downslope, Pacific Palisades, California. (U. S. Geological Survey)

## PHYSICAL PRINCIPLES AFFECTING SLOPES

The physical principles affecting slopes include: (1) the Earth's gravity and the relative effectiveness of the two components of gravity on a slope, (2) the strength and cohesion of the materials forming the slope, and (3) factors that affect the material in such a way as to change the stability of the slope.

### Components of Gravity on a Slope

The Earth's *gravity* is the inward-acting force that tends to pull all particles toward the center of the Earth. This force acts along lines that are at right angles to the Earth's horizon (Figure 11.2). On a slope the Earth's gravity (line *AD* in Figure 11.3) can be resolved into two components. One component acts along the slope and down the slope. This is the *slope-parallel component* (line *AC* in Figure 11.3); because the slope-parallel component tends to pull material downslope, we refer to it as the *pulling force.* The other component of gravity acts at right angles to the slope and against the slope (line *AB* in Figure 11.3). This is the *slope-normal component.* Because the slope-normal component tends to push the material against the slope and thus to prevent movement, we will call it the *resisting force.*

These two components of gravity always act upon the material forming a slope. What happens depends on the relative size of the pulling force compared with the resisting force and on the inherent strength of the material underlying the slope.

**Influences on the pulling force** The chief variable that affects the pulling force is the *angle of slope.* The intensity of the pulling force increases as the slope angle becomes steeper (Figure 11.4).

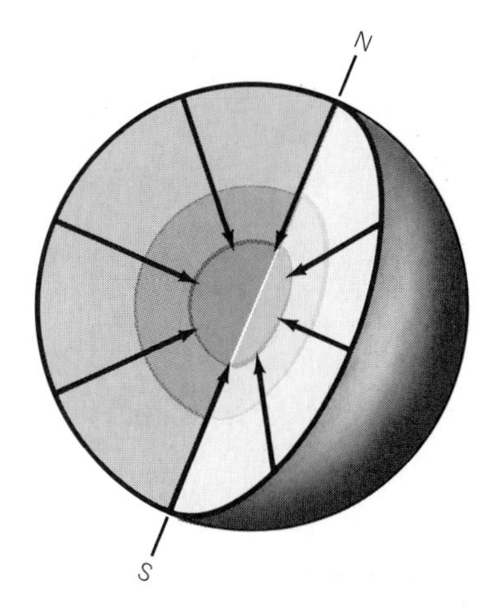


**11.2 The Earth's gravity tends to pull all objects toward the center of the Earth,** acting along lines shown by arrows in this cutaway sketch.

**11.4 As slope angle changes, the relative values of the two components of the Earth's gravity shift.** On gentle slopes, the slope-normal component, or resisting force (= $F_r$), acting against the slope, is larger. As the slope angle increases, the slope-parallel component, or pulling force (= $F_p$). acting along the slope, becomes larger.

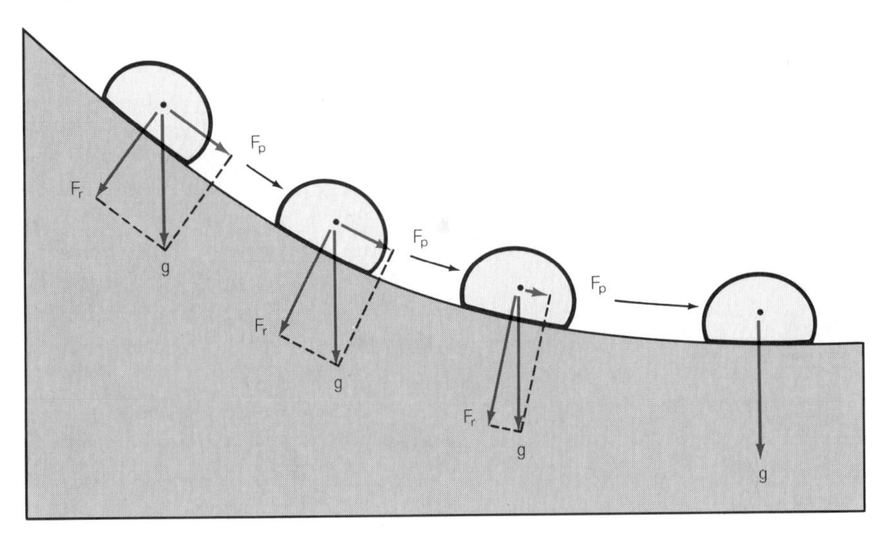


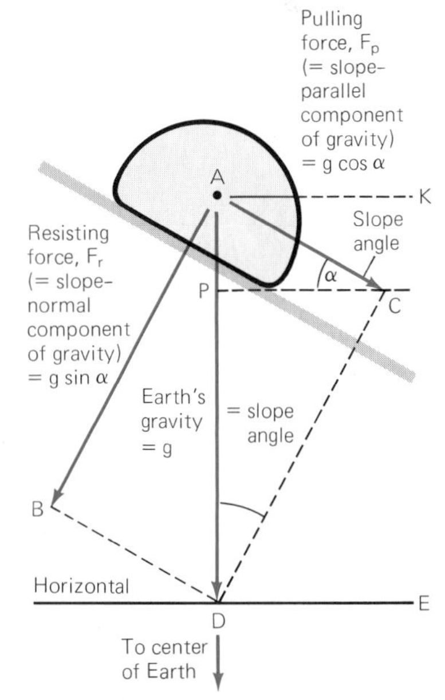


**11.3 Contrasting components of the Earth's gravity on a slope.** Slope angle, ∠*ACP*, equals angle *PDC*. Further explanation in text.

Other factors that can increase the pulling force include addition of weight to the upper part of the slope or removal of material from the toe of the slope. Mass added upslope increases the tendency for the rest of the material to move downslope. Taking away material decreases lateral stability.

**Influences on the resisting force** The intensity of the resisting force changes with angle of slope, but in an inverse relationship. As the slope becomes steeper, the resisting force decreases. On a horizontal surface, the resisting force is equal to the total gravitational pull; the downslope pulling force is zero (see Figure 11.4).

The resisting force can also vary according to changes of conditions that affect the strength of the material underlying the slope. In order to understand this relationship, we must discuss the strength of materials.

## Materials Forming Slopes

**Strength of materials** The term *strength of materials* refers to a material's ability to resist deformation.

A fundamental concept in understanding strength of materials is that external conditions, called *forces,* establish internal responses, called *stresses.* Stresses are of three kinds: (1) *compressive stresses,* which tend to push the material closer together; (2) *tensile stresses,* which tend to pull the material apart; and (3) *shearing stresses,* which tend to make the material slip sideways along a series of parallel planes (Figure 11.5). Within a body subjected to deforming forces, all three kinds of stresses exist simultaneously. A material's reaction to deforming forces depends in part on its elastic properties. Two important aspects of the elastic properties of a solid are changes of shape and changes of volume (see Appendix D and Box D.1 on elastic properties of solids). A third aspect relates to breaking or fracturing (named rupture in engineering usage). The upper limit of strength of a material is its *elastic limit,* that is, the value of the stress to which a solid can be subjected and still recover its original shape or its original volume when the forces that created the stresses have been removed.

Stress in excess of the elastic limit creates permanent deformation. Such deformation implies that the solid has failed. It may do so either by plastic flow or by rupture. Ductile substances fail by plastic flow, whereas brittle substances fail by rupture.

**11.5 Slippage of slabs along parallel shearing planes,** an effect of the dominance of the pulling force, can be compared to the sliding that takes place between the individual cards in a deck of playing cards.

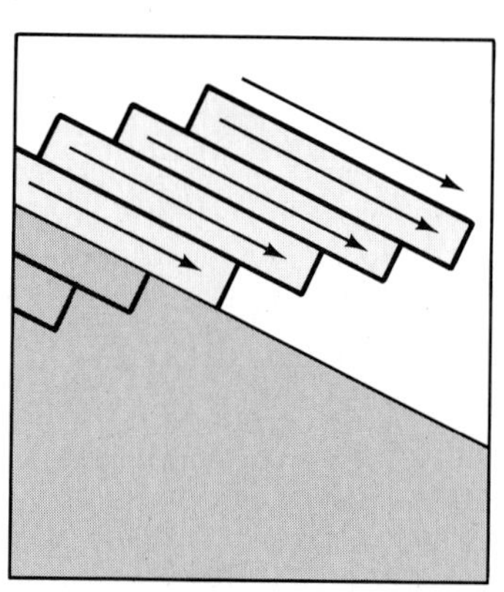

**Cohesion and cohesionless materials; angle of repose** The term *cohesion* refers to the ability of a body of particles to stick together. The building-block materials of solid rocks stick together because the growing crystal lattices are interlocked or because loose particles are bound together by the precipitation of a cement. The cohesion of a body of regolith or of sediment is many times less than that of a body of solid rock. Nevertheless, some kinds of noncemented regolith or sediment do possess cohesion. One kind of cohesion depends on sizes of the particles and another, on the surface tension of water.

Fine-grained regolith or sediment, in which the sizes of the particles are about $\frac{1}{16}$ of a millimeter and smaller (silt- and clay sizes), possess what is known as *electrostatic cohesion.* This comes about because in these tiny sizes, the amount of surface area relative to the mass increases.

**11.6 Natural vertical slope** formed in **loess,** a cohesive sediment. (G. R. Roberts)

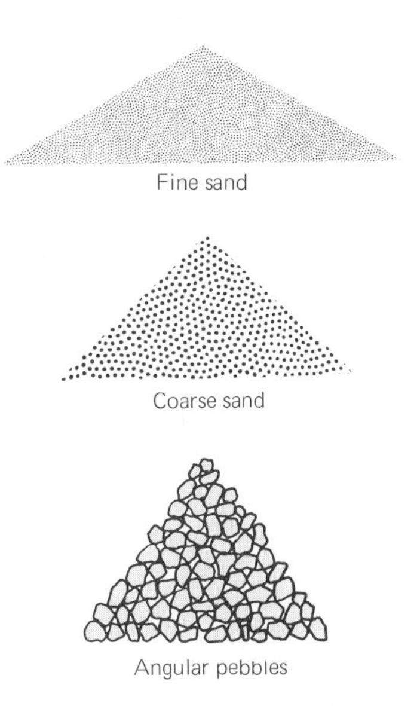


**11.7 Angles of repose** shown by profiles and sections through cones of sediment of contrasting particle sizes.

In addition, because of the nature of the crystal lattices of clay minerals, the ionic charges are not uniformly distributed.

The surface tension of water in moist sand tends to hold the sand particles together. This effect is possible only in sand-size sediments. It depends upon the fact that the pore spaces are so small that capillary water can be held within them. Such cohesion in sand is *capillary cohesion.*

Cohesive regolith or sediment, whether based on electrostatic forces or surface tension of water, tends to stand in vertical slopes (Figure 11.6).

A body of cohesionless particles tends to form a slope lying between 30 and 40 degrees. Such an angle assumed by a body of cohesionless particles is known as the *angle of repose* (Figure 11.7). The angle of repose is the natural slope which results from the interplay of the gravity components acting against the slope and along the slope (see Figure 11.3). Cohesionless particles subjected only to the normal influences of the two components of gravity do not accumulate at any angle that is steeper than the angle of repose. Instead, they will spontaneously readjust by downslope movement (Figure 11.8).

It is important to realize that although cohesionless particles do not accumulate at angles *steeper* than their angle of repose, once they have formed such an angle, they may be subjected to factors that reduce their resistance to flow and thus may move downslope. The angle might better be named the "angle of temporary repose." We illustrate by considering the effects of water and fluid pressure.

**11.8 Sand cones at lower ends of small channels in a sandy slope.** Diameter of lens cap is 50 millimeters. South bank of St. Mary's River, Florida. (J. E. Sanders)

## Factors Affecting Stability

**Abundance of intersitial water** The effect of water on slope materials depends on how abundant the water is. Slopes are most likely to fail during, or shortly after, periods of heavy rain.

The surface tension of thin films of capillary water bind small particles together. If the spaces between particles become completely filled with water, the surface-tension effect is lost, cohesion is reduced, and the

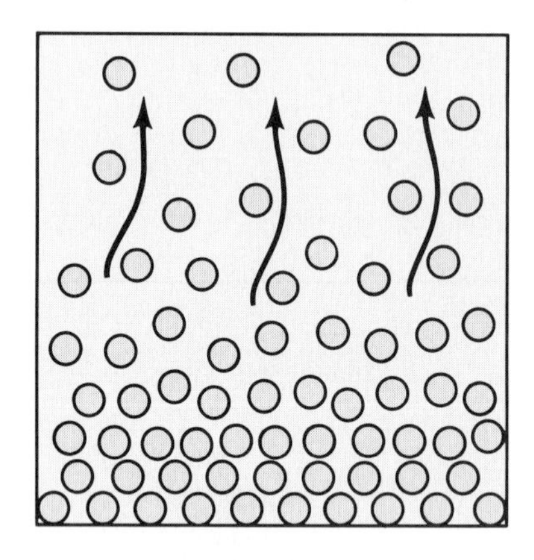

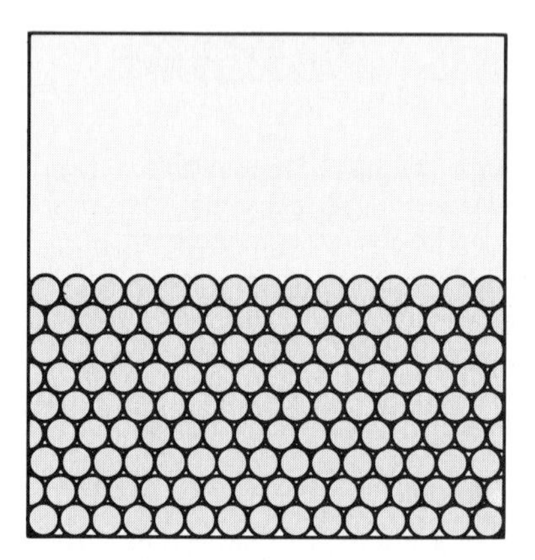

**11.9 Flow of interstitial fluid at high pore pressure,** in the pore spaces within a body of particles that touch one another. Fluid that flows outward or upward (as here), can separate the particles, and thus cause them to occupy an increased volume (or, in technical terms, to be dilated). Once the particles are no longer in contact, they can flow as a liquid.

particles can move more easily. Thus, a film of water promotes stability, but a full load of water reduces stability.

The saturation of a slope by heavy rains or seepage adds considerable weight to the mass of debris. This extra weight increases both the pulling force and the resisting force, but the effect on the pulling force is greater because water has no shearing strength and thus contributes nothing but its weight to the resisting force. When the water becomes so abundant that its weight increases the pressure within the pore spaces, the pressurized water tends to "float" adjacent particles. This *pore pressure* pushes the particles apart, and thus overcomes the resisting force (Figure 11.9).

**Liquefaction** A body of cohesionless particles can be turned into a liquidlike mass by processes other than upward flow of interstitial water. The general name for the change of stable solidlike, cohesionless particles to a liquidlike condition is *liquefaction.* If this condition sets in abruptly, the process is *spontaneous liquefaction.* Cohesionless sediment can be spontaneously liquified by rapid jolts, such as earthquakes or the shocks of a pile driver, or by changes that follow the rapid fall of the level of underground water. The best-known flows of this kind have occurred in The Netherlands on gentle sandy slopes along the coast; between 1881 and 1946, 229 such flows were reported.

**Changes in water level** The stability of a slope may change greatly if the water level adjacent to it falls by more than a few meters per day. This may happen to a river after a flood or around a storage reservoir whose level is being lowered. In both cases, a volume of regolith that had been under water is suddenly above water level. Unless water can escape from the regolith almost as fast as the level of the river or reservoir drops, the trapped water exerts tremendous sideward force on the slope near the dropping water level. The force exerted on particles of regolith by the outward-flowing interstitial water is *seepage pressure,* which acts in the direction of the flow and increases in proportion to the speed of seepage. Both seepage pressure and speed are greatest at the foot of the slope. If these two factors become great enough, the slope may fail. Once the lower part of a slope has failed, the upper part soon follows. Rapid drawdown is a common cause of slope failures, especially in sediments of a size somewhere between sand and clay.

**Presence of ice** By the mere act of freezing, which causes water to expand and lift the particles, ice can overcome the resisting force. This lifting away from the slope overcomes the resisting force and the particles finally let go when the ice melts. Ice becomes slippery only when it is covered by a thin film of water, and ice crystals in regolith therefore act as a cement rather than a lubricant.

**Effects of seismic waves** During an earthquake, seismic waves passing through regolith physically lift the particles, thus overcoming the effects of the resisting force. Once the particles have started to move, their downslope motion pushes them away from the slope, creating a self-perpetuating effect that is transferred to other particles along the way (Figure 11.10).

**Changes in toes of slopes** Any change at the base of a slope usually affects what happens higher up the slope. Removal of support at the toe

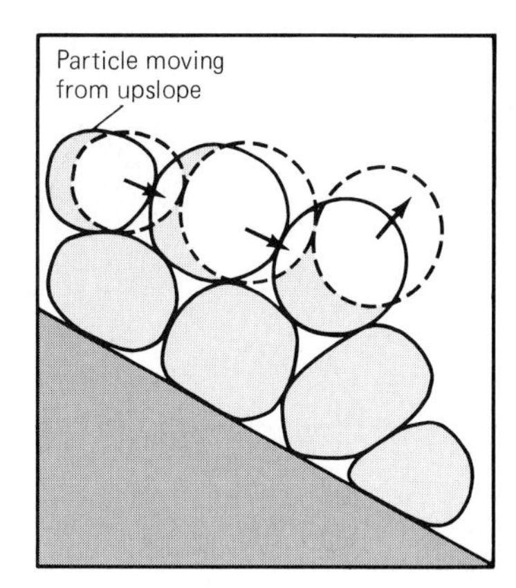


**11.10 Dilatant effect of particles moving along a slope.** As particles are struck by other particles moving from upslope, they tend to bump into other particles and to be deflected away from the slope. When many particles have been deflected, the former rigid framework of particles in contact is destroyed. At that point, the now-dispersed and dilated particles can move down even the slightest slope. Experiments have proved that shearing of particles parallel to a slope creates a dispersive "pressure" of particles being deflected away from the slope.

of a slope in construction projects commonly causes a slope to fail (Figure 11.11). The same effect is produced where a stream undermines its banks or a sheet of water takes away sand from the top of a beach. Before it was removed, the material at the toe of the slope contributed to the slope's stability. After it has been taken away, however, the slope fails.

## GRAVITY-TRANSPORT PROCESSES

We can understand slopes and downslope movements by classifying materials and transport processes into two groups each. The categories of slope-forming materials are: (1) hard bedrock cut by intersecting joints and bedding planes and (2) soft bedrock and regolith. The transport processes may be divided into: (1) rapid, that is, transport at a rate faster than 0.5 meter per week, and (2) slow, less than 0.5 meter per week. (The various kinds of downslope gravity-transport processes are summarized by the block diagrams on pages 268 and 269.)

### Rapid Downslope Movements Involving Jointed, Hard Bedrock

Nearly all bedrock at the Earth's surface lies in what can be considered to be a *zone of fracture* where intersecting cracks and partings along bedding surfaces have cut the rock into innumerable blocks of various sizes (Figure 11.12). This contrasts with a deeper-lying zone of flow, where at depths of 20 kilometers or more, the weight of the overlying rock has pushed the material together and has eliminated all openings along joints.

**11.11 Removal of material from the toe of a slope,** as in a roadcut, creates instability and invites slope failure.

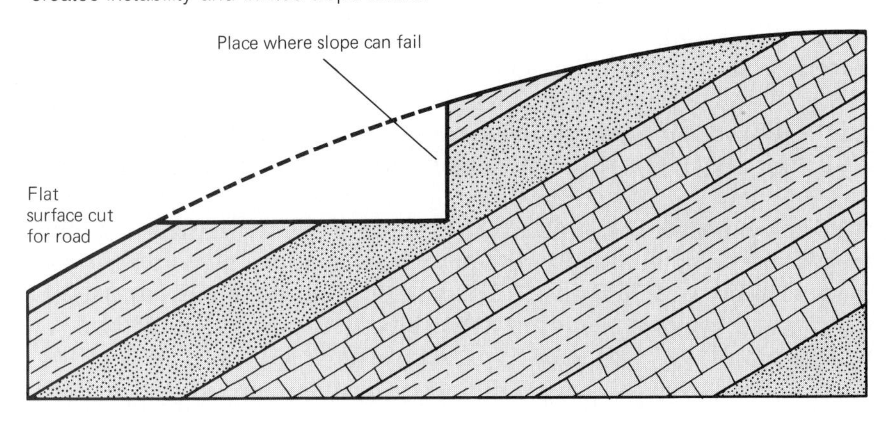


**11.12 Intersecting joints in a body of bedrock** at the Earth's surface cuts the otherwise-solid rock into many blocks that fall down and collect at the bases of the slopes. Schematic profile and section.

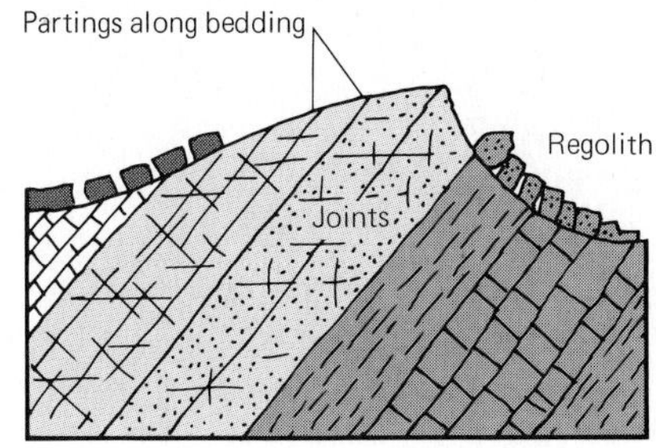

**Figure 11.13 Result of a rock slide along a steeply inclined joint in massive sandstone,** near Standing Cow Ruin, Canyon de Chelly, Arizona. (U. S. National Park Service)

The rapid gravity-transport processes can involve single pieces of rock or great bodies of bedrock, which, after they have broken loose, become "instant regolith." Such processes include: (1) rock fall, (2) rock slide, and (3) rock avalanche.

**Rock fall** A common event on cliffs and steep slopes underlain by bedrock is the relatively free fall of detached individual blocks broken from the bedrock. Such movement of individual blocks is named *rock fall* (see Figure 2.17). The fallen rocks may be of any size. They can fall from a cliff, a steep slope, a cave, or an arch. The rock may fall directly down or make a series of bounds and rebounds over other rocks or regolith on a steep slope. Little or no interaction among the falling fragments takes place. Rock falls are common along the head walls of canyons and along rocky cliffs at the shore. Such falls may be caused by the wedging of jointed rocks by tree roots or ice.

**Rock slide** Some movements involve not single pieces, but many pieces broken loose from solid masses of bedrock. The term *rock slide* describes the downward and usually rapid movement of many newly detached segments of former bedrock. The sliding takes place along some parting surface, as along a joint, along a fault, or between beds (Figure 11.13). Large, natural rock slides occur frequently in such rugged areas as the Rocky Mountains. Smaller, but more frequent rock slides are produced by such human activities as the excavation and undercutting of rocky slopes to build roads and railroads. An accumulation of rock fragments that have fallen from above, either individually or in groups, is a *talus* (Figure 11.14). The angular blocks in a talus are known as *sliderock.*

**11.14 Sliderock accumulating in fanlike talus cones at foot of steep slope** underlain by massive igneous rock in northern Afghanistan. (United Nations)

**Rock avalanche** The most awesome variety of downhill movement is the *avalanche.* Materials varying from pure ice and snow to rocky debris move at speeds more than 100 kilometers per hour. A *rock avalanche* is defined as the extremely rapid downslope transport of a large body of rock debris.

One of the most spectacular avalanches of modern times crashed down

(a)

**11.15 Deposits of a rock avalanche from Sherman Peak,** Alaska, triggered by the Prince William Sound earthquake on 27 March 1964, cover part of Sherman Glacier. (Photo by Austin Post, University of Washington and U. S. Geological Survey)

Schematized profile and section drawn along main path of avalanche from Sherman Peak to Sherman Glacier shows small clump of trees still standing on the west side of Spur Ridge. The avalanche literally flew over the tops of these trees. (Plotted by J. E. Sanders from data in R. L. Shreve, 1966, Sherman landslide (*sic*): Science, v. 154, p. 1639–1643, Figure 2, p. 1640)

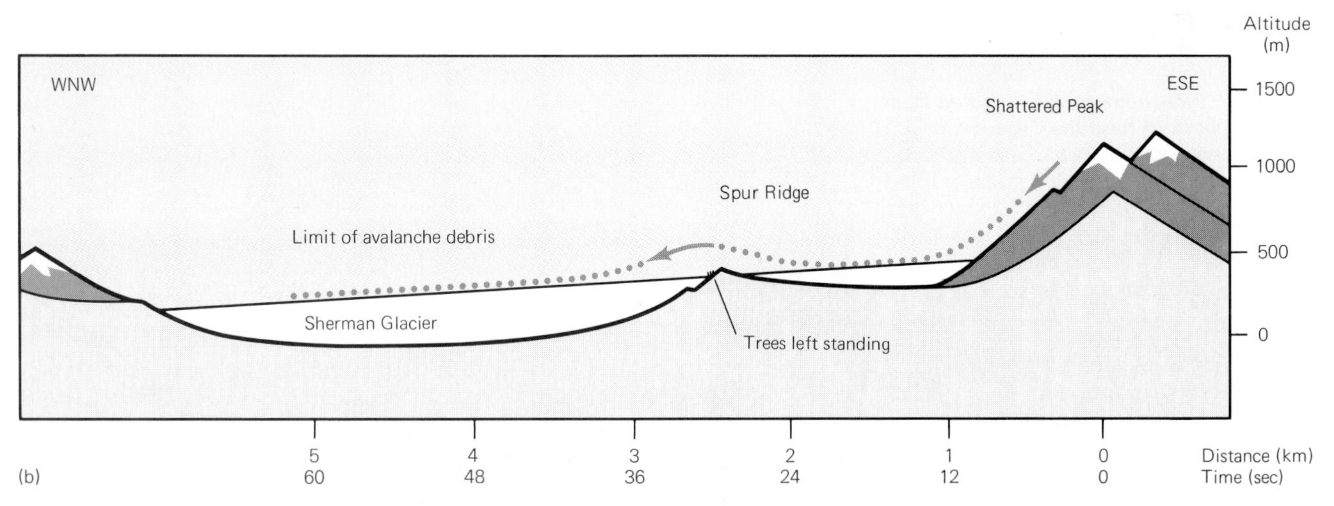


(b)

from the high Alaskan coast ranges and spread across the Sherman Glacier (Figure 11.15). The solid bedrock was shaken loose on 27 March 1964, during the Prince William Sound earthquake, whose source was hundreds of kilometers away. The avalanche debris literally flew over the tops of trees growing in a low hollow along its course. The avalanche finally came to rest as a blanket of debris one to three meters thick.

Avalanches typically accompany earthquakes. A great avalanche set off

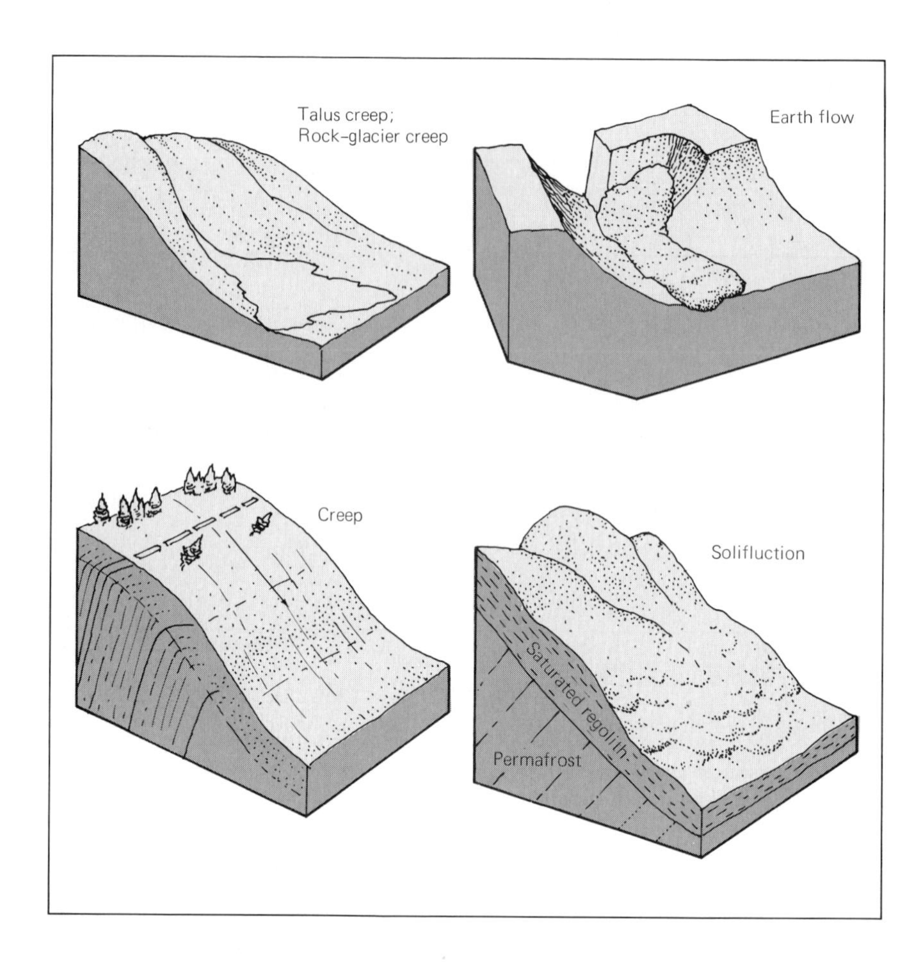


by the Peru earthquake of 1970 left a huge deposit in a steep mountain valley (Figure 11.16).

**11.16 Avalanche debris triggered from north peak of Nevados Huascaran,** Peru by earthquake of 31 May 1970, includes everything from mud (beneath cracked surface at lower right) to giant boulders weighing an estimated 700 tons. (U. S. Geological Survey)

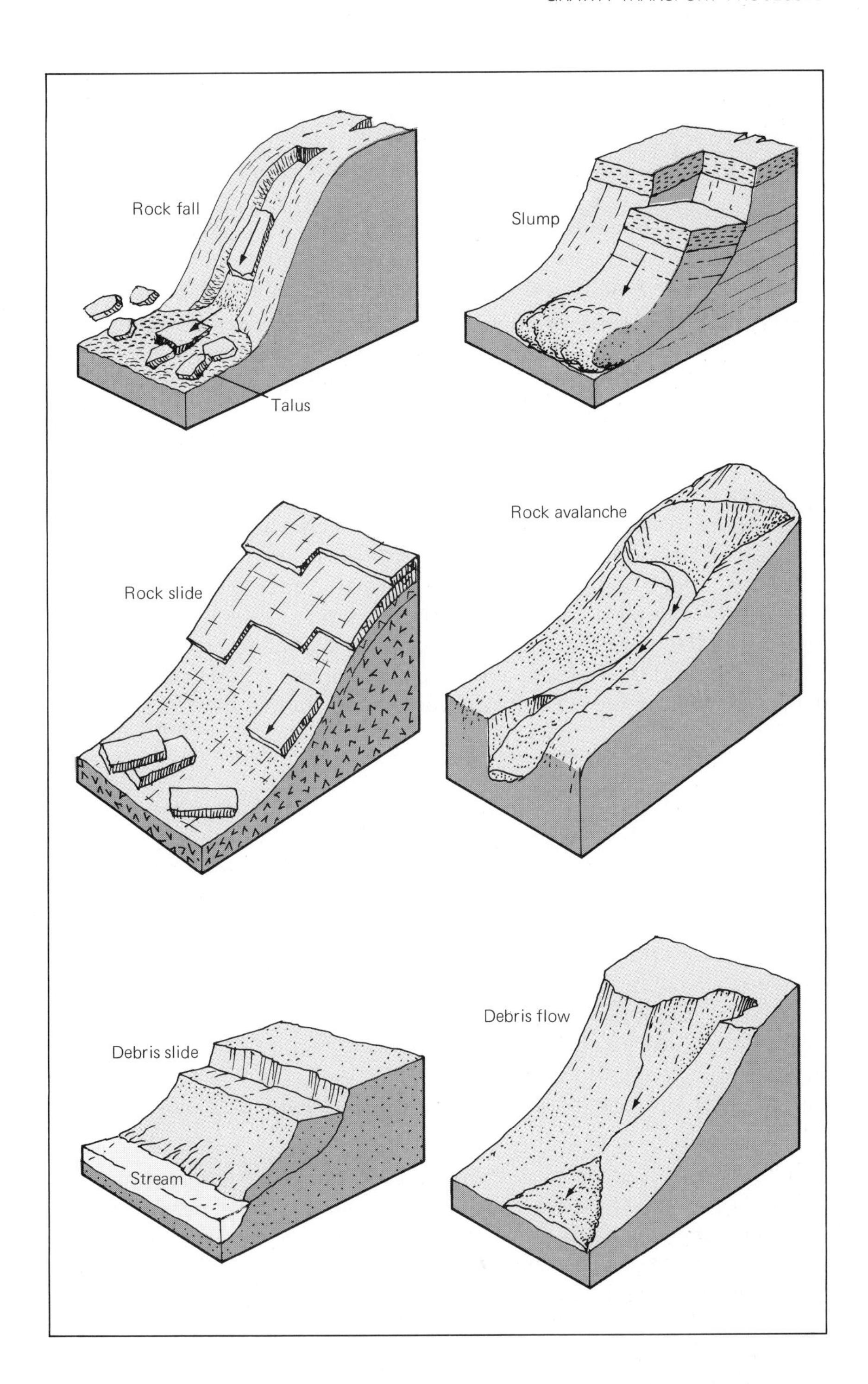


## Rapid Downslope Movement Involving Soft Bedrock or Regolith

Slopes underlain by soft bedrock or by regolith lack the block-making characteristics of slopes underlain by hard bedrock. The rapid processes involve the shifting of bodies of soft bedrock or the shifting of sand or clay along well-defined surfaces. The speed of such shifting may not equal that of rock slides or avalanches, but is generally considered to be rapid rather than slow.

**Slump** When a mass of soft bedrock or regolith breaks loose along a distinct surface of failure, it moves downward and rotates so that the formerly horizontal top of the mass dips toward the surface of failure. Such blocklike failure along a distinct surface is *slump*. Once started, such a moving mass that began as a unit may lose its internal structure and continue as a chaotic, tonguelike flow (see Plate 11, p. 259). A slumped mass usually does not travel very far or spectacularly fast. Nevertheless, slumping is one of the chief problems facing a highway engineer. Slumping takes place along a failure surface that is typically spoon shaped. If a slump affects the steep edge of a flat-topped plateau, what was initially a horizontal surface commonly tilts backward into the hillside. At the bottom, the slumped mass bulges upward. Because the entire mass often slumps as a single unit, plants and even houses on the former horizontal surface may move downslope and tilt intact. In mountainous areas, lakes commonly form in these sunken regions.

Slumping usually occurs in regions where massive sedimentary strata (such as sandstones and limestones) or sheets of volcanic rock lie above weak formations of shale or clay. Such weak formations tend to erode away uniformly, and the resistant cap rock fails as large blocks break loose along fractures or along slump surfaces. The landscape of the Snake River Plains and the Columbia Plateau is broken by many small, steplike slumps where weak rock has eroded beneath the harder sheets of extrusive basalt.

**Debris slide** Bodies of sand commonly slump along steep surfaces because particles have been moved away from the base — as along the bank of a stream or a beach being eroded by thin sheets of water flowing along the shore. As soon as the body of sand breaks loose, its internal structure disappears and it becomes a *debris slide*. Many debris slides, for example, those that occur along undercut banks of streams, are so small that they are not recorded in detail. During high water along a river such as the Mississippi, the collapse of sections of an undercut bank may be heard many times a day. Debris collapses along the Amazon River have been known to continue for several hours, affecting sections of bank more than a kilometer long and creating large waves.

**Debris flow** Wherever dry regolith on a slope becomes fully saturated, it is likely to begin to flow as a viscous, pasty mass. The general name for such movement is *debris flow*. The lower ends of slumped masses of clay-rich regolith or soft shale commonly lose their internal structure and become chaotic liquidlike flows. If fine particles exceed 50 percent by volume the term *mudflow* is appropriate. Debris flows are common in three places: (1) at the toes of slumped masses, (2) in stream valleys in dry regions subject to occasional torrential downpours, and (3) on volcanoes, where tephra and rainfall can mix together.

Mudflows are capable of transporting complex mixtures of rock debris, including boulders several meters in diameter. They are common where little or no vegetation anchors the soil, where slopes are moderately steep, where periods of intense rainfall alternate with dry periods, and where a deep layer of regolith contains enough clay or silt to aid in lubrication. Low-viscosity mudflows have been known to move so fast that they overtook railroad trains.

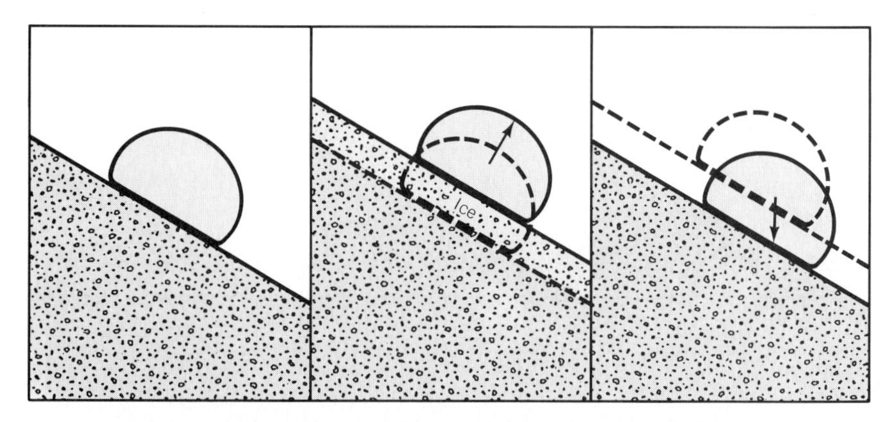


**11.17 Repeated cycles of freezing and thawing plus gravity combine to shift pebble downslope.** In left panel, pebble rests on slope. In center panel, ice crystals have formed and have pushed the surface material outward parallel to itself and have moved the pebble upward and outward in a direction normal to the slope. In right panel, after thaw, gravity has pulled the pebble vertically downward to a new position slightly downslope of the pebble's location in left panel.

A mudflow of volcanic origin accompanied the famous eruption of Mt. Somma, the ancestor of modern Vesuvius in Italy (see Figure 5.20). Heavy rains soaked the newly fallen fine tephra on the slopes of the volcano, converting the formerly dry tephra into a heavy unstable mass that began to move downslope rapidly. The mass, still hot when it eventually reached the Roman city of Herculaneum, at the base of Mt. Vesuvius, buried the city completely, just as the fallout of tephra from the atmosphere had snuffed out Pompeii.

### Slow Downslope Movement Involving Loose Blocks of Hard Rock

Once blocks of hard rock have fallen from a cliff, they may continue to move downslope. Two processes are talus creep and rock-glacier creep.

**Talus creep** As long as a talus is fed by a continuing supply of rocks from above, it will maintain an angle of about 40 degrees. If the addition of new rock should cease, however, the talus will gradually flatten out as the rocks shift farther downhill by *talus creep.*

One of the most important causes of creep is the growth of ice crystals. As crystals enlarge, they grow perpendicularly away from a slope, pushing particles of rock debris ahead of them. When the crystals melt, the particles settle vertically downward for a net downhill movement (Figure 11.17), or they even topple downslope for a meter or more. Talus creep is most rapid in cold regions where the rocks are moved by the alternating expansion and thawing of ice between the fragments.

**Rock-glacier creep** In some alpine and arctic mountains, tongue-shaped masses of large rocks form features called *rock glaciers* (Figure 11.18). Some of the world's most impressive rock glaciers are in Alaska, where single masses may measure more than a kilometer long and half a kilometer wide.

**11.18 Rock glacier** on Atlin Mountain, southern British Columbia. (Richard Wright/National Film Board of Canada.)

The origin of rock glaciers is still not well understood. Many geologists think that interstitial ice is an important factor. This ice could be the remnants of the compacted snow that forms true glaciers. Others think the ice of rock glaciers derives not only from compacted snow but also from meltwater, rain, and rising groundwater. Still others think that such ice is not needed to cause rock glaciers to flow.

### Slow Downslope Movement Involving Fine-Grained Regolith

Slow movement (less than 0.5 m/week) occurs on almost every moderate or steep slope composed of fine-grained regolith, and greatly affects agriculture and soil fertility by removing topsoil faster than it can be replaced by natural soil-forming processes. Slow movements include creep, solifluction, and earth flow.

**Creep** The most widespread type of slow gravity transport is *creep,* the imperceptible downslope movement of surface soil or rock debris. (The term *creep* is also used for slow movement on faults; see Chapter 17.) Creep affects all sizes of rock debris, from fine sands and silts to boulders tens of meters across. The debris may be dry or fully saturated with water. Although creep appears to be continuous, as the name implies, it is in reality a never-ceasing series of minute individual movements.

Normally, creep operates just in the top few meters of the regolith, but the mechanism is so effective that it carries vegetation and human-made objects with it. In fact, the best evidences for creep can be seen on almost any steep slope in the form of tilted fence posts, tilted monuments, and tilted utility poles; displaced walls or foundations; misaligned roads; and inclined trees (Figure 11.19). Near-surface vertical strata of sedimentary rock may be tipped over so that they are inclined downhill. Perhaps the most striking demonstration of creep are tree trunks which may bend uphill as they strive to counteract the downhill tilting.

**Solifluction** A special type of creep, which occurs in high latitudes or at high altitudes where the ground freezes deeply, is *solifluction* (from the

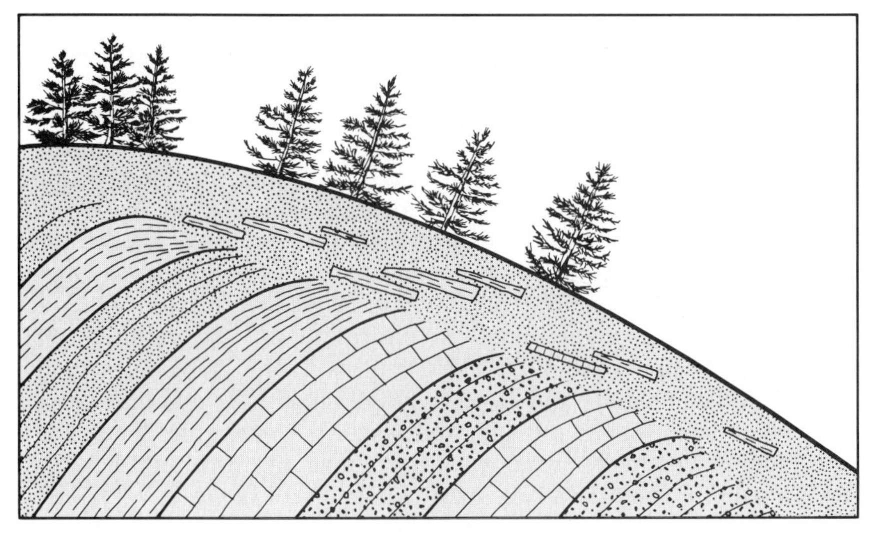

(a)

(b)

**11.19 Effects of creep.**
(a) Trees tipped over downslope and angle of inclination of sedimentary strata changed by creep of loose material downslope. Schematic profile and section.
(b) Curved tree trunks formed where upward vertical growth has counteracted the downhill tilting near the roots. Field observations have shown that tree roots creep downhill as a unit and are not strung out in an uphill direction.

Latin for "flowing soil"). Solifluction is typical of subpolar areas where permanently frozen regolith, or permafrost, is covered by a shallow layer that thaws in summer. This active layer, a couple of meters thick, consists of tundra peat, rock debris, and other weathering products. Because water permeates the layer in summer, but cannot penetrate the zone of permafrost below, the surface mass may flow along slopes as gentle as two or three degrees. The movement of solifluction is slower and more continuous than mudflow, and it may affect whole slopes rather than just local areas.

Solifluction is not altogether restricted to subpolar regions. It may occur any place where loose regolith overlies packed clay or impermeable bedrock. The water cannot escape downward, and the effect is the same as if permafrost were below.

**Earth flow** When the regolith on a steep, sod-covered slope becomes saturated with water, it may sag downward, without breaking the sod cover, to form a series of irregular terraces. This process is known as an *earth flow,* and its effects are seen most commonly in humid climate zones after a heavy rain, where the regolith has flowed down to form a bulging toe with large, smooth wrinkles. An earth flow may be confined to an area as small as a few square meters or it may cover many acres. Some earth flows are shallow and affect only the regolith; others may extend much deeper, especially if the bedrock consists of unstable clay or shale or of deeply weathered igneous rock. Millions of tons of bedrock have been moved in huge, mudlike, plastic deep earth flows.

## BEHAVIOR OF SLOPES AND USE OF THE LAND

From what we have read in this chapter, it should be obvious that any slope is subject to change; clearly no one is justified in assuming that a slope is never going to move. In this final section we review areas of hazardous slope conditions, summarize some engineering aspects of slope behavior, and explore the possibilities for predicting slope failures.

### Areas of Hazardous Slope Conditions

Although any slope is a site of potential movement, some slopes are definitely more hazardous than are others. Any steep slope, whether underlain by bedrock or by regolith, can be the site of a disastrous rock avalanche or other rapid downslope transport of material.

Slopes in earthquake areas are particularly prone to sudden and unexpected failures. The passage of seismic waves through the ground places slopes in extra jeopardy; the waves temporarily reduce the strength of the material and this enables the pulling force to take over unopposed. During the great 1964 earthquake at Prince William Sound, Alaska, seismic waves triggered slope failures that damaged many structures and temporarily crippled the economy of the whole state. At Valdez and Seward, spontaneous liquefaction and slumping below sea level were responsible for carrying away the waterfronts of both towns. In the giant slump and earth flow at Turnagain Heights, Anchorage, Alaska, houses and trees that had stood as much as 20 meters above sea

**11.20 Flow of regolith and houses into the sea** as a result of slope failures during Prince William Sound earthquake of 27 March 1964. (United Press International)

level were swept down and out to sea (Figure 11.20). In one case, material moved more than one kilometer across an intertidal mudflat and into the ocean.

The steep upper slopes on major volcanic composite cones can serve as "launching pads," so to speak, of potentially damaging volcanic debris flows. The "witches brew" for such flows typically is prepared during a major eruption, when rapidly rising hot air currents generate thunderstorms and the rainfall mixes with the water from the volcanic activity and with the tephra to create a pasty mixture that can flow readily.

Slopes underlain by regolith consisting of mixed particle sizes can create special hazards. Examples are numerous in areas that have been influenced by former glaciers; the glacier-deposited sediment typically includes particles whose sizes range from clay to boulders. Water trickling down the surfaces of slopes underlain by such mixed sizes tends to wash away the fine particles. Eventually, the larger particles are undermined. In this way, boulders may be released to crash down on what lies at the base of the slope.

Regions where former sea floors have been greatly elevated in comparatively recent geologic time, as in California, Italy, and New Zealand, may display high hills and steep slopes underlain by soft shale bedrock. Slopes composed of such material are especially prone to move after heavy rainfalls.

Other damage-prone areas are located along the banks of rivers and the shores of lakes and the sea where the toes of slopes can be undercut. An additional factor on such slopes is water that may flow over the surface or seep out from within. Studies of coastal cliffs in Boston Harbor, Massachusetts, have shown that the rate of retreat correlates with times of heavy rainfall.

## Morphologic Effects of Slope Failures

Slope failures affect the land in several different ways. On the upper parts, where the slopes have "let go," various scars may be present

**11.21 Slope failure along valley of Gros Ventre River,** Wyoming. (a) Sketch map of the locality (b) Conditions before the slope failed. (c) Conditions after the slope failed. (a, b, and c from W. C. Alden, 1928, Landslide and flood at Gros Ventre, Wyoming: American Institute of Mining Engineers, Transactions, v. 76, p. 347–359, Figure 1, p. 347 and Figure 2, p. 348) (d) View east from an airplane showing treeless scar (at center) where the slope failed and lake formed where debris blocked the river (left). (John S. Shelton) (Austin Post/University of Washington and USGS)

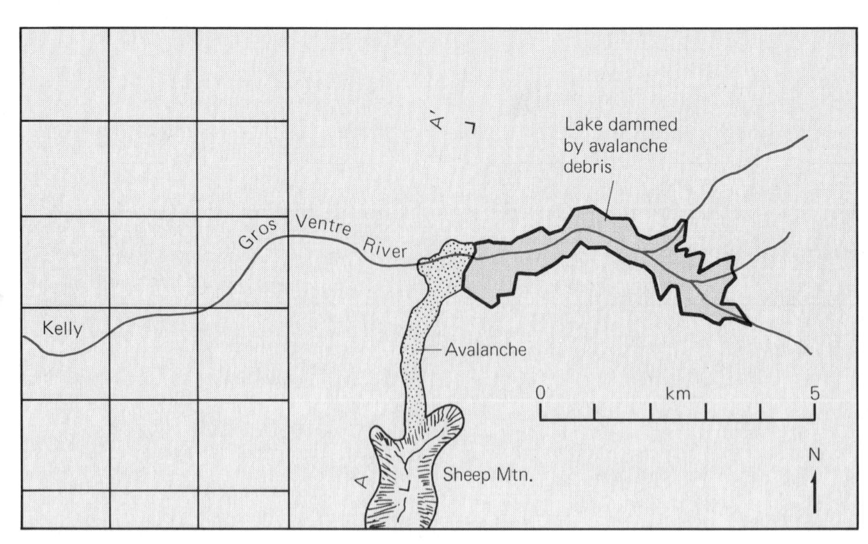


(a)

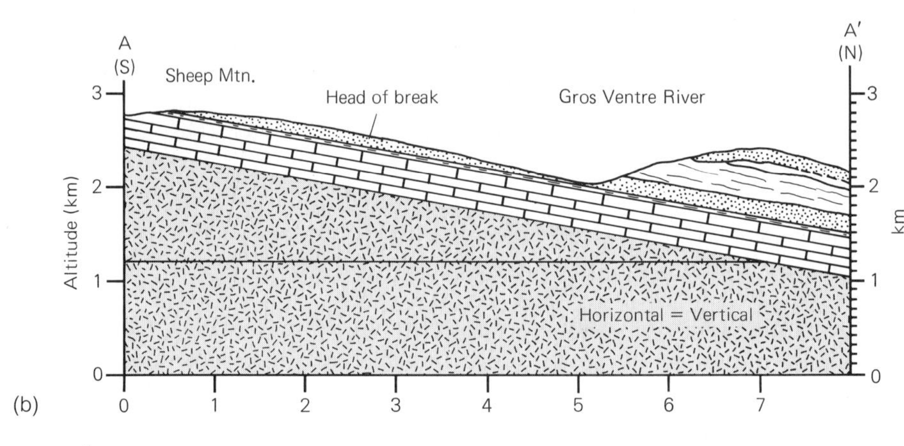


(b)

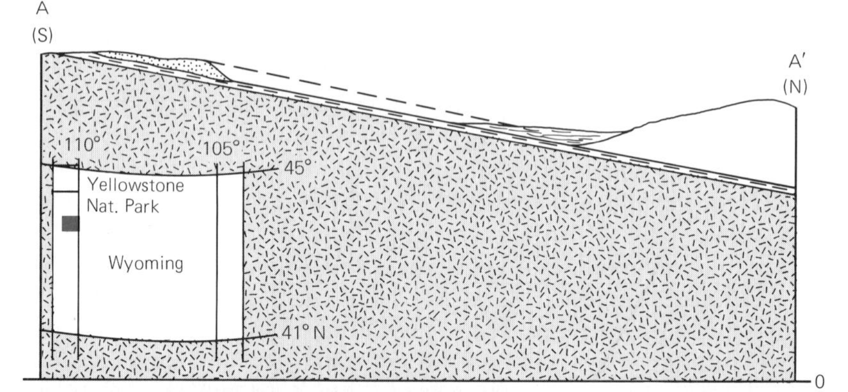


(c)

(d)

where vegetation has not been reestablished. Curved bowllike features mark surfaces along which slumping took place. Maps prepared in a study of slope conditions in California showed that about 25 percent of the area studied displayed indications of former slope failures.

Gravity transport of material down steep valley sides typically results in the formation of earth-fill dams. One of the most intensively studied slope failures in the United States disrupted the Gros Ventre River valley in Wyoming in 1925 (Figure 11.21). Some 38 million cubic meters of rocky debris broke loose from one side of the valley and traveled about 600 meters down a slope inclined from 18 to 21 degrees. The sliding mass lunged across the valley, climbed 100 meters up the opposite slope, and then settled back, forming a dam 70 to 75 meters high and more than half a kilometer long. The river backed up in a lake nearly eight kilometers long. In 1927 heavy spring runoff washed away 15 meters of the dam, causing extensive damage downstream and killing several persons in the town of Kelly, Wyoming.

A general term for all sediments that have been transported down subaerial slopes by gravity is *colluvium*. Such material consists of mixtures of coarse and fine particles whose degree of chemical weathering may vary greatly. The fine particles may show the effects of intense chemical weathering, but the coarser particles may be fresh and not weathered.

## Engineers' Manipulation of Slopes

Wherever roads, buildings, or other structures are built in the vicinity of slopes, the builders must be aware of the potential danger of slope failure. If they are not aware of the hazards of slope failures, then their projects may result in large damages and incur high costs. Highways, railroads, and pipelines are examples of projects in which excavations on slopes are made or new slopes are created by removal of bedrock or of regolith.

Large areas of farmlands have been destroyed by slope failures in Switzerland, Czechoslovakia, and Quebec, Canada. In Oregon a single slope failure removed 24 houses from a city. A slope failure along Lake Michigan cost a railroad a quarter of a million dollars.

To try to anchor slopes and thus to forestall failure, engineers and engineering geologists have used many techniques. One of the commonest and most effective approaches is to reduce the water content of the regolith underlying a slope. High pore pressure and excess water tends to fluidize the regolith and thus to destroy its strength. Lowering the pore pressure of the interstitial water by large-scale drainage or drying promotes capillary cohesion which tends to stabilize the regolith. Groundwater can be controlled by building deep drainage galleries or by drilling special drainage holes from which water can be pumped out of the regolith. Tile-lined ditches or concrete-lined ditches may be used to keep water from flowing over the surface of a slope.

The stability of a slope can sometimes be enhanced by removing the load at the head of the slope, by adding material to the toe, or both. This technique can reduce the pulling force along a slope without expensive construction.

If slopes are adjacent to railroads, highways, or buildings, it may be necessary to build retaining walls to protect against slope failures. To be effective, such walls must be up to several meters thick. Less-expensive

**11.22 Steel-mesh curtains hung on face of deep roadcut** to prevent falling rocks from bouncing away from the face and onto the road. Interstate 80, Englewood, New Jersey. (J. E. Sanders)

structures that provide some slope stability include rock bolts (huge bolts or rods sunk through bedrock) and pilings (large wooden or steel poles that are pounded into the regolith).

A large, flexible, steel-mesh "curtain" has been hung from the steep vertical roadcuts where Interstate 80 cuts through the Palisades sill in Englewood, New Jersey (Figure 11.22). The mesh prevents individual rocks from bouncing off the vertical face of the cut and out onto the highway.

Finally, vegetation can increase the stability of a slope. Plants such as alders, poplars, willows, birches, and other trees that draw water rapidly from the regolith send down extensive roots that serve to anchor the slope. In regions where rainfall triggers many slope failures, extensive cutting of trees only invites problems.

Once a slope has started to fail, nothing in the way of forestalling total slope failure can be done, unless the movement is slow. The principal job of the engineering geologist is to prevent a recurrence of old failures and to anticipate future failures. Preventive techniques address essentially the same kinds of problems as the engineering solutions described above. Many other methods of preventing slopes from failing have been tried, particularly in emergency situations such as leakage of water into a slope or overloading by unusually heavy rains. One is the artificial hardening of the regolith by the injection of chemicals or cement. Another means of increasing regolith's resistance to shearing is electro-osmosis. In this process, pore water is induced to flow outward by passing an electric current between electrodes driven into the ground. Highway cuts in nonsorted glacial sediments have been successfully stabilized by armoring them with a "pavement" of large rocks.

### Predicting Future Slope Failures

Slopes may appear to fail without warning. The material may rumble downhill so suddenly that anyone unfortunate enough to be trapped in its path usually cannot escape. In reality, however, the only slopes that truly fail without warning are those triggered by earthquakes or by spontaneous liquefaction. In all others, progressive deformation of the regolith or bedrock or downhill movement of the entire surface precedes the actual moment of rapid failure. Rather than saying a slope fails without warning, it is more accurate to say that casual observers fail to detect the warning signs. The massive slope failure at Goldau, Switzerland, in 1806, took the villagers by surprise. But, several hours before the event, horses and cattle became restless and bees deserted their hives.

The accounts of unusual animal behavior prior to the 1806 slope failure at Goldau suggest the possibility of using piezoelectric effects (Chapter 4) to give last-minute warnings. The theory behind such a procedure is that deformation of quartz lattices in the regolith or in the bedrock underlying a slope about to fail causes these lattices to store electric charges. When the deforming forces on the lattices are released, as they would be just before massive movement begins, the electric charges are abruptly discharged from the lattices. The discharge may fill the air with static electricity or create unusual local magnetic fields, conditions which may contribute to the unusual animal behavior. Instruments, not yet built, should be able to register such changes, or to detect some other phenomenon that affects the animals, and thus could give warnings to enable people to move to safe places in time. These ideas are so new that they have not yet been tested.

# CHAPTER REVIEW

1. The surface of the Earth consists of low-lying, nearly horizontal plains, typically underlain by sediment, and inclined areas, or *slopes,* which may be underlain by regolith or bedrock. Gravity is what makes slopes change.
2. The study of the landforms and the changes they undergo is *geomorphology.*
3. The physical principles affecting slopes include (1) the Earth's gravity and the relative effectiveness of the two components of gravity on a slope, and (2) the strength of the materials underlying the slope and the ways in which the strength of the materials can change as a result of responses to the slope components of gravity and to other factors.
4. The Earth's *gravity* is the inward-acting force that tends to pull all particles toward the center of the Earth. It can be resolved into two components that act upon the material forming a slope: the *slope-parallel component,* or the *pulling force,* tends to pull material downslope; the *slope-normal component,* or the *resisting force,* tends to push the material against the slope and thus prevent movement. The chief variable that affects the pulling force is the *angle of slope.*
5. External conditions, or *forces,* establish internal responses, or *stresses.* Three kinds of stresses are *compressive, tensile,* and *shearing.*
6. The upper limit of strength of a material is its *elastic limit.* Stress in excess of the elastic limit creates permanent deformation.
7. *Cohesion* refers to the ability of a body of particles to stick together. Coarse particles that are not affected by electrostatic forces, and have pore spaces too large for effective capillary action are *cohesionless.*
8. Pore pressure pushes particles apart, and thus overcomes the resisting force.
9. The change from otherwise-stable cohesionless particles to a liquidlike condition is *liquefaction.* If this condition sets in abruptly, the process is *spontaneous liquefaction.*
10. A rapid drop in the water level adjacent to a slope may create *seepage pressure,* the force exerted on the soil particles by the outward-flowing water.
11. Other factors affecting the stability of slopes are the presence of ice, seismic waves, and changes at the toes of slopes.
12. Slope-forming materials may be classified as hard bedrock cut by intersecting joints and bedding planes, and soft bedrock and regolith.
13. Downslope movements can be either rapid or slow. The rapid movements are *rock fall, rock slide, rock ava-*

*lanche, slump, debris slide,* and *debris flow.* The slow movements are *talus creep, rock-glacier creep, creep, solifluction,* and *earth flow.*

14. Precarious slope conditions may be costly and dangerous, and engineers are constantly studying the possibilities of predicting and preventing future slope failures.

## QUESTIONS

1. Draw a simple sketch showing how the Earth's gravity on a slope can be resolved into two components. Name the components.
2. Explain how a stable slope can be made unstable by changing the values of the components of gravity.
3. Define *capillary cohesion.* How does a slope underlain by material displaying capillary cohesion differ from a slope underlain by *cohesionless material?*
4. Define *angle of repose.* Is it possible for a slope to fail if the slope angle is less than the angle of repose? Explain.
5. What is *pore pressure? Liquefaction? Seepage pressure?*
6. Describe the effects of saturation of a slope during a heavy rainfall.
7. Describe how removal of the toe of a slope may affect the slope.
8. Describe three fast gravity-transport processes that affect slopes underlain by jointed, hard bedrock.
9. Explain the circumstances by which the Sherman Glacier, Alaska, acquired a coating of regolith in March 1964.
10. Describe *slump, debris slide,* and *debris flow.* Which of these activities is likely to be found on a volcanic cone? Why?
11. Define *talus.* Explain the process of *talus creep.*
12. Compare *creep, solifluction,* and *earth flow.*
13. Discuss some of the hazardous slope conditions that accompany earthquakes and volcanic activity.
14. Define *colluvium.*
15. Discuss some engineering problems associated with slopes and some proposed solutions to these problems.
16. Explain how piezoelectricity may be useful in predicting slope failures.

## RECOMMENDED READING

Browning, J. M., 1973, Catastrophic Rock Slide, Mount Huascaran, North-central Peru, May 31, 1970: *American Association of Petroleum Geologists, Bulletin,* v. 57, no. 7, p. 1335–1341.

Bunting, B. T., 1961, The Role of Seepage Moisture in Soil Formation, Slope Development, and Stream Initiation: *American Journal of Science,* v. 259, no. 7, p. 503–518.

Crandell, D. R., and Mullineaux, D. R., 1967, Volcanic Hazards at Mount Rainier, Washington: *U.S. Geological Survey, Bulletin 1238,* 26 p.

Hsü, K. J., 1975, Catastrophic Debris Streams (Sturzstroms) Generated by Rockfalls: *Geological Society of America, Bulletin,* v. 86, no. 1, p. 129–140.

Johnson, A. M., 1970, *Physical Processes in Geology:* San Francisco, Freeman, Cooper, and Co., 577 p.

Matthes, G. H., 1953, Quicksand: *Scientific American,* v. 188, no. 6, p. 97–102.

Nilsen, T. H., and Turner, Barbara, 1976, Influence of Rainfall and Ancient Landslide Deposits on Recent Landslides (1950–1971) in Urban Areas of Contra Costa County: *U.S. Geological Survey, Bulletin 1388,* 18 p.

Schuster, R. L., and Krizek, R. J., *eds.,* 1979, *Landslides: Analysis and Control:* Washington, D.C., U.S. Transportation Research Board, TRB Special Report 176, 485 p.

Wahrhaftig, Clive, and Cox, Alan, 1959, Rock Glaciers in the Alaskan Range: *Geological Society of America, Bulletin,* v. 70, no. 4, p. 383–436.

# CHAPTER 12 RUNNING WATER AS A GEOLOGIC AGENT

PHYSICAL ASPECTS OF FLOWING WATER
STREAMS AS SELF-ADJUSTING TRANSPORT SYSTEMS
STREAMS AS AGENTS OF LANDSCAPE EVOLUTION
STREAMS AS ENVIRONMENTS FOR DEPOSITING SEDIMENTS
STREAM RESPONSES TO MAJOR GEOLOGIC CHANGES

PLATE 12
Narrow V-shaped valley that has been deepened by upstream retreat of this waterfall, Yellowstone River in Yellowstone National Park, Wyoming. (George Grant, U. S. National Park Service)

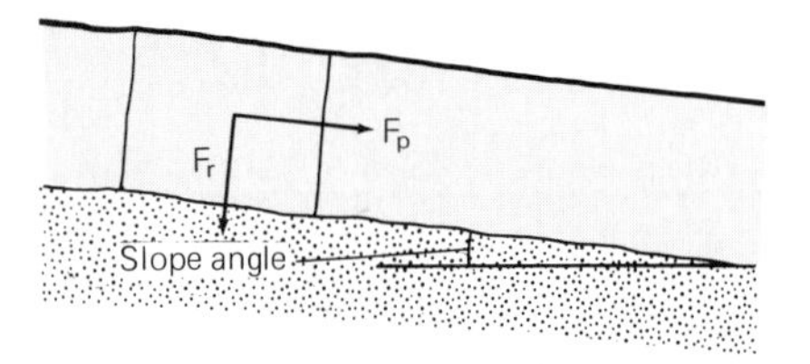


**12.1 Water in a stream being sheared downslope** by the pulling force of gravity $(F_p)$ and held against the stream bed by the resisting force $(F_r)$.

IN THIS CHAPTER, WE EXAMINE THE MANY WAYS IN WHICH RUNNING water serves as a geologic agent. Our method is to begin with some physical aspects of flowing water. We then take up four major geologic topics: (1) streams as self-adjusting transport systems; (2) streams as agents of landscape evolution; (3) streams as environments for depositing sediments; and (4) stream responses to major geologic changes.

## PHYSICAL ASPECTS OF FLOWING WATER

The behavior of water on a slope is governed by the same components of the Earth's gravity that were discussed in Chapter 11. The water is subject to the shearing of the pulling force, or slope-parallel component of gravity, and is held against the slope by the resisting force, or slope-normal component of gravity (Figure 12.1; see also Figure 11.3). A flowing body of water moving down a slope responds to the pulling force in two ways. The fluid itself responds to shearing within the water by forming one of two characteristic kinds of flow and the moving flow exerts a shearing drag on the surface or bed over which it flows.

### Shearing Within the Water: Laminar Flow and Turbulent Flow

Masses of water respond to shearing by flowing in one of two ways. At extremely low shearing stresses, which cause the water to flow very slowly (less than 1 mm/sec), parallel layers of water shear past one another as smooth surfaces that conform to the shape of the bed upon which the water flows. This is *laminar flow* (Figure 12.2; compare with Figure 11.5). The flow in natural streams is almost never laminar.

**12.2 Laminar flow** is characterized by smooth layers of water flowing one over another, either as planes (shown here) or as concordant curved surfaces.

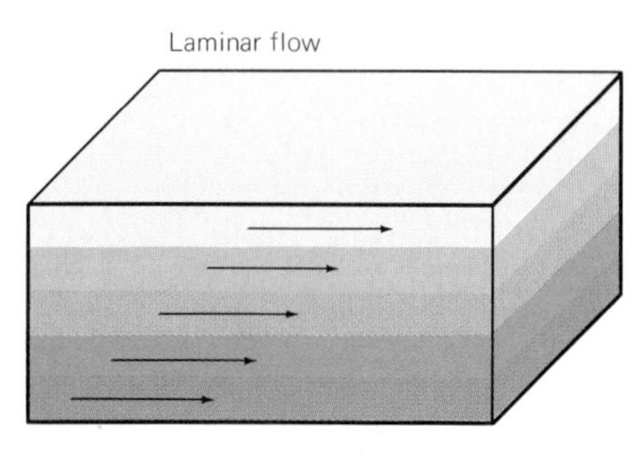


At most intensities of shearing found in nature, the flow of water is characterized by a variety of vortices and eddies that continually form and disappear. This is *turbulent flow*. Turbulence is a fluid's unique way of responding to intense shearing. If one could trace the path of an individual water molecule within a turbulent current, one would see a looping, forward-and-backward, corkscrew kind of motion (Figure 12.3).

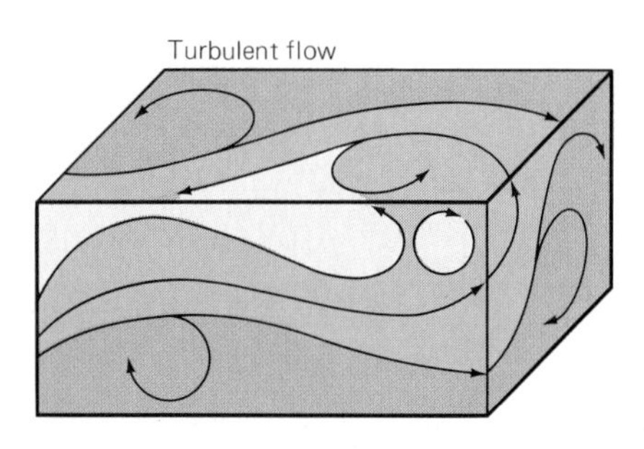


**12.3 Turbulent flow** is characterized by complex, curved flow paths. Schematic block diagram.

The law of continuity within the flowing water requires that the total amount of fluid flowing upward must equal the total amount of fluid flowing downward. The upward movement of flow paths within turbulence provides the force for lifting and suspending sediment. The fact that a turbulent current is able to transport sediment means that the total area of upward-flowing water exceeds the total area of the downward-flowing water. This implies that a generalized circulation pattern becomes established within the water in which broad areas of upward flow are counterbalanced by narrow areas of downward flow.

### Shearing Effects of Water Flowing over Various Substrates

The effects of water flowing on a substrate differ depending on the kind of substrate. We consider two contrasting cases: cohesionless sediments and bedrock.

**Effect over sediments** The effect of water shearing over sediments varies through a considerable range. What happens depends on several variables, including the speed of the flow, the depth of the flow, and the

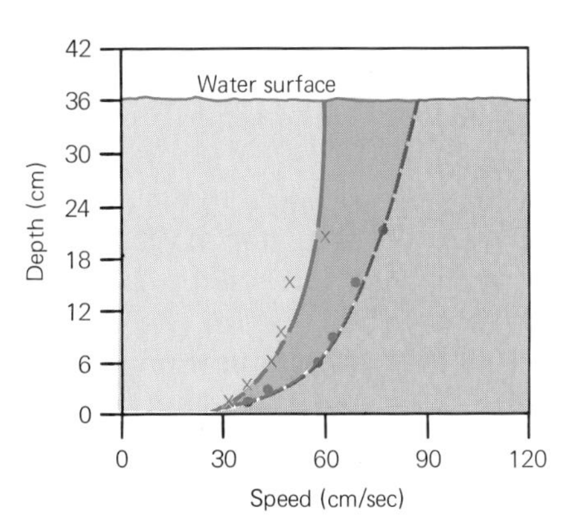


**12.4 Speeds of water at various depths of a stream** flowing over contrasting kinds of sediment on different slopes. Because the water flows faster at successively greater heights above the base of the stream, the water is exerting progressively greater shearing stresses. (Replotted by J. E. Sanders; data in L. B. Leopold, M. G. Wolman, and J. P. Miller, 1964, Fluvial processes in geomorphology: San Francisco and London, W. H. Freeman and Co., 522p., Figure 6.1, p. 154)

**12.5 Speeds of water at various depths of a stream** flowing over contrasting kinds of sediment on different slopes. The same data as in Figure 12.4 but with depth plotted on a $\log_{10}$ scale. Slopes of lines give intensities of shearing; lower slopes mean more intense shearing. (Replotted by J. E. Sanders; data in L. B. Leopold, M. G. Wolman, and J. P. Miller, 1964, reference cited in caption of Figure 12.4, Figure 6.2, p. 155)

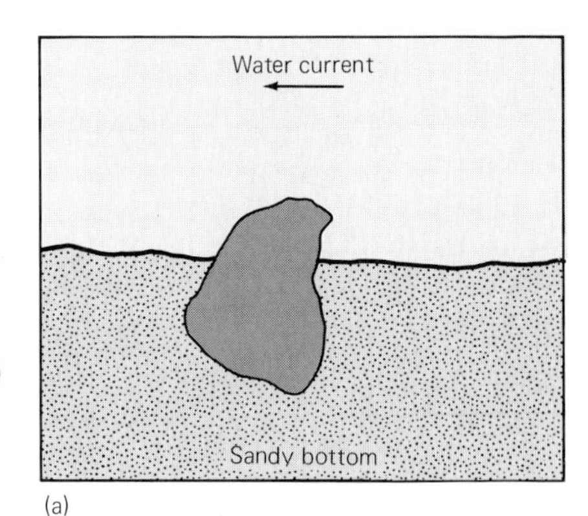

(a)

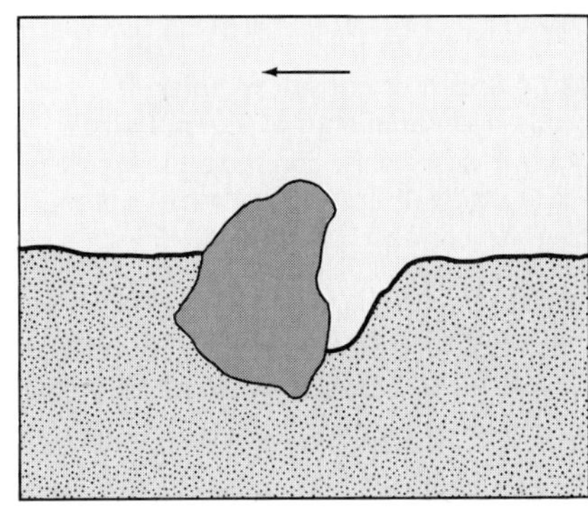
(b)

sizes of the sediment particles.

The speed of the flow governs the intensity of shearing. This relationship is evident from a graph of the change of speed with depth in a current (Figures 12.4, 12.5). The intensity of shearing is determined by the rate of change of speed between successive depths within the flow. If two flows are the same depth, but one flows faster than the other, then the faster current exerts a greater shearing stress on the bed of the flow. The depth of flow is important in ways that are not fully understood. Some of these are related to the sizes of the sediment.

The sizes of the sediment particles determine the characteristics of the surface over which the water flows, known as *roughness,* and also determine whether the sediment will be cohesive or cohesionless (see Chapter 11). Large particles create a rougher surface over which the water must flow and also project farther upward into the flow than small particles. Where a few large particles project above a sandy surface, the current tends to bury them. Eddies and vortices on the upcurrent side of the projecting object cause the water to scour a hole that may eventually become large enough so that when the large particle has been undermined, it will sink out of sight (Figure 12.6).

Particles smaller than 0.125 millimeter tend to present a smooth surface to a current of water and also to display cohesion. Particles larger than

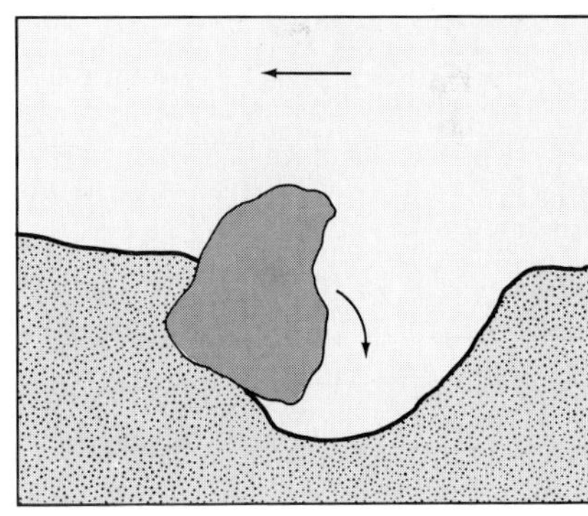
(c)

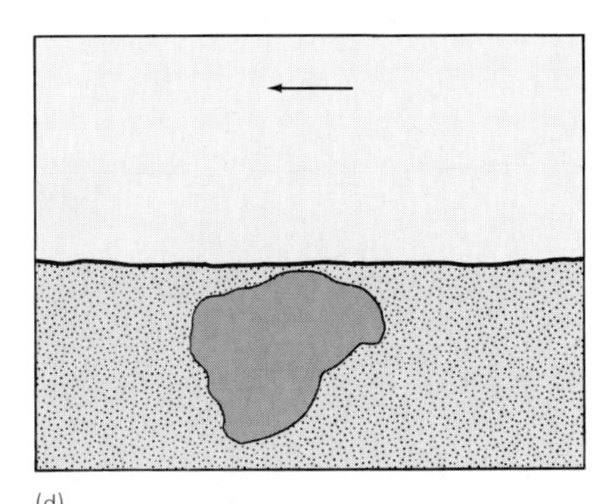
(d)

**12.6 Undermining and sinking of pebble on a sandy stream bed** (a). Scour of hole on upstream side of pebble (b) causes pebble to tip over upstream into the hole (c) and then to be buried by the sand (d). By such scouring and burial, a current can decrease the roughness of its bed and thus can flow more efficiently.

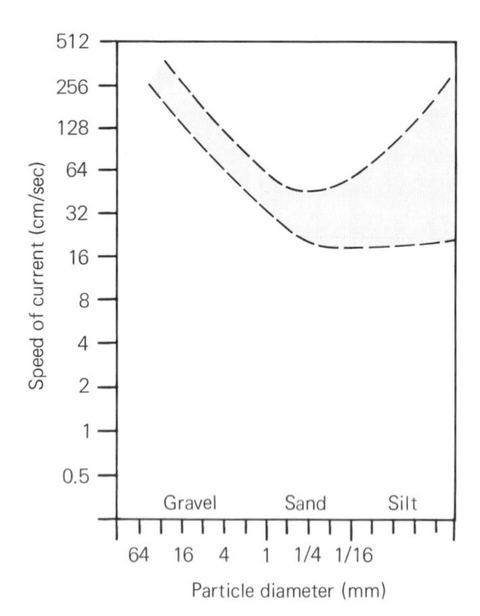


**12.7 Speed of current required to erode bed sediments of various sizes.** Notice that a swifter current is required to erode cohesive silt than to erode coarser cohesionless sand. (After G. M. Friedman and J. E. Sanders, 1978, Principles of sedimentology: New York-Santa Barbara-Chichester. Brisbane-Toronto, John Wiley and Sons, 792p., Figure 4-26, p. 110)

0.125 millimeter are usually cohesionless. If the particles are cohesionless, then a direct relationship exists between the sizes of the particles and the speed of a current required to erode them. With cohesive particles, however, the opposite is true. A swifter current is required to erode the finer particles (Figure 12.7).

Water moves sediment in two contrasting ways: (1) within the current and (2) along the base of the current. Finer particles are moved within the current. They become entrained within the flow and are wafted about by the irregular flow paths within the turbulent current (Figure 12.8). Such particles can be supported by the threads of the turbulent flow only if the upward speeds of the water exceed the settling speeds of the particles. The settling speed of a sediment particle is governed by its diameter, shape, and density. A graph of settling speeds of spheres of quartz of various diameters is shown in Figure 12.9. Particles supported by the upward components of a turbulent flow constitute the *suspended load.*

Particles that do not become caught up within the turbulent eddies can be transported along the bottom of the current. They constitute the *bed load;* such particles are rolled, dragged, or bounced along. Without being in the suspended load, a particle may rise a few centimeters to as much as 30 centimeters above the bottom, travel downstream some distance, and then be pulled back by gravity (see Figure 12.8). Such movement is known as *saltation* (Latin for "leap"). A collective name for all processes of bed-load transport is *traction.*

The separation of suspended load and bed load is not always distinct. As conditions of flow change, particles of the bed load may suddenly become suspended or vice versa. Particles of sand, gravel, and small cobbles tend to move as bed load, either individually or in groups.

**12.8 Paths of particles in the two kinds of sediment loads of a stream** shown by schematic longitudinal profile and section through the water. Fine particles in suspension follow irregular, curved, complex flow paths within turbulent eddies. Coarser particles in bed load roll or bounce along the bottom, following paths that are nearly straight when viewed from above (that is, in the plane of the profile). In the vertical plane shown by this section, saltating particles in bed load follow curved trajectories shown by overlapping circles.

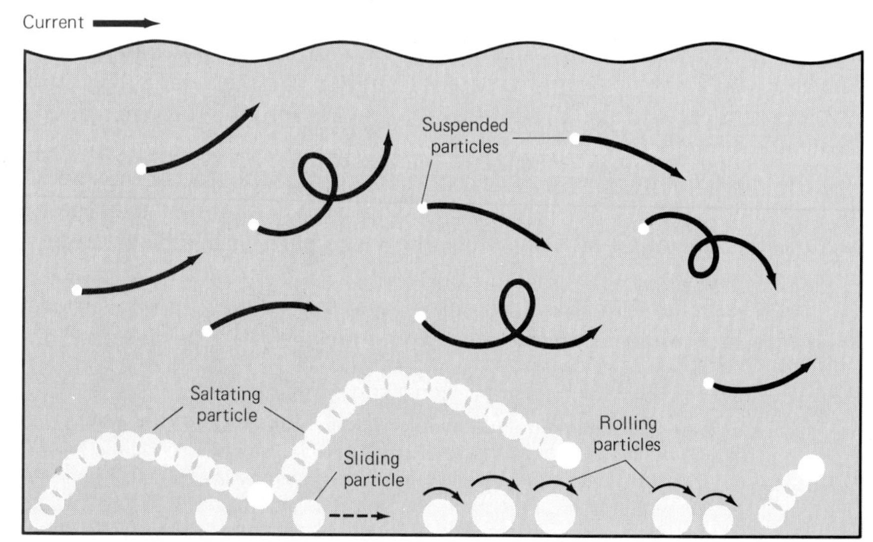

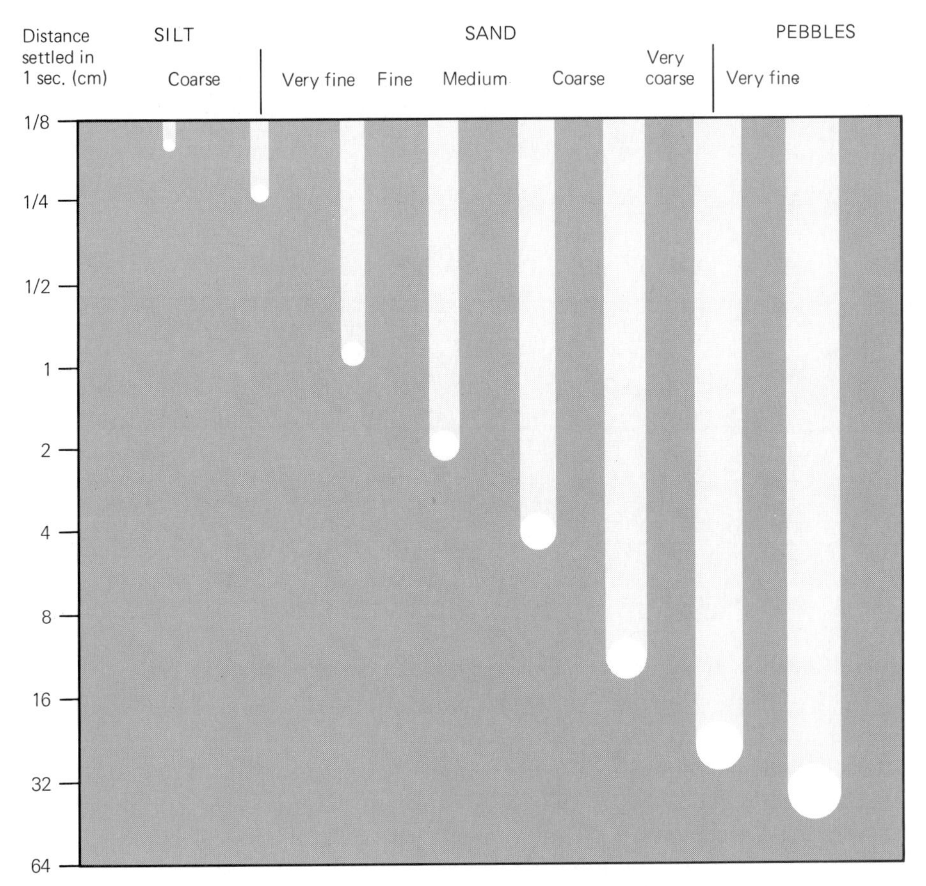


**12.9 Settling speeds of quartz spheres of various diameters** in still fresh water shown by lengths of lines measured down from top of graph (to represent vertical distances the individual spheres would settle in one second). Sizes of particles schematic; both settling distances and particles sizes plotted on $\log_2$ scales. (Data from R. J. Gibbs, M. D. Matthews, and D. A. Link, 1971, The relationship between sphere size and settling velocity *(sic):* Journal of Sedimentary Petrology, v. 41, no. 1, p. 7–18, Table 4, p. 13–16)

**Effect over bedrock** Where water shears over bedrock, it is not possible for the water to fashion the surface on which it flows into any kind of bed forms, as with sand. The current can smooth the rock and erode out rounded depressions. If the current transports bed-load sediment across bedrock or if a body of liquefied sediment is flowing at the base of the current where it crosses bedrock, then the effect is to abrade and thus lower the surface of the bedrock.

## STREAMS AS SELF-ADJUSTING TRANSPORT SYSTEMS

One of the truly remarkable examples in nature of a system with a series of built-in mechanisms for self adjustment is a stream network. A stream system transports various materials — water and its physical load of sediment particles and its chemical load of ions in solution. As the quantities of these materials change, the system makes adjustments. Because all the factors interact, it is difficult to organize a coherent discussion of them by proceeding one at a time. Moreover, it should be remembered that the factors do not work in isolation.

### Discharge of Water

The volume of water flowing through a cross section of a stream channel during a given time is its *discharge* (Figure 12.10). Sediment and ions

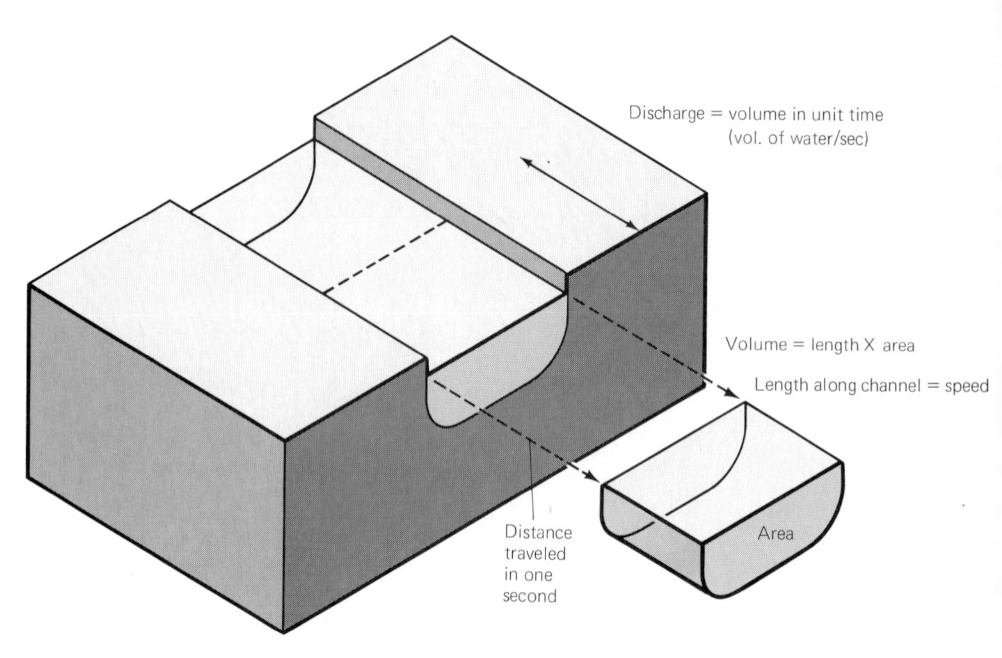


**12.10 Discharge of a stream** shown schematically by removing the quantity of water that has flowed along it in one second. All the parcels of water in a stream are lined up like cars in a bumper-to-bumper traffic jam. A parcel of water cannot move until a corresponding parcel of water downstream from it has moved onward. As each parcel passes down the channel, its place is taken by another parcel from upstream. The volume of water discharged *(Q)* equals the cross-sectional area (width, *w*, times depth, *d*) times the speed of the current *(c)*, or in mathematical notation: $Q = w \cdot d \cdot c$.

carried and deposited by a stream may also be referred to as stream discharges. When the term discharge is meant to refer to the volume of water, it is used by itself. When other kinds of discharge are meant, an appropriate modifier is added.

The subject of stream discharge involves many subthemes. We will try to deal with them by examining here where the water comes from and the causes of variation at a single station. In a later section, we discuss downstream changes.

**Supply of water to streams** Water is delivered to a stream as a more-or-less steady supply and as intermittent large quantities. The *base flow* is the usually steady supply of water that keeps a stream flowing between rains. If the base-flow water becomes insufficient or intermittent, then the stream ceases to flow as a "permanent stream" and becomes an *intermittent stream,* one that flows only now and then.

In order to visualize how the base flow affects stream behavior, it is necessary to go all the way down the stream to its mouth and work upstream. The level of the water at the mouth of a stream is a *base level* that serves as a check valve on the stream. From base level upstream, the base-flow water is backed up continuously, as are the cars in a bumper-to-bumper traffic jam at rush hour. The upstream water locked into the base flow cannot flow any faster than the water downstream, just as a car in the middle of a traffic jam cannot be driven any faster than the cars ahead of it. Another way to visualize the base flow is to think of it as a great conveyor belt or a moving walkway that moves all that it contains as a unit.

Any water other than the base flow usually represents "extra" water that comes from surface sources. A large quantity of "extra" water typically forms a *flood bulge,* a long wave of water that travels downstream at its own speed, independently of and added to the speed of the base flow.

The moving wave of a flood bulge is a *kinematic wave,* a wave that travels along with the water "particles" within it. If the flood water did not make such a moving wave, it could not travel any faster than the base flow. In terms of the conveyor-belt analogy, a flood bulge can be compared to a person walking on a conveyor belt in the same direction as it is traveling. The total speed of movement is speed of walking plus the speed of the belt.

The independence of the base flow and a flood bulge is shown in Figure 12.11. The base flow exists before a flood bulge arrives and is still present after a flood bulge has moved out of a given part of the stream.

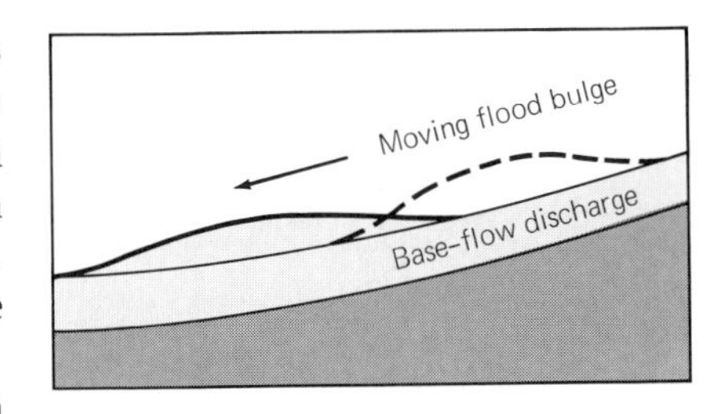


**12.11 Base flow and flood bulge,** schematic longitudinal profile of a stream. The moving flood bulge is known as *kinematic wave,* but the relationships of the moving bulge to the base-flow water are not known.

**Causes of discharge variation at a single station** Variation in stream flow results from changes in the supply of water within the drainage area. Such changes can be caused by the seasonal distribution of rainfall, by the seasonal melting of snow and ice, by the erratic distribution of larger-than-usual rainstorms, by cyclic variation in rainfall during periods longer than one year, and by the release of water from dams either in a controlled way or by the bursting of a dam and sudden discharge of the water.

## The Water's Loads

The water of a stream contains two kinds of loads that are controlled by the products of weathering: (1) the physical load consisting of sediment and (2) the chemical load consisting of ions in solution.

**Sediment load** As mentioned previously, the physical load of a stream is carried as suspended load or bed load. The suspended load comes from sheet flows and rivulets draining across regolith that lacks a cover of plants and from erosion along the sides of the channel. When suspended particles are abundant, they cause the water to appear murky and turbid.

In general, the amount of the suspended load increases as discharge increases (Figure 12.12), but similar increases in discharge are not necessarily accompanied by similar increases in suspended load. For example, the floods in August 1955 in Connecticut caused much water to flow in the rivers, but because of the cover of vegetation, very little gully erosion took place. As a result, the amount of suspended sediment was not large and the peak of discharge of suspended sediment came many hours before the peak of water discharge.

The bed load is not easy to measure exactly; even estimates may not be very accurate. Investigators generally agree that bed load makes up about 10 percent of the total solid load, but in fast-moving rivers may exceed 50 percent.

Once in motion, large particles move faster than smaller ones, and rounded particles move more easily than angular particles or those having flat shapes. The largest particle a stream can move is a measure of the stream's *competence.* The total quantity of sediment a stream is able to transport is its *capacity.*

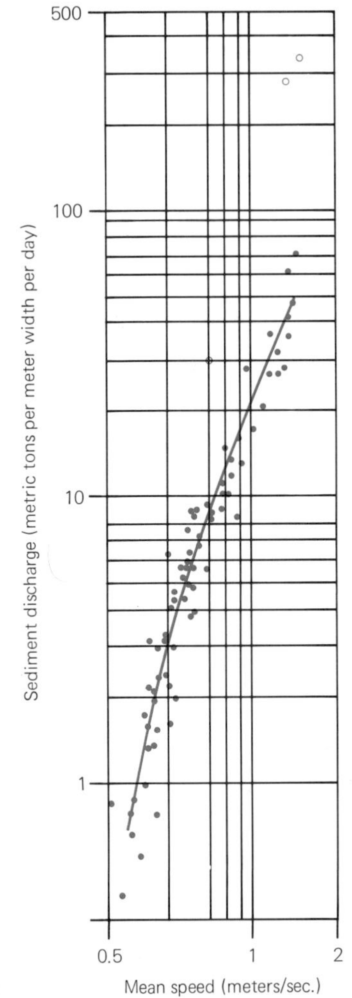


**12.12 Increase of suspended sediment with increasing discharge,** Powder River, Arvada, Wyoming. Both discharge and weight of suspended sediment are plotted on $\log_{10}$ scales. (Data from Luna B. Leopold and Thomas Mattock, 1953, The hydraulic geometry of stream channels and some physiographic implications: U. S. Geological Survey, Professional Paper 252, 37p., Figure 13, p. 20)

Particles in the bed load collide repeatedly. These collisions may break particles into smaller fragments, dislodge them, or abrade them. The effectiveness of abrasion is apparent to anyone who has noticed the smooth, rounded pebbles along a stream (see Figure 8.10) or a beach.

**Dissolved load** The chemical load carried by a stream consists of ions in solution. In some cases, the weight of the dissolved load may be as much or more than the solid, sediment load. Each year, an estimated 4000 million tons of soluble matter is transported from the continents to the oceans. The loads of selected rivers are shown in Table 12.1.

## Dependent Variables Involved in the Hydraulic Geometry of a Stream

The *hydraulic geometry of a stream* is an expression for the important variables in a stream system and how they interact. The independent variable is discharge. The dependent variables are shape and depth of the channel, speed of the flow, slope of the water surface, and slope of the stream bed, or gradient. We define each of these dependent variables first and then show how they interact as the discharge changes.

**Shape and depth of the channel** The channel of a stream may be pictured as a long, narrow trough. What affects the conditions of the flowing water is the shape of the channel's cross section. The depth divided by the width is a measure known as the *form ratio.* The form ratio of a river that is 1 meter deep and 100 meters wide is 1/100. The form ratio during the summer of the Columbia River near the Canadian border is 1/19, whereas that of the Platte River near Duncan, Nebraska, is 1/160.

**Speed of the flow** The speed of flow varies in different parts of the channel. The location of maximum speed of flow depends upon the channel shape, roughness, and sinuosity. The point where the current flows fastest usually lies near the center of the channel and below the surface, about a quarter of the way to the bottom (Figure 12.13). Speed of flow decreases toward the sides of the channel and the bottom. One of the greatest speeds measured by the U.S. Geological Survey was 6.6 meters per second in a rocky gorge of the Potomac River during the flood

**12.13 Speeds of flow of water in various parts of a channel** shown by transverse profile and section. Lines connect points of equal speed that was measured in centimeters per second. Based on current meters placed at 10-meter intervals across the River Klaralven, southern Sweden, and at 0.5-meter intervals of depth. (Åke Sundborg, 1956, The River Klarälven, a study of fluvial processes: Geografiska Annaler, v. 38, no. 2–3, p. 127–316, Figure 55, p. 283)

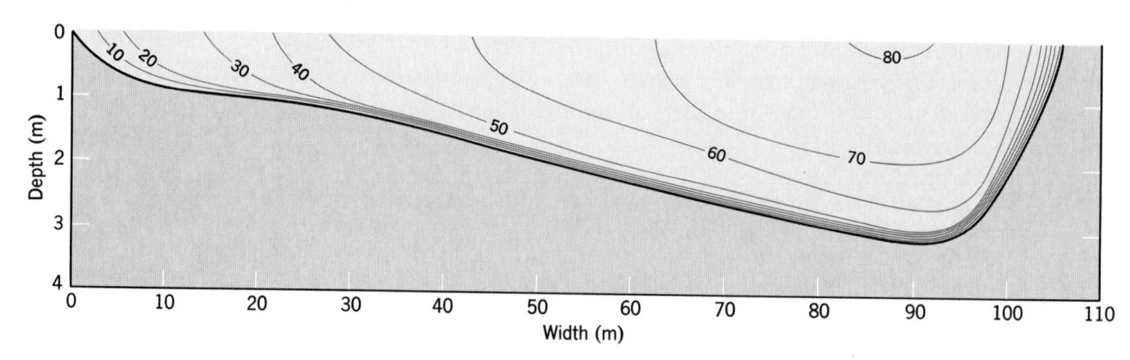

of March 1936. The highest natural speed for running water is about 9 meters per second (32 km/hr).

**Slope of water surface** The loss of height within a unit distance in the downstream direction is the slope of the water surface. The steeper this slope, the faster the water flows.

**Gradient and longitudinal profile** The slope on which a stream flows is its *gradient*. The gradients of almost all streams are steeper near their sources than they are farther downstream. The aggregate of all the local gradients constitutes the *longitudinal profile* of a stream. The profile forms a curve with rapidly increasing steepness toward the headwaters (Figure 12.14). The main reason for this shape is that, in general, discharge increases downstream. As the volume of water increases, the shape of the channel becomes more efficient, and thus the larger amounts of water can flow faster on a gentler slope than can the smaller amounts of water in localities farther upstream where the gradient is steeper.

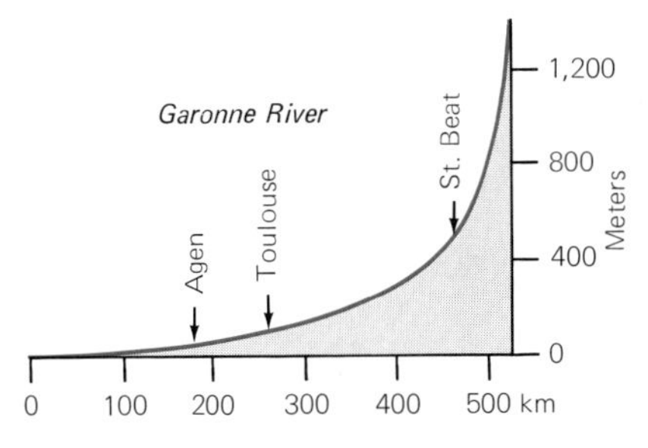


**12.14 Longitudinal profile of a stream** is concave up, that is, steeper in headwater areas and gentler in downstream areas. Garonne River, France. (Emmanuel de Martonne, 1951, Traité de géographie physique, neuviéme éd.: Paris, Librarie Armand Colin, v. 2, Le relief du sol, p. 499–1057; Figure 213, p. 552, from l'Atlas de France, Plate 21)

## Responses to Changes in Discharge; The Concept of a Graded Stream

The importance of discharge in stream behavior can be illustrated by considering how the shape and depth of the channel, speed of the flow, and gradient change as discharge fluctuates. As discharge increases when a flood bulge arrives, the altitude of the water surface is elevated, the level of a sediment bed is usually lowered (thus, the depth increases in two ways), the width of the flowing water may increase, and the speed of the flow increases (Figure 12.15).

The increase in depth that results from the lowering of a stream's bed takes place because bed sediment is dispersed and dilated. This stirring

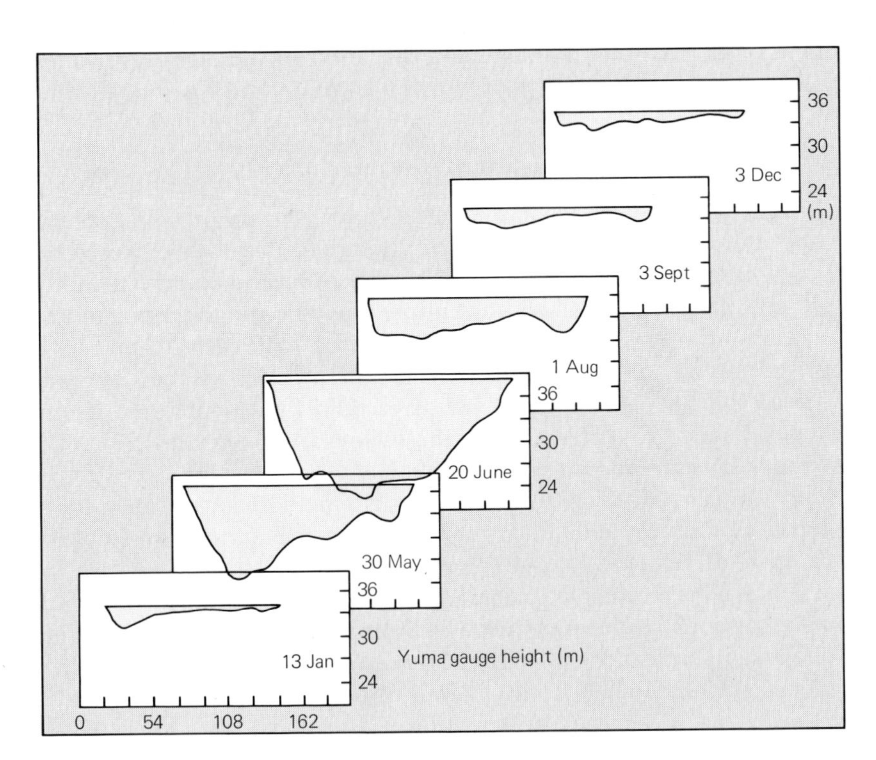


**12.15 Effects of 1912 flood in Colorado River,** profile at Yuma, Arizona on days shown. Both increasing height of water surface and lowering of elevation of stream bed can be referenced against the identical profile panels. (After G. M. Friedman and J. E. Sanders, 1978, reference cited in caption of Figure 12.7, Figure 8-42, p. 224, based on data of E. W. Lane and W. M. Borland, 1953, River-bed scour during floods: American Society of Civil Engineers, Proceedings, v. 79, p. 254–1 to 254–14, Figure 4, p. 1076)

up of the bed sediment is a cleansing mechanism. The stirred-up sediment may or may not be transported very far downstream. In either event, as soon as the discharge decreases, the dispersed sediment is usually redeposited on the bed. As a result, the level of the bed after the flood may be the same as it was before the flood. To a casual observer, it may appear that the flood made no changes.

If discharge increases downstream, then speed also increases downstream. It is easy to see that rivers grow wider and deeper downstream, but it is not so apparent that in a downstream direction speed increases as well. A wild, leaping mountain stream spends much of its energy creating turbulent eddies with almost as much backward as forward motion. For example, in a downstream direction speed increases from the tiny Owl Creek in Wyoming (0.6 m/sec) to the Bighorn River in Montana (0.75 m/sec) to the Mississippi River itself at Vicksburg, Mississippi (1.5 m/sec).

The interaction of all the factors just discussed establishes a condition of dynamic balance which characterizes a *graded stream.* Any change in the loads, the total material a river carries, causes it to make a corresponding change in gradient. If the river cannot move its load beyond a certain point on the profile, it will increase its gradient. The gradient can be steepened by depositing some of its load at that point, building up the channel bed and creating a steeper slope below the point. At the same time, this buildup decreases the slope above the point, so that the river continues the process in the upstream direction until the entire slope has been adjusted.

## STREAMS AS AGENTS OF LANDSCAPE EVOLUTION

In this section, we examine how streams and associated activities erode bedrock. We start with the origin of valleys and their relationship to streams, then consider drainage patterns, and conclude with some ideas about cycles and rates of stream erosion through long periods of time.

### Origin of Valleys; Relationship to Streams

**Playfair's law** Before the nineteenth century, the origin of valleys was a subject of vigorous debate. One group argued that valleys were great cracks where the Earth's crust had been split open and that rivers merely found the ready-made valleys and simply flowed through them. Another group argued that valleys had been scoured by swift currents on the sea floor at times when the ocean had covered all the lands, as they imagined had been the case, for example, during the Old Testament flood. A third viewpoint stated that rivers carve the valleys in which they flow. This third idea became known as Playfair's law.

John Playfair, (1748–1819) was an English mathematician and a friend of James Hutton. In 1802 Playfair, writing on the formation of valleys, restated Hutton's idea that streams do not simply occupy valleys, but create them. This principle, *Playfair's law,* is now generally accepted, with some exceptions, and forms the basis for most modern studies of rivers.

The ancient volcanoes of central France (see Figure 5.4) provided the evidence which allowed Charles Lyell to confirm Playfair's law and to

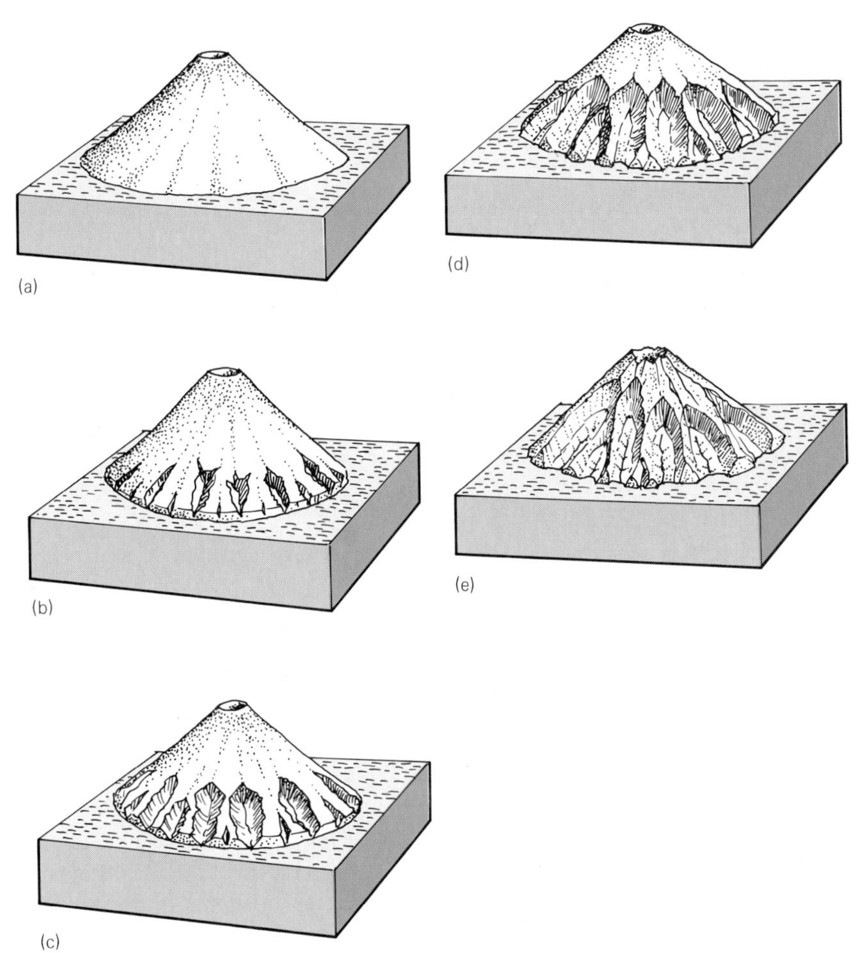


**12.16 Erosion of inactive volcanic cone,** Hawaii, by streams and downslope gravity-transport processes. Steep-sided gorges from when waterfalls retreat rapidly upstream. Later, side slopes are lowered and valleys widen. (Based on Gordon A. Macdonald and Agatin T. Abbott, 1970, Volcanoes in the sea: Honolulu, University of Hawaii Press, 441p., Figure 113, p. 175, and photographs in Chapter 9, Stream erosion, p. 161–183)

refute the argument that valleys on land had been carved by strong currents at the bottom of the sea. A basaltic lava flow from the two tephra cones, Puy de Lassolas and Puy de la Vache, flowed down the Veyre valley and blocked off a tributary, forming Lac d'Aydat. Since then, a narrow gorge many meters deep has been cut through the basalt. Lyell argued that this gorge could have been cut only by the modern stream and could not possibly have been eroded by currents at the bottom of a sea. The loose tephra cones were the clinchers. Any hypothetical valley-carving sea that covered the area after the lava had been erupted and had cooled to form the basalt would surely have destroyed these cones, yet there they stand. Since Lyell's time, the age of the lava flow has been fixed by radiocarbon dating at 7650 ± 350 years.

**Headward erosion** The formation of a valley begins soon after "new" land is raised above sea level. The sequence has been studied in Hawaii on inactive volcanic cones (Figure 12.16). Rain collects in cracks and flows downhill. Tiny rivulets form and lead into larger trickles of water which come together in small streamlets. Eventually, at the lowest level, the streamlets come together in a master stream, which cuts downward rapidly.

A stream erodes bedrock chiefly by drilling, sawing, and quarrying.

**12.17 Potholes in limestone bed of Catskill Creek,** northwest of Catskill, New York. (J. E. Sanders)

The drilling and sawing take place at waterfalls. The falling water drills *potholes* (Figure 12.17), which form easily in weak rock. The sawing takes place as streams erode along cracks in the bedrock. Swiftly flowing water can quarry out loose blocks bounded by cracks or bedding-plane partings.

Erosion of the edge of a waterfall causes the upstream retreat of the waterfall, leaving behind a steep-sided gorge (Figure 12.18). After this gorge has formed by headward erosion, the main work of the stream itself is virtually completed. Other processes become involved in further activities on the side slopes.

**Factors shaping valley-side slopes** Although, according to Playfair's law, rivers are responsible for carving valleys, the rivers themselves do not erode the entire volume of sediment removed from the cross profile of a valley. Most of the sediment travels from the valley slopes to the stream by other processes (Figure 12.19).

The exact behavior of valley-side slopes has been the subject of some controversy. It is still not agreed what happens after the original slot has been cut. Everyone agrees that the next stage is the formation of a V-shaped cross section, which is typical of stream valleys (Figure 12.20). One opinion is that the V-shaped cross section gradually widens like a book being opened and laid flat (Figure 12.21a). The other viewpoint, which has gained support from recent field observations, is that the sides of the V retreat more or less parallel to each other (Figure 12.21b).

The important processes operating on valley-side slopes are the downslope gravity-transport processes, raindrop impact, and the flow of water in sheets and small gullies.

After rain has begun to fall, soil pores may become clogged, a process known as surface sealing. Once the surface of the soil has been sealed, puddles form and the accumulated water begins to flow in rather uniform thin layers known as *sheet flows.* The thin sheets of water exert a drag on loose particles. Depending upon their speeds of flow, the sheets pick up

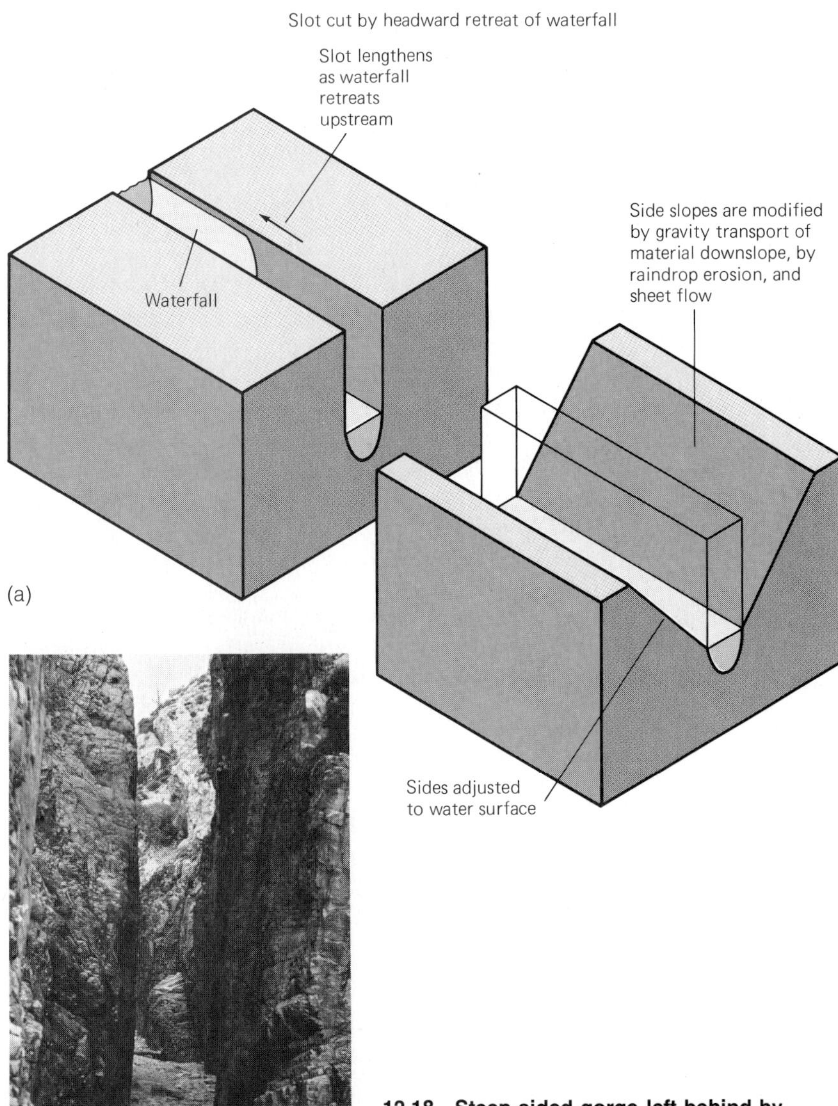


**12.19 Characteristic V-shaped stream-valley** transverse profile results after side slopes have been lowered by various downslope gravity-transport processes; schematic block diagram.

**12.18 Steep-sided gorge left behind by upstream retreat of a waterfall.** (a) Slot-like gorge is a direct result of stream erosion; schematic block diagram. (b) Steep-walled gorge carved by stream in massive bedrock, Tamina "canyon," northeastern Switzerland. (Swiss National Tourist Office)

particles ranging in size from fine clay to sand. Sheet flows may also carry off valuable organic matter that has been dissolved from soil aggregates by rain splash. Sheet flows and rain splash are most severe on steep slopes where the water moves swiftly. During heavy rainfalls on steep slopes, sheet flows become more localized and eventually begin to carve countless tiny channels in the soil called *rills*. If rills continue unchecked, they may grow into even-deeper channels called *gullies*. In turn, gullies may form a landscape of closely spaced V-shaped valleys known as *badlands* (Figure 12.22).

**12.20 Typical V-shaped transverse profile of a stream valley,** north shore of Long Island, view N toward Long Island Sound, Wildwood State Park, Wading River, New York. (J. E. Sanders)

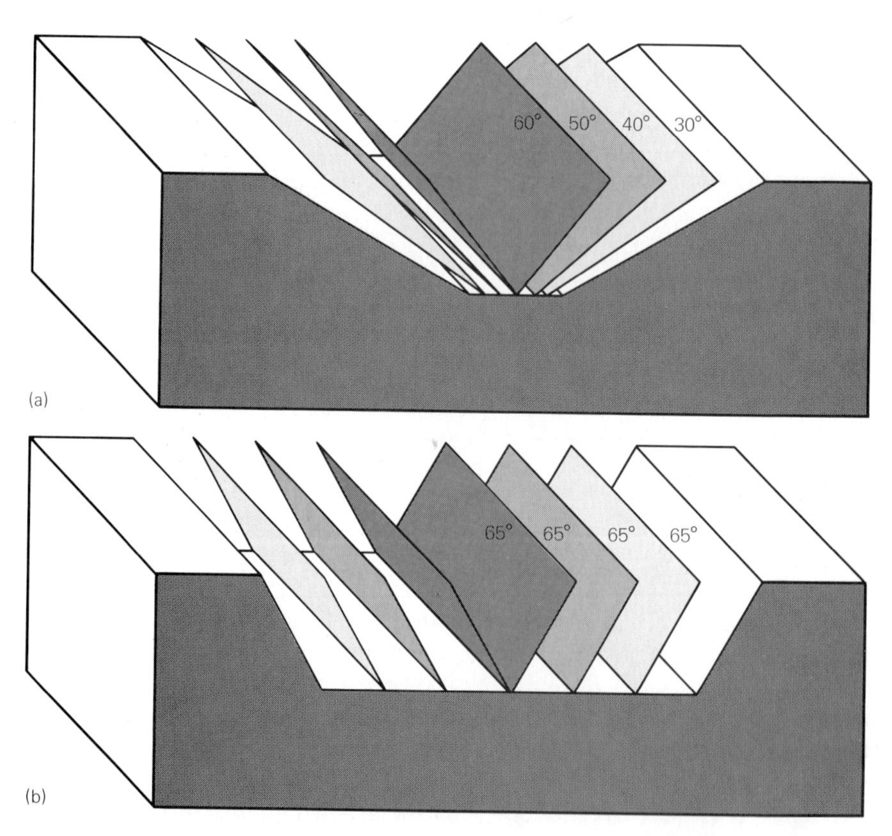


**12.21 Contrasting ideas about behavior through time of valley-side slopes,** schematic block diagrams. (a) Concept that slopes are gradually lowered. (b) Concept that valleys widen as slopes retreat parallel to themselves.

**12.22 Badlands created by stream erosion** in an arid region, Death Valley, California. (U. S. National Park Service)

### Patterns of Valleys; Drainage Patterns

A master stream and its tributaries make complex drainage patterns. These patterns depend upon the amount of water and the kinds of bedrock in the drainage basin. Where water is plentiful, the chief drainage patterns are dendritic, trellis-rectangular, radial, and annular (Figure 12.23).

### The Concept of the Erosion Cycle

The concept of the erosion cycle, widely popularized by William Morris Davis (1850–1934) in the nineteenth century, dominated geologic thinking for 50 years. Many doubts are now being raised about the validity of the cycle concept. We review the ideas and evaluate their present status.

The *erosion cycle* is thought to begin with the uplift of a land mass, ranging from an island to a mountain range to an entire continent. The cycle is presumed to end when this uplifted area has been eroded away and all that is left is an almost-flat plain, known as a *peneplain,* lying near base level. An erosion cycle takes so long that no one has ever seen a complete cycle. Geologists try to piece together the parts of the cycle by examining different regions in the modern world and by looking at the geologic record.

The idealized erosion cycle is based on the presumption that nothing interrupts the regime of erosion. Actually, many interruptions can take place.

**Stages in the erosion cycle: youth, maturity, and old age** The three stages in the Davis erosion cycle have been named youth, maturity, and old age. In the youthful stage of the cycle, streams erode headward actively. Valley walls are steep, even vertical (Figure 12.24a); waterfalls

**12.23 Drainage patterns** shown by schematic maps. (a) *Dendritic pattern* results where underlying materials erode uniformly. (b) *Trellis pattern* indicates long, narrow belts of rock that erode unequally. The straight master streams flow in belts of weak rocks, as along faults or the tilted edges of strata. Tributaries enter master streams at right angles. (c) *Rectangular pattern* in which master streams make many right-angle bends and tributaries enter master streams at right angles. (d) *Radial pattern* develops where streams flow away from a high central point, as on a volcanic cone or a circular dome. (e) *Annular pattern* appears where master streams flow along curved belts of weak rock among the gently dipping strata of a large dome.

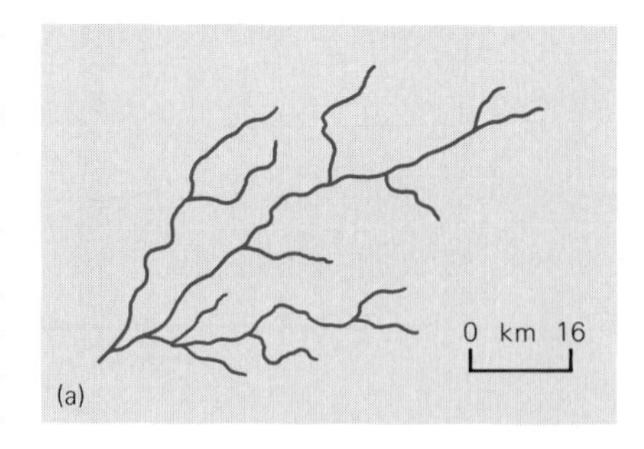


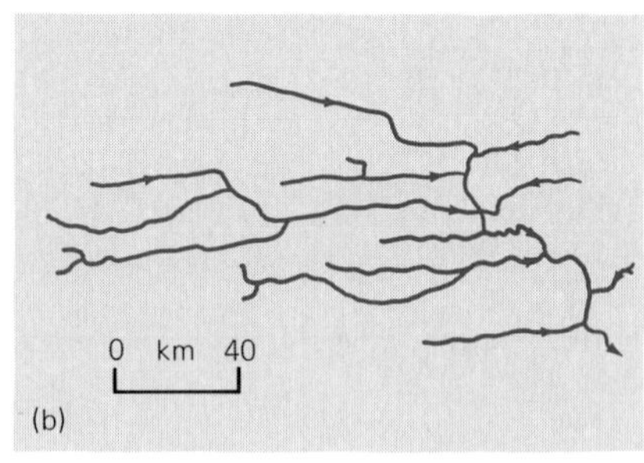


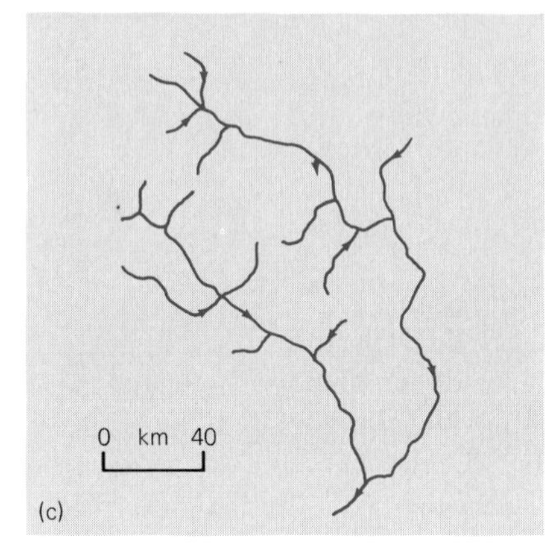


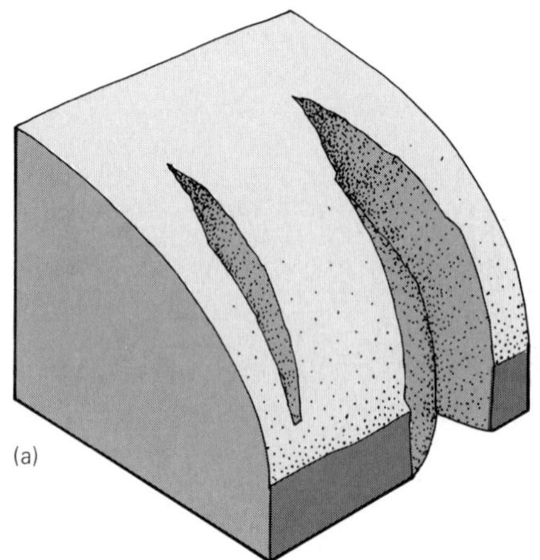


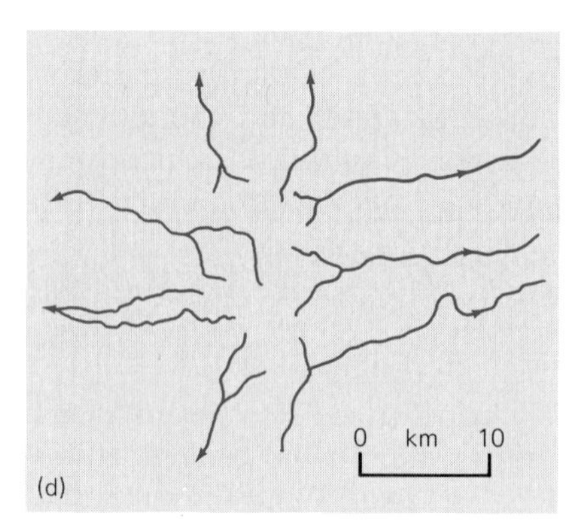


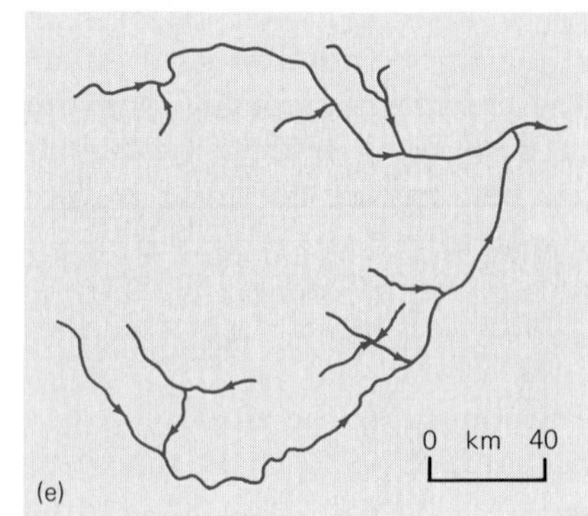


and rapids are numerous; and large areas in between streams have not yet been dissected. Lakes and swamps may be numerous.

The stage of maturity is achieved when the young valleys have been cut into the uplifted block, new networks of tributaries have formed, and all valleys show typical V-shaped profiles. In the mature stage, the network of streams extends itself completely across the area (Figures 12.24b and 12.25). Flood plains appear on valley floors.

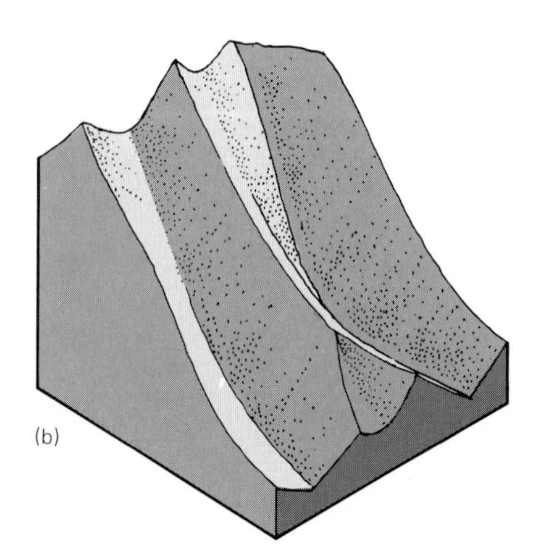


**12.24 Cycle of erosion of a landmass by streams and by downslope gravity-transport processes;** stages of youth and maturity shown by schematic block diagrams. (a) Youthful stage: narrow, steep-sided streams; areas between streams not yet dissected. (b) Mature stage: stream network extends to all parts of the region; divides are narrow.

**12.25 Mature landscape** dissected by V-shaped stream valleys; image of area near Sandy Hook, Kentucky, made from above by side-scanning radar. Width of photograph is 10 kilometers. (Autometric Division of Raytheon Company, courtesy of U. S. Army Engineers, Topographic Laboratories, Fort Belvoir, Virginia)

The stages of youth and maturity can be illustrated by many modern examples, but the stage of old age is where doubts about the Davis cycle have been raised. The old-age stage is supposed to be a surface of low relief across which streams meander. Many alleged modern examples have proved to be alluvial plains that appear the same regardless of their ages. From the geologic record, we know that ancient peneplains did exist. But no peneplains have been proved in the modern world. Therefore, the details of the shift from the stage of maturity to that of old age are conjectural and controversial. Part of this problem is related to the history of valley-side slopes, shown in Figure 12.21.

Despite the problems about the final stages in the erosion of a landmass, the cycle concept is useful in that it synthesizes how stream-eroded landscapes change with time. With new attention being given to the behavior of slopes, we can expect that the problems about the old-age part of the cycle eventually may be resolved.

**Rates of erosion** Erosion can be measured in two ways: (1) by finding the rate at which the land surface is lowered and (2) by measuring the amount of sediment loads and dissolved loads in streams. The easiest method is to measure the stream loads and then calculate rates of the lowering of the land surface based on amounts of material in the streams. The amount of sediment supplied per unit area is known as the *sediment yield* (Figure 12.26). At present, running water removes soil from the surface of the United States at an average rate of 1 centimeter per 300 years.

**12.26 Sediment yields from lands used in various ways.** Compiled from various sources.

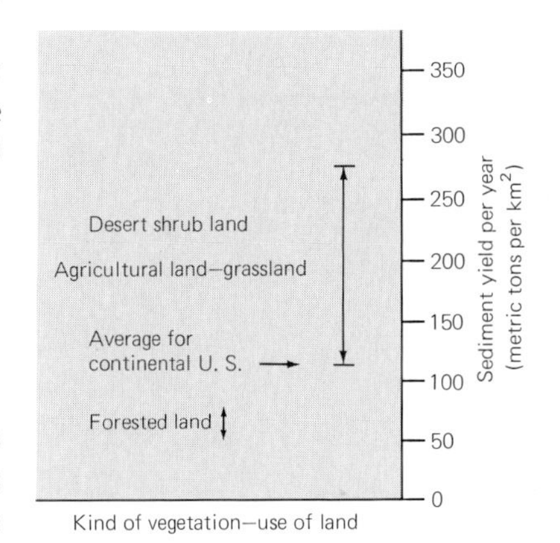


## STREAMS AS ENVIRONMENTS FOR DEPOSITING SEDIMENTS

A geologic agent that erodes and transports material can also function as a mechanism for depositing sediment. All stream-deposited sediments are known as *alluvium.* Streams erode materials from regions that are

**12.27 Fans,** Death Valley, California. (U. S. National Park Service)

being uplifted and transport these materials to regions that are subsiding. Here, streams deposit their alluvium. Four environmental settings from upstream areas to the mouths of streams in which streams may deposit sediment are: (1) fans, (2) areas having braided channels, (3) lowlands in which the channels form great sweeping curves (flood-plain rivers), and (4) deltas.

### Fans

Where a stream leaves a bedrock valley in a highland block and enters an open lowland in which its channel is free to change directions, it deposits *fans,* low, fan-shaped cones of alluvial sands and gravels ranging from a few meters to many kilometers in extent (Figure 12.27). A stream flowing across a fan is able to maintain its course, temporarily at least, by depositing sediment and thus building upward as the fan subsides. Eventually, however, the built-up channel will stand higher than its surroundings. This is a situation that is not stable; it is subject to quick changes. The stream may break out of its built-up channel and start a new course across one of the lower parts of the fan. The stream then builds upward in this new course, breaks out, and occupies another lower position. By such repeated shifting, the stream builds up all parts of the fan. In the Los Angeles basin, the communities of Burbank, Glendale, Montrose, and Pasadena, built along fans from the San Gabriel Mountains, are repeatedly subjected to flash floods such as those that built the fans in the first place.

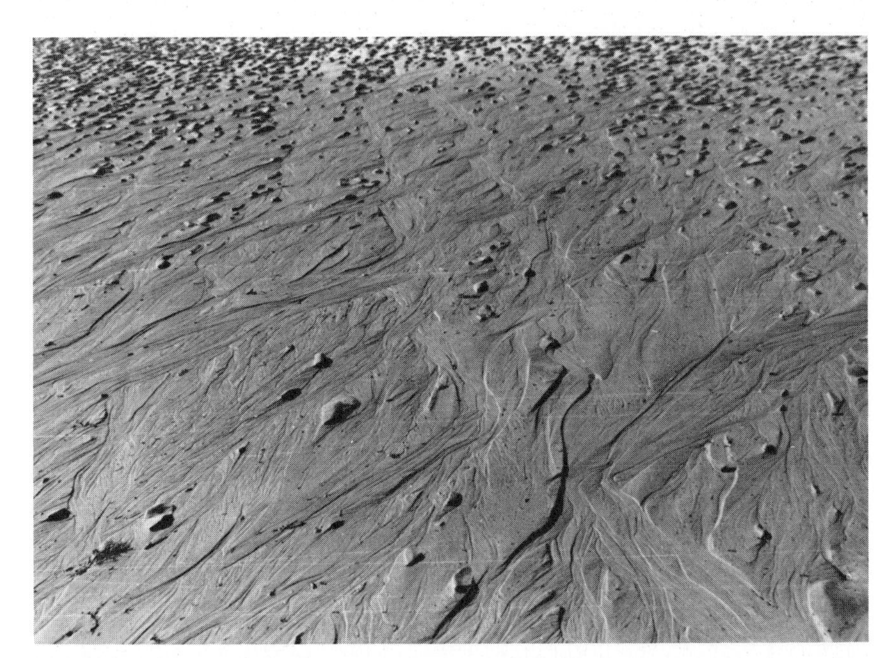

(a)

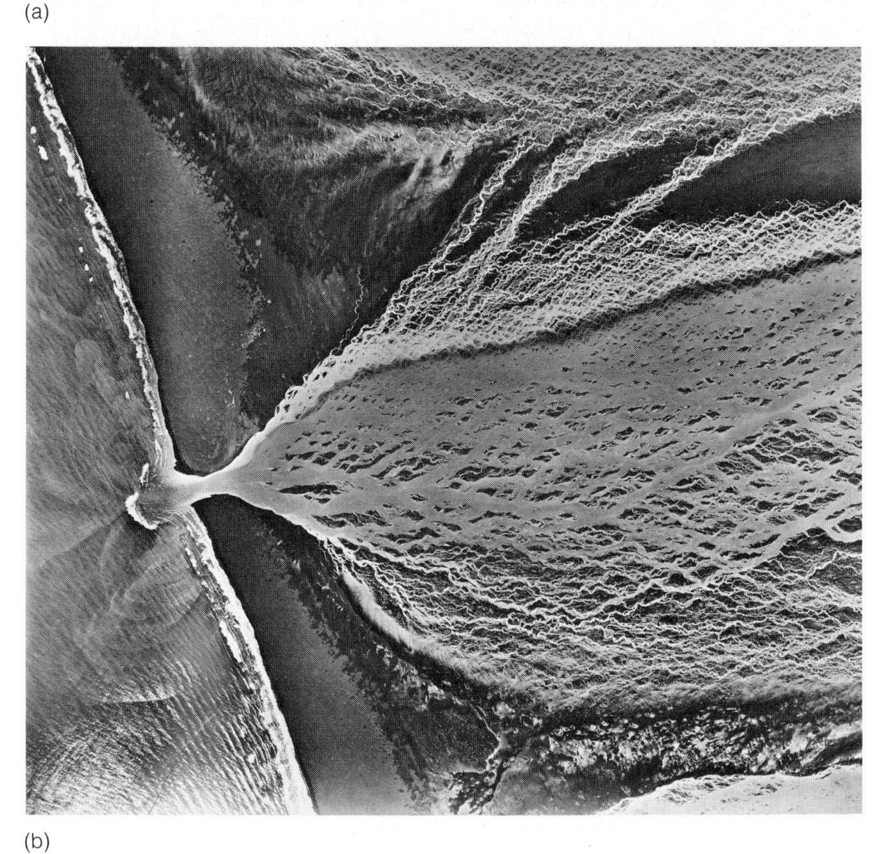

(b)

**12.28 Braided streams** at various scales. (a) Tiny braided channels formed on a sandy intertidal flat by water draining away during a falling tide. Five Islands, Nova Scotia, Canada. (J. E. Sanders) (b) Giant network of braided streams from a melting glacier, SE coast of Iceland, viewed from an airplane. (Copyright Icelandic Geodetic Survey)

## Areas Having Braided Channels

In streams with highly variable discharge and easily erodible banks, the channel may become interspersed with bars and islands, forming a pattern named a *braided stream* (Figure 12.28). The process of building up the channel in this way is *aggradation;* the opposite process, the deepening of a stream channel, is *degradation.* Braided channels typically are broad and shallow. A braided pattern commonly forms in response to an excess load of sand-size or coarser sediment during a flood. The stream spreads the excess material widely across the channel bed and, as part of its self-adjusting ability, tries to increase its gradient. When the flood waters have receded, the stream deposits its load on bars, islands, and in the bed. Channels of braided streams are often seen in arid regions where flash floods may overwhelm a channel with eroded rock waste. Braided streams are typical patterns of streams draining away from melting glaciers.

### Flood-Plain Rivers

The three major features of flood-plain rivers are (1) the channel and its bordering low ridges composed of sediment; (2) the great curves, named meanders, and associated features; and (3) the wide, low-lying flood plains.

**Channels and natural levees** A profile through a flood-plain river shows a well-defined *channel.* Along the banks of the channel are the highest points on the flood plain, *natural levees,* broad, low ridges of fine sand or silt that has been deposited by repeated overflow (Figure 12.29). Levees range in width from about 20 meters on small streams to several kilometers on the lower Mississippi. Heights range from about 10 centimeters along small streams to several tens of meters along the Mississippi.

As flood waters rise over the banks at the sides of the channel, the speed of the water drops abruptly. At lower speeds, the water can no longer carry all of its load, so it deposits sediment on the natural levees. The largest particles settle first; suspended smaller particles, such as fine sand and silt, may be transported long distances from the channel. Hence, the banks themselves receive the thickest deposits of sand and other coarse sediment, and thus are built upward above the adjacent flood plain.

**Meanders and point bars** It is probable that the only truly straight rivers in the world are those following faults. Almost certainly, no river is straight for a distance greater than 10 times its width. Most stream channels are parts of great sweeping curves known as *meanders* (Figure

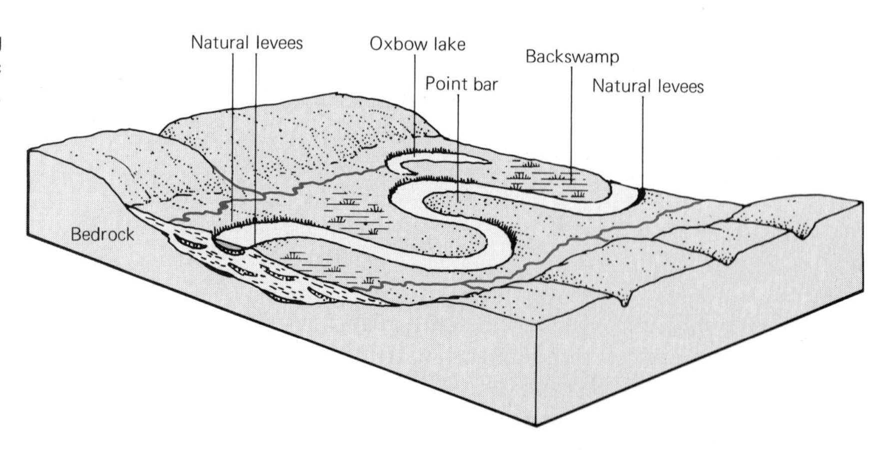


**12.29 Features of a large meandering river and adjacent flood plain,** schematic block diagram. Further explanation in text.

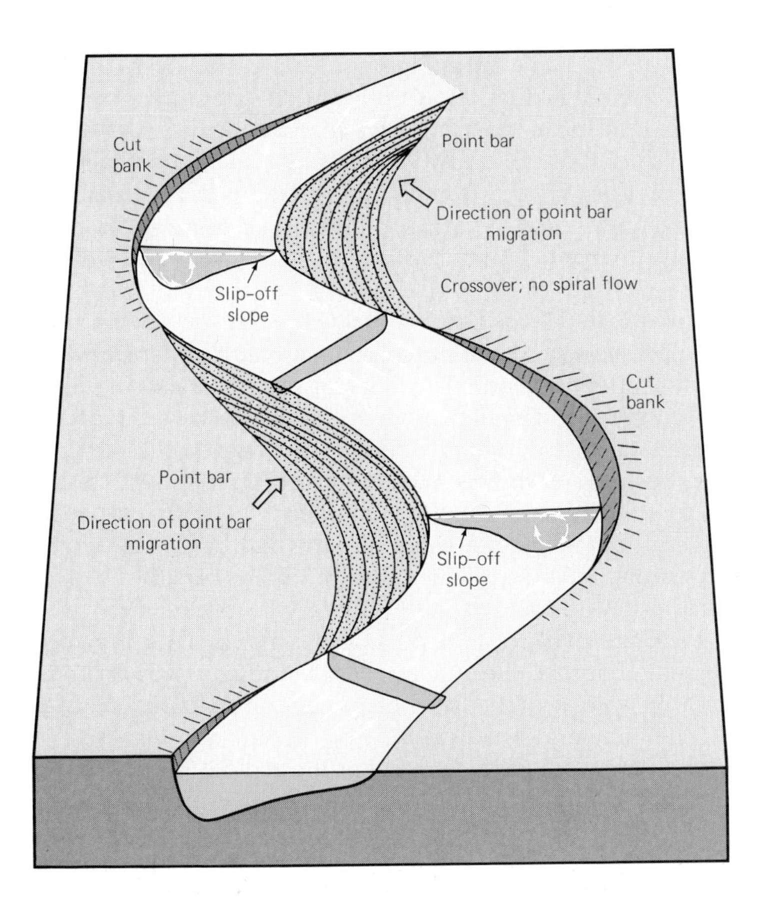


(a)

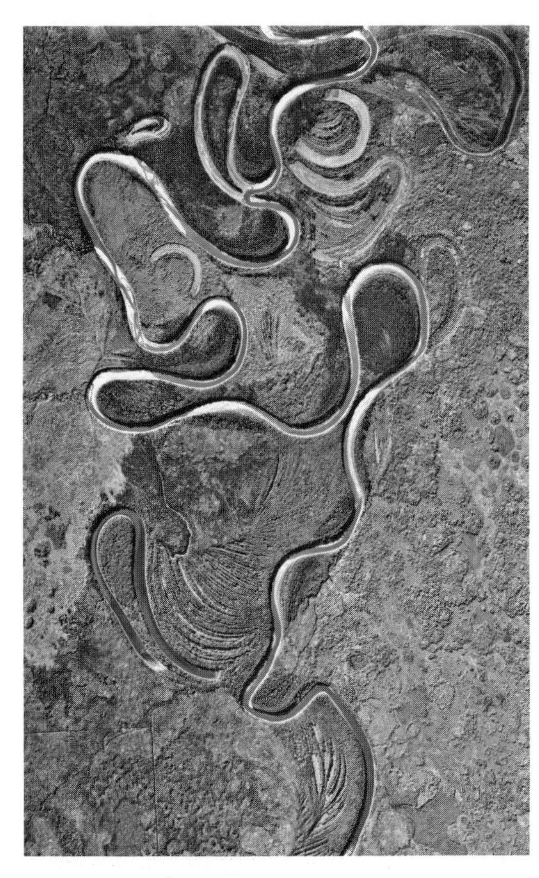

(b)

**12.30 Stream meanders.** (a) Mechanics of meandering. Heaping up of water at surface on the outsides (cut banks) of curves sets up return flows along the bottom away from the cut banks toward the opposite sides, the point bars or slipoff slopes. (After G. M. Friedman and J. E. Sanders, 1978, reference cited in caption of Figure 12.7, Figure 8–39, p. 221, based on J. F. Friedkin, 1945, A laboratory study of the meandering of alluvial rivers: Vicksburg, Mississippi, U. S. Army Corps of Engineers, Mississippi River Commission, Waterways Experiment Station, 40p., Plates 1–5) (b) Meanders and cutoffs, including many ox-bow lakes. Photograph from an airplane looking vertically downward. Hay River, NW Alberta, Canada. (Geological Survey of Canada)

12.30), the most efficient route for a river to follow. Water meanders because its surface does not remain level and because of large eddies within the flow.

Meanders form most commonly in loose alluvium which erodes easily. The alluvium is eroded from the outside portion of a bend. As the water turns, it is "banked" as though it were following a bobsled course. The surface water flows toward the outside of the bend. The banked-up surface water causes the water along the bottom to flow toward the opposite bank. Material eroded from the outsides of bends is deposited on the insides of bends, forming *point bars* (see Figure 12.30).

The whole system of meanders erodes its way through the valley-bottom sediments like a snake crawling through the grass. As long as all sediments in the path of the meanders are being eroded at a uniform rate, the overall shapes of the meanders do not change; all the meanders move

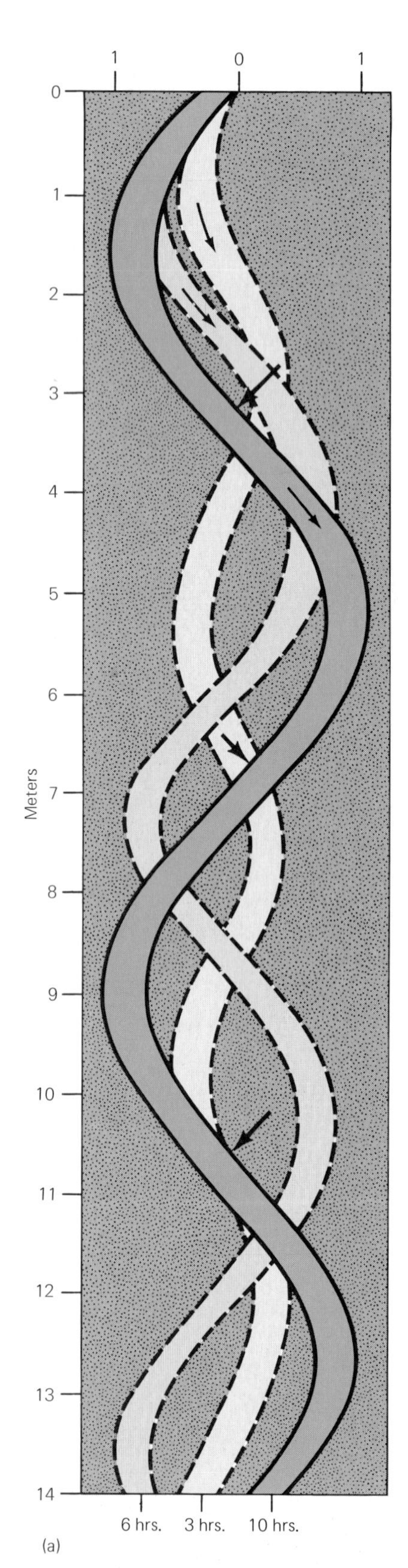


down the valley at the same rate (Figure 12.31a). However, typical features of valley-bottom sediments are thick bodies of tough clay that were deposited in flood-plain lakes. Nearly a century ago, Colonel James B. Eads (1820–1887) of the U.S. Army Corps of Engineers prowled over kilometers of Mississippi River bottom using a diving bell. He remarked that where the river bed consisted of clay, it resisted erosion so strongly that it might as well have been made of marble.

**Flood plains; ox-bows** The low areas outside the channel and natural levees form the *flood plain.* Many features of flood plains result from the migration of a meandering channel. If one set of meanders overtakes another meander whose downvalley progress has been slowed, the next meander upstream from the halted one cuts through and isolates a crescent-shaped section which becomes an *ox-bow lake* (Figure 12.31b). During floods, the radius of curvature of a meander is larger than normal. Larger discharge produces larger meanders. Thus, during floods, channels may be cut more or less straight across the point bars.

Flood plains are sites of swamps and temporary lakes where clay and organic matter can accumulate. If flood-plain basins dry out between floods, the fine sediments become mud cracked. During the 1952 flooding of the Missouri River, areas of the flood plain near Kansas City received as much as 15 centimeters of sediment. Flood-plain deposits rich in organic matter are "instant" soils. Because of these soils, many ancient civilizations, in Egypt, India, China, and other countries, concentrated their farms on flood plains.

**Strata deposited by migrating channels** Another effect of channel migration is to deposit distinctive sediments beneath the point bar. These sediments are as thick as the depth of the channel in a flood. At the bottom of the channel are waterlogged tree trunks, bones of animals, and large clay pebbles. With time and the addition of heat, these logs may turn into coal and cause uranium minerals to be deposited in their vicinity (see Chapter 8). During and after World War II, much research was done on stream-channel sediments as the basis for locating uranium deposits.

## Deltas

When a stream flows into a lake or the ocean, its speed is checked, and it deposits its sediment. The body of stream-laid sediment deposited at the mouth of a river is a *delta.* The name delta (the Greek letter Δ) was assigned to such sediment deposits because of the triangular shape of the deposits at the mouth of the Nile River (Figure 12.32). However, a delta may form not only delta-shaped patterns, but also radial, arc-shaped, or "bird-foot" patterns (see Figure 3.5g, h).

**12.31 Meanders in uniform and nonuniform valley-floor sediments.** (Left page) Downvalley migration of meanders, based on constant discharge of experimental stream flowing over sand. (After J. F. Friedkin, 1945, reference cited in caption of Figure 12.30, Plate 25) (Right page) Clay plug, deposited in ox-bow lake, prevents meander that encounters the resistant clay from further downvalley migration. Upstream meanders, not thus stopped, overtake the stopped meander and cut off one of its loops, forming a new ox-bow lake. (Schematic map of stream)

**12.32 Nile delta,** as photographed looking eastward from Gemini IV, orbit 12, on 4 June 1965, J. A. McDivit and E. H. White aboard, from about 180 kilometers height. Black areas are densely vegetated; light-colored areas lack vegetation. (NASA)

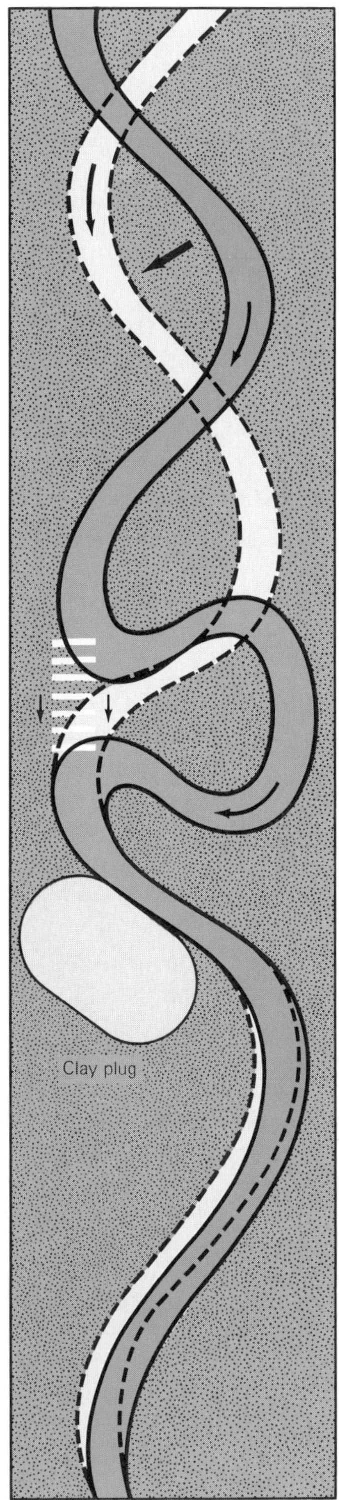


## STREAM RESPONSES TO MAJOR GEOLOGIC CHANGES

As we have seen, a given stream tends to be adjusted to a particular base level; to a particular range of variation of discharge; and, within these limits, to erode uplifted areas and to deposit in areas that are subsiding. Thus, major geologic changes affecting any of these factors can cause a stream's behavior to shift. Such changes include: changes in base level, changes in discharge, regional uplift, local uplift, and the building of natural dams.

### Changes in Base Level

Because base level acts as a check valve for the base flow of a stream, any change in base level sets off some kind of response by the stream. If base level drops, the lower end of the stream flows with renewed vigor and begins to erode through sediments previously deposited. The result may be a *stream terrace,* a flat, benchlike area formed when a stream has cut downward through its flat valley floor. Not uncommonly a river may be flanked by a series of terraces (Figure 12.33). If base level is raised, the lower end of a stream becomes flooded and the speed of flow in the lower reaches is diminished. As a result, the stream deposits sediment. Such sediment may fill the valley partially or completely. If a valley becomes completely filled with sediment, it disappears as part of the landscape. Its presence beneath the filling sediment may be found only by means of borings or other techniques for testing the subsurface materials (see Chapter 19).

**12.33 Sets of dissected terraces,** Rangitikei River, North Island, New Zealand. (S. N. Beatus, New Zealand Geological Survey)

## Changes in Discharge

Because discharge is such a fundamental factor in stream behavior, it follows that any departure from the usual range of variation of discharge will bring about changes. A great decrease in discharge simply diminishes stream activity. A great increase in discharge, followed by a later decrease, may cause erosion of bedrock and the deposition of large quantities of coarse sediment.

## Regional Uplift

Regional uplift that does not take place faster than an established stream can erode downward causes a stream to become deeply incised. Meandering valleys cut into bedrock (Figure 12.34) are considered to be the products of regional uplift at a rate that could be matched by active downcutting.

**12.34 Incised meanders,** Green River Dinosaur National Monument, Utah. (U. S. National Park Service)

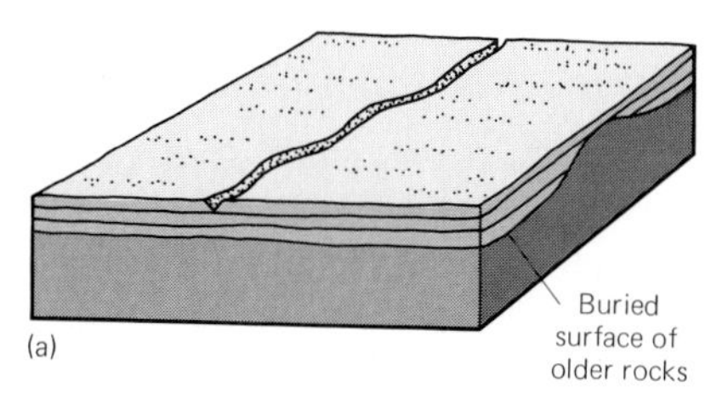


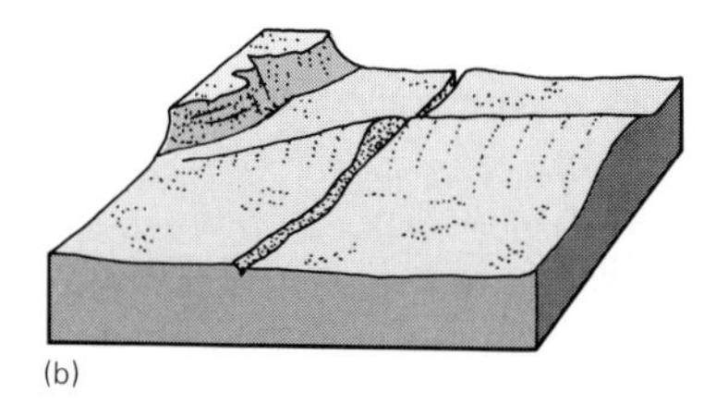


**12.35 Superposed stream,** stages in development shown by schematic block diagrams. (a) Stream starts to flow on nearly horizontal strata which bury an old landscape that includes an elongate, rounded ridge. (b) After much uplift and erosion, the stream has cut vertically downward and has eroded a steep gorge through the now-uncovered ridge. Nearly all traces of the covering strata have been removed.

In other cases of regional uplift, a stream that is flowing in some kind of flat-lying covering layer may erode through to an underlying block in which the geologic structure is complex. As a result of simple downcutting in the course that it had established on the covering layer, the stream's position with respect to the structure in the underlying block may become anomalous. For example, the stream may cut through a ridge that it would otherwise flow around. A stream that has eroded into a complex block from an overlying covering layer is said to have been *superposed* (Figure 12.35). In some cases, streams may be superposed from the top of a large glacier, and regional uplift may not have been involved.

### Local Uplift

If a narrow, local segment of the Earth's crust is uplifted across the path of a stream at a rate that does not exceed the rate of downcutting by the stream, then the stream may be able to maintain its course across the rising area. After a time, the stream may have eroded a gorge through a high ridge that the stream would otherwise not flow across. In this case, because the stream is older than the ridge, the stream is designated as an *antecedent stream* (Figure 12.36).

### Natural Dams

The existence of natural dams creates lakes in stream valleys. Eventually, gaps may be eroded through these dams and the lakes drained. While the lakes are in existence, however, fine-grained sediments may be deposited within them. Therefore, geologic evidence for the former existence of a natural dam consists of a gap eroded through the dam and of the kinds of sediments deposited in quiet water. The rapid draining of the dammed-up water may create great floods (see Figures 3.7, 3.8).

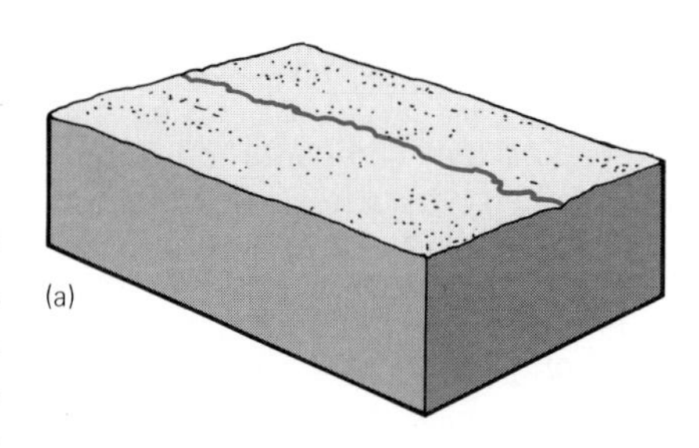


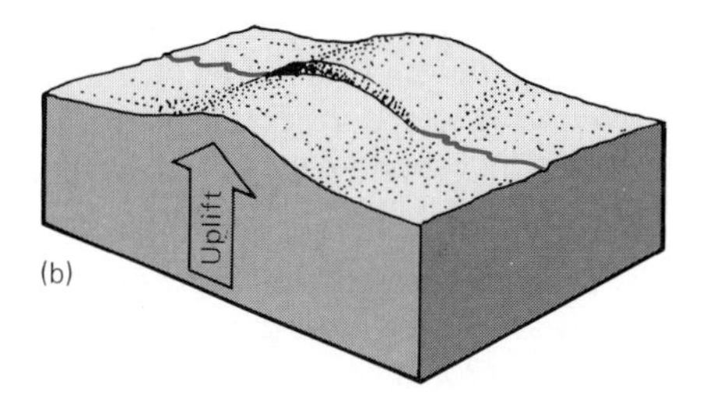


**12.36 Antecedent stream,** stages in development shown by schematic block diagrams. (a) Stream flowing across surface before anticline is elevated athwart its path. (b) Anticline is elevated, but the rate of uplift is matched by the rate of stream downcutting. As a result, the stream maintains its course and erodes a steep gorge across the anticline.

# CHAPTER REVIEW

1. A flowing body of water moving down a slope is subject to the shearing of the pulling force. Shearing within the fluid itself results in characteristic kinds of flow, and the moving flow exerts a shearing drag on the surface or bed over which it flows.

2. In response to shearing, masses of water move by either laminar flow or turbulent flow. *Laminar flow* takes place only in smooth, straight channels as parallel layers of water shear past one another. Far more common is *turbulent flow,* characterized by a variety of vortices and eddies that are continually forming and disappearing.

3. The effect of water shearing over sediments depends on several variables, including the speed of the flow, the depth of the flow, and the sizes of the sediment particles.

4. Particles supported within the current by the upward components of a turbulent flow constitute the *suspended load.* Particles that are rolled, dragged, or bounced along the base of the current constitute the *bed load.*

5. The volume of water flowing through a cross section of a stream channel during a given time is its *discharge.*

6. The *base flow* is the water that keeps a stream flowing between rains. The "extra" water that enters a stream rapidly forms a *flood bulge,* a long wave of water that travels downstream independently of the base flow.

7. Variation in stream flow results from variation in the supply of water within the drainage area. Such variation is caused by the seasonal distribution of rainfall, the seasonal melting of snow and ice, the erratic distribution of larger-than-usual rainstorms, and cyclic variation in rainfall during periods longer than one year.

8. The slope on which a stream flows is its *gradient.* The aggregate of all the local gradients constitutes the *longitudinal profile* of a stream. The interaction of all the factors that enter into the definition of the hydraulic geometry of a stream establishes a condition of dynamic balance which characterizes a *graded stream.*

9. *Playfair's law* states the principle that streams do not simply occupy valleys, but create them. The main work of rivers in creating valleys is the *headward erosion* of the stream bed. Other processes involved in *changing the valley-side slopes* to complete the profile of the valley are the downslope gravity-transport processes, raindrop impact, and the flow of water in sheets and small gullies.

10. A master stream and its tributaries make complex drainage patterns. These patterns depend upon the amount of water and the kinds of bedrock in the drainage basin. Where water is plentiful, the chief drainage patterns are *dendritic, trellis-rectangular, radial,* and *annular.*

11. An idealized *erosion cycle* is thought to begin with the uplift of a large land mass. The cycle ends when this uplifted area has been eroded away and all that is left is an almost-flat *peneplain* lying near base level. The cycle can be broken down into three stages defined on the basis of the results of stream erosion: *youth, maturity,* and *old age.*

12. Erosion can be measured by finding the rate at which the land surface is lowered, and by measuring the amount of sediment and dissolved loads in streams. The amount of sediment supplied per unit area is the *sediment yield.*

13. All stream deposits are known collectively as *alluvium.* Four environmental settings in which streams deposit sediment are: (1) *fans,* (2) areas having *braided channels,* (3) *flood-plain rivers,* and (4) *deltas.*

14. Some features of flood-plain rivers are *natural levees, meanders* and *point bars,* and *ox-bow lakes.*

15. When a stream flows into a lake or the ocean, its speed is checked, and it deposits its sediment. The body of stream-laid sediment deposited at the mouth of a river is a *delta.*

16. A stream tends to be adjusted to a particular base level; to a particular range of variation of discharge; and, within these limits, tends to erode uplifted areas and to deposit in areas that are subsiding. Thus, major geologic changes affecting any of these factors can cause a stream's behavior to shift.

## QUESTIONS

1. Name the two ways in which a body of water moving down a slope responds to the pulling force (slope-parallel component of the Earth's gravity).

2. Compare and contrast *laminar flow* and *turbulent flow.*

3. Define *suspended load* and *bed load.* In which load would one most likely find a silt-size particle? A gravel-size particle? Why?

4. What is the *discharge* of a stream? Discuss the sources of a stream's discharge and summarize the factors that cause discharge to vary.

5. Define *stream gradient.* Sketch a typical longitudinal profile of a stream.

6. What is a *graded stream?*

7. What are the two chief mechanisms by which rivers create valleys? Explain how each works.
8. List and sketch the four major kinds of drainage networks in areas where water is plentiful.
9. What is the *erosion cycle?* Name and characterize the stages into which this cycle has been subdivided.
10. Define *sediment yield* and explain the factors that affect sediment yield.
11. What is *alluvium?* Name and describe four environmental settings in which alluvium may be found.
12. Define *flood plain* and list three features one would expect to find on a flood plain.
13. What is a *braided stream?* What conditions are thought to be responsible for a stream's becoming braided?
14. Define *delta* and cite locations where one would expect to find deltas.

**RECOMMENDED READING**

Bloom, A. L., 1978, *Geomorphology; a systematic analysis of Late Cenozoic land forms:* Englewood Cliffs, New Jersey, Prentice-Hall, Inc., 152 p.

Friedman, G. M., and Sanders, J. E., 1978, *Principles of Sedimentology:* New York, John Wiley and Sons, 792 p.

Leopold, L. B., and Langbein, W. B., 1966, River Meanders: *Scientific American,* v. 214, no. 6, p. 60–73.

Leopold, L. B., Wolman, M. G., and Miller, J. P., 1964, *Fluvial Processes in Geomorphology:* San Francisco and New York, W. H. Freeman and Co., 522 p.

Shimer, J. A., 1959, *This Sculptured Earth: The Landscape of America:* New York, Columbia University Press, 255 p.

Wright, L. D., and Coleman, J. M., 1973, Variations in Morphology of Major River Deltas as Functions of Ocean Wave (*sic*) and River Discharge (*sic*) Regimes: *American Association of Petroleum Geologists, Bulletin,* v. 57, no. 2, p. 370–398.

# CHAPTER 13 GLACIERS AND GLACIATION



PLATE 13
Terminations of four valley glaciers, three of which have built prominent end-moraine ridges. Glacier at lower right has nearly blocked the valley, enabling a small proglacial lake to form. Eastern Baffin Island, Canada. (National Air Photo Library, Canada)

GLACIERS ARE ONE OF THE MOST POWERFUL AGENTS SHAPING THE surface of the Earth. They carve bedrock and are prime creators of sediments. In rasping motions comparable to the massive downslope movements described in Chapter 11, glaciers create "instant regolith." The sediments deposited by glaciers are quite unlike any other kind and form distinctive shapes and landforms.

The deposits of the most recent glaciers still form the regolith in many parts of the world. But in the ancient bedrock, still-older glacial deposits have been lithified; they indicate that glacial episodes took place hundreds of millions of years ago.

## PHYSICAL ASPECTS OF GLACIERS

A *glacier* is a flowing mass of ice that formed by the recrystallization of snow, is powered by gravity, and has flowed outward beyond the snowline (Figure 13.1). We first examine two physical aspects of glaciers — the origin of glacial ice and the mechanics and rates of glacial movement.

**13.1 Snowline crossing small modern glacier.** Where the surface appears smooth, snow is present. The deep cracks that appear in the lower part of the glacier extend beneath the snow, making glacier crossing extremely hazardous. North face of the Bernini Group, Alps, Switzerland. (Mario Fantin, Photo Researchers, Inc.)

## Snowfall and the Origin of Glacial Ice

Each year, at least a little snow falls on each continent. By the end of winter much or all of this snow usually has disappeared. For example, studies in the Antarctic, begun during the International Geophysical Year (1958), have shown that the average annual fresh snowfall at the South Pole is only about 15 centimeters — about equivalent to the amount of precipitation that falls on the barren Australian desert. (The only major land mass lacking glaciers is the continent of Australia.) Each winter parts of New England may accumulate nearly three meters of snow, yet in summer the land is green with vegetation. Snow lingers through the year only under certain climatic conditions and above certain altitudes. Such areas are said to be above the *snowline.* Glaciers can form above the snowline if snowfields form and in them firn, and finally, glacier ice are created.

**Snowfields** Large areas of snow that last from one winter to the next are *snowfields.* The snow in a snowfield may be thinly coated by ice, but beneath this surface layer, it remains powdery and porous. The existence of snowfields is a first requirement for a glacier; more fresh snow must fall each year than is lost to melting.

**Firn** Snow crystals that melt and refreeze into granular particles become *firn* (density at least 0.55 grams per cubic centimeter). Firn is not compact enough to prevent the passage of air and water.

**Glacial ice** As firn layers build up year after year, the increasing weight of the added layers affects the melting-and-refreezing cycle. The weight of the overburden forces out more and more of the trapped air. When the ice mass has attained a density of 0.84 grams per cubic centimeter, has become impermeable to air, and has begun to move under the pressure of its own great weight, it has become *glacial ice.*

## Mechanics and Rates of Glacial Movements

Glaciers are not static ice masses, but are in constant flux. They respond continuously to the snow that accumulates and melts, to rises and falls of temperature, and to the changing seasons.

A glacier, by definition, always flows internally. If its motion ceases, it is no longer a glacier, and is nothing more than a large block of ice. One component of the internal motion is the flow away from a point of accumulation of new snow. However, the flow away from the zone of accumulation does *not* necessarily mean that the outermost margin of the ice sheet will advance. If melting at the margin takes place faster than the flow from the zone of accumulation, outward, then the outermost margin will *retreat.* Response to melting is a second component of motion. Retreat does not mean that the glacier has actually shifted into reverse gear and moved backward, but rather that the outermost edge of the glacier is receding faster than the ice is flowing outward. Thus, the glacier's margin is a dynamic response to the rate of supply of new snow and the rate of melting.

Glaciers flow in two modes: (1) normal or nonsurging and (2) surging.

Nonsurging glacial movement is thought to consist of two different mechanisms: (1) slippage or sliding along discrete shear planes (see Figure 11.5) and (2) plastic flow from deformation of ice crystals. Both of these mechanisms are related to the fact that the ice in the lower parts of the glacier is under enormous pressure. The factor responsible for making glaciers move is the component of gravity acting along a slope (see Figure 11.3).

**Sliding** Although the ice tends to slip along many internal shear planes, the weight of overlying ice tends to concentrate most of this motion along a single surface near the base of the glacier. Movement on the bottommost layer is facilitated by the presence of water. This water can come from melting related to air temperature or from the effects of great pressure.

**Plastic flow** Solid ice can flow because it is crystalline (see Figure 4.8b), and crystals can be deformed. The long axes of the ice crystals in the lower parts of the glacier become aligned so that they are roughly parallel to the surface of the glacier. When pressure from above and gravity from below interact, the sheets within the glacier are set in motion.

**The zones of fracture and flow** Glacial ice under low pressure is a brittle solid that fractures readily. Glacial ice under high pressure can flow plastically. The effect of the increase in pressure downward in a glacier is to divide the ice into two zones: (1) an upper zone of fracture and (2) a lower zone of flow.

The most distinctive features of the zone of fracture are *crevasses,* deep (as much as 20 to 30 meters) fissures in the upper zone of the glacier (see Plate 13), just as are the rocks in the lithosphere. Glacier ice compares to deformed metamorphic rocks. Crevasses are created by the differing rates and mechanics of flow of the surface and the interior parts of a glacier.

Crevasses are much more common in fast-moving valley glaciers of mountain chains than in slow-moving continental ice sheets. Many crevasses are created because of the drag between moving ice and stationary valley walls.

In the zone of flow, crevasses become closed, and layers of sediment become intricately contorted (Figure 13.2). Both on a small scale and on a large scale, features found in the zone of flow in a glacier resemble the features found in metamorphic rocks in the interiors of mountain chains (see Chapters 16 and 18).

**13.2 Intricately contorted layers of sediment** (black) at terminus of glacier that has nearly filled a small proglacial lake with icebergs. Atlantic Ocean is in lower part of view. (Copyright by Icelandic Geodetic Survey)

**Rates of movement of nonsurging and surging glaciers** The first attempts to measure glacial movement were made by glaciologists who drove a line of stakes across the ice and then observed how the stakes moved with time (Figure 13.3). On the basis of his studies of glacial movement in the Alps in 1840, the British geologist J. D. Forbes (1809–1868) was able to make a remarkable forecast. In 1820, three French mountaineers had fallen to their deaths from Mt. Blanc, and their corpses had been buried in the snowfield atop a valley glacier. Noting the point where the three bodies had fallen into the snowfield and calculating the mean rate of flow of the glacier, Professor Forbes estimated that the corpses would reach the plain below and would be released from the ice in 1860. The frozen bodies were discovered in 1863.

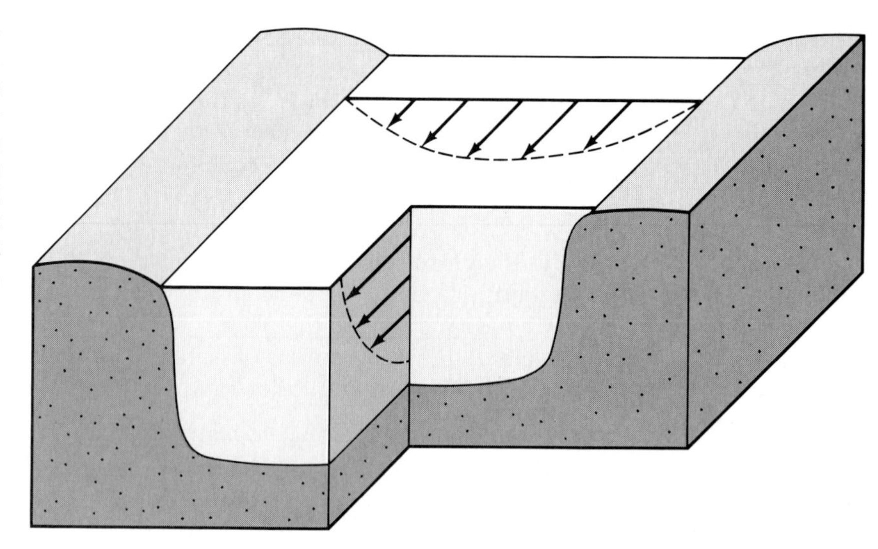

**13.3 Flow of a valley glacier** proved by lines of stakes and bent-over pipes, schematic block diagram. Deflections of stakes driven in a line across the top of the glacier prove that the glacier flows fastest in the center and slowest at the sides, as in a river. Bending over of pipes inserted into holes drilled into glaciers at different depths prove that the upper levels of a glacier flow faster than the lower levels.

The typical rate of glacial flow is 15 to 60 centimeters per day. Where large outlet glaciers in Greenland squeeze through narrow mountain passes, their speeds have been clocked at better than 30 meters per day.

Unusual conditions have produced some rapid if short-lived gallops called *surges*. In the mid-1930s, the Black Rapids Glacier in Alaska suddenly began a six-month surge. The glacier advanced almost 7 kilometers: its average speed was 35 meters a day, and its maximum speed was 65 meters in 24 hours. It finally stopped, just short of a hunting lodge and just before it obliterated the only road link between Fairbanks and the "lower 48" states. The conditions that produce such surges are still not completely known.

**13.4 Small valley glaciers,** some of which (lower right) are eroding their cirques into arêtes. Axel Heiberg Island, Northwest Territories, Canada. (Canada Department of Energy, Mines, and Resources)

## MODERN GLACIERS: KINDS AND LOCATIONS

The three chief kinds of modern glaciers, in order of increasing size, are: (1) valley (Alpine) glaciers, (2) piedmont glaciers, and (3) ice sheets (continental glaciers).

### Valley Glaciers

As their name implies, *valley* (or *Alpine*) *glaciers* are ice masses located on all the world's mountains whose peaks lie above the snowline. A valley glacier, which flows from high altitudes down valleys (Figure 13.4), can be identified from afar by its distinctive tonguelike appearance. A valley glacier can be said to resemble a spoonful of topping syrup flowing partway down the ice cream in a sundae. The "syrup" flows as long as the glacier receives an adequate supply of snow from the mountains above.

Large parts of some valley glaciers lie well below the snowline, and thus appear not to be supplied by direct snowfall. Such a glacier can be fed by avalanches of snow from valley-side slopes higher up the mountain (Figure 13.5). This movement of snow from the valley-side slopes is analogous to the movement of sediment to streams down valley-side slopes by gravity-transport processes.

### Piedmont Glaciers

A *piedmont glacier* is a glacier formed by the flowing together of two or more valley glaciers (Figure 13.6; see also Figure 13.2, upper right). A piedmont glacier is formed by ice that is "third-hand." Snow falls first on the peak; after it has become ice, it flows down into valleys; and, finally, it is extruded onto flatlands below. If a valley glacier can be compared to one spoonful of syrup, then we can visualize a piedmont glacier as consisting of the running together of several spoonfuls of syrup, that is, the syrup (ice) has flowed all the way down the ice cream (valley) to form a pool (piedmont glacier) below.

Some piedmont glaciers are enormous. In Alaska, the Malaspina Glacier is the product of several separate piedmonts that have joined together to cover an area larger than Rhode Island. In theory, a piedmont glacier such as the Malaspina can build upward until its top eventually lies higher than the snowline. Should this happen — as it may have in Antarctica and Greenland — then that part of the glacier above the snowline would receive its snow "first-hand," and thus would grow considerably more than would a piedmont glacier being nourished entirely "third-hand."

**13.5 Snow avalanches** are major sources of raw materials to form new ice in valley glaciers. Compare Figure 11.15. View from airplane of lobate margin of 1952 avalanche. (Copyright, Photo Library, FAO, Rome, Italy)

### Ice Sheets (Continental glaciers)

The greatest glaciers in the world today all fall into the category of *ice sheets.* To complete the syrup analogy, the formation of ice sheets is analogous to spilling the whole can of syrup onto a flat table top. The syrup would tend to flow outward in all directions, forming a relatively smooth surface. The ice naturally seeks downhill slopes, but when the margin of an ice sheet encounters an incline whose top is lower than the top of the lens-like glacier, the ice will actually travel upslope. The center of a sheet always remains higher than the sides.

**13.6 A piedmont glacier.** View from airplane of Malaspina Glacier, St. Elias Mountains, Alaska. (Austin Post, University of Washington and U. S. Geological Survey)

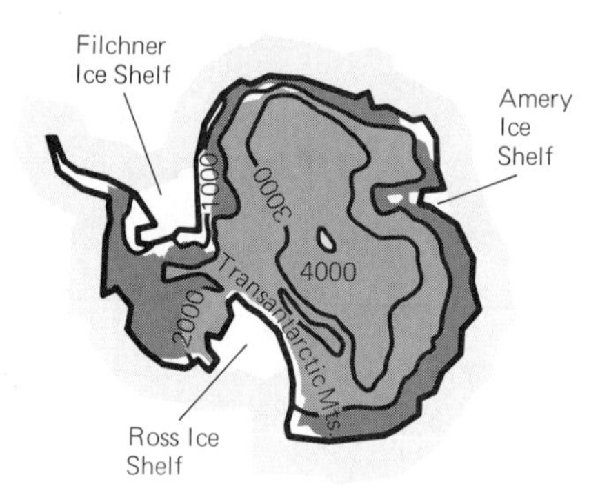


**13.7 Map of Antarctica** showing altitude on top of ice sheet. Because ice extends to sea level and below, the altitude of the top is an approximate indication of the minimum thickness of glacier ice. Contours in meters above sea level. (American Geographical Society)

An ice sheet is directly nourished by snowfall. As snow accumulates on the surface, the ice below begins to move outward and downward. In addition, the tremendous weight of the ice sheet causes the edges to spread outward.

Existing glaciers occupy 16 million square kilometers, roughly 10 percent of the Earth's surface area. Of this, some 12.5 million square kilometers are located in Antarctica. The giant Antarctic Ice Sheet may be as much as 10 million years old; parts of it are at least 4000 meters thick (Figure 13.7). The sheer weight of the ice has depressed the surface of the continental lithosphere more than a kilometer below sea level. Today, however, the land is slowly rising. This probably indicates that in the past, the ice sheet was even thicker than it is now. Scientists from many nations maintain scientific outposts on the Antarctic Ice Sheet, but the interior of that mighty glacier remains practically unknown.

The second-largest modern glacier is the Greenland Ice Sheet, which covers almost 2 million square kilometers (Figure 13.8). Scientists have been probing Greenland's ice sheet for three decades. Considerable climatic information has been gathered, but the formation and development of the glacier itself remain virtually as mysterious as the aspects of the Antarctic glacier. Other ice sheets are found in Iceland, Spitsbergen, Novaya Zemlya in the Soviet Union, and the ice caps that cover parts of Canada and Alaska.

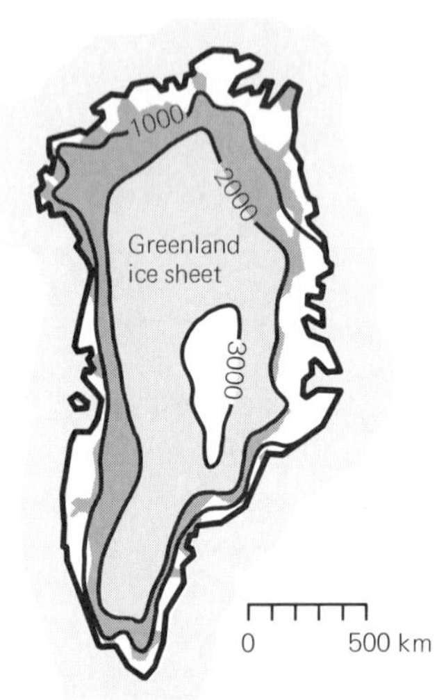


**13.8 Map of Greenland,** showing altitude of top of the ice sheet. Contours in meters above sea level. (U. S. Army, Cold Regions Laboratory)

## MODERN GLACIERS AND PLEISTOCENE GLACIATION

Nearly everyone who lives in the northern United States or in northwestern Europe is familiar with the concept that great glaciers once

**13.9 The last days of the glacial age,** as interpreted in this painting of the region near Lucerne, Switzerland, as the artist visualized it appeared about 30,000 years ago. Conceptualization by W. Amrein, Prof. Albert Heim, and Ernst Hodel. (Swiss National Tourist Office)

covered the areas where they live (Figure 13.9). So much has been said and written about the "ice ages" and so much speculating has been done about whether or not a new ice age may be at hand that we may tend to take such subjects for granted. But, ideas about the way glaciers work and the geologic proofs that former glaciers were much more extensive than today's glaciers are relatively recent developments.

In this section, we examine the origin of the concept of ice ages and the distinctive deposits and erosive effects of modern glaciers that provided the basis for the ice-ages concepts. Because many effects of Pleistocene glaciers are still so fresh, we discuss examples of them along with those of modern glaciers.

### Origin of the Concept of Ice Ages

Early in the nineteenth century, many Europeans were struggling to get accustomed to the idea that some of their lands had been recently covered by red-hot lava from geologically young volcanoes that were no longer active. All of a sudden, they began to hear talk that much of northern Europe had been recently subjected to widespread *glaciation,* the advance and retreat of a glacier over an area.

Although Jean Louis Rodolphe Agassiz (1807–1873) was not the first person to propose the idea that glaciers once covered much of Europe, he was most responsible for winning acceptance of the ideas which became known as the glacial theory. Originally, Agassiz was skeptical about the entire notion. In fact, in 1836, he made a field trip through the upper Rhône Valley just to disprove the glacial theory. As a result of this trip, just the opposite occurred. Agassiz became an enthusiastic convert and a hard-working champion of the glacial theory. He traveled across Europe and America arguing for his views. Today, Agassiz's views form the basis of an entire branch of geology known as glacial geology.

### Processes of Glacial Erosion

No agent of erosion on Earth is more potent than an advancing glacier. A glacier that may be 2000 to 4000 meters thick can pull loose large chunks of bedrock, grind solid rocks into fine fragments, and sculpt the land within a few centuries. To achieve comparable results, wind and water would be required to work for millions of years.

**Quarrying** In places where the ice is moving slowly, a glacier can attach itself to surfaces of bedrock. Water formed from pressure melting or from warm air temperatures fills cracks in bedrock. When this water freezes, it forms a solid jacket of ice around a block of bedrock. As the ice eventually flows away, it pulls out the blocks that have been jacketed by ice (Figure 13.10). Such removal of large blocks of bedrock is known as *quarrying* or *plucking*.

**Abrasion** As a glacier moves, it acquires a load of rock particles, chiefly in its basal parts and, in a valley glacier, also along the sides. It grinds these particles even finer and at the same time uses the load of rock particles to abrade the underlying bedrock.

The fine chips and powder created by abrasion of embedded rocks against one another and against solid bedrock constitute *rock flour.*

**13.10 Quarrying of blocks and rasping by a glacier** form a rôche moutonné with gentle slope toward the direction from which the glacier flowed and a steep, irregular slope on the side toward which the ice flowed and thus pulled away from the rock. Schematic profile and section.

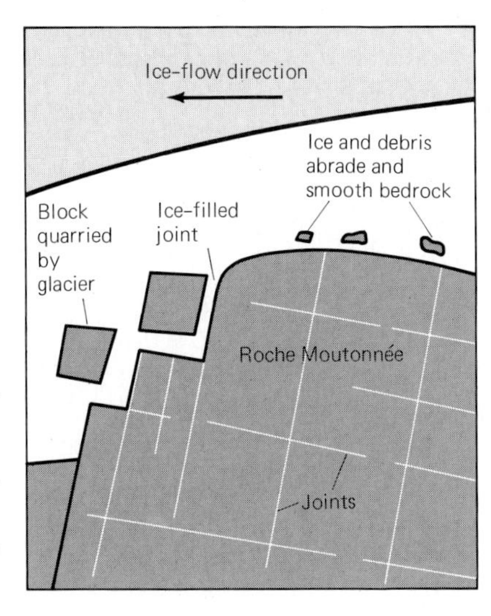

**13.11 Cirques** seen from an airplane, Wind River Mountains, Wyoming. (Austin Post, University of Washington and U. S. Geological Survey)

Because of their tiny sizes, the particles of rock flour are not easily observed, but they become conspicuous where they leave the glacier in a stream of meltwater. The rock flour makes the water appear milky, and when the milky stream enters a lake, the effect is to color the water. The pale blue of the water of Lake Louise in Canada and the cerulean hues of the water of Kashmir's Dal Lake are examples. Water charged with rock flour from a glacier is evidently good for people's health. Remote peoples who drink such water live long and healthy lives.

### Evidence of Glacial Erosion: Landforms Eroded in Bedrock

Glaciers erode distinctive landforms in bedrock. These include: (1) cirques; (2) arêtes, cols, and horns; (3) roches moutonnées; and (4) U-shaped and hanging valleys and fjords. In addition, glaciers cut striae and grooves into bedrock surfaces and leave many bedrock surfaces smoothed and polished.

**Cirques** The most common mountain landscape feature created by glaciation is a *cirque* (SEERK), a steep-walled, bowl-shaped niche in which a valley glacier originated (Figure 13.11). A cirque, which begins at the head of a valley formed by other processes, is the area above the snowline that nourishes a valley glacier. When the outermost margin of a valley glacier is being melted back faster than the ice is flowing outward, the cirque is the place where the last traces of ice remain.

Because the ice stays in a cirque longer than anywhere else, erosion is deepest there. The shape of the glacier resembles an old-fashioned keyhole, with the cirque forming the round part on top. Cirques are rimmed on three sides by a mountain; the fourth side is the downhill opening through which the glacier descends. Typically, the remaining mountain rim towers high above the bottom of a cirque because centuries of frost wedging enable the glacier to quarry deeply into the rock. The walls of a cirque tend to be more or less smooth, because continued abrasion wears away most of the high spots.

**Arêtes, cols, and horns** Rarely does only one valley glacier exist in a mountainous area. Commonly, separate glaciers, each fed from snowfall at a peak, will move down different sides of the crest of the mountain chain. Eventually, each glacier forms its own cirque. Adjacent cirques may grow until they are separated by only a relatively thin, vertical, rocky remnant. The remnant is called an arête (ah-RET; from the French word for "stop").

Eventually, gaps are eroded in the wall-like arête. A gap in an arête between two cirques is a *col.*

A mountain pinnacle may be left after the wearing away of several arêtes. A pinnacle which overlooks three or more cirques, is called a *horn.* Two of the world's best-known peaks are horns: Mt. Everest, which is bordered by four mammoth valley glaciers, and the Matterhorn in the Alps (Figure 13.12).

**13.12 Matterhorn,** a horn formed by glacial erosion from three sides. Near Zermatt, Valais, Switzerland. (Agi Castelli)

**13.13 U-shaped valley** formed by flow of a valley glacier, now melted. Berner Oberland, Interlaken, Switzerland. (Swiss National Tourist Office)

**Roches moutonnées** The combined action of abrasion, smoothing, and rounding on one side (the side up which the ice flowed) and quarrying on the opposite side (downflow side) creates asymmetrical rock hillocks known as *roches moutonnées* (ROASH mootahn-AY; see Figure 13.10).

**U-shaped valleys, hanging valleys, and fjords** A valley glacier characteristically alters the V-shaped profile created by running water (see Figure 12.20). The result is a U-shaped profile, with steeper sides and a rounder bottom (Figure 13.13).

**13.14 Fjord** at the head of Milford Sound, New Zealand. (New Zealand Consulate General, New York)

A main-valley glacier may be fed by tributary glaciers. Each tributary glacier has its own head or source and finds its own route until it joins the main tongue. Because these feeders are by definition smaller, they pack less erosive force than the main-valley glacier. They do not erode as deeply, so that they often empty not at the floor of the main-valley glacier, but at a higher level. After the main-valley glacier has melted away, the bottom of the main-valley may be several hundred meters lower than the bottoms of its tributaries. These higher tributaries, aptly called *hanging valleys* (see Figure 13.13), are common in the European Alps.

Near a coastline, glaciers can create U-shaped valleys that may be flooded by the sea after the glaciers have melted. Such glaciated valleys occupied by an arm of the sea are *fjords* (Figure 13.14). Some fjords extend as far as 190 kilometers inland.

**13.15 Bedrock surface smoothed, grooved, and striated** by a glacier whose nonsorted deposits are exposed in section at right. East Berlin, Connecticut; bedrock is Lower Jurassic siltstone. (J. E. Sanders)

**Striae and grooves** The rasping, abrasive effect of glaciers on massive solid bedrock is to smooth and polish the surface and to sculpt it into large-scale rounded forms and grooves (Figures 13.15 and 13.16). No other geologic agent creates such features. A thick ice sheet tends to flow regionally in a single direction that crosses hills and valleys as if they did not exist. The direction of flow of a thinner sheet or of a valley glacier is influenced by local morphology. The result of one-directional flow is the rasping out of tiny linear grooves, named *striae,* or of larger troughs or deep grooves. The striae and grooves are parallel to the direction of the glacier's flow.

### Glacial Sediments

Glacial debris, collectively named *drift,* is strewn in characteristic deposits all along the glacier's path. Drift is short for "northern drift," a nineteenth-century term applied in Europe to the material transported southward from Scandinavia. When the term was first applied, geologists thought that the sediment had been rafted by icebergs. The sediments deposited directly by a glacier include erratics and till. Other glacier-controlled sediments are not deposited directly by glaciers, but accumulate in bodies of water closely related to a glacier.

**13.16 Glacial grooves at base of glacier,** Pumori South Glacier, Himalayas. (Fritz Muller, Copyright Swiss Foundation for Alpine Research, Zurich)

**13.17 Till,** consisting of light-colored particles set randomly in dark-colored matrix that is nonstratified and poorly sorted. Montauk Point, Long Island, New York. (Robert LaFleur, Rensselaer Polytechnic Institute)

**Sediments deposited directly from a glacier: till and erratics** The sediment deposited directly by a glacier is *till.* A glacier cannot separate its debris by size. Thus, till is a nonstratified, nonsorted heterogeneous collection of debris (Figure 13.17). The range of particle sizes in till can be from microscopic rock flour up to immense boulders. Glaciers, which possess great powers for transporting large rock fragments long distances, can deposit debris on bedrock that is not like the transported fragments. A rock fragment which differs from the underlying bedrock is an *erratic.* If the erratic rests on a polished, striated rock pavement, then almost certainly that erratic was deposited by a glacier. The finding of distinctive rocks from the Alps in parts of Switzerland where no such bedrock exists was important evidence in support of the idea of "ice ages."

Within a given glacier the distance an erratic may travel depends on rock composition. Limestone and other easily weathered rocks can survive only short trips. Granite can move long distances. Plymouth Rock is a granite boulder that may have been swept from New Hampshire to its resting place near Boston. In the nineteenth century the discovery of 11 diamonds scattered in Wisconsin, Indiana, and Ohio pointed to the existence of a rich lode somewhere in Canada. The precise location, however, is not known. Diamonds are the most resistant of minerals, so that these may well have traveled thousands of kilometers. Prospectors are still searching.

**Sediments deposited by meltwater: outwash and varved lake deposits** When overall melting exceeds outflow the glacier retreats or wastes away. The margin of a glacier may retreat under one of two conditions: (1) melting back while the ice continues to flow actively, or (2) as large masses of stagnant ice. In either case, water is plentiful. The water may flow away from the melting ice in great streams that deposit stratified sediment derived from the glacier. Or, the meltwater may form lakes in which the melted-out sediments accumulate. Thus arise, respectively, outwash and varved lake deposits.

**13.18 Proglacial lakes** being fed by meltwater from valley glaciers three of which have built prominent end moraines. Eastern Baffin Island, Canada, viewed obliquely from an airplane. (National Air Photo Library, Canada)

The body of the glacier possesses no mechanism for sorting its sediment load by size, but meltwater from the ice mass can organize glacially transported sediment to a certain degree. The sediments, chiefly sand and gravel, deposited by streams flowing from a glacier are called *outwash*. Another name for all glacially derived sediments deposited by streams is *stratified drift*. A meltwater stream flowing away from a glacier deposits larger and heavier particles first, nearer the glacier, and washes the smaller and lighter material farther away.

Lakes that receive meltwater from a glacier are named *proglacial lakes* (Figure 13.18; see also Figure 13.2). Such lakes are frozen over during a large part of the year and are free of ice only briefly, perhaps for a month or more in the summer. The supply of sediment to a proglacial lake is strictly seasonal. During the summer, meltwater streams actively transport sediment, but during the winter, the supply is cut off as the streams freeze solid. In summer, rock flour becomes widely dispersed in the lake water, and fine sand and silt may flow along the bottom. In winter, the flow stops and the fine particles of last summer's suspended rock flour slowly settle to the bottom according to size, largest first. Because of the long settling period, the winter-time deposit is always graded. Summertime sediment, though variable, is coarser than the wintertime deposit, and overlies the previous winter's layer along a sharp contact. Summer sediment may be laminated, possibly even micro-rippled, and may contain several graded layers. Summer sediment grades upward into the following winter's deposit.

In 1879, Gerard De Geer (1858–1943), a Swedish geologist, decided that this cycle of sedimentation in proglacial lakes would leave distinctive deposits. His pioneering research confirmed the existence of the seasonal stratification in proglacial lake sediments. He proposed the name *varve* for the sediment deposited in a year (Figure 13.19), and early in the twentieth century, he counted the varves in Scandinavia like tree rings to tell time. He was thus able to date the melting of the last Würm glacier in Northern Europe fairly accurately at about 10,000 B.C.

## Evidence of Glacial Deposition: Landforms Deposited by Glaciers or Related to Glaciers

Glaciers shape their sediments into distinctive features. Parts of the landscape in many areas were formed by Pleistocene glaciers. Flowing ice is responsible for moraine ridges and drumlins. Meltwater, with or without blocks of stagnant ice, deposits eskers and kames. Collapse of sediment over buried blocks of ice creates kettles.

**Moraine ridges** A large body of drift (consisting of till, stratified drift, or both) that has been shaped into a rounded ridge is a *moraine*. Several

**13.19 Varved sediments** deposited in proglacial lake of Late Quaternary age, Columbia River Valley, Ferry County, Washington. Light-colored sediment is coarse silt and very fine sand; dark-colored sediment is clay. (F. O. Jones, U. S. Geological Survey)

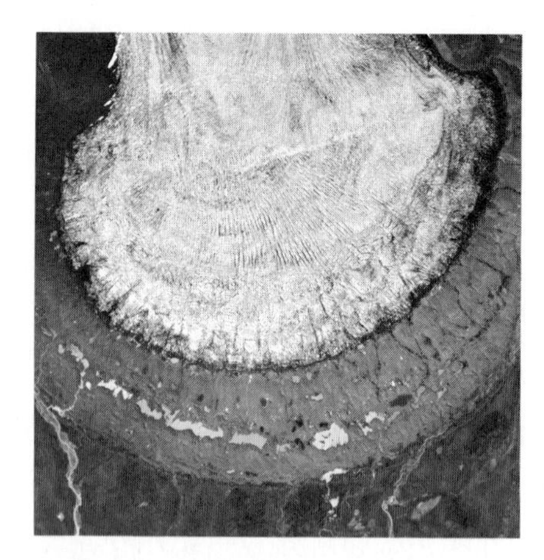

**13.20 Terminal moraine** at end of valley glacier, Iceland. Pattern of radial crevasses resulted from spreading into semicircular outline after ice that had flowed out the end of the valley (see Figure 13.4 and 13.18). The outermost ridge is the terminal moraine; between it and the edge of the glacier are many small lakes, probably kettles formed by collapse of sediment deposited above isolated blocks of ice. (Copyright Icelandic Geodetic Survey)

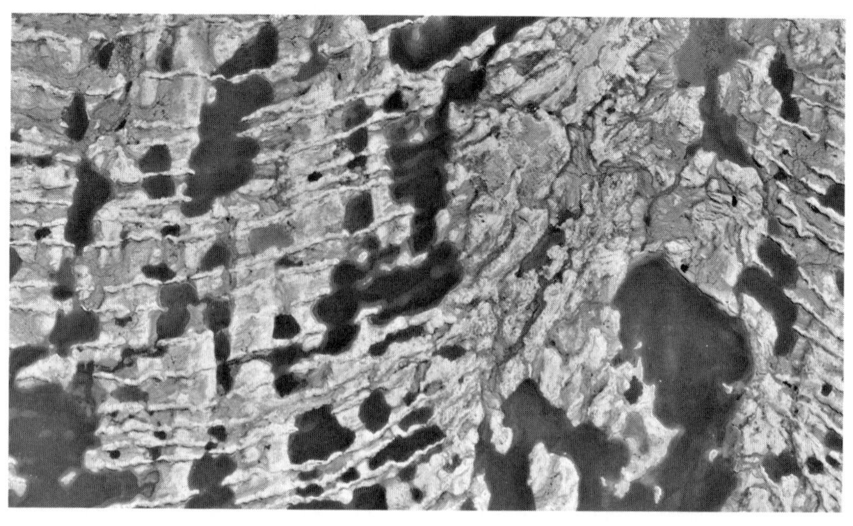

**13.21 Recessional moraines** deposited by two retreating glacier lobes. Dark areas are lakes. Western Quebec, Canada, about 80 kilometers E of Hudson Bay. Width of view is 15 kilometers. (Geological Survey of Canada)

varieties are recognized. At the outer margin of a glacier that has reached its maximum extent, the ice pushes up debris into a ridge whose trend follows the edge of the ice. This ridge is known as a *terminal moraine* (Figure 13.20). A single glacier deposits only one terminal moraine — at the point of its greatest extent. During its retreat, it may deposit many other morainic ridges along its margins at places where the rate of retreat is temporarily slowed. These ridges are known as *recessional moraines* or *end moraines* (Figure 13.21).

A *lateral moraine* is formed along the nonleading edges of a valley glacier. A *medial moraine,* the product of converging lateral moraines that flow together where two valley glaciers meet, occurs atop and within the glacier itself (Figure 13.22).

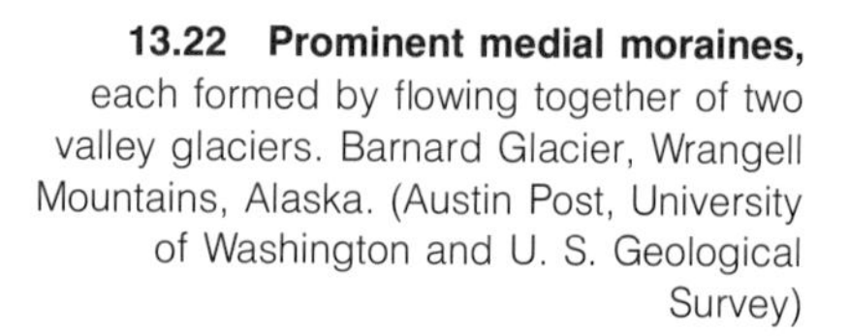

**13.22 Prominent medial moraines,** each formed by flowing together of two valley glaciers. Barnard Glacier, Wrangell Mountains, Alaska. (Austin Post, University of Washington and U. S. Geological Survey)

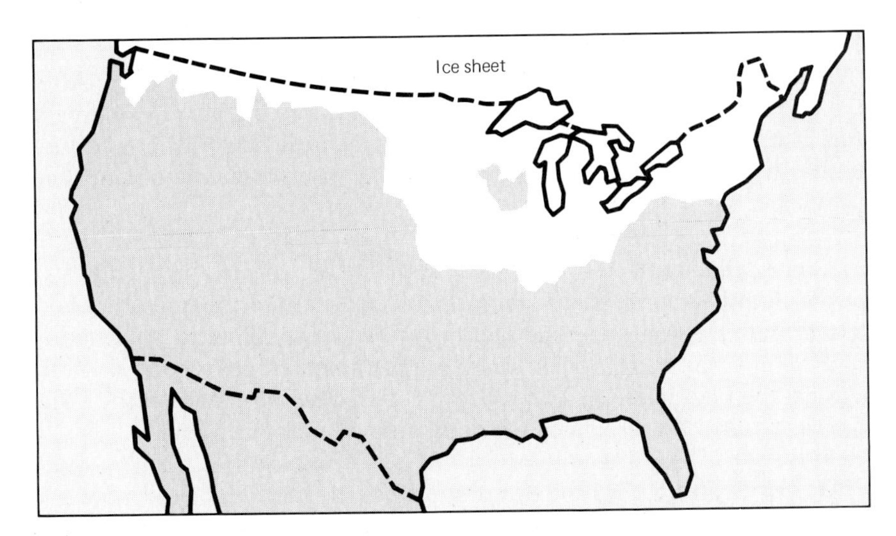


13.23 **Limits of Quaternary glacial ice** in continental United States shown by combining outermost points reached by various individual glacial advances. Local ice caps in mountain regions farther south have been omitted. (Based on Glacial Map of North America, Geological Society of America, 1945, R. F. Flint, Chairman)

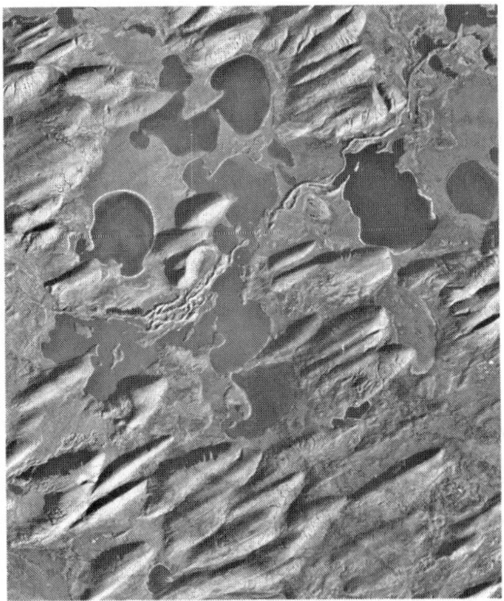

13.24 **Swarm of drumlins** seen from above, northern Saskatchewan, Canada, deposited by Pleistocene continental glacier that flowed from NE (upper right) to SW. Lakes (dark areas) are about 1 kilometer in diameter. Prominent esker extends from upper right to center of view. (Geological Survey of Canada, Canadian Government copyright)

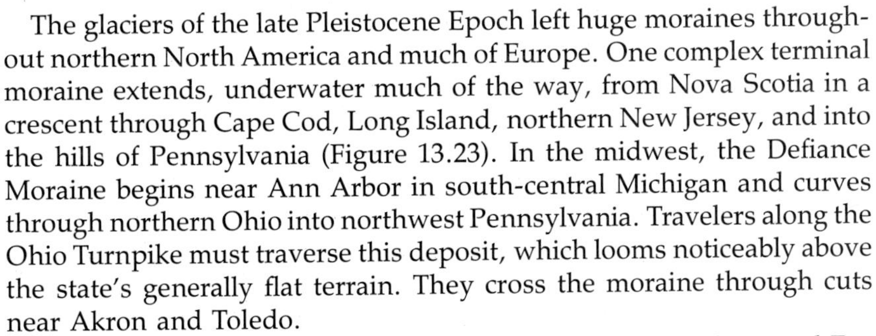

The glaciers of the late Pleistocene Epoch left huge moraines throughout northern North America and much of Europe. One complex terminal moraine extends, underwater much of the way, from Nova Scotia in a crescent through Cape Cod, Long Island, northern New Jersey, and into the hills of Pennsylvania (Figure 13.23). In the midwest, the Defiance Moraine begins near Ann Arbor in south-central Michigan and curves through northern Ohio into northwest Pennsylvania. Travelers along the Ohio Turnpike must traverse this deposit, which looms noticeably above the state's generally flat terrain. They cross the moraine through cuts near Akron and Toledo.

There are equally impressive moraines throughout north-central Europe. Much of Denmark is crossed by ridges deposited by successive glaciers whose margins extended along what is now the eastern edge of the North Sea.

**Drumlins** In contrast to a moraine, which can be considered the peripheral leavings of a glacier, a *drumlin* consists of till that a glacier has fashioned into a streamlined shape. Characteristically, a drumlin is eroded into an elongated shape with its steepest slope facing the direction from which the glacier originated. The lowest end points to the direction in which the ice mass moved (Figure 13.24).

Drumlins range in height from 15 to 60 meters and in length, up to a kilometer or more. Parts of Canada and the states of Massachusetts, New York, and Wisconsin are dotted with drumlins. In the central Hudson Valley and northeastward into New England, apple orchards have been planted on drumlins.

**Eskers** An *esker* (Figure 13.25) is a long, winding, continuous embankment composed of a layer of gravel beneath a low, narrow mound of sand and silt, that can stretch for more than 60 kilometers. An esker is a distinctive deposit made by a stream of meltwater beneath a large block

13.25 **An esker,** a sinuous body of sediment deposited by a stream flowing beneath a stagnant mass of ice. Width of view is about 2 kilometers. (Austin Post, University of Washington and U. S. Geological Survey)

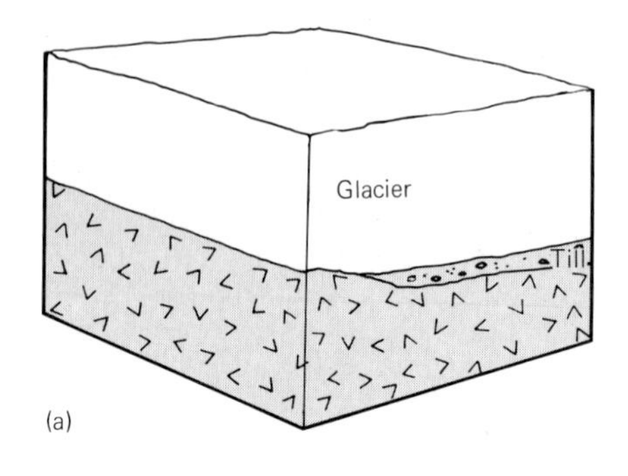


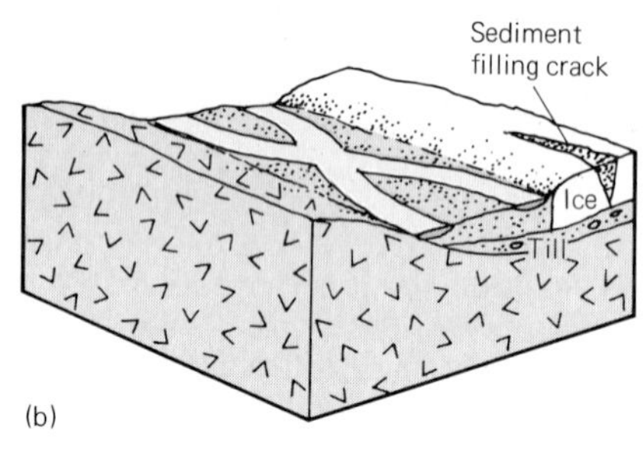


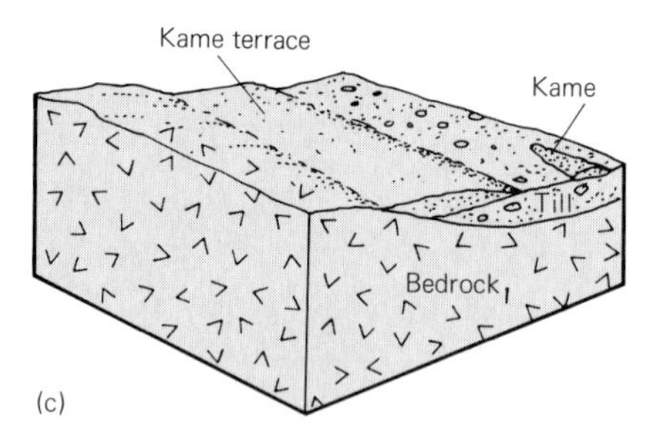


**13.26 Origin of kame terraces and kames,** schematic block diagram. (a) Braided stream flows in "valley" between bedrock wall and stagnant block of ice from former glacier and deposits a body of sediments, one side of which is in direct contact with the ice. (b) After ice has melted, the stream-laid sediments form a terrace along the margin of the valley. Sediments that were deposited from above into fissures in ice remain as a small raised knoll (a kame).

of stagnant ice. First, the stream erodes a sinuous tunnel. Then, the stream partially or fully fills this tunnel with meltwater deposits. After the ice has melted away, these deposits are left as an embankment standing above the surrounding lowlands. An esker would not be preserved if the ice began to flow as a glacier again after the stream had deposited the sediment.

**Kames and kame terraces** Kames and kame terraces are bodies of outwash built against stagnant blocks of ice. These sediment bodies take two forms: (1) short, steep-sided mounds (*kames*), or (2) terracelike bodies built against the margin of a former glacier (*kame terrace*) on one side and against the bedrock walls of valleys that had been glaciated on the other (Figure 13.26). Many kames are remnants of former eskers.

**Kettles** Scattered within the outwash or the till deposited near the edge of a glacier may be large blocks of ice. These blocks may be buried and slowly melt away. As the ice disappears, the overlying surface is let down and, if no more sediment is added, a closed depression, known as a *kettle,* forms above the former block of ice (see Figure 13.20). Kettles may fill with water and become small lakes. A subsequent deposition of sediment may turn the kettle lakes into swamps.

## INFERRED GLACIAL DEPOSITS IN ANCIENT BEDROCK

Once they had become aware of the distinctive geologic work of glaciers, geologists began to study the evidence related to former glacial episodes. As we have just seen, evidence in the regolith and landscapes for many Pleistocene glaciations (during the last 2.5 million years or so) is widespread. What is more, in the ancient bedrock, geologists have found diagnostic deposits left by glaciers hundreds of millions of years ago. The evidence for pre-Pleistocene glaciers consists only of distinctive kinds of rock; the former glacial landscapes have long since been eroded away.

Till that has been cemented so that it forms part of the bedrock of an area is a sedimentary rock named *tillite.* Ancient tillites, containing erratics and resting on striated and polished rock surfaces, have been found in rocks of Permian, Devonian, Ordovician, early Cambrian, and older Precambrian ages.

Late in the Paleozoic Era, more than 230 million years ago, glaciers again advanced northward and southward from the poles. During this time, when the Earth's surface is thought to have been a single large landmass, the ice reached not only areas that are parts of the present North America and Europe, but also parts of Africa, Australia, the subcontinent of India, and South America.

Tillite and related glacial deposits of Ordovician age covered what is now the Sahara. Tillites of Precambrian age are known from Australia, Utah, Canada, Scotland, and Scandinavia.

## OTHER GEOLOGIC ASPECTS OF GLACIERS AND GLACIATION

Other geologic aspects of glaciers and glaciation include use of layers of snow and ice as historical archives of atmospheric fallout and of ancient climates, the relationship of glaciers to sea level, erosion by glaciers, and the mechanics of flow of solids.

### Historical Archives of Atmospheric Fallout and of Ancient Climates

A glacier and its accumulated cap of snow and ice hold many clues to what was in the atmosphere while the snow was falling and to the climate that prevailed while the ice formed.

The firn layers, and any materials which settle on the surface of the snow-ice, form annual deposits that enable glaciologists to establish the age of the ice by simply counting the layers. A vivid example is the material thrown into the air by the 1912 eruption of Mt. Katmai on the Alaskan peninsula. A thin layer of this tephra has been found in all glaciers in western North America and even within the Greenland Glacier 3200 kilometers away. Proof indicating the record of the pollution of the atmosphere by lead smelting and the burning of tetraethyl lead in gasoline has been found in large specimens dug out of snowfields on Greenland and Antarctica. In addition, the firn strata of all glaciers may contain anything that can blow onto them between snowfalls, including pollen, spores, and insects.

The proportion of oxygen isotopes in the ice of the world's glaciers provides clues about ancient temperatures. When water falls as precipitation, most of the oxygen atoms are the common isotope $^{16}O$. However, some ions of the isotope $^{18}O$ are always present, and the warmer the temperature at the time of precipitation, the greater the concentration of $^{18}O$. A mass spectrometer can determine the ratio of the two isotopes of oxygen. Using this ratio, scientists can infer ancient temperatures (Figure 13.27).

### The Relationship of Glaciers to Sea Level

Glaciers are closely related to the hydrologic cycle and the world's climate. The water stored on land in glaciers is water that has been temporarily removed from active participation in the hydrologic cycle.

When the world's climate warms even slightly, nearly all modern glaciers retreat. When the climate cools, most glaciers advance. Because the water to make glaciers comes ultimately from evaporation of sea water, a great advance of glaciers results in a lowering of sea level. When the glaciers melt, the water returns to the oceans and causes sea level to

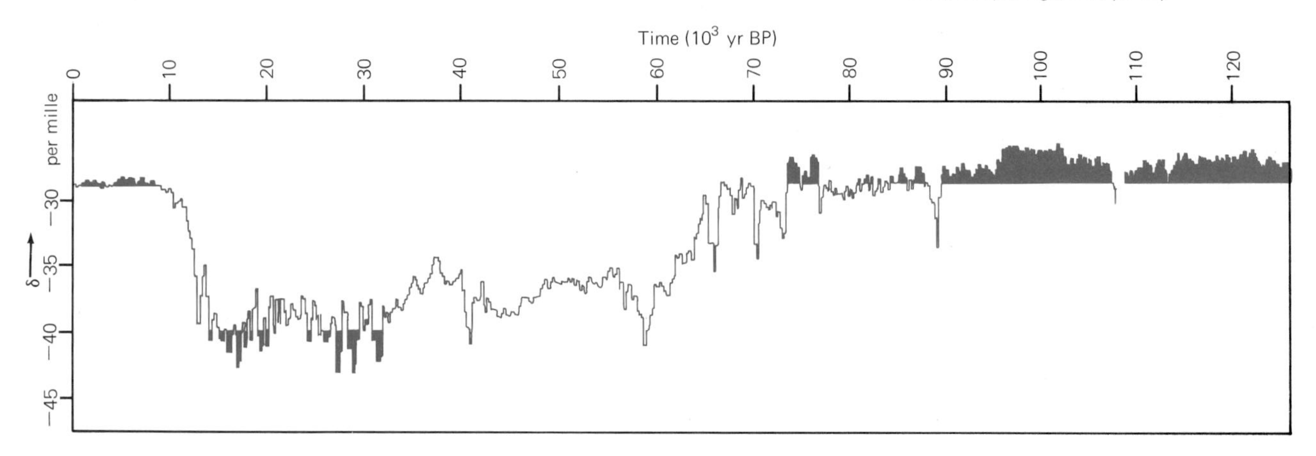


**13.27 Variation in oxygen isotopes** with depth in boring, 1390 meters long, from the top of the glacier to bedrock, through the North Greenland Ice Sheet at Camp Century, 225 kilometers E of Thule reflects fluctuations in climate. Samples were not dated by counting annual layers in the snow and ice, but rather by applying a mathematical model of ice flow and location of the core on the ice sheet. Synthesis of two harmonics, 78 and 191 years ± 5 percent, based on samples representing 10 to 20 years' worth of ice. (Willi Dansgaard, S. J. Johnsen, H. B. Clausen, and C. C. Langway, Jr., 1971, Climatic record revealed by the Camp Century ice core, p. 37–56 *in* Turekian, K. K., *ed.*, Late Cenozoic glacial ages: New Haven, Connecticut, Yale University Press, 606p., Figure 3, p. 43)

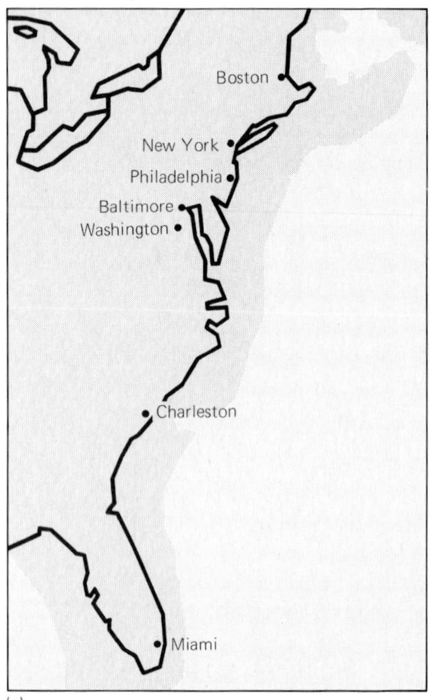


(a)

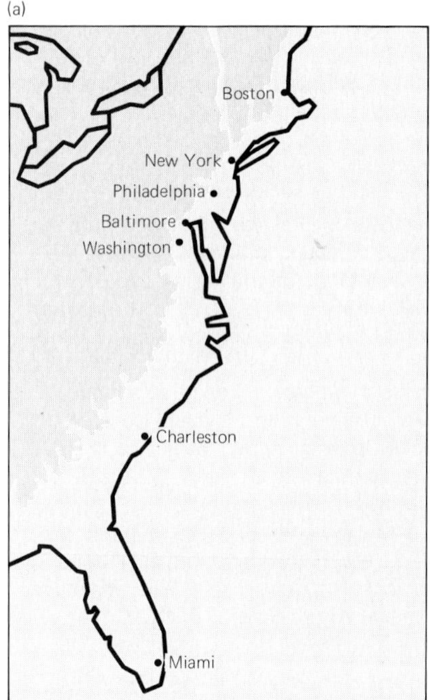


(b)

**13.28 Worldwide changes of sea level** accompanied Quaternary climatic oscillations, shown by maps of shorelines, eastern United States. (a) Shoreline with sea level about 100 meters lower than now, as during an ice age. (b) Shoreline with sea level about 100 meters higher than now, a predicted result if all the world's present glaciers were to melt and the water were to flow into the oceans.

rise. During times when the climate was generally cooler than it is now, glaciers covered millions of square kilometers of the land areas in the Northern Hemisphere and world sea level dropped more than 150 meters (Figure 13.28). During intervening times of warmer climate, the great glaciers melted and sea level rose to where it is now, and even slightly higher.

Depending on the estimate, the ice of modern glaciers now contains between 2.5 and 25 million cubic kilometers of water. Should all the glaciers melt, world sea level would rise by an estimated 20 to 60 meters, more than enough to flood most of our great cities. Pleistocene glaciers may have covered as much as 32 percent of the Earth's land areas (Figure 13.29).

### Mechanics of Flow of Solids

The physics and geophysics of glaciers have yielded data useful in understanding the behavior of rocks. Glacial ice flows and creates patterns that are similar to those created by certain processes of rock metamorphism (compare Figure 13.2 with Plate 9). The weight of a glacier creates a load on the Earth's lithosphere. The downward movements of the lithosphere under the weight of a fully developed glacier contrast with the upward movements that take place after the glacier has melted (see Chapter 18).

## THEORIES ABOUT CLIMATE

Much current interest and considerable research effort are centered on the question of what controls the Earth's climate. This question is

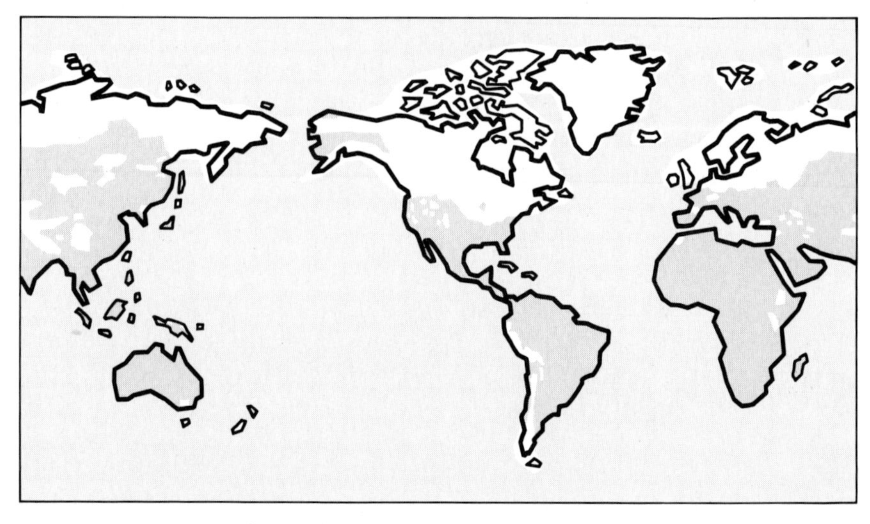

**13.29 Generalized extent of Quaternary continental glaciers** during ice ages.

important not only to geologists trying to understand the Earth's history, particularly during the ice ages, but also to farmers, economists, and government officials concerned with strategic planning. Theories about climate can be arranged in two groups: (1) astronomic theories, based on factors external to the Earth, and (2) terrestrial theories, based on factors on Earth.

### Astronomic Theories

Theories based on external factors, known as *astronomic theories,* contend that climate changes as a result of variations in the Sun's output of energy or in how this energy is received on Earth. An important characteristic of astronomic theories is that they are based on variations that recur with exact precision and have always happened.

One of the most significant developments concerning the astronomic theories is the recent realization that the Solar System functions as a unit and that the motions of all the planets cause slight variations in the pull of gravity that are felt on all parts of the system. These seem to affect the Sun's output, and this variation, in turn, affects the energy received by the planets.

At least three astronomic factors affect how the Earth relates to solar energy. These are: (1) ellipticity (amount of departure from a circle) of the Earth's orbit; (2) angle of tilt of the Earth's axis; and (3) direction of tilt of the Earth's axis.

Because the Earth's orbit is an ellipse, the distance from the Earth to the Sun varies during the year. In addition, the amount of the eccentricity of the ellipse (Figure 13.30) changes on a cycle of 102,000 years. At present,

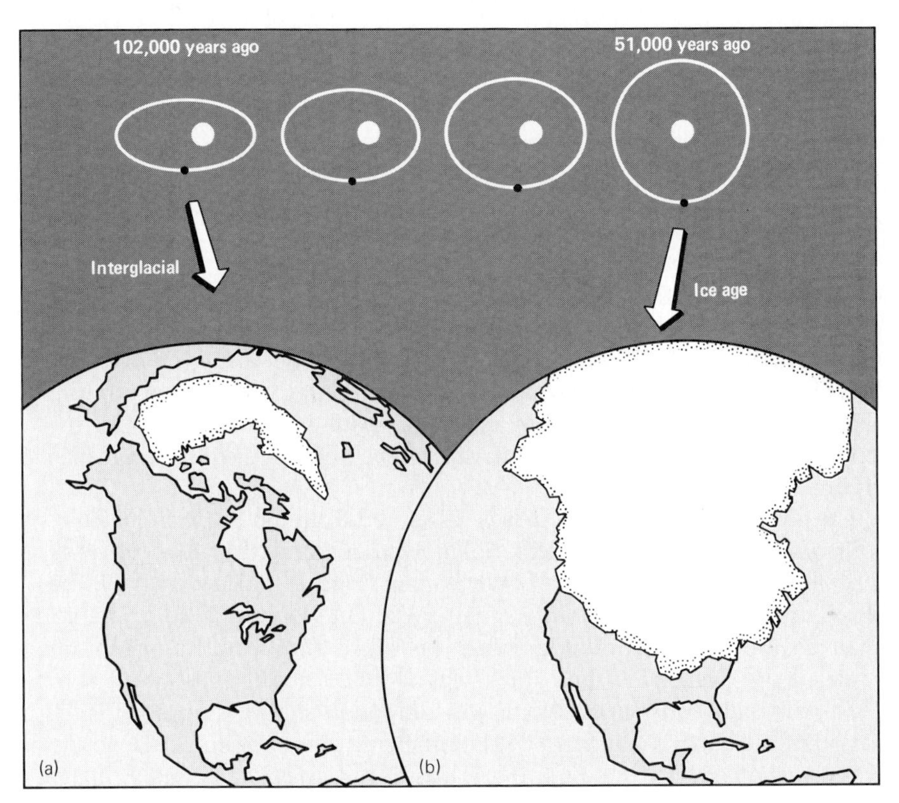


**13.30 Changing ellipticity of Earth's orbit** results from variation in length of minor axis of the ellipse; the length of the major axis remains constant. The ellipticity has been enormously exaggerated to emphasize the relationships. (a) The present ellipticity is near its mean value of about 2 percent, but is decreasing. At times of greatest ellipticity (maximum eccentricity), the length of the minor axis is at its minimum, and the Earth is closest to the Sun a condition causing warmth and general melting of glaciers. The last time of maximum orbital eccentricity (about 4 percent) was attained about 120,000 years ago, during the Last Interglacial age. (b) When the orbit is nearly circular, as it was about 40,000 years ago during an ice age, the Earth-Sun distance is at a maximum, a condition causing cold climate and growth of glaciers.

**13.31 Tilt of the Earth's axis of rotation.** (a) Tilt, measured between the Earth's pole and the Mean Solar Pole (see Figure 1.8). (b) Oblignity of the Ecliptic, the smaller component of tilt, shown by inclination of Ecliptic Pole with respect to Mean Solar Pole. See Figure 1.11 for larger component of tilt. Further explanation in text.

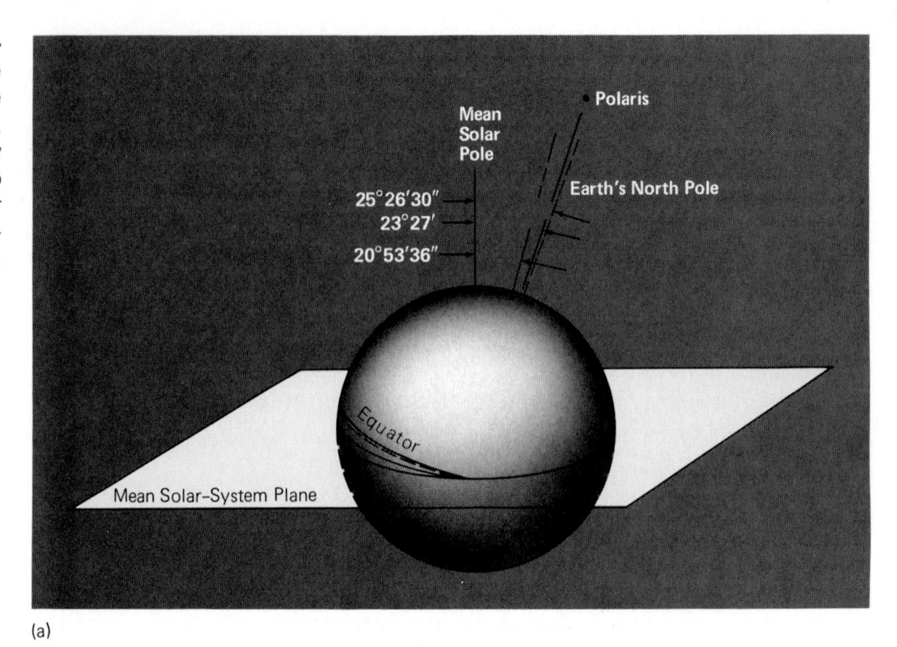


(a)

90°
Mean Solar Pole
1°36'
Ecliptic Pole
Plane of the Ecliptic
Mean Solar-System Plane
Plane of the Ecliptic
1°36'
1°36'

(b)

the eccentricity is 1.67 percent, nearly its mean value, and is decreasing. The ellipticity of the nearly circular orbit is 0.5 percent. The maximum ellipticity is 4.1 percent. These changes result from variations in the length of the minor axis; the length of the major axis does not change (see Figure 1.3).

The tilt of the Earth's polar axis is measured from the Mean Solar Pole, a line perpendicular to the Mean Solar-System Plane, the mean plane of the orbit of the planets of the Solar System (Figure 13.31a). In 1979, the tilt of the Earth's axis from the Mean Solar Pole was 23°26'31". The tilt changes about 0.4 seconds per year; it is decreasing from its maximum value of the present cycle, which was attained about 6000 years ago.

The tilt angle consists of two major components: (a) the angle of inclination of the Earth's axis from the Ecliptic Pole (line perpendicular to the Plane of the Ecliptic) (see Figure 1.11), and (b) the *obliquity of the Ecliptic,*

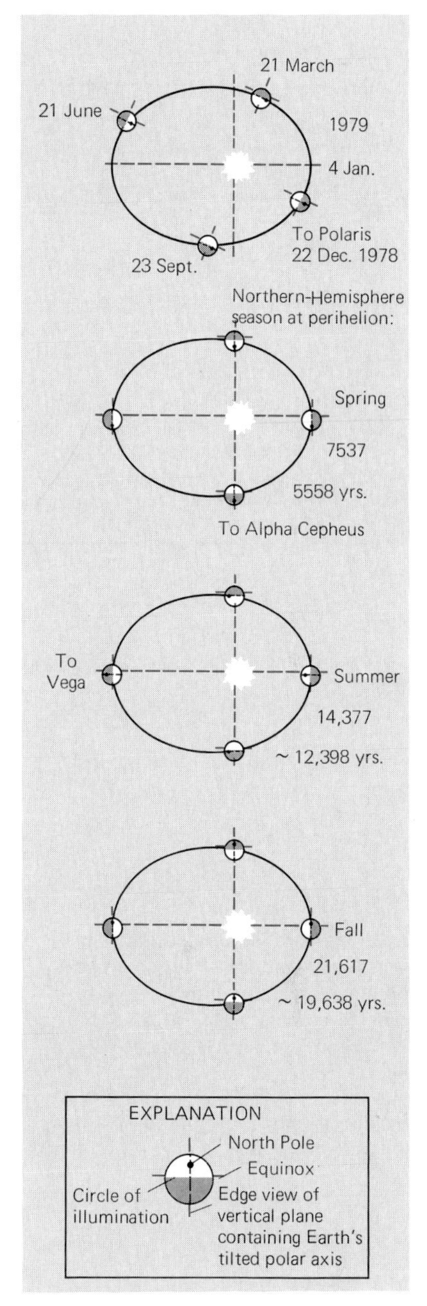


**13.32 Effect of precession through time,** schematic maps of Earth and Sun viewed from above the Mean Solar-System Plane. Orbital path shown as a single, much-exaggerated ellipse (changes in orbital ellipticity through time not shown). The Northern Hemisphere attains maximum warmth when Northern-Hemisphere summer coincides with the point of closest approach to the Sun, as predicted for the year 14,377 (12,399 years into the future from 1978).

which is the angle between the Mean Solar-System Plane and the Plane of the Ecliptic (Figure 13.31b). The angle of inclination of the Earth's pole with respect to the Ecliptic Pole ranges between a maximum of 23°50′30″ and a minimum of 22°29′36″, the mean is 23°10′00″. The cycle is 17,280 years, which is two-thirds of a complete precession cycle of 25,920 years. The obliquity is 1°36′ above and below the Mean Solar-System Plane in a cycle of 362,880 years, which is 14 full precession cycles.

The direction of inclination of the Earth's axis and the point of closest approach of the Earth to the Sun in the yearly orbit affect the amount of contrast between winter and summer temperatures in the Northern and Southern hemispheres. At present, the minimum Earth-Sun separation, hence the time when the intensity of solar energy reaches its maximum, is early in January, at a time when the North Pole is pointed away from the Sun. Thus, the time of maximum solar energy coincides with Northern-Hemisphere winter and Southern-Hemisphere summer. Because precession is causing the vertical plane containing the Earth's axis to move away from a vertical plane through the principal axis of the orbital ellipse, Southern-Hemisphere summers are becoming cooler (Figure 13.32).

The first attempt to calculate the effect on the Earth's climate of these orbital changes and axial variations was made by Milutin Milanković (1879–1958), a Serbian (thus now Yugoslavian) astronomer. Climate curves based on various dated specimens clearly show a cyclic variation that coincides closely with the curve drawn by Milanković (Figure 13.33).

## Terrestrial Theories

Climate theories based on factors on Earth are known collectively as *terrestrial theories.* These theories maintain that climate changes because of variations in the Earth's magnetic field, in amounts of volcanic dust in atmosphere, and in tectonic activity that can change the altitudes of continents, configurations of lands and seas and circulation within the seas, and locations of the pole of rotation.

**Variations in the Earth's magnetic field** The magnetic field extends far out into space and interacts with incoming solar energy (see Figure 1.16). The magnetic field serves as a filter, trapping some energy waves (and particles) and allowing other energy waves and particles to pass. It has been suggested that a change in the Earth's magnetic field can result in a change in solar energy received and hence in a change of climate.

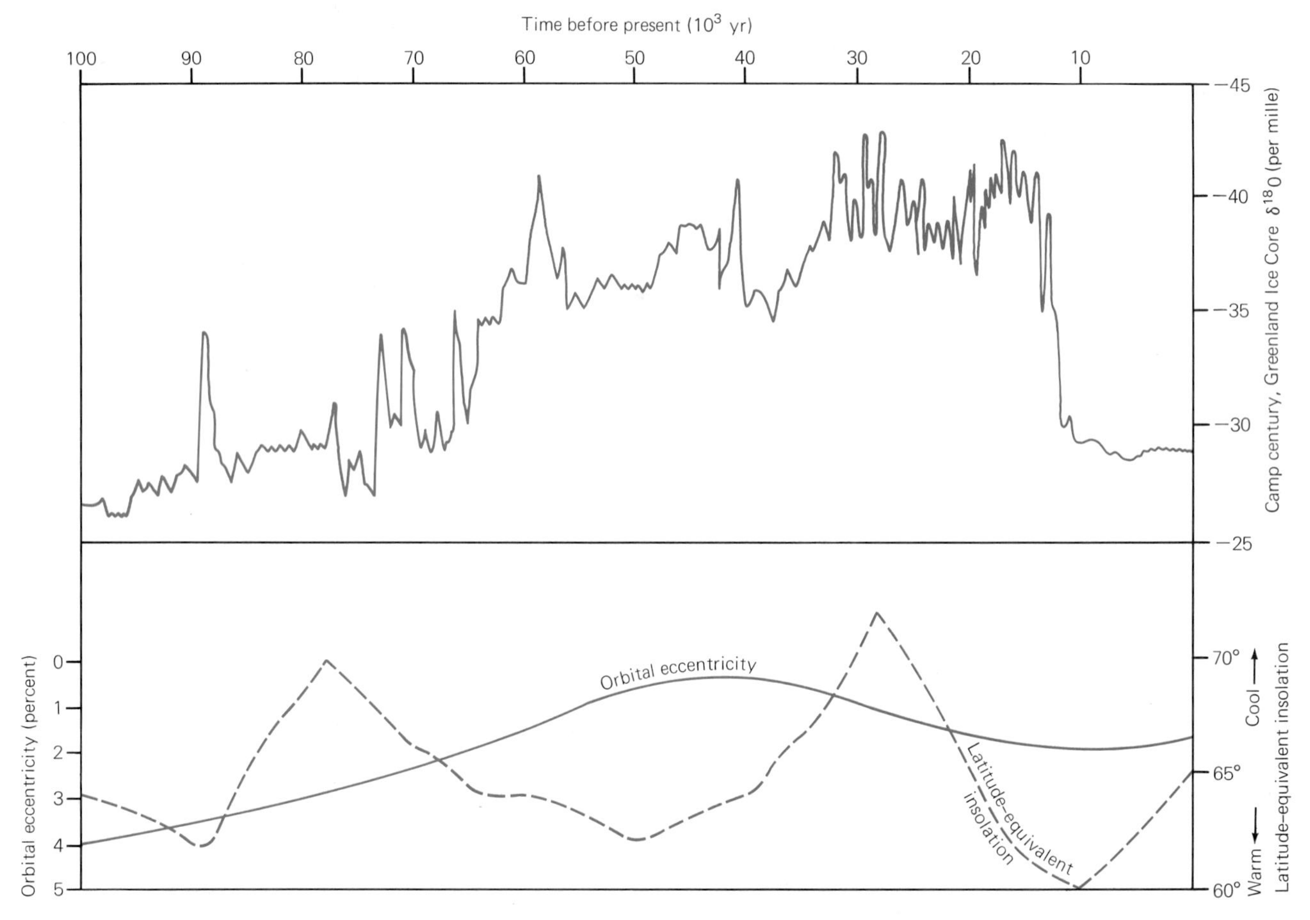


**13.33 Ice-core isotope curve from North Greenland ice sheet compared with Milanković curve** expressed as latitude-equivalent insolation. (Greenland isotopic curve from Willi Dansgaard, S. J. Johnsen, H. B. Clausen, and C. C. Langway, Jr., 1971, reference cited in caption to Figure 13.27, Figure 9, p. 52; Milanković curve from A. J. J. van Woerkom, 1961, The astronomical theory of climate changes, p. 147–157 *in* Shapley, Harlow, *ed.*, Climatic change. Evidence, causes, and effects: Cambridge, Massachusetts, Harvard University Press, 318p., Figure 4, p. 156)

**Atmospheric conditions; variations in volcanic dust** The Earth's atmosphere responds dynamically to solar energy. What is more, the quantity of volcanic dust in the stratosphere is known to vary and also to be effective in blocking incoming solar energy. After major volcanic eruptions — Skaptar Jøkull on Iceland in 1683, Tomboro in Indonesia in 1815, Krakatoa in 1883, Mt. Pelée in the West Indies in 1902 — vast amounts of tephra entered the atmosphere and drifted with high-altitude winds around the Earth. These particles not only created spectacular red sunsets, but decreased the amount of radiation entering the atmosphere lowering atmospheric temperatures.

**Variations resulting from tectonic activity** Tectonic activity can affect the climate in several ways. No matter how warm or cold it is worldwide, any mountain high enough to project above the local snowline can create glaciers. At present, the altitudes of continental masses are much greater than they were earlier in the Cenozoic Era. This change resulted from post-Miocene uplift. Throughout geologic history, times of high continents have been times of generally cooler climates.

The existence and shapes of oceans greatly influence climate and these aspects of oceans can be modified by plate movements. Thus, the continued opening of the Atlantic Ocean during the Cenozoic, the shift of the North Pole into the Arctic Ocean basin, and the closing of the Isthmus of Panama brought about the present arrangement by which water warmed in the tropics flows far to the north and warms northern Europe (see Figure 15.23).

As a result of movements of lithosphere plates, the North Pole lies within the Arctic Ocean basin. In addition, a large high area in the bottom restricts the flow of water from the North Atlantic into the Arctic Ocean. Because pack ice covers most of the Arctic Ocean, winds blowing off this ocean toward the land are dry and the surrounding lands are cold, polar deserts. As sea level continues to rise, however, more and more warm water comes in from the North Atlantic. Eventually, goes one argument, the pack ice could melt. If so, then winds could pick up moisture, and when they blow over the land, they would drop the moisture as snow and glaciers could grow. As glaciers grow, they lower sea level. This would cut off the Arctic Ocean from the North Atlantic. Hence, the Arctic pack ice would return, the glaciers would be starved of new snow, and they would eventually melt. As they melted, sea level would rise, and eventually warm water from the North Atlantic would melt the Arctic pack ice, and the glaciers would expand. If such events took place, they would be a neat mechanism for starting and stopping glaciers.

### Explanation of Glacial and Interglacial Climate Changes

From all the evidence that has been compiled about climates, it seems clear that some combination of astronomic factors and terrestrial factors must be just right to cause glacial climates and interglacial climates to alternate repeatedly.

The astronomic causes have been established, but if they were the only factors, then glaciers would have been advancing and retreating throughout geologic history, not merely during the last 10 million years or so, and in larger cycles that involve several hundred million years.

## THE GREAT LAKES AND THEIR GEOLOGIC HISTORY

In north-central North America many large freshwater lakes occupy basins that have resulted from the effects of Pleistocene glaciers on the preglacial landscape. The modern lakes are one of a series of lakes that were formed as each glacier retreated from the area. In this section we describe the modern Great Lakes, discuss their Pleistocene ancestors, and discuss their present conditons.

### Modern Great Lakes

The interconnecting bodies of fresh water called the Great Lakes stretch from Minnesota to New York, border eight states and two Canadian provinces, and contain a volume of 23,000 cubic kilometers of water. These lakes constitute far and away the largest such network in the world. Their total area is 248,000 square kilometers. The five individual lakes, and their worldwide ranks in size, are: Superior (1), Huron (4), Michigan (5), Erie (11), and Ontario (13).

Although the Great Lakes basin amounts to only 3.5 percent of the land of the United States, almost 15 percent of the population lives within the basin, as do about 33 percent of all Canadians. This concentration of population is a testimony to the importance of fresh water. Water from the lakes directly and indirectly nourishes some of the most fertile farmlands in the continent. And what once seemed to be an inexhaustible supply of fresh water fed the heavy industries in lakefront cities ranging from Rochester, New York, to Milwaukee, Wisconsin. The lakes themselves served as the medium for low-cost shipping of goods.

### The Ancestral Great Lakes

When a huge glacier melts and retreats, it leaves a terminal moraine and possibly one or many recessional moraines behind as distinctive landmarks. If the plowed depressed ground left by the vanished ice mass then fills with water, the waves rework the till at the shores into beaches. Glaciologists have found a number of such beaches in the Great Lakes basin. Through various dating techniques, scientists have proved that these features were created at different times. Experts now calculate that at least seven major lakes successively covered the area occupied by the present five bodies.

The last ice mass to cover the Great Lakes basin arrived during the late Wisconsin Stage, starting 20,000 years ago. Moving south-southwest over soft shale bedrock that had already been broken and shaped by previous glaciers, this ice sheet did the final landscaping. The water-filled depressions that are today's five lakes were scoured to their present contours by this last glacier.

When this last glacier retreated north, it left its load, derived from the fine crushing of shale, over much of the upper midwest. That crushed debris became the rich topsoil that now graces the farmlands of the region. Geologists have determined that the bedrock underlying the soil is rugged and hilly, but few traces of these irregularities remain in the flattened landscape. As the glacier retreated, its shifting margin formed a massive dam of ice. As meltwater and natural precipitation gathered in the depressions the ice left behind, the water collected and gradually found its way to the present perimeters of the Great Lakes (Figure 13.34).

By the carbon-dating of vegetation known to have been killed by the advance of the last glacier, scientists have established that glaciation reached its maximum extent about 14,000 years ago. This ice mass did not retreat completely out of the Great Lakes basin until about 10,000 years ago, and still lingered in northern Canada until about 6000 years ago.

### Present Conditions in Great Lakes

The Great Lakes were once vibrant with populations of fish and other aquatic life, but since the turn of the century they have deteriorated markedly. Today, these populations have been greatly reduced. There are even large "dead" spots in the middle of Lakes Erie and Ontario where dissolved oxygen is not abundant enough to support fish. Sectors of the shorelines of the various bodies are unfit for human bathing, and only the water of Lake Superior is free enough from bacteria to be drunk, but even it may contain harmful asbestos fibers. Bottom sediments and fish of Lake Michigan contain PCBs.

The culprits are human beings. The large cities and factories built near the lakes to tap the seemingly endless supply of fresh water have in turn discharged a continual flow of toxic effluents back into the lakes. So heavy has this pollution been at times that the Cuyahoga River has

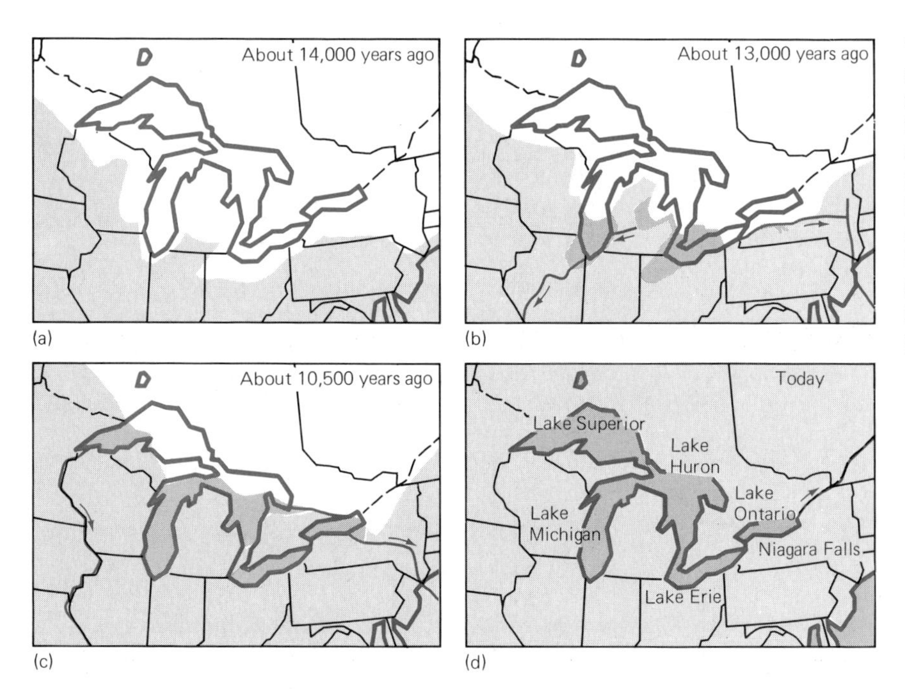


**13.34 Evolution of the Great Lakes** shown by maps in which geologic evidence has been used to reconstruct locations of the margins of former glaciers and extent of ancient lakes. (Based on F. B. Taylor *in* Leverett, Frank, and Taylor, F. B., 1915, The Pleistocene of Indiana and Michigan and the history of The Great Lakes: U.S. Geological Survey, Monograph 53, 529 p.; (b), from Plate 17, facing p. 392; (c), from Plate 19, facing p. 400)

actually caught fire from shore to shore. The results of pollution have been the loss of valuable plant life (except for the oxygen-depleting algae that thrive on pollution) and an ensuing loss of fish species. Lake trout and other fish species have been further reduced by the migration through the St. Lawrence Seaway of parasitic sea lampreys from the Atlantic Ocean.

By the late 1960s, conditions were so bad that the public began to apply heavy pressure to industries and cities to filter their wastes before dumping them into the lakes. The cleanup was complicated by interstate and international quarreling, but by now Americans and Canadians alike have recognized that without the Great Lakes, wholesale changes would be necessary in the lives of millions of people. Today there is hope that the Great Lakes will survive to greet the next glacier, the successor of the former glaciers that created them.

# CHAPTER REVIEW

1. A *glacier* is a flowing mass of ice that formed by the recrystallization of snow, is powered by gravity, and has flowed outward beyond the snowline.

2. A mass of snow goes through a series of changes before it becomes a glacier. These changes result in the formation of *snowfields, firn,* and *glacial ice.*

3. Glaciers advance and retreat almost daily. A glacier always flows internally; if its motion ceases, it is still ice, but by definition it is no longer a glacier. A glacier is said to retreat if melting at the front margin takes place faster than the outward flow of the ice from a point of accumulation of new snow.

4. The typical rate of normal or nonsurging flow in modern glacials is about 15 to 60 centimeters per day; short-lived *surges* have moved glaciers as much as an average of 35 meters per day. Nonsurging glacial movement is thought to consist of *slippage* or *sliding* along discrete shear planes and *plastic flow* from deformation of ice crystals.

5. Geologists recognize three chief kinds of modern glaciers: *valley* or *Alpine glaciers, piedmont glaciers,* and *ice sheets* (continental glaciers).

6. The results of *glaciation* (the advance and retreat of a glacier over an area) include *erosion* and *deposition.* An advancing glacier is the most potent agent of erosion on Earth. Processes of erosion include *quarrying* (or *plucking*) and *abrasion.* Distinctive landforms eroded in

bedrock by glaciers include: *cirques; arêtes, cols,* and *horns; roches moutonnées; U-shaped valleys, hanging valleys* and *fjords;* and *striae* and *grooves.*

7. Glacial drift, the regolith and rock debris torn from the terrain, is strewn in characteristic deposits all along the glacier's path. Some kinds of glacial sediments are *erratics, till, outwash,* and *varved lake deposits.* Depositional landforms include *moraines, drumlins, eskers, kames* and *kame terraces,* and *kettles.*

8. The most recent episode of widespread glaciation took place less than 20,000 years ago. In North America, Pleistocene glaciers covered all of Canada, the upper tier of the Plains States, the midwest, and all of the northeast coast. The evidence for pre-Pleistocene glaciers consists of distinctive kinds of bedrock. The glacial landscapes formed by these very ancient glaciers have long since been eroded away.

9. The accumulation of layers of firn on the surface of glaciers is important as a record of what was in the atmosphere and of climate during the history of the glacier.

10. Glaciers are closely related to the hydrologic cycle and the world's climate. If all the present glaciers were to melt, the world sea level would rise 20 to 60 meters, enough to inundate most major cities. The chief existing ice sheets are on Antarctica and Greenland.

11. Astronomic theories about climate change deal with changes in solar radiation and changing ellipticity of the Earth's orbit, angle of tilt of the Earth's axis, and direction of tilt of the Earth's axis. Terrestrial theories are based on changes in the Earth's magnetic field, variations in atmospheric conditions and volcanic activity, tectonic changes related to elevation of continents, and shapes of the oceans.

12. The Great Lakes have resulted from the effects of Pleistocene glaciers on the preglacial landscape. The modern lakes together are one of a series of lakes that were formed as the last glacier retreated from the area.

## QUESTIONS

1. Define *glacier.* List the steps involved in the origin of a glacier.
2. Explain the mechanisms by which glaciers move. Cite typical rates of advance of glaciers.
3. Does the rate of advance of the front edge of a glacier always equal the rate of movement of the ice? Explain.
4. List, briefly describe, and give a modern example of the three major kinds of glaciers.
5. What are the two chief processes of erosion by glaciers? Explain how these processes are related to the distinctive landforms eroded in bedrock by glaciers.
6. Compare *drift, erratics, till, outwash,* and varved proglacial lake sediments.
7. List and describe by annotated sketches six depositional landforms created by glaciers.
8. Discuss the evidence that can be used to infer the direction of flow of a now-vanished glacier.
9. Trace the origin of the idea that the latest parts of the Earth's history included ice ages.
10. Compare and contrast the geologic evidence for inferring the former existence of glaciers of Pleistocene age and of pre-Pleistocene age.
11. Explain the relationships among glaciers, the hydrologic cycle, the world's climate, and sea level.
12. List some of the *external factors* that are thought to affect the Earth's climate.
13. List and explain three *factors on Earth* that can affect climate.
14. Defend or attack the concept that both astronomic factors and terrestrial factors contributed to Pleistocene changes of climate.
15. Explain the relationship between the Great Lakes and Pleistocene glaciers.

## RECOMMENDED READING

Andrews, J. T., 1975, *Glacial Systems:* North Scituate, Massachusetts, Duxbury Press, 191 p.

Denton, George, and Porter, S. C., 1970, Neoglaciation: *Scientific American,* v. 222, no. 6, p. 101–110.

Imbrie, John, and Imbrie, K. P., 1979, *Ice Ages. Solving the Mystery:* Short Hills, New Jersey, Enslow Publishers, 224 p.

Kennett, J. P., and Thunnell, R. C., 1975, Global Increase in Quaternary Explosive Volcanism: *Science,* v. 187, p. 497–507.

Knight, C., and Knight, N., 1973, Snow Crystals: *Scientific American,* v. 228, no. 1, p. 100–107.

Kukla, G. J., 1976, Around the Ice-age World: Reconstruction of Landscape and Climate from 18,000 Years Ago: *Natural History,* v. 85, no. 4, p. 56–61.

Ninkovich, Dragoslav, and Donn, W. L., 1976, Explosive Cenozoic Volcanism and Climatic Implications: *Science,* v. 194, p. 899–906.

Post, Austin, and LaChapelle, E. R., 1971, *Glacier Ice:* Seattle and London, University of Washington Press, 110 p.

# CHAPTER 14 DESERTS AND THE WIND

PLATE 14
Two common effects of wind in a desert: sand dunes (background) and pebble-strewn surface formed by concentration and interlocking of particles that could not be blown away (foreground, beneath plants). Great Sand Dunes National Monument, Colorado. (U. S. National Park Service)

THE WORD DESERT IS DERIVED FROM THE LATIN WORD FOR forsaken or not joined together. The archaic definition was an abandoned or uninhabited place. A more modern definition describes deserts as regions characterized by evaporation that greatly exceeds precipitation.

In the modern world, areas classified as "dry" cover more land area than do areas of any other kind. About one-third of all land area is desert (Figure 14.1), an impressive amount, considering that two-thirds of the Earth's surface is covered by water. Most land classified as deserts receives less than 25 centimeters of rainfall per year (Figure 14.2); extremely arid deserts receive less than 5 centimeters of rainfall per year.

In this chapter we study deserts of the hot, dry regions of low latitudes as special effects of extreme climatic conditions. (Deserts set apart by distinctive features of low-temperature conditions are not discussed.) As far as we can tell from examining modern deserts and comparing their products with the geologic record, deserts must have been present since early in the Earth's history. Determining the distribution of ancient deserts is very important in connection with reconstructing the ancient positions of the continents. We will also deal with how the atmosphere transports sediments. The importance of such transport is emphasized by

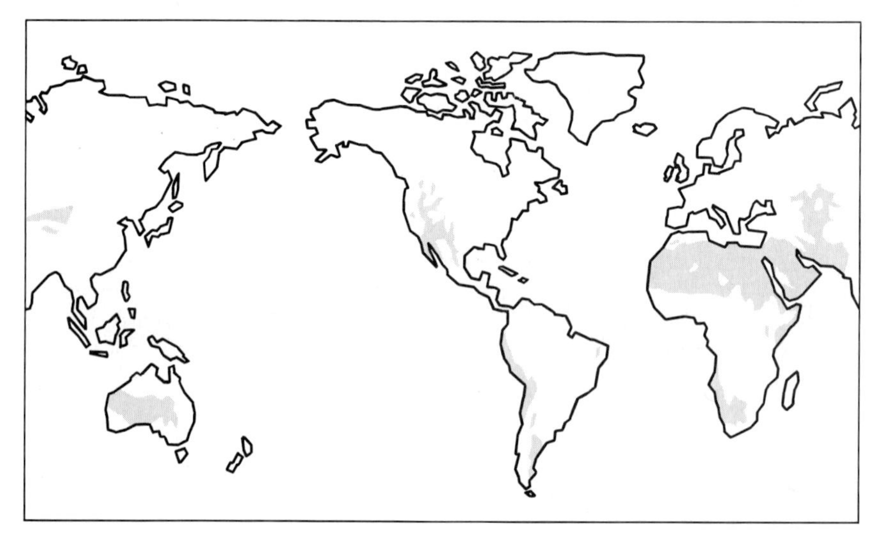

**14.1 The distribution of deserts of the hot variety,** shown by gray areas on world map.

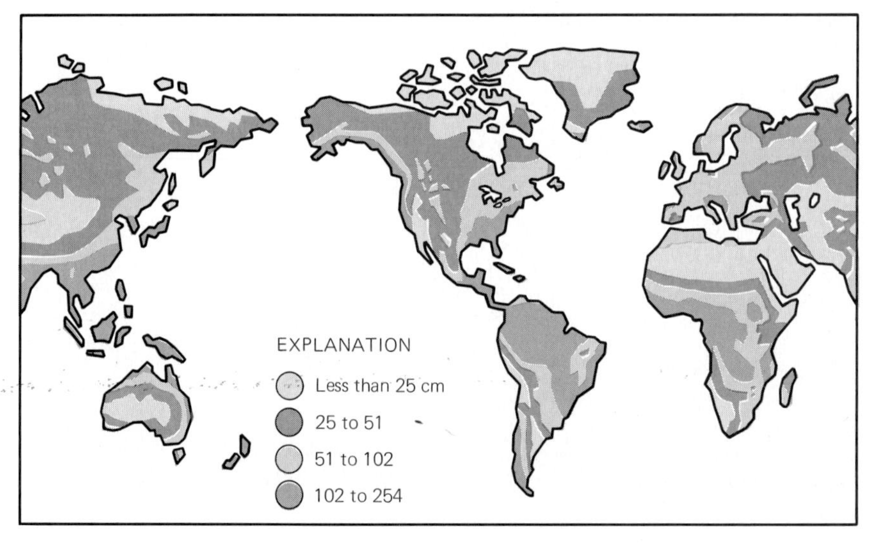


**14.2 Annual precipitation** shown on same world base map as in Figure 14.1. Notice the close proximity of hot deserts to areas of great rainfall and also the locations of cold, polar areas that receive negligible precipitation.

the fact that many of our present soils have formed by the weathering of ancient windblown dust. Movement of sediment in the atmosphere has given us important clues to the distribution of tephra as it is related to ancient volcanoes and to the fallout patterns of the radioactive products of atmospheric nuclear-weapons tests.

Our approach to the material of this chapter is to review the causes and distribution of modern deserts, to study desert sediments in terms of landforms and processes, to summarize the cycle of erosion in deserts, to look at desert lakes, and finally, to see how the wind transports sediments and shapes desert landscapes.

## MODERN DESERTS: DISTRIBUTION AND ENVIRONMENTAL CONDITIONS

### Types and Distribution of Modern Deserts

Most modern deserts can be arranged in four groups according to causes and locations: (1) subtropical deserts, (2) continental-interior deserts, (3) rainshadow deserts, and (4) coastal deserts. Semiarid regions commonly associated with deserts are steppes.

Most of the world's deserts lie in two broad belts of high pressure that trend east-west and encircle the globe. One lies north of the Equator and the other, south of the Equator (Figure 14.3). The location of deserts in these belts is controlled by the global air circulation system.

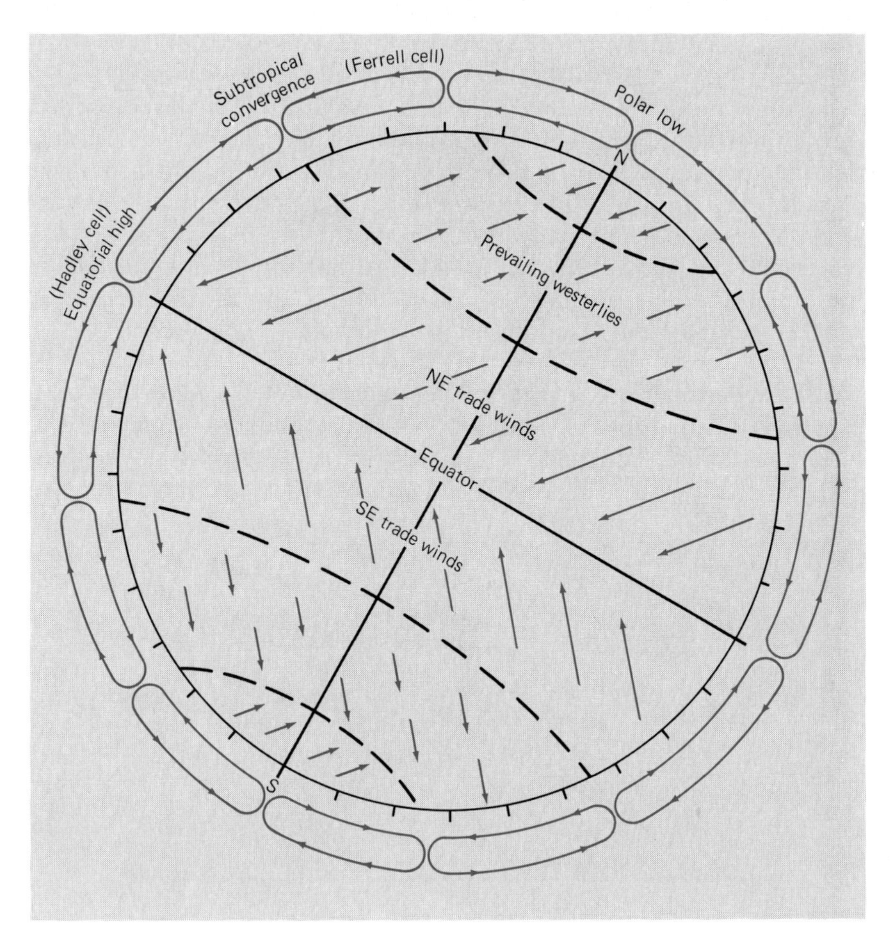


**14.3 Generalized large-scale circulation of Earth's atmosphere** that results from uneven distribution of solar energy and from the Earth's rotation. Schematic projection of Earth and lower part of the atmosphere. Further explanation in text.

**Subtropical deserts** Deserts created primarily by the cool, dry, descending air just outside the tropics are *subtropical deserts.* The greatest deserts on Earth belong to a vast subtropical swath of arid land extending for 8000 kilometers from the west coast of North Africa to Pakistan and western India. At its eastern end, this belt of deserts is broken by the monsoon region of Asia. Another subtropical desert covering nearly 1.3 million square kilometers is centered at 30 degrees north latitude in the southwestern United States and northern Mexico. This desert, too, stops toward the east where humid air moves northward from the Gulf of Mexico. The Kalahari Desert occupies most of southwest Africa (Namibia); a surrounding semiarid area extends north and east. More than 80 percent of Australia is arid or semiarid, for it is situated in (and helps to maintain) one of the world's most persistent zones of dry, descending air.

**Continental-interior deserts** Some deserts form in regions so far from the sea that little moisture ever reaches them. Such *continental-interior deserts,* where air is heated and dried in the summer and cooled in the winter, occupy much of central Asia. The Gobi Desert and the Takla Makan Desert of western China are among the harshest environments on Earth. Stable, dry cells of high-pressure air are trapped by adjacent highlands. Only polar air interrupts the cold monotony — with air that is equally dry and colder still.

**Rainshadow deserts** A landform barrier, such as a mountain range, may block and deflect air currents. Air that encounters a mountain range rises and is cooled. Because cool air cannot hold as much water vapor as warm air, the water vapor condenses and falls as rain or snow.

In the western United States, the Cascades and Sierra Nevada deflect upward the humid winds blowing inland from the Pacific Ocean. Most of this precipitation falls on the western flanks of the mountain ranges, feeding coastal rivers. By the time the winds have reached eastern Oregon, eastern California, Nevada, and Utah, little moisture is left. The result is a *rainshadow desert* (Figure 14.4). Other rainshadow deserts on the dry side of the American cordillera include the Sonoran Desert in Mexico and the Patagonian desert in Argentina.

**Coastal deserts** Some of the world's driest deserts lie along the coasts of Africa, North America, and South America, where currents of cold water create dry "rainshadows" on adjacent land. Winds blowing onshore from cold water are cooled and cannot hold much moisture. This

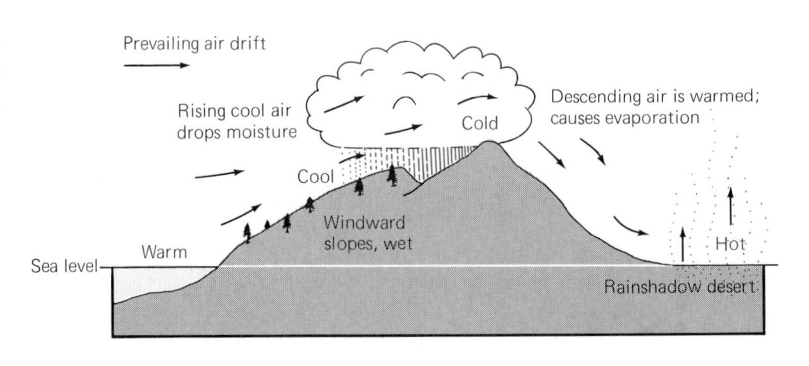


**14.4 Rainshadow desert** on lee side of coastal mountain range, schematic profile and section.

dense, dry air not only brings no rain of its own, but also blocks lighter, moist air from entering.

The Humboldt Current is responsible for the Atacama-Peruvian Desert in Chile and Peru, the world's smallest desert system. Although the coast here is usually foggy, this desert is the world's driest; it averages less than 1.5 centimeters of rain per year. In Southwest Africa, the Namib Desert, the coastal extension of the Kalahari, lies adjacent to the cold Benguela Current. The cold California Current dries southern California and Baja California.

### Steppes

Most true deserts are surrounded by semiarid regions called *steppes.* In steppes, average rainfall is 25 to 50 centimeters per year (see Figure 14.2).

As patterns of rainfall change, the border between steppe and desert shifts widely. The effect of drought in a former steppe area is far reaching. Animals congregate near the waterholes and destroy all the vegetation. Eventually, they perish for lack of food and water.

The largest steppes are in Asia. They are located in the Soviet Union bordering the Gobi and Takla Makan continental-interior deserts. The Great Plains of the United States and the prairies of Canada are steppes that have been converted into rich agricultural areas; they are the "breadbaskets of the world". Despite the agricultural successes of recent years, these areas are always on the margins of their water supplies. In drought years, the crops suffer.

### Environmental Conditions in Deserts

In addition to lack of rainfall and great evaporation, the desert environment is characterized by large temperature ranges and distinctive vegetation.

**Evaporation** As we have seen, the most significant feature of deserts is their general lack of water. The relationship between precipitation and evaporation typically is defined in terms of *potential evaporation,* the amount of water that would evaporate if it were present to do so. Thus, deserts include regions where abundant rain may fall for a few weeks or days, but where it is dry the rest of the year. One weather station in the Kalahari Desert in Africa recorded no rain at all for 15 consecutive years. The lowest mean annual precipitation in the world (0.4 mm) occurs at Dakla, Egypt. In Australia potential evaporation may exceed precipitation by 10 to 1, in the Sahara by 100 to 1.

**Temperature** Despite the fact that no desert occurs within 15 degrees of the Equator, deserts are the world's hottest places. The reason is that desert air holds little moisture to form clouds and thus to block incoming solar radiation. In cloudy humid regions only about 40 percent of incoming solar radiation reaches the Earth's surface; in some deserts as much as 90 percent of the Sun's energy reaches the ground. The lack of surface and atmospheric moisture also allows temperatures to fluctuate widely from day to night and from summer to winter. A peak summer temperature of 58°C has been reported in el Azizia, Libya, the hottest place on Earth. The ground surface is up to 10°C hotter. California's

**14.5 Sparsely vegetated desert landscape** in Libya; most plants visible here have been planted for the purpose of keeping the sand in place. F. A. O. photo, Photo library, FAO, Rome, Italy)

Death Valley is almost as hot. With little moisture in the air to retain warmth, the temperature may also drop sharply as insolation decreases at night or in the winter. One station in Libya recorded a nighttime temperature of −9°C, and the midwinter mean in the Gobi Desert is below −12°C. Daily and annual temperature fluctuations of 60°C are on record for West Africa and Central Australia.

**Vegetation** Sparse desert vegetation is adapted to make maximum use of scanty rainfall and groundwater (Figure 14.5). Perennial plants are spaced widely, leaving room for extensive root systems. This thin plant cover offers the surface little protection from winds, heating and cooling, and rapid erosion by the little rain that does fall. *Xerophytes* (from the Greek *xeros,* "dry," and *phyton,* "plant") have evolved ingenious means of surviving in deserts (Figure 14.6). Foliage is leathery and waxy in order to slow evapotranspiration. Root systems may be shallow to trap any rain that falls or deep to seek moisture far underground. The creosote bush combines shallow roots with a deep taproot; salt cedar, mesquite, and greasewood may extend roots 6 meters or more below the surface. Cactus plants concentrate root systems near the surface to trap rainfall before the water can seep beyond reach, while their thick trunks store water to be doled out sparingly between rains.

## THE GEOLOGIC CYCLE IN DESERTS: SEDIMENTS AND THE CYCLE OF EROSION

The operation of the geologic cycle in deserts is drastically affected by the climate, particularly by the way water behaves, and by the lack of vegetation. In this section, we examine characteristics of running water

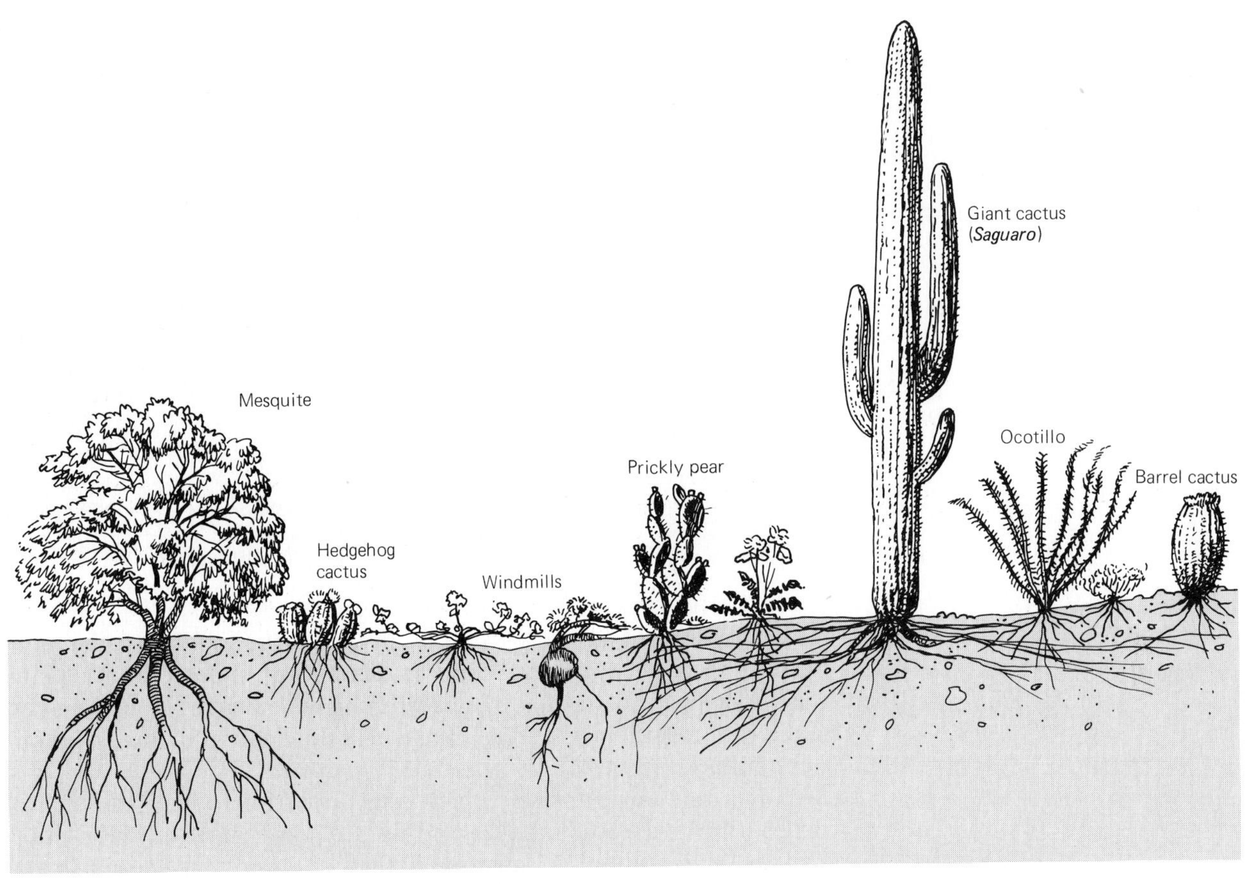


**14.6 Desert plants and their root systems.** (Time-Life Books, Deserts)

and drainage in deserts, how the operation of the geologic cycle in deserts affects the sediments, and the cycle of erosion in deserts.

## Running Water and Interior Drainage

When rain does come to the desert, it often comes in brief, high-intensity "cloudbursts" which may dump an entire year's precipitation in an hour. Once water is on the ground, its great transport capacity makes it more important than wind in shaping desert landscapes. On gentle slopes, water from a cloudburst characteristically moves as *sheetwash,* or sheet floods, a broad surface runoff laden with suspended sediment. Sheetwash collects rock debris into broad, thin alluvial deposits. Where slopes are steeper, the surface is quickly scarred by gullies that develop into steep-sided, flat-bottomed valleys (Figure 14.7). These features are the *arroyos* and *box canyons* typical of the American southwest and the *wadis* of the Arabian desert and the Sahara.

Cloudbursts may cause flash floods that within a minute or two can fill an arroyo with a fast-moving mixture of water and sediment 3 meters or more deep. Such muddy waters typically form mudflows. Flash floods and mudflows may do their erosive work for only a few hours, yet they can move tremendous quantities of material, including large boulders

**14.7 Box canyon,** showing typical vertical sides and flat, sediment-covered floor. Dry River, Martinique, the same valley down which the nuées ardentes from Mt. Pelée traveled from the vent to the sea (see Figs. 5.26 and 5.27). (Embassy of France in New York City, Press and Information Office)

and other large objects (Figure 14.8). In arid regions, hillsides commonly become intricately dissected into badlands (see Figure 12.22) or intersecting gullies and arroyos. Where running water leaves its channels and spreads over flatlands, the resulting sheet floods form broad and prominent fans.

Only a few desert rivers, which originate in mountains of abundant water supply, such as the Nile and the Colorado, flow the year around and are large enough to traverse deserts and reach the sea. Far more common are rivers that drain into lowlands or closed basins within the desert and commonly disappear by evaporation and infiltration into the thirsty soil. Such rivers constitute *interior drainage* (Figure 14.9). Groundwater, always deep in deserts, is nearest the surface under these streams. In the Sahara and several other deserts deep sandstone beds may bring artesian water to deserts from the foothills of distant moun-

**14.8 Overturned bus, part of a debris flow** about 5 meters thick that overwhelmed the site of former Plaza de Armas in town of Yungay, Peru. (U.S. Geological Survey).

**14.9 Interior drainage basin** illustrated by the termination of the Armargosa River, Death Valley, California. (Bill Belknap, Rapho/Photo Researchers, Inc.)

tains. Here, the exposed tilted edges of the sandstones can catch abundant rainfall (see Figure 10.11). An *oasis* is a spot where the desert surface intersects the water table. Where streams change into interior drainage, the only reliable sources of water are from aquifers that must be conserved carefully.

### Desert Sediments: Processes and Landforms

Deserts are dominated by bare rock surfaces and by great bodies of sediment. Because of the rare, but intense periods of rainfall, the episodes of transport of sediment by water are brief but vigorous. Moreover, because of interior drainage, desert highlands tend to be buried in their own debris. This situation produces distinctive processes that affect sediments; in turn, the sediments as well as the bedrock of desert areas are fashioned into distinctive landforms.

During desert cloudbursts, mountain streams fill quickly with water that flows vigorously enough to sweep sand, pebbles, and even rocks down mountain slopes. As these streams leave the mountains, they deposit their sediment loads on fans (see Figure 12.27).

As long as nearby mountains are high enough to furnish erosion products, sediment continues to be shed into such basins. In some of the enclosed basins of the southwestern United States, sediment is hundreds of meters thick (see Figure 8.20).

**Features of desert basins** Blocks of elevated bedrock in deserts commonly form highlands that yield abundant sediment to an adjacent lowland. The lowland subsides and thus becomes a basin that is kept supplied with sediments. A desert basin surrounded by eroding mountains is known as a *bolson,* the Spanish word for purse. Distinctive features along the margins of bolsons are composed of coarse sediment. Toward the center of a bolson, finer sediments are typical. Commonly, neighboring fans begin to overlap and coalesce. A slope, underlain by the sediments of coalesced fans is a *bajada* (ba-HA-da).

Farther from the mountains, beyond fans and bajadas, the slope of a

**14.10 Playa at center of large bolson,** Australia, as viewed from a low-flying airplane. (G. R. Roberts)

bolson becomes flatter. A sun-baked flat underlain by clays and salts in a bolson is a *playa* (PLY-ya). Sudden rains may produce enough water to create a desert lake, or *playa lake,* at the center of a bolson. Playa lakes may cover large areas with water to depths of a few meters for brief periods. Their beds are normally encrusted with evaporate minerals or covered by clay and fine sand (Figure 14.10).

## Cycle of Erosion in Deserts

The erosion of desert bedrock highlands is thought to follow a predictable series of stages that form the desert erosion cycle, which is analogous to, but not the same as the erosion cycle in humid areas. The control of the cycle of erosion in a humid area is sea level, the ultimate base level. In deserts having interior drainage, the surface of sediments in a filling bolson is the local base level. As the bolson fills with sediments, the base level moves upward.

Between the foot of many desert mountain blocks and the neighboring bajadas, one typically finds a gently sloping bedrock surface called a *pediment.* In cross section a pediment, extending from either side of a mountain range, resembles the triangular gable for which it is named. Pediments, bare or thinly covered with sediment, are the dominant erosional features of many desert landscapes (Figure 14.11a). When fully developed, pediments are thought to correspond to the peneplains of humid regions.

The extent of pediments provides a basis for recognizing three stages in the erosion cycle of a desert mountain block. As in the erosion cycle in humid climates, the cycle in deserts has been divided into youth, maturity, and old age.

In the youthful stage, pediments have not yet appeared or are very narrow. Wide, well-developed pediments characterize the mature stage. The lower portions of pediments have been buried under a bajada. At the same time, the upper edges have enlarged upward and have encroached

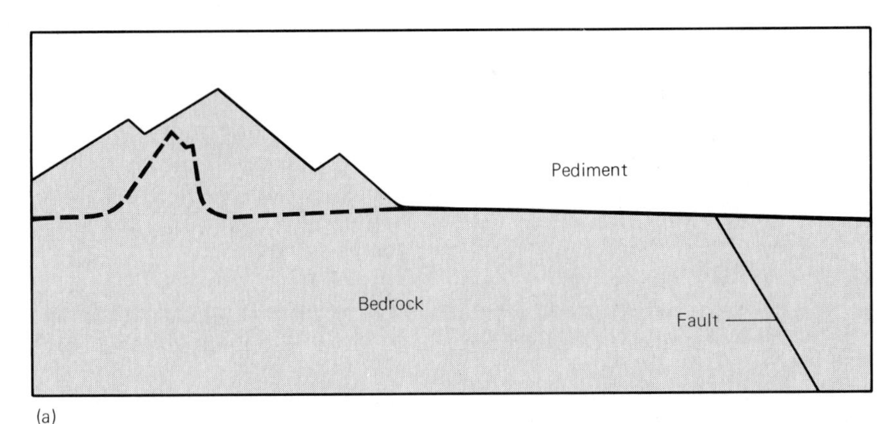


(a)

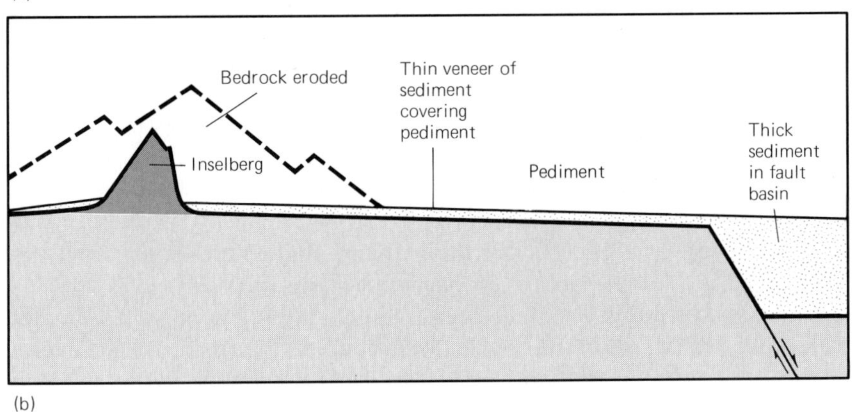


(b)

**14.11 Erosion of bedrock and retreat of slope of mountain block creates a wide pediment.** (a) After front of mountain has retreated half way back from the bounding fault (right). (b) Non-eroded remnants of the mountain block are named *inselbergs*. Schematic profiles and sections.

on the talus slopes or cliff faces. In the stage of old age, pediments on both sides of a mountain range meet, opening broad passes (Figure 14.11b).

## DESERT LAKES

Although most surface water in deserts disappears into sediments, under certain conditions, permanent lakes form as dominant features of desert landscapes. We will discuss some modern desert lakes and then examine the evidence that the ancestors of some desert lakes were much larger than the lakes of today.

### Modern Desert Lakes

When a playa lake is replenished by stream flow faster than the water can drain away, either through surface streams or by infiltration, permanent desert lakes form. These lakes range in size from modest ponds to Utah's Great Salt Lake, whose surface area during the past century has ranged between 2430 and 6150 square kilometers.

The waters in almost all modern desert lakes are brackish, or saline, to varying degrees. The brackishness is largely a result of evaporation: as the lake water continually evaporates in the desert heat, ions, chiefly of potassium, sodium, calcium, and magnesium, dissolved from rocks, are left behind. In dry years, the salinity of the water in Great Salt Lake may

**14.12 Vast field of algal stromatolites,** built by algae in the shallow margins of Great Salt Lake, Utah, and exposed for this view because the water level dropped. (Courtesy Saunders Brine Shrimp Company, Utah, and Utah Geological and Mineral Survey)

reach as much as 270 parts per thousand — eight times more than the normal salinity of the ocean (35 parts per thousand).

Desert lakes are extremely sensitive to slight changes in climate, especially to changes in the amount of rainfall. Droughts may last several years; as the dry years continue, lake levels drop steadily, leaving an ever-widening belt of halite or other evaporate minerals and laminated algal deposits, named algal stromatolites (Figure 14.12) around the shores. Nonmarine evaporites, such as borax and gypsum, are mined in ancient lake beds in the Mojave Desert in California; potash is taken from the salt flats of Utah and the Dead Sea.

## Pluvial Lakes

That former desert lakes of great size existed previously in some desert basins is shown by deltas, beaches, "sea" cliffs, and other shoreline features now high above the water levels of modern desert lakes (Figure 14.13). These larger ancestors of desert lakes are called *pluvial lakes,* because they were fed by abundant rainfall in the past. Because in some places, old shorelines are cut on sediments deposited by glaciers (Chapter 13), many geologists believe that the maximum sizes of pluvial lakes coincided with periods of glaciation. Other geologists have argued that ice ages are accompanied by periods of great aridity in areas marginal to the glaciers.

In the western United States, spectacular remnants of ancient pluvial lakes have been found (Figure 14.14). Great Salt Lake and the Bonneville Salt Flats around it are the remains of Lake Bonneville, a giant pluvial lake which rivaled today's Lake Michigan in size. The water level of Lake Bonneville was as much as 300 meters above that of its shrunken descendant.

Probably the world's largest pluvial lake once spread across the lowlands that now surround the Caspian Sea, connecting it with the Aral Sea and the Black Sea. Archeologists have found some of the earliest traces of agriculture in this region, known popularly as the Fertile Crescent.

**14.13 Shorelines of Ancient Lake Bonneville,** Utah, appearing as terraces high on the sides of the mountain slopes that enclosed the basin. (American Airlines)

Neolithic man first tamed cattle in the Fertile Crescent, and the increase in grazing and browsing by large herds helped to denude the soil and promote erosion. Since then, the great lake has shrunk until all that remains is the Caspian Sea, and even that body of water continues tc dwindle.

In second place behind water as a geologic agent in deserts is the wind. The wind is an everyday factor in deserts, but the total work of the wind is not as great as the geologic work of occasional spurts of water.

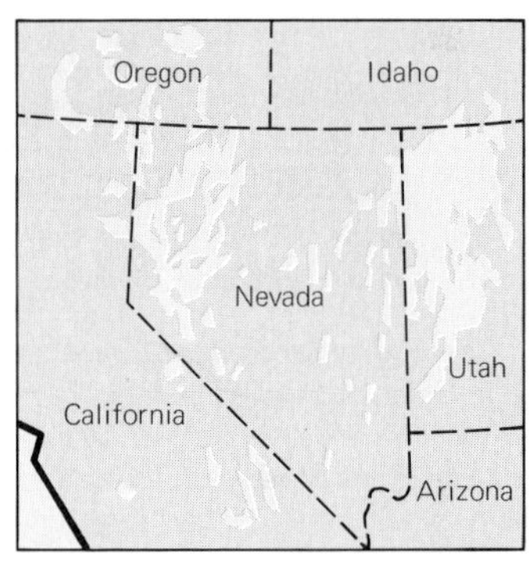


**14.14 Extents of pluvial lakes of Pleistocene age in western United States,** as reconstructed from studies of ancient lake deposits and shoreline features. (U. S. Geological Survey)

## THE GEOLOGIC WORK OF THE WIND

We discuss the geologic work of the wind under three main headings: (1) mechanics of wind transport and sediments deposited by the wind; (2) features formed by wind erosion of regolith; and (3) the features formed in rock as a result of abrasion by wind-transported particles.

### Mechanics of Wind Transport and Wind-Deposited Sediments

The mechanics of sediment transport and deposition by the wind compare closely with those of a stream of water. The wind can carry fine sediment as suspended load and can push sand along as bed load. The deposits of wind-suspended sediment consist largely of silt; the deposits of the wind's bed load consist of various kinds of sand dunes. Sediments transported and deposited by the wind are collectively designated as *eolian sediments*.

**The wind's suspended load** As with a turbulent stream of water, the speeds of the upward flows within a turbulent wind may exceed the settling speeds of silt particles and clay flakes smaller than 0.06 millimeters in diameter. As a result, the wind can suspend these particles. An initial disturbance, such as the impact of an animal's hoof or a powerful downdraft in the wind, may start the sediment particles moving.

**14.15 Leading edge of dust storm** arriving in Union County, New Mexico on 21 May 1937. Before this cloud arrived, only gentle breezes blew, but in less than 15 minutes, the winds reached hurricane force for 30 minutes as the front of the cloud passed. Curved dark indentations along margin of cloud are places where air from in front of the cloud is drawn in to mix with the dust. (Al Carter, U. S. Department of Agriculture)

*Dust storms* sweep finer material to great heights and over great distances (Figure 14.15). During the 1930s, dust storms sometimes dimmed the midday sun over cities of the eastern seaboard and swept dust from Colorado into the North Atlantic Ocean. Although the individual suspended particles are much lighter than sand particles driven along the ground, the total weight of fine particles entrained during a dust storm may be far greater than that of sand in a sandstorm. Winds during the dust-bowl years raised dust-filled clouds as much as three kilometers high, holding many thousands of tons per cubic kilometer. Little wonder that vast areas of the Great Plains were denuded and whole houses buried by settling dust. Only the termination of the sustained drought quenched the dust storms that threatened to convert wheatfields and grazing lands into a desert (Figure 14.16).

Evidence of enormous ancient dust storms has been left in the form of extensive deposits of fine-grained, wind-blown silt, named *loess* (see Figure 11.6a). Deposits of loess typically are nonstratified and consist chiefly of silt but also contain fine sand and clay. Loess varies in color from reddish-brown through yellow to gray. The minerals in loess consist mostly of quartz, feldspars, and carbonates. The grains are angular and are packed irregularly. This gives loess high porosity, usually 60 percent or more. Loess is an excellent moisture-retaining base for fertile loam. Most of the world's grain is cultivated where wild grasses once trapped windblown dust.

Deposits of loess are widespread. They form a blanket over the underlying soil to depths from a few meters to more than 30 meters.

**14.16 Abandoned farm,** a result of the drought and formation of the "Dust Bowl" in the 1930s. Near Guyman, Oklahoma. (B. C. McLean, U. S. Department of Agriculture)

Large areas of the central United States in the Mississippi and Missouri basins (including most of Illinois, Iowa, Nebraska, and large parts of South Dakota, Kansas, and Missouri) have been coated by loess deposits 1.5 to 24 meters thick. Much of this loess in the central United States was derived from dried-out, fine-grained outwash deposited by meltwater streams draining southward from the melting Pleistocene glaciers (Figure 14.17).

An elongated belt of loess reaches from western Europe through the Ukraine and Russian steppes into northern China. The world's thickest deposits cover thousands of square kilometers of Shensi Province in China to a depth of 60 meters or more. This thick loess is thought to have blown there from the Gobi Desert. Similarly, the Argentine deposits are believed to have come from the Patagonian Desert. Although many loess deposits can be traced to sources nearby, loess may come from distant sources. For example, the loess covering the plain of central Hungary has been derived from the Sahara. Study of the Hungarian loess prompted the concept of *high-altitude suspension* of windblown material.

According to this idea, *low-altitude suspension* involves only the lowest part of the atmosphere in a zone 3 to 5 kilometers thick. Most loess is transported by this means. By contrast, high-altitude suspension involves the upper atmosphere's jet streams, which are localized currents of air, having a tunnellike appearance, within which wind speeds blow at several hundreds of kilometers per hour. Such high-speed winds are capable of keeping fine sand in suspension. The distinctive feature of a high-altitude suspension is that once the material has been boosted high enough to be transported in the jet stream, very little of it may fall out while it is being transported hundreds or even thousands of kilometers.

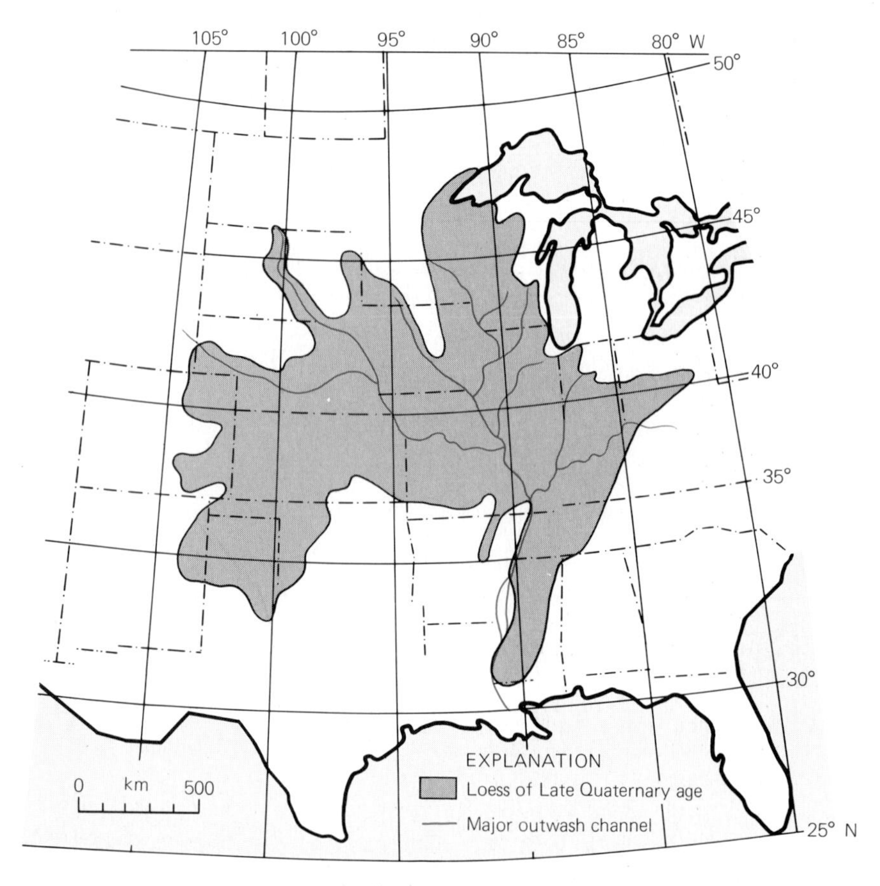


**14.17 Loess in relation to Pleistocene outwash channels** that drained south from melting continental glacier, shown on map of central United States. (After Robert V. Ruhe, 1965, Quaternary Paleopedology, p. 755–764 *in* Wright, H. E., Jr., and Frey, D. G., *eds.*, The Quaternary of the United States: Princeton, New Jersey, Princeton University Press, 922 p., Figure 2, p. 757)

Eventually, in a zone of downdraft, some or all of the material may return to the Earth's surface (Figure 14.18). Thus, the material from the Sahara was transported to Hungary without any fallout upon the intervening territory.

**14.18 Low-altitude suspension compared with high-altitude suspension,** schematic profile through lower part of Earth's atmosphere. Further explanation in text. (G. M. Friedman and J. E. Sanders, 1978, Principles of sedimentology: New York-Santa Barbara-Chichester-Brisbane-Toronto, John Wiley and Sons, 792 p., Figure 4-19, p. 103, based on L. Moldvay, 1961, on the laws governing sedimentation from eolian suspensions: Univ. Szeged, Acta Mineralogica-Petrographica, v. 14, p. 75–109)

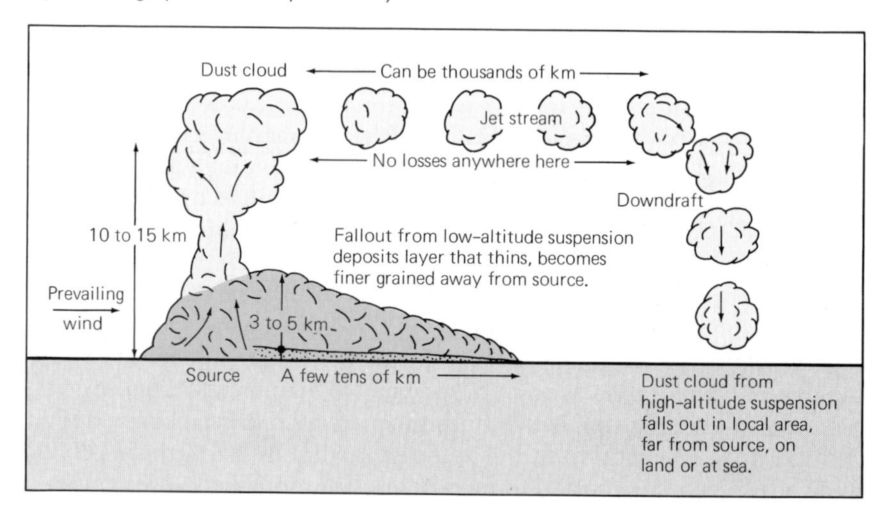

Similarly, on March 29, 1947, Hekla in Iceland erupted explosively, sending some tephra to heights of nine kilometers. This high-flying tephra cloud traveled southward 880 kilometers, changed course, and moved nearly 3000 kilometers northeastward. It landed in Finland 51 hours later. Its mean speed for the trip was 75 kilometers per hour.

**The wind's bed load** Most of the particles in the wind's bed load move by *saltation,* a series of elastic bounces and impacts that rarely carry them more than 10 centimeters above the ground (Figure 14.19). The maintenance of any saltation at all requires a wind of at least 16 kilometers per hour. Even much more powerful winds cannot cause saltating sand grains to bounce more than a few meters above the surface. Particles within this moving carpet of sand collide with one another, thus promoting rounding of the particles.

The wind is affected by the roughness of the surface over which it blows. A very thin layer of air, about $\frac{1}{30}$ the diameter of the mean height of the features on the surface, lies just above the surface. Within this thin layer, wind speed is zero. Thus, large rocks and boulders tend to break up the airflow and cause saltating sand to accumulate on the ground around them.

Large areas of shifting eolian sand, in various distinctive landforms, cover between one-quarter and one-third of all deserts. Virtually all active or moving eolian sand occurs not in isolated dunes but in vast, migrating *ergs,* or sand seas, greater than 125 square kilometers in area.

Despite the violence and seeming chaos of desert sandstorms, the wind does not drop its sand in random fashion. Rather, the sand gathers in patterns of surprising consistency.

Winds resemble streams in depositing their largest particles first as their speeds decrease. The dynamics of air flow and turbulence are extremely complicated, but clearly a rock, shrub, or other obstacle that causes deposition can begin an accumulation — a "sand shadow" — that encourages further growth. Passing air currents tend to eddy into this shadow and there slow down, adding to the deposit. This irregularity alters the windstream still further; it gives rise to repeated zones of

**14.19 Saltating sand particles being blown by a wind** from left to right. At tops of their trajectories, the wind drives the particles forward; gravity then pulls the particles back to the ground, where they strike other particles, spattering them upward and pushing them forward along the ground. (After R. A. Bagnold, 1941, The physics of blown sand and desert dunes: London, Metheun and Co., Ltd., 265 p., Figure 5, p. 17; Plate 2, opposite p. 34, Figure 10, 12, 36; and Figure 19, p. 62)

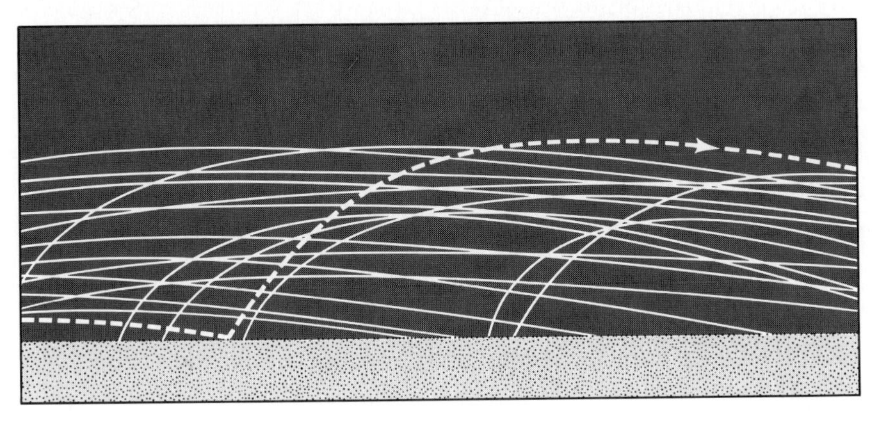

**14.20** Large dunes, covered by small-scale ripples (about 8 cm from crest to crest), Death Valley, California. (U. S. National Park Service)

deposition that range from ripples a few millimeters high to great piles as high as 200 meters (Figure 14.20). Such deposits soon become independent of the obstacle that caused them originally. A *sand dune* may be defined as a mobile heap of sand independent of either ground form or any fixed obstruction to the wind.

The longitudinal profiles through sand dunes are characteristic; the windward face is gentle and the lee slope is steeper; the maximum angle of repose is approximately 34 degrees. Dunes show many of the same patterns found among the bed forms on stream beds described in Chapter 12. Blown sand migrates up to the crest (whose height may depend on wind speed, size of sand particles, degree of sorting, and other factors) and falls over into the *slip face.* A net downwind transfer of sand takes place; hence, the dune moves bodily forward in the direction of the prevailing wind (see Figure 14.20). A migrating dune's advance (some travel as much as 50 meters per year) can be halted and the dune stabilized by the growth of hardy grasses whose windbreaking effect and roots hold the surface sand in place.

The common kinds of dunes are barchan dunes, longitudinal dunes, transverse dunes, and U-shaped (parabolic) dunes (Figure 14.21). Dunes with distinctive crescent shapes (in plan view) are *barchan dunes.* The horns of the crescent curve downwind from the main body. Barchans are most common in deserts with steady, moderate winds and limited sand cover. Crosswinds may cause the elongation of one horn, and in deeper sand barchans may coalesce into curved networks called *star dunes.*

In deserts where winds blow regularly enough to establish great cylindrical flow cells, *longitudinal dunes* are common. The long axes of such dunes are aligned parallel to the general direction of the wind. Called *seif dunes* by Arabs, longitudinal dunes reach lengths of almost 100 kilometers and heights of 90 meters or more. As the wind speed varies in the cylindrical flow cells, the cross currents build a string (or several strings) of shifting slip faces along the spine of a seif dune.

Dunes that are linear, usually short, and aligned at right angles to the prevailing wind are *transverse dunes.* This dune form is most often seen along coastlines where steady winds blow; if vegetation interrupts their formation, the shapes of transverse dunes may be irregular.

A dune which resembles an elongated, reversed barchan is a *U-shaped dune (parabolic dune),* with horns extending upwind instead of downwind. Parabolic dunes are sometimes found among beach dunes migrating inland, and may form as sand drifts behind a gap in an obstructing ridge. Wind funneled through such a gap spreads out and slows enough to drop the sand it carries (just as a stream may deposit its load if the channel volume is suddenly increased). This sand accumulates in a U-shaped, or parabolic, curve just beyond the gap. Later, a parabolic dune may detach itself from its site of formation and migrate independently.

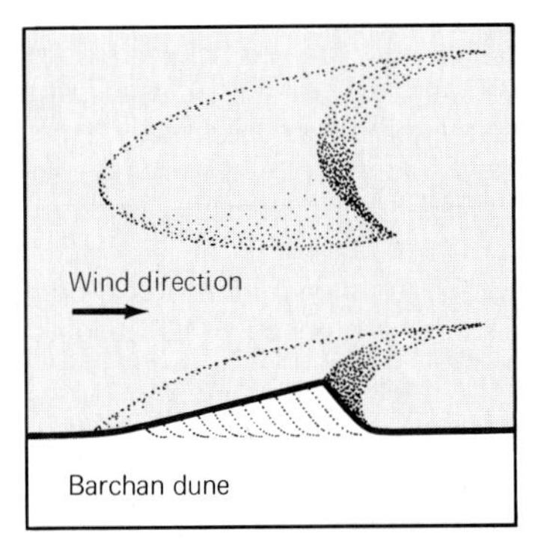


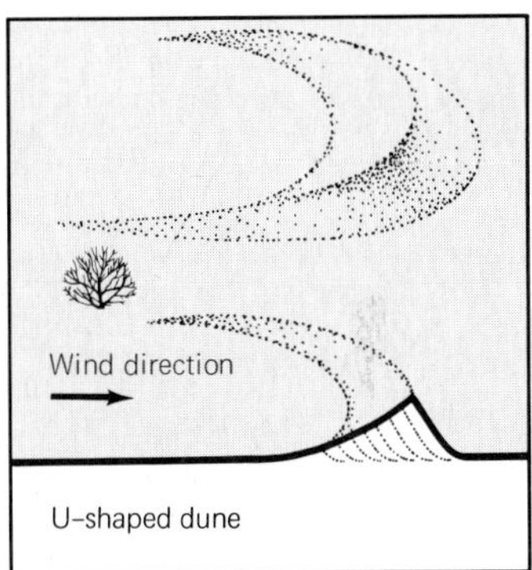


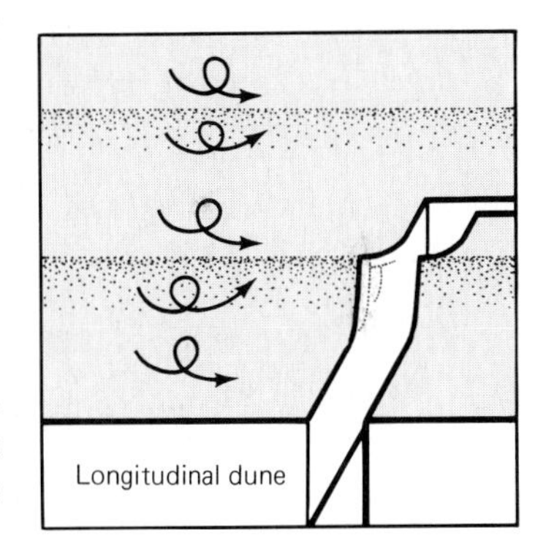


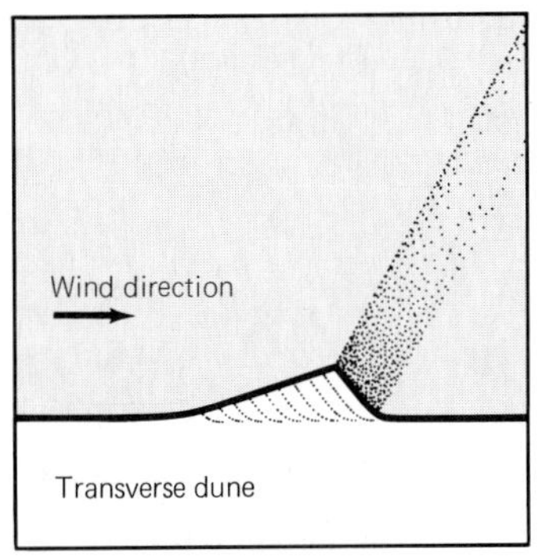


**14.21 Major kinds of dunes,** schematic block diagrams. Arrows indicate direction of wind. Further explanation in text. Dotted lines on vertical faces suggest cross strata formed by shifting of dunes.

## Wind Erosion of Loose Sediment Particles

Wind is capable of picking up and transporting loose sediment particles. The wind's effectiveness in picking up sediment from a particular area depends on speed, direction, and duration. Another factor is the amount of moisture it contains, for a wind accompanied by rainfall picks up little sediment. Finally, wind erosion is affected by the plant cover of an area.

**Deflation** The picking up and carrying away of dry, weathered, granular material, usually ranging in size downward from coarse sand to the finest dust is *deflation.* Where the surfaces of broad desert areas have been lowered uniformly, the effects of deflation are difficult to measure. Where deflation has been locally retarded by moisture, vegetation, or exposed bedrock, however, contrasts can be striking. Under certain conditions, deflation may lift away a meter of topsoil in a single year, especially when plowed soil has been dried during a drought. Many midwestern American farms were stripped bare of topsoil in the dust bowl years of the 1930s (see Figure 14.16).

When semiarid grassland becomes desert, the change is accompanied by the deflation of fine materials which were bound by the fine root systems of the grasses. This process is particularly harmful to cropland, because the first materials to blow away are those which alone could encourage new plant growth upon the return of the rains. Extensive

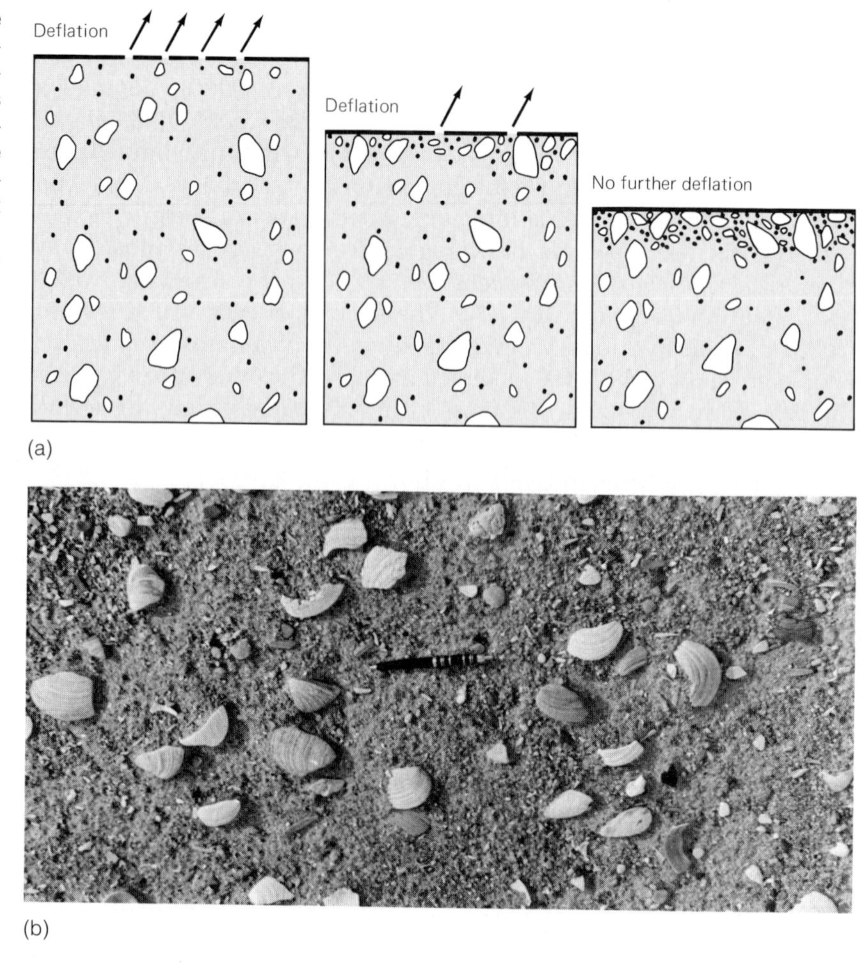


**14.22 Deflation armor.** (a) Successive stages in development, starting with pebbly sand or shelly sand, schematic profiles and sections. As the wind blows away the finer particles, the coarser materials are left behind. Eventually, these become packed together to form a continuous layer that protects the sediment below from further erosion. (b) Photograph of deflation lag formed mostly of broken shells on a beach. (J. E. Sanders)

deflation is halted only when the surface approaches the water table (where moisture keeps the sediment saturated) or when the remaining pebbles and larger stones (*lag gravel*) accumulate so thickly that they protect the sand and dust beneath. Such an accumulation is called *deflation armor* (Figure 14.22); when the stones are close and tightly fitted, it may be given the name of *desert pavement*. However, a desert pavement need not necessarily be a deflation armor. Sheets of running water can also erode away fine particles from a mixed sediment and leave behind a lag of pebbles and larger sizes.

### Features Formed in Rock by Abrasion

Rock masses are shaped, polished, and etched by the natural sandblasting of wind-transported sediment. Wind abrasion grinds away at isolated rock masses, sculpting slim-legged pedestals, hollows, arches, and other weird shapes characteristic of desert landscapes (Figure 14.23). *Ventifacts* (from the Latin for "made by wind") are pebbles and larger rocks with smooth facets abraded and polished by wind action. Most ventifacts display two or three facets, but changes in wind direction or shifting of the rock may cause an even-greater number of facets to appear.

(a)

(b)

(c)

**14.23 Features formed by wind abrasion.** (a) Pedestal, base sandblasted by saltating windblown particles. Death Valley, California. (American Airlines)
(b) Wind-formed arch in sandstone, Monument Valley, Arizona. (American Airlines)
(c) Ventifacts, each showing several facets abraded by windblown sand. (G. R. Roberts)

# CHAPTER REVIEW

1. A *desert* is a region characterized by low rainfall and evaporation that greatly exceeds precipitation. Most deserts receive less than 25 centimeters of rainfall per year. About one-third of all land area is desert.

2. According to their causes and locations, most modern deserts can be arranged in four groups: *subtropical, continental-interior, rainshadow,* and *coastal.*

3. Most true deserts are surrounded by semi-arid regions called *steppes* within which the yearly rainfall averages 25 to 50 centimeters.

4. Apart from lack of rainfall and great evaporation, the desert environment is characterized by a large temperature range and distinctive vegetation. The lack of surface and atmospheric moisture allows temperatures to

reach the highest levels on Earth and to fluctuate widely; clouds, which block solar radiation, do not form and there is not enough moisture to retain heat through the night and cold seasons. Sparse vegetation is adapted to make maximum use of scanty rainfall and groundwater.

5. Rain often comes to the desert in brief high-intensity cloudbursts, which may dump an entire year's precipitation in an hour. On gentle slopes, water from a cloudburst characteristically moves as *sheetwash*. Water running down steep slopes forms features named *arroyos, box canyons,* or *wadis*. Many desert regions display *interior drainage,* rivers that do not traverse the desert, but rather drain into lowlands or disappear within the desert through evaporation and infiltration.
6. Desert sediments, as well as desert bedrock, are fashioned into distinctive landforms, including *bolsons, bajadas, pediments,* and *playas*.
7. Although most surface water in deserts disappears into sediments, under certain conditions permanent lakes form as dominant features of the desert landscape. Geologic evidence indicates that ancient *pluvial lakes* were much larger than modern desert lakes.
8. The geologic work of the wind can be discussed under three main headings: (1) mechanics of wind transport, and sediments deposited by the wind; (2) features formed by wind erosion of the regolith; and (3) the features formed in rock as a result of abrasion by wind-transported particles.
9. Sediments transported and deposited by the wind are collectively designated as *eolian sediments*. These include the fine-grained suspended load deposited as wind-blown silt (*loess*), and bed-load material that is deposited in great bodies of sand which typically form dunes.
10. The common kinds of dunes are *barchan, longitudinal, transverse,* and *U-shaped* (also known as *parabolic*).
11. The picking up and transport of material by the wind is *deflation,* a process that can destroy plowed fields. Deflation stops when the water table is exposed or when a layer of coarse particles (a *deflation armor*) too coarse for the wind to transport forms by the loss of fine particles.
12. The process by which windblown sand and dust wear away rock surfaces is *wind abrasion*. *Ventifacts* are created by wind abrasion.

## QUESTIONS

1. Define *desert*. What proportion of the world's land area would be classified as desert?
2. List and describe the geologic circumstances associated with four kinds of modern deserts.
3. Why are deserts hot in the daytime yet cold at night?
4. What are *steppes?* How are steppes related to deserts?
5. Describe desert rainfall and its effects.
6. What is meant by *interior drainage?* Compare the interior drainage of a desert with the drainage in a region having karst morphology.
7. Define *bolson, bajada,* and *playa*. Summarize how each of these features behaves during a rainstorm and during a long dry spell.
8. Draw an annotated schematic profile showing a *pediment* and an *inselberg*.
9. Explain how the cycle of erosion is thought to operate in deserts. Compare the cycle of erosion in a desert region with that in a region where rainfall is plentiful.
10. What is a *pluvial lake?* Are pluvial lakes related to lakes in modern deserts? Explain.
11. Define *eolian sediments*. How does eolian sediment compare with alluvium?
12. What is *loess?* Compare loess deposited from a low-altitude suspension with loess deposited from a high-altitude suspension.
13. Explain the mechanisms by which the wind transports sand.
14. List the major kinds of dunes and describe each by means of an annotated sketch showing wind direction.
15. What is *wind abrasion?* How are *ventifacts* related to wind abrasion?

## RECOMMENDED READING

Abbey, Edward, 1968, *Desert Solitaire. A Season in the Wilderness:* New York, McGraw-Hill Book Co., Inc., 269 p.

Bagnold, R. A., 1941 (reprinted 1954), *The Physics of Blown Sand and Desert Dunes:* London, Methuen and Co., Ltd., 265 p.

Coffey, Marilyn, 1978, The Dust Storms: *Natural History,* v. 87, no. 2, p. 72–83.

Garbell, M. A., 1963, The Sea that Spills into a Desert: *Scientific American,* v. 209, no. 2, p. 94–100.

Leighton, M. M., and Willman, H. B., 1950, Loess Formations of the Mississippi Valley: *Journal of Geology,* v. 58, no. 6, p. 599–623.

Mabbutt, J. A., 1977 *Desert Landforms:* Cambridge, Massachusetts, The MIT Press, 340 p.

McKee, E. D., *ed.*, 1979, A Study of Global Sand Seas: *U.S. Geological Survey, Professional Paper 1052,* 429 p.

Norris, R. M., and Norris, K. S., 1961, Algodones Dunes of Southeastern California: *Geological Society of America, Bulletin,* v. 72, no. 4, p. 605–620.

Prospero, J. M., 1979, Dust from the Sahara: *Natural History,* v. 88, no. 5, p. 54–61.

Solbrig, O. T., and Orians, G. H., 1977, The Adaptive Characteristics of Desert Plants: *American Scientist,* v. 65, no. 4, p. 412–421.

# CHAPTER 15 OCEANS AND SHORELINES



PLATE 15

Numerous coastal features, Tomales Bay, California, where San Andreas fault goes out to sea. Dune-covered spit (lower left) contrasts with eroding bedrock cliffs facing open Pacific Ocean on headland at lower right. (Pacific Aerial Survey)

THE EARTH IS NOT KNOWN AS THE "WATER PLANET" for nothing. Among the planets of our Solar System the Earth is unique in possessing a continuous body of salt water, the oceans. The oceans cover almost 71 percent of the Earth's surface. The distribution of lands and seas over the Earth is asymmetric; land is concentrated in the Northern Hemisphere (Figure 15.1). Ocean water is distributed in four deep-ocean basins: (1) Pacific (as large as all the others put together), (2) Atlantic, (3) Indian, and (4) Arctic. The southern ocean, known as the Antarctic Ocean, does not occupy a separate basin, but a depression that is formed by southward extensions of the Pacific, Atlantic, and Indian ocean basins. Smaller, partially isolated bodies of ocean water are *seas*. A few examples are the Caribbean Sea, the Mediterranean Sea, and the South China Sea. Still-smaller arms of the oceans are gulfs, bays, channels, and straits.

The total surface area of the oceans is about 358 million square kilometers. As we noted previously, the average depth of the oceans is about 4 kilometers. The total volume of seawater is 1350 cubic kilometers — 18 times the volume of all the land lying above sea level.

The oceans affect every aspect of human life. They are a vast part of the biosphere and are the habitat for hordes of organisms. The interactions between ocean water and the atmosphere affect the weather; both are responsive to the forces that determine the Earth's climate. The oceans are the ultimate reservoir of the water that circulates in the hydrologic cycle.

**15.1 Uneven distribution of lands and seas** shown on circular map projections of the Earth and as pie diagrams. (a) Pacific and Indian Oceans dominate the water hemisphere. (b) The small proportion of land shown as gray area above water (blue). (c) Eurasia, Africa, and North America dominate the land hemisphere. (d) The proportion of land is about half the total area.

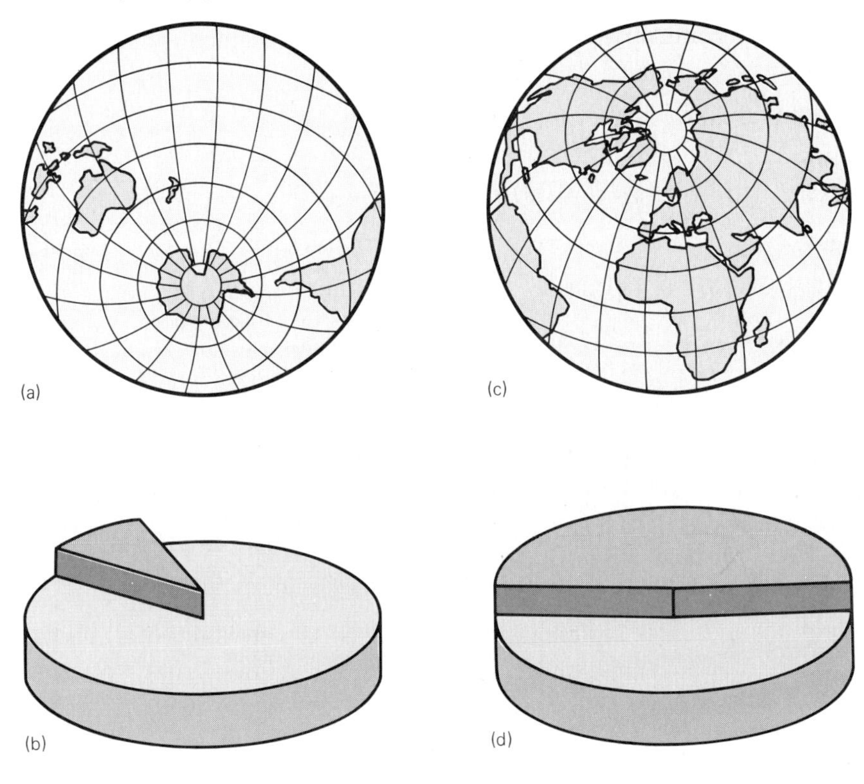

## WAVES, TIDES, AND OTHER COASTAL PROCESSES

Our discussion of coastal processes includes wave action; short-term changes and long-term changes of water level; and various other coastal processes, such as currents at mouths of rivers, activities of coastal organisms, action of shore ice, and chemical precipitation in marginal evaporitic flats.

### Wave Action

In order to understand how waves affect shorelines, we need to answer some questions: How do waves originate? How far from shore do they influence the bottom? What bottom features are caused by waves?

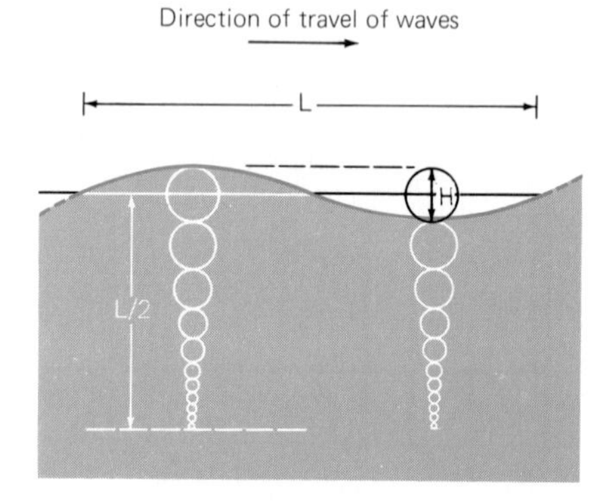


**15.2 Wave definitions** shown on schematic profile with large vertical exaggeration. See also Table 15.1.

**Wave mechanics** Simple waves that cross the water surface without being influenced by the bottom are called *deep-water waves.* If waves of only one size are present, their profiles will resemble sine curves (Figure 15.2). The definitions and standard letter symbols of various characteristics of waves of all kinds are shown in Table 15.1.

The separate droplets of water affected by a deep-water wave do not move forward, but merely rotate in a circle that lies in a vertical plane whose edge is parallel to the direction the waves are traveling. After each complete turn the droplets return almost to their original locations The entire mass of water does not move forward, but rather only the *wave form* passes along the water surface. As we will see in the next section, most waves are generated by the wind, and wind also pushes water along as surface currents.

At the surface, the diameters of the circular orbits are equal to wave height, $H$. As long as the orbits are circular, the waves are *deep-water waves,* that is, waves traveling in water deeper than half their wavelengths. In going downward from the surface, the diameters of the orbits

**TABLE 15.1**
TERMS AND SYMBOLS USED IN DESCRIBING WAVES

| Applications | Term | Definition | Symbol |
|---|---|---|---|
| Applicable to all waves | Wavelength | Horizontal distance between successive wave crests or between successive wave troughs | $L$ |
| | Wave height | Vertical distance from bottom of a trough to the top of a crest | $H$ |
| | Wave period | Time required for a wave to advance a distance equal to one wavelength | $T$ |
| | Wave celerity (or speed; incorrectly referred to as "velocity") | Distance traveled by a wave in a unit of time | $c$ |
| Special terms for water waves | Depth | Vertical distance from a water surface lacking waves to the bottom | $h$ |
| | Wave steepness | Ratio of wave height to wavelength | $H/L$ |

**15.3 Storm at sea in North Atlantic,** viewed from Apollo 7 spacecraft. Because the winds blow around the center of the storm and the diameter of this storm is approximately 200 km, the maximum fetch for winds from any direction within it is one quarter of the circumference, or $\frac{\pi \cdot 200}{4} = 157$ kilometers. (National Aeronautics and Space Administration)

decrease rapidly. Therefore, where water is deeper than half the wavelength ($L/2$), the waves do not disturb the bottom. At a depth of $L/2$, the circular motion is negligible.

**Sizes and origins of waves** Most water waves result from wind action. A great storm crossing the open sea churns the water almost as if it were a gigantic eggbeater with a diameter of several hundred kilometers (Figure 15.3). A wind can transfer its kinetic energy to the water surface and thus create waves ranging in size from tiny ripples to mountainous seas. How this energy is transferred is not fully understood, but it is thought to involve fluctuations of friction and pressure.

Three important factors influence the origin of wind-generated waves: (1) wind speed, (2) wind duration, and (3) length of the stretch of open water across which the wind from a particular direction can flow, known as *fetch*. The periods of waves generated directly by winds range from fractions of a second to about 10 seconds.

Winds not only generate storm waves, but given sufficient time and fetch, they also prevent storm waves from growing beyond a maximum size by blowing their tops off. Waves being actively blown by the wind are steep and irregular; they are named *sea waves*, or just *seas*. In the open sea, the sizes of sea waves depend on the speed and duration of the wind. In the open sea, the periods of sea waves range from about three to seven seconds; periods increase with the speed of the wind. Wind-generated waves radiate outward in all directions from storm centers. In doing so, the waves mutually interact, with the result that some waves combine and grow larger, whereas others conflict and become smaller or obliterate one another altogether. Once these waves have traveled beyond the places where the wind actively blows the water, they become transformed into longer, lower, and more regular waves than sea waves. Such transformed sea waves are *swells*, which transfer energy from a storm at sea to distant shores. The periods of most swells range between 6 and 12 seconds, but some are as large as 20 or 22 seconds.

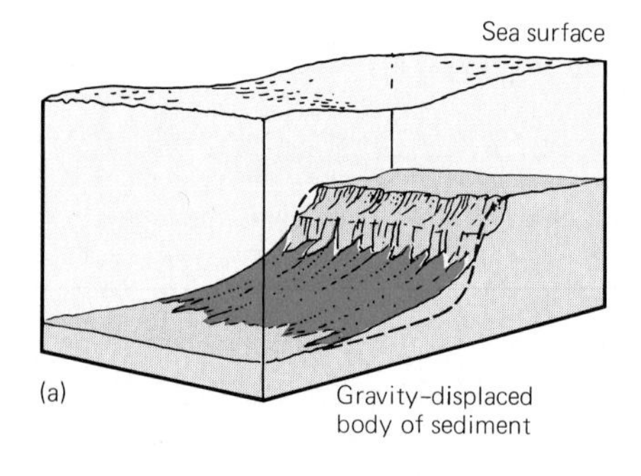


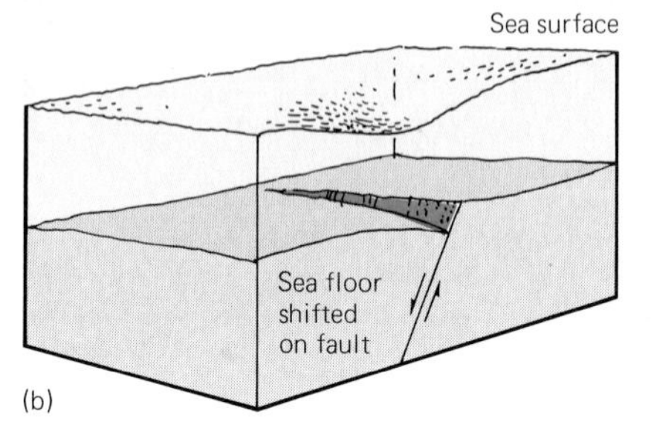


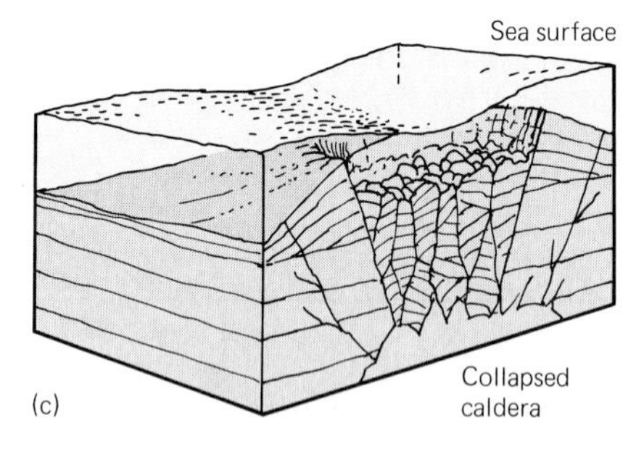


**15.4 Abrupt displacement of the sea floor generates tsunami,** schematic block diagrams. (a) Failure of underwater slope and sudden shifting of bottom material. (b) Fault. (c) Collapse of a caldera.

Some large waves are not generated by the wind. Tsunami are the result of a sudden displacement of the sea floor. This type of disturbance produces a series of long, low waves having periods of 12 to 15 minutes, wavelengths of 100 to 200 kilometers, and wave heights in the open sea of only a meter or so. Because of their enormous wavelengths, tsunami are scarcely visible in the open ocean and are never deep-water waves. They are a special kind of *very-shallow-water wave,* that is, a wave in which the wavelength is more than 20 times the depth and whose speed is a function of water depth. The speeds of tsunami in the Pacific Ocean range from 720 to 800 kilometers per hour.

The kinds of displacements of the bottom capable of generating tsunami are: (1) shifting of a body of sediment, (2) faulting, and (3) collapse of a volcanic caldera (Figure 15.4). Nearly all tsunami originate in zones of active earthquakes beneath deep-ocean trenches. The close relationship between earthquakes and tsunami is the basis of a reliable tsunami-warning system (see Chapter 17). After the large Chilean earthquake in 1960, a tsunami was predicted for Hilo, Hawaii, 15 hours after the earthquake. The tsunami arrived at the predicted time.

## Effects of Waves Approaching the Shore

As deep-water waves approach the shore, they eventually reach a depth equal to half their wavelengths, where they begin to "feel bottom." While they are still in deep water, the speeds of the waves are governed by their wavelengths: the longer the waves, the faster they travel. As waves begin to interact with the bottom, three things happen: (1) the waves are subject to shoaling transformations; (2) the oscillating bottom water transports sediment; and (3) the directions of the wave crests are changed so that the waves tend to become parallel to the shore, a process known as wave refraction.

**Shoaling transformations** As soon as the waves begin to drag bottom, they start a series of systematic changes known as *shoaling transformations.* These changes are scarcely noticeable at first, but by the time the depth is about $\frac{1}{8}$ of the wavelengths ($L/8$), the shapes of the waves and the kinds of motion they impart to the water change significantly. The symmetrical crests and troughs of deep-water waves become narrow, steep crests and wide, flat troughs. The water particles now travel in ellipses instead of circles (Figure 15.5). As water depths decrease, the ellipses become flatter and flatter. Over impermeable bottoms, the ellipses eventually give way to a simple back-and-forth oscillation parallel to the bottom. If the steepness of the waves (height/length ratio) is less than 0.039, then a depth of $L/20$ will be reached before the waves break. At that depth the waves become very-shallow-water waves, whose speeds are governed by the square root of the Earth's gravitational acceleration times the depth of water. Waves traveling between water depth of $L/8$ and $L/20$ are *shallow-water waves.*

**Sediment transport by shoaling waves** Where shallow-water waves travel over permeable bottom materials, such as cohesionless sand or gravel, the elliptical-oscillatory motions may be intense enough to transport sediment. The water motions beneath shallow-water waves are both cyclical and asymmetric. The cycle can be visualized in four parts: (1)

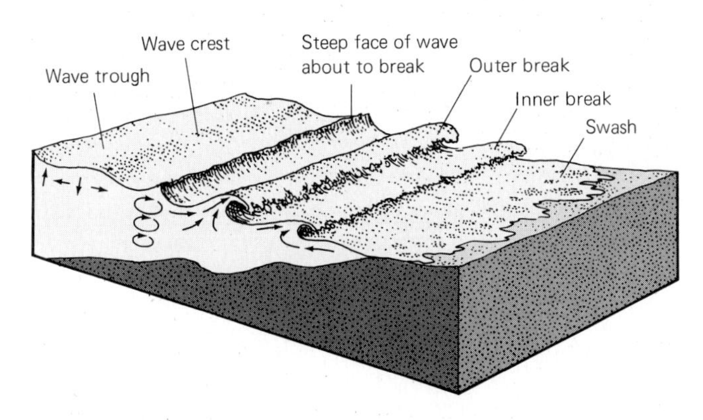


**15.5 Shoaling transformations and breaking waves,** schematic block diagram near a shoreline. Low-rounded deep-water waves (at left; small arrows indicate direction of movement of the water as the wave form passes) become steep-crested shallow-water waves which eventually break as they approach shore. Orbits of deep-water waves are circles (not shown); orbits of shallow-water waves are ellipses. Alternating surges of bottom water, toward land beneath wave crests and away from the land beneath wave troughs, transport sediment and create small-scale ripples in sand. The long axes of the ripples are at right angles to the direction of travel of the water-surface waves.

upward as wave crests approach, (2) landward beneath the crests, (3) downward after the crests have passed, and (4) seaward beneath the troughs (see Figure 15.5). The upward pulse of bottom water before the landward surge beneath wave crests creates a pressure burst that affects both the bottom water and the interstitial water in the bottom sediments. As a result, the sediment commonly appears to "explode" upward and is then driven toward shore. This upward pulse can be compared with throwing a tennis ball up into the air before serving it. By contrast, just before the seaward surge beneath the troughs, the bottom water is pushed downward. This tends to drive the particles against the bottom just before the water surges seaward. Consequently, many particles that surge toward shore beneath the wave crests are deposited and do not move outward with the seaward surge beneath the wave troughs. The result of many such cycles is to drive toward shore most of the large particles that can be moved.

**Wave refraction** Starting at a depth of $L/2$, a wave that approaches a coast straight on begins to "feel bottom" simultaneously along its entire crest. If a wave approaches a coast obliquely, only a small segment feels bottom at any one time. Vertically above the point where each segment starts to feel bottom, shoaling transformations begin. Wavelength decreases, height increases, and speed is reduced. As a result, the crestlines of obliquely approaching waves swing around and tend to become parallel to shore. The process by which the direction of a series of waves, moving in shallow water at an angle to the shoreline, is changed so that waves tend to become parallel to the shore is known as *wave refraction* (Figure 15.6)

**15.6 Refracting very-shallow-water waves** displayed in oblique view from an airplane late in the day with the Sun at a low angle. Democrat Point, SW end of Fire Island, Long Island, New York. (Bruce Caplan)

**Breakers, surf, and longshore drift** As waves move into shallow water, they undergo shoaling transformations that affect their shapes. At a depth equal to $2H$ the wave profile becomes very peaked and asymmetrical; its front is steep (see Figure 15.5) and its seaward side slopes more gently. When depth equals $1.28H$, the wave form is converted into a *breaker,* a wave that is collapsing. Beneath a breaker, the motion of the water is very turbulent and is predominantly upward. As each wave crest collapses, water from both adjacent troughs flows toward the *breaker zone,* the zone where waves collapse. The water from many breaking waves flowing toward the lines of breakers builds a bar beneath each. Under some wave conditions, such breaker bars migrate landward. Under other wave conditions, the breaker bars migrate seaward.

**15.7 Breaking wave and smaller waves formed from it,** Pacific coast of central California. In foreground, seaward-flowing backwash is colliding with arriving wave of translation (the surf). (J. E. Sanders, 1960)

(a)

**15.8 Tidal range of about 15 meters.** at Parrsboro, Nova Scotia, in Minas Basin, an embayment at the NE end of the Bay of Fundy. (a) At high tide, Parrsboro Harbor is full of water; ocean-going freighters can navigate alongside the dock. (b) At low tide, the only water in the harbor is a small stream that flows between the dock and the breakwater and lighthouse (left). When the tide is out and it is resting on special timbers placed beside the dock, an ocean-going ship experiences a few hours' worth of "free" drydock. (Stan Frank and J. E. Sanders, 1966)

Shoreward of each breaker bar is a trough where the water deepens. When breakers surge into the deeper water of this trough, they are reorganized into smaller waves of oscillation (Figure 15.7). Where the smaller waves collapse, a line of secondary breakers forms and under it, another breaker bar. When the water becomes so shallow that no new waves of oscillation can form shoreward of the breaker line, the breakers are transformed into *waves of translation,* waves that displace the water masses within the moving crests. Waves of translation landward of the breakers and seaward of the backwash constitute the *surf.* This mass of water surges toward the land and flows up the gently sloping shore as *swash,* returning as *backwash.*

## Changes of Water Level

Because the level of the water serves as a base level for streams (Chapter 12) and also is the level at which so many activities are concentrated, it is obvious that any change in water level is an important event. Some changes of water level are tidal oscillations that take place once or twice every day (Figure 15.8). Other changes of water level result from storms and last for only a day or two. Still other changes of water level are long-term events related to local tectonic elevation or subsidence of the land or to various factors that control the worldwide supply of sea water or the volumes of the ocean basins.

**Mechanics of tides and tidal oscillation** The *tide* is a rhythmic rise and fall of water level. The vertical difference in altitude of the water surface between the level of a high water and that of the low water immediately preceding or following is the *tidal amplitude* (or, *tidal range*) of that tide.

(b)

The astronomic aspects of tides include the gravitational pulls on the Earth and the Earth's water by the Moon, the Sun, and the other planets of the Solar System; the orbiting of the Earth by the Moon; and the spinning of the Earth on its polar axis.

The phenomenon of lunar tides results from the effects of the Moon's gravitational pull and the motions of the Earth and the Moon. The gravitational pull of the Moon on the Earth bulges out the Earth's ocean water (Figure 15.9). The chief factor in the lunar tide is the distance between the Earth and the Moon. The distance changes with the Moon's elliptical movement around the Earth. The actual tide at any particular location is also a function of the Earth's rotation.

The Sun and the other planets also exert gravitational pulls on the Earth. In comparison to the Moon's pull, their effect is small, but they do exist. The tidal bulge caused by the Sun's pull accounts for the tides in Tahiti, an island located where the water within the lunar tidal bulge coincides with the shape of the solid Earth. On Tahiti, the tide rises when the Sun comes up and drops when the Sun sets.

**15.9 Bulges of water that form the tides,** schematic profile through Earth and Moon. Moon's gravitational attraction bulges out the Earth's water on side toward the Moon. Movement of center of Earth around the Earth-Moon barycenter (not shown) creates a centrifugal force away from the Moon that creates a second bulge on the side of the Earth away from the Moon. The effect of this barycenter-related centrifugal force (not to be confused with the centrifugal force of the Earth's spinning on its polar axis) creates a bulge that is smaller than the bulge nearest the Moon.

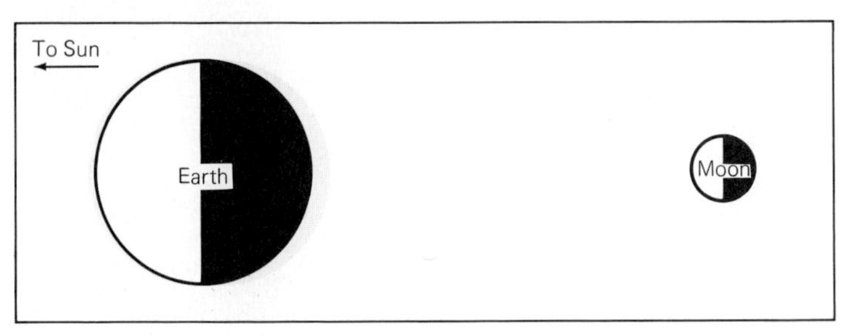

The solar tide regularly adds to or subtracts from the lunar tide, depending on where the Moon is with respect to the Sun. At full Moon and new Moon, the Sun, the Earth, and the Moon lie along a line. This alignment causes *spring tides,* which are tidal amplitudes larger than normal (Figure 15.10). When the Moon is in its quarter phases, the lunar tide and the solar tide tend to cancel each other. Thus, the tidal range is smaller than normal, creating *neap tides.* Figure 15.11 shows a simplified representation of the Earth, the Moon, and the Sun in relation to spring tides and neap tides.

**15.10 Monthly variation in levels of tide,** New York harbor, September 1950 and phases of the Moon. Spring tides are at New Moon or Full Moon and neap tides, at quarter phases. The greatest spring tides are recorded when New Moon and the Moon's closest approach to the Earth coincide. On this graph, the maximum spring tides coincide with Full Moon. This happened because the Moon's closest approach to the Earth came at Full Moon. In the panel at left, a line has been drawn through the center of the Earth, the Earth-Moon barycenter, and the center of the Moon to emphasize the orbit of the center of the Earth around the Earth-Moon barycenter. Dots mark daily advances of the Earth-Moon barycenter in the elliptical orbit of the Earth-Moon pair around the Solar-System barycenter. (G. M. Friedman and J. E. Sanders, 1978, Principles of Sedimentology: New York-Santa Barbara-Chichester-Brisbane-Toronto, John Wiley and Sons, 792 p., Figure E-2, p. 539)

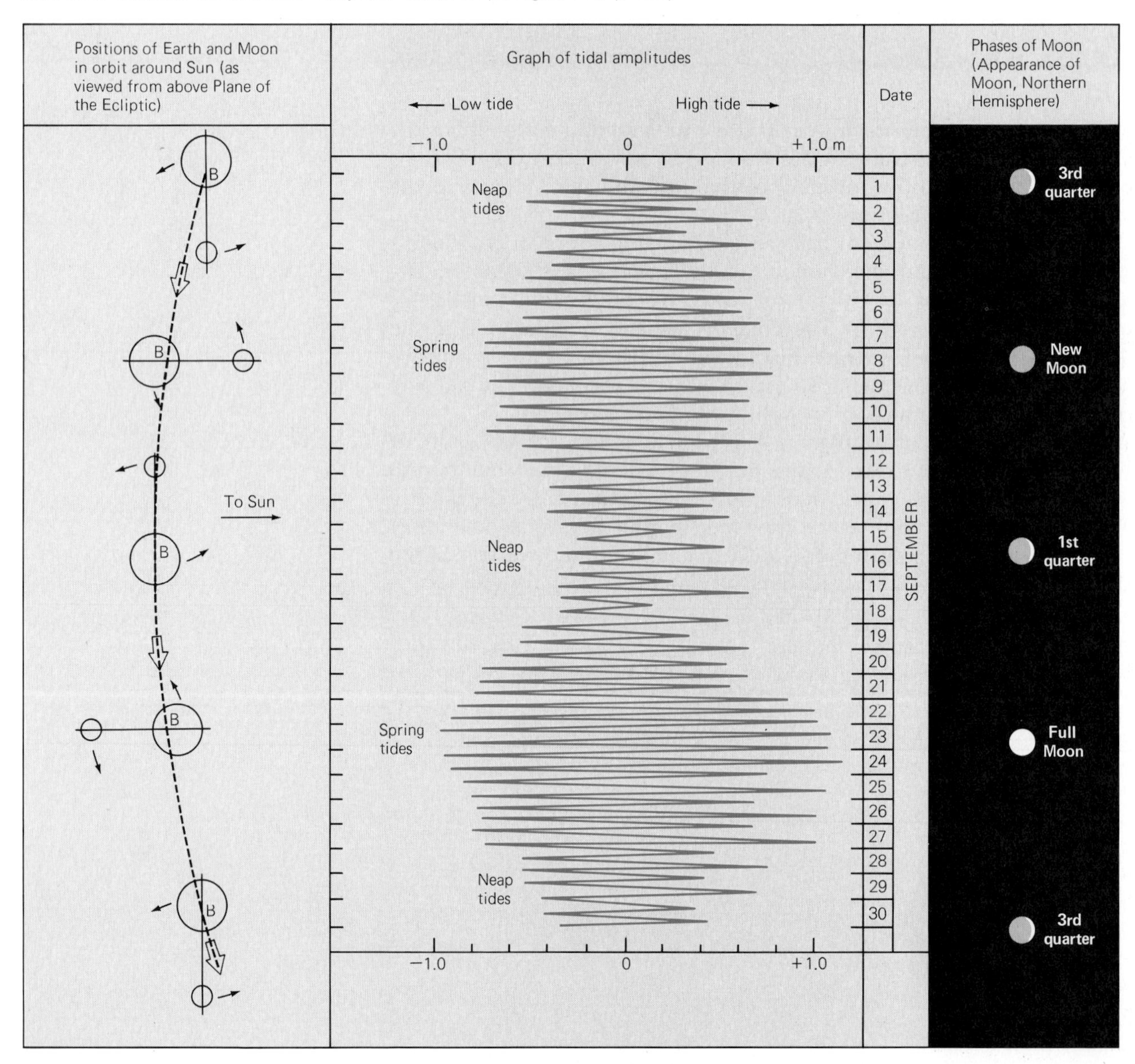

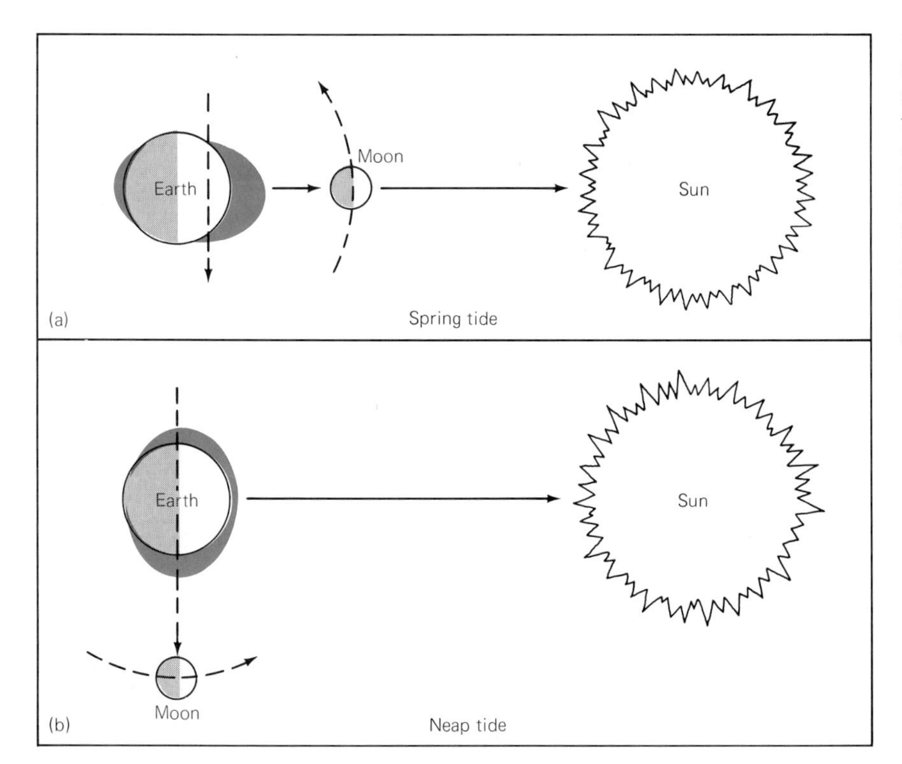


**15.11 Alignment of Earth, Moon, and Sun and corresponding tidal phases,** schematic views from above the Plane of the Ecliptic. (a) Earth, Moon, and Sun are aligned at spring tides, either with the Moon on the side of the Earth toward the Sun (New Moon) as shown here, or on the side of the Earth opposite the Sun (Full Moon, not shown). (b) When the Earth, Moon, and Sun are aligned at right angles (Moon in quarter phase), the tidal ranges are least (neap tides).

As the tide rises and falls, it creates horizontal tidal currents. These range from barely perceptible currents to rushing streams that make navigation difficult or dangerous. Tides flow into and out of San Francisco Bay at speeds of about 8 kilometers per hour and through the entrance to the Minas Basin, Nova Scotia, at 18 kilometers per hour. A rising tide is known as a *flood tide;* associated currents are flood-tidal currents. A falling tide is an *ebb tide;* currents related to a falling tide are ebb-tidal currents. In between are times of slack water, named *high-water* or *flood slack* and *low-water* or *ebb slack,* respectively.

**Deleveling** The term *deleveling* refers to long-term changes of the relative position of a water surface against the land. These changes can come about either because the land shifts up or down, the water level changes for some reason, or both the land and the water shift. A relative rise of water level against the land is called *submergence,* and a relative drop of water level against the land is *emergence.*

## Other Coastal Processes

Other important coastal processes that affect the geologic features formed within the water include currents at the mouths of rivers, diverse activities of coastal organisms, the action of shore ice, and chemical precipitation of evaporites in marginal flats.

**Currents at mouths of rivers** At the mouth of a river, various currents are established that depend on the relative densities of the river water and the water into which it flows. Where a river enters the sea, whose density is 1.025 grams per cubic centimeter, the river's water-plus-

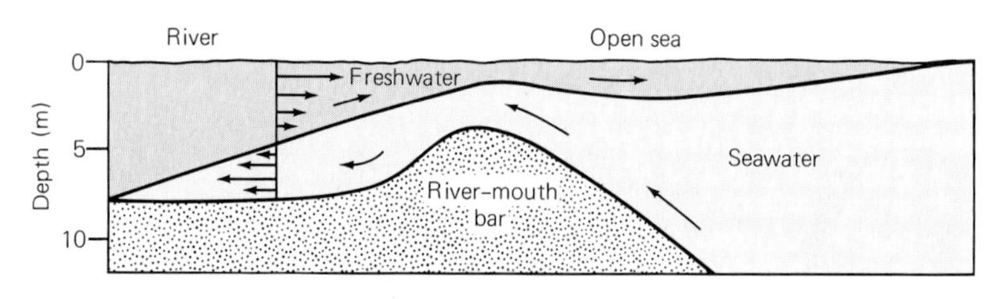


**15.12 Currents at the mouth of a river entering the sea,** schematic profile and section in vertical plane along the channel. Downriver transport of bedload sediment stops where the fresh water meets the tapering edge of the saltwater wedge and thus ceases to be in contact with the bottom. During a flood, the extra-thick column of fresh water drives out the saltwater wedge and enables bedload sediment to be transported out of the river.

sediment mixture spreads out over the seawater because the density of the river mixture is almost never greater than the density of seawater. This spreading of the plume of fresh water creates two wedges, an upper wedge of fresh water that tapers seaward and a lower wedge of salt water that tapers landward (Figure 15.12). Within the wedge of fresh water, a current flows outward. Within the wedge of salt water, a current flows inward, or landward. Along the boundary between the two wedges, complex mixing takes place as internal waves form and break.

**Activities of coastal organisms** Coastal organisms, including many kinds of invertebrate animals and various plants, are capable of leaving distinct impacts on shores. The activities of organisms range from boring and excavating bedrock to cause *bioerosion* of bedrock coasts; to the sediment trapping of saltwater grasses, algal mats, and mangrove roots; to the burrowing by crabs, worms, and crustaceans; and to the massive rock-hard reefs built by organisms, such as coralline algae and reef-building corals, which secrete calcium carbonate (Figure 15.13).

**Action of shore ice** Along coasts subjected to wintertime freezing, slabs of ice may pile up on the shore. The ice may incorporate sediment within itself and transport this sediment away from the shore. When the ice eventually melts, the sediment is dropped onto the bottom.

Wind blowing against great slabs of floes of ice may drive the floes ashore with considerable force. The edge of such an ice floe acts like a giant bulldozer and heaps up ice-push ridges.

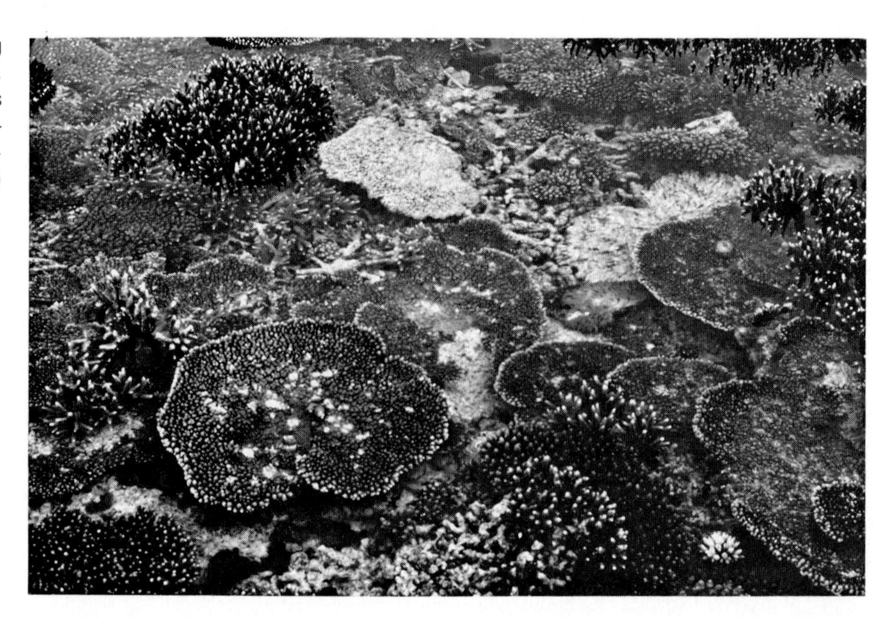

**15.13 Coral-reef community** consisting of crowded-together colonies of corals. Diameter of coral clump in upper left is about 1 meter. Great Barrier Reef, Australia. (G. M. Friedman, Rensselaer Polytechnic Institute)

**Chemical precipitation in marginal evaporitic flats** One of the most significant discoveries about coastal processes made in recent years concerns the chemical reactions that take place in porous, permeable sediments along shores in hot, dry climates where evaporation greatly exceeds precipitation. In such places, the salinity of the interstitital water becomes greatly elevated to the point where various evaporite minerals are precipitated. These include halite, anhydrite, and dolomite. Such marginal evaporitic flats have been named *sabkhas,* after the Arab usage in the Middle East. Coastal sabkhas are widespread features around desert shores of both lakes and the sea.

## MAJOR COASTAL FEATURES

The major coastal features we describe are those made on bedrock coasts, those formed where waves predominate (beaches and related bodies of coastal sediment), those formed where sediment is being actively supplied by rivers or by dry offshore winds, and those built by tropical reefs.

### Bedrock Coasts

Where bedrock and the sea are in contact, the dominant geologic process is erosion. The effects of wave action and abundant salt-water spray are added to the usual activities of weathering and downslope gravity-transport processes. By removing fallen products transported by gravity, waves maintain steep slopes and thus encourage further instability of slopes.

The particular effects of the waves involve the impact of water, compression of air in joints as water crashes into the cliffs, and impacts of hurled rocks. Storm waves can drive boulders toward shore with such great intensity that the boulders can richochet off the bottom (or off another boulder) and fly out of the water as if they were mortar shells. One large boulder, weighing about 60 kilograms, dropped through the roof of the lighthouse keeper's cottage on Tillamook Rock, off the Oregon coast a few kilometers south of the mouth of the Columbia River. How the boulder was propelled is not known, but the result was astonishing. The house was built on top of a cliff about 40 meters above sea level!

Erosion of bedrock by waves creates complex landforms (Figure 15.14).

**15.14 Stack, sea arch, and cliff,** coastal landforms eroded in bedrock by waves. New Brunswick coast of Canada. (Mark R. Chartrand, III)

**15.15 Effects of bioerosion,** limestone along north coast of Jamaica. Both the pitted and jagged surface at upper right and the flat terrace being washed by the breaking wave at center are the work of organisms. (J. E. Sanders)

Ultimately, these are destroyed and the result is a wave-cut bench. In tropical regions, where limestone coasts are common, various organisms are effective agents of erosion (Figure 15.15).

## Wave-Dominated Coasts: Beaches, Spits, Barriers, and Beach Plains

Where any kind of cohesionless sediments are available at the shore, wave action arranges and rearranges the sediments into the characteristic forms of beaches, spits, and barriers.

**Beaches** A *beach* is a body of nonconsolidated sand-size or coarser sediment along the shore of a body of water (lake, bay, sea, or ocean). The sediment forming beaches can consist of nearly any kind of particle, but most beaches consist of sand-size quartz (Figure 15.16). The inner limit of a beach is a boundary on which all agree; it is the upper limit of wave action or the place where the material changes (from beach sediment to bedrock, for example). The outer limit of a beach has been variously defined. Outer boundaries that have been proposed include the low-water line, seaward limit of wave action, and the outermost line of breakers. Because we think the distinctive features of beaches result from the effects of breaking waves, we prefer the boundary based on the outermost breakers. In the open sea this boundary lies at a depth that changes with the sizes of the waves, and during great storms may extend down to 10 meters or so.

(a)

(b)

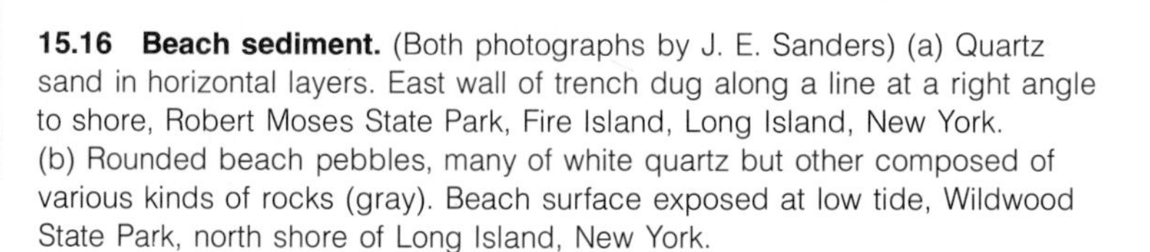

**15.16 Beach sediment.** (Both photographs by J. E. Sanders) (a) Quartz sand in horizontal layers. East wall of trench dug along a line at a right angle to shore, Robert Moses State Park, Fire Island, Long Island, New York. (b) Rounded beach pebbles, many of white quartz but other composed of various kinds of rocks (gray). Beach surface exposed at low tide, Wildwood State Park, north shore of Long Island, New York.

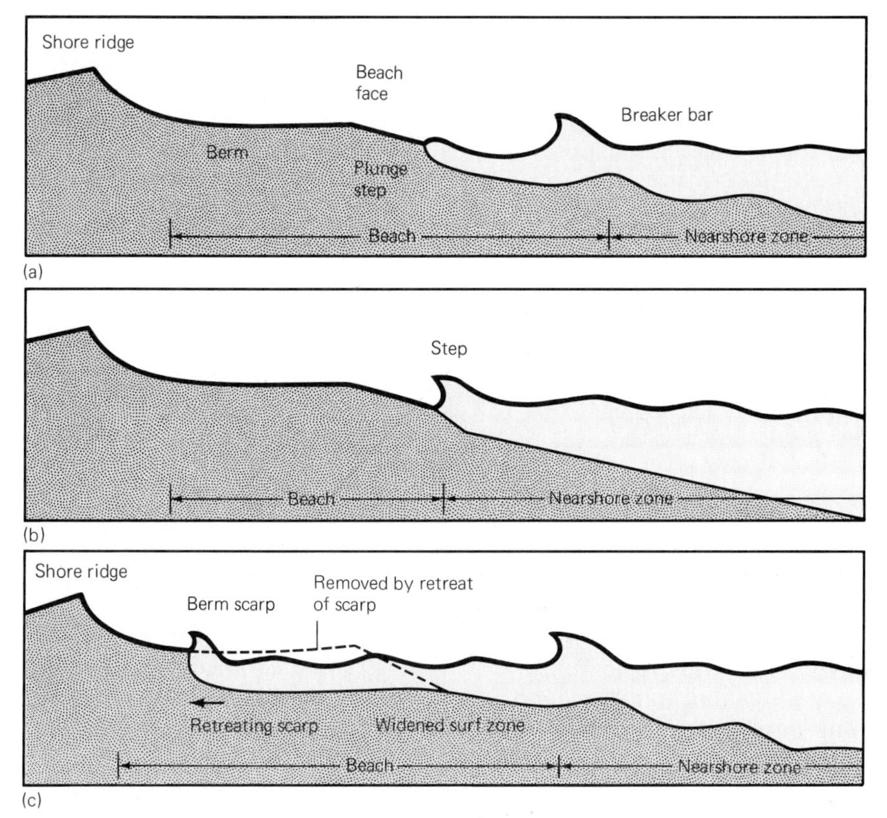


**15.17 Beach profiles through contrasting kinds of beaches.** Schematic sketches showing views at right angles to shore. (a) Wide berm formed as a result of seaward accretion of beach face by sediment transported through the breaker zone. (b) The same, except for absence of breaker bar; waves break at plunge step. (c) Berm scarp formed by undercutting and collapse.

A typical marine beach consists of two chief levels: (1) a lower level, submerged except at low tide, and (2) an upper level, known as a *berm,* which is exposed to the air except during occasional times of high water. In between these two levels is a comparatively even, regular slope, the *beach face* (Figure 15.17).

Sediment may be moved along the shore in several ways. Oblique approach of waves creates a parabolic, zig-zag motion of water and sediment on the beach face. This motion consists of oblique swash and straight-down backwash which transports sediment parallel to shore in a process named *beach drifting.* During particularly vigorous beach drifting, each wave can shift particles of sediment along the beach face by as much as several meters.

Oblique approach also creates currents that transport sand parallel to shore at many levels. All such transport of sediment parallel to shore is known collectively as *longshore drift* (or *longshore transport*).

On most beaches, long periods of berm growth are separated by short episodes of berm erosion. As sediment is brought through the breaker zone by the waves and added to the beach face, the berm widens. Berms are destroyed by thin sheets of water that undercut the sand and cause it to collapse along a vertical scarp (Figure 15.17c). As the tide rises, such scarps retreat landward. During a few hours, scarp retreat can wipe out a berm that may have been building for many months. However, a scarped beach eventually builds back; the waves reconstruct another berm that lasts until the next episode of scarp retreat.

**Spits and barriers** Longshore transport of sediment can extend one or both ends of a beach. A small sediment peninsula connected to land or a

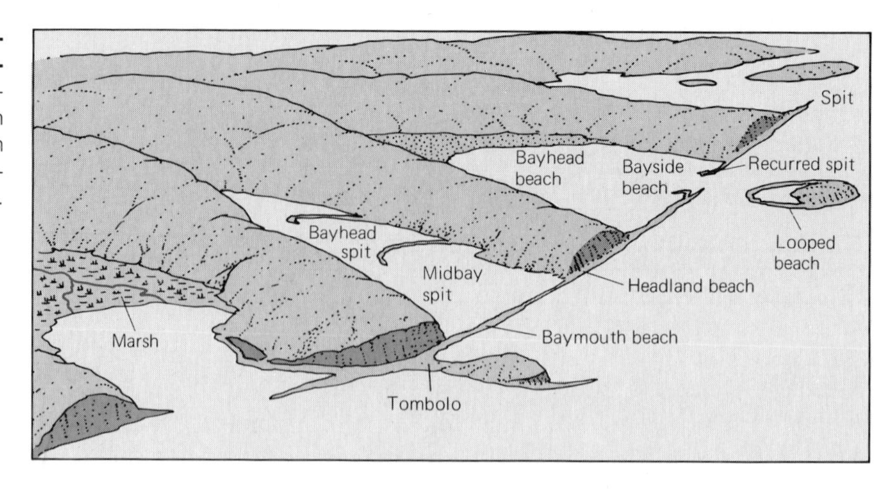


**15.18 Kinds of beaches based on relationships to mainland and coastal islands.** Beaches in bays are named according to their positions near the mouth of the bay along the side of the bay, in the middle of the bay, or near the landward (head) end of the bay.

large island at one end and terminating in open water at the other is a *spit* (Figure 15.18).

Refraction of waves around an island causes sand to accumulate between the island and the mainland or between adjacent islands. Beaches connecting an island to a mainland or to another island are *tombolos* (see Figure 15.18).

By longshore transport and spit growth, by the bulldozing action of long-period large swells, or by both processes, waves build *barriers,* long, narrow strips of sediment along coast (Figure 15.19). Between a barrier and the mainland (or larger island) are shallow bodies of water that are named *lagoons* if they are elongated parallel to shore and *bays* if their shapes are irregular.

**Beach plains** Where beach sediment is abundant, a mainland beach may grow seaward by adding on layers of sediment. After the shore has been pushed outward, a **beach plain** is the result.

### Tropical Shorelines: Reefs

Very distinctive features of tropical coasts are *reefs*. Modern reefs are built by corals that are hosts to unicellular algae, and hence require light for vigorous growth. Surrounding most reefs are great bodies of calcium-carbonate skeletal debris. The skeletal material was secreted originally

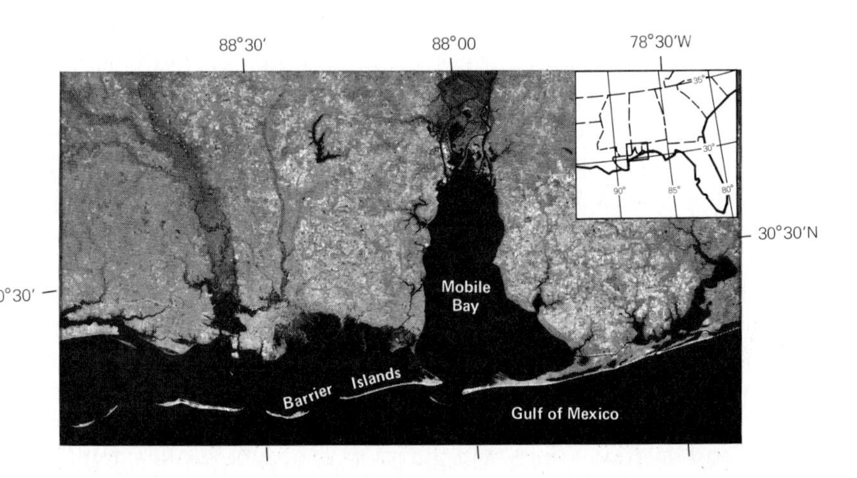


**15.19 Barrier islands and barrier spit** (white), northern Gulf of Mexico. In this view, water shows as black. The large embayment at right of center is Mobile Bay, Alabama. Satellite image, 17 November 1973, multispectral scanner band 6. (National Aeronautics and Space Administration)

not only by the organisms that form the reef but also by those that live on or near the reefs. Reefs can grow upward as sea level rises, provided that the rate of rise is not excessive.

### Classification of Coasts

*Coasts* (the general area where land and a body of standing water meet) can be classified on the basis of (1) the composition and morphology of the land area next to the body of water, both of which are functions of the previous geologic history of the area; (2) the kinds of coastal processes at work, which are functions of the present situation and are largely influenced by the climate; and (3) the stability of the coast, that is, whether the shoreline is shifting.

The morphology of the land may consist of landscapes eroded in bedrock by rivers or by glaciers, of bedrock displaying prominent tectonic features, or of bedrock of various other categories. It may consist of volcanic cones or of coastal plains. It may be shaped of regolith that has been deposited in a previous geologic setting, such as till and well-sorted sands, or of sediment actively building outward, as on deltas, delta-marginal plains, fans, beach plains, and encroaching dunes. The land may also consist of recently emerged lake bottoms or sea floors.

The coastal processes may be chiefly physical, such as those related to wave action, tides, or ice floes; biologic, such as marshes, mangrove thickets, reefs, and bioerosional; or chemical, as on evaporitic flats. The resulting coastal features can be organized into various groups, as shown in Figure 15.20.

## MOVEMENTS OF OCEAN WATER

Currents are created by the flow of water between areas of different surface levels or different densities. The waters of the oceans are constantly exposed to conditions that cause such differences and therefore are constantly moving. The oceans are: (1) subjected to uneven heating by the Sun; (2) blown by the winds; (3) evaporated more in some places than in others; (4) pulled by the gravitational attraction of the Moon, the Sun, and the other planets in the Solar System; and (5) affected by the Earth's rotation.

### Surface Currents

The waters at or near the surface of the oceans are set in motion by the interplay of climate zones, major wind belts, and the rotation of the Earth. The Sun is responsible for creating the Earth's climate belts and its major wind belts. The rotation of the Earth is responsible for the Coriolis effect. Together, these factors create the major surface currents in all oceans.

**The driving force: major wind belts related to climate zones** Intense heating by the Sun in the equatorial regions creates uniform masses of warm tropical seawater, whose temperature is one of the world's great constants (about 28°C). The heated equatorial air rises,

Group 1: Submerged coasts

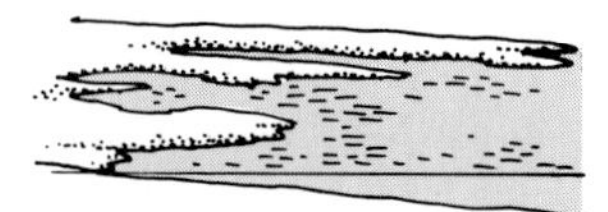

(a) Coastal plain dissected by rivers

(b) Massive bedrock dissected by rivers

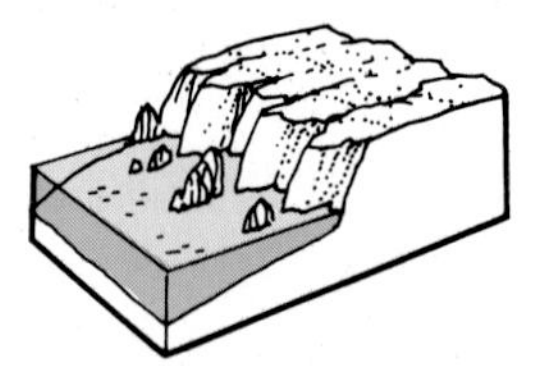

(c) Cliff partly eroded by waves

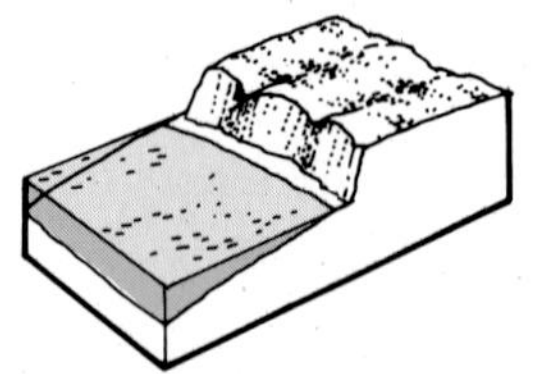

(d) Cliffed shore straightened by wave erosion

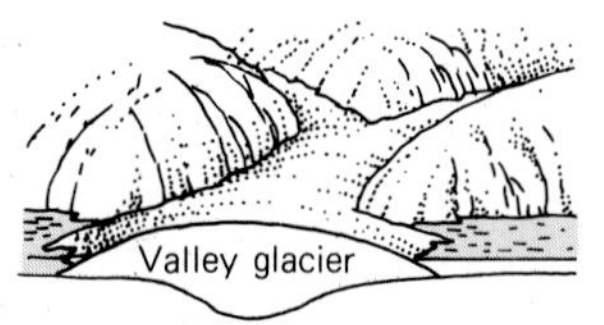


(e) Mountains; valley glacier entering sea

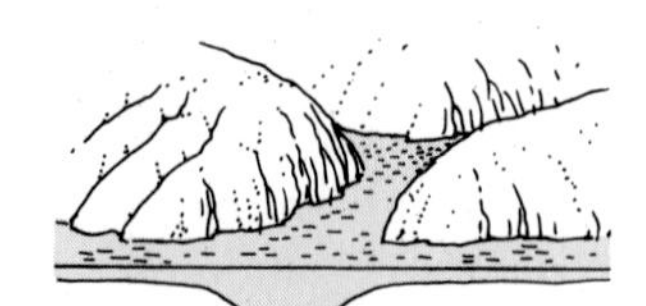

(f) Fjord; formed from (e) after glacier melted and sea entered

Group 2: Coasts related to tectonic features

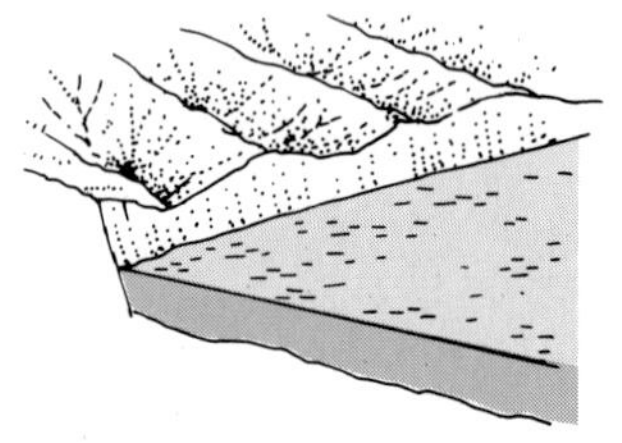

(g) Fault forms coastal cliff

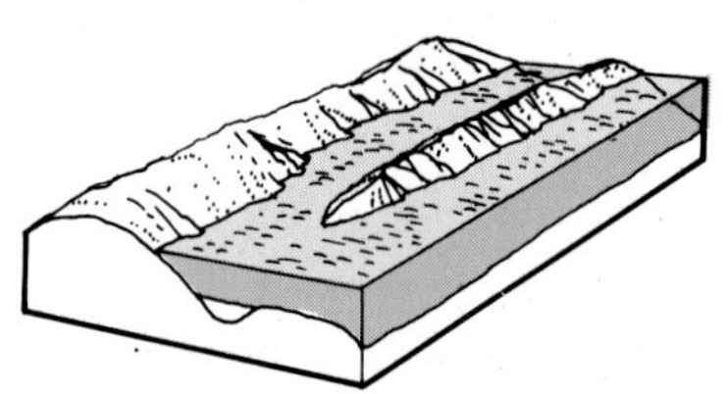

(h) Young folds trending parallel to the coast

Group 3: Volcanic coasts

(i) Circular volcanic features

Group 4: Coasts where sediment is abundant

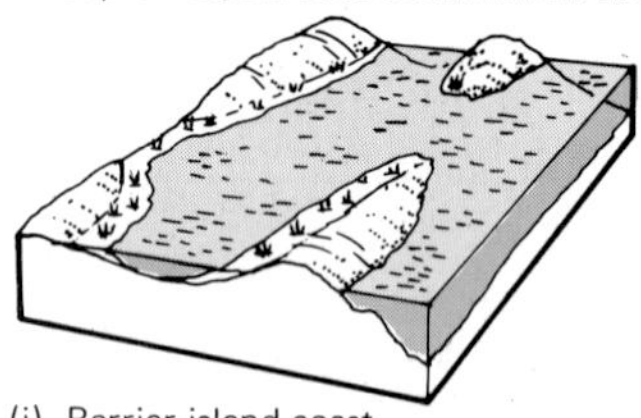

(j) Barrier-island coast

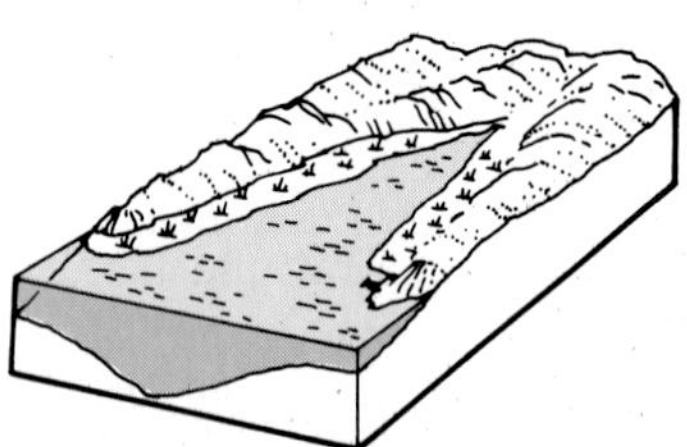

(k) Estuary coast

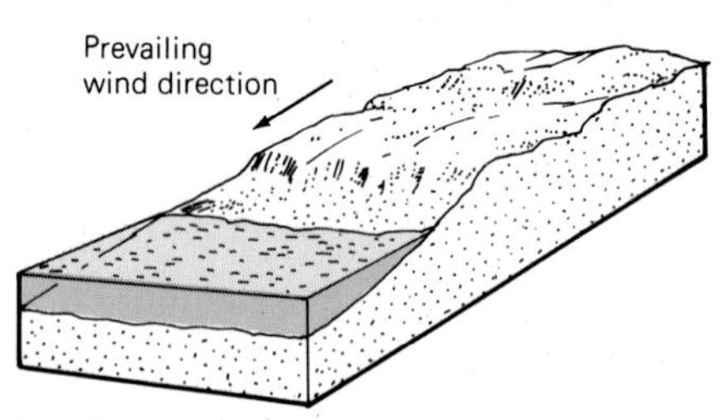


(m) Coast where windblown sand is pushing back the sea

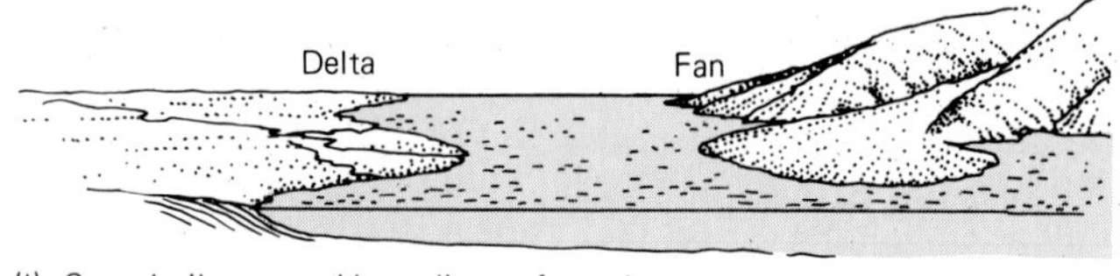


(l) Coast built outward by sediment from rivers

Group 5: Tropical coasts

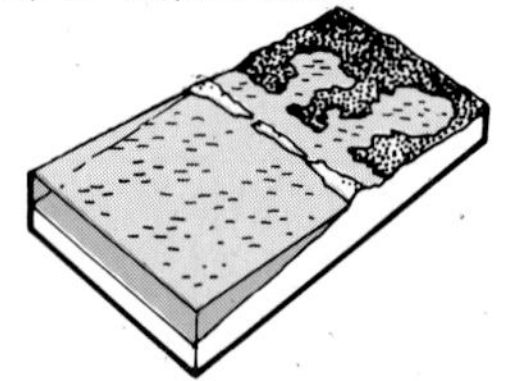

(n) Barrier reef and mangrove swamp

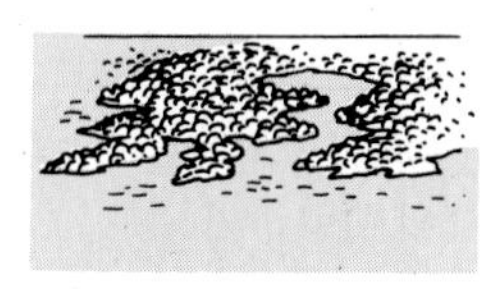

(o) Mangrove swamp

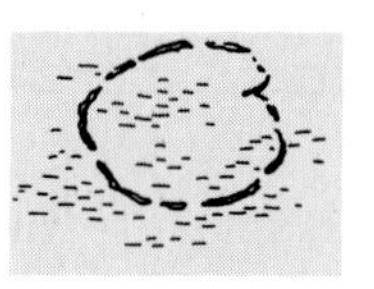

(p) Coral atoll

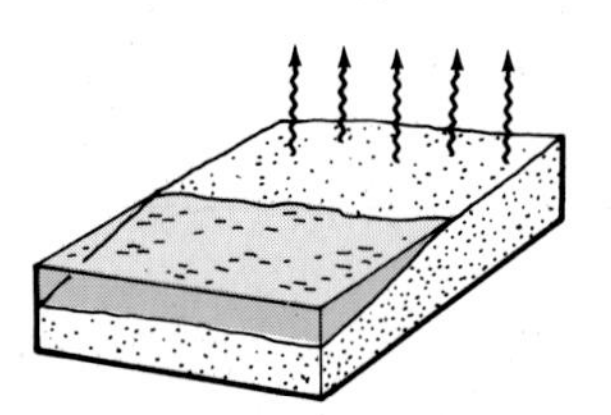

(q) Marginal evaporitic flat **(sabkha)**

**15.20 Kinds of coasts as related to the composition and morphology of the land and to the kinds of marine activities and their duration.**

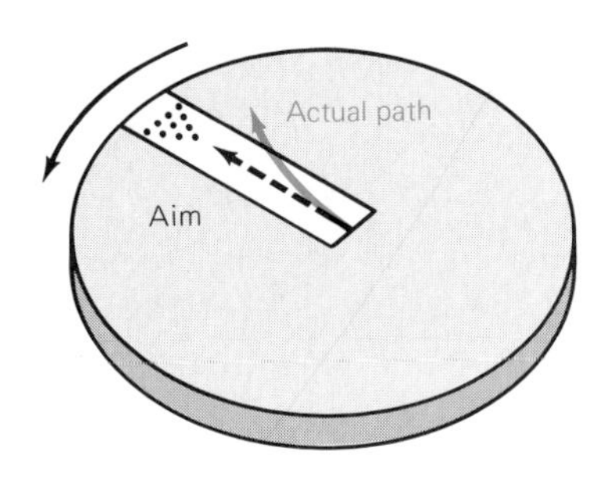


**15.21 Coriolis effect shown by the way in which a giant turntable deflects a bowling ball.** On a turntable rotating counterclockwise, as shown here, a ball aimed directly at the head pin will always roll into the right-hand gutter. If turntable were rotated clockwise, a ball aimed at the head pin would always roll into the left-hand gutter.

drops its moisture, flows away from the Equator toward both the north and the south, and descends in regions of persistent high pressure. Here, the descending air is heated and takes up moisture from the Earth's surface; it creates belts of deserts on the continents and zones of maximum evaporation at sea. Some of the descending air returns to the Equator, and some of it moves toward the poles. The winds blowing over the surface of the oceans drive the water in great surface currents and also create waves.

**The diverting force: the Coriolis effect** Any motion on the surface of an object that is rotating, such as the wind or an ocean current on the surface of the Earth, is subjected to a systematic deflection known as the *Coriolis effect.* The deflection of moving objects with respect to a rotating frame of reference was first described by the French mathematician, Gaspard Gustave de Coriolis (1792–1843). Two important points to be remembered about the Coriolis effect are: (1) the relationship between the direction of rotation and the sense of deflection and (2) the direction of rotation of the Earth.

Coriolis found that wherever the sense of rotation of the frame of reference is counterclockwise, all moving objects are deflected toward their right, as seen in the direction they are moving (Figure 15.21). By contrast, if the frame of reference rotates clockwise, then the deflection is to the left.

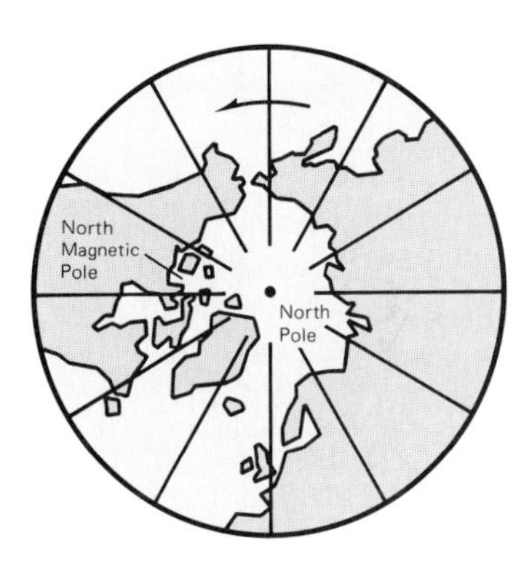


(a)

To apply the Coriolis effect to the Earth, we must know the direction of the Earth's rotation. As viewed from above the North Pole, the Earth rotates counterclockwise (Figure 15.22a). However, when viewed from above the South Pole, the same rotation of the Earth appears to be in a clockwise direction (Figures 15.22b and 15.22c). You can confirm this by spinning a small globe so that its rotation is counterclockwise when viewed from above the globe's north pole. While the globe is still spinning, view it from above its south pole. How would you describe this south-pole view of its rotation?

Because the Earth rotates, neither the winds nor water currents move in simple north-to-south or south-to-north directions. The solid Earth spins independently of the water or air above it. The movement caused by rotation is greatest at the Equator and decreases regularly toward the poles. Accordingly, anything that moves is deflected, and the deflection is greater in equatorial latitudes than in polar latitudes.

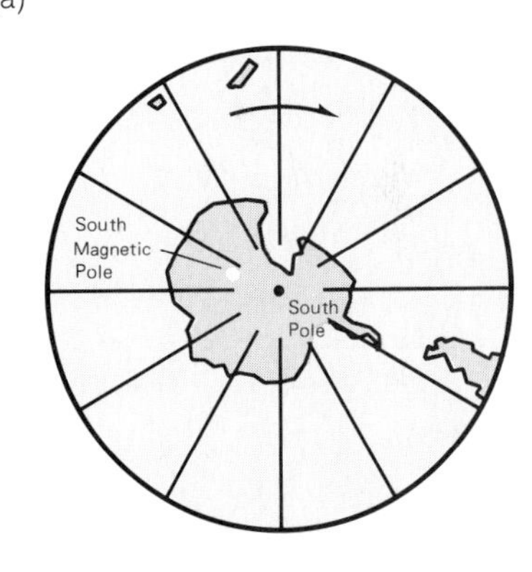


(b)

**The resultant effects: major surface currents** The major surface currents of the oceans involve nine major systems: four great spirals or *gyres* (JHY-ers), two in the Northern Hemisphere and two in the Southern

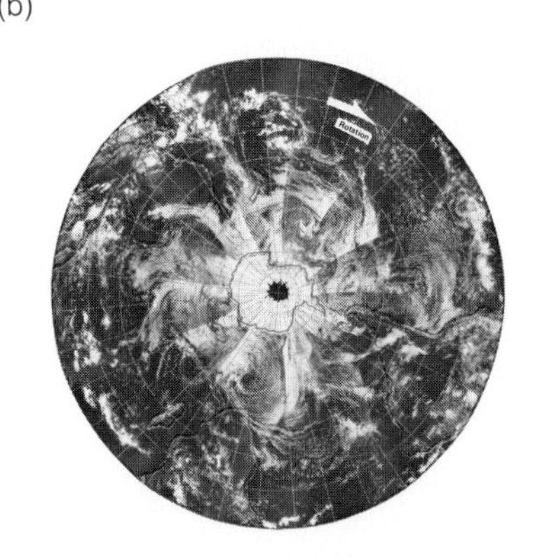

**15.22 The Earth rotates in only one direction,** but the sense of that rotation depends on where one views it. (a) From above the North Pole, the sense of rotation is counterclockwise. (b) From above the South Pole, the sense of rotation is clockwise. (c) Composite satellite photograph looking directly down at the South Pole shows clouds in streamers curved as a result of the Earth's clockwise rotation. (U. S. Department of Commerce, National Ocean and Atmospheric Administration) (c)

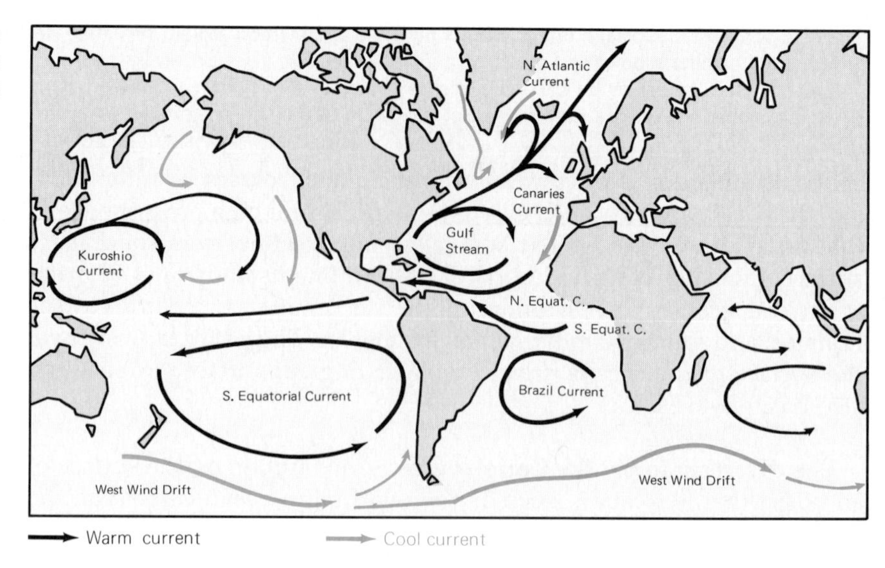


**15.23 Major surface currents of the oceans.** (U. S. Navy Oceanographic Office)

Hemisphere; four lesser gyres; and a single round-the-world current in the Antarctic Ocean (Figure 15.23).

### Density Currents and Deep Circulation: Turbidity Currents

Seawater moves not only as waves, but also as complex currents. The entire body of ocean water is a dynamic system whose movements have many analogies to those of the atmosphere. Water moves along various paths in response to gravity acting on internal differences of density and to the Coriolis effect.

At the same time as the great wind-driven currents move slowly through shallow surface waters, the deeper waters circulate in response to gravity acting on water masses differing in density and to the effects of the Earth's rotation. Dense water sinks, displacing less-dense water. The result is a *density current,* a localized current, flowing because it consists of fluid denser than that of the body of fluid through which it moves.* In polar seas, cold air chills surface water, causing it to become denser and to sink.

If the sinking water is denser than all the other water, it reaches the bottom and flows along the sea floor as a density current. Because of the Earth's rotation, however, these currents do not flow as broad sheets along the deepest parts of the ocean basins. Instead, as a result of the Coriolis effect and their speeds, the waters are banked up along the margins of ocean basins and follow submarine contours as flat, fast-moving tongues. These *contour currents* flow southward along the western margin of the North Atlantic and northward along the western margin of the South Atlantic. Measured speeds of these contour currents range from 5 to 25 centimeters per second, fast enough to transport almost all sizes of deep-ocean sediment particles.

*During the Second World War, German submarines turned off their motors and glided silently through the Strait of Gibraltar. The dense, salty water of the Mediterranean Sea sinks and produces a subsurface current that carried the subs westward. The lighter, less-salty water from the Atlantic Ocean produces an eastward current at a shallower depth that carried the subs inward.

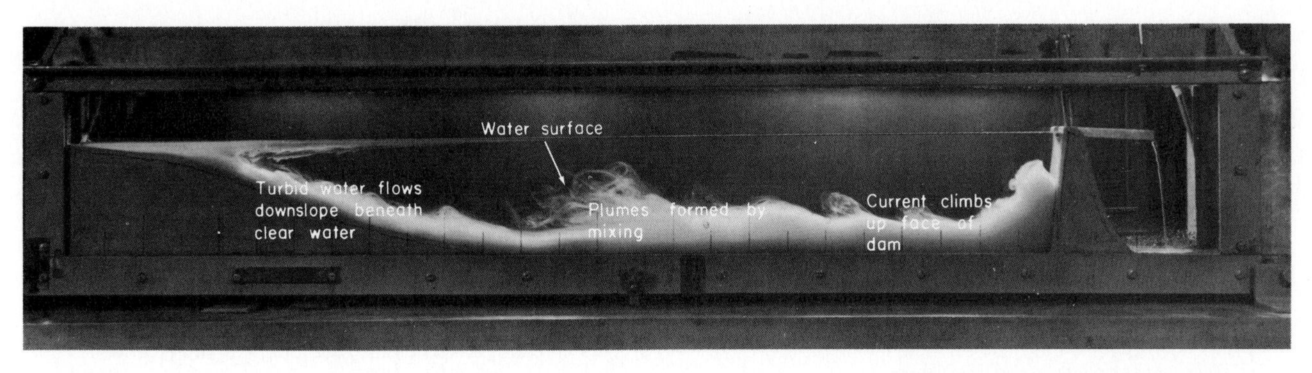


**15.24 Laboratory turbidity current.** (California Institute of Technology)

*Turbidity currents* are a variety of density current in which the excess density that causes the current to flow is contributed by sediment in suspension (Figure 15.24).

## MAJOR FEATURES OF OCEAN BASINS

Distinctive features of the ocean basins were discovered in the 1950s and 1960s; many of these features were important in developing the ideas about plate tectonics. The two contrasting regions of the ocean basins are the basin margins and the basin floors (end papers).

### Features of Basin Margins

Based on abundance of sediment and tectonic setting, all the thousands of kilometers of ocean-basin margins can be placed in one of three major varieties: (1) deep-sea trenches, formed where sediment is not abundant; (2) continental shelf, continental slope, and continental rise, formed in nontropical regions where sediment is abundant; and (3) steep escarpments, formed in some tropical regions where sediment is abundant.

The trenches are thought to characterize plate boundaries where a continent and deep-sea basin are converging. Sediment is abundant along what are named trailing edges of continents, where no significant movement between continental crust and ocean crust is in progress.

**Converging margins of continents** A typical continental margin that is thought to be converging with the adjacent deep-sea floor is that off eastern Asia. Here, a shallow sea next to the continent is hemmed in by festoons of islands, next to which is a deep-sea trench. Farther seaward is the deep-sea floor. A *sea-floor trench* is an elongated, narrow, steep-sided depression, generally deeper than the adjacent sea floor by 2000 meters or more, extending parallel to the margin of an ocean basin. In most trenches, water depths exceed 8000 meters. The deepest known trench is the Mariana Trench in the northwestern Pacific Ocean, which is almost 11,000 meters deep.

**Trailing edges of continents** A typical trailing edge of a continent is the eastern margin of North America. Sediment has built out a wide *continental shelf,* which passes seaward into a *continental slope,* defined as the relatively steep (3° to 6°) slope that lies seaward of the continental shelf. Next seaward is the *continental rise,* the gentle slope with gradient

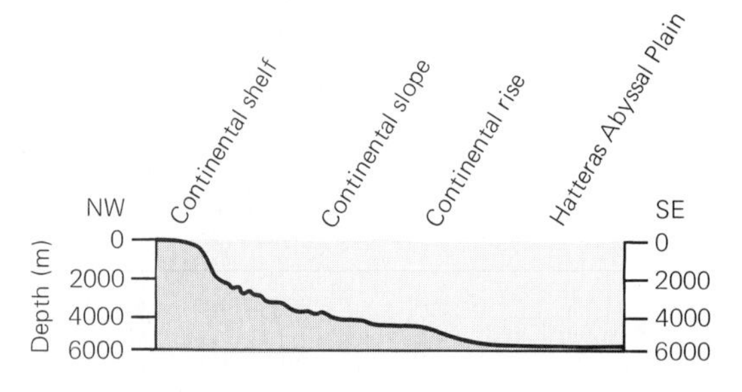


**15.25 Continental rise,** profile of bottom southeastward from Norfolk, Virginia to the Hatteras Abyssal Plain, a horizontal distance of 600 kilometers (from the outer edge of the continental shelf to the abyssal plain). (Bruce C. Heezen, Marie Tharp, and Maurice Ewing, 1956 , The floors of the oceans. I, The North Atlantic: Geological Society of America, Special Paper 65, 122 p., 30 pls., Plate 24, profile W18)

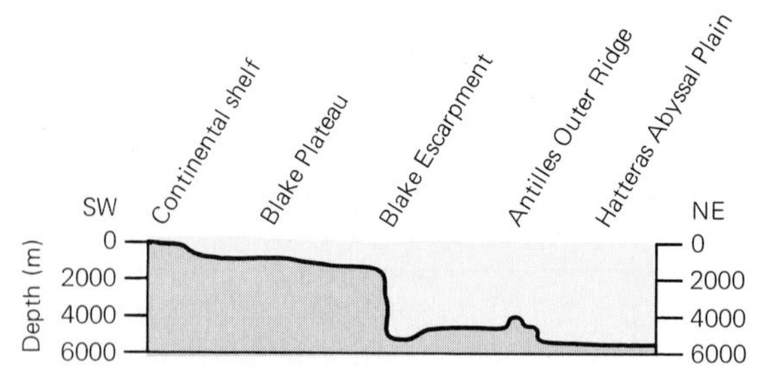


**15.26 Blake Plateau and Blake Escarpment,** displayed in exaggerated profile extending from continental shelf to the Hatteras Abyssal Plain. (Bruce C. Heezen, Marie Tharp, and Maurice Ewing, 1956, reference cited in caption to Figure 15.25, Plate 24, profile W24)

between 1:100 and 1:700 that lies seaward of the continental slope (Figure 15.25). The closest feature on land to a continental rise is the great plains of central United States, a gently sloping body of sediment that extends from the Rocky Mountains nearly to the Mississippi River.

Continental rises are swept by bottom-following currents driven by the general ocean circulation. In some places, these currents are deflected away from the continent. In such places, they have built *outer ridges,* long ridges of sediment that extend upward 200 to 2000 meters above the adjacent sea floor.

In many tropical regions, where carbonate sediments are forming limestones beneath the sea, the sea floor is a virtual cliff. Such steep slopes have been named *escarpments.* An example is the Blake Escarpment, which forms the seaward margin of the Blake Plateau (Figure 15.26).

## Features of Basin Floors

The floors of the ocean basins consist of low areas and higher ridges. Low areas are known collectively as the *abyssal floor.* The higher areas, known as oceanic rises and ridges, with rough morphology are classified as distinct features apart from the abyssal floor.

The abyssal floor consists of smooth areas underlain by sediment and rough areas underlain by rock. A smooth area may be an *abyssal plain* (Figure 15.27), a flat part of the abyssal floor, underlain by sediment having an imperceptible slope of less than 1:1000, or an *abyssal fan,* a fanlike accumulation of sediment at the mouth of a submarine canyon. The rough parts of the abyssal floor are grouped collectively as *abyssal-hills provinces,* parts of the floor consisting almost completely of irregular rocky hills a few hundred meters to a few kilometers wide and having relief of 50 to 100 meters. Abyssal-hills provinces are widespread on the floor of the Pacific basin; their aggregate area exceeds that of any other relief feature on Earth.

The higher parts of the ocean basin floors consist of *oceanic rises and ridges,* continuous rocky ridges on the ocean floor many hundreds to a few thousands of kilometers wide, whose relief is 600 meters or more (see end papers). These ridges have been cut into countless blocks by many parallel fractures. Continuous seismic profiles have shown that the

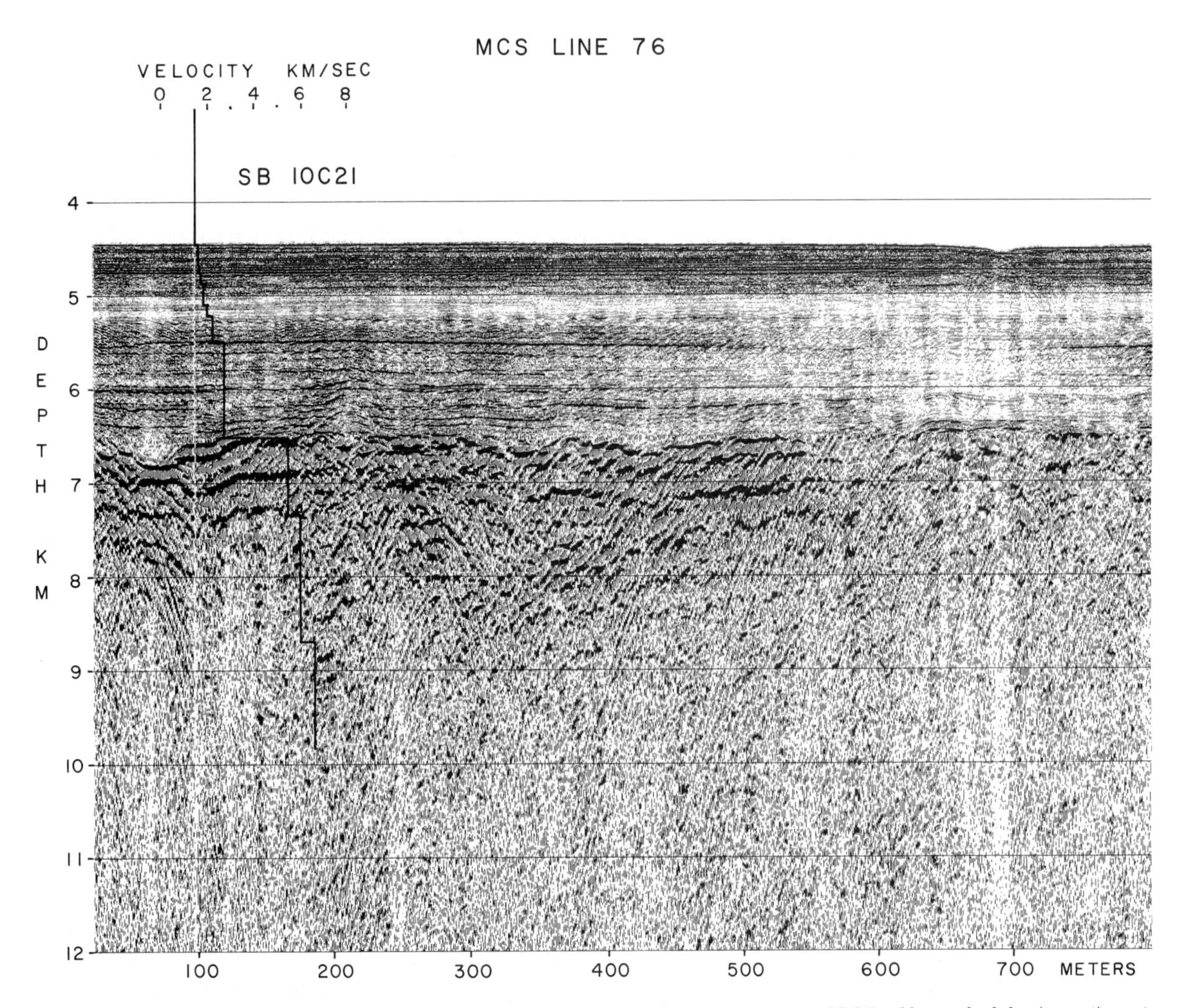


**15.27 Abyssal plain,** in northwestern Atlantic Ocean near Deep-Sea Drilling Project Site 106 (latitude 36°20′ N, longitude 69°30′ W), displayed on a record from a modern multi-channel continuous seismic reflection profiler. DSDP Hole 106 penetrated 350 meters of interbedded sand and silt of Quarternary age, about 600 meters of clay ranging in age from Pliocene to Medial Miocene, and stopped in Eocene silicified claystone at a subbottom depth of 1030 meters. Mafic oceanic crustal rock, not quite fully penetrated on this record, begins between 6 and 7 on the depth scale, SB10C21 designates the sonobuoy used to determine the speed of P waves. (Lamont-Doherty Geological Observatory of Columbia University, courtesy of Paul Stoffa)

rough, rocky bottom characteristic of oceanic rises and ridges (on a large scale) and of abyssal-hills provinces (on a smaller scale) extends beneath the sediment underlying abyssal plains.

Many conical edifices built by volcanoes occur on the sea floor. Some project above sea level as islands.

## BASIN-FLOOR SEDIMENTS

We begin our presentation of basin sediments with the kinds of material and the processes by which those materials are deposited. Then, we present a classification based on the materials and processes. We close with a summary of the chronology of deep-sea sediments.

### Kinds of Material

The first systematic studies of deep-sea sediments, made in the nineteenth century, showed that deep-sea sediments come from three principal sources: (1) the lands, (2) the ocean water itself, and (3) to a much smaller extent, outer space. The lands contribute detrital particles (*ter-*

*rigenous sediment*) as well as tephra and ions in solution. Some particles of terrigenous sediment are large enough to be seen with the naked eye, but others are so small that their study requires special instruments and techniques. The ocean water contributes particles consisting chiefly of the hard parts of tiny planktonic organisms, notably the one-celled plants, diatoms and coccoliths, and the one-celled animals, Foraminifera and Radiolaria. Contributions from outer space consist of tiny spherules of nickel-iron and of glassy materials (*tektites*).

### Processes of Basin-Floor Sedimentation

During the nineteenth-century, geologists supposed that no currents flow in the deep sea. They thought that the only sediment came from an "eternal snowfall" of tiny skeletal debris, which accumulated slowly and constantly to build on the bottom a layer of sediment that was both nonstratified and uniformly thick in all localities. Oceanographic expeditions completed since about 1950 have shown that these concepts are not correct. We now know that processes of deep-sea sedimentation are complex and that the deep-sea floor is swept by many kinds of currents. What is even more remarkable, we now know that sediment deposited on submarine slopes is subject to displacement by gravity and as a result, can be spread widely by gravity-transport processes down slopes and across basin floors beyond the toes of the slopes. We also know that within the sediments, many kinds of strata are present, that the thickness of deep-sea sediments varies from place to place, and that in many places sediments are not present. Both physical processes and chemical processes operate.

### Deep-Sea Sediments

Basin-floor sediments, especially deep-sea sediments, are classfied according to the kinds of material composing them and by the processes involved in their deposition. In the sediments, a general progression can be seen starting at the shore and moving away from shore. This progression is from terrigenous sediments, or in many tropical regions from shallow-water carbonate sediments composed of the calcareous skeletal remains of reefs and other invertebrates, to pelagic sediments, in which these land-marginal sediments form only a minor proportion or are absent and the material consists of skeletal remains of microorganisms.

The major groups of deep-sea sediments are the following (Figure 15.28). (1) *Terrigenous sediments* are detrital particles deposited by vertical fallout. Large quantities of fine-grained terrigenous mud are present in many basins along the ocean margins as well as on the upper parts of continental slopes and continental rises. (2) *Gravity-displaced sediments* are a large variety of materials that have been displaced down an underwater slope by any one of numerous gravity-transport processes, including turbidity currents, liquefied cohesionless-particle flows, slumps, and debris flows. (3) *Glacial-marine sediments* are deposited on the sea floor when icebergs melt; the individual objects dropped from floating ice are known as *dropstones*. (4) *Pelagic sediments* are open-sea deposits containing predominantly skeletal remains of microorganisms and clays or products derived from clays. Pelagic sediments may be classified as *ooze*,

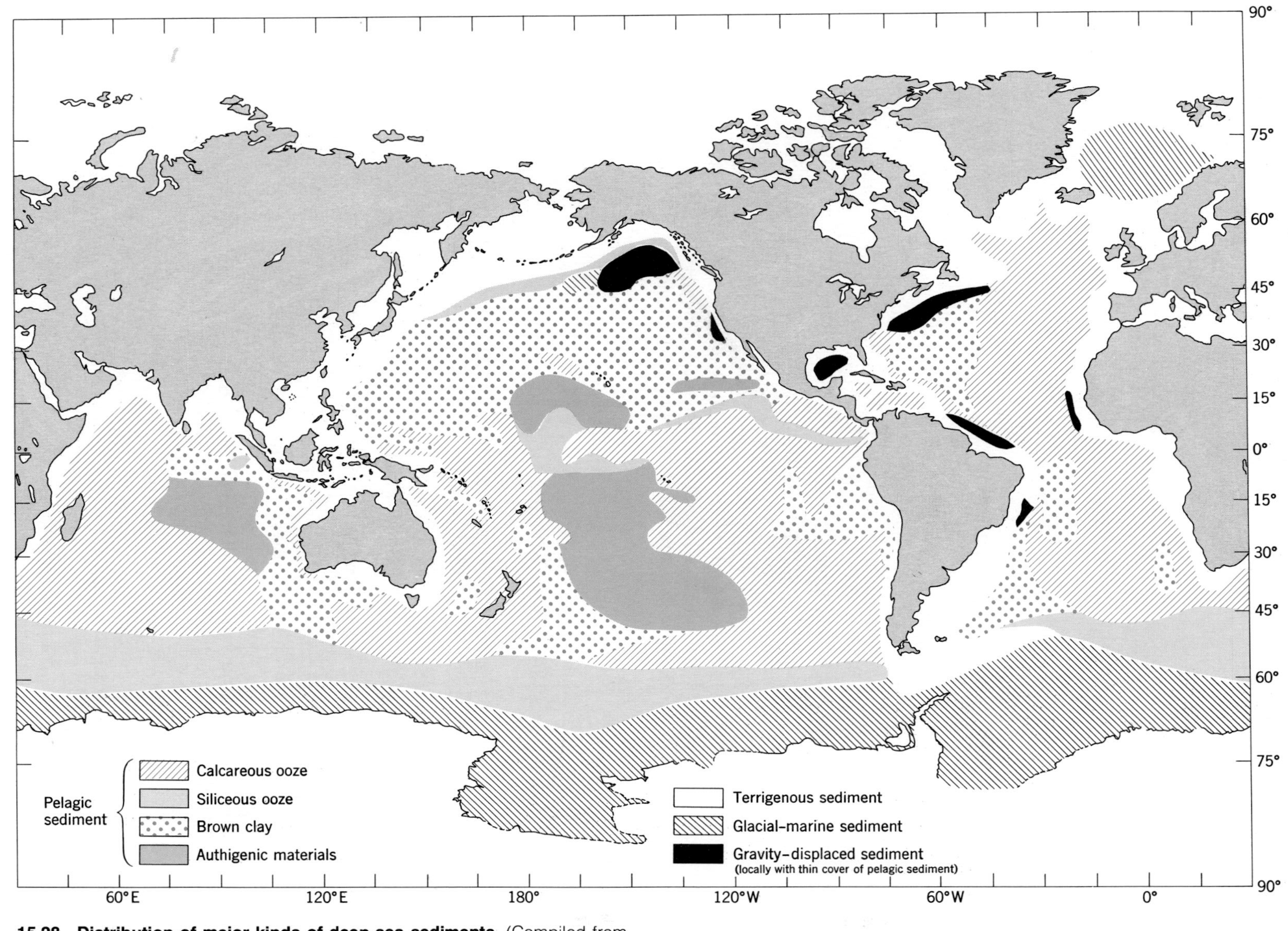


**15.28 Distribution of major kinds of deep-sea sediments.** (Compiled from various sources)

consisting of more than 30 percent of skeletal remains of planktonic microorganisms, or *brown clay,* containing less than 30 percent of such skeletal remains. (5) *Tephra* deposits qualify for recognition as a distinct group of deep-sea sediments because of their volcanic origin and composition. (6) *Authigenic sediments* are sedimentary deposits formed in place, not from physically transported material, but rather consisting of minerals that crystallized out of seawater. (7) *Euxinic sediments* are black muds deposited in basins where the bottom waters become stagnant and hydrogen sulfide destroys all organisms except the anaerobic bacteria. (8) *Open-water evaporites* include those evaporite minerals that are precipitated from the open waters of a basin having restricted circulation.

### Chronology of Deep-Sea Sediments

Close studies of the uppermost 20 meters of deep-sea sediments in representative cores have made it possible to construct a composite record of many changes of climate, probably extending far back into the Pleistocene Epoch. The record shows that the Pleistocene climatic changes were worldwide. During glacial ages, cool climates seem to have prevailed simultaneously in both hemispheres. Likewise, the warmer climates of the interglacial ages apparently existed everywhere at about the same time.

The earlier expectation of finding in deep-sea sediments a continuous record of events from early in the history of the Earth down to the present has not been realized. Cores from holes drilled by the *Glomar Challenger* provide a continuous record that extends only as far back as Late Jurassic time. The lack of older sedimentary strata is taken as evidence that the ocean basins are no older than the Jurassic.

## SPECIAL ASPECTS OF DEEP BASINS AND DEEP-BASIN SEDIMENTS

Two special aspects of deep basins are the origin of the underwater canyons at their margins and of the fans and plains on their floors.

**15.29 Submarine canyon,** transverse profile sketched from observations made by diver-geologists off California. (Robert F. Dill, Caribbean Marine Laboratory, Fairleigh-Dickinson University)

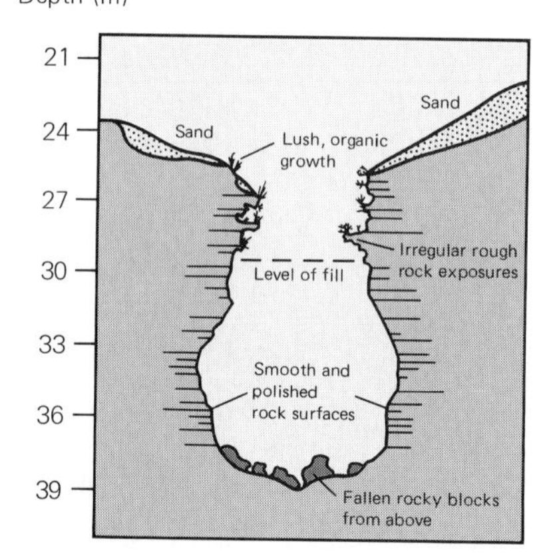


### Origin of Submarine Canyons

The outer slopes of all continental masses are cut by enormous *submarine canyons.* These sinuous, V-shaped valleys have a variable number of tributaries, cross all or part of a continental shelf, and extend down continental slopes. They may be hundreds of kilometers long. The sides of these valleys slope steeply, and their narrow floors, some lying as much as 1.5 kilometers below the valley rims, are inclined seaward as much as 80 meters per kilometer and extend down to 3.6 kilometers below sea level. The walls of many canyons consist of rock (Figure 15.29). Some canyons have been partly filled with unconsolidated sediments. Some canyons, such as the Hudson and Congo, appear to be seaward extension of land rivers; others are not obviously connected with land rivers.

For more than a century, submarine canyons have engaged the interest of geologists, who have proposed many hypotheses of origin for these tremendous valleys. Most of their ideas are based on the premise that the canyons originated within a geologically short time interval. One group of hypotheses invokes processes that work on land, another group involves submarine processes.

Direct observations in the upper parts of canyons have disclosed that at least six processes are actively eroding today's canyons: (1) creeping sediment, (2) progressive slumping, (3) flowing sand, (4) turbidity currents (see Figure 15.24; only a few small ones actually have been seen), (5) bottom currents created by tides, and (6) bottom currents created by internal water movements. Submarine mass movement of sediment is abrading bedrock walls of some canyons, much as a valley glacier does.

Evidently, parts of some canyons originated by several processes acting through long periods of time. Their headward parts may have originated as land valleys, but they subsided below sea level and were then modified by submarine processes. Parts of many canyons have never been above sea level. Some canyons may have been "left behind" as sediment accumulated at their margins and thus the walls of the canyons grew upward.

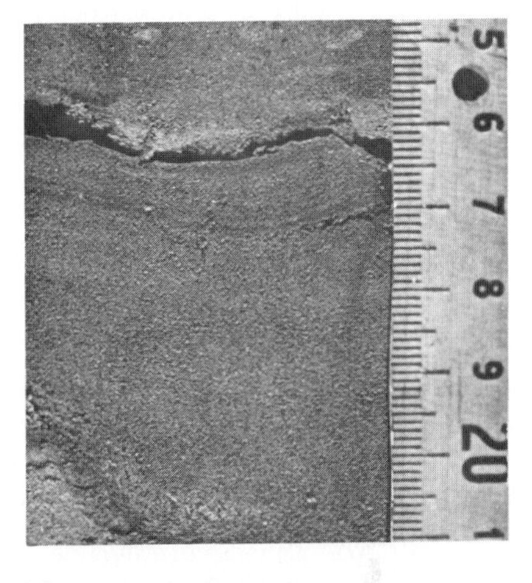

(a)

### Origin of Basin-Floor Fans and Basin-Floor Plains

Basin-floor *fans,* bodies of sediment with fanlike forms, lie opposite the mouths of many submarine canyons. If these lie in abyssal depths, they are named abyssal fans. The Hudson abyssal fan, at a depth of about 4.5 kilometers, was first cored in 1951. Since then, cores have been raised from other abyssal fans. Coarse gravel displaced from shallow water has been cored at depths of 3.3 kilometers at a distance of 220 kilometers from the head of Monterey Canyon off California.

Basin-floor *plains,* which are not restricted to the vicinity of submarine canyons but are spread widely on deep-basin floors, are the flat tops of bodies of sediment much more extensive than those constituting basin-floor fans. As with the fans, the term abyssal plains is used if the water is deep enough to rank as abyssal. Cores show that most abyssal plains on the North Atlantic floor are underlain by alternating layers of graded terrigenous sediment and nongraded pelagic sediment (Figure 15.30). The graded layers are thought to have been deposited by turbidity currents generated on continental slopes. Continuous seismic profiles show that in most, the subbottom sediments contain many widespread parallel layers having contrasting physical properties (see Figure 15.24).

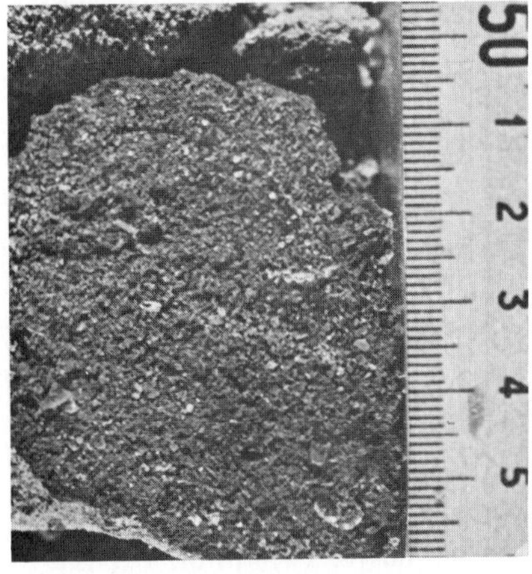

(b)

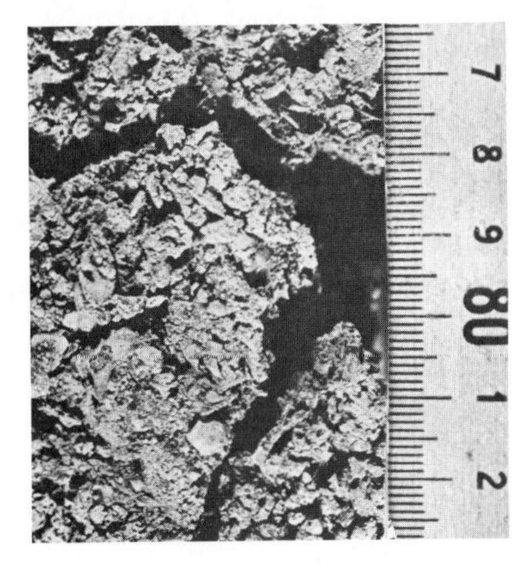

(c)

**15.30 Graded layer of deep-sea sediment** raised from Hatteras Abyssal Plain, western Atlantic. Three segments of a lingle layer more than a meter thick that occupies the second meter of the core (measuring down from the top). Hence, centimeter marks on meter stick at right should read 120 centimeters below top instead of 20 centimeters; the 50 should be 150 centimeters; and the 80, 180 centimeters. (Lamont-Doherty Gelogical Observatory of Columbia University, courtesy E. D. Schneider)

# CHAPTER REVIEW

1. Among the planets of our Solar System the Earth is unique in possessing a continuous body of salt water, the *oceans*. The oceans cover almost 71 percent of the Earth's surface in four deep-ocean basins, the Pacific, Atlantic, Indian, and Arctic.

2. Terms used to describe all kinds of waves are *wavelength, wave height, wave period,* and *wave celerity* (or speed). Additional terms for water waves only are *depth* and *steepness*.

3. Most water waves result from wind action. The most important factors influencing the origin of wind-generated waves are *wind speed, wind duration,* and *fetch*.

4. *Sea waves* are steep and irregular waves being actively blown by the wind. *Swells* are long, low, and regular former sea waves that have traveled beyond the places where the wind actively blows the water.

5. Long, low waves generated by the sudden displacement of the sea floor are *tsunami*.

6. As waves begin to interact with the bottom, three things happen: the waves are subject to *shoaling transformations;* the oscillating bottom water transports sediment; and wave refraction occurs, the process by which the direction of a series of waves is changed so that the waves tend to become parallel to the shore.

7. A *breaker* is a wave that is collapsing. As each wave crest collapses, water from both adjacent troughs flows toward the *breaker zone,* the zone where waves collapse.

8. *Waves of translation* are waves that displace the water masses within the moving crests. Waves of translation landward of the breakers and seaward of the backwash constitute the *surf*.

9. The *tide* is a rhythmic rise and fall of water level. A rising tide is a *flood tide,* and a falling tide is an *ebb tide*.

10. *Deleveling* refers to long-term changes of the relative position of a water surface against the land. A relative rise of water level against the land is *submergence,* and a relative drop of water level is *emergence*.

11. Other important coastal processes that affect the geologic features formed within the water include currents at the mouths of rivers, diverse activities of coastal organisms, the action of shore ice, and chemical precipitation of evaporites in marginal flats.

12. Major features of *bedrock coasts* are formed by wave action, saltwater spray, the usual weathering and downslope gravity-transport processes, and by the erosive activities of organisms.

13. Wave action arranges and rearranges cohesionless sediments available at the shore, into the characteristic forms of *beaches, spits,* and *barriers*. A beach is a body of nonconsolidated sand-size or coarser sediment along the shore of a lake or ocean.

14. Very distinctive features of tropical coasts that are not influenced by large rivers are *reefs,* which are built by corals that are hosts to unicellular algae.

15. The major factors involved in the classification of coasts are the composition and morphology of the land; the coastal processes at work; and whether the shoreline is shifting.

16. The waters of the oceans are constantly moving because of many factors that create changes of water level and/or density that cause currents to flow.

17. Because the Earth is rotating, the winds and ocean currents are subjected to a systematic deflection known as the *Coriolis effect*.

18. Seawater moves in major surface currents and also in *density currents* and *contour currents* beneath the surface.

19. Ocean-basin margins can be classified into three major types: (1) *sea-floor trenches,* occurring where sediment is not present and thought to characterize converging margins of continents; (2) *continental shelves, continental slopes,* and *continental rises,* occurring in nontropical regions where sediment is abundant and characteristic of trailing edges of continents; and (3) steep *escarpments,* occurring in tropical regions where sediment is abundant in the form of limestone-forming carbonates.

20. The floors of the ocean basins consist of low areas known as the *abyssal floor,* and higher parts are called *oceanic rises* and *ridges*.

21. Deep-sea sediments come from three principal sources, the land, the ocean water itself, and outer space. The major groups of deep-sea sediments are *terrigenous sediments, gravity-displaced sediments, glacial-marine sediments, pelagic sediments, tephra, authigenic sediments, euxinic sediments,* and *open-water evaporites*.

22. Special features of deep basins are *submarine canyons,* valleys cut through the outer slopes of continental masses; *basin-floor fans,* fanlike forms of sediment at the mouths of submarine canyons; and *basin-floor plains,* extensive, flat-topped bodies of sediment.

## QUESTIONS

1. List and discuss the three main factors affecting the origin of wind-generated waves on the surface of a body of water.
2. Define *sea waves* and describe their characteristics.
3. What are *swells?* Describe the characteristics of swells and indicate where one would expect to find swells.
4. Define *tsunami* and list their chief characteristics. How do tsunami originate?
5. What are *shoaling transformations?* What are the effects of these transformations?
6. Define *breakers* and describe the water motions and bottom features associated with breakers.
7. What is a *wave of translation?* Compare waves of translation with *waves of oscillation*. Where are each of these kinds of waves likely to be found?
8. Discuss the relationships among *surf, swash,* and *backwash*.
9. What factors cause the *tide?* What is a *flood tide?* An *ebb tide?*
10. Define *submergence,* and *emergence*.
11. List and explain four geologic processes other than waves and changes of level that affect coasts.
12. What is a *beach?*
13. Define *spit, barrier,* and *beach plain*.
14. What is a *reef?* Where are modern reefs found?
15. List the factors involved in the classification of coasts. Name and briefly describe the major kinds of coasts.
16. What are the chief two factors that cause ocean currents to flow?
17. What is the *Coriolis effect?* Explain how the Coriolis effect operates in both the Northern Hemisphere and Southern Hemisphere.
18. Define *density current, contour current,* and *turbidity current*.
19. List and describe the three major kinds of features found at ocean-basin margins.
20. Compare the features of the trailing edge of a continental plate with those of a converging margin.
21. List and describe four major features of deep-ocean basins.
22. What are the chief constituents of deep-sea sediment? Where do those materials come from?
23. List the eight major kinds of deep-sea sediments and briefly indicate the processes involved in their accumulation.
24. Define *submarine canyons* and list the processes that are thought to cause these canyons.
25. Compare *basin-floor fans* and *basin-floor plains* and discuss their origins.

## RECOMMENDED READING

Bascom, Willard, 1980, *Waves and Beaches. The Dynamics of the Ocean Surface* (revised, updated, and enlarged ed.): Garden City, New York, Anchor Books, Doubleday and Co., Inc., 366 p. (paperback)

Bernstein, Joseph, 1954, Tsunamis (*sic*): *Scientific American,* v. 191, no. 2, p. 60–64.

Damuth, J. E., and Kumar, Naresh, 1975, Amazon Cone: Morphology, Sediments, Age, and Growth Pattern: *Geological Society of America, Bulletin,* v. 86, no. 4, p. 863–878.

Davies, T. A., 1976, Five Hundred and Seventy Three Holes in the Bottom of the Sea—Some Results from Seven Years of Deep-sea Drilling: *Journal of Geological Education,* v. 24, no. 5, p. 142–155.

Gross, M. G., 1977, Oceanography. *A View of the Earth,* 2nd ed.: Englewood Cliffs, New Jersey, Prentice-Hall, Inc., 497 p.

Heirtzler, J. R., Taylor, P. T., Ballard, R. D., and Houghton, R. L., 1977, A Visit to the New England Seamounts: *American Scientist,* v. 65, no. 4, p. 466–472.

Ingle, R. M., 1954, The Life of an Estuary: *Scientific American,* v. 190, no. 5, p. 64–69.

Kuenen, Ph. H., and Migliorini, C. I., 1950, Turbidity Currents as a Cause of Graded Bedding: *Journal of Geology,* v. 58, no. 1, p. 91–127.

Menard, H. W., 1964, *Marine Geology of the Pacific:* New York, McGraw-Hill Book Co., Inc., 271 p.

Menard, H. W., and Frazer, J. W., 1978, Manganese Nodules on the Sea Floor: Inverse Correlation Between Grade and Abundance: *Science,* v. 199, p. 969–971.

# CHAPTER 16 GEOLOGIC STRUCTURES AND MOUNTAINS

PLATE 16
Results of great displacement along the Glarus horizontal overthrust (outcrop trace marked by dark line at base of vertical cliff), central Alps, Switzerland. (Heena Pon)

IN THE NEXT THREE CHAPTERS, WE WILL STUDY IMPORTANT TOPICS related to the origin of geologic structures and the deformation of the lithosphere, the geologic dating of structural features, the origin of earthquakes, the use of seismic waves to infer the interior of the Earth, and tectonic concepts about how all these things take place. Our plan is to describe geologic structures and kinds of mountains in this chapter. Familiarity with kinds of faults provides a basis for understanding the discussion of earthquakes, in Chapter 17. Similarly, once we have developed concepts of how we can use seismic waves to interpret the interior structure of the Earth, it is possible to discuss the tectonic concepts that have been proposed to explain such phenomena as the origin and history of mountains and to grasp the many new ideas about the concept of plate tectonics. The subjects of plate tectonics and its possible relationship to mountain building, are discussed in Chapter 18.

Geologic structures are formed by many kinds of processes, which may be grouped under two major headings: (1) *tectonic structures,* formed by large-scale motions of the lithosphere, and (2) *nontectonic structures,* formed by processes other than large-scale movements of the lithosphere and often involving sediments before they have been lithified. Some examples of nontectonic structures are deformation by glaciers (see Chapter 13), deformation of sediment by currents, gravity deformation of sediments, diapirs, and deformed evaporites (see Chapter 8).

In this chapter, we will concentrate on tectonic structures. In Chapter 18, we relate tectonic structures to concepts of plate tectonics.

## TECTONIC STRUCTURES

### Basic Geologic Structural Features

Geologic structural features are usually defined by the ways they have affected sedimentary strata. Always implied is the concept that the strata were deposited in horizontal positions without interference and were deformed afterward.

**Anticlines and synclines** Strata that have been crowded together form wavelike arrangements known as *folds.* Two types of folds are archlike upfolds or *anticlines* and troughlike downfolds or *synclines* (see Figure 6). The characteristic feature of anticlines is that the *oldest* strata form the core of the fold and the successively younger strata, dipping away, appear on the flanks. (See Box 16.1 for an explanation of *dip* and *strike,* terms used to describe the orientation of strata.) In synclines, the *youngest* strata form the core of the fold and the successively older strata, dipping inward, appear on the flanks. Figure 16.1 explains the various parts of folds and shows the chief varieties.

**Joints** *Joints* are simple fractures or cracks along which the opposite sides have not shifted relative to each other (see Figures 5 and 7.4). Joints, typically form *sets* of parallel cracks. The intersections of two or more sets of joints outline blocks of rock.

**Faults** The term fault and many of the basic concepts of faulting have been adopted from terms used by miners. *Faults* are fractures along

**BOX 16.1**

# Attitude of strata; strike and dip

The term ***attitude*** is used by geologists to refer to the spatial orientation of layers, of fractures, or of contacts.

Some strata are horizontal. The attitude of these is specified by stating in words that they are horizontal. The symbol for horizontal strata is ⊕.

If strata are not horizontal, their attitude is specified by giving the measurements of two angles. The two angles used by geologists to express the attitudes of strata are named *strike* and *dip*. In order to understand what the angles mean, it is necessary to visualize how the intersecting planes that form the angles are defined.

The angle involved in the concept of strike is a compass bearing, that is, it is an angle measured in the horizontal plane referenced to true north as determined by a compass (see Appendix F). In the United States, compass bearings for strike are measured in the quadrants east or west of magnetic north, using angles starting with 0 for true north. Thus the value of the *strike angle* ranges from 0° to 90° East or from 0° to 90° West.

Of the two lines needed to define strike angle, therefore, one is magnetic north. The other line, called the *strike line,* is more difficult to visualize. It is a horizontal line formed by the intersection between a horizontal plane and the dipping plane being measured (Sketch A). One place that is easy to visualize is a strike line formed where a dipping layer enters a body of water; it is the shoreline along the dipping surface. The strike angle, then, is the compass bearing of the strike line. When the term strike is used alone, it refers to the strike angle (which, after all, cannot be measured until the strike line has been identified). In summary, *strike* may be defined as the compass bearing of the horizontal line formed by the intersection with the horizontal plane of the nonhorizontal plane being studied.

Dip is the angle measured in the vertical plane, down from horizontal, that is, horizontal is set at 0°and the angles increase to vertical, which is 90°. The first of the two lines that define the dip angle is the intersection of a vertical plane at right angles to strike with the horizontal plane (defines the horizontal line); the second is the intersection with the dipping plane (defines the nonhorizontal line). The dip angle is shown in Sketch B.

For each strike, two directions having the same amount of dip are possible (Sketch B). In order to specify the single correct plane, one must indicate both an angle and a direction for the dip. For example, if the strike is N45°E, and the dip angle is 30°, the dip could be toward the northwest or the southeast. Thus, a complete strike and dip is: N45°E, 30°SE.

Geologists measure strike and dip in the field using a special geologic compass, named a Brunton compass (or simply Brunton for short). The sides of a Brunton compass case have been carefully machined so that they are parallel. The face of a Brunton differs from regular compasses in that the positions of east and west are reversed. This is done so that when the sides of the compass are parallel to the strike line and the needle has swung to north, the point of the needle reads the correct strike angle (Sketch C). A Brunton includes a small clinometer that can be used to measure dip angles.

Line of strike
Map symbol
30
30° – Angle of dip

(A)

Map notation for strike and dip is a T-shaped symbol with a long crossbar and short stem. The crossbar is drawn with a protractor using true north as 0. Because the crossbar has been drawn to show its correct strike angle, one need not indicate the number of degrees of the angle (Sketch D). The stem points in the correct dip direction, and a number at the end of the stem indicates the amount of dip angle in degrees. The symbol for vertical strata (a dip of 90°) is an elongated plus sign, with the longer line showing the strike.

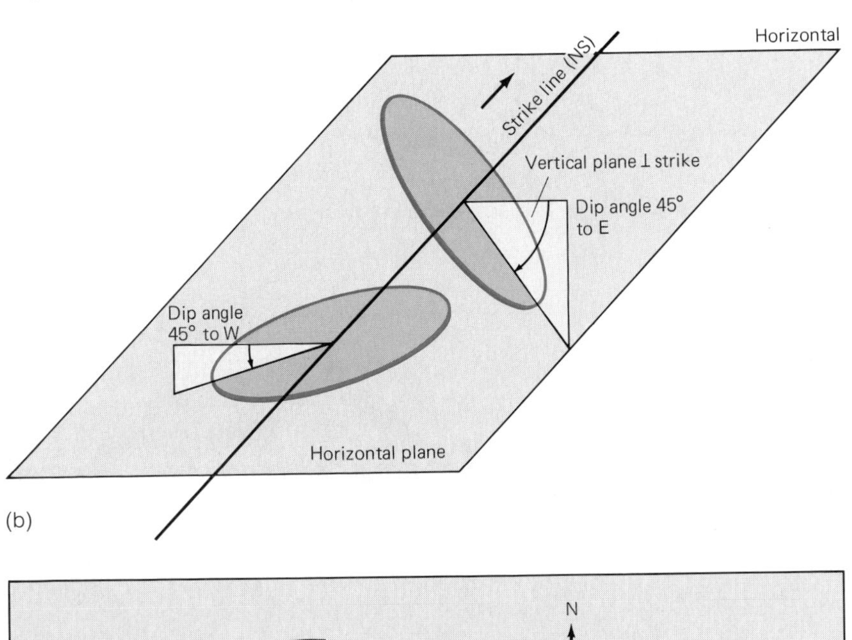


(b)

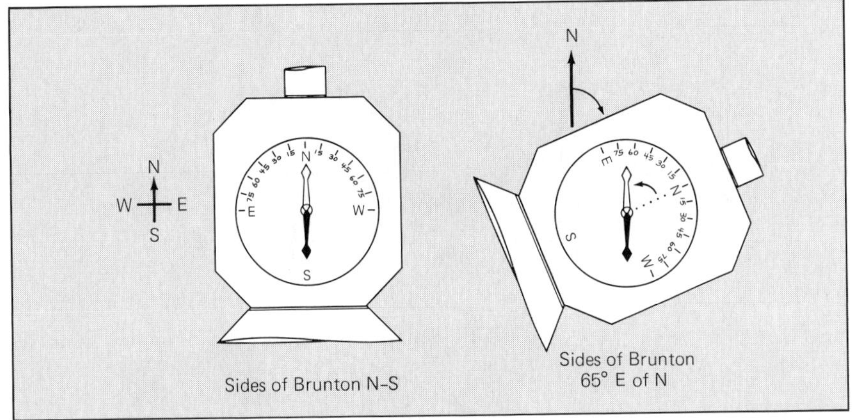


(c)

|  |  |  |  |
|---|---|---|---|
| ⊢25 | Strike N-S, dip 25° E | 25⊣ | Strike N-S, dip 25° W |
| | | 40 | Strike NE, dip 40° NW |
| 45 | Strike NE, dip 45° SE | | |
| 10 | Strike E-W, dip 10°N | 10 | Strike EW, dip 10° S |
| 70 | Strike NW, dip 70° NE | 70 | Strike NW, dip 70° SW |
| + | Strike EW, vertical dip | + | Strike NS, vertical dip |

(d)

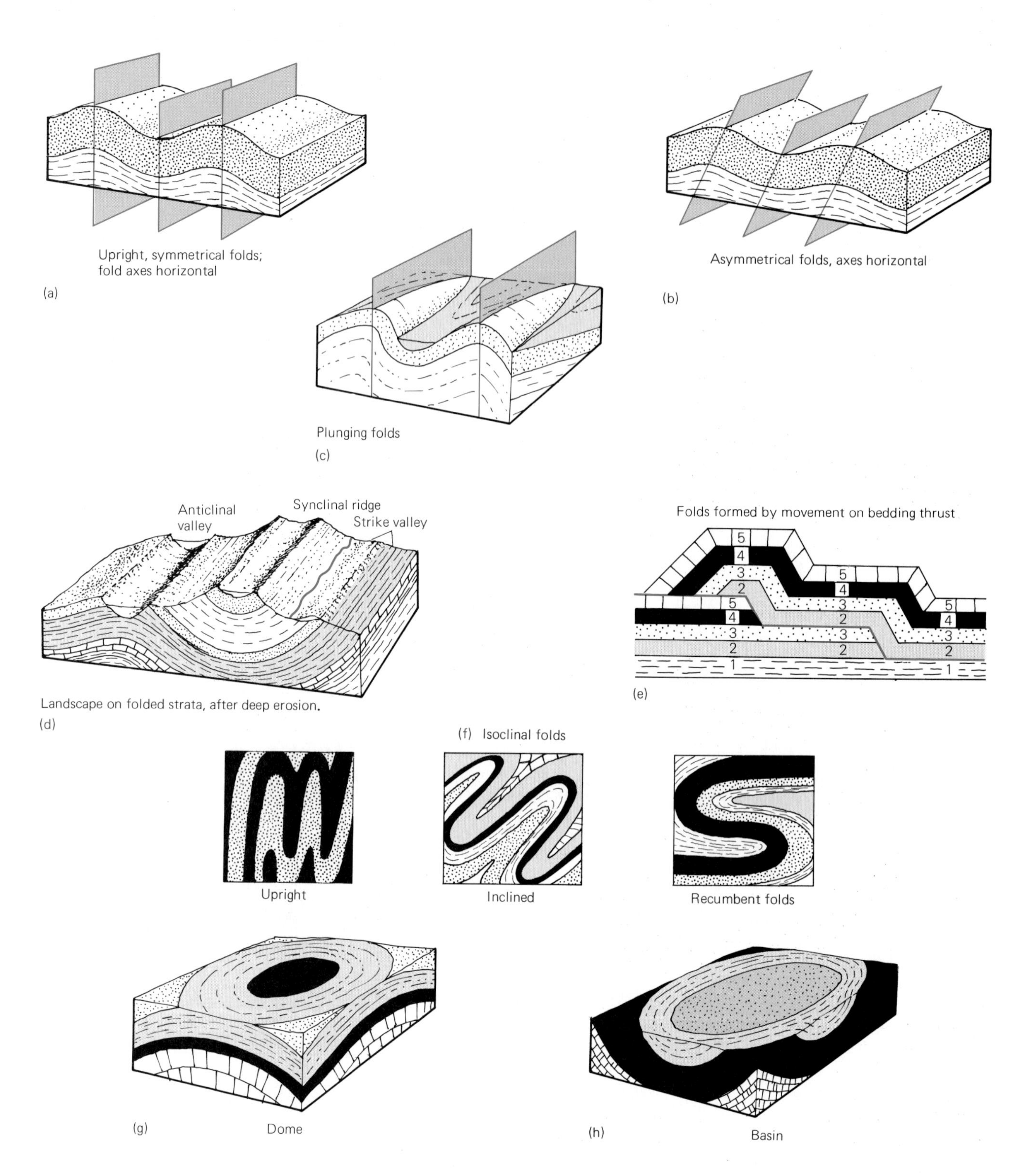


which opposite sides have been displaced in a direction parallel to the fracture (see Figure 5). The blocks on opposite sides of a nonvertical fault are the *hanging-wall block* and the *footwall block;* by definition, a nonvertical fault dips beneath the hanging-wall block (Figure 16.2). Figure 16.3

**16.1 Principal kinds of folds,** schematic block diagrams and sections. (a) Symmetrical anticlines and synclines with *axial planes* (the axial plane is the plane passing through points where strata bend around) vertical and *fold axes* (the fold axis is the median line around which the strata of a fold curve) horizontal. The maximum inclinations of the strata on both limbs of symmetrical folds are equal. (b) Asymmetrical folds; the maximum inclination of the strata on one limb exceeds that of the strata of the other limb. (c) Plunging folds; axial planes are vertical, but fold axis are not horizontal. (d) Landscape on folded strata after deep erosion is the inverse of that on new folds. A valley has been eroded along the axis of an anticline and a ridge has formed near the axis of a syncline. A valley, eroded in dipping strata, whose trend is parallel to the strike of the strata is a *strike valley.* (e) Folds formed by movement on bedding thrust that cuts across strata at steep angle in two places, but is elsewhere parallel to the beds. The inclination of the strata on the upper block has formed either because the strata are dipping back into the thrust surface or are parallel to the places where the thrust cuts across the beds at a high angle. (f) Isoclinal folds with limbs parallel. In upright isoclinal folds (1), axial planes are vertical. In inclined isoclinal folds (2), axial planes are neither horizontal nor vertical. Isoclinal folds with horizontal axial planes are *recumbent folds* (3). The inclined isoclinal folds (2) display Z-shaped patterns, whereas the recumbent folds (3) are S-shaped folds. (g) In a dome, the strata dip away from the center where the oldest strata are exposed; successively younger strata, dip outward in all directions, form concentric belts around the oldest strata. (h) In a basin, the strata dip inward; the youngest strata are preserved at the center and older strata form concentric circular belts around the youngest strata.

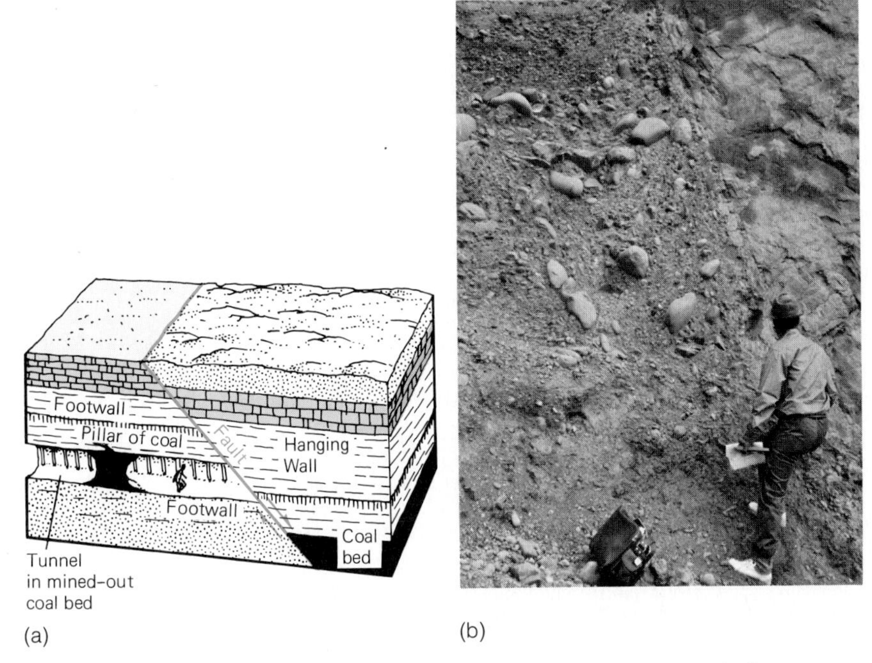


**16.2 Steep faults that are not vertical** (high-angle faults). (a) A nonvertical fault dips under the hanging-wall block (or simply hanging wall, in the miners' usage, where the term originated). The basis for this usage is the situation that prevails where a tunnel intersects a fault. The hanging wall looms over the miner's head, whereas the miner can walk on the footwall (or footwall block of geologic usage). Schematic. (b) Steep fault in weakly lithified sedimentary strata, Hanging-wall block is at right; footwall block, at left. Small syncline on footwall block suggests that the hanging-wall block moved relatively upward, making this a reverse fault (see Figure 16.3c). (R. W. Fairbridge, Columbia University)

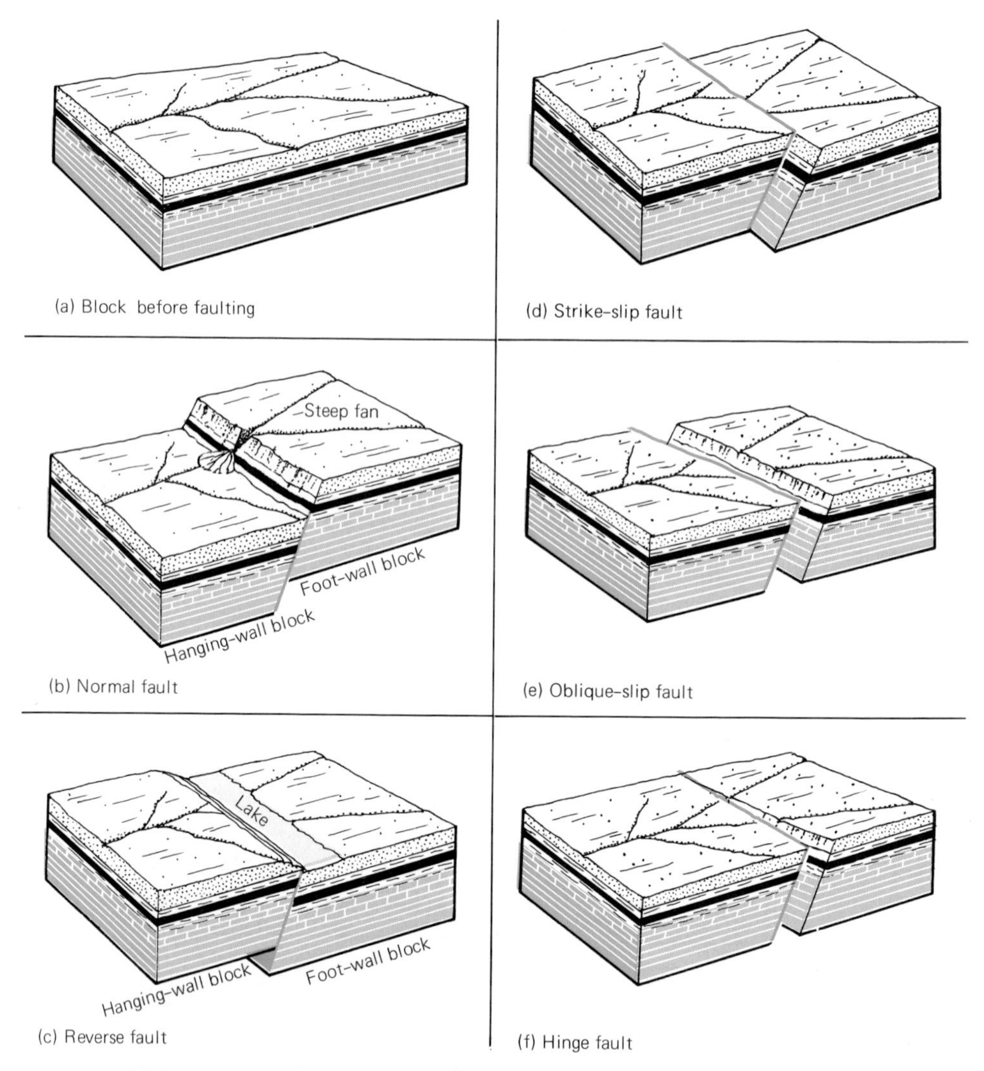


**16.3 Principal kinds of high-angle faults,** schematic blocks. (a) Block before faulting, showing strata and drainage from left to right. (b) *Normal fault,* a high-angle fault (usually along which the hanging-wall block has been displaced relatively downward, directly down the dip of the fault. (c) *Reverse fault,* a high-angle fault along which the hanging-wall block has been displaced relatively upward, directly up the dip of the fault. (d) *Strike-slip fault,* a fault along which displacement has been horizontal. Many strike-slip faults are vertical; shown here is a *right-lateral* strike-slip fault, the direction being determined by facing the fault and finding out whether the block across the fault has moved to the left (defines left-lateral fault) or, as here, to the *right* (defines right-lateral fault). (e) *Oblique-slip fault,* a fault along which the displacement has been neither horizontal nor vertical, but includes two components, one horizontal and the other vertical. (f) *Hinge fault,* a fault along which the displacement changes noticeably along strike, such that the fault ends at a definite point in one direction and its displacement increases in the other direction.

shows various kinds of faults. We discuss the origins and patterns of faults further in the following section.

## Tensional Structures: Patterns of Normal Faults

Within a part of the lithosphere being deformed, tensional forces may be applied on a regional scale or on a local scale. Whatever the cause or the scale, the immediate effect on the rocks is to pull them apart and normal faults form. Dikes may be intruded into sets of normal faults or the open cracks may be filled with mineral deposits known as veins. Parallel sets of normal faults give rise to rift valleys and grabens. Other sets of normal faults form radial patterns, concentric patterns, or en-echelon patterns.

**Rift valleys and grabens** The effect of tensional forces applied on a regional scale typically is to form sets of more-or-less parallel normal faults. Between pairs of such faults a large block of the lithosphere may subside to form a graben. A rift valley is a special case of a graben in which the present relief of the land surface has been caused by the fault displacement (Figure 16.4a). In some cases, the subsiding block accumulates sediment and its surface relief tends to be hidden (Figure 16.4b). Notable rift structures in the world include the Rhine graben in Europe, the Andean graben in South America, the Ottawa-Bonnecherre graben in Canada, the central graben of Iceland, the system of rift valleys in eastern Africa, and the axial rift valley in the crest of the mid-Atlantic ridge and the extensions of this feature into other oceans (see Figure 1.20).

**Radial patterns of faults and dikes** Where a cylindrical pluton or cylindrical diapir punches its way upward, it typically arches up the overlying strata and may cause them to break into sets of normal faults having radial patterns. Typical radial patterns of dikes are associated with volcanoes or stocks. Sets of radial normal faults usually lacking intruded dikes are commonly seen above salt diapirs.

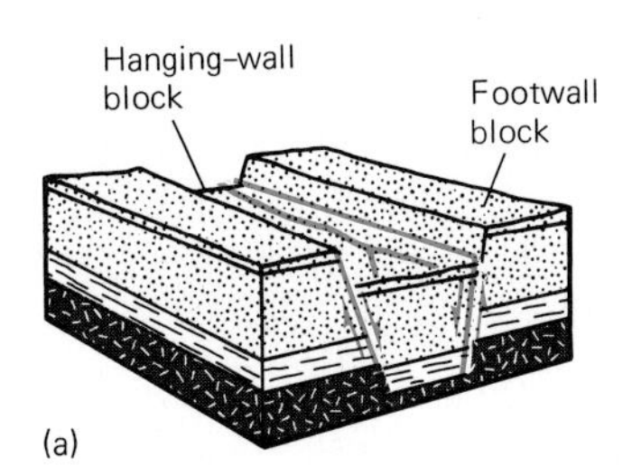


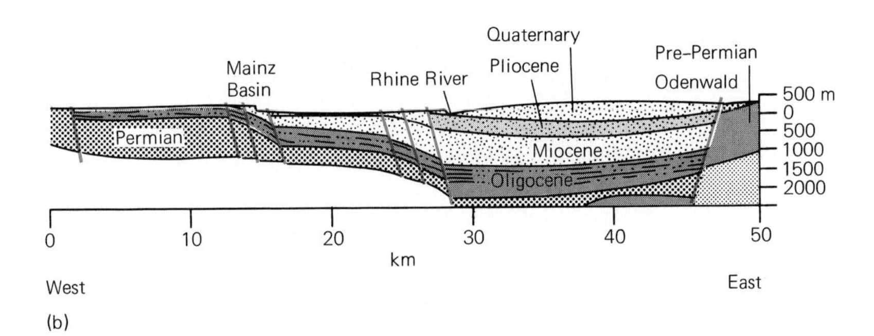


**16.4 Rift valley and graben.** (a) Rift valley, where the graben forms a low-lying block between two highlands on the footwall blocks. (b) The Rhine graben, in which sediment has buried a former rift valley. (Rhine graben after Paul Dorn, 1960, Geologie von Mitteleuropa (2, neubearbeitete und stark verbesserte Anflage); Stuttgart, E. Schweizerbart'sche Verlagsbuchhandlung (Nägele und Obermiller), 474 p.; Figure 97, p. 324)

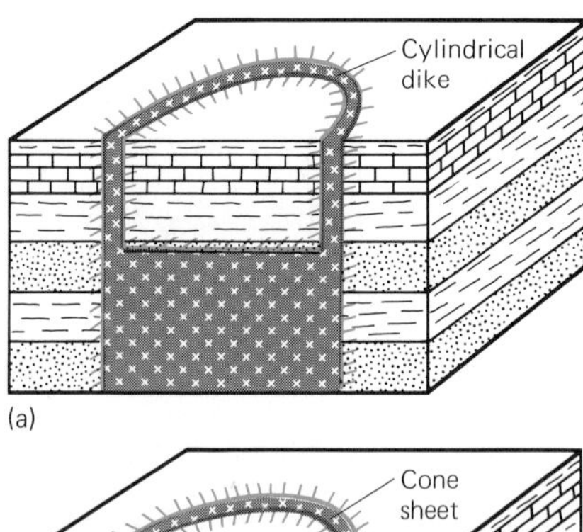


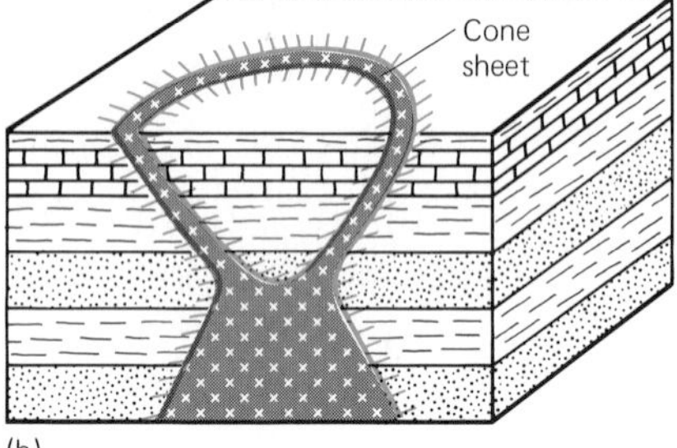


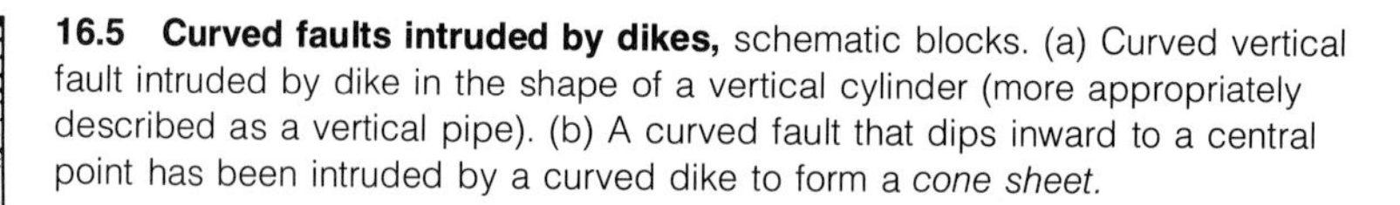
**16.5 Curved faults intruded by dikes,** schematic blocks. (a) Curved vertical fault intruded by dike in the shape of a vertical cylinder (more appropriately described as a vertical pipe). (b) A curved fault that dips inward to a central point has been intruded by a curved dike to form a *cone sheet.*

**Concentric patterns of faults and dikes** The collapse of a caldera typically forms a circular closed depression bounded by one or more curved normal faults (see Figures 5.19, 5.22, 5.24 and 5.30). Parallel to and concentric with the circular wall of the caldera may be other curved normal faults, into which curved dikes may have been intruded. The curved dikes may be essentially vertical and thus *cylindrical* or they may be inclined inward to form funnel-shaped *cone sheets* (Figure 16.5).

**En-echelon patterns** An en-echelon pattern refers to a series of short, parallel, overlapping features whose ends lie along a series of two other parallel lines that are not parallel to the short features (Figure 16.6). En-echelon cracks or normal faults, with or without fillings of vein minerals or of dikes, exist on all scales ranging from tens of centimeters to hundreds of kilometers. En-echelon patterns form in several ways. The en-echelon faults in north-central Oklahoma that trend northeast-southwest have been found by drilling to be surface expressions of larger deeper-lying faults that trend north-south. The patterns of en-echelon gash fractures shown in Figure 16.6 or of the en-echelon crevasses at the margins of valley glaciers shown in Figure 13.1 are related to shearing couples, as explained in the next section.

## Shearing Couples and Strike-Slip Faults

One of the major kinds of deformation of the lithosphere is the application of a shearing couple of regional extent. A *shearing couple* is a pair of forces that are parallel but act in opposite directions. The effects of shearing couples are usually expressed in the form of strike-slip faults. The sense of shearing of a couple is described in the same way as is the displacement along a strike-slip fault (see Figure 16.3d). Thus, a shearing couple is described as being right lateral or left lateral (Figure 16.7).

In the lithosphere, application of large-scale shearing couples in which the planes of maximum shearing stresses are vertical may be responsible for forming three contrasting kinds of structures: (1) an en-echelon set of vertical fractures, some of which may be open; (2) folds having vertical axial planes and axes that are parallel among themselves but perpendicular to the set of vertical open fractures; and (3) one or more strike-slip faults along a direction that is parallel to the plane of maximum shearing stresses and diagonal to the en-echelon sets of open fractures and the parallel folds (Figure 16.8).

Strike-slip faults first became widely noticed by geologists as a result of the devastating San Francisco earthquake of 1906 (see Chapter 17). One

**16.6 *En-echelon fractures* filled with white quartz.** These fractures were formerly open; the rock was pulled apart at right angles to the widest parts of the quartz filling and simultaneously squeezed along a direction parallel to the lengths of the fractures. This could have been the result of a left-lateral couple operating in a direction parallel to the hammer handle, or of a left-lateral couple acting at right angles to the hammer handle. (Precambrian quartzite, Baraboo Range, Wisconsin, J. E. Sanders)

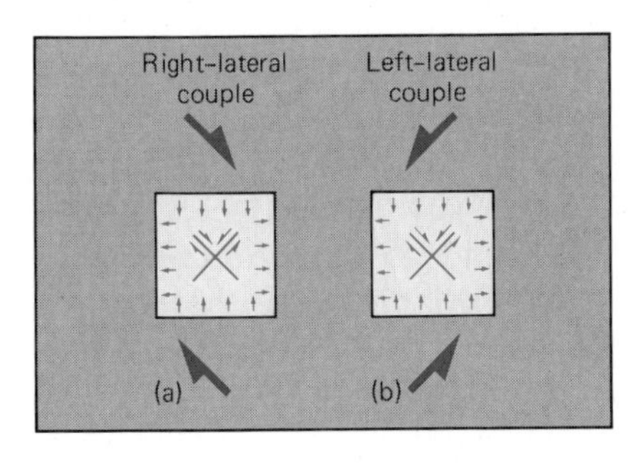


**16.7 Effects of shearing couples applied to cubes,** schematic views from above. (a) Right-lateral couple applied along a line that trends NW-SE creates maximum tensile stresses in an E-W direction and maximum compressive stresses in a N-S direction; X-shaped lines indicate vertical planes along which shearing stresses reach maximum values. (b) Left-lateral shearing couple applied along a line trending NE-SW. Internal stresses as in (a).

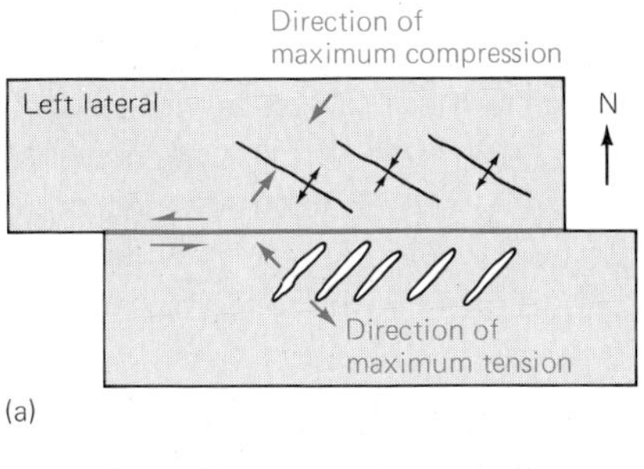


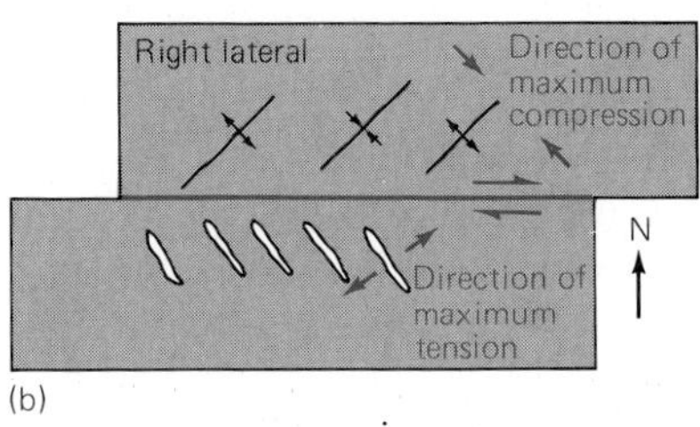


**16.8 Structural features formed along strike-slip faults,** schematic maps with folds and tensional fractures shown on one fault block only (in nature, both folds and tensional fractures would be found on both fault blocks). (a) Left-lateral strike-slip fault trending E-W may form a series of *en-echelon* anticlines and synclines with axes oriented NW-SE and a series of *en-echelon* fractures (compare Figure 16.6) trending NE-SW. (b) Right-lateral strike-slip fault trending E-W may form a series of *en-echelon* folds with axes oriented NE-SW and a series of *en-echelon* fractures trending NW-SE.

of the world's largest strike-slip faults, the San Andreas fault, extends across California for about 800 kilometers (see Figure 17.22). Other notable strike-slip faults have been found in China (Figure 16.9) and Scotland.

**16.9 Major left-lateral strike-slip fault,** western China, trends E-W and offsets a mountain range (dark areas), possibly as much as 40 km (white arrows). (National Aeronautics and Space Administration)

### Transform Faults

A special kind of strike-slip fault, named a *transform fault,* is defined as a fault that ends where expansion takes place and new material is added, such as on the fractured crust of a lava flow (Plate 1) or between parts of the lithosphere where new material is being added from the mantle, as at the centers of mid-oceanic ridges. Most of the large fracture zones that have been discovered on the deep-sea floor are transform faults. We discuss transform faults further in Chapter 18.

### Effects of Faulting on Adjacent Materials

The materials adjacent to a fault may or may not have been much changed by the effects of the fault movement. In some cases, the materials adjacent to a fault have been displaced, but not otherwise changed. Examples include faults in sediments (Figure 16.10). Some great overthrusts are remarkably clean cut and do not seem to have affected the adjacent rocks noticeably. In other cases, however, faults may create sets of striae and grooves, may break the rock into pieces and even grind down the pieces into small bits, or may orient the platy minerals so that rock cleavage results.

**Slickensides** The term *slickensides* refers to sets of parallel striae that are found in rock next to faults (Figure 16.11). Slickensides may occur in the minerals of rocks. In some cases, slickensides are not present in the minerals of the rock, but are present only in minerals such as calcite or quartz that were deposited in veins and later deformed. Other slickensides are in clays. The striae of slickensides mark the line of last movement on the fault; the tiny step-like tension fractures at right angles to the striae enable one to infer the direction of movement.

**Fault breccia, fault gouge, and mylonite** Where faults cut brittle materials and the moving fault blocks are pressed together under great

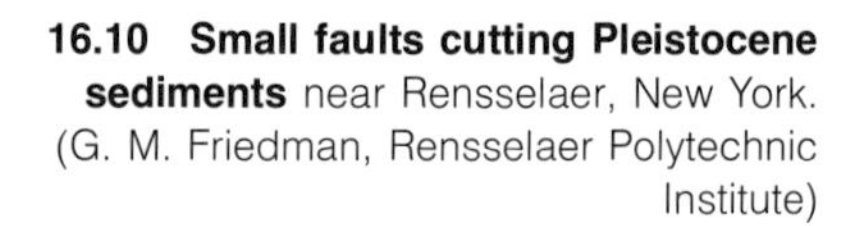

**16.10 Small faults cutting Pleistocene sediments** near Rensselaer, New York. (G. M. Friedman, Rensselaer Polytechnic Institute)

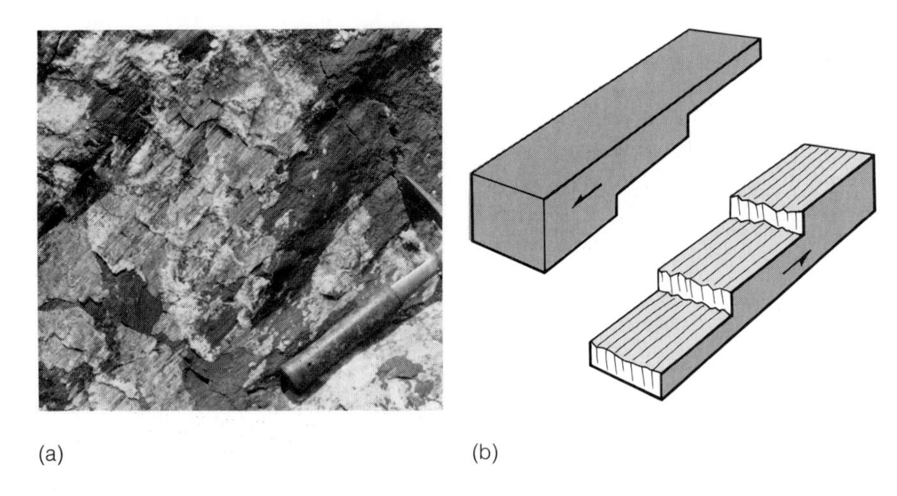

(a) (b)

**16.11 Slickensides.** (a) Slickensides in calcite deposited along a fault cutting Talcott Basalt (Upper Jurassic Newark Group), Meriden, Connecticut. Block shown moved diagonally toward upper right corner with respect to block that has been eroded away. (J. E. Sanders) (b) Use of slickensides to infer fault motion, schematic blocks. The offset of the fault's most recent motion is parallel to the striae (parallel lines along the treads of the tiny "stairsteps" shown here) and in a direction that causes the blocks to separate along the vertical surfaces at right angles to the striae (the risers of the tiny "stairsteps"). The sense of motion between the two blocks is left lateral.

pressure, fault motion may create a body of broken rock having angular pieces, named *fault breccia,* (Figure 16.12) or various ground-up rock material, named *fault gouge* and *mylonite* (see Chapter 9).

A fault breccia is usually accompanied by a network of pore spaces through which subsurface solutions can pass. As a result, the spaces within a fault breccia commonly are the sites where various minerals are deposited. By contrast, fault gouge tends to lack such openings. Zones of fault gouge typically serve as barriers to the flow of subsurface fluids.

**Cleavage** Where fine-grained sedimentary materials are adjacent to a fault, the effect of motion may be to align the platy minerals. This alignment may give rise to *slaty cleavage* (see Appendix Figure C.13).

### Patterns of Folds

Folds vary from tiny isoclinal features within a metamorphic rock (Plate 9) to structures so large that one is not immediately aware that they are present. Three patterns of folds are concentric folds not involving basement rock, folds involving basement rock, and similar folds.

**Concentric folds not involving basement rocks** In concentric folds, all the layers are bent around the fold axis, and internal adjustments are made by slippage along bedding surfaces. In this way, slickensides may form with their striae oriented perpendicular to the strike of the fold axis (or axial plane). The overall shapes of the folds depend upon the possibilities for motion downward. If the folds are underlain by a rigid basement that does not permit motion downward, then anticlines are free to grow upward, resisted only by the force of gravity acting on the mass of the material involved, but the synclines cannot develop downward. The resulting folds are a series of flat-bottomed synclines and narrowly curved anticlines (Figure 16.13a). Equal curvature of both anticlines and synclines is possible only where the synclines can move downward at the same time that the anticlines are growing upward (Figure 16.13b).

In concentric folds, the limbs are squeezed and space tends to be formed in the crestal areas. Such spaces are best developed in anticlines, where the strata may separate and form arcuate openings.

**16.12 Fault breccia** whose former open spaces have been filled by copper-sulfide minerals. (J. E. Sanders)

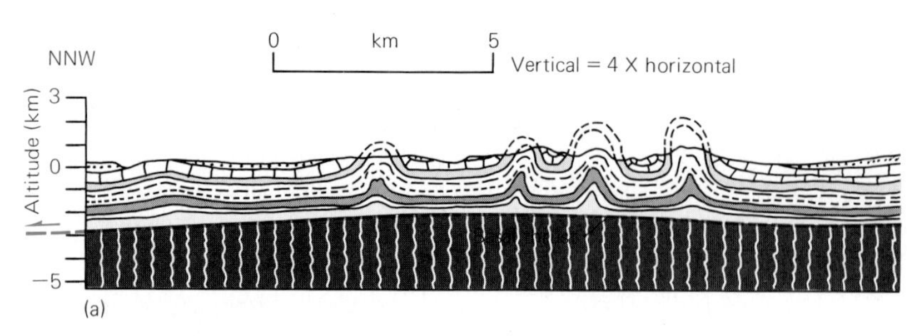


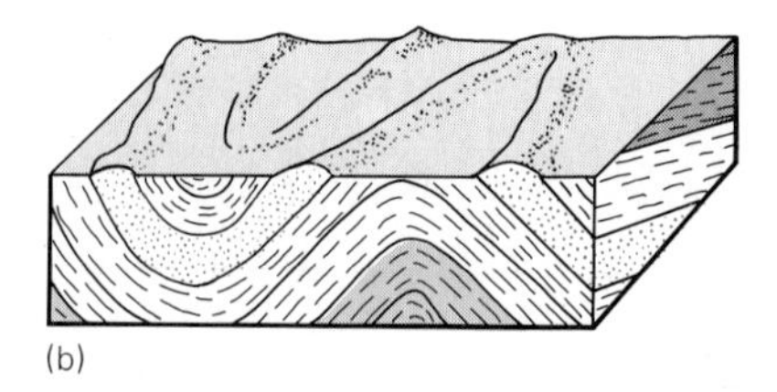


**16.13 Contrasting fold patterns as functions of basement behavior.** (a) Basement does not yield downward; as a result, strata slip along a basal thrust. Anticlines grow upward, but synclines cannot shift downward, but are expressed as flat-bottomed areas between anticlines. Profile and section through Jura chain, Switzerland. (b) Plunging anticlines and synclines of comparable curvature, a product of folding where the underlying basement yields downward during deformation. Schematic block.

**Folds involving basement rocks** Where the body of rocks being deformed is thick enough to include both basement and a covering of stratified materials, then the strata may be comparatively passive and the basement rocks may be actively motivating the growth of the folds. A typical setting for such folds is the zone of rocks affected by a strike-slip couple, as along a strike-slip fault.

**Similar folds** The folds designated as *similar folds* result from flowage of materials along a series of parallel planes, as in a glacier (see Figure 13.2). In similar folds, the flow planes become the axial planes; the original plane surface assumes the shape of one or more folds not so much by flexing as a result of squeezing from the sides, but by slippage as in a deck of cards (see Figure 11.5).

## Overthrusts

As a first approximation, an *overthrust* can be considered as nothing more than a reverse fault having a low angle of inclination. Indeed, many overthrusts that formed by shearing across fold axes fit this concept (Figure 16.14). However, some great overthrusts clearly have formed without breaking across a fold, have had the fault appear at the Earth's surface, and the upper block has continued to move along this surface, either on the sea floor or on land.

Movement on some overthrusts of this kind has been as great as 200 kilometers and more. Along many such overthrusts, great bodies of flat-lying strata have been shifted over other flat-lying strata, along what are known as *bedding thrusts* (see Figure 16.1). Along a few great overthrusts, basement rocks have been transported over strata.

As some great overthrusts have moved forward, they have shed down their frontal slopes sedimentary debris eroded from the moving upper block. By continued movement the upper block has overridden this debris. The action is somewhat comparable to what happens at the front of a flow of block lava (see Figure 5.5c). Blocks that tumble off the front of the lava flow are then covered by the flow as it advances over them.

After the strata have been displaced on a great overthrust, further deformation may take place. The overthrust may be folded, overturned, and even displaced by other faults, including other overthrusts.

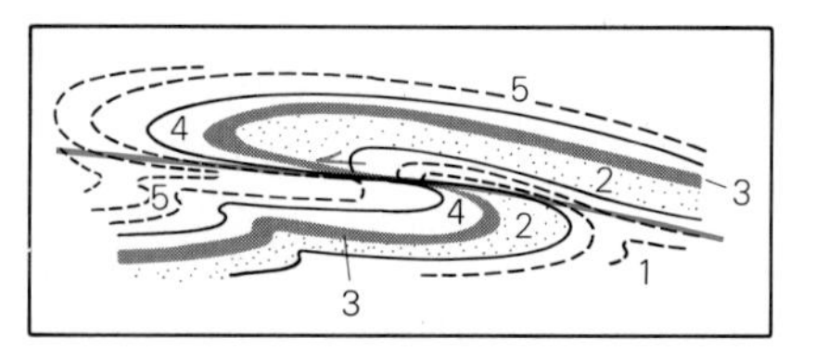


**16.14 Overthrust formed by breaking across limb of fold.** Schematic section.

## TECTONIC INFLUENCES ON SEDIMENTATION

Tectonic movements of the lithosphere affect sediments in several ways. The speed and duration of uplift of an area determine to a large extent the coarseness of the sedimentary debris eroded from the uplifted area. The relationship between rate of subsidence and rate of supply of sediment governs the depth of water in the basin where sediment accumulates. Growing faults or folds can affect drainage patterns on land and movements of sediment under water.

### Abundant Coarse Sediment

It is axiomatic in geology that if strata consist of detrital sediment, then some area must have been elevated and eroded. The kind of sediment will depend, first of all, on the kind of materials forming the elevated tract. Thus, if a body of ancient limestones is elevated, the detritus will consist of pieces of limestone. Similarly, if a granite is elevated, the possibility exists for particles of granite to be in the detritus. We have to say "possibility," because granite contains abundant feldspar and, depending on the conditions of weathering, the feldspar may or may not survive. In turn, the conditions of weathering can be affected by the rate and extent of elevation. If rapid elevation creates steep relief, then the feldspar probably will survive.

The coarseness of the detritus depends on many factors, only one of which is relief. However, once a supply of coarse sediment has been established, then its continuation nearly always implies that tectonic uplift persisted.

### Thick Shallow-Water Strata

One of the important controls on the kind of sediment deposited in a body of water is the depth. Among other things, depth influences the distribution of bottom-dwelling invertebrates. Numerous examples exist in the geologic record in which organisms that are confined to shallow depths are distributed through a much greater thickness of strata than the depth range in the sea spanned by the living organisms. Thus, the distribution of certain shallow-water organisms that live in water only a few meters deep through a stratigraphic thickness of more than 10 kilometers beneath the Bahama Bank implies that the lithosphere there subsided through 10 kilometers at a rate that could be matched by the accumulation of sediment (Figure 16.15).

### Formation of Deep-Water Basins

Where the surface of the lithosphere subsides faster than sediments can accumulate, the water simply becomes deeper. At a later time, sediments may fill the basin. If they do, then the lower parts of the filling sediments will consist of deep-water sediments and the upper parts, of shallow-water sediments. An example is the Ventura Basin, California, which subsided rapidly during much of the Cenozoic Era and formed a depression in which the depths reached 2000 meters. The Pliocene strata that filled this basin are 3000 meters thick and consist of deep-water deposits. During the Pleistocene Epoch, the water became progressively

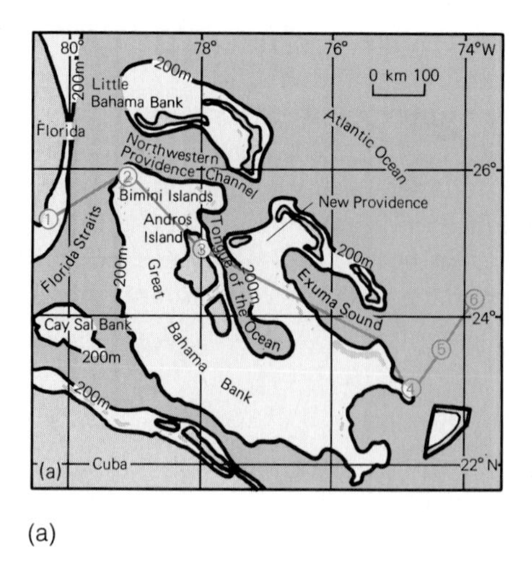


(a)

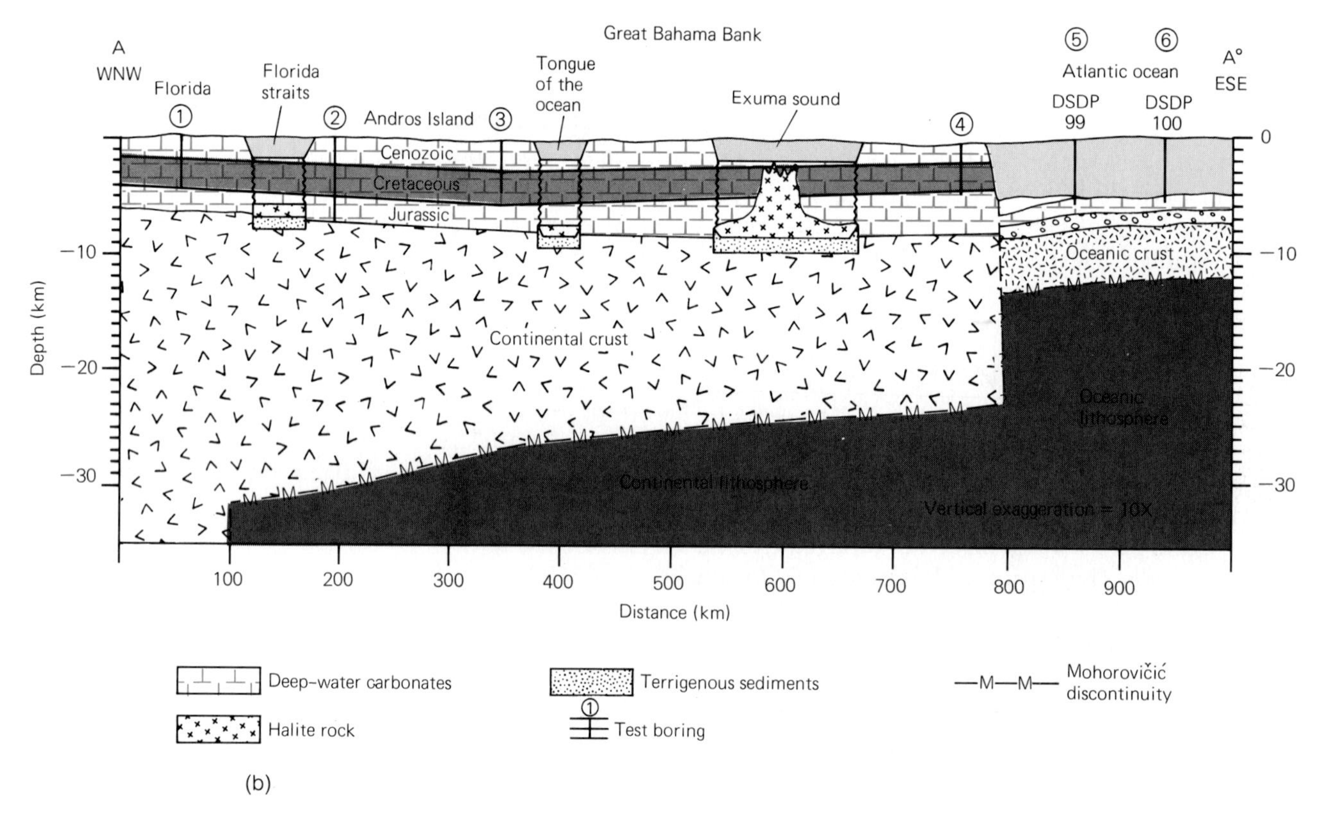


(b)

**16.15 Results of subsidence, The Bahamas.** (a) Map of the Bahamas, a series of low islands and adjoining areas of shallow water (depth usually < 10 meters), into which project various deep channels, extensions of the adjacent Atlantic Ocean. (Gerald M. Friedman and John E. Sanders, 1978, Principles of sedimentology: New York-Santa Barbara-Chichester-Brisbane-Toronto, John Wiley and Sons, 792 p., Figure 12-25, p. 371) (b) Subsurface relationships beneath the Bahamas and vicinity, based on results of petroleum-exploratory drilling (holes 1 to 4), scientific holes drilled by the *Glomar Challenger* (holes 99 and 100), and geophysical measurements and interpretations. The total subsidence since the Jurassic Period has been 8 to 10 kilometers, at an average rate of about 0.05 mm per year (about 5 centimeters per 1000 years). Since the layer of salt was deposited beneath the modern deep-water areas, deep-water carbonate sediments have accumulated in these areas, whereas beneath present shallow-water areas and adjoining banks and islands, shallow-water carbonate sediments have thickened. The salt beneath Exuma Sound has penetrated upward as one or more diapirs. The detailed relationships between oceanic structure and the structure beneath the Bahamas are not known, but are suggestive of a fault. (Redrawn from Henry T. Mullins, George W. Lynts, A. Conrad Neumann, and Mahlon M. Ball, 1978, Characteristics of deep Bahama channels in relation to hydrocarbon potential: American Association of Petroleum Geologists, Bulletin, v. 62, no. 4, p. 693–704, Figure 2, p. 696)

shallower and eventually, the basin became part of the land. The Pleistocene strata are about 1800 meters thick.

### Growth Structures

The term *growth structures* designates features of local scale that affect the thickness and characteristics of sediments being deposited. The term refers to the recording of the growth of the structures by the sediments as contrasted with sedimentary strata that are deposited as horizontal sheets and afterwards are deformed. The effects of growth structures differ, depending upon whether the sediments near the structures are being deposited by rivers flowing parallel to the long axes of the structures or are accumulating on the sea floor.

## DISTRIBUTION OF TECTONIC STRUCTURES: TECTONIC PROVINCES

In the modern world, it is possible to divide the lithosphere into two contrasting areas: (1) stable, broad expanses, usually in the interiors of continents, where modern tectonic activity is negligible, and where not much tectonic activity has taken place during the past hundred million years or more, and (2) unstable, elongate zones where tectonic activity is in progress or has taken place during the Cenozoic Era. The unstable areas of lithosphere can be further subdivided into orogenic belts and fracture zones. This recognition of this arrangement contributed to the fundamental ideas of plate tectonics which state that tectonic activity is negligible in the interiors of plates and is concentrated at plate boundaries.

### Stable Regions

The parts of the lithosphere that have been tectonically stable in the later parts of geologic history can be identified by their thin veneer of horizontal strata resting on basement rocks (Figure 16.16). In these broad areas, the surface of the lithosphere has moved downward and upward more or less as a single unit. The strata that accumulated may have been cut by faults or gently folded or warped on a broad scale, but they have not been closely folded nor extensively involved with overthrusts.

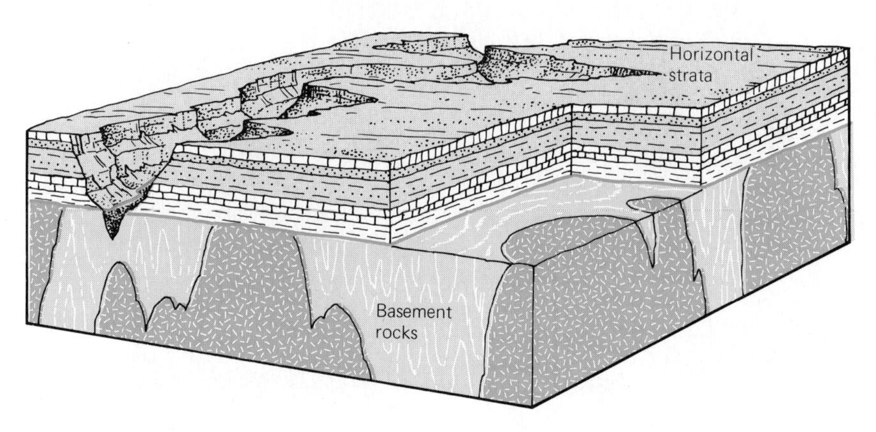


**16.16 Tectonically stable area,** schematic block diagram. Although the thickness of the strata means that the block formerly subsided, and the erosion of the river valley implies vertical uplift, the nonfaulted horizontal strata indicate tectonic stability. The underlying basement rock, however, is the product of an ancient belt of vigorous tectonic activity.

## Unstable Regions

The parts of the lithosphere that have not been tectonically stable in the later parts of geologic history are comparatively narrow, elongate belts in which the strata typically are unusually thick and have been extensively deformed. In such belts, the strata may have been closely folded and overthrust and may display the effects of great heating (orogenic belts) or they may have been extensively fractured to form rift valleys or great blocks that have been displaced laterally.

**Orogenic belts** Literally defined, an *orogenic belt* is a mountain belt; the term *orogeny* refers to the birth of mountains, which will be discussed in Chapter 18. Here, we want to discuss orogenic belts in terms of their structural features.

Orogenic belts are characterized by their thick strata and extensive development of geologic structures. All kinds of anticlines and synclines are present (Figure 16.17); these usually are accompanied by great overthrusts, which may or may not be related to broken folds.

In the interior parts of orogenic belts, the strata have been metamorphosed on a regional scale (see Chapter 9); vast batholiths of granitic rocks may be present. The overall impression one gets from the structures of orogenic belts is that the deformation resulted from the coming together of great parts of the lithosphere and that the materials being deformed were pliable and ductile. Also, the metamorphic cores of orogenic belts are new additions to the world's supply of basement rocks.

**Fracture zones: rift valleys** The fracture zones in which the lithosphere breaks along great faults and then rift valleys or grabens form are products of extension or of pulling apart (see Plate 1). If the conditions that create a rift valley continue for tens of millions of years or longer, than a new ocean may form. Material from the mantle fills in where the lithosphere has broken apart. This kind of activity is explained further in Chapter 18.

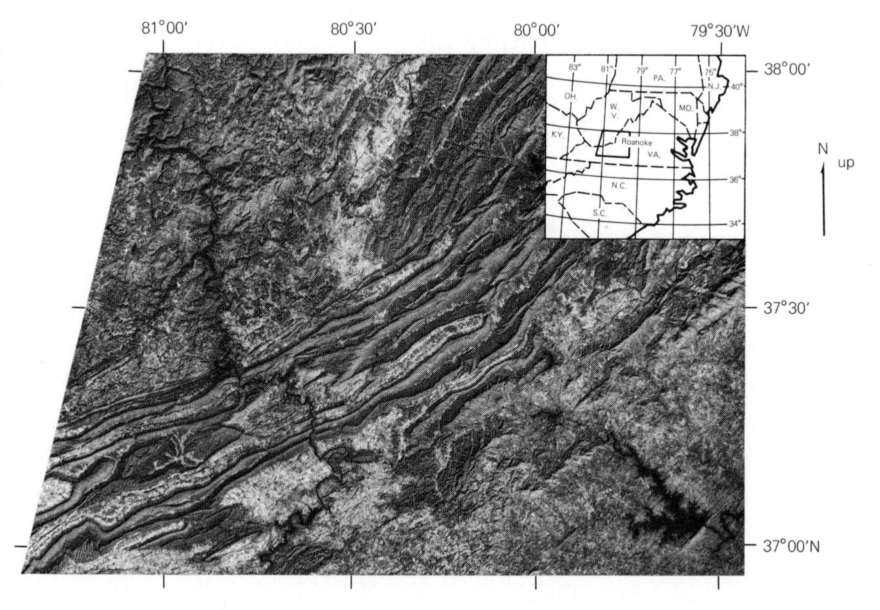


**16.17 Appalachian orogenic belt;** numerous large folds displayed by tilted edges of resistant strata. Satellite image on 26 October 1973 of central southwestern Virginia and adjacent parts of southern West Virginia. New River flows in incised meanders across the folded strata. Roanoke River and James River (at right) flow southeastward. (NASA)

**Fracture zones: blocks that are offset laterally** Along some great fracture zones, the predominant direction of motion is lateral and strike-slip offset takes place. Great zones of strike-slip faults on continents or of transform faults on the deep-sea floor are examples (Figure 16.18).

In contrast to orogenic belts, the style of deformation in fracture zones reflects a brittle condition. Rocks are broken, either cleanly where they are being pulled apart or with accompanying zones of fault breccia, fault gouge, or mylonite where the blocks are shifting while being pressed together.

In closing this section on distribution of tectonic structures, we emphasize that the tectonic conditions described for a given area apply only within a specified period of geologic time. At earlier times, other tectonic conditions may have prevailed. Thus, a broad, stable region under present conditions may contain one or more ancient, inactive orogenic belts and may have been crossed by many now-inactive fracture zones.

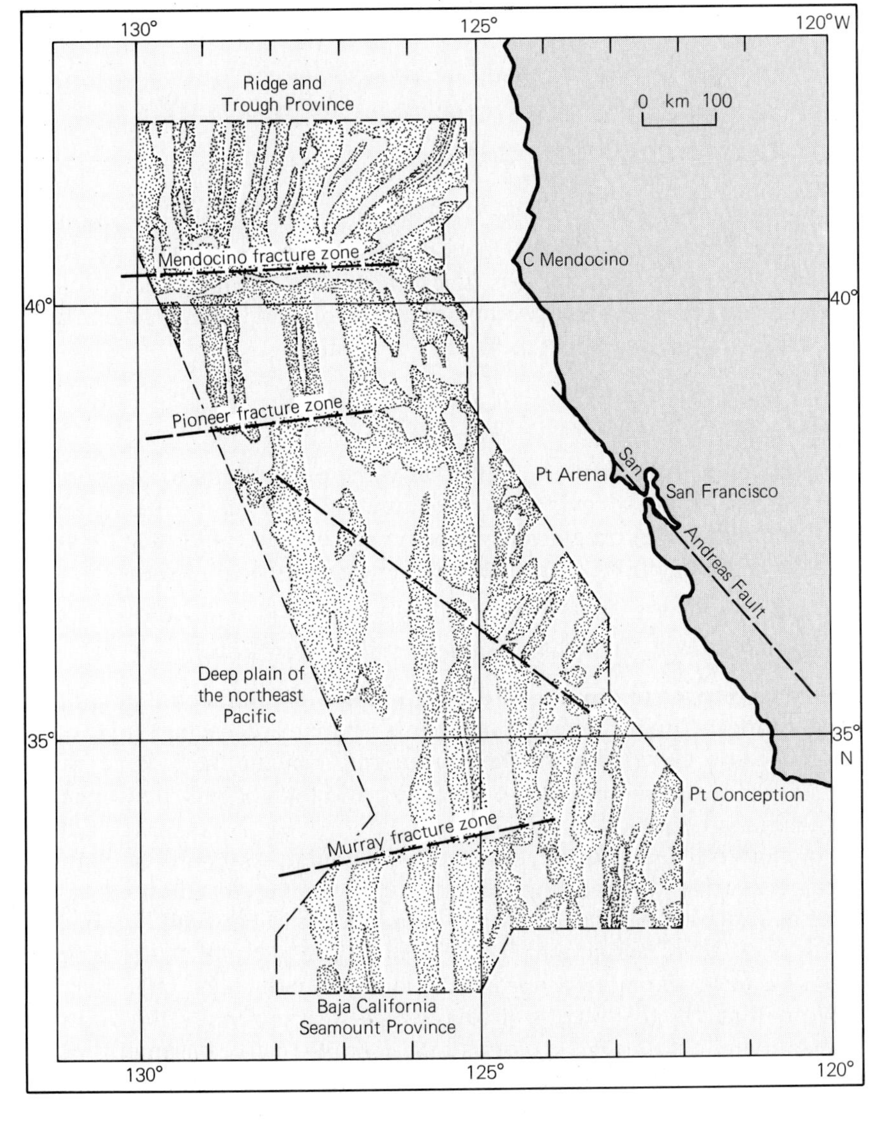


**16.18 Fracture zones on floor of Pacific Ocean,** shown on first modern magnetic survey. The discovery of these remarkable linear patterns of magnetic anomalies began a new era in the study of the sea floor and of tectonics (see Chapter 18). (Arthur D. Mason and Ronald G. Raff, 1961, Magnetic survey off the west coast of North America, 32°N Latitude to 42°N Latitude: Geological Society of America Bulletin, v. 72, p. 1259–1266, Figure 1, p. 1260)

(a)

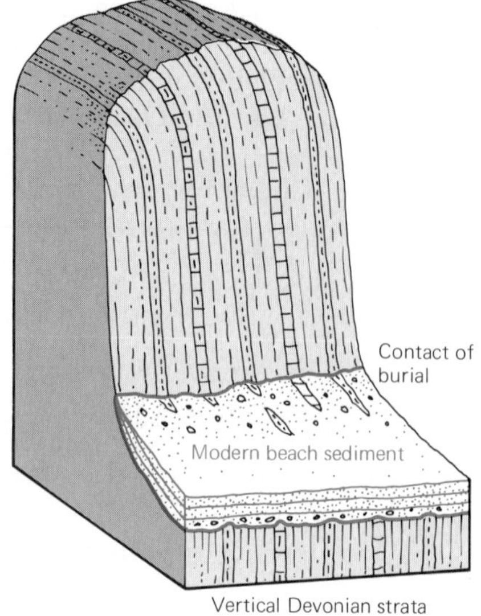


(b)

(c)

**16.19 Relationship of angular unconformity.** (a) Vertical Devonian strata being buried by modern sediments, Mt. Joli, Percé, Quebec. (J. E. Sanders, August 1965) (b) Block diagram extending from area of photograph of Figure 16.19a outward to show horizontal strata of modern marine sediments burying vertical Devonian strata. (c) Silurian limestones, dipping steeply to left, resting with angular unconformity on nearly vertical Ordovician grayworkes and siltstones. Roadcut, N.Y. 32, W. of Rip Van Winkle bridge, Catskill, N.Y. (Thomas H. Sanders)

## DATING TECTONIC MOVEMENTS

Even if one is not able to explain how a given tectonic feature came into existence, it is usually possible to assemble geologic evidence that can be used to determine more or less when the tectonic feature formed. The evidence can be grouped as sedimentary, as related to geologic contacts of burial, and as related to radioactive minerals.

### Sedimentary Evidence for Tectonic Movements

As explained in the section relating tectonic movements to sedimentation, several lines of evidence in sediments may show that tectonic movements were in progress while sediment was accumulating. In such cases, the dating of the sediments dates the time of tectonic movements.

### Geologic Contacts of Burial: Concept of Unconformity

The continuity of an episode of deposition can be interrupted in various ways connected with tectonic movements. In many cases, parts of the sea floor become elevated onto the continents, remain there for a time, subside to become submerged once again, and then begin a new episode of sedimentation.

For example, in Figure 16.19, vertical Devonian strata are overlain by modern beach sediment. The relationship of the beach sediments to the vertical strata is one of *unconformity* (a relationship between two bodies along a contact of burial, which corresponds to a gap in the geologic record). Because the nearly horizontal modern sediments bury vertical Devonian strata, this section displays *angular unconformity* the younger strata are not parallel to the older strata. If strata overlie basement rocks, the relationship is one of nonconformity (see Figure 16.16).

Not all strata deposited on the sea floor become folded when they are elevated onto the continents. In broad, stable areas, vertical movements take place with little disturbance of the horizontal strata. Along some contacts of burial, the older eroded strata are parallel to the younger covering strata. An unconformable relationship between parallel groups of marine strata that are separated by an erosion surface is one of *disconformity*. The term disconformity has been applied also in a broader sense to refer to the situation in which any kind of parallel strata include gaps marking times when no strata accumulated.

The age of tectonic movements is determined by finding the youngest strata that have been affected by a particular event and by seeking the oldest covering strata that were deposited later on.

### Radioactive Dating of Tectonic Movements

Tectonic movements sometimes involve the formation of new minerals or the recrystallization of old minerals. In either case, the minerals, such as micas, contain radioactive isotopes, and thus the time when the new lattices grew is a time of new accumulation of decay products. A radioactive date on appropriate specimens, therefore, can be the basis for determining when a tectonic event took place.

Radioactive minerals that were deeply buried may have been so hot that the daughter-isotope gases escaped continuously from the crystal lattices. However, when the deeply buried rocks were tectonically elevated into a cooler region, the gaseous decay products could remain within the lattices. In such a case, rocks of widely different geologic ages, say from Precambrian to Devonian, would all show about the same radiometric age that was clearly not related to an episode of thermal metamorphism. For example, all the dates might be about 180 million years, which would be in the Jurassic Period. The Jurassic dates, therefore, would mark the tectonic event of moving upward into a cooler environment and *not* the dates of the rock units.

**16.20 The world's mountains** represented as the only remaining land areas if sea level were to rise by 1000 meters. (a) Main area of the Western Hemisphere is the Cordillera extending from Alaska to the southern tip of South America. (b) Chief areas in Eastern Hemisphere are the Alpine-Himalayan belt, parts of Africa, and various island peaks.

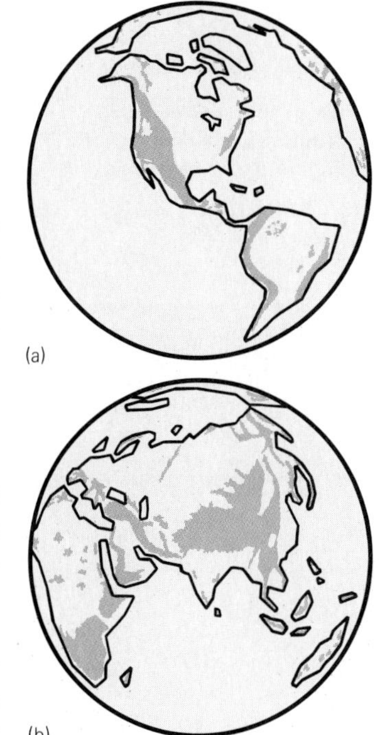


## MOUNTAINS

Mountains are major relief features of continents. The term *mountain* has been applied to any isolated feature that stands 400 meters or more higher than its surroundings and to various complex groups of ridges that form enormous areas (Figure 16.20). In analyzing mountains, geologists make distinctions based on geologic structure.

### Kinds of Mountains Based on Geologic Structures

Using geologic structure as a basis for classification, one can recognize six major kinds of mountains: (1) volcanic cones, (2) plateaus, (3) large domes exposing basement rocks, (4) fold mountains, (5) complex mountains, and (6) fault-block mountains.

**Volcanic cones** As discussed in Chapter 5, volcanic products may accumulate to build cones or groups of cones having local relief in excess of 400 meters. On the basis of their relief, large volcanic cones clearly

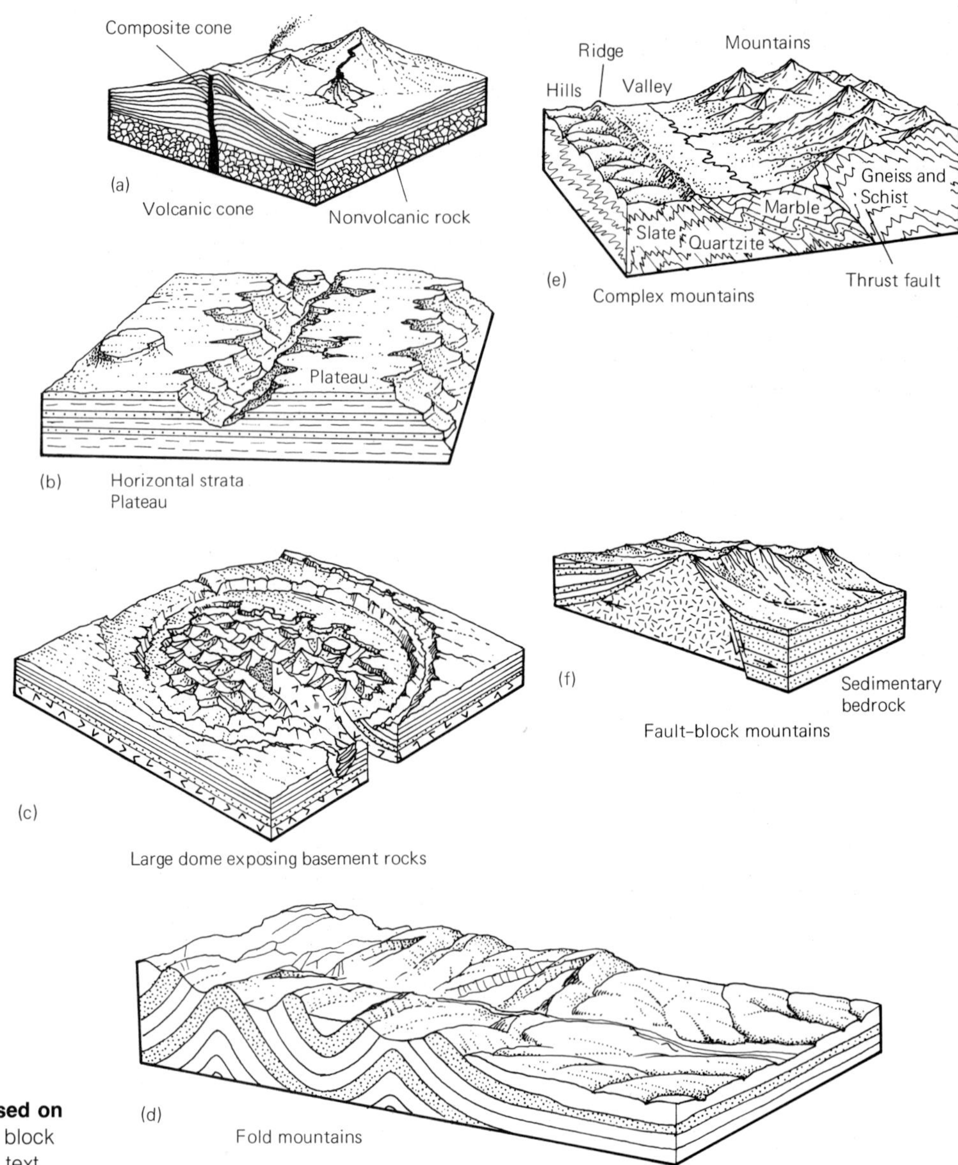


**16.21 Kinds of mountains based on geologic structures,** schematic block diagrams. Further explanation in text.

qualify as mountains (Figure 16.21a). Some volcanic cones that have built upward from the floor of the Pacific Ocean constitute some of the largest individual morphologic features on the Earth (see Figure 5.13).

**Plateaus** A *plateau* is a deeply dissected area in which the stratified rocks, either sedimentary or volcanic, are essentially horizontal (Figures 16.21 and 16.22). Many plateaus have been named mountains, for example, the Catskills in New York State. The high-standing Catskills and low-lying adjacent Hudson Valley provide an instructive example showing the independence of relief and geologic structure. The local relief in the Catskill Mountains is more than 1000 meters. Nevertheless, the strata underlying the Catskills are essentially horizontal. By contrast, the rocks beneath the Hudson Valley display many of the structural characteristics of an orogenic belt, such as overturned folds and over-

**16.22 Plateau,** viewed obliquely from an airplane. Mountainous relief in a cold climate results in capping of snow, S coast of Devon Island, Northwest Territories, Canada, looking E. Nearly horizontal Cambrian strata overlie Precambrian basement rocks (exposed at lower left). (Canadian Government Copyright, National Air Photo Reproduction Centre, Canadian Department of Energy, Mines, and Resources)

thrusts. Despite such crumpled structures, the local relief within the Hudson Valley is never more than a few hundred meters.

**Large domes exposing basement rocks** Mountains formed by *large domes exposing basement rocks* result from great vertical uplift of a cylindrical segment of the lithosphere and later differential erosion (Figure 16.21c). The basement rocks that become elevated in the cores of these great domes generally are much more resistant to erosion than are the sedimentary strata which originally covered the ancient rocks. After great erosion, the resistant cores remain as mountains, whereas the land surface underlain by the covering strata is lowered considerably. Examples include the Black Hills in South Dakota, the Ozarks in Missouri, and the Adirondacks in New York State.

**Fold mountains** When geologists refer to *fold mountains* they mean mountain belts whose major relief is related to the eroded parts of extensive groups of folds (Figure 16.21d).

The earliest geologic observers noticed small folds that could be seen, for example, in a coastal cliff (Figure 16.23), but, the idea that the strata of an entire mountain chain had been folded on a large scale did not occur to them. That notion was first proposed about the middle of the nineteenth century, when the Appalachian chain in Pennsylvania was being mapped. In eastern Pennsylvania, eroded edges of inclined strata form parts of enormous parallel folds that are hundreds of kilometers long and up to tens of kilometers wide (Figure 16.24). These folds have elevated and depressed the strata by thousands of meters. The present-day surface of the Earth is the result of great erosion that took place after folding. This part of the Appalachians has been named the Valley and Ridge Province.

**16.23 Folds exposed in coastal cliff.** Lulworth, Dorset, England. (G. R. Roberts)

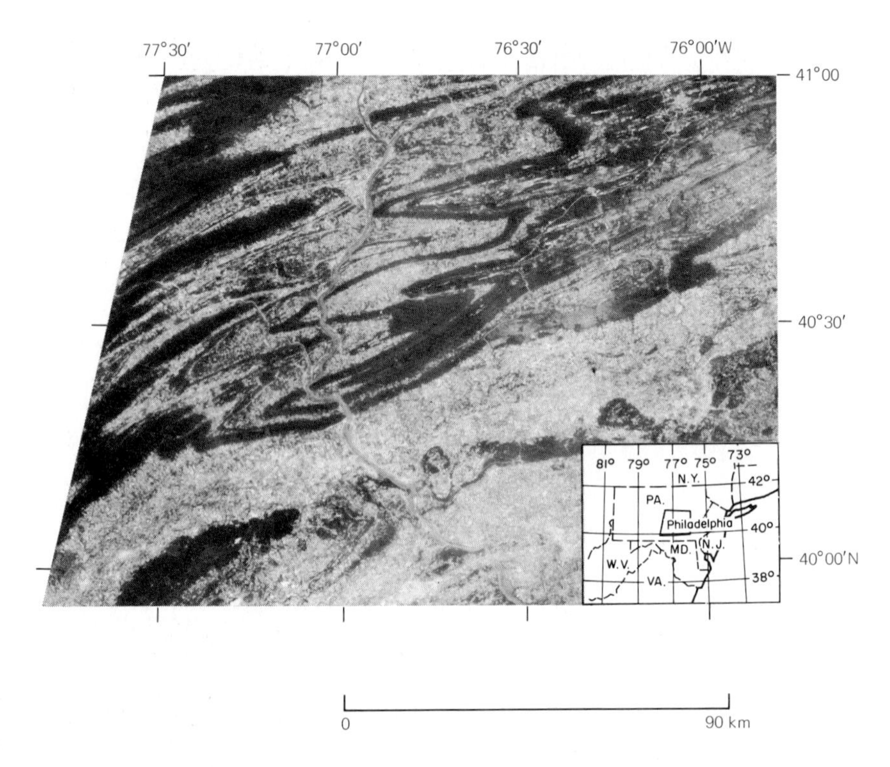


**16.24 Fold mountains,** satellite image on 8 July 1973 of Appalachian Valley and Ridge province in eastern Pennsylvania north and east of Harrisburg. Susquehanna River flows S at left of center of view. Black areas are tree-covered ridges formed by the tilted edges of resistant sandstone strata on limbs and noses of eroded plunging folds. (National Aeronautics and Space Administration)

**16.25 Complex mountains,** the Swiss Alps. The three ridges in the foreground are underlain by Mesozoic marine strata that form parts of great thrust sheets. The jagged peaks on the skyline are composed of Paleozoic granitic gneisses. Peak at extreme left is the Jungfrau. (Swiss National Tourist Office)

Other fold mountains besides the Valley and Ridge Province include the Zagros chain in Iran and the Jura in France and Switzerland. In young mountain ranges, many anticlines form elongated hills and synclines form valleys. In old fold chains, this relationship between folds and the landscape may have been reversed. After much differential erosion, relief is controlled by the resistance of the strata. In such mountains, the axes of anticlines may underlie valleys, whereas the rocks in the troughs of synclines cap high ridges (see Figure 16.1d).

**Complex mountains** Mountainous relief that results from the elevation and differential erosion of orogenic belts having complex internal structures, including folds, great thrusts, batholiths, and belts of regionally metamorphosed rocks, are *complex mountains* (Figure 16.21e). Nearly all orogenic belts contain at least some parts that would be classed as complex mountains. These parts usually are in the interior of the chain. Examples include the Caledonian chain of Wales, Scotland, and Norway; the interior parts of the Alps (Figure 16.25); the interior parts of the Himalayas; and parts of the Appalachians in eastern Canada, New England, and the Great Smokies in Tennessee and North Carolina.

**Fault-block mountains** Faults may raise and lower any parts of the Earth's surface. Commonly, however, orogenic belts become broken by networks of faults. A typical sequence would be a newly deformed orogenic belt becomes a fold-mountain chain or a complex-mountain chain; the mountain chain is eroded to a peneplain; shortly afterward, the peneplain is elevated and the rocks are broken into a series of fault blocks, some of which may become fault-block mountains (Figure 16.26). The mountainous relief of *fault-block mountains* results from vertical elevation and possibly also from differential erosion, as in domal uplifts (see Figure 16.21f).

Fault blocks may be tilted because they were elevated much more on one fault than on an adjacent fault. The great Sierra Nevada of eastern California is an example of an enormous tilted fault-block range (Figure 16.27). The steep east-facing escarpment contrasts with the gentle slope toward the west. Numerous other tilted fault-block ranges are found in southern Nevada and western Utah, in the region known as the Basin and Range Province. This region is part of an extensive orogenic belt whose age spans the Paleozoic and Mesozoic eras. The belt was subjected to several orogenies, the terminal one about 60 to 70 million years ago. The fault-block ranges are only a few million years old; some of them are still being actively elevated.

In Chapter 18 we will discuss some of the ways that mountains have been formed according to the theory of plate tectonics.

**16.26 Fault-block mountain** is being actively eroded and the debris is accumulating as fans, desert of Chile. The fault is the linear feature passing the apex of each of the two fans. The fault-block mountain's elevated, dissected area lies just to the left of the fault. Vertical air view on 14 April 1955. (Army Map Service photograph, courtesy U. S. Geological Survey)

**16.27 Sierra Nevada, California,** a tilted fault-block mountain. (Roland van Huene, U.S. Geological Survey)

# CHAPTER REVIEW

1. *Tectonic structures* are formed by large-scale motions of the lithosphere. *Nontectonic structures* are formed by processes other than large-scale movements. Nontectonic structures resulting from some kind of disturbance of sediments that takes place before the sediments became lithified are called *sedimentary structures.*

2. Geologic structures are usually defined by the ways they have affected sedimentary strata. Implied throughout is the concept that the strata were deposited in a horizontal arrangement without interference and then were deformed.

3. The term *attitude* refers to the spatial orientation of strata or some other feature, such as a fracture or a geologic contact between two different kinds of rocks. *Strike* is the compass bearing of the line formed by the intersection of the nonhorizontal surface and a horizontal plane measured east or west of true north. *Dip* is the angle, in the vertical plane at right angles to the strike, of the nonhorizontal surface and the horizontal plane, measured down from the horizontal; the maximum dip angle is 90°.

4. Strata that have been crowded together form wavelike arrangements known as folds. Two types of folds are *anticlines,* archlike upfolds, and *synclines,* troughlike downfolds.

5. *Joints* are fractures along which the opposite sides have not been displaced. Joints typically form sets of parallel cracks.

6. Fractures along which the sides have been displaced in a direction parallel to the fracture are *faults.*

7. Tectonic movements of the lithosphere affect sediments in several ways. The speed and duration of uplift of an area determine the coarseness of the sedimentary debris eroded from it; the relationship between rate of subsidence and rate of supply of sediment governs the depth of water in the basin where sediment accumulates; growing faults or folds can affect drainage patterns on land and movements of sediment under water.

8. It is possible to divide the lithosphere into areas that are tectonically stable and areas that are tectonically unstable.

9. Stable areas are broad expanses, usually in the interior of continents. Unstable areas can be further subdivided into *orogenic belts* (mountain belts), *fracture zones* in which rift valleys are formed, and fracture zones in which blocks are offset laterally.

10. Tectonic movements can be dated on the basis of sedimentary evidence, relationships along a contact of burial, and radioactivity of mineral specimens.

11. A *mountain* is any isolated feature that stands 400 meters or more higher than its surroundings. There are six major kinds of mountains: volcanic cones, plateaus, large domes exposing basement rocks, fold mountains, complex mountains, and fault-block mountains.

## QUESTIONS

1. Define *attitude, strike,* and *dip.*
2. What are *folds?* Name the two major contrasting kinds of folds and describe each by means of an annotated sketch.
3. Compare *joints* and *faults.*
4. Define *tectonic structures.*
5. List the common patterns of normal faults and briefly discuss their conditions of origin.
6. Define *shearing couple.* How are shearing couples named? What structural features are formed by shearing couples?
7. What is a *transform fault?*
8. List some of the effects of fault movement on materials adjacent to the fault.
9. What are three major patterns of folds?
10. Define *overthrust* and *bedding thrust.*
11. Discuss ways in which tectonic movements affect sediments. What are *growth structures?*
12. What is a *tectonic province?* Compare *stable areas* with *unstable areas.*
13. Discuss the geologic evidence for dating tectonic movements. Define *unconformity, angular unconformity,* and *disconformity.*
14. Define *mountain.* List six major kinds of mountains based on geologic structure.

## RECOMMENDED READING

Bain, G. W., and Beebe, J. H., 1954, Scale Model (*sic*) Reproduction of Tension Faults: *American Journal of Science,* v. 252, no. 12, p. 745–754.

Bruce, C. H., 1973, Pressured Shale and Related Sediment Deformation: Mechanism for Development of Regional Contemporaneous Faults: *American Association of Petroleum Geologists, Bulletin,* v. 57, no. 5, p. 878–886.

Demy, J. W., 1974, Turkey's North Anatolian Fault: A Comparison with the San Andreas Fault: *U.S. Geological Survey, Earthquake Information, Bulletin,* v. 6, no. 3, p. 12–16.

Engelder, Terry, 1974, Cataclasis and the Generation of Fault Gouge: *Geological Society of America, Bulletin,* v. 85, no. 10, p. 1515–1522.

Engelder, Terry, and Engelder, Richard, 1977, Fossil Distortion and Décollement Thrusting of the Appalachian Plateau: *Geology,* v. 5, no. 8, p. 457–460.

Harris, L. D., and Milici, R. C., 1977, Characteristics of Thin-skinned Style of Deformation in the Southern Appalachians, and Potential Hydrocarbon Traps: *U.S. Geological Survey, Professional Paper 1018,* 40 p.

Hobbs, B. E., Means, W. D., and Williams, P. F., 1976, *An Outline of Structural Geology:* New York, John Wiley and Sons, 571 p.

Kupfer, D. H., 1962, Structure of Morton Salt Company Mine, Weeks Island Salt Dome, Louisiana: *American Association of Petroleum Geologists, Bulletin,* v. 46, no. 8, p. 1460–1467.

# CHAPTER 17
# EARTHQUAKES, SEISMOLOGY, AND THE EARTH'S INTERIOR

EARTHQUAKES AND SEISMIC WAVES
SOME HISTORIC EARTHQUAKES
SEISMOLOGY AND THE PROPERTIES OF SEISMIC WAVES
THE DISTRIBUTION OF EARTHQUAKE EPICENTERS
THE INTERIOR OF THE EARTH
FORECASTING AND PREDICTING FUTURE EARTHQUAKES
EARTHQUAKES AND HUMAN ACTIVITES

PLATE 17
Earthquake devastation at Lioni, Italy. The quake, which coincided with a near-perigee full Moon, struck at 0734 local time on Sunday, 23 November 1980, killed more than 3000 people, and left an estimated 200,000 homeless. The calculated Richter magnitude was 6.8. (United Press International)

EARLY JAPANESE PEOPLE THOUGHT THAT EARTHQUAKES WERE THE result of sudden movements of a giant spider that carried the Earth on its back. Mongolian mythology recorded that the huge, but occasionally unbalanced, supporter of the Earth was a giant mole. Greek and Roman scholars, considerably more scientific, ascribed earthquakes to air escaping from underground caverns or to the collapse of vast subterranean cavities. In the Middle Ages, earthquakes were often thought to be divine punishment inflicted upon cities for the sins of the inhabitants.

European scholars began a more scientific treatment of earthquakes after the great tremor that devastated Lisbon, Portugal, in 1755. The systematic study of earthquakes in the United States was given its impetus by the 1906 San Francisco earthquake, a disaster that killed almost 700 people and reduced much of the city to ruins. The fault movement responsible for this earthquake left surface indications that were carefully studied by geologists, who produced the first fully documented accounts of the effects of horizontal fault displacement on the Earth's surface.

## EARTHQUAKES AND SEISMIC WAVES

### Basic Definitions

The Earth is not a static body, but is changing, moving, and full of energy. When parts of the Earth move, rocks are deformed. The changes in shape or volume which result from deformation are known as *strain*. When accumulated *strain* is suddenly released, it creates kinetic energy which travels through the Earth as *seismic waves* (from the Greek *seismos*, "shock" or "earthquake"). Seismic waves are also called *tremors*, if gentle, and *shocks*, if powerful. The study of seismic waves and earthquakes is *seismology*.

According to modern geophysicists, what we call *earthquakes* are trains of energetic seismic waves (movements that we can feel) and principal events in what may be a prolonged sequence of seismic waves. Such a sequence of waves may begin with precursory tremors that become increasingly large; these are known as *foreshocks*. The sequence usually ends with waves that begin to fade after the shock of the earthquake itself; these are known as *aftershocks*.

Many, but by no means all, earthquakes are related to motion along faults. At times, the opposite sides of a fault become locked together and both bend as further motion takes place. Eventually, the fault becomes unlocked and snaps out of its bent position (Figure 17.1). Such motion creates a large earthquake. Slow, gradual movements on faults, named *fault creep*, are accompanied by small tremors, but not by large earthquakes.

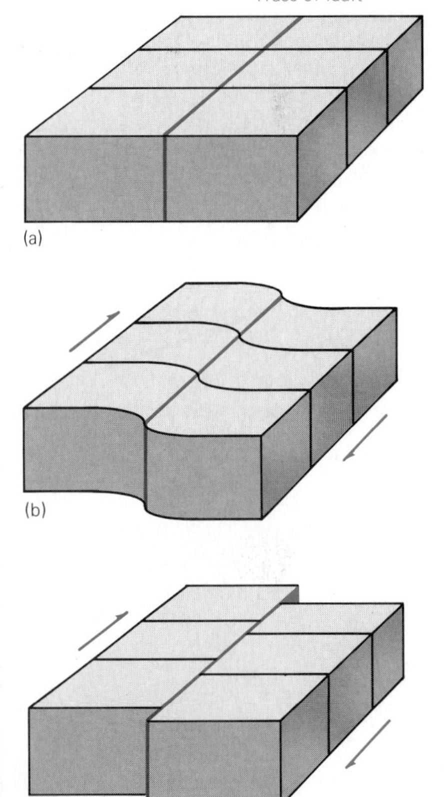


**17.1 Fault movement and earthquakes, elastic-rebound concept.** (a) Strike-slip fault with two reference lines drawn across it at right angles. (b) Right-lateral shifting of blocks adjacent to locked part of fault that does not slip; deformation is entirely by bending. (c) Eventually, the fault becomes unlocked, the blocks shift abruptly, creating an earthquake and offsetting the bent lines. When the displacement changed from bending to fault slippage, the reference lines straightened out but became offset as shown. (Based on H. F. Reid, 1911, The elastic-rebound theory of earthquakes: California University at Berkeley, Department of Geology *Bulletin*, v. 6, p. 413–443, Figure 2, p. 420, and San Andreas fault, California)

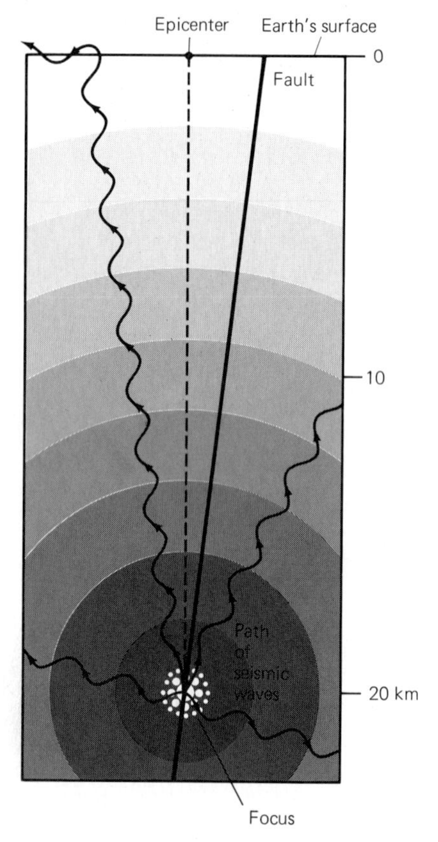


**17.2 Energy released at earthquake focus,** 20 kilometers beneath the surface, radiates outward in all directions (with fronts of waves shown by arcs of circles). Paths of body waves in four selected directions shown. Body waves that reach the surface (they arrive first at the epicenter) change kind and direction; they travel *along the surface* as surface waves (top left).

The place where the energy is first released is the *focus,* and the place on the surface of the Earth directly above the focus is the *epicenter* (Figure 17.2). The depth of the focus of an earthquake can vary from approximately 5 to 700 kilometers. Systematic mapping of these depths of focus shows a distribution throughout the world that appears to be related to the location of plates (see Figures 1.23 and 1.24).

## Measuring the Magnitude of Earthquakes

The energy released during a great earthquake (such as the 1906 California earthquake) has been compared to the force of 100,000 atomic bombs. However, to seismologists the most meaningful way to describe earthquakes is not in terms of atomic bombs or tons of TNT, but in terms of ground motion and energy released. The first crude instrument for recording ground motion caused by seismic waves was used in Italy in 1841. Modern descendants of this instrument, called *seismographs* (Figure 17.3) are still being refined, but the principle remains the same. The rec-

**17.3 An array of three seismographs,** including one for recording seismic waves in the vertical plane (upper right) and two for recording horizontal motions aligned at right angles (foreground, with slanted windows) resting on a concrete pier anchored on bedrock at Palisades, New York. The recorder being examined by the man at left is for display only; the functioning recorders, one drum for each of the instruments shown, are in an adjacent dark room. The key link between the inertia weights of the seismographs on the concrete pier and the recording drums is a galvanometer (of which four are shown here; they are the small vertical cylindrical objects between the seismograph instrument cases), an electrical device that sends signals from the interia weights to a movable mirror that reflects a narrow beam of light onto the light-sensitive film wrapped around the enclosed recording drums. The film records, which are changed daily, must be developed photographically before they can be studied. (Lamont-Doherty Geological Observatory of Columbia University).

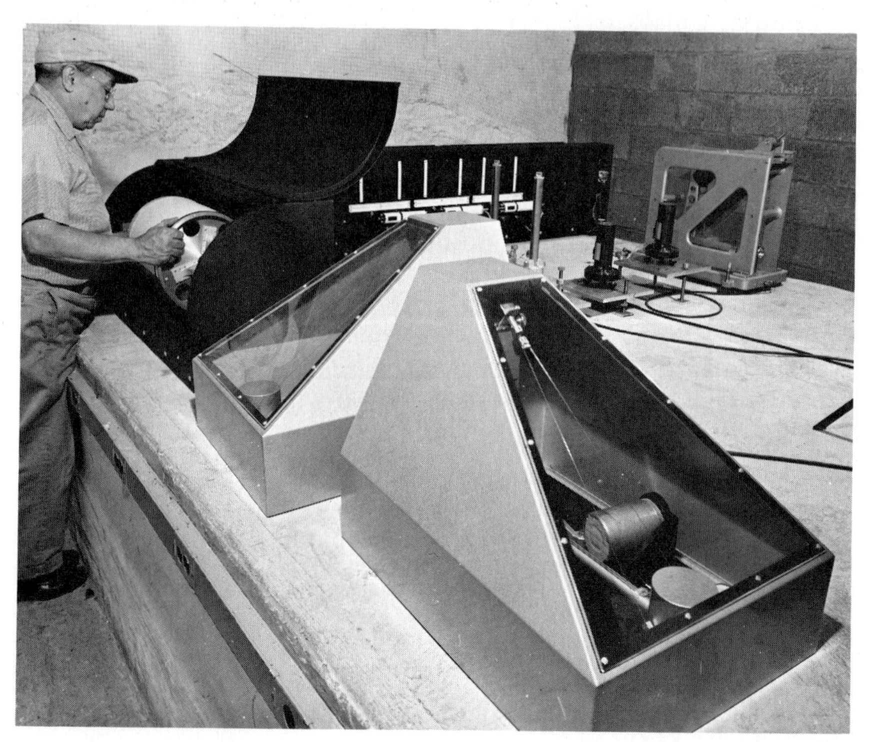

ord of seismic waves made by a seismograph is a *seismogram* (Figure 17.4).

The amount of energy released is rated by a scale devised in 1935 by Charles F. Richter (1900–      ) of the California Institute of Technology (Table 17.1). The *Richter Magnitude Scale* is an order-of-magnitude scale based on logarithms to the base 10. Thus, the amount of energy released in an earthquake of magnitude 6 is 10 times greater than a magnitude 5 earthquake. An earthquake of less than magnitude 2 is usually not felt by humans; any earthquake of magnitude 6 or greater is considered a major quake.

Early seismologists used a scale developed in the 1880s which defines earthquakes by a subjective assessment of damage and other observable effects. The scale was revised in 1902 by the Italian seismologist Giuseppe Mercalli (1850–1914) and then modified again in 1931. The modern version, the *Modified Mercalli Scale,* runs from Roman numeral I through XII and is still sometimes used along with the Richter scale (see Table 17.1).

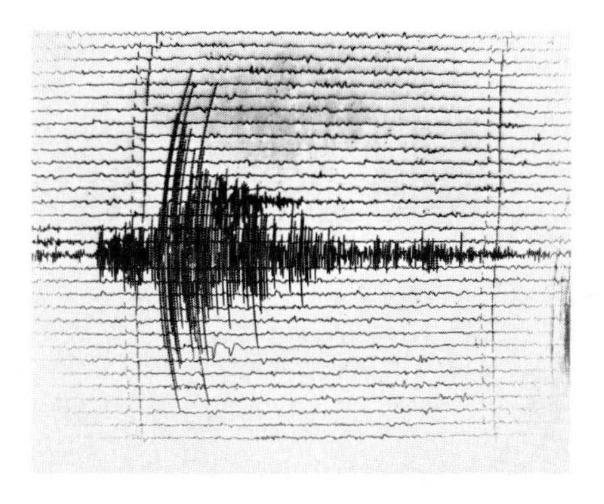

**17.4 Seismogram** of a typical high-frequency earthquake recorded in the summit region of Kilauea Volcano. The predominant frequency of this event is about 10 cycles per second and its duration about 1½ minutes. Similar earthquakes related to times of volcanic activity occur in swarms of many hundreds per day. (Hawaiian Volcano Observatory, U.S. Geological Survey).

**TABLE 17.1**
MODIFIED MERCALLI SCALE OF EARTHQUAKE INTENSITIES WITH CORRESPONDING RICHTER MAGNITUDES

| Modified Mercalli Intensity Scale | | Corresponding Richter Magnitude | Number per Year |
|---|---|---|---|
| I | Detected only by seismographs | Less than 3.4 | 800,000 |
| II | Felt only by a few people, usually on upper floors of buildings | 3.5 | 30,000 |
| III | Felt by people at rest; duration long enough to be estimated; vibrations similar to passing truck | 4.2 | |
| IV | Shaking noticed by some people outdoors and many indoors; crockery rattles, standing vehicles rock | 4.3 | 4,800 |
| V | Felt by nearly everyone, many sleepers awakened; shaking of furniture and beds, buildings and other tall objects may be disturbed, pendulum clocks may stop | 4.8 | |
| VI | Felt by all; windows may break, plaster may crack, chandeliers swing, heavy furniture may move; damage slight | 4.9–5.4 | 1,400 |
| VII | General alarm, everyone runs outdoors; noticed by people driving cars; little damage to well-constructed buildings. | 5.5–6.1 | 500 |
| VIII | Panel walls thrown out of frame structures; fall of chimneys, factory stacks, monuments, walls; heavy furniture overturned; sand and mud ejected in small amounts. | 6.2 | 100 |
| IX | Buildings shifted off foundations, underground pipes broken; ground cracked conspicuously. | 6.9 | |
| X | Most stone buildings destroyed, bridges and solid wooden buildings badly damaged, railway lines bent; slope failure on steep slopes, water splashed over banks | 7.0–7.3 | 15 |
| XI | Few buildings remain standing, bridges destroyed, underground pipes completely out of service; broad fissures in ground, earth slumps and land slips in soft ground | 7.4–8.1 | 4 |
| XII | Damage total, objects thrown into air; waves seen on ground surfaces, lines of sight and level distorted | Greater than 8.1 | 1 every 5 to 10 years |

### Effects of Earthquakes

Most earthquakes do only nuisance harm, such as stopping pendulum clocks, setting off burglar alarms and rattling tableware. But every few years a "great" earthquake — greater than Richter magnitude 8.0 — happens near enough to a population center to wreak widespread devastation (Table 17.2).

In addition to the damage done by the quake itself, further devastation may result from other natural events triggered by the earthquake. Such events include slope failures, tsunami, and fires. During the 1964 Alaska earthquake, shock-induced slope failures caused major damage in Anchorage and other areas, and tsunami accounted for more damage in coastal regions than did actual shocks.

An indirect, but severe effect of earthquakes is fire. Perhaps the most fearsome example of the destructive power of an earthquake fire in the United States followed the 1906 San Francisco earthquake (Figure 17.5). Although the main shaking of the earthquake was all over in less than a minute, fires started by hot overturned stoves, bared electric wires, broken gas pipes, and crumbled chimneys. These events started fires that lasted for days. Fire-fighting efforts were crippled because movement along the San Andreas fault broke water mains and leveled pumping stations.

**TABLE 17.2**
SOME MAJOR WORLDWIDE EARTHQUAKES

| Date | Location | Death Toll | Richter Magnitude |
|---|---|---|---|
| 1456 | Naples, Italy | 30,000 | |
| 1556 | Shensi, China | 830,000 | |
| 1716 | Algieria | 20,000 | |
| 1755 | Lisbon, Portugal | 30,000+ | |
| 1759 | Baalbek, Lebanon | 20,000 | |
| 1783 | Calabria, Italy | 50,000 | |
| 1891 | Mino-Owari, Japan | 7,300 | |
| 1899 | Yakutat, Alaska | | 8.6 |
| 1905 | Kangra, India | 19,000 | 8.6 |
| 1906 | Andes of Colombia/Ecuador | | 8.6 |
| 1906 | Valparaiso, Chile | 1,500 | 8.4 |
| 1908 | Messina, Italy | 120,000 | 7.5 |
| 1911 | Tien Shan, China | | 8.4 |
| 1915 | Avezzano, Italy | 30,000 | |
| 1920 | Kansu, China | 180,000 | 8.5 |
| 1923 | Kwanto, Japan | 140,000 | 8.2 |
| 1933 | Sanriku, Japan | 3,000 | 8.5 |
| 1939 | Concepción, Chile | 25,000 | 8.3 |
| 1950 | North Assam, India | 1,500 | 8.6 |
| 1960 | Chile (three major shocks) | 10,000 | 8.3–8.9 |
| 1964 | Prince William Sound, Alaska | 130 | 8.6 |
| 1970 | Peru | 66,000 | 7.8 |
| 1972 | Managua, Nicaragua | 10,000 | 5.7 |
| 1974 | Pakistan | 5,200 | 6.3 |
| 1975 | Ankara, Turkey | 2,400 | 6.8 |
| 1975 | Hilo, Hawaii | 1 | 7.3 |
| 1976 | Guatemala | 22,000 | 7.9 |
| 1976 | Tangshan, China | 700,000 | 7.6 |
| 1977 | Vrancea, Romania | 2,000 | 7.2 |
| 1978 | Tabas, Iran | 25,000+ | 7.7 |

**17.5 Earthquake and fiery aftermath** totally devastated most of San Francisco, California on 18 April 1906. (California Historical Society)

## SOME HISTORIC EARTHQUAKES

Our examples of historic earthquakes are Lisbon, 1755; New Madrid, Missouri, 1811–1812; and Prince William Sound, Alaska, 1964.

### Lisbon, Portugal, 1755

The destructive earthquake that struck Lisbon on 1 November 1755 probably shook people's ideas as much as it did buildings. This was the first great earthquake to strike Europe in hundreds of years.

The earthquake came as a series of three great shocks. The first shocks lasted six or seven minutes — a long time by usual earthquake standards. Within a few minutes the shaking destroyed all large buildings and ruined half of the houses in the city. Because it was All-Saints' Day, thousands of worshippers had gathered in the churches. Of the 30,000 known fatalities, many were killed when the churches collapsed.

In the second series of shocks the massive new stone quay along the Tagus River where hundreds of people had sought safety, sank into the river.

Not long afterward a tsunami generated by the earthquake roared up the river, smashed the waterfront, and devastated everything in its path for as much as a kilometer inland. Other tsunami reached Ireland and even Antigua in the West Indies, 5600 kilometers away on the opposite side of the Atlantic Ocean.

The leader of Portugal at the time, the Marquess of Pombal (1699–1782), distributed and collected questionnaires to record eyewitness accounts and to provide material for cataloging and classifying the effects of the earthquake on various kinds of buildings. The written records kept by Portugese priests are still preserved today; they represent the first systematic attempt to document the effects of an earthquake.

### New Madrid, Missouri, 1811–1812

Just after two o'clock in the morning of 16 December 1811, an earthquake centered near New Madrid, Missouri, caused damage throughout much of the Mississippi River Valley and shook the Earth's crust with such violence that people were awakened in cities as far away as Pittsburgh and Norfolk (Figure 17.6).

The three largest shocks in the series were felt in Quebec, on the Canadian Atlantic coast, and in the Rocky Mountains. The shaking continued on and off through March 1812, and aftershocks were still felt five years later.

In comparison with many other famous earthquakes, the quake at New Madrid was neither one of the most dramatic nor the most destructive. Its magnitude was one of the largest, however, and this earthquake serves as an important reminder that earthquakes can happen, and have

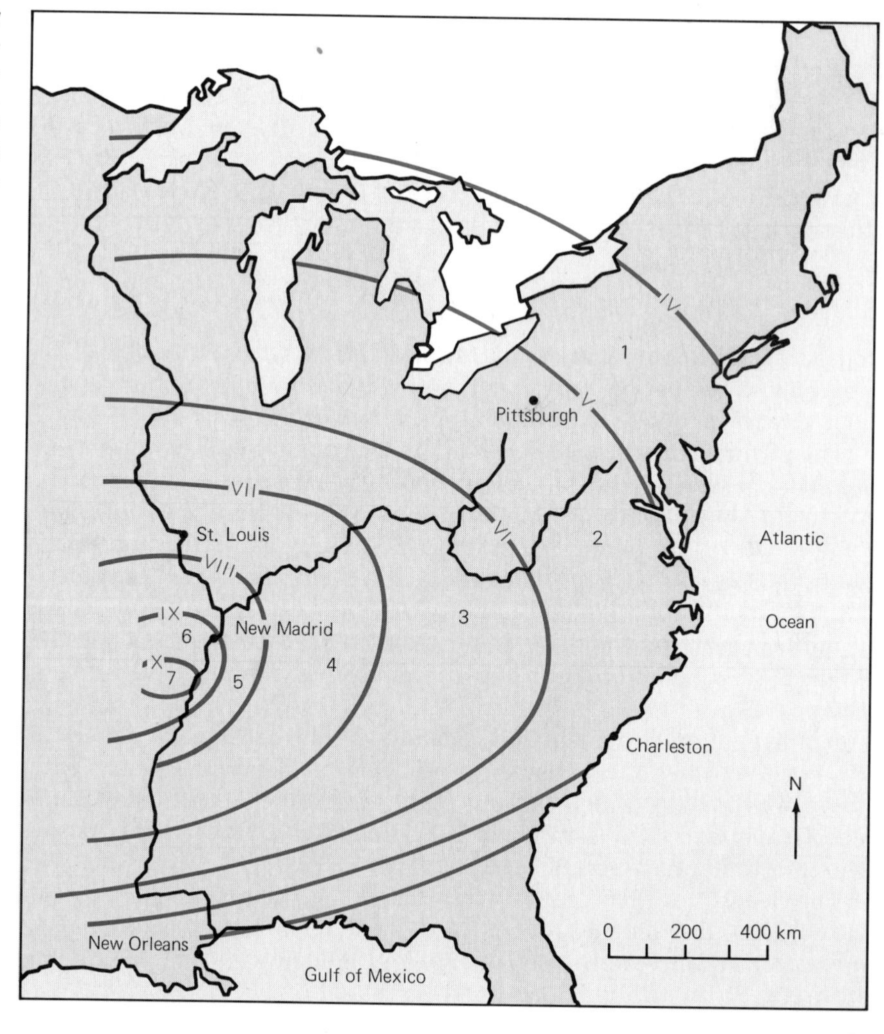


**17.6 Areas influenced by the New Madrid earthquake,** a long series of tremors that began on 16 December 1811. The lines on the map are *isoseismals*, lines that connect areas subjected to equal intensity of wave action from a given earthquake.

**17.7 Destruction in downtown Anchorage,** Alaska, Fourth Avenue near C Street, a result of ground shifting during the Prince William Sound earthquake of 27 March 1964. (U. S. Geological Survey)

happened, in the United States in areas other than the west coast and Alaska. Another important point to note about the New Madrid earthquake is that its epicenter is situated far from any recognized plate boundary.

### Prince William Sound, Alaska, 1964

The greatest earthquake of modern times jolted Alaska on 27 March 1964, killing 114 people in Alaska and causing $350 million worth of damage. The Richter magnitude of the quake was 8.6 probably the largest of this century and one of the largest in history. Severe slope failures and slumps destroyed business and residential areas in the principal city of Anchorage (Figure 17.7). Railroad lines were twisted like string, highways crumpled, gas lines ruptured, and the airport control tower collapsed. Soft, watery clay beneath the Turnagain Heights section of Anchorage failed and carried houses partway down a mudflat toward Cook Inlet (see Figure 11.20). The main tremor lasted three to four minutes; it caused significant damage to the ground and to structures throughout an area of 130,000 square kilometers.

Most of the seismic activity took place along the Denali fault zone. To the west of this zone of activity the surface within an area of about 78,000 square kilometers was lowered as much as 2 meters; marine shellfish are now growing on submerged spruce trees. To the east, an area of as much as 128,000 square kilometers was raised, in some places as much as 10 meters. Docks were elevated beyond the reach, except at the highest tides, of the boats that formerly used them.

Abrupt displacement of the sea floor generated tsunami. The waves wiped out whole fishing fleets, harbor facilities, and canneries. Tsunami destroyed or damaged the waterfronts of Seward, Whittier, Homer,

**17.8 Damage by tsunami,** Kodiak, Alaska, a destructive side effect of the Prince William Sound earthquake, 27 March 1964. (Official U. S. Navy photograph)

Cordova, and Kodiak (Figure 17.8) and caused damage as far away as parts of the coasts of British Columbia, Washington, Oregon, and California (Figure 17.9). Small waves were recorded even in Hawaii, Japan, and Antarctica.

## SEISMOLOGY AND THE PROPERTIES OF SEISMIC WAVES

Now that we have looked at some examples of earthquakes and their effects, we will examine properties of seismic waves and the types of information that can be derived from their behavior. Then we will discuss the scientific uses of seismic waves in mapping the distribution of earthquake epicenters, in analyzing the interior of the Earth, and in predicting earthquakes.

### Kinds of Seismic Waves

The properties of seismic waves that race through and around the Earth during an earthquake depend on two factors: (1) the kind of wave, and (2) the elastic properties of the material through which the waves travel. We describe four kinds of seismic waves that are subdivided into two groups according to their travel paths and into three groups according to their appearance on seismograms. These three seismogram-grouped waves have been labelled *P* waves, *S* waves, and *L* waves (Box 17.1).

Waves designated as *P waves,* or *compressional waves,* by seismologists are equivalent to sound waves. Such waves alternately push and pull particles of rock or other material through which they travel. For this reason, *P* waves are sometimes called "push-pull" waves. The back-and-forth motion is along the travel paths of the waves. The *P* is an abbreviation for "primary." These are the fastest waves, and thus are the first to be recorded on a seismogram.

*S waves,* or *shear waves,* tend to displace particles at right angles to the direction in which the waves travel. In crude diagram form, *S* waves

**17.9 Travel-time map for tsunami** generated by Prince William Sound earthquake, 27 March 1964. The average speed of the tsunami that traveled the 1600 kilometers to the N end of Vancouver Island, Canada, in four hours was 400 kilometers per hour. (National Oceanographic and Atmospheric Administration)

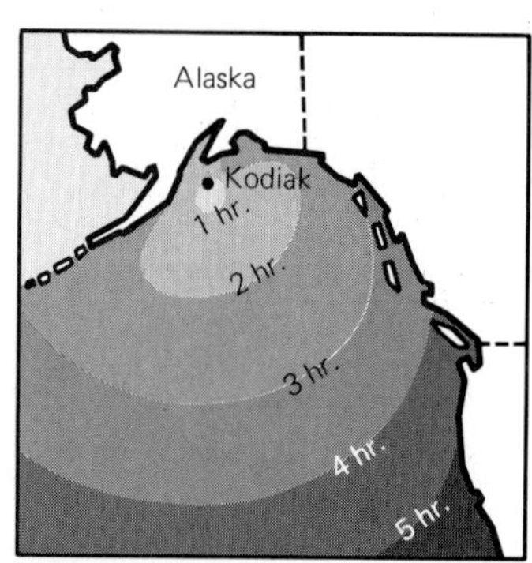

resemble a rope that is fixed at one end and shaken at the other. Because they travel through the main body of the Earth, *P* and *S* waves are also known as *body waves.*

*L waves,* or *long waves,* travel along the surface of the Earth. Two kinds are recognized, Love waves and Rayleigh waves. They are named after the scientists who discovered them, the British mathematician A. E. H. Love (1863–1940) and the British physicist Lord Rayleigh (John William Strutt, 1842–1919). *L* waves cause the strongest ground motion — a back-and-forth horizontal oscillation. Rayleigh waves are short and steep; they accomplish most of the damage to buildings.

## The Scientific Uses of Seismic Waves

If earthquakes generated only one kind of seismic wave or if the different kinds of seismic waves all traveled with the same speeds, seismologists could only record their occurrence. However, as we have seen, at least four kinds of seismic waves are known, and the behaviors of these waves are influenced by the kinds of materials they pass through. Therefore, seismologists can make careful records of these waves and their properties. From analysis of the records, seismologists can locate earthquake epicenters, determine if parts of the Earth's interior are liquid or solid, and infer the densities of rocks.

**Locating earthquake epicenters** In locating epicenters of earthquakes, seismologists use a procedure similar to one used in estimating how far away a bolt of lightning has struck. The lightning creates two kinds of waves, light and sound (thunder), which travel outward from their point of origin at different speeds. Because sound waves travel slower than light waves, one experiences a time lag between the flash of lightning and the sound of thunder. The farther away the point where the lightning strikes the ground, the greater the gap between the light and the sound. (The lightning-thunder time lag is about 3 sec/km).

The difference in speeds between *P* and *S* waves is the basis for locating earthquake epicenters (Figure 17.10). *P* waves travel about 1.75 times as fast as *S* waves. The time interval between the arrival of the first *P* waves and the arrival of the first *S* waves is a direct function of distance from the recording station to epicenter. For example, if the first *P* waves arrive 2 minutes, 41 seconds before the first *S* waves, seismologists know that the distance from the station to the epicenter is 1600 kilometers; if the difference is 6 minutes, 27 seconds, the station-to-epicenter distance is 4800 kilometers. (See these specific time intervals and their corresponding distances marked on Figure 17.10.)

To determine the exact location of an epicenter information from three stations is required. The station-to-epicenter distance defines the radius of a circle which can be drawn with a station at the center. One station can determine only that the epicenter lies somewhere along its circle. Circles drawn around two stations intersect at two points, either of which could be the epicenter. A circle around a third station will pass through only one of the two points and the point of intersection of the three circles is the epicenter (Figure 17.11).

**17.10 Speeds of two kinds of earthquake waves** shown on a graph of time *versus* distance (expressed both as degress of arc along the Earth's surface as measured by the angle between lines drawn from the center of the Earth to two points on the Earth's surface; and as linear distances, kilometer scale below, mile scale at top).

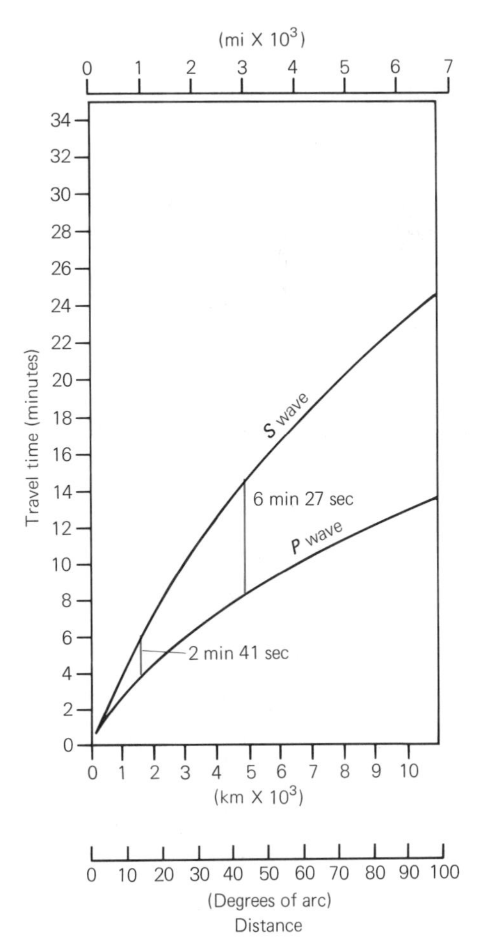

**BOX 17.1**

# Seismic waves

Seismic waves are so important for our understanding of the interior of the Earth that it is essential to understand the mechanics of the various waves and how they can be used for interpreting the properties of the materials they pass through. Many different names are used; here, we attempt to present the information about the waves and how they are named. To start with, we note that a synonym for seismic waves is *elastic waves,* so named because the waves are related to the elastic properties of matter (see Appendix D). Waves are organized into two groups depending on *where* they travel. Thus, any waves traveling within the Earth are *body waves,* whereas any waves moving along the surface of the Earth are *surface waves.* In an earthquake, only body waves leave the focus. After the body waves have reached the surface, they create the motions that give rise to the surface waves.

Another kind of subdivision of seismic waves is based on how they look on seismograms. In this kind of usage, the two chief factors in giving names have to do with the times of arrival and the sizes of the traces. Thus, the first groups of waves to be recorded from an earthquake are the primary waves, named *P waves.* A little later, another group arrives; these are the secondary waves, named *S waves.* Finally, the wave traces become very confused from the effects of large waves; they are named *L waves.*

Yet-another two-group arrangement of waves is based on the relationship between the directions the waves are traveling and the directions of motion of the material as the waves pass. Motion along the direction of travel defines *longitudinal waves;* motion at right angles to direction of travel constitutes the basis for *transverse waves.*

The seismic *P* waves are mechanically equivalent to sound waves, that is, they pass through solids, liquids, and gases by setting up back-and-forth movements of compression and dilation in the direction of travel (Sketch A). The effect resembles the motion of an accordian as its ends are alternately pushed together and pulled apart. Another name for sound waves is compressional waves. Because of the back-and-forth motion along the line of travel, they are also longitudinal waves. Seismic *P* waves can be easily remembered by thinking of them as push-pull waves. This uses a word starting in *P*, for *P* waves. (Ignore the other first letters, such as *E* for elastic, *L* for longitudinal, *C* for compressional, *S* for sound, *B* for body.)

Seismic *S* waves cause material through which they pass to be sheared back and forth along lines at right angles to the direction of wave travel (Sketch B). Thus, they are named *shear waves.* (Remember the *S* for shear waves as well as secondary; ignore the *T* for transverse, *E* for elastic, and *B* for body.) Shear waves can pass through solids only. Fluids lack the shearing resistance necessary to transmit *S* waves. Seismic *S* waves are usually recorded on seismograms from both the horizontal seismographs and the vertical seismographs. This means that the plane of oscillation of the particles as the waves pass is neither horizontal nor vertical, but oblique, as shown in Sketch B.

Seismic *L* waves consist of two important kinds; both are surface waves (omit the *S* for surface; remember the *L* for large). Transverse motion in the horizontal plane (Sketch C) characterizes *Love waves* (named for A. E. H. Love, a British mathematician who discovered them). Love waves can be visualized as shear waves in the horizontal plane. During an earthquake Love waves are felt as a gentle side-to-side swaying. (Add another *L* for Love.)

The second kind of seismic *L* waves cause the land surface to roll up and down, as water waves on the sea. The motion is in the vertical plane whose edge is the line along which the waves travel. The motion is an ellipse whose vertical axis is 1.2 times its horizontal axis (Sketch D). These are the *Rayleigh waves,* named for the British physicist Lord Rayleigh (John William Strutt, 1842–1919). The short, choppy Rayleigh waves cause extensive damages to buildings during an earthquake. (Omit the *R* for Rayleigh, the *E* for elastic, and the *T* for transverse.)

**Determining state of matter** *P* waves pass through matter in all three of its states: solid, liquid, and gas, but *S* waves can travel only in solids. If *P* waves from an earthquake, but not *S* waves, are recorded at a given station, we know that a gas-filled or liquid-filled space lies between the focus and the seismograph.

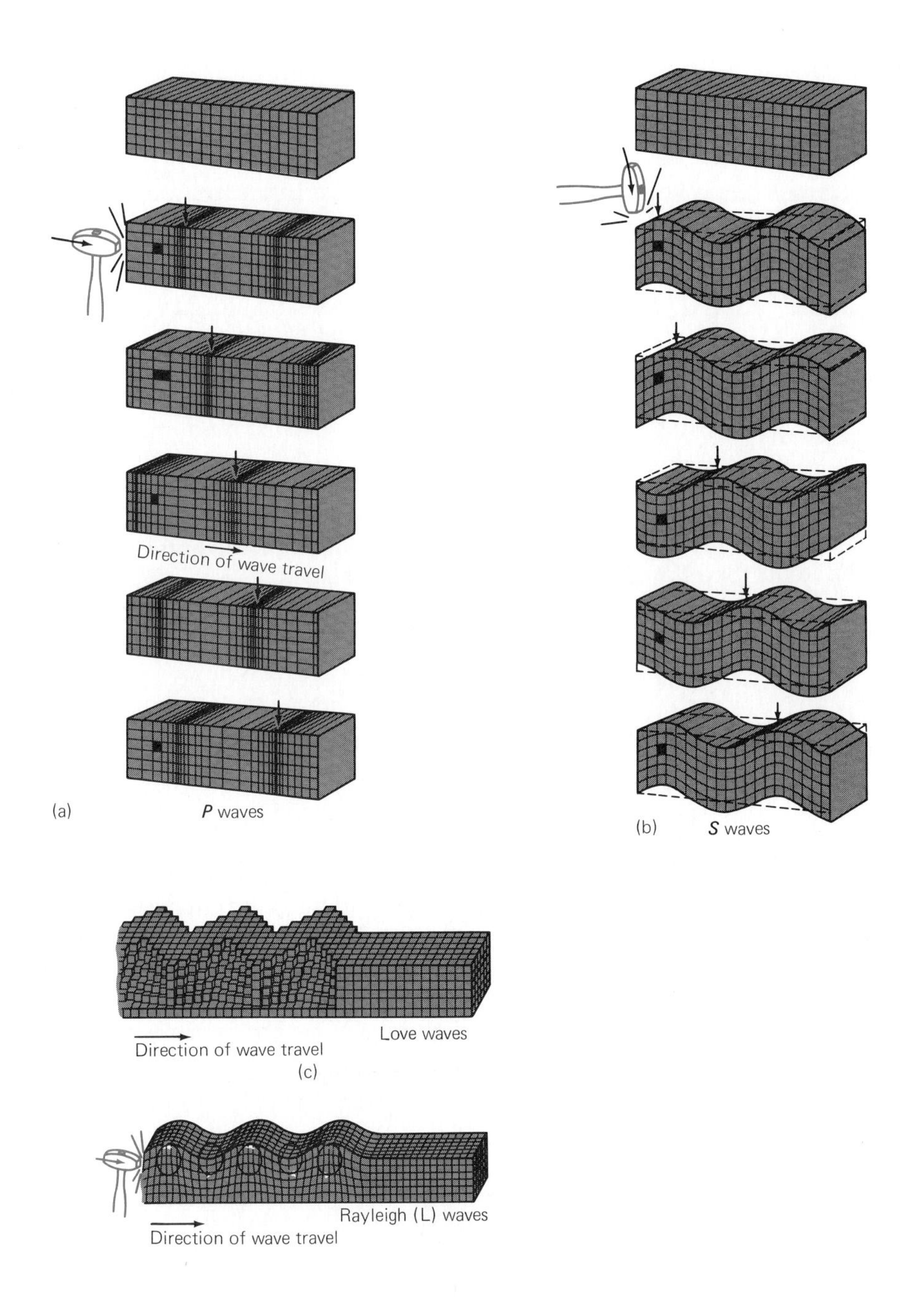


**Determining densities of materials** The speeds of both $P$ and $S$ waves are determined by the kinds of material through which they travel. In general, the speeds of body waves increase with the density of the material. If seismographs at two separate stations receive $P$ waves which have traveled at different speeds from a common focus, then we can

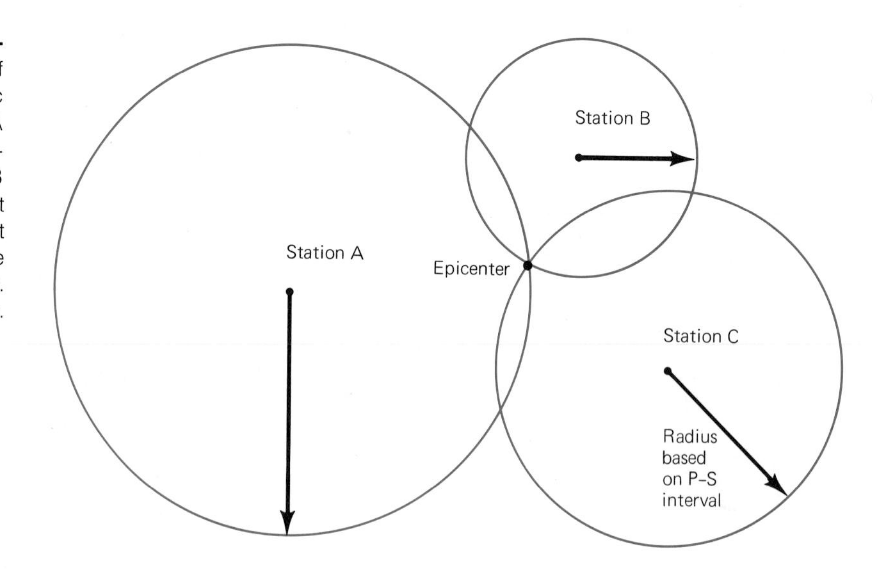


**17.11 Locating an earthquake epicenter** by finding the point of intersection of three circles drawn around three seismic recording stations. Circles from stations A and B intersect at two points, the epicenter and point X. Circles from stations B and C intersect at the epicenter and point Z. Circles from Stations A and C intersect at the epicenter and point Y. Only at the epicenter do the three circles intersect. Schematic.

conclude that the density of the material along the two routes must be different. (See Box D.1 in Appendix D.)

## THE DISTRIBUTION OF EARTHQUAKE EPICENTERS

An organization of stations around the world, known as the World Wide Standardized Seismograph Network (WWSSN), was established in 1961 by the U.S. Coast and Geodetic Survey. (All United States governmental involvement in the study of earthquakes was subsequently transferred to the U.S. Geological Survey.) One of the prime reasons for the development of the WWSSN was to monitor underground nuclear blasts throughout the world. Presently the WWSN has 116 stations in 61 countries.

The first worldwide network of seismograph stations for locating epicenters was set up more than 75 years ago by an enterprising Scot named John Milne (1850–1913). He took advantage of the global extent of the British Empire to place simple measuring devices at strategic locations throughout the colonies.

In the early days of seismology, scientists felt satisfied if they could plot an earthquake epicenter to within 150 kilometers. Nowadays, it is possible to locate earthquake epicenters within 8 kilometers.

The plotting of epicenters has shown concentrations of long, linear belts, with relatively quiet areas in between. The three worldwide belts are the Circum-Pacific belt, the Alpine-Himalayan belt, and the mid-oceanic-ridge belt. Figure 17.12a shows seismic belts and epicenters of earthquakes of all depths, and Figure 17.12b illustrates the locations of earthquakes with foci deeper than 100 kilometers. The first map corresponds with all kinds of margins of lithospheric plates, whereas the second map shows seismic belts that correspond only with the locations of sea-floor trenches, where lithosphere plates are thought to be colliding.

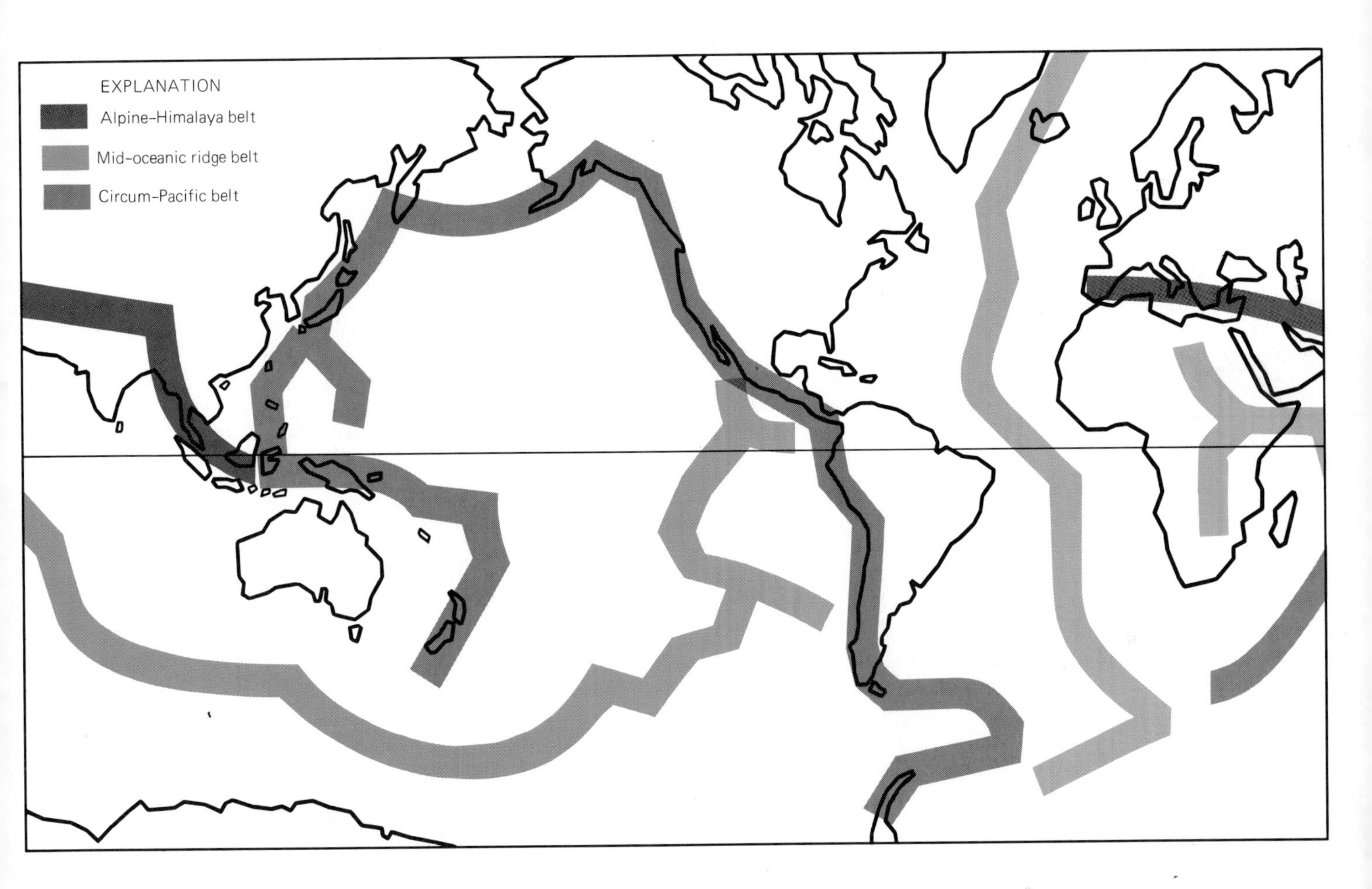


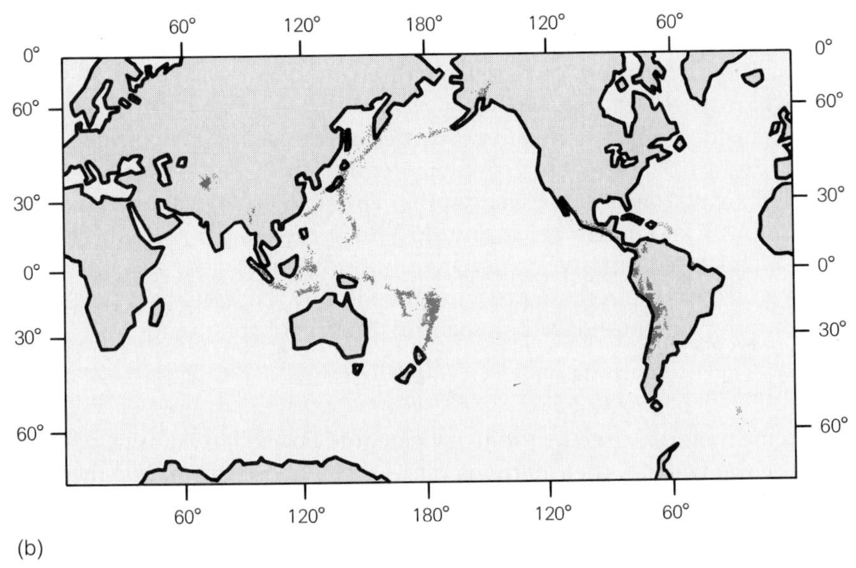


(b)

**17.12 Worldwide belts where earthquake epicenters cluster.** (a) Belts based on locations of epicenters from earthquakes of all depths. (b) Locations of earthquakes from foci deeper than 100 kilometers. (After Muawia Barazangi and James Dorman, 1969, Worldwide seismicity maps compiled from ESSA, Coast and Geodetic Survey, Epicenter data, 1961–1967: Seismological Society of America, Bulletin, v. 59, no. 1, p. 369–380, (a), Plate 1; (b), Plate 2)

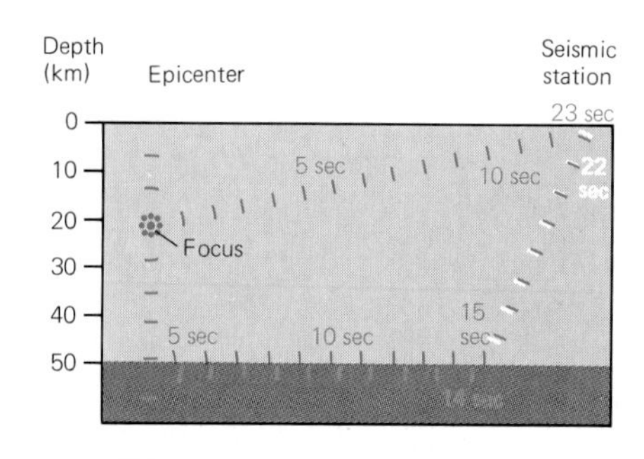


**17.13 Basis for Mohorovičić discontinuity,** schematic profile and section through outer part of Earth from an earthquake focus 20 km beneath the surface to a nearby seismic station. P waves (wave fronts shown by short blue arcs of concentric circles with spacing proportional to speeds of travel) moving through crust at 7 kilometers per second reach three points by direct paths: (1) epicenter, in just under 3 seconds; (2) base of crust, in a little more than 4 seconds; and (3) seismic station in 13 seconds. Energy from P waves that reach the base of the crust divides. Some forms P waves that remain within the crust and travel along its base at 7 kilometers per second, and then up to the seismic station, arriving there about 23 seconds after the start of the earthquake. Other energy penetrates into the mantle, and becomes P waves that travel at 8.2 kilometers per second (short white lines). During each second of time, therefore, the mantle P waves gain 1.2 kilometers with respect to the crustal P waves. Thus, after 9 seconds of travel, the mantle P waves have spanned $9 \times 8.2 = 73.8$ kilometers, whereas the crustal P waves have covered only $9 \times 7 = 63$ kilometers. The difference, 10.8 kilometers, is slightly more than 1 second worth of travel time. In traveling from the base of the crust to the seismic station, both groups of P waves progress at the same speed (7 kilometers per second). Because the P waves that traversed the mantle traveled faster than the P waves that remained within the crust, the station records two arrivals of P waves that followed indirect paths, in this case, separated by a gap of 1.3 seconds. The size of the time gap depends on the depth of focus of the earthquake, the distance from the focus to the recording station, the speeds of travel of P waves in the crust and in the mantle, and on the thickness of the crust.

## THE INTERIOR OF THE EARTH

Analysis of the behavior of seismic waves as they pass through the Earth from the focus to seismograph stations has yielded basic information about the architecture of the Earth. In this section, we discuss the internal structure of the Earth and how it has been inferred.

### The Mohorovičić Discontinuity; The Crust-Mantle Boundary

The *Mohorovičić discontinuity* is a seismic interface between the crust and the mantle. This interface was discovered in 1909 by a Yugoslav geophysicist named Andrija Mohorovičić (1857–1936). His study of earthquakes in the Balkan Peninsula provided evidence for the concept that below depths of a few tens of kilometers, both *P* waves and *S* waves traveled faster than at shallower depths. The speeds of *P* waves increased from about 7 kilometers per second to 8.2 kilometers per second (Figure 17.13). This discontinuity, named in honor of Mohorovičić, was found to be a worldwide phenomenon and eventually became the basis for determining the boundary between the crust and the mantle.

### The Mantle

The same methods used by Mohorovičić could be used to examine deeper parts of the Earth. Other methods for learning about the mantle include examination of exotic blocks thrown out of volcanoes and study of the depth distribution of earthquake foci. Blocks thrown out of volcanoes indicate that the mantle contains mafic silicate rocks including garnets. These materials are under great pressure and are mostly in the solid state. Around the Pacific Ocean, earthquake foci are concentrated along planes that dip 45 degrees away from the oceans. Such inclined zones of earthquake foci have been named *Benioff zones* (Figure 17.14).

When variations in the properties of the upper mantle were first reported, almost no one accepted the results. A zone was found in which

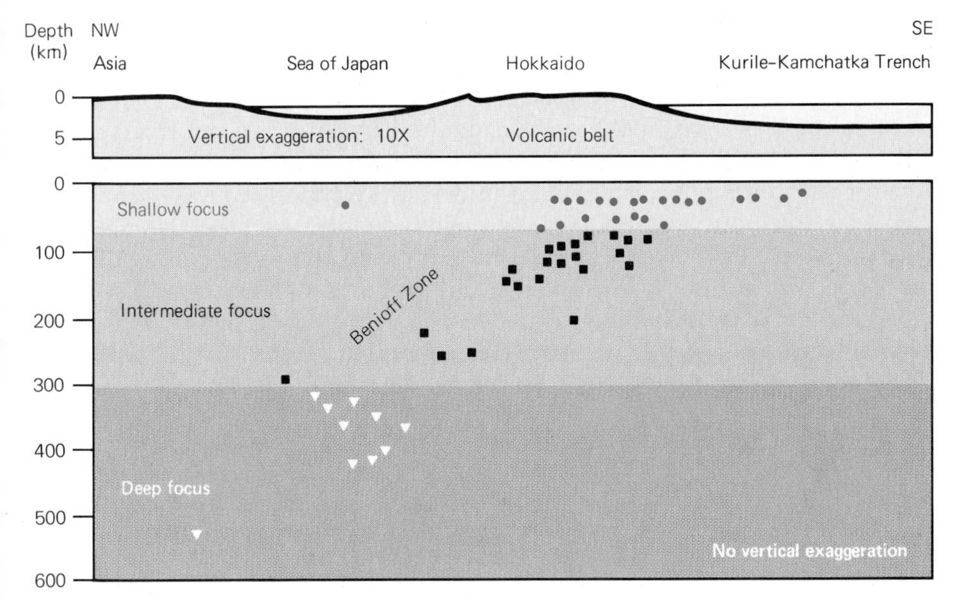


**17.14 Benioff zone beneath Japan and vicinity,** profile and section. (a) Exaggerated profile showing morphology of surface. (b) Benioff zone, dipping away from the deep-ocean basin, extends from trench to a depth of 600 kilometers, to a point well beneath the mainland of Asia. (Beno Gutenberg and Charles F. Richter, 1954, *Seismicity of the Earth and Associated Phenomena:* Princeton, New Jersey, Princeton University Press, 310 p. Reprinted 1965 New York and London, Hafner Publishing Co., Figure 6, p. 29)

the speeds of the *S* and *P* waves did not increase regularly with depth, as they should if the properties of the mantle are changing gradually. Instead, in one zone the speeds *decreased* before they resumed their usual pattern of increase with depth. This zone lies in the depth range of about 100 to 250 kilometers (Figure 17.15). Together, the mantle and crust above this zone where the speeds of seismic waves reverse constitute the *lithosphere*. The reversal of speeds of seismic waves is taken to indicate the existence of a zone of low strength and easy flow, which has been named the *asthenosphere*.

Another boundary within the mantle lies at a depth of 700 kilometers. This marks the lower limit of deep-focus earthquakes. The depth of 700 kilometers divides the upper mantle from the lower mantle.

Within the mantle, no other pronounced seismic irregularities deeper than 700 kilometers have been found. The next seismic boundary is a remarkable interface that is encountered where the *S* waves vanish and the speeds and directions of the *P* waves change notably. The effect of

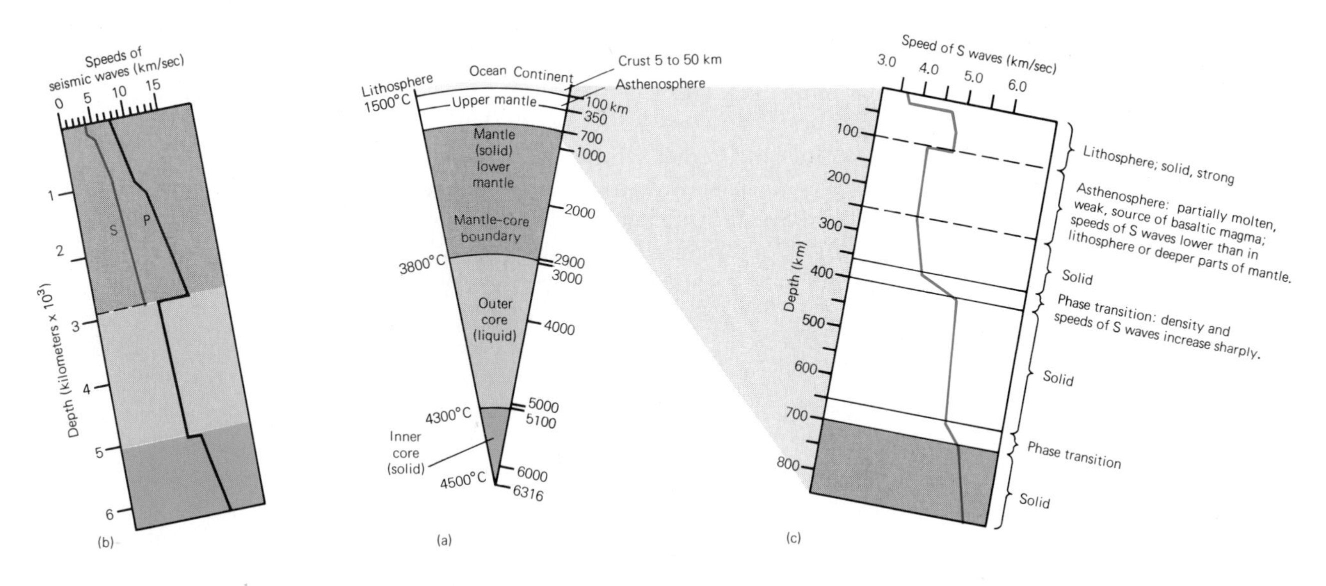


**17.15 Variations of speeds of seismic waves and inferred internal structure of the Earth.** (a) Pie diagram showing internal structure of the Earth and estimated temperatures. (b) S waves stop at a depth of 2900 km, the level at which the solid lower mantle overlies the fluid outer core. (c) Changes of speed of S waves with depth in upper mantle. (Depths are for typical oceanic lithosphere and oceanic asthenosphere; continental values for both are thicker.)

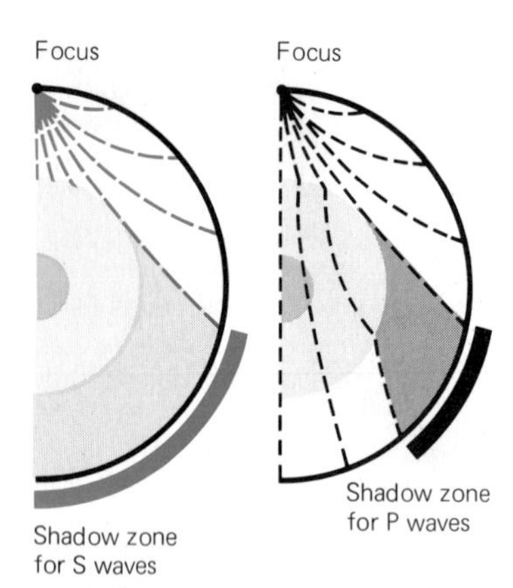


**17.16 Shadow zones created by effect of Earth's core on paths of seismic waves from distant sources,** schematic sections showing half of the Earth. Paths of waves body (dashed lines) are curved because the elastic properties of the Earth change with depth. The fluid outer core stops S waves (left). P waves (right) can pass through the fluid part of the Earth's core, but their paths are bent.

this disappearance of the *S* waves is to create a "shadow zone" for each earthquake, within which no *S* waves and no direct arrivals of the *P* waves are recorded (Figure 17.16).

These changes in seismic-wave behavior indicate that the shadow zone is caused by a pronounced seismic interface at a depth of 2900 kilometers. Above this interface the material transmits both *S* waves and *P* waves and therefore, this material must be solid Below this interface the material does not transmit *S* waves. The only materials we know at the surface of the Earth that do not transmit *S* waves are fluids (liquids or gases).

The *mantle* is then defined as the mostly solid portion of the Earth between the crust and a depth of 2900 kilometers. The seismic interface at 2900 kilometers is known as the *mantle-core boundary*.

### The Core

The Earth's *core* is the smooth, dense sphere that forms the innermost portion of the Earth (see Figure 17.15a). Because *S* waves do not propagate through the core, at least part of it must be molten. By the speeds of *P* waves and laboratory evidence, we infer that the core consists largely of iron, that contains some oxygen to explain its density. The temperature is probably about 4500°C, and the density of the material about 13.5 times that of water.

Movements of the liquid outer core are thought to generate the Earth's magnetic field. We infer that the liquid in the core moves in complex ways that are not altogether understood. Factors that could make the liquid part of the core move are: (1) the Earth's rotation, (2) thermal convection currents, and (3) possibly other sources of energy. The liquid, which is thought to be molten iron, is an electrical conductor. The flow of an electrical conductor is capable of creating a magnetic field.

## FORECASTING AND PREDICTING FUTURE EARTHQUAKES

A distinction is made between forecasting and predicting earthquakes. A forecast includes only the place and possibly the magnitude, whereas a prediction must include place, Richter magnitude, *and* time. In trying to increase their ability to predict, seismologists are concentrating on strike-slip faults to find out if the effects of the tremendous accumulation of strain on rocks can be measured. If many sets of measurements through time show when a critical point has been reached, then it may be possible to predict when the snapping point will occur.

Seismologists have searched for some predictable relationship between strained rocks and some measurable property. These indicators have included changes of speeds of seismic waves, changes in the tilt of the ground, and various other properties.

### Changes in Speeds of Seismic Waves

One kind of precursor that can be observed and that has proved to be a valuable indicator in some localities is change in speeds of seismic waves (Figure 17.17). The first observation of these changes came from Russian seismologists recently working the Garm region of the Soviet Republic of Tadzhik in central Asia. In examining the seismograms of medium-sized tremors coming from a single, known source, the Russians noticed that for several months before a medium-intensity earthquake, the speeds of the *P* waves consistently dropped. They were able to notice this because on their seismograms, the *P–S* interval was changing. Ordinarily, the *P–S* interval would not change, but in this case, the strain on the rocks had built up enough to slow down the *P* waves. The drop in speed was followed by a return to normal shortly before an earthquake took place. American scientists soon found comparable peculiarities in some of their own data and proposed explanations to account for these peculiarities.

What do we know about rocks being strained that might account for this remarkable change in the speeds of *P* waves? Laboratory tests indicate that when a rock is strained to the point of breaking, just before it actually breaks, it *dilates*. Under strain, a network of microfractures develops, and these fractures increase the volume of the specimen. This dilatancy is thought to be part of the explanation for the changes in speeds of the *P* waves.

According to the *rock-dilatancy theory*, proposed by Yash Aggarwal, Christopher Scholz, and Lynn Sykes at Columbia University's Lamont–Doherty Geological Observatory, freshly microfractured and dilated rock differs from nearby crustal rock and the difference affects the speed of *P* waves as follows. As discussed in Chapter 10, most bedrock is saturated with groundwater. However, microfractures can create new pore space in rocks faster than groundwater can seep in to saturate the pores. The density of a body of microfractured rock having pores not filled by groundwater is less than the density of ordinary water-saturated rock. The speeds of *P* waves are related to rock density (see Appendix D). Therefore, a lowering of the density slows down the *P* waves. Eventually, groundwater from adjacent bodies of rock seeps into the pores of the microfractured rock. After such seepage, the speeds of *P* waves return to normal. In addition, this new water arriving in the microfractured rock may raise the pore pressure beyond its original value. The increased pore pressure decreases the frictional resistance along the fault and this decrease helps to trigger an earthquake.

Several successful predictions of small quakes have already been made by monitoring the speeds of *P* waves. The first successful prediction was made in 1973 in New York State. On 1 August, a prediction was made that a quake of magnitude 2.5 or more would strike near Blue Mountain Lake in the Adirondack region "in a couple of days." Two days later, the quake came at magnitude 2.5.

## Changes in the Tilt of the Ground

It is still too early to judge whether rock dilatancy is a sufficiently widespread phenomenon to be used as a universal earthquake predictor. For example, rock dilatancy does not seem to apply to deep-focus tremors, and its general applicability has been challenged. Nevertheless, for shallow-focus quakes, dilantancy does seem to explain several important precursors. In addition to causing the speeds of *P* waves to change, the dilatancy of a body of rock would cause the ground to be uplifted and tilted. Tilting of the ground has been studied in detail in California and Japan and usually has been found to be a reliable indicator of a forthcoming quake (see Figure 17.17). Just before the quake occurs,

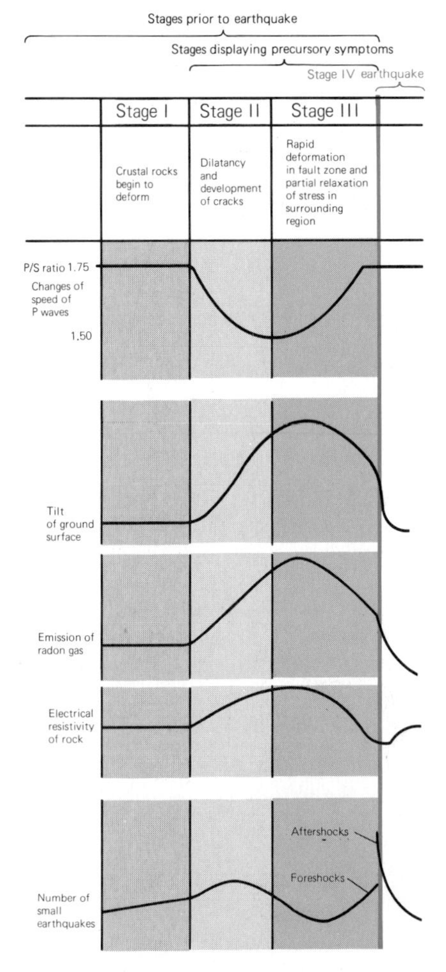


**17.17 Changes of speeds of seismic waves and other precursory phenomena of earthquakes,** classified into stages based on dilatancy. Top curve records changing speeds of P waves with respect to S waves from a known source (to eliminate the distance factor in the P minus S interval). The P/S ratio decreases by 10 to 20 percent in dilated rocks and then returns to normal just before an earthquake. The patterns of the curves for ground tilt, radon emission, and electrical resistivity are comparable in that each builds to a maximum in stage III and decreases to or toward normal just before the earthquake. The bottom curve, for microearthquakes, shows a maximum in stage II, a minimum in stage III, and then a rapid buildup in the form of foreshocks just before the earthquake. (Christopher H. Scholz, Lynn R. Sykes, and Yash P. Aggarwal, 1973, The physical basis for earthquake prediction: Science, v. 181, p. 803–807, Figure 3, p. 806)

the tiltmeters record rapid changes. In the part of the San Andreas fault near San Francisco that has been locked for many years, a closely controlled network of triangulation stations exists. These points are surveyed by instruments that use laser beams; thus, accuracy within millimeters is possible. Early in 1970, many of the points of the triangulation network seemed to be moving away from one another. What is more, it is now evident that substantial ground uplift and tilt is occurring farther south.

### Seismic Gaps Along Active Faults

On 29 November 1978, a major earthquake (Richter magnitude 7.9) struck southern Mexico. This earthquake was noteworthy not only because it was one of the strongest in the last 25 years, but also because its epicenter and magnitude had been forecast almost exactly by scientists at the University of Texas.

The method used to forecast the 29 November earthquake involved a two-step process. First, the investigators checked the seismic history of the area to determine the location and frequency of former seismic activity. Then the established map of epicenters was examined for *seismic gaps,* areas where there was a break in the chain of epicenters. Such a gap may occur where an active fault has become locked and is temporarily not undergoing creep in which both sides along the fault shift regularly by small amounts. The area of locking — the seismic gap — is considered an excellent candidate for future seismic activity. Once such a gap has been located, scientists try to monitor the region carefully for tremors that may foretell an impending earthquake.

Earthquakes had occurred at the edges of the seismic gap in 1965 and 1968. Their unusual seismic activity had started less than two years before the quakes actually struck. With this in mind, the University of Texas scientists were concerned by a suspicious quiescent stage in the area of the seismic gap in 1977. (Such a decrease in normal seismic activity usually precedes a significant increase in seismic activity.) The scientists recommended that the 150-kilometer long seismic gap be monitored in great detail. However, their request was rejected and their earthquake warning was ignored by government officials.

### Southern California's Palmdale Bulge

Palmdale, California, is situated 60 kilometers northeast of Los Angeles, about 110 kilometers from the point where the Garlock fault meets the San Andreas fault. Palmdale is becoming famous, because it lies directly in the middle of an uplift, the "Palmdale Bulge," that covers about 12,000 square kilometers (Figure 17.18). The maximum uplift has been 25 centimeters. The uplift, which started in 1960 and began to be publicized in 1975, is thought to be related to movement along the San Andreas fault. Some geologists predict that the steady shifting of the plates along the San Andreas fault indicates an impending earthquake. Recent satellite measurements have shown that the plates are moving almost twice as fast as previously thought (Figure 17.19). Until the new laser measuring technique had been developed and used, the plates on either side of the San Andreas fault were thought to be moving past each other at the rate

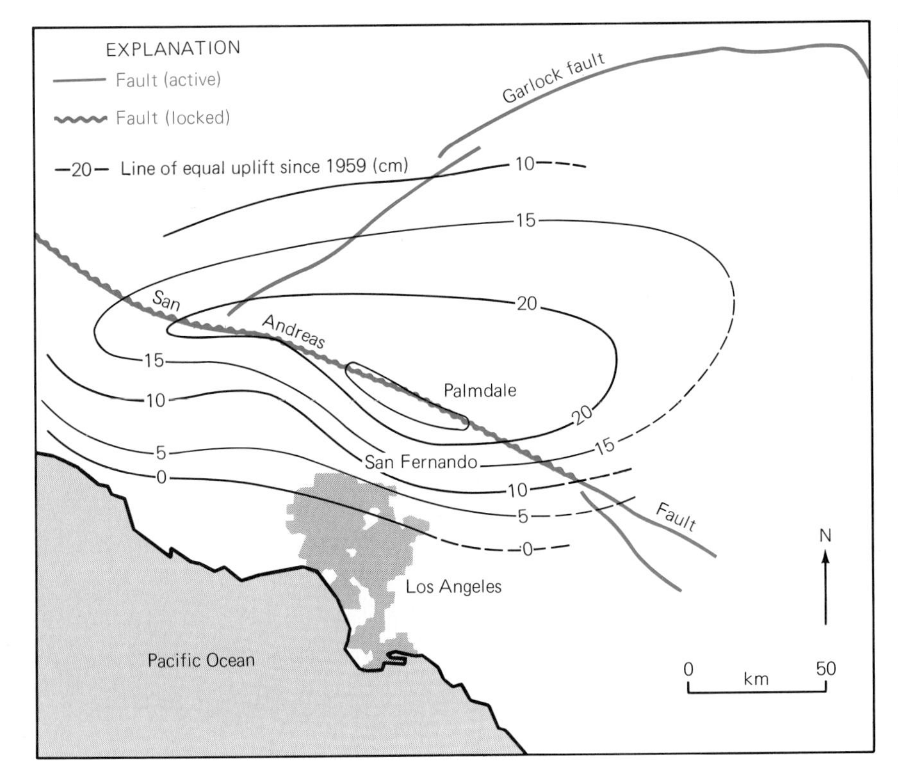


**17.18 The Palmdale bulge,** an area of recently elevated ground along a locked part of the San Andreas fault, southern California. Lines show areas of inferred equal uplift since 1959; each line represents a vertical change of 0.05 meters (5 centimeters, or about 2 inches). (U. S. Geological Survey; the basis for this interpretation of the Palmdale bulge has recently been questioned)

of about 5 centimeters a year. Now, it appears that a more accurate rate is 9 centimeters a year.

What does it all mean? No one can say with certainty that the Palmdale Bulge is a definite precursor of a major earthquake. What can be done is to try to relate this California uplift to similar uplifts that are known to have preceded significant earthquakes. Comparable cycles of uplift and collapse have been recorded before the 1964 Alaska earthquake (Richter magnitude 8.6), the 1960 Chile earthquake (Richter magnitude 8.3), and the 1971 San Fernando Valley, California, earthquake (Richter magnitude 6.5). But, similar uplifts have also been noted in areas where no subsequent seismic activity occurred. Scientists are doing all that they can do at this time: watching and waiting. Some geologists have been willing to make one prediction, however: because the Palmdale Bulge has been rising and subsiding for such a long time, if an earthquake *does* strike the Palmdale area, it may very well be one of the largest in history (Figure 17.20).

### "The Great Tokai Earthquake"

As we have seen, when lithospheric plates push past each other, as along the San Andreas fault, the eventual release of the built-up strain may cause an earthquake. Although such an earthquake may be of Richter magnitude 7 or more, it probably will be less severe than an earthquake caused by the action of three plates grinding against one another that drives one of the plates below the surface, a process called *subduction* (see Chapter 18). The subduction of plates causes the Earth's crust to be

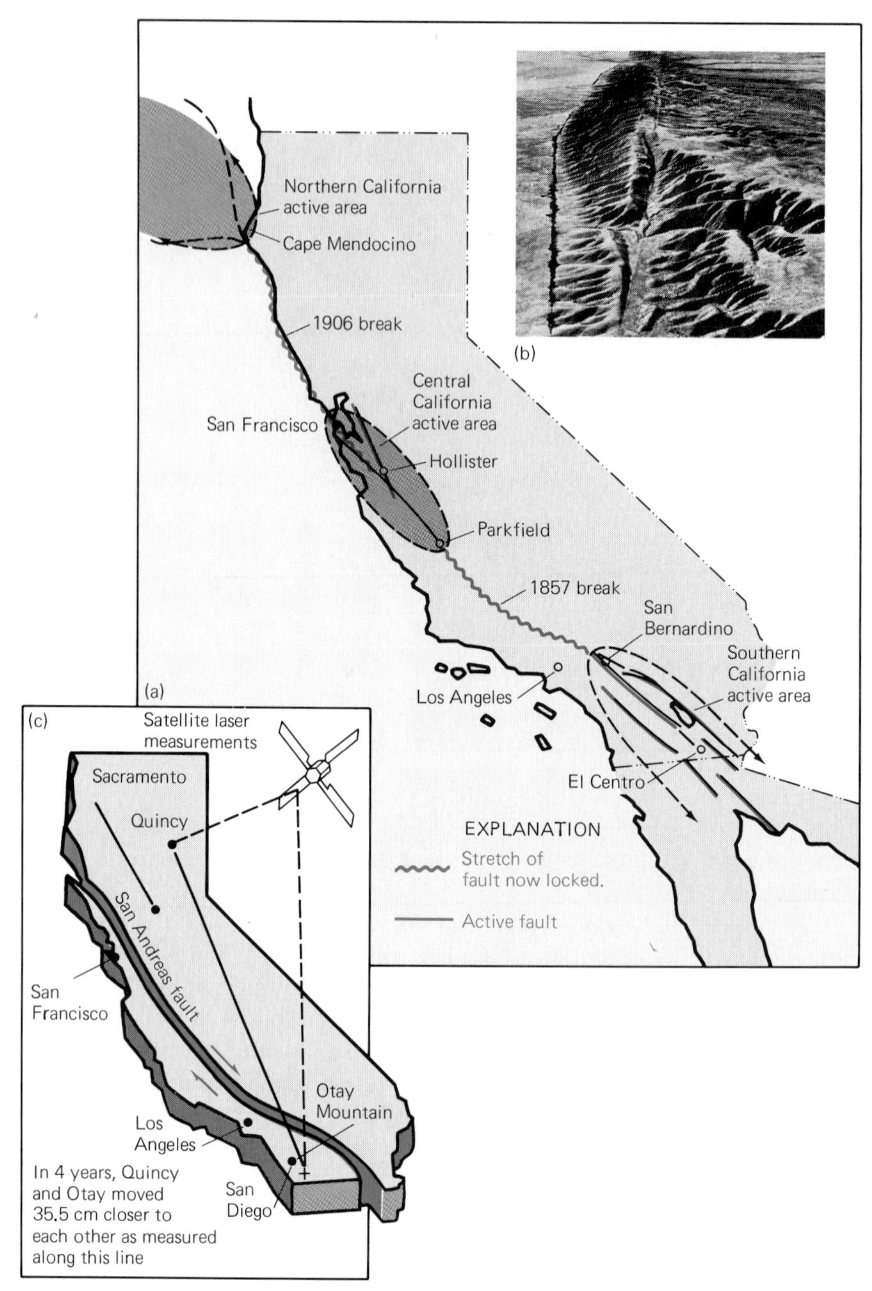


**17.19 San Andreas fault and other major faults in California.** (a) Location map showing active zones (inside dashed lines), locked areas, and selected earthquake locations. (U. S. Geological Survey) (b) View northwestward along the San Andreas fault in 1965 shows that a stream (lower center) has been offset. Carrizo Plain at left; Elkhorn Plain at right. (R. E. Wallace, U. S. Geological Survey). (c) Results of laser measurement of distance indicate that southern California has been compressed along a north-south direction much more than anyone had supposed. (Based on data of U. S. Geological survey)

compressed and great stresses to build up. If the surface of the Earth eventually springs upward, an earthquake will probably be triggered. Such an earthquake is likely to be deeper and spread out farther than one caused by slipping on a fault that is confined to shallow depths.

Such a subduction-caused earthquake has been "predicted" for Shizuoka City in the Tokai region of Japan, an area about 150 kilometers southwest of Tokyo that includes Mt. Fuji (Figure 17.21). Japanese scientists have already named the expected quake "The Great Tokai Earthquake." The land in the Tokai region has subsided about 30 centimeters during this century, and the lack of recent local earthquakes of

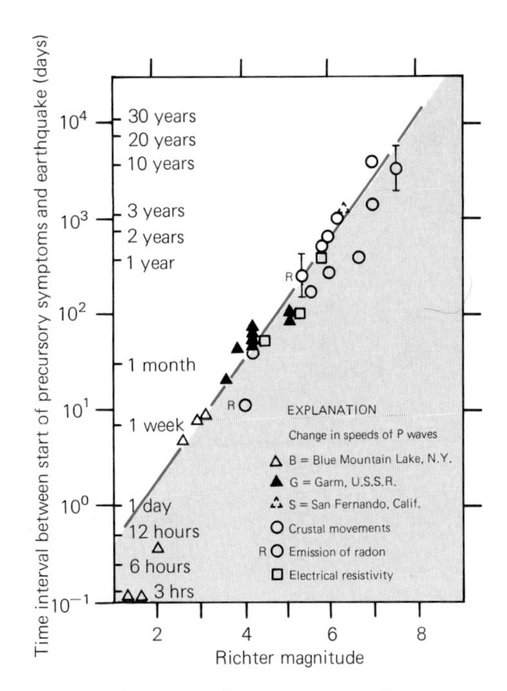


**17.20 Time intervals between start of earthquake-precursory phenomena and earthquake plotted against Richter magnitude of earthquake** shows the basic dilemma surrounding earthquake prediction. The time lag between start of symptoms and a small earthquake is only a few days. By contrast, the time delay involved in a major earthquake exceeds 10 years. If the Palmdale bulge proves to be a valid precursor of an earthquake in southern California, then the 20 years of time between 1959 and 1979 implies that the Richter magnitude will exceed 7.6. Every day's delay increases the magnitude of the forthcoming earthquake, which had not yet happened in late 1980 (Christopher H. Scholz, Lynn R. Sykes, and Yash P. Aggarwal, 1973, reference cited in caption of Figure 17.17, Figure 8, p. 808)

high magnitude has caused Japanese geologists to expect an imminent release of the "locked" plates. In the middle of the nineteenth century the same area was struck by two major earthquakes. Every 100 years or so, a series of quakes jars the nearby boundary between the Asian and Philippine plates. In January 1978, an earthquake of magnitude 7.0 hit the Izu Peninsula — the peninsula is also in the Tokai region, just across Suruga Bay from Shizuoka City.

In addition to land subsidence and tilting, the Tokai region has exhibited other classic signs of impending seismic activity, including rock dilatancy, changes in underground pools of liquid, compression of the crust, acceleration of land movement, and peculiar animal behavior. Some scientists believe that the Izu quake was a precursor of "The Great Tokai Earthquake," which could exceed a magnitude of 8.0 and would be powerful enough to cause major damage in Yokohama and Tokyo, where 14.5 million people reside.

## EARTHQUAKES AND HUMAN ACTIVITIES

In the remotest parts of the Himalayas, the modern-day natives will not allow geologists or paleontologists to hammer on the bedrock for fear of

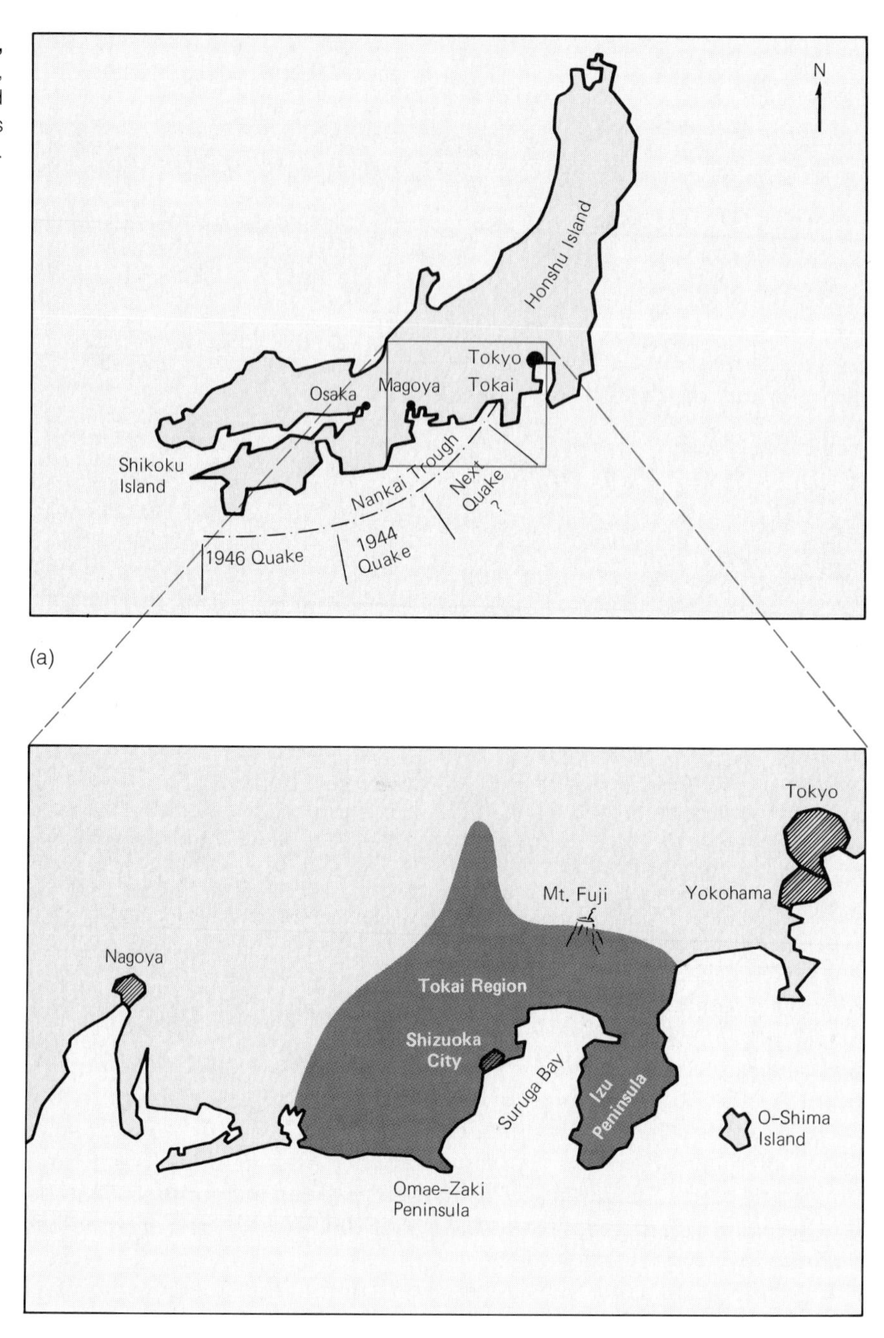


**17.21 Location maps of Tokai region, Japan.** (a) Index map of Honshu Island, showing location of larger map and epicentral areas of previous earthquakes in Nankai Trough. (b) Tokai region.

arousing the sleeping gods who are thought to live within the Earth and to make earthquakes by shaking the ground in anger when their slumbers are disturbed. In some other areas of the world, supposedly more advanced, attitudes exist about earthquakes that do not differ greatly from those of the Himalayan natives.

People have known for some time how to construct buildings that can withstand the vibrations of an earthquake, and yet some recent United States earthquakes have caused millions (or even billions) of dollars' worth of damage to structures.

**17.22 Ruins of Tokyo,** Japan (Byobashiku, one of the business quarters), a result of the 1923 earthquake and fire. (UPI)

In 1921, the great American architect Frank Lloyd Wright (1869–1959) designed Tokyo's Imperial Hotel. The sprawling hotel was located in an earthquake-prone area, but Wright boasted that his hotel, designed with a flexible foundation, could survive any earthquake. The test came quickly. In September, 1923, a great earthquake leveled both Yokohama and Tokyo, reducing most parts of both cities to heaps of rubble (Figure 17.22). Initial reports to Wright in Los Angeles declared that his building had been destroyed along with all the others. But Wright confidently laughed at these reports. A week later the accurate news finally arrived in a cable from Tokyo:

HOTEL STANDS UNDAMAGED AS MONUMENT TO YOUR GENIUS. HUNDREDS OF HOMELESS PROVIDED BY PERFECTLY MAINTAINED SERVICE. CONGRATULATIONS. OKURA.

Obviously, lessons can be learned from the past. We can also heed the scientific knowledge of the present. We can try to coexist with earthquakes or we can ignore them. Let us consider briefly the repercussions of these alternatives.

### Recognizing the Potential Danger

From a human standpoint, the most important effect of earthquakes is destruction of structures and loss of life. With respect to loss of life, the United States has been relatively lucky. Although earthquakes during the past century have caused some $2 billion worth of property damage, only about 1500 Americans have died. (Compare American fatalities with some of the figures listed in Table 17.2).

The great San Francisco earthquake of 1906 occurred early in the morning when few people were exposed to the danger of falling buildings; fewer than 700 persons died. A severe earthquake at Long Beach,

**TABLE 17.3**
DESTRUCTIVE CALIFORNIA EARTHQUAKES

| Date | Location | Richter Magnitude | Death Toll | Damage |
|---|---|---|---|---|
| 1812 | San Juan Capistrano | 7–8 | 50+ | |
| 1857 | Fort Tejon | 8+ | 1 | |
| 1868 | Hayward | 7? | 30 | $350,000 |
| 1872 | Owens Valley | 8+ | 27 | $250,000 |
| 1899 | San Jacinto | 7? | 6 | |
| 1906 | San Francisco | 8.3 | 700 | $1 billion |
| 1915 | Imperial Valley | 6–7 | 9 | $1 million |
| 1925 | Santa Barbara | 6.3 | 13 | $8 million |
| 1933 | Long Beach | 6.3 | 115 | $40 million |
| 1940 | Imperial Valley (El Centro) | 7.1 | 9 | $6 million |
| 1952 | Tehachapi (Kern County) | 7.7 | 14 | $60 million |
| 1954 | Eureka | 6.5 | 1 | $2.1 million |
| 1971 | San Fernando Valley | 6.5 | 65 | $550 million |

*Source:* 1972 Britannica Yearbook of Science and the Future. (Copyright 1971 by Encyclopaedia Britannica, Inc.)

California, on 10 March 1933, destroyed many school buildings, but school was not in session. The waterfront of Seward, Alaska, was completely destroyed by tsunami following the 1964 earthquake; a festival scheduled to take place on the waterfront had been cancelled shortly before the quake.

By contrast, other countries have suffered enormous losses of life, as well as property damage, and seismologists warn that the luck of the United States cannot continue indefinitely (Table 17.3). Based on records of previous earthquakes, government seismologists have prepared earthquake-hazard maps of the United States (Figure 17.23).

**Building regulations and common sense** What an earthquake does to a building is to give it a lot of energy it does not need and usually is not built to accommodate. A building is first hit by *P* waves, followed

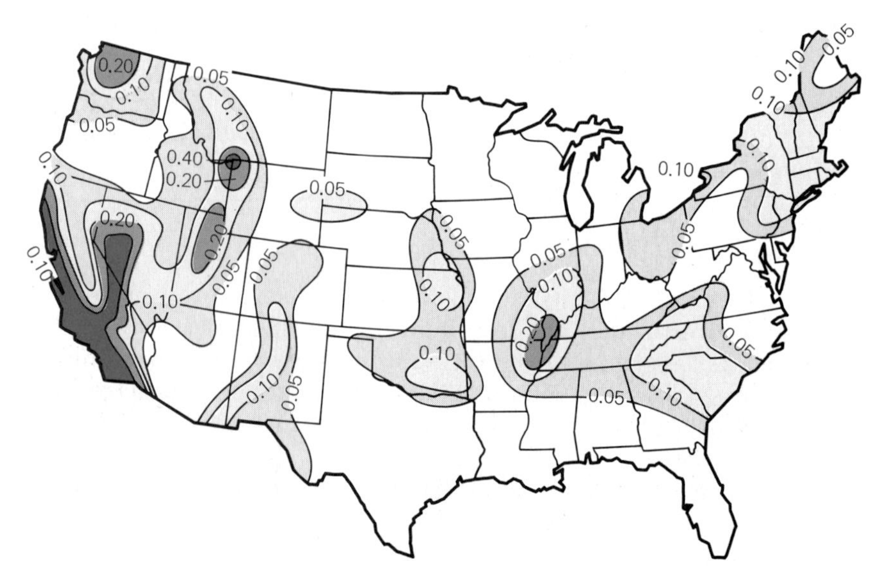


**17.23 Earthquake-risk map for the conterminous United States.** The lines show the maximum estimated accelerations to be expected with a probability of 10 percent (one chance in ten) from earthquake-related seismic waves, expressed as proportions of the Earth's gravitational acceleration (9.8 meters per second per second). Thus, 0.10 means an acceleration of 10 percent of the Earth's gravitational acceleration or 0.98 meters per second per second. (U. S. Geological Survey)

seconds or minutes later by the *S* waves, leading some people to think there have been two earthquakes. Finally, the *L* waves arrive. The long-period Love waves shift the ground from side to side, but may do little or no damage. The most damaging seismic waves are other surface waves, the Rayleigh waves. These are steep and "choppy" and come close together.

A building that cannot absorb the energy of these arriving waves disintegrates. Earthquake engineers attempt to design buildings that can absorb excess energy without falling down, as a tree absorbs energy from the wind without toppling. Because they provide a treelike capacity to sway, stretch, or vibrate instead of breaking, steel beams are the best reinforcement for buildings. Brittle materials, such as brick, concrete, glass, and adobe, are inflexible and tend to crack easily during earthquakes.

If we know how to construct buildings that can resist earthquake damage, why do buildings keep falling down? We know that the closer a structure is to a fault, the more likely it is to be damaged during an earthquake. Unfortunately, building regulations have been so lax that typically this warning is ignored. Near San Francisco Bay, for example, many structures, including a Mormon temple, several freeway overpasses, the new automated BART subway line, and the football stadium of the University of California at Berkeley, have been built on or adjacent to a major active fault zone, the Hayward Fault, which is parallel to the San Andreas Fault. As many as 50 new housing developments are built each year across known fault zones. According to official surveys, about 50 schools and hospitals occupy sites subject to severe earthquake damage. In the summer of 1975, city officials in Los Angeles approved the idea for construction of an underground mass-transit subway system, possibly extending all the way from the San Fernando Valley to downtown Los Angeles. Considering the severity of the 1971 San Fernando Valley earthquake (Figure 17.24) if it actually gets underway, this new underground construction will probably be observed with great interest by scientists and city residents alike.

**17.24 Collapsed freeway overpass,** a connector between Foothill Boulevard and the Golden State Freeway, an effect of the 1971 San Fernando Valley earthquake, southern California. (U. S. Geological Survey)

**Table 17.4**
WHAT TO DO WHEN AN EARTHQUAKE STRIKES

When an earthquake strikes, for a minute or two the solid Earth may pitch and roll like the deck of a ship. The motion is frightening, but unless it shakes something down on you, it is harmless. Keep calm and ride it out. Your chances of survival are good if you know how to act.

**During the Shaking:**

1. If indoors, stay indoors. Hide under sturdy furniture. Stay near the center of the building. Stay away from glass.
2. Don't use candles, matches, or other open flames.
3. Don't run through or near buildings where there is danger of falling debris.
4. If outside, stay in the open away from buildings and utility wires.
5. If in a moving car, stop but stay inside.

**After the Shaking:**

1. Check utilities. If water pipes are damaged or electrical wires are shorting, turn off at primary control point. If gas leakage is detected, shut off at main valve, open windows, leave house, report to authorities, and stay away until utility officials say it is safe.
2. Turn on transistor radio or television for emergency bulletins.
3. Stay out of damaged buildings; aftershocks can shake them down.

**Tsunami: The Deadly Companion of Some Earthquakes**

Tsunami are a danger on our western and Alaskan coastlines and on Pacific islands. Generated by undersea earthquakes, these great seawaves can be a quake's worst killer. Sooner or later, tsunami visit every coastline in the Pacific, so these cautions apply to you if you live in any Pacific coastal area.

1. If there is an earthquake in your area, leave low-lying coastal sections for high ground.
2. Tsunami come in series; the first wave may not be the largest.
3. Never go to the beach to watch for a tsunami.
4. Stay out of all danger areas until an all-clear is issued by competent authority.
5. Stay tuned to radio or television during a tsunami emergency. Bulletins issued through local public safety agencies and National Oceanic and Atmospheric Administration (NOAA) offices can help you save your life.

*Source:* U.S. Department of Commerce, National Oceanic and Atmospheric Administration. (Copies of this table in a convenient card format are available for 10 cents, or $7.50/hundred, from the Superintendent of Documents, U.S. Government Printing Office, Washington, D.C. 20402.)

Scientists have recently warned that a repetition of the 1906 San Francisco earthquake today could cause tens of thousands of deaths and billions of dollars of damage. They indicate further that such potential losses could be reduced through improved engineering and construction practices and through more judicious land utilization.

The 4000-year-old Code of Hammurabi states: "If a builder build a house for a man and do (*sic*) not make its construction firm, and the house which he has built collapses and causes the death of the owner of the house, that builder should be put to death." Few people today would advocate such a powerful judgment. Nevertheless, in one way or another, when we build without regard for the lessons of the past, we continue to kill people.

# CHAPTER REVIEW

1. *Seismic waves* are special forms by which kinetic energy is transmitted inside the Earth and along the rocky parts of its surface. Seismic waves are created when strain is released by abrupt movement along *faults*. Parts of some faults occasionally become locked and further movement bends rocks out of shape. Seismic waves and tremors are created when the rocks eventually snap out of their bent positions. When we can feel these tremors, we call them *earthquakes*.

2. The *focus* of an earthquake is the location within the Earth at which the energy is first released. The *epicenter* is the place on the surface of the Earth directly above the focus. A *seismograph* is an instrument which records seismic waves passing through the Earth. The *Richter Magnitude Scale* and the *Modified Mercalli Scale* are two methods used to express earthquake intensity.

3. Every few years a "great" earthquake (more than Richter magnitude 7.0) happens near enough to a population center to wreak widespread devastation. Earthquakes can also cause other, equally harmful events such as slope failures, tsunami, and fires. The 1964 Prince William Sound earthquake in Alaska registered a Richter magnitude of 8.6 — probably the largest of the century and one of the largest in history.

4. The seismic waves that race through and around the Earth during an earthquake are complex. The most common types have been labelled *P waves* (compressional waves), *S waves* (shear waves), and *L waves* (long waves). The variety of *L waves* named *Love waves* causes the strongest ground motion during an earthquake. Other, "choppy" surface waves, named *Rayleigh waves* accomplish most of the damage.

5. By analyzing the behavior of different kinds of seismic waves, seismologists can locate earthquake epicenters, determine if parts of the Earth's interior are liquid or solid, and infer densities of rocks and the architecture of the Earth.

6. Currently, it is possible to locate earthquake epicenters within 8 kilometers. The plotting of epicenters has shown that many epicenters are concentrated in long, linear belts, with relatively quiet areas in between. The three worldwide belts are the *Circum-Pacific belt*, the *Alpine-Himalayan belt*, and the *mid-oceanic-ridge belt*.

7. The *Mohorovičić discontinuity* is a seismic interface between the crust and the mantle. The mantle and crust above the zone where the speed of seismic waves reverses constitute the lithosphere. The reversal of speeds of seismic waves is taken to indicate the existence of a zone of low strength and easy flow, which has been named the asthenosphere. The limit of deep-focus earthquakes is marked at a depth of 700 kilometers.

8. The *mantle* is the mostly solid portion of the Earth between the crust and a depth of 2900 kilometers. The seismic interface at 2900 kilometers is the *mantle-core boundary*. The *core* is the smooth, dense sphere that forms the innermost portion of the Earth.

9. Changes resulting from the effects of rock dilatancy in speeds of seismic waves, changes in the tilt of the ground, and areas of seismic gaps are likely indicators of future seismic activity.

## QUESTIONS

1. What are *seismic waves?* How are seismic waves created?
2. Define *focus, epicenter, seismograph*.
3. List the important physical characteristics of *P* waves, *S* waves, Love waves, and Rayleigh waves.
4. Compare the *Richter Magnitude Scale* and the *Modified Mercalli Scale* for evaluating earthquakes.
5. Explain how seismologists determine the location of the epicenter of an earthquake.
6. Summarize the worldwide distribution of earthquake epicenters.
7. Explain the seismic evidence for inferring that the outer part of the Earth's core consists of a liquid.
8. What is the *Mohorovičić discontinuity?*
9. Explain *rock dilatancy*. How is rock dilatancy thought to be related to the speeds of seismic waves?
10. Discuss the problem of predicting earthquakes.
11. Identify the *Palmdale bulge* and *seismic gap*.
12. Which California city, Los Angeles or San Francisco, is likely to experience a major earthquake before the other? Why?
13. Discuss some of the natural side effects that may accompany an earthquake.
14. Are the epicenters of all earthquakes located at plate boundaries? Explain.

**RECOMMENDED READING**


Bolt, B. A., 1978, *Earthquakes: a primer:* San Francisco, W. H. Freeman and Co., 241 p. (paperback)

Brown, B. W., and Brown, W. R., 1974, *Historical Catastrophes: Earthquakes:* Reading, Massachusetts, Addison-Wesley Pub. Co., Inc., 191 p.

Duke, C. M., 1960, The Chilean Earthquakes of May 1960: *Science,* v. 132, p. 1797–1802.

Johnston, M. J. S., and Mortenson, C. E., 1974, Tilt Precursors Before Earthquakes on the San Andreas Fault, California: *Science,* v. 186, p. 1031–1034.

Kerr, R. A., 1978, Earthquakes: Prediction Proving Elusive: *Science,* v. 200, p. 419–421 (research news).

Lowman, Paul, Wilkes, Kathleen, and Ridky, R. W., 1978, Earthquakes and Plate Boundaries: *Journal of Geological Education,* v. 26, no. 2, p. 69–72.

Page, R. A., 1980, Estimating Earthquake Potential: *Earthquake Information Bulletin,* v. 12, no. 1, p. 16–25.

Penick, James, Jr., 1976, *The New Madrid Earthquakes of 1811–1812:* Columbia, Missouri, University of Missouri Press, 181 p.

Richter, C. F., 1969, Earthquakes: *Natural History,* v. 78, no. 10, p. 36–45.

Roberts, Elliott, 1967, *Volcanoes and Earthquakes:* New York, Pyramid Publications, 171 p. (paperback)

Wallace, R. E., 1970, Earthquake Recurrence Intervals on the San Andreas Fault: *Geological Society of America, Bulletin,* v. 81, no. 10, p. 2875–2889.

Wyllie, P. J., 1975, The Earth's Mantle: *Scientific American,* v. 232, no. 3, p. 50–57.

GEOPHYSICAL CHARACTERISTICS OF THE LITHOSPHERE
VERTICAL DISPLACEMENTS: THE LITHOSPHERE'S RESPONSE TO NATURAL LOADS
VERTICAL DISPLACEMENTS: SUBSIDENCE, EPEIROGENY, AND ISOSTASY
VERTICAL DISPLACEMENTS AND HORIZONTAL DISPLACEMENTS: OROGENY
HORIZONTAL DISPLACEMENTS: CONTINENTAL DRIFT AND POLAR WANDERING
PLATE TECTONICS

# CHAPTER 18 DYNAMICS OF THE LITHOSPHERE, OROGENIC BELTS, AND PLATE TECTONICS

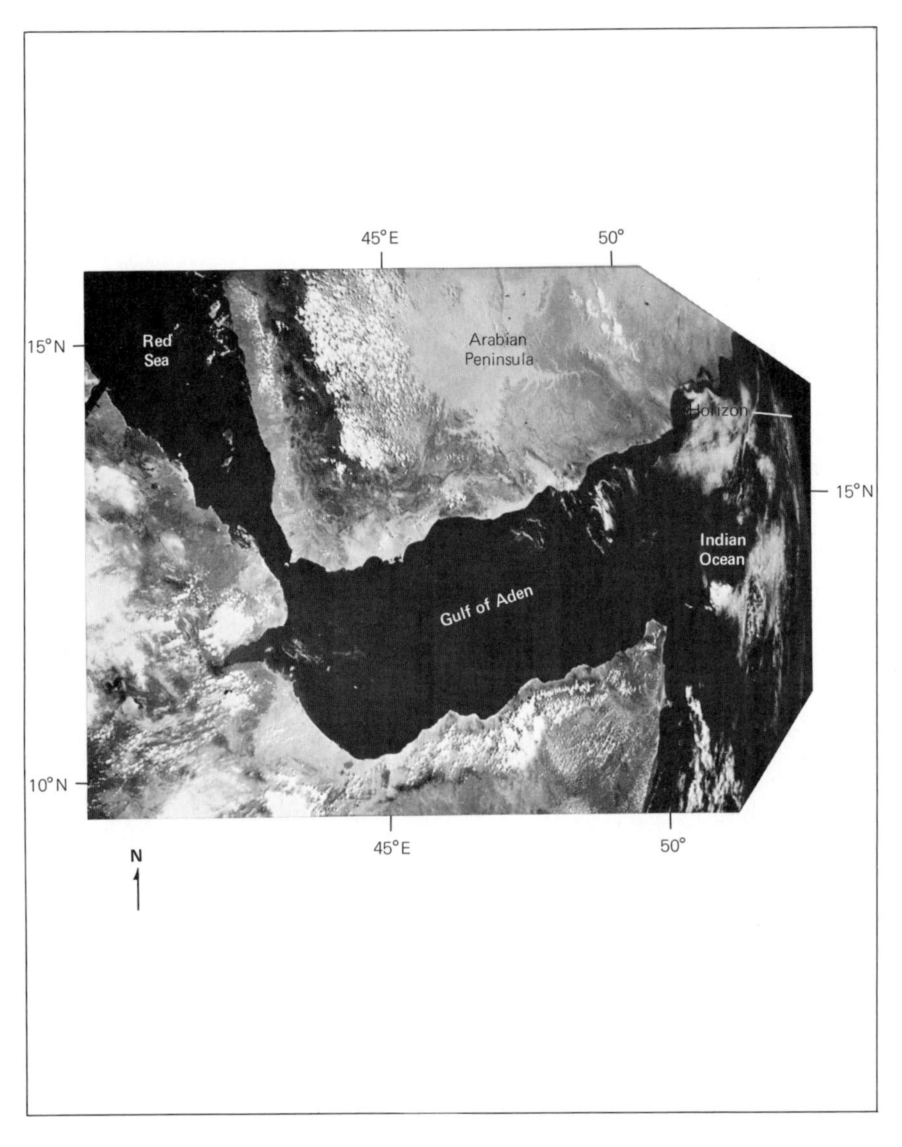


PLATE 18
Results of plate tectonics in Middle East. Red Sea and Gulf of Aden are thought to have formed by the spreading of the Arabian Peninsula away from Africa. View from Gemini XI spacecraft on 15 September 1966 with camera pointing northeastward along Gulf of Aden toward Indian Ocean. The far horizon of Indian Ocean at right shows Earth's curvature. (National Aeronautical and Space Administration)

NEARLY EVERY FIRST-ORDER MORPHOLOGIC FEATURE AT THE surface of the Earth and most geologic structures have formed because of some dynamic activity of the lithosphere, all of which are included in the term *tectonics*. One of the most challenging assignments facing a geologist is to be able to assemble and interpret the large-scale fabric of the lithosphere that has resulted from such tectonic activities.

Modern concepts emphasize the importance of dealing with the entire lithosphere, but even this new realization of the importance of the lithosphere may not be adequate. After all, the lithosphere is but a trifling part of the whole Earth, and the properties of the lithosphere may be determined by what happens at still-deeper levels within the mantle.

In this chapter, we summarize the geophysical properties of the lithosphere and the underlying part of the mantle. We discuss vertical displacements caused by natural loads and by more complex internal mechanisms. The subject of combined vertical and horizontal displacements serves as a basis for presenting the dynamic history of orogenic belts and the origin of complex mountains. We also relate horizontal displacements to ideas about continental drift. All of these topics lead us to the concepts of plate tectonics. We trace the origin of these concepts, summarize some of the major features that are explained by the concepts of plate tectonics, and compare some of the predictions made by these concepts with field evidence.

## GEOPHYSICAL CHARACTERISTICS OF THE LITHOSPHERE

Knowledge of the structure and physical properties of the lithosphere is necessary in order to understand its large-scale dynamics. Because we know how different properties affect the speeds of seismic waves, local gravity measurements, local magnetic fields, and the amount of heat flowing upward to the Earth's surface, we can use measurements of these factors to perform a geophysical survey of the lithosphere. Many of these properties are closely interrelated; they are functions of the kind of rock

**18.1 Geophysical profiles through three contrasting kinds of lithosphere.** In stable areas, whether on a continent or in a deep-ocean basin, heat flow is only about 1 HFU (= heat-flow unit; which is 1 microcalorie, or $10^{-6}$ calorie, per square centimeter per second) and the speed of P waves beneath the Mohorovičić discontinuity is 8.2 or 8.3 kilometers per second. In unstable areas of all kinds, heat flow ranges up to 8 HFU or higher and the speeds of P waves beneath the Mohorovičić discontinuity are less than 7.6 kilometers per second. (Compiled from various sources, chiefly Brune)

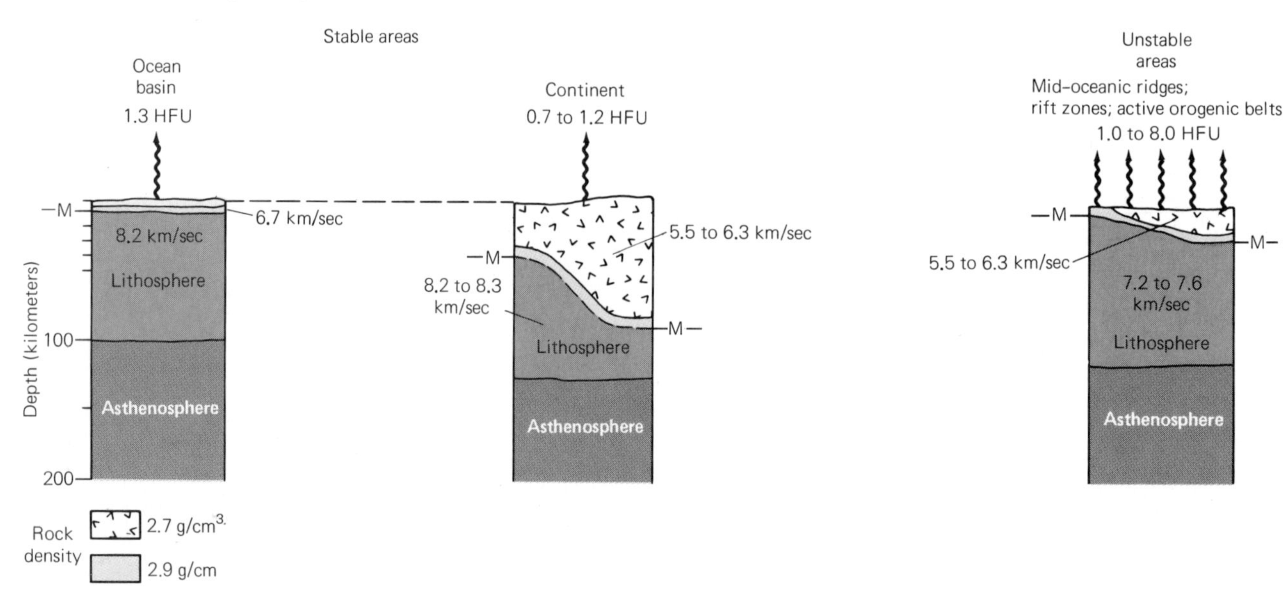

material and how much heat is flowing through it.

In analyzing thousands of determinations made in all parts of the world, geophysicists have found that the measurements can be organized into three groups. One cluster of closely similar results comes from parts of the deep-ocean basins that are not near an active mid-oceanic ridge or a trench. A second cluster of comparable results characterizes broad parts of continents. The third cluster comes from long, linear belts where active movements are in progress — from such locations as mid-oceanic ridges, trenches, rift valleys, and active orogenic belts. The first two are considered to be representative of stable lithosphere and the third, of unstable lithosphere. A summary of the characteristics of these three areas based on densities of rocks, speeds of seismic *P* waves, thicknesses of the various layers, and amounts of heat flowing upward is shown in Figure 18.1.

## VERTICAL DISPLACEMENTS: THE LITHOSPHERE'S RESPONSE TO NATURAL LOADS

An important way to begin the dynamic analysis of the lithosphere is to measure its response to natural loads. The level of the lithosphere shifts upward as load is decreased and downward as load is increased. For example, the atmosphere's weight varies with altitude and movements of storms and other air masses and the level of the lithosphere has been found to shift upward slightly as air pressure decreases and to sink downward when it increases. Similar changes have been observed in response to the Moon's gravitational pull.

The lithosphere sags under the weight of water in large lakes and rebounds to its previous level if the water is removed. As the lithosphere rebounds, shoreline features formed by the presence of water may be elevated in the shape of a dome. Accumulation of sediment, as in deltas, and volcanic material also has a noticeable effect on the surface of the lithosphere. The load of glacial ice is great enough to cause the lithosphere to sag (Figure 18.2). As with the draining of a lake, the melting of a glacier

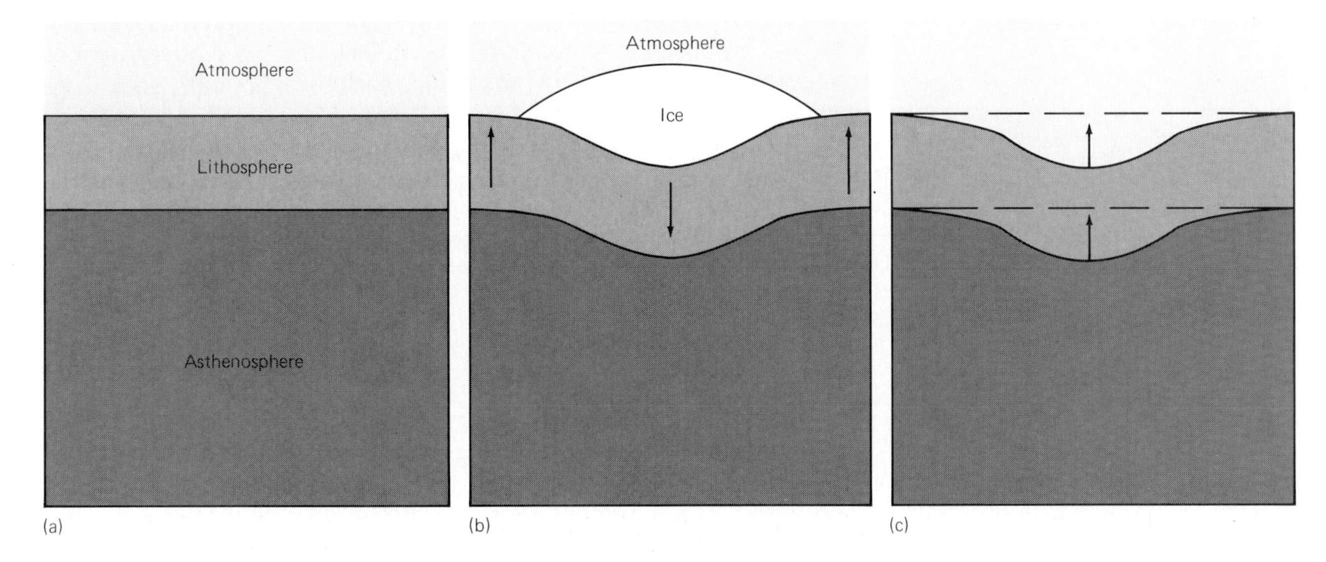


**18.2 Effect of continental glacier on the lithosphere,** schematic profiles and sections with thickness of ice greatly exaggerated with respect to thickness of lithosphere. (a) Condition before glacier forms. (b) Lithosphere sags under weight of ice, which may represent a load of about 360 kilograms per square centimeter. (c) Glacier has melted, but lithosphere has not yet returned to its preglacial level (dashed lines).

removes the load from the lithosphere. After the load has been removed, the surface of the lithosphere rebounds to its preglacial level, deforming shoreline features that may have been built or carved while the lithosphere sagged.

All these natural loads demonstrate that the lithosphere, although seemingly solid, is far from being rigid on a large scale. In fact, these loads indicate that the asthenosphere, the layer which supports the lithosphere, yields easily. Another remarkable characteristic of the lithosphere is its ability to rebound when a load has been removed. This implies that part of the asthenosphere can move aside, yet can return after the load has been taken away. In the next section, we examine larger-scale and longer-term vertical displacements.

## VERTICAL DISPLACEMENTS: SUBSIDENCE, EPEIROGENY, AND ISOSTASY

In many instances, the surface of the lithosphere seems to have moved upward or downward without being subjected to the kinds of external loads discussed above. We must conclude, therefore, that some kind of internal mechanism operates to lower or raise the surface of broad areas. Much of the geologic record and many of the Earth's modern surface features have come about because of these vertical movements. The downward movements are named *subsidence.* Vertical uplift that takes place without crumpling the strata is *epeirogeny.* We first assemble evidence that proves that vertical displacement has indeed happened. We begin with subsidence, then illustrate uplift, and finally show that subsidence has been followed by uplift.

### Proofs of Subsidence

The first geologist to prove that a large part of the Earth's surface subsided as a unit was Charles Darwin (1809–1882), author of the epoch-making book, *The Origin of Species.* Darwin's explorations ashore during the voyage of H.M.S. *Beagle,* turned up evidence that during the Cenozoic Era, much of South America had first subsided and later had been elevated. After he left South America, Darwin explored Pacific coral atolls, and proposed the hypothesis that atolls had formed because corals were capable of growing upward while the foundations beneath them subsided (Figure 18.3). Our proofs of subsidence include drowned features made by waves, submerged landscapes, and thick shallow-water strata. In all cases where submergence is used to indicate that the

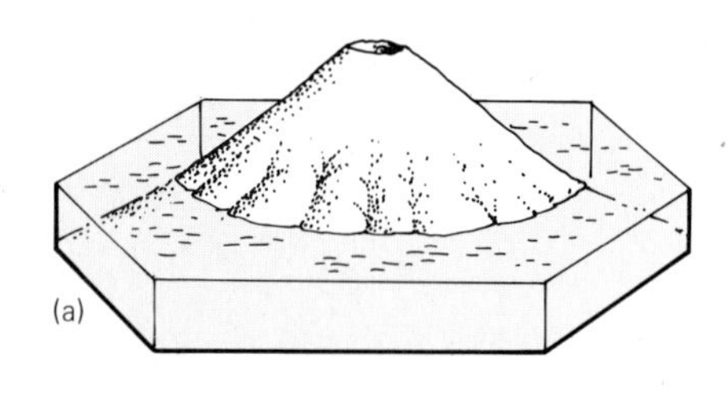


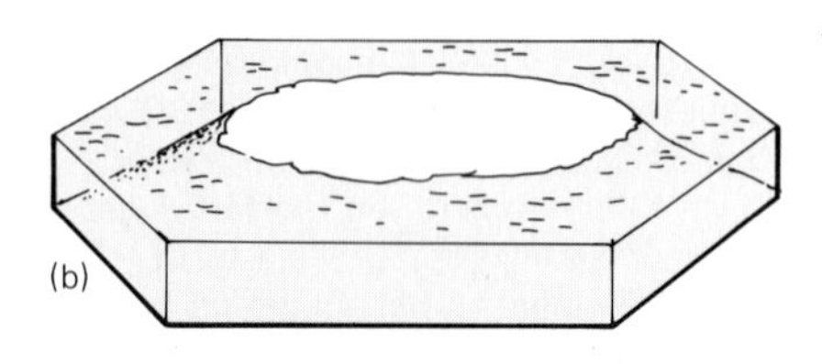


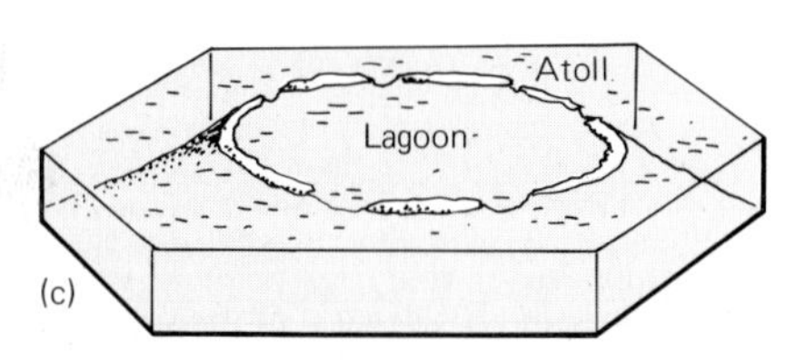


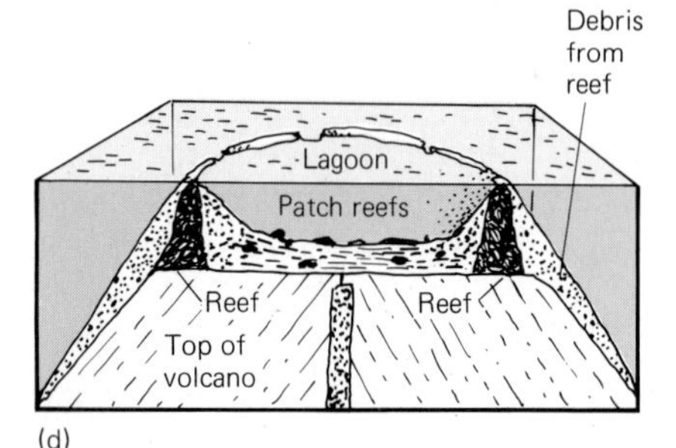


**18.3 Volcanic cone subsides and corals grow upward to form an atoll.** (a) Conical volcanic island becomes inactive and corals grow around its shore, as in many modern examples that formed the basis of Charles Darwin's concept of the origin of atolls by reef growth on subsiding volcanic cones. (b) Volcanic peak is eroded to sea level by waves before corals become established. In this case, corals start to grow on a nearly flat circular base, a possibility not considered by Darwin. (c) Truncated cone subsides and coral grow upward creating an atoll with a shallow lagoon. (d) Further subsidence creates atoll with deep central lagoon having an essentially flat floor. Coral growth around an intact volcanic cone that subsided would result in an atoll having a central shoal, the apex of the cone. No such shoals have been found.

lithosphere has subsided, it is necessary for one to be able to show that the effect has not been caused by changing amounts of water, which may not be related to movement of the lithosphere.

**Drowned wave-built features: shorelines and guyots** Many features that are fashioned at the shore of a lake or the sea become submerged and thus appear as terracelike features on echograms. The waves quickly plane off the tops of many volcanic cones, reducing them from islands to shoals (see Chapter 5). The planed-off top of a volcanic cone is usually taken to be a former sea-level position. Flat-topped conical features named *guyots* (Figure 18.4) were first discovered by Harry H. Hess (1903–1969) on the floor of the Pacific Ocean and later found by others elsewhere.

**Submerged landscapes** Various distinctive landscape features have been found on the sea floor. These include river valleys, karst features such as caves and sinkholes, soils, and forests.

**Thick shallow-water strata** Various kinds of sedimentary strata are deposited under the control of the level of a lake or the sea or are governed by depth of water. If the bottom of the sea subsides at a rate that can be matched by the deposition of these strata, then the result will be a thick body of strata, all having shallow-water characteristics. Many examples have been proved by drilling. These include Pacific coral atolls (see Figure 18.3); broad continental terraces, as off the east coast of the United States; the Bahama Banks; and the Gulf of Mexico.

## Proofs of Uplift

Uplift on a broad scale is shown by various elevated shoreline features, by the deformation of coastal-plain strata adjacent to New England, by

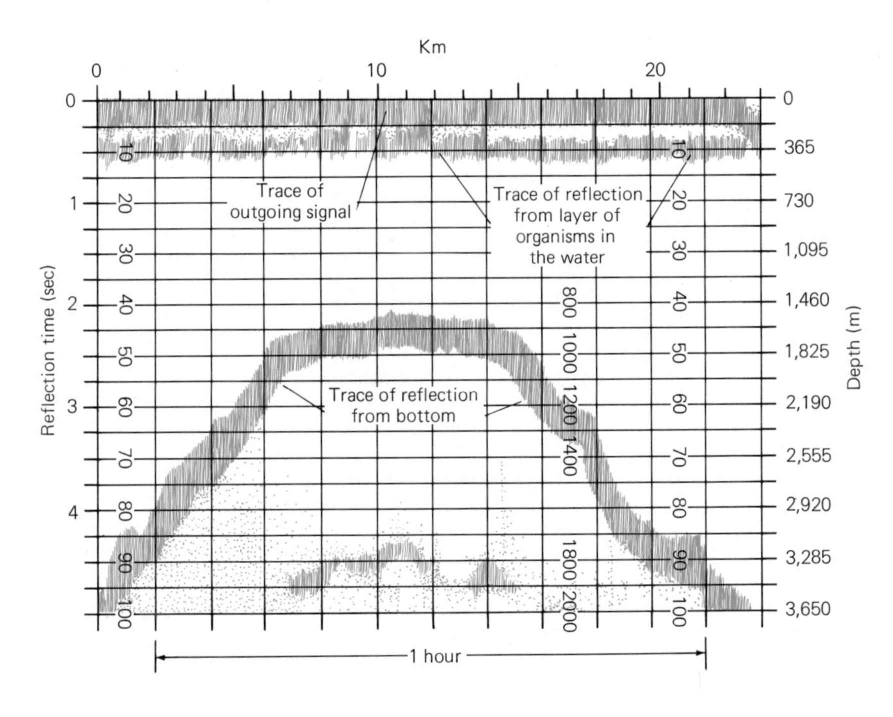


**18.4 Guyot,** profile displayed on drawing of an echogram having vertical exaggeration of about 5X. Guyot in Gulf of Alaska, about 151.5°W and 52.5°N. (After Henry W. Menard and Robert S. Dietz, 1951, Submarine geology of the Gulf of Alaska: Geological Society of America, Bulletin. v. 62, no. 10. p. 1263–1285, Guyot GA-12, plate 3)

exposure of deep-sea sediments on land, and by sheets of coarse sediment. We have previously mentioned elevated shorelines in connection with the loads on the lithosphere. Other elevated shorelines, such as those in California, on Barbados, in New Guinea, and on other islands in the Pacific area, have resulted from tectonic uplift.

In the New England states of northeastern United States, the Cenozoic history has been complicated by the advances and retreats of Pleistocene glaciers. However, despite this complication, the deformation of the nearly horizontal coastal-plain strata and erosion of deep valleys clearly indicate that the region has been elevated. The uplift evidently began late in the Miocene Epoch and continued during the Pliocene Epoch. By the time the first Pleistocene glaciers had arrived, many major valleys, including what is now Long Island Sound, had already been eroded.

Some of the sediments deposited in the deep-sea basins are not found elsewhere. The most distinctive kinds are those that were deposited in water deeper than the carbonate-compensation depth; the typical example is brown clay. Deep-ocean type of brown clay has been found on Barbados and on some islands in Indonesia. Uplift in excess of four kilometers is implied.

One of the consequences of the tectonic elevation of a large landmass is the formation of great quantities of coarse sediment. Long after the elevated tract has been eroded away, indirect evidence of its former existence is provided by the sediment shed from it and deposited in areas that subsided.

### Proofs of Subsidence Followed by Uplift

In many parts of continents, the exposed geologic record contains proofs that the region subsided and accumulated sedimentary strata, and afterward these strata were elevated and eroded. An example of such proofs is the exposure at the surface of materials that have been subjected to effects of burial, such as shales, altered organic remains, and regionally metamorphosed rocks.

The Colorado Plateau is an area of nearly half a million square kilometers in southwestern United States that has been deeply dissected by the Colorado River and its tributaries. One of the spectacular wonders of the world, the Grand Canyon (see Figure 3.9), lies in the southwestern part of the plateau country. Strata of Paleozoic, Mesozoic, and early Cenozoic ages, in many places still essentially horizontal, are exposed. Two episodes in the plateau's tectonic history are implied. The first episode was a long period of subsidence (3.5 km during about 500 million years, for an average rate of 0.007 mm/yr) during which the strata accumulated. The second episode was a period of rapid uplift (4 km in less than 20 million years, for an average rate of 0.2 mm/yr, nearly 30 times the average rate of subsidence).

The presence at the modern land surface of broad areas of mica schists and other rocks formed by large-scale metamorphism deep within the lithosphere implies both downward and later upward movements. The downward movement of subsidence is implied by the body of sediments that form the parent rocks later metamorphosed. The metamorphism itself may have involved additional subsidence. Finally, in order to bring to the surface such materials that form only deep down in the lithosphere, a great period of elevation and erosion must take place.

### The Concept of Isostasy

The condition of dynamic balance between the lithosphere and the asthenosphere is known as *isostasy* (from the Greek for "equal standing"). The term implies that the Earth's gravity causes segments of the lithosphere to shift upward or downward according to their densities, almost as if they were following Archimedes' principle of buoyancy. We use the term "buoyancy" somewhat advisedly, because the asthenosphere transmits seismic *S* waves and thus we know that it must be solid, not liquid. Nevertheless, the strength of the asthenosphere is so small that it flows plastically and thus displays certain liquidlike features.

In the early days of the twentieth century, isostasy was explained in terms of two layers, the *sial* and the *sima,* separated by the Mohorovičić discontinuity (see Figure 17.13). According to this interpretation, the sial was thought to be a comparatively rigid layer, the crust, which "floated" on a denser, yielding layer, the sima. The concept of plate tectonics now indicates that the balancing movements of isostasy take place at the base of the lithosphere, which lies far below the Mohorovičić discontinuity. Under the new idea, the actual weights of columns of material extending from the Earth's surface to some depth in the asthenosphere are equal (Figure 18.5). Vertical changes are thought to take place because the lithosphere is balanced by the asthenosphere and also undergoes certain changes. We examine these changes in the next section.

### Possible Subsurface Changes That Affect Isostasy

Any changes in the thickness of the lithosphere, in the densities of the rocks composing it, or in the situation within the asthenosphere can affect isostasy and thus can cause the surface of the lithosphere to be displaced vertically. These changes include erosion of felsic crust, tectonic thickening, regional metamorphism and thermal density changes, subsurface phase changes, and flow of the asthenosphere.

If a region such as New England, composed of felsic continental rocks and being elevated is persistently eroded, it is possible to strip away all the continental crust. If this were to happen, then as soon as whatever subsurface forces causing the uplift cease to operate, and isostasy takes over, then the area will subside, possibly to oceanic depths (see Figure 18.5 g-i).

One of the consequences of great overthrusts and close folding is to thicken the sedimentary strata. After this thickening has taken place and the forces responsible for it have relaxed, then establishment of isostasy will cause the area where the strata were tectonically thickened to rise (Figure 18.6).

One of the chief effects of regional metamorphism is to increase the densities of rock materials. Other density changes can be caused by changes in temperature as a result of simple thermal expansion or thermal contraction without any changes in mineral composition. Heating rock materials causes them to expand and to become less dense. Both of these effects would tend to elevate the surface.

An important change in the mantle that can affect the lithosphere is the hydration of olivine to form serpentine (a hydrated magnesium silicate), which takes place at a temperature of 500°C. This reaction is reversible and is accompanied by a change in volume of 25 percent. The volume

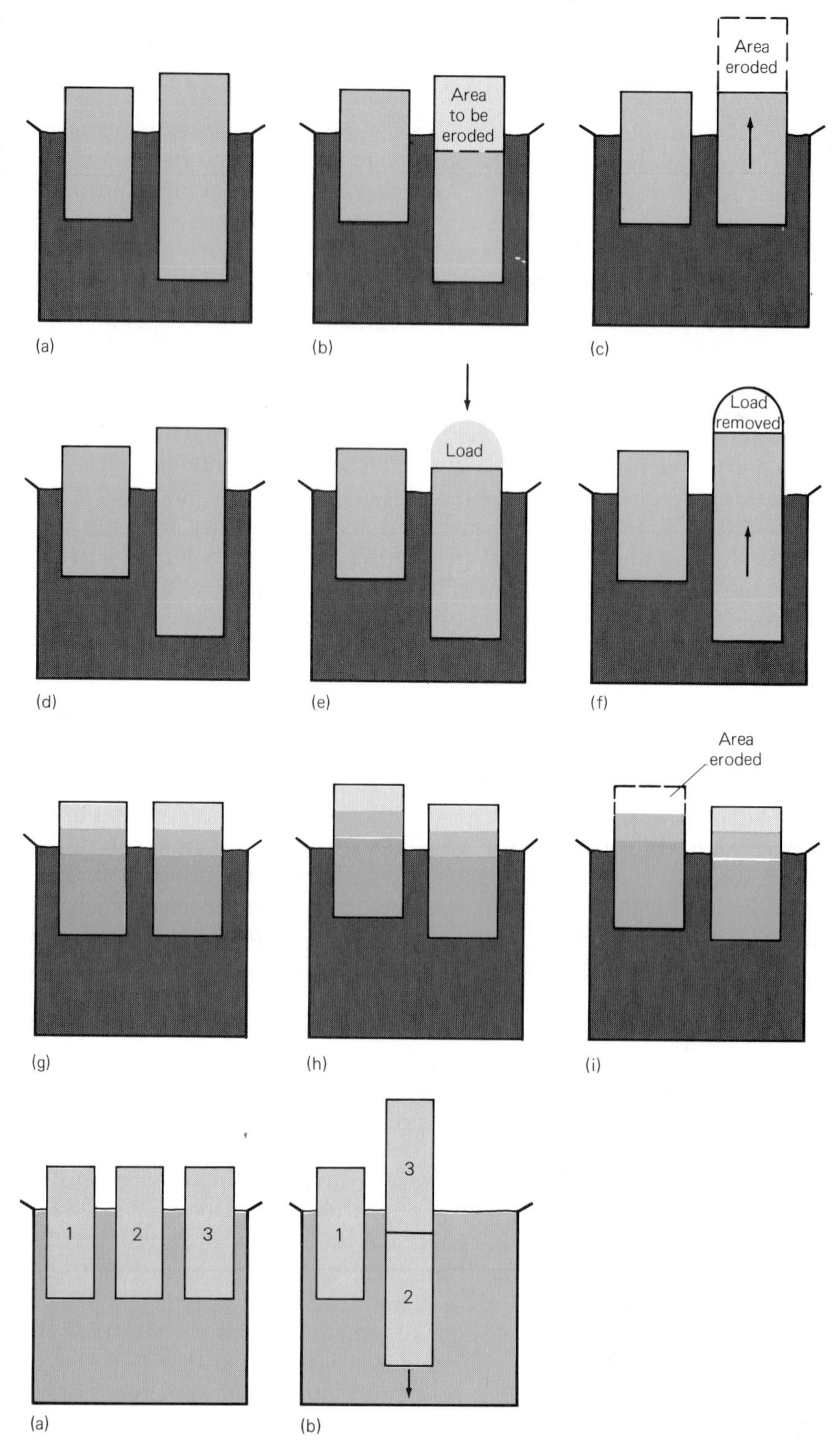


**18.5 Concept of isostasy** illustrated by wooden blocks floating in water to simulate relationship between lithosphere and asthenosphere. Top row: blocks of equal density but of unequal thickness. (a) Top of thicker block rises higher and its base extends deeper than corresponding parts of thinner block. (b) Top of thicker block removed. (c) After the thickness of both blocks has become the same, their tops and bottoms attain identical levels. Middle row: effects of adding and later removing a load. (d) Blocks before load is added. (e) Load added depresses the loaded block whose base reaches a new depth, governed by thickness and density of the load compared with thickness and density of the block. (f) After load has been removed, the unloaded block rises to its former level (compare Figure 18.3). Bottom row: effects of composite blocks having layers of differing densities. (g) Identical blocks; levels of bases and tops coincide. (h) Left block elevated by force from below; top eroded. (i) After erosion and cessation of elevating force, left block sinks to restore balance. Top of left block stands lower than top of right block; base of left block stands higher than base of right block.

**18.6 Effects of tectonic thickening,** blocks of equal thickness and density. (a) Condition before overthrusting. (b) Overthrust thickens the crust. Thickened block sinks and its top is elevated.

increases when serpentine is formed and decreases when olivine crystallizes. This means that heating serpentine enough to form olivine would cause the Earth's surface to subside. And, hydrating and cooling olivine to make serpentine would cause the Earth's surface to be elevated.

From our review of the response of the lithosphere to natural loads, it is clear that the asthenosphere is able to flow aside when the base of the lithosphere is pressed downward. No one knows for sure if the asthenosphere is capable of flowing on its own, so to speak. Perhaps the asthenosphere flows via a slow-moving convection current. Wherever the asthenosphere could flow away laterally, the lithosphere would sink downward. This would cause its surface to subside. Similarly, if moving asthenosphere material could pile up anywhere, the effect would be to elevate the overlying lithosphere.

## VERTICAL DISPLACEMENTS AND HORIZONTAL DISPLACEMENTS: OROGENY

In Chapter 16 we defined various kinds of mountains based on their geologic structures. We will now discuss orogeny or origin of the kinds of mountains that are related to orogenic belts. We begin in this section by summarizing the horizontal displacements. Then we discuss the horizontal displacements and vertical displacements that are parts of the six stages of the dynamic history of an orogenic belt.

### Horizontal Displacements Implied by Fold Belts and Great Overthrusts

In complete contrast to the kinds of vertical displacements summarized in previous sections are the vast horizontal displacements that are implied by fold belts and great overthrusts. Horizontal displacements on the order of 100 to 200 kilometers have been determined. One of the remarkable features of such long-distance overthrusts is that deep-water strata from the former ocean side of the orogenic belt are thrust over shallow-water strata on the side toward the interior of the continent.

In some orogenic belts, ocean-floor lithosphere, in slabs as much as 10 kilometers thick, has been displaced over shallow-water sediments that were deposited on felsic continental crust. In any orogenic belt, this time of thrusting is an event of great significance. Examples are known from all parts of the world. In North America during the Paleozoic Era, such thrusting took place at different times in the three marginal geosynclines: the Appalachian on the east (Figure 18.7), the Ouachita on the south, and the Cordilleran on the west.

**18.7 Great overthrusts in southern Appalachians,** as inferred from recently surveyed deep seismic-reflection profiles, shown on slightly schematic transverse geologic section. If this section is correct, then the displacement on the Great Smokies overthrust has been 200 kilometers. (Modified from F. A. Cook, L. D. Brown, and J. E. Oliver, 1980, The southern Appalachians and the growth of continents: Scientific American, v. 243, no. 4, p. 156-157, 159-168, figures on p. 164 and 165; and U.S. Geological Survey)

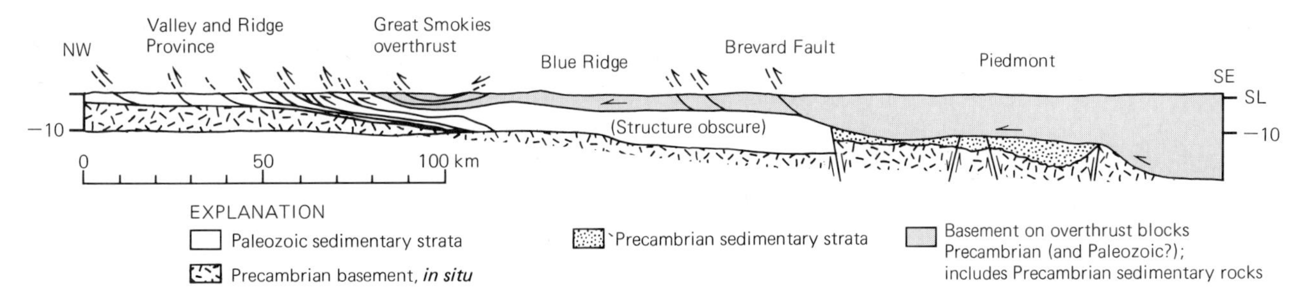

During the Ordovician Period in the Appalachian geosyncline, a body of dark-colored, fine-grained, deep-water strata of Cambrian and Ordovician ages measuring several thousand kilometers long, hundreds of kilometers wide, and thousands of meters thick, traveled as much as 150 kilometers westward across the sea floor and came to rest above shallow-water carbonate rocks, also of Cambrian and Ordovician ages. In New York state, the transport was from east of what is now the Green Mountains to the Taconic Upland.

During the Pennsylvanian Period, comparable activity took place in what is now southwestern Arkansas and southeastern Oklahoma. Once again, dark-colored shales and siltstones, deposited in deep water, were thrust over shallow-water strata. The direction of displacement was northward and westward. During the Mississippian Period, deep-water Paleozoic shales were displaced eastward from what is now western Nevada onto a shallow-water platform.

The mechanics of such great thrusts is one of the mysteries of tectonics.

## Dynamic History of an Orogenic Belt

The dynamic history of an orogenic belt involves both kinds of displacement, vertical and horizontal. We can recognize six stages: (1) geosynclinal stage of accumulation of strata; (2) terminal-orogeny stage; (3) initial elevation, usually including a post-orogenic stage of block faulting and vertical movements; (4) long-continued erosion; (5) peneplanation; and (6) recycling.

**Geosynclinal stage** The history of an orogenic belt begins with the formation of a *geosyncline,* an elongated part of the Earth's crust that subsides beneath sea level and accumulates sediment. If the region that subsides is at the margin of a continent, the strata accumulate at the continental margin, with the continental mass on one side and an island arc or a deep-ocean basin on the other. Alternatively, if the subsided area lies within a continental mass, the geosyncline is a true trough with land bordering it on both sides.

The detailed history of the crust during the geosynclinal stage can be read from the kinds of strata present, their thicknesses, and the sizes of their particles. In some geosynclines, subsidence was matched exactly by accumulation of sediment. The water started shallow, and even though the bottom may have continued to subside and sink through thousands of meters, sediment kept accumulating so that the water remained shallow. For example, this kind of subsidence and accumulation of limestones and carbonate sediments has been going on in the Bahamas for the last hundred million years. Exploratory borings indicate that shallow-water limestones and dolostones and evaporites deposited on intertidal flats are 5500 meters thick (see Figure 16.15). If that much sinking had taken place without a corresponding accumulation of strata, then the water would have become 5500 meters deep.

In some geosynclines deep water evidently did form. Later, the deep area filled with sediments. The oldest sediments were deposited in deep water and the younger ones, in shallower and shallower depths.

In some geosynclines, folds and faults were active while the strata were accumulating. The growing folds affected the thickness of the strata, which became thinned over crests of folds and thickened in troughs of synclines. Similar changes may have taken place on fault blocks.

**Terminal-orogeny stage** While the various sedimentary strata of a future mountain chain are accumulating in a geosyncline, some of them may be deformed locally. However, such local deformation usually is overshadowed by what happens during the grand climax, the *terminal orogeny,* which involves deformation, metamorphism, and plutonism of the geosynclinal belt. Great thrusting takes place. The usual arrangement is that deep-water strata are thrust over the shallow-water strata. Orogeny typically includes great thermal events — regional metamorphism, formation of granite batholiths (see Figure 6.16), and the concentration and deposition of many metal deposits. While the strata are being subjected to high temperatures and pressures, they may recrystallize as they are being deformed.

**Initial-elevation stage** Shortly after all the deformation and heating that accompany the terminal orogeny have taken place, the orogenic belt typically becomes greatly elevated and a mountain range is born. While all the vertical movements are in progress, local fault-bounded basins may form. In these basins, thick bodies of sediments and some volcanic rocks may accumulate. Rapid uplift and subsidence may create thick bodies of nonmarine fan deposits along the margins of the basins (see Figure 8.20).

**Long-continued-erosion stage** After the stage of initial elevation and possible formation of basins within the rising chain, the dominant process becomes erosion and further vertical adjustments related to isostasy (see Figure 18.5). In order to wear down the relief of the mountain chain, vast quantities of rock must be eroded. The interface at the base of the lithosphere and top of the asthenosphere is a dynamic surface, much akin to the boundary between fresh water and salt water under a place such as Long Island, New York. Erosional removal of rock from a mountain chain is analogous to pumping down the groundwater table. As the water table is lowered, the level of the salt water in the subsurface rises. Similarly, as erosion removes material from the mountains, isostasy brings up the base of the lithosphere and this contributes more material to be eroded.

**Peneplanation stage** After much erosion, the relief of the former mountain chain may disappear. On such a surface of low relief, a new suite of sedimentary strata may accumulate.

Figure 18.8 emphasizes how the fabric of an orogenic belt remains the same, whereas the surface expression of the belt may change with time. The sequence shown begins with mountains that are eroded to a peneplain. By later subsidence, the peneplained orogenic belt is covered with younger strata. Now, the orogenic belt is completely invisible on the Earth's surface; nonetheless, its original fabric remains. Later, the buried

**18.8 Structural fabric in rocks of an orogenic belt persists** despite changes in elevation of the land surface. (a) Typical structural features of an orogenic belt; folds, overthrust, and granite batholith, eroded to a peneplane. (b) Peneplain subsides and is buried by marine strata. (c) Later domal uplift tilts the covering strata. Erosion removes covering strata in center of dome, unroofing the buried orogenic belt and possibly forming a domal mountain (see Figures 16.21c and 3.15).

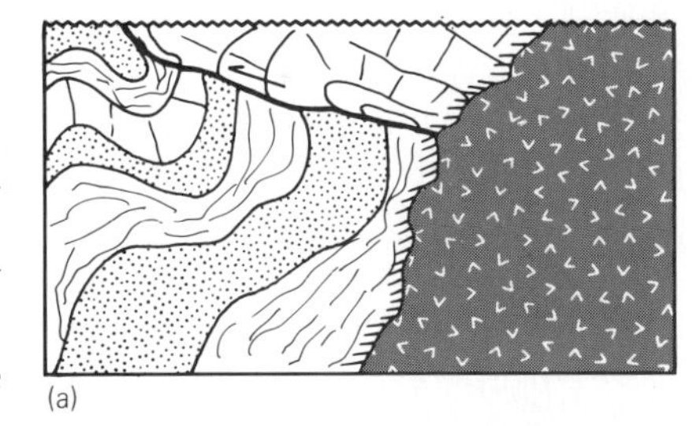
(a)

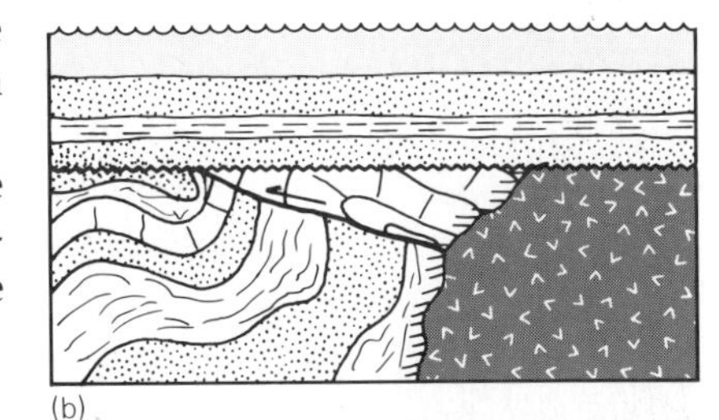
(b)

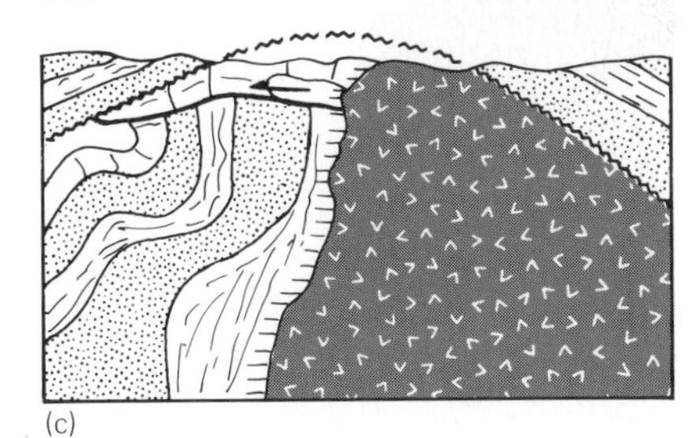
(c)

orogenic belt may come to the surface and form mountains once again. This time, the mountains may be of the kind related to fault blocks or a great domal uplift exposing the basement rocks.

**Recycling stage** After a mountain chain has been worn down to a peneplain, has subsided, and has been covered with sediment, the worn-down mountains may simply become part of the fabric of the stable part of a continent. However, it may start to go through the stages of its history again. If it does, we can say that it has been *recycled.*

In terms of geologic time, mountains are short lived. They are built up, eroded away, and recycled. In the process, they help to keep the continents alive by providing solid mass faster than erosion can erase it. But as mountains come and go, the fabrics of the rocks in an orogenic belt remain basically untouched, much the way your genes do not change because your appendix has been removed or because your nose has been reshaped.

It may help you to remember the dynamic history of orogenic belts if we use an analogy with human lives. If, for the purposes of our analogy, we are permitted to include the idea of human reincarnation, the comparative stages are shown in Table 18.1.

Without the existence of a geosyncline, the process of "gestation" and orogeny could not happen. Geosynclines are likely to appear on the sea floor, so paradoxically, the highest points in the world (mountains) begin their lives at the lowest points in the world (the bottom of the ocean), where sediments accumulate.

## HORIZONTAL DISPLACEMENTS: CONTINENTAL DRIFT AND POLAR WANDERING

**18.9 Alfred Wegener,** the German meteorologist and geophysicist who devoted his life to what was a nearly futile attempt to convince geologists that continents had drifted. (Wide World Photos)

Because the evidence for vertical displacement of the lithosphere is so convincing and has been known for so long, it was only natural that some geologists would wonder whether large-scale horizontal motions were possible also. Two varieties of large-scale horizontal displacements include continental drift and polar wandering.

### Modern Progress of the Concept of Continental Drift

A young German meteorologist named Alfred Wegener (1880–1930) (Figure 18.9) became interested in the idea that continents might have moved on a global scale. In 1912, Wegener published his first essays on

**TABLE 18.1**
A COMPARISON OF THE HISTORY OF OROGENIC BELTS AND HUMAN LIVES

| Orogenic Belts and Related Mountains | Human Lives |
|---|---|
| Geosyncline (initial subsidence) | Conception |
| Accumulation of thick sediments | Gestation |
| Orogeny | Birth |
| Initial elevation | Youth |
| Long-continued erosion | Maturity |
| Peneplain | Old age/Death |
| Recycling | Reincarnation |

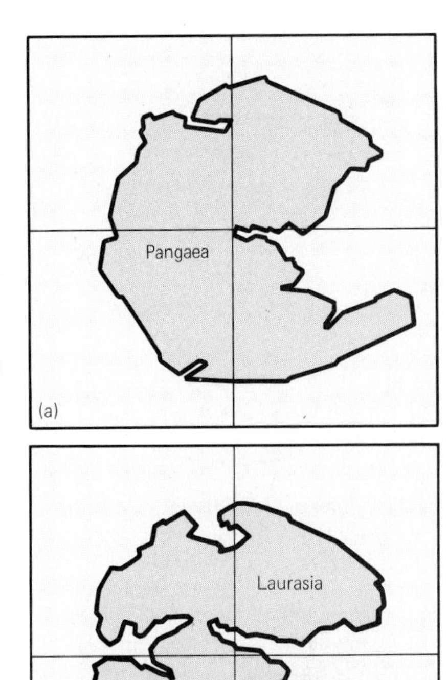


**18.10 Pre-drift reconstructions of continents.** (a) Pangaea, as reconstructed by Wegener. (b) Reconstruction according to Alexander DuToit, involving two major continental masses, Laurasia on the north and Gondwanaland on the south.

continental drift; in 1924, an English translation of his now-famous book, *The Origin of the Continents and Oceans,* was published. This book brought Wegener's theory to the attention of the English-speaking world and initiated a long series of debates and criticisms.

In Wegener's own words, the exact fit of the Atlantic coastlines of South America and Africa

> *"was the starting point of a new conception of the nature of the earth's crust and of the movements occurring therein; this new idea is called the theory of the displacement of continents, or, more shortly, the displacement theory, since its most prominent component is the assumption of great horizontal drifting movements which the continental blocks underwent in the course of geological time and which presumably continue even today."*

Wegener asserted that many surprising and apparently disconnected facts from geology, climatology, and paleontology could be explained by only one idea: that the continents had, until about 200 million years ago, been united in a single land mass which has since broken apart. He called his supercontinent *Pangaea* (Figure 18.10), or "all world."

Whereas Wegener's compilation of the evidence for continental drift was impressive, it was largely circumstantial. Wegener believed that the continents had moved, but he was not able to propose a satisfactory mechanism which could move them. The final impasse resulted when Wegener suggested that the ocean floor would have to be younger than the continents. This view was just the opposite of what nearly all geologists believed, namely, that the ocean floor was as old as the continents. As a result, the subject of continental drift did not progress very far. Wegener's arguments were ignored by most geologists and many of the important scientific questions that he raised were not answered.

**18.11 Remanent magnetism in detrital sedimentary deposits,** product of alignment of magnetic particles during settling. Polarity of specimens collected at high latitudes is indicated by upward- or downward direction north-seeking poles point. In the Earth's magnetic field today, such poles point down in the Northern Hemisphere and up in the Southern Hemisphere. (Schematic)

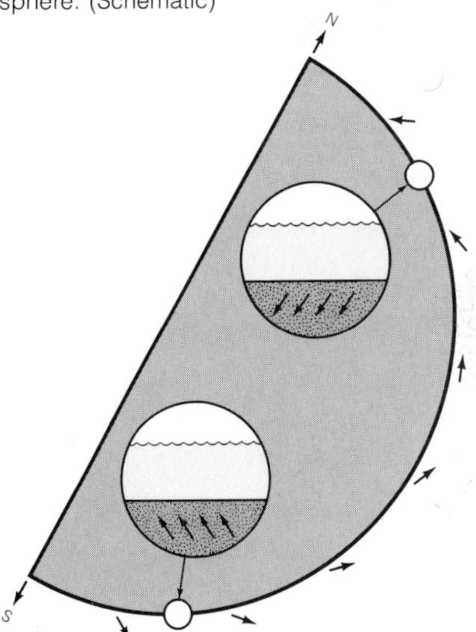


## Evidence Supporting Continental Drift

**Remanent magnetism** Early in the 1950s the science of paleomagnetism was born. As explained previously, igneous rocks preserve records of the magnetic field at the time they cooled, and sedimentary particles may be aligned during settling (Figure 18.11). Rocks preserving evidence of ancient magnetic fields are said to possess *remanent magnetism.*

In Great Britain the remanent magnetism of the Triassic red sandstones was determined in 1953. The results defined a Triassic Magnetic North Pole located 30 degrees away from the present pole and a latitude for England near the ancient Equator. The best explanation of these results was the interpretation that since the Triassic Period, about 200 million years ago, England had rotated clockwise by 30 degrees and had moved northward thousands of kilometers.

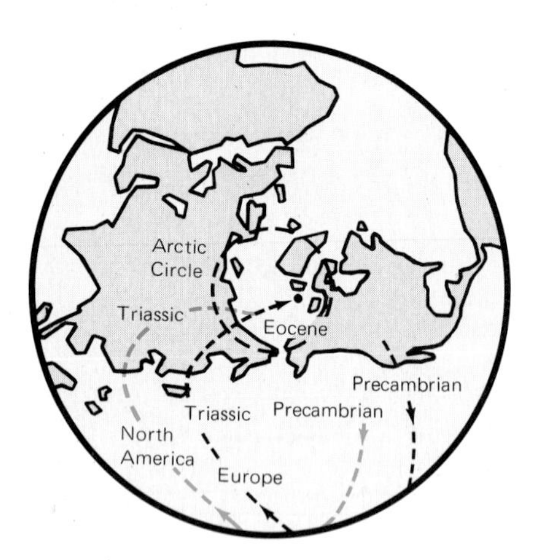


**18.12 Polar-wandering paths,** Europe and North America, Precambrian to Cenozoic. (Based on John W. Northrop, III, and Arthur A. Meyerhoff, 1963, Validity of polar and continental movement (*sic*) hypotheses based on paleomagnetic studies: American Association of Petroleum Geologists, Bulletin, v. 47, no. 4, p. 575–585, Figure 1, p. 576, after Allan Cox and R. R. Doell, 1960)

**Polar wandering** Studies of the remanent magnetism of rocks of many ages from many locations indicated that the Magnetic North Pole seemed to have moved from place to place through time, a motion called *polar wandering* (Figure 18.12). Evidence that the magnetic poles had wandered was used to revive interest in Wegener's idea that the continents had drifted. This new evidence from geophysics removed continental drift from its position at the fringes of science and made the concept into a serious subject of discussion by leading scientists. However, research about the sea floor then being done seemed to indicate that continental drift as visualized by Wegener and others was impossible.

**Impact of early research on the sea floor** The first results from the deep-sea floor that were related to the concepts of continental drift involved the shape of the bottom, which was being revealed by continuously recording precision echo sounders; the kinds of sediments, which were being sampled by poston corers; and by seismic experiments, which were indicating the thickness and characteristics of the sediments. Although the early results from the sea floor were opposed to continental drift, as more information was compiled from the sea floor, the situation changed.

Wegener and others imagined that the continents had plowed through the sea-floor sediments and left smooth tracks behind. If, indeed, the continents had plowed through deep-sea sediments as suggested, then those sediments should have been disturbed. Maurice Ewing (1906–1975) and his associates showed that the deep-sea sediments were not disturbed; they found that many parts of the deep-sea floor are rocky and jagged, not smooth as would have been the case had the sea floor consisted of "tracks" left beneath drifting continents.

As soon as the median rift valley and generally rocky, fractured appearance of the mid-Atlantic Ridge had been revealed by many records from echo sounders, Bruce C. Heezen (1924–1976) suggested that the continents might have separated because the sea floor had expanded. He presented this idea to a symposium held in France, and because his paper was published in French, many English-speaking scientists did not read it. But not long afterward, a rapid-fire chain of events led to our present concepts of plate tectonics.

## PLATE TECTONICS

In discussing plate tectonics, we will trace the origin of the concept, review definitions of plates and plate boundaries, and summarize some explanations and predictions that grow out of plate-tectonic theory.

### Origin of the Concept of Plate Tectonics

The concept of plate tectonics was born by the union of two hypotheses: (1) Harry H. Hess proposed the idea that the sea floor could move; and (2) Frederick J. Vine (1939– ) and Drummond Matthews (1931– ) proposed the idea that if the sea floor did move, it would lock in a record of the history of the Earth's magnetic field.

**Harry Hess and the moving sea floor** In a paper published in 1962, Harry H. Hess (Figure 18.13) argued that the deep-sea floor had been, and still is, moving. Furthermore, Hess proposed that the deep-sea floor is locked tightly to the mantle and thus the ocean floor could move along with the mantle without disturbing the deep-sea sediments (Figure 18.14). In other words, Hess suggested that instead of the continents plowing through the ocean floor, as Wegener and others had supposed, the ocean floor was moving at a deeper level, carrying not only the deep-sea sediments, but also the underlying oceanic crust and continents with it. Hess wrote that along the vast mid-oceanic ridge system, mantle rock is converted into new oceanic crustal rock. He thought this conversion took place at a temperature of 500°C. Hess reasoned that the ocean floor was moving away from the mid-oceanic ridges and disappearing at the deep-sea trenches.

**18.13 Harry H. Hess,** originator of the hypothesis of sea-floor spreading by movement of thick plates along a zone of weakness in the upper mantle. (Orren Jack Turner)

A few geologists have proposed that another possible way to explain the effects of continuous conversion of mantle material into new oceanic crust and to explain the separation of the continents is to suppose that the Earth is expanding. However, most geologists do not accept the notion of an expanding Earth because convincing proof has not yet been presented.

At the rate of movement that Hess proposed, the ocean floor at any point on the Earth could not be older than about 260 million years — brand new in comparison with continental rocks as old as 3.8 billion years. Such a young age for the oceans was consistent with the ages of deep-sea sediments. No deep-sea sediments older than about 180 million years have been found.

Hess also rejuvenated the idea, originally proposed in 1928 by Arthur Holmes (1890–1965), that the driving force for moving the continents consisted of convection currents within the mantle (Figure 18.15). Convection currents occur in fluids, when the fluids are heated unevenly. The heated fluid rises from the heat source and flows laterally; as the material cools, it flows downward, where it can be heated and then rise again. When Holmes first published his idea, convection was thought to be restricted to fluids and most geologists considered that Holmes' scheme of convection currents within the Earth's solid mantle was outrageous. But by the early 1960s, when the evidence in support of continental displacement was becoming stronger, the idea of convection currents in the mantle was not thought to be outrageous after all. Figure

**18.14 The Earth's oceanic crust attached firmly to the underlying mantle,** according to concept proposed in 1965 by Harry H. Hess. Schematic profile and section. (Harry H. Hess, 1965, Mid-oceanic ridges and the tectonics of the sea-floor (*sic*), p. 317–322 *in* Whittard, W. F., and Bradshaw, R., *eds.*, Submarine geology and geophysics: Colston Research Society Symposium, 17th, University of Bristol, 5–9 April 1965, Proceedings, London, Butterworths, 461 p., 39 pls., Figure 123, p. 324)

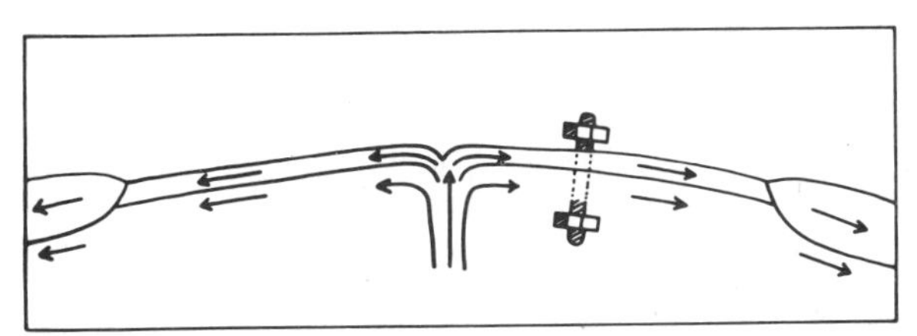

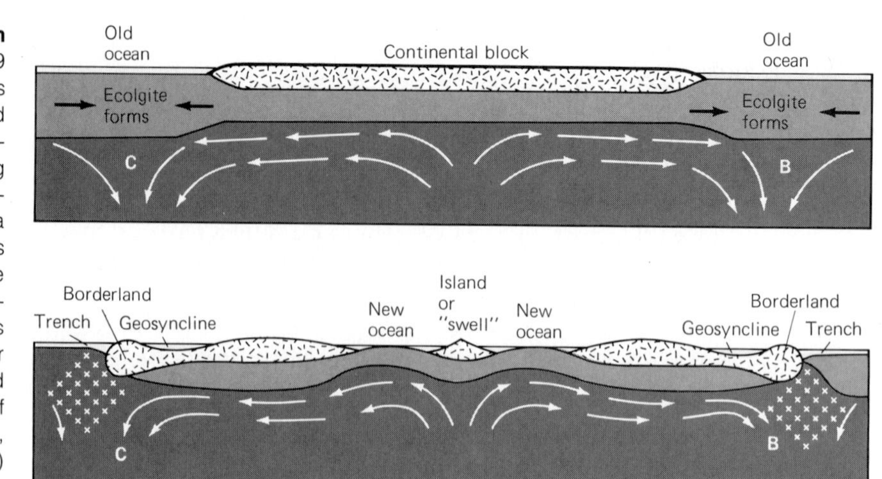


**18.15 Effects of convection currents in the mantle,** a concept proposed in 1929 by Arthur Holmes. (a) Convection currents rise beneath a continent and descend beneath an old ocean basin. (b) Over rising current, continent splits apart, forming new oceans and leaving behind a remnant of continental material to form a median ridge or island. This profile was intended to explain the mid-Atlantic Ridge and was drawn before geophysical surveys proved that the crust beneath this ridge is oceanic, not continental. (Arthur Holmes, 1928–1931, Radioactivity and Earth movements: Geological Society of Glasgow, Transactions, v. 18, p. 575–583, Figures 2 and 3, p. 579)

18.16 shows some proposed models of convection currents in the Earth's interior.

In 1961, Robert S. Dietz (1914– ) proposed the name *sea-floor spreading* for the outward movement of new sea-floor crust away from a mid-oceanic ridge.

**The Vine-Matthews hypothesis: the "magnetic tape recorder"** The polarity of the Earth's magnetic field reverses itself every half million years or so. The phenomenon of *polarity reversal* was first noticed in 1906 by a Frenchman, Bernard Brunhes (1867–1910), who thought at first that his instruments were playing tricks on him. In the 1960s, however, researchers on remanent magnetism not only confirmed the information

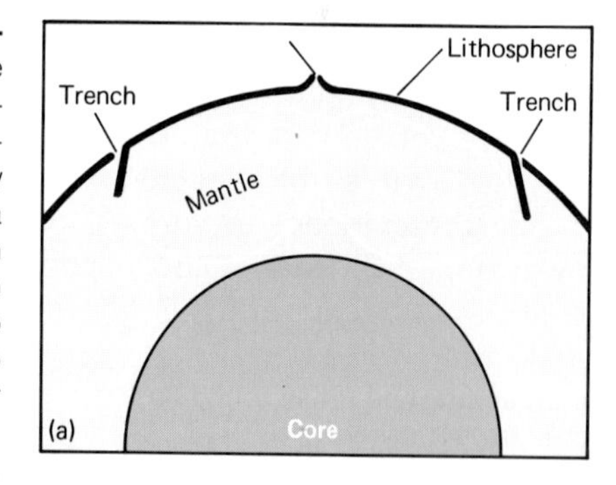


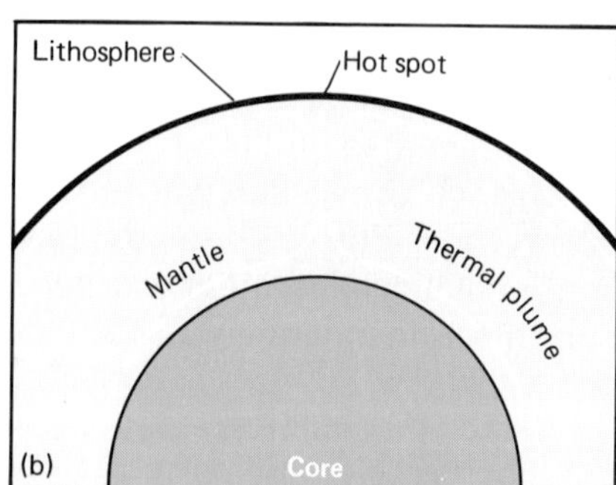


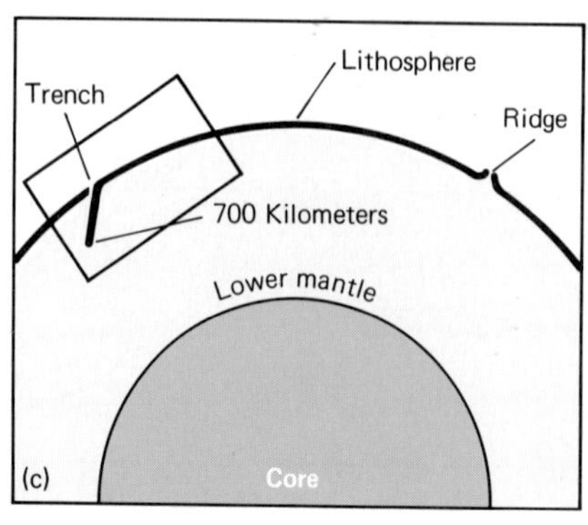


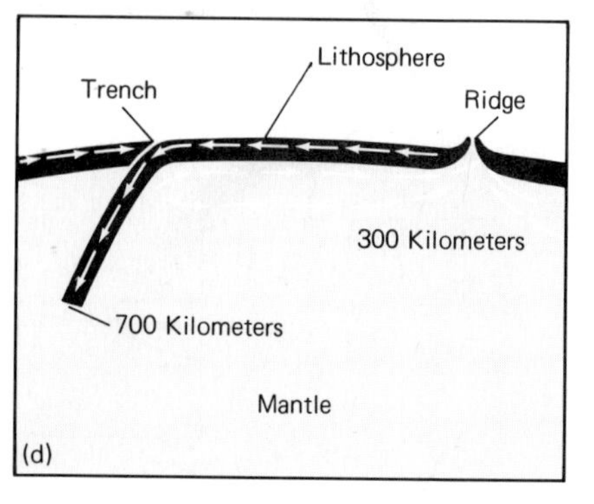


**18.16 Various concepts of mantle convection.** Convection from outer core through entire mantle: (a) Upward convection currents beneath mid-oceanic spreading ridges and concentrated return flow beneath trenches. (b) Upward motion in a few cylindrical plumes; downward return flows dispersed widely. Convection within the mantle: (c) Convection confined to asthenosphere. (d) Convection confined to top 300 kilometers of mantle. (After Peter J. Wyllie, 1975, The Earth's mantle: Scientific American, v. 232, no. 3, p. 50–57, 60–63, figures on p. 63)

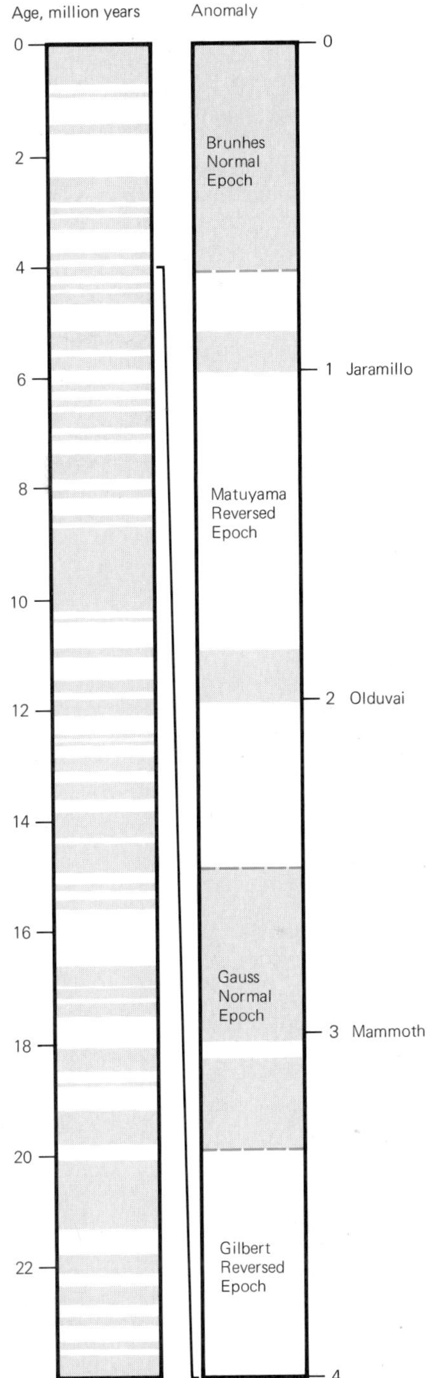


**18.17 Magnetic-polarity time scale,** based on synthesis of paleomagnetic information from continental volcanic rocks and deep-sea sediments and on paleontologic analyses of various fossils. (Courtesy W.F.B. Ryan, Lamont-Doherty Geological observatory of Columbia University)

that polarity reversals had taken place, but began to establish time scales for the magnetic reversals (Figure 18.17).

Studies in the 1960s with a sensitive magnetometer towed behind a ship recorded a baffling consistent pattern of magnetic stripes on many parts of the deep-sea floor. Half the stripes were *positive anomalies,* that is, zones of sea floor with magnetic strengths higher than the Earth's average value, and the other half were *negative anomalies,* with readings lower than the adjacent stripes. Drawn on a map, the stripes made a pattern as distinctive as the flank of a zebra (Figure 18.18). The explanation proposed for this pattern was that the positive anomalies are produced because the polarity of the magnetic particles in the oceanic crustal rocks is the same as that of the Earth's present magnetic field. Negative anomalies were inferred to be present over rocks having magnetite with polarity opposite to that of the Earth's (Figure 18.19).

In 1962, Frederick J. Vine and Drummond Matthews proposed that the magnetic anomalies around the mid-oceanic ridge are symmetrical. In 1963, they further proposed that if the sea floor had been spreading continuously and magnetic polarity reversals had occurred only occasionally, then a magnetometer survey directly over a ridge crest should reveal a pattern of anomaly stripes extending symmetrically away from the ridge in both directions. The width of the stripes should equal the rate of sea-floor spreading multiplied by the time between reversals (Figure 18.20).

## Definitions of Plates and Plate Boundaries

Lithosphere plates were defined on the basis of three kinds of motion: (1) spreading apart, with the formation of new oceanic crust; (2) coming together, with the presumed destruction of oceanic crust; and (3) slipping

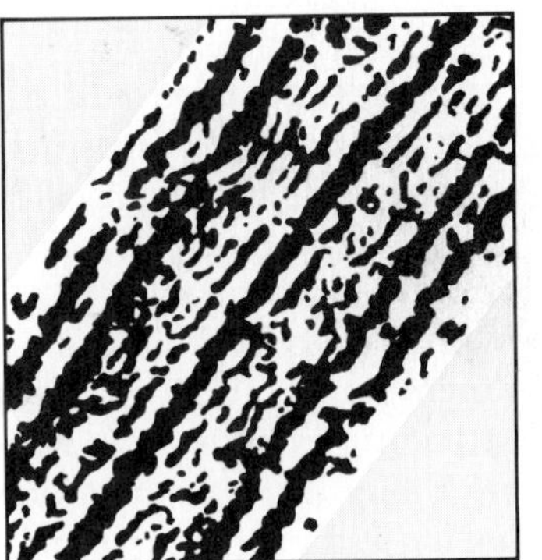

**18.18 Map of magnetic anomalies,** Reykjanes Ridge, SW of Iceland, where pattern of symmetry is best displayed. Black areas, positive anomalies; white areas, negative anomalies. (After James R. Heirtzler, Xavier LePichon, and J. Gregory Baron, 1966, Magnetic anomalies over the Reykjanes ridge: Deep-Sea Research, v. 13, no. 3, p. 427–443, Figure 1, p. 428 and Figure 7, p. 435)

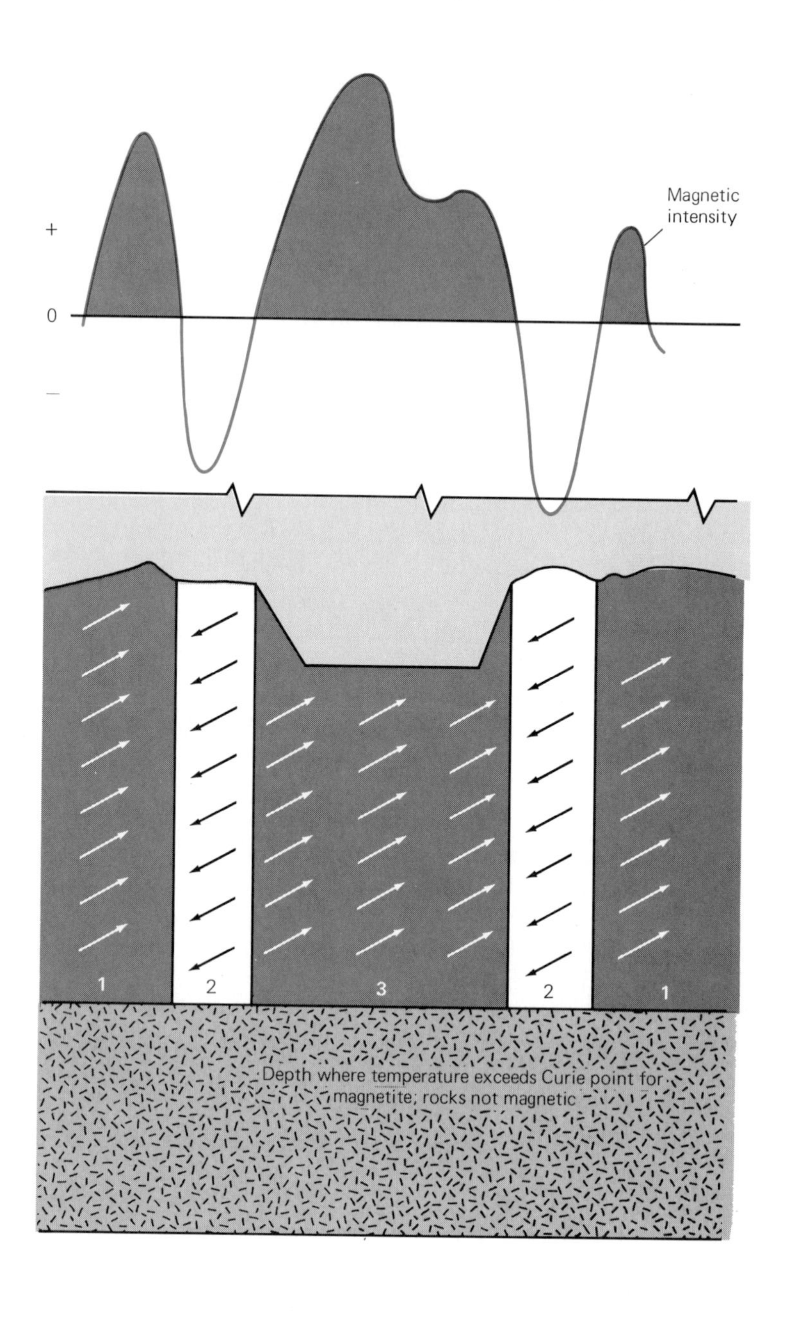


**18.19 Vine-Matthews hypothesis for explaining sea-floor magnetic anomalies,** schematic profile and section. Arrows show polarities of magnetic minerals in layer of pillowed basalt, about 0.5 kilometers thick. Positive anomalies are inferred to characterize rocks in which the remanent polarity matches the Earth's present polarity (arrows pointing up). Negative anomalies are thought to coincide with belts of rock in which the remanent polarities are opposite to that of the Earth's present polarity (arrows pointing downward). Numbers designate ages of rocks; 1 is oldest, 3, youngest.

past each other, in transcurrent motion, with crust along the faults being neither created nor destroyed.

**Spreading ridges and rift valleys** Following the idea, proposed by Hess, the mid-oceanic ridges are thought to be examples of modern *divergent* or *spreading boundaries* between two lithosphere plates (Figure 18.21). Here, as mentioned, underlying mantle material is thought to be forming new oceanic crust.

According to the ideas proposed by John F. Dewey (1937–     ) and Kevin Burke (1929–     ), the first step in the formation of a new spreading ridge is an isolated domal uplift (Figure 18.22a). Such domes are thought to lie above *mantle plumes,* cylindrical parts of the mantle having higher-than-normal heat flow. As the dome is elevated, it frac-

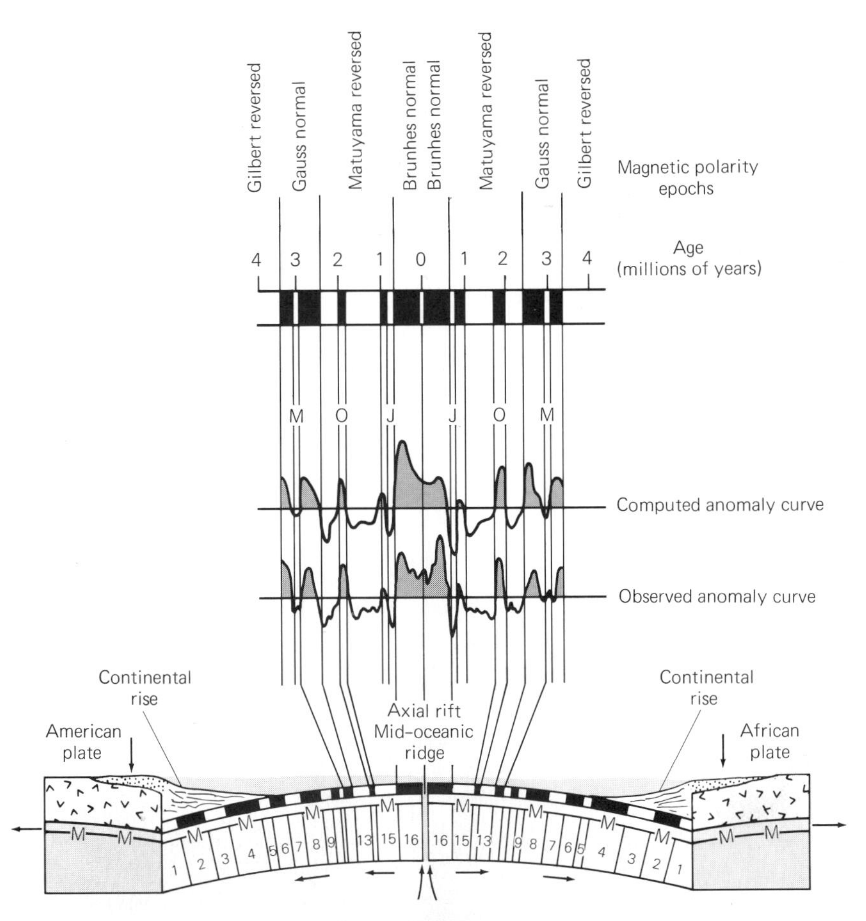


**18.20 Variable widths of magnetic stripes,** a function of rate of sea-floor spreading and duration of polarity episodes. Schematic profile and section. Rock segments are numbered by ages; 1 is oldest and 16, youngest. Numbers at top designate ages in millions of years. Anomaly curve displays intensity of the Earth's magnetic field. Positive values (above horizontal line) are readings that were higher than expected. Negative values (below horizontal line) are readings that were less than expected (assuming uniform magnetic properties in the oceanic crustal rocks).

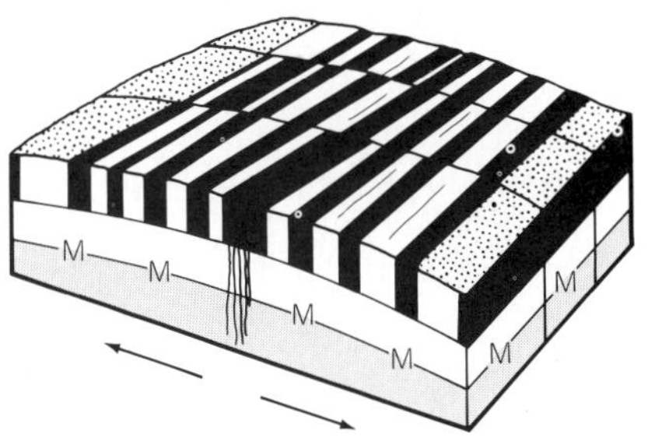

**18.21 Divergent plate boundary,** schematic block.

**18.22 Inferred surface expression of mantle plume lithosphere that is not being displaced across the plume.**
(a) Uplift of dome having diameter of 100 to 200 kilometers or so; no fractures. (b) Three cracks form at 120-degree angles, at first confined within the circumference of the dome. (c) Lithosphere plates begin to spread away from hot spot along two of the three fractures. (d) Junction of the Red Sea, Gulf of Aden, and the African rift system. (After Kevin C. A. Burke and John F. Dewey, 1973, Plume-generated triple junctions: key indicators in applying plate tectonics to old rocks: Journal of Geology, v. 81, no. 4, 406–433, Figure 2, p. 408) See also Plate 18.

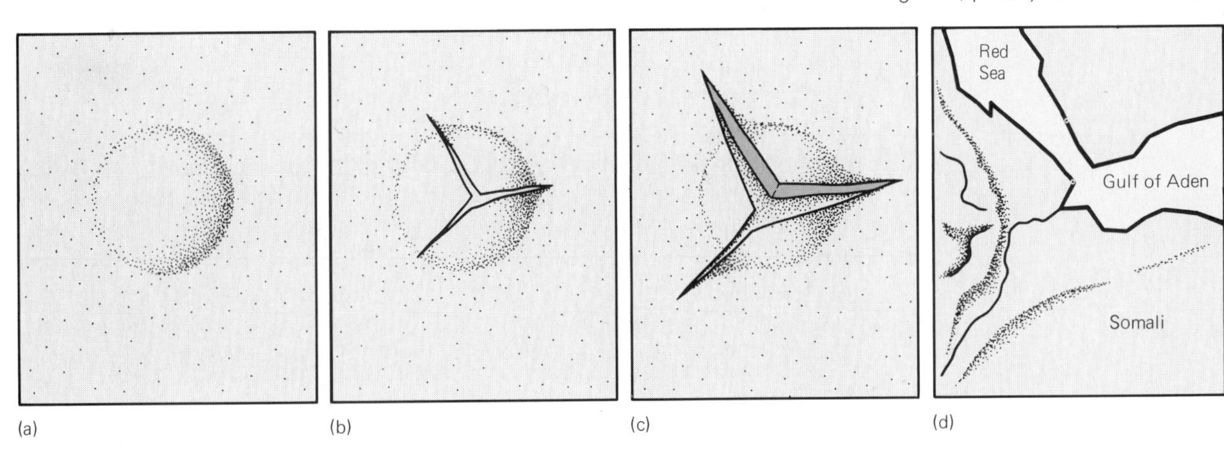

tures along three cracks arranged at angles of 120 degrees (Figure 18.22b).

If spreading begins from these fractures, then new oceans may open along all three. Alternatively, new oceans may open along only one or two of the fractures (Figure 18.22c). An example in the modern world where this activity is inferred to have occurred is the junction of the Red Sea, the Gulf of Aden, and the African rift at Afar, Ethiopia (Figures 18.22d and Plate 18).

**Benioff zones, deep-sea trenches, and subduction zones** The opposite of a divergent plate boundary is a *convergent plate boundary,* where two plates are inferred to collide (Figure 18.23). Benioff zones are places where earthquake foci of intermediate and deep varieties are concentrated along inclined zones that dip away from the oceans (see Figure 17.14). New, detailed studies of seismograms have indicated that the first-motion direction implied by earthquakes originating in Benioff zones is lithosphere above the Benioff zone moving relatively upward over the block below this zone.

It was obvious from the time when Benioff zones were first described that these zones are tectonic features of the greatest significance. From the first formulations of the concepts of plate tectonics, Benioff zones have been regarded as the places where oceanic crust disappears beneath continental crust. The downward movement of oceanic crust has been named *subduction,* and the places where it is supposed to happen,

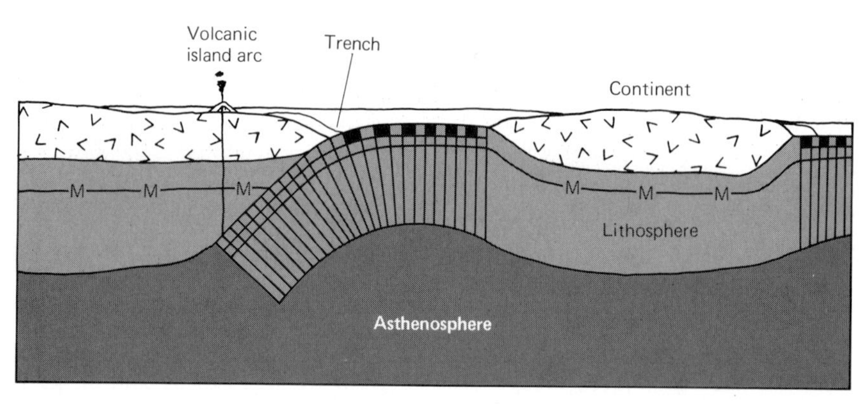


(a)

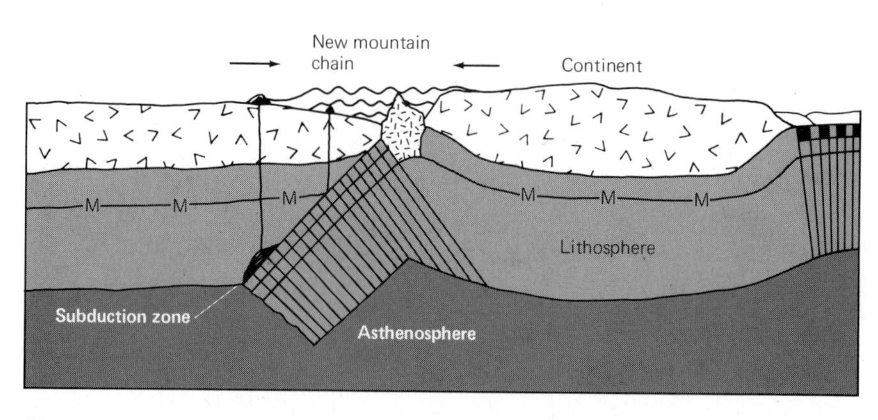


(b)

**18.23 Convergent plate boundary,** schematic profiles and sections. (a) Oceanic lithosphere, "consumed" at a subduction zone, is recycled back into the mantle; without a spreading ridge, the ocean shown presumably will disappear.

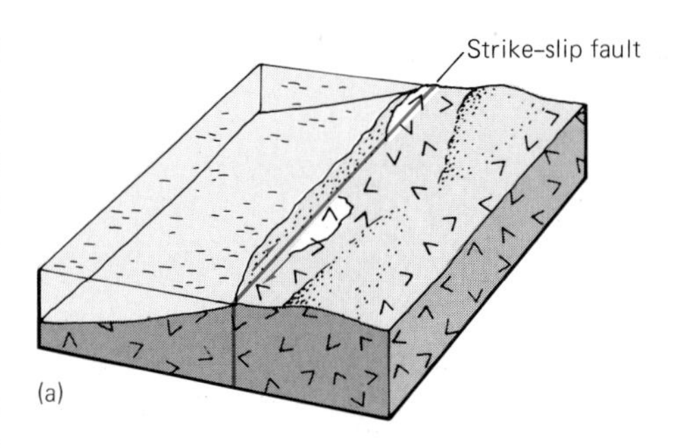


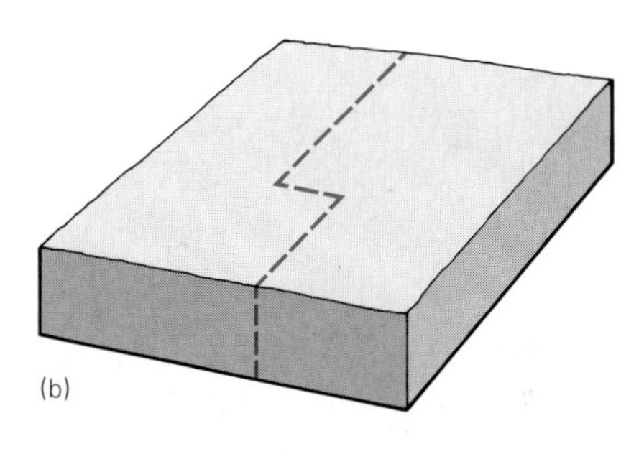


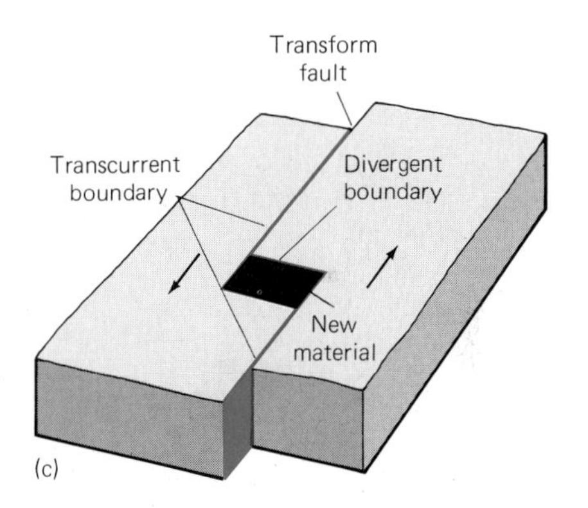


**18.24 Transcurrent plate boundaries,** schematic blocks. (a) Right-lateral strike-slip fault, no mid-oceanic ridge is involved. Transcurrent boundary involving transform fault and spreading center where new material is added: (b) Reference block before faulting and spreading. (c) Left-lateral movement on transform faults form the boundaries at ends of a segment of a spreading center.

beneath the deep-ocean trenches, are called *subduction zones* (see Figure 18.23). The concept of subduction is perhaps one of the most dramatic aspects of the whole idea of plate tectonics. Subduction zones are visualized as nature's ultimate disposal systems. Whatever has accumulated on top of the ocean crust, pelagic sediments or even entire volcanoes, is presumed to disappear down the trenches by subduction. It has even been seriously suggested that we dump all our garbage, radioactive wastes, and toxic chemicals down the trenches, where the stuff will be subducted into oblivion.

**Zones of transcurrent movement** Along a *transcurrent plate boundary,* two plates shift laterally, one past the other along a great transcurrent fault, either a strike-slip fault or a transform fault (Figure 18.24a). At plate boundaries that are strike-slip faults, no new material is added nor is any destroyed. The action is simply one of sideways shifting. An example of a transcurrent plate boundary at a strike-slip fault is thought to be the San Andreas fault in California (see Figure 17.19). This fault is regarded as the transcurrent boundary between the North American Plate and the Pacific Plate (see Figure 1.14).

At plate boundaries that are transform faults, the boundary is a zig-zag pattern that alternates between transcurrent and divergent. This follows from the nature of transform faults, which are connected to a spreading ridge (Figure 18.24b, c). An example of a plate boundary that includes transform faults is along the mid-Atlantic Ridge between the African Plate and the South American Plate.

## Plate-Tectonic Explanations and Predictions

The concept of plate tectonics is regarded by some enthusiasts as being the ultimate explanation of nearly all geologic phenomena. In the following sections, we consider only a few central hypotheses.

**Plate tectonics and continental drift** According to the theory of plate tectonics, continental drift is a normal part of the Earth's behavior; the great lithosphere plates that compose the outermost shell of the Earth are always moving. Some continents are split apart as new oceans open. Other continents collide as ocean basins disappear. As an illustration of how continental drift and plate-tectonic theory are related, let us return to Wegener's example of Pangaea and examine it in light of modern concepts.

From the dates of the oldest sediments and fossils in the Atlantic Ocean, geologists have inferred that Pangaea began to break apart during Late Jurassic time (about 180 to 200 million years ago). This is the same break-up date assigned by Wegener. As Europe slowly moved away from the eastern seaboard of North America, Africa was also beginning to split

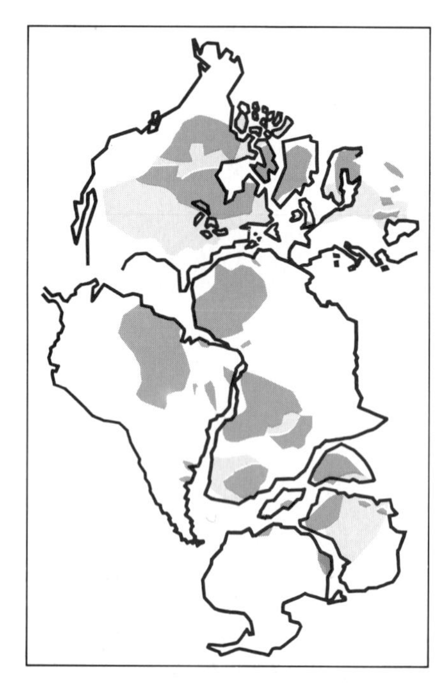

**18.25 Parts of Pangaea bordering future Atlantic Ocean** restored as they are thought to have been before spreading and separation. Gray areas indicate exposed Upper Precambrian rocks. (Based on Patrick M. Hurley and John R. Rand, 1969, Pre-drift continental nuclei: Science, v. 164, p. 1229–1242, Figure 8, p. 1237)

off from South America (Figure 18.25). The South Atlantic began to open about 135 million years ago — much later than the North Atlantic.

At about the same time that the North Atlantic began to open, a seam is thought to have formed between Africa and the rest of Pangaea. Another oceanic ridge developed between the tip of South America and Antarctica. A short distance to the northeast, this ridge branches and divides India from Antarctica and Africa (Figure 18.26). India began its long northward journey, which was to end in collision with Asia 80 million years and 7000 kilometers later, deforming the crust to form the Himalayan mountain complex (Figure 18.27).

The last of the great landmasses to split apart is thought to have been the combined Australia and Antarctica. These two continents remained joined until perhaps 40 million years ago. Antarctica is inferred to have moved southward. Australia set off northward at high speed, traveling more than its own length in only a few tens of millions of years.

Meanwhile, in the Pacific, along the huge East Pacific Ridge, the direction of spreading changed from east-west (as implied by the great

**18.26 Inferred position of India with respect to Africa-Arabia** and **Antarctica-Australia early in the Cenozoic Era.** (a) About 75 million years ago, India had separated from the adjacent continents. (b) About 35 million years ago, both India and one of its spreading ridges had shifted to the northeast

fracture zones) to roughly northwest-southeast (as implied by the rows of volcanic islands, such as the Hawaiian chain). Such changes in direction of inferred spreading are not yet well explained.

**Predicted future plate movements** The continued northwesterly movement of the Pacific Plate would bring Los Angeles abreast of San Francisco in about 10 million years (Figure 18.28). In about 60 million years, at present rates of spreading, Los Angeles will reach the Aleutian Trench, where the Pacific Plate is presumed to be consumed by being recycled downward into the mantle.

Some geologists predict that even greater changes than those discussed above will affect Africa and the Mediterranean. They expect that as Africa rotates counterclockwise, the Red Sea and the Gulf of Aden will widen. At the same time, they foresee that the western Mediterranean will swing shut, closing off the Strait Gibraltar. Finally, they predict that the African Rift Valleys will widen and become new oceans, thus splitting Africa lengthwise.

and Australia had separated from Antarctica. Numbers indicate ages of sea floor (in millions of years), based on magnetic-anomaly patterns. (After John B. Sclater, and Daniel B. McKenzie 1973. The evolution of the Indian Ocean: Scientific American, v. 228, no. 5, p. 62–72, figures on p. 71 and 72)

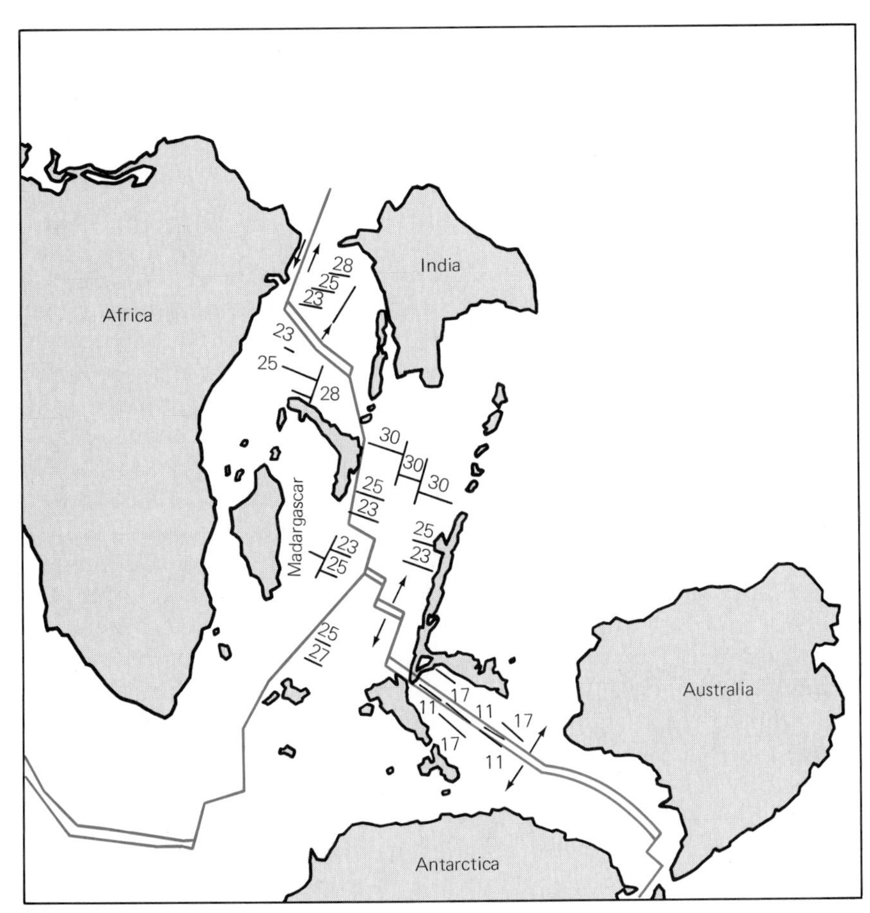


(b)

**18.27 The Himalayan Mountains,** strata that were deformed and forced upward—possible consequences of the collision between India and Asia, according to the concept of plate tectonics. (United Nations)

**Predicted ages of sea-floor crust and sediments** One of the attractive features of the Vine-Matthews hypothesis, now incorporated within the concepts of plate tectonics, is the use of magnetic anomalies on the deep-sea floor to predict the age of the crust underlying the sea floor at any point. The belts of magnetic anomalies are simply counted outward, and assuming that each anomaly represents a corresponding time of magnetic-polarity reversal and that spreading has never stopped, the ages of the sea-floor crust can be assigned by matching the number of the anomaly with the time scale of magnetic-polarity reversals (see Figure 18.19). It is important to emphasize that this procedure absolutely depends on the notion that spreading is continuous and that no anomalies are skipped.

Rarely are geologists able to decide unambiguously which anomaly is which; many of them look more or less alike. Therefore, anomalies can be used for certain dating only if *all are present* and thus simple counting can be the basis for converting anomaly to geologic age. After counting and matching anomalies, geologists made maps of the predicted age of the crust underlying various parts of the ocean floors (back end papers). Although anomaly counting seems to give reliable ages in the main ocean basins, it has failed utterly in the Red Sea. According to the anomaly curves, the Red Sea was considered to be 2 million years old. This age was widely believed until the *Glomar Challenger* drilled a hole that penetrated Miocene evaporites 15 million years old.

This chain of events in the Red Sea illustrates what the British biologist Thomas Huxley (1825–1895) termed a "scientific tragedy": "a beautiful theory killed by an ugly fact."

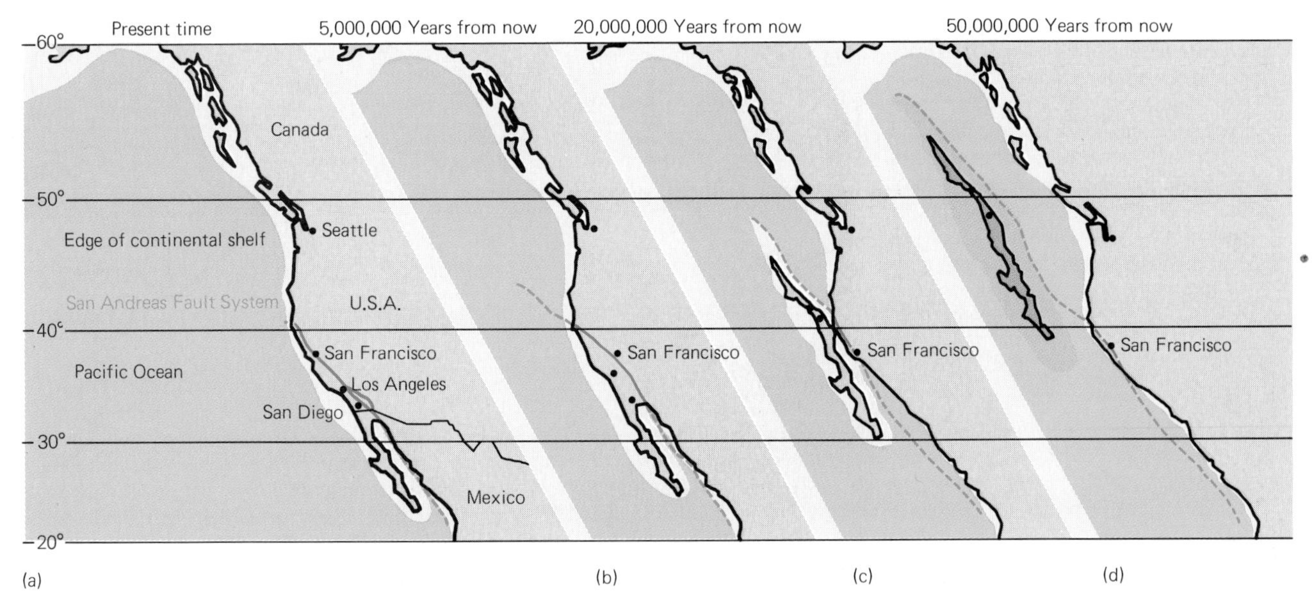


**18.28 Predicted effects of further motion on San Andreas fault,** California, schematic maps based on persistence without significant change of present boundaries between land and water. (a) Present configuration. (b) After 250 kilometers of right-lateral movement, the entrance to the Gulf of California has widened and the entrance to San Francisco Bay has been blocked. (c) After another 750 kilometers of motion, central Baja California is abreast of San Francisco, and a large "Gulf of Oregon" has formed. (d) After 1500 kilometers of additional shifting, for a total shift of 2500 kilometers, a new island occupies parts of the Gulf of Alaska, and the two previous gulfs have vanished. (Conjectural, based on the concept of continuous movement at a rate of 5 centimeters per year and consumption of Pacific crust at the Aleutian trench)

**Mantle hot spots and rows of volcanic islands** Within ocean basins are many rows of volcanic islands that become systematically younger in one direction. An example is the Hawaiian chain, which begins on the northwest with the atoll of Midway, and ends on the southeast with the modern volcano Kilauea, on the island of Hawaii (Figure 18.29).

According to plate-tectonic explanation, the Pacific Plate is moving northwestward over a fixed plume of heat rising through the underlying mantle. Where the plate is over the plume, a volcano is active. As the plate moves away, the volcanic cone is cut off from the rising heat and becomes inactive. Ultimately, the top of the inactive cone is eroded. Moreover, as it moves down the slope of the dome created by the hot spot, the island may be submerged. If this view is correct, then the row of islands marks the track of the Pacific Plate for the past few tens of millions of years.

If the Pacific Plate is moving across a fixed hot spot in the mantle, then one cannot suppose that movements of the mantle provide the force for driving the plate. The concept of islands formed by motions of the lithosphere over a fixed hot spot in the mantle implies that the plate is moving independently of the mantle. This is the exact opposite of the Hess view that the sea-floor crust is "bolted" to the underlying mantle (see Figure 18.14). Clearly, the theoreticians cannot "have it both ways." Both ideas could be wrong, but only one can be correct. If the moving-island-row concept is correct, then some mechanism other than large-scale mantle convection must be sought for driving the plates. If the plates are "locked" to the underlying mantle, then the moving-island-row theory needs to be modified. This difficult dilemma will be resolved only when the mechanism of plate motion has been fully understood.

**The Wilson cycle: disappearing and new oceans** One of the logical consequences of the concept of plate tectonics is that oceans can be both created and destroyed. Creation results from spreading away from a

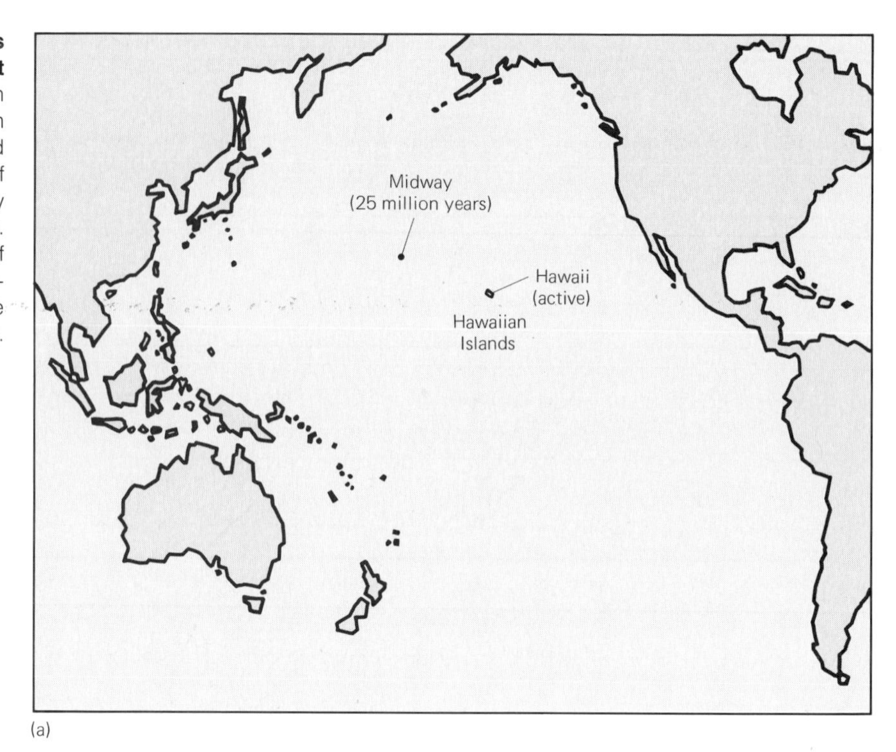


(a)

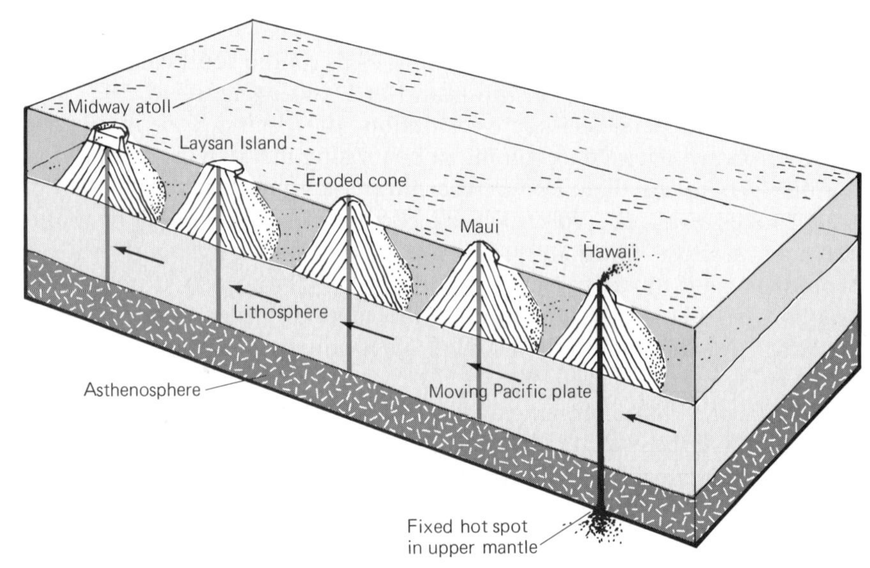


**18.29 Hawaiian Islands interpreted as plate-movement track over a fixed hot spot in the upper mantle.** (a) Location map, showing ends of Hawaiian chain and their ages. (b) Volcanic cones and conduits move away from source of magma in the asthenosphere. The only active volcano is on the island of Hawaii. Progressive erosion and subsidence of inactive cones yields as an ultimate product a coral atoll, as at Midway (compare Figure 18.4).

mid-oceanic ridge; destruction is what presumably happens when an ocean is consumed in a subduction zone. By supposing that sites of spreading and of consumption can be shifted now and then, one can visualize how an ocean could appear, be consumed, and then a new ocean can be formed on the site of the first ocean. This concept was formulated by J. Tuzo Wilson (1908– ) for the Atlantic Ocean; subsequently the progression of events was named a *Wilson cycle.* As applied to the Atlantic Ocean area, the cycle involves three stages. In Stage 1 (Figure

18.30a, b, c) a new ocean forms by spreading away from a mid-oceanic ridge. This spreading is matched by lithosphere consumption at some remote location, so that the western boundary of the growing ocean coincides with the trailing edge of a plate, as the North American plate is behaving at present. In Stage 2 (Figure 18.30d, e), the sites of spreading and of subduction change. Now, the formerly active mid-oceanic ridge becomes inert; a subduction zone appears at the ocean's western boundary, where formerly there was a passive, trailing margin; and a new spreading center becomes active somewhere else. With time, the now-inert ocean basin is totally consumed, and formerly separated continents are joined together. In Stage 3 (Figure 18.30f), spreading takes up again where it started, and a new ocean forms, with its western boundary being a passive margin, as in Stage 1.

**Plate tectonics and the dynamic history of orogenic belts** According to the concepts of plate tectonics, orogenic belts such as the Appalachians are born at continental margins by the operation of Wilson cycles. During the first stage of such a cycle, a passive continental margin subsides and traps the sediment eroded from the interior of the continent to build a continental terrace. A modern example is thought to be the great continental terrace along the eastern margin of North America (Figure 18.31). A comparable terrace was built along the eastern margin of the North American continent during the Cambrian and Ordovician periods. At that time, an ocean is thought to have stood to the east, where the Atlantic Ocean is now situated. If this ocean were spreading during the Cambrian and Ordovican periods, then the accompanying plate consumption must have been going on in the Pacific region, not along the eastern margin of North America.

Proponents of plate tectonics suggest that orogeny takes place when a continent or an island arc appears at a subduction zone. Subduction is thought to be something that affects only oceanic lithosphere. Because of its buoyant capping of felsic continental crust, continental lithosphere is thought not to be capable of undergoing subduction. Instead, when a continent arrives at a subduction zone, continental collision takes place, and such a collision is what is thought to cause orogeny.

Referring back to the Appalachian area, the plate-tectonic explanation is that subduction began in the Appalachians during the Ordovician Period. At this time, the whole pattern of the Earth's major interior motions is thought to have changed. The former passive continental margin became a volcanic island arc above a subduction zone; the former ocean stopped spreading; but elsewhere, ocean-ridge spreading became active, and motion away from it toward the new subduction zone caused the formerly spreading, but now-inactive ocean to disappear. After this inactive ocean had been consumed, Europe appeared from over the eastern horizon, and a great orogeny took place, in medial Devonian time. This orogeny was powered by the collision of Europe against North America. Once joined, these two continental masses remained together until the Jurassic Period, when the present cycle of spreading from the mid-Atlantic Ridge began, and Europe and North America began to separate.

The plate-tectonics explanation of some existing mountain chains is as follows. The Himalayas were formed when the free-moving India slammed northward into the Asian mainland and pushed the mountains upward as India crunched beneath the surface (see Figure 18.27). Africa

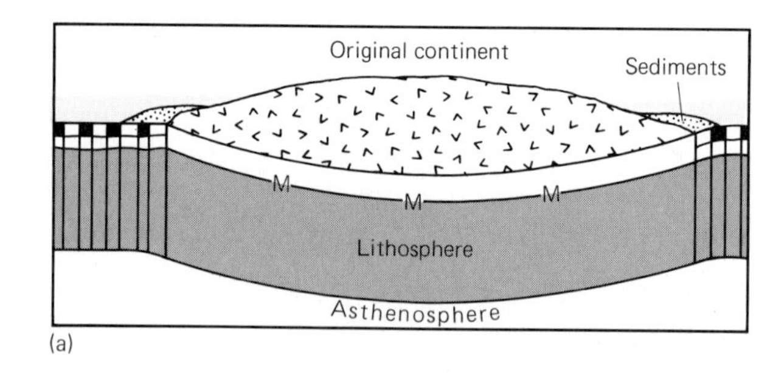


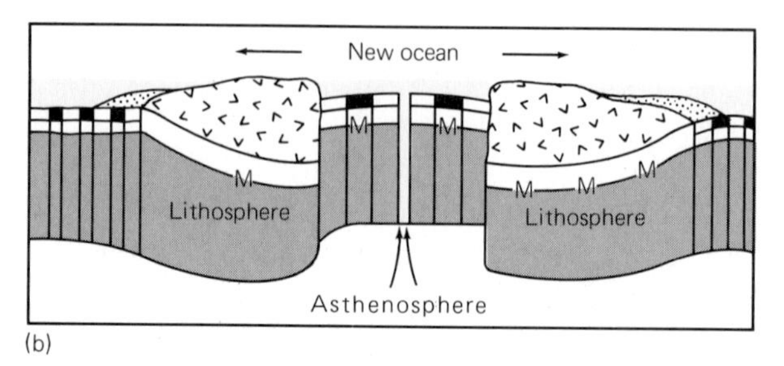


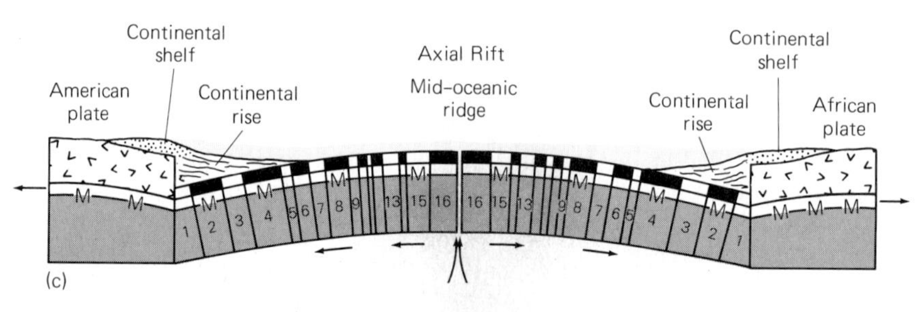


**18.30 Wilson cycle as applied to areas near the Atlantic Ocean.** (a) Continent bordered on each side by an old ocean (compare Figure 18.15a). (b) New ocean forms where former continent splits apart (compare Figure 18.15b). (c) Fully developed new ocean; spreading in this ocean is not matched by loss of this ocean's lithosphere at a subduction zone at its margins, but rather by subduction elsewhere (not shown). At trailing edge of continent, sediments build continental terrace and continental rise. (d) Totally new spreading pattern begins; active ocean becomes an old ocean. (No such old oceans have been found in the modern world; hence, the rest of the cycle depends on what is considered to be reasonable conjecture, not on specific analogy.) Crust of old ocean begins to disappear at newly established subduction zone and accompanying volcanic island arc. (e) Continent-continent collision deforms strata, creates orogeny with accompanying overthrusts and batholiths. (f) Continent is split and new ocean starts the cycle again, as in (b). (After R. S. Dietz, 1972, Geosynclines, mountains, and continent-building: Scientific American, v. 226, no. 3, p. 30–38; figures on p. 37)

and Europe revolved in a motion that pushed Italy into central Europe and created the Alps, also building up the Matterhorn from a former piece of Italy. The collision of Arabia and Eurasia moving in opposite circular paths forced the uplifting of the Zagros Mountains in Iran.

**Plate tectonics and igneous activity** The concepts of plate tectonics can be used to explain the major sites of igneous activity. Mafic magmas, formed by partial melting of "normal" upper-mantle material, rise to the surface along mid-oceanic ridges or along deep fractures that extend through the lithosphere. By contrast, felsic magmas are formed only where thick bodies of sediment are subjected to conditions where the

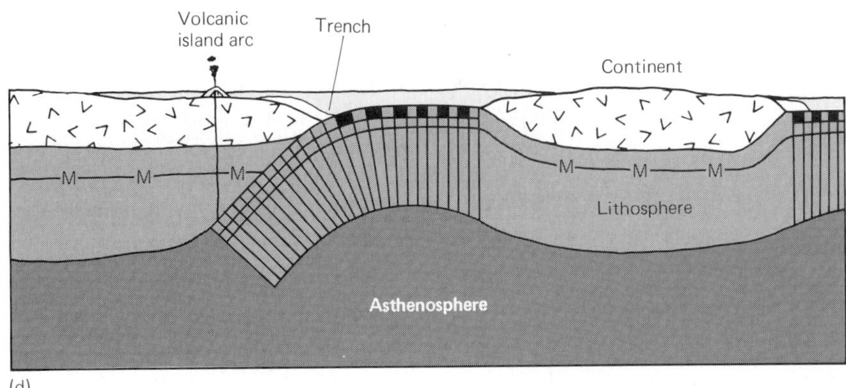


(d)

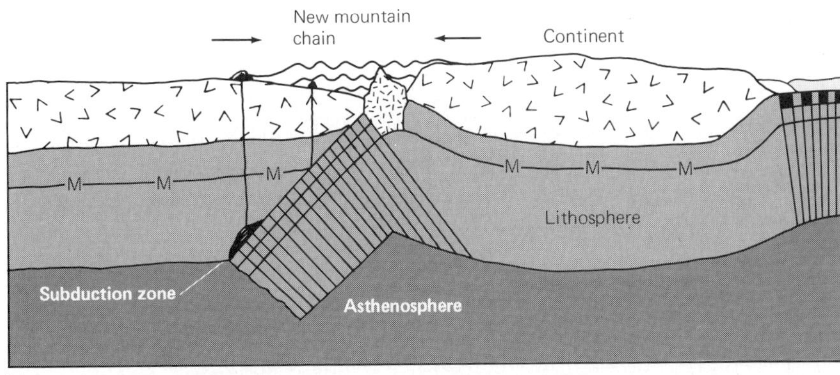


(e)

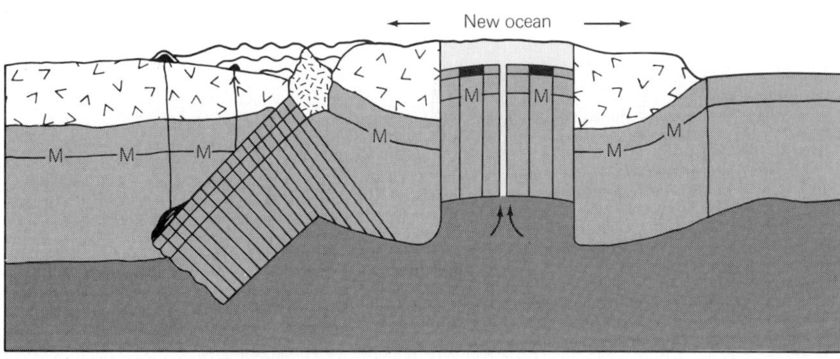


(f)

material can melt. Intermediate magmas are thought to form only at subduction zones, where mixtures of deep-ocean sediments and oceanic crust are melted en route to being consumed by disappearing down a subduction zone.

**Plate tectonics and natural resources** Orogenic belts are the preferred geologic habitats of plutons composed of granitic rocks or of intermediate rocks. Thus, orogenic belts are places where one would start to search for primary deposits of tin or tungsten that may have been deposited with a

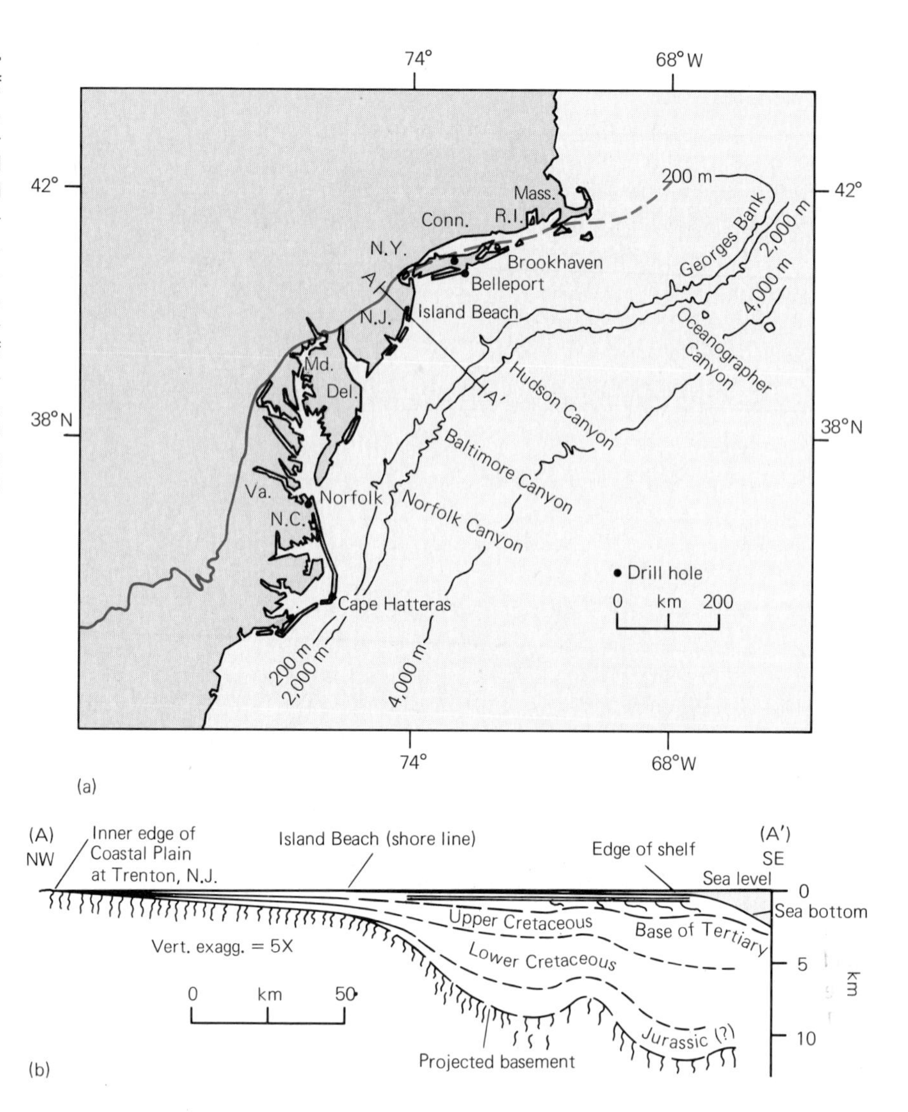


**18.31 Continental terrace along eastern North America.** (a) Index map of northeastern conterminous United States, showing base of coastal-plain strata, location of profile and section, and selected onshore deep test borings. (b) Profile and section, based on continuous seismic profiles and a few test borings. (After J. P. Minard, W. J. Perry, E. G. A. Weed, E. C. Rhodehamel, E. I. Robbins, and R. B. Mixon, 1974. Preliminary report on geology along Atlantic continental margin of northeastern United States: American Association of Petroleum Geologists, Bulletin, v. 58, no. 6, part II, p. 1169–1178, Figure 1, p. 1170, Figure 2, p. 1171, and Figure 4, p. 1172)

granite, for example, and for many kinds of hydrothermal deposits that may be associated with felsic or intermediate rocks. These include copper-impregnated porphyries, deposits of gold, silver, lead, zinc, and many other metals.

Geologic studies have shown that in an orogenic belt, the usual sequence of events is: (1) accumulation of strata, (2) deformation of the strata, including folding and thrusting, (3) emplacement of felsic or intermediate plutons, (4) movement along high-angle faults, and (5) deposition of ore minerals.

In a few orogenic belts, great overthrusts have involved oceanic crust. As a result, sheets and slabs of mafic and ultramafic oceanic crust and associated deep-sea sediments have been emplaced above sedimentary strata deposited in shallow water. Within these bodies of ultramafic rock may be various ores as explained in a following paragraph.

One of the logical extensions of the concept of plate tectonics was to determine if ore deposits are related systematically to plate boundaries, to the interiors of plates, or to such features as rift valleys that cross plates. Another possibility is to see if ore deposits are related to any particular tectonic stage such as a special time during the opening of a new ocean. These questions are just now being addressed. One of the most successful results has been the discovery of numerous additional copper-impregnated prophry ore bodies along what are inferred to be convergent plate margins in the southwestern Pacific-Indonesian area, and new understanding of massive sulfide deposits associated with submarine volcanic rocks.

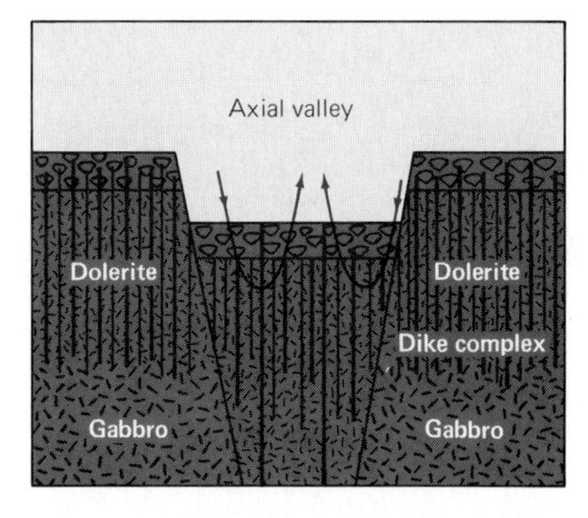


**18.32 Hydrothermal circulation in axial valley of mid-oceanic ridge** (blue arrows); ocean water descends through pores in volcanic rock and reacts with the hot basalt and underlying rocks. The water is responsible for altering the silicate minerals to hydrated forms and for depositing metallic sulfide minerals that may be abundant enough to be ore deposits. (After Enrico Bonatti, J. Honnorez, P. Kirst, and F. Radicati, 1975, reference cited in caption of Figure 18.16, Figure 11, p. 73; and Maaten J. deWit and Charles R. Stern, 1976, A model for ocean-floor meta morphism, seismic layering and magnetism: Nature, v. 264, p. 615–619, Figure 2, p. 616)

Many ore deposits of metallic minerals are associated with mafic igneous rocks that formed at a spreading ridge on the sea floor. Detailed studies of heat flow on the sea floor have shown that the cold ocean-bottom sea water circulates downward through the top part of the oceanic mafic crustal rocks and is heated there (Figure 18.32). Later, the heated water rises through the rocks and is discharged into the sea again. The effect is to dissolve and re-precipitate ions and to alter the rocks. Many of the precipitates include copper and manganese minerals. The massive layered copper sulfides of the Troodos massif, on Cyprus, and elsewhere, are thought to have formed in this way.

Deeper down within the oceanic crust and in the uppermost part of the mantle, massive gabbro and associated ultramafic rocks are present (see Figure 18.32). Within these are found crystal cumulates, which commonly include chromite, nickel, platinum, and magnetite.

## Some Dissenting Opinions About Plate Tectonics

The hypothesis of plate tectonics has been adopted by an overwhelming number of geologists. Nevertheless, some important questions about tectonics still have not been answered. Several dissenting ideas have been expressed concerning various aspects of plate tectonics.

The chief difficulty centers around the concept of subduction. The notion that subduction takes place is considered to be necessary to dispose of all the material created at spreading ridges on an Earth whose diameter does not change. However, S. W. Carey (1911–    ) and others, including the late Arthur Holmes, insist that the Earth may be expanding. If, indeed, the Earth is expanding, then the argument in favor of subduction becomes much less compelling.

Several objections to the ideas of plate tectonics have been voiced by A. A. Meyerhoff (1928–    ) and Howard A. Meyerhoff (1899–    ), based on field evidence and other geologic considerations. For example, the Meyerhoffs contend that on Iceland, a land that is presumably being split apart by spreading from the mid-Atlantic Ridge, folds are present. These and other features suggest to the Meyerhoffs that Iceland has been pushed together in its past and has not been subjected only to pulling apart, as predicted by the concept of plate tectonics.

The Soviet academician V. V. Beloussov (1907–    ) has raised questions about the importance of vertical movements, especially in the interiors of continents. The concepts of plate tectonics do not deal with vertical movements at such places. Beloussov also asks how can the Red Sea be spreading obliquely to the local magnetic anomalies.

The concept that deep-sea sediments should be piling up in trenches and there be offscraped as the crust beneath them enters a subduction has been questioned by David W. Scholl (1934–    ) and M. S. Marlow

(1945– ). They have published seismic reflection profiles that show the sediments of some trenches are flat lying and not crumpled as one would expect them to be according to the hypothesis of subduction.

A final point to be considered concerns the evidence from the analysis of seismic *S* waves that the substructure of continents extends to depths of approximately 400 kilometers. Thomas H. Jordan (1948– ) has pointed out that the thick continental underpinnings seem to have traveled along with the continents. If this idea is correct, then the simplified notion that the continents move on an asthenosphere at the same depths as the oceanic crust needs to be changed. Jordan advocates use of the term *tectosphere* to refer to the zone of motion of both continents and oceans. Jordan's idea makes considerable sense; the implication of his analysis is that a Benioff zone is a normal boundary between continents and oceans and may be more a zone of great temperature changes than of subductionlike motion (see Figure 1.14).

Other questions need to be considered, and probably the most critical issue involves the driving mechanism needed for extensive plate movement. Even if it *does* happen, no one is really sure *how.* And the arguments will continue until an acceptable reconciliation is reached. In the meantime, the hypothesis of plate tectonics will continue to fascinate and inspire scientists throughout the world.

# CHAPTER REVIEW

1. Nearly every large-scale feature at the surface of the Earth has formed because of some dynamic activity of the lithosphere, the process known as *tectonics.*
2. Physical properties of the lithosphere and underlying parts of the mantle affect tectonic activities. Using measurements of rock density, speeds of seismic waves, thicknesses of various layers, and amounts of heat flowing upward, it is possible to recognize three kinds of lithosphere: stable areas of deep-ocean basins not near a mid-oceanic ridge or a trench; stable areas of broad parts of continents; and unstable linear belts that are active tectonically, including *mid-oceanic ridges, trenches, rift valleys,* and *active orogenic belts.*
3. The lithosphere is not rigid but responds to natural loads by shifting downward with an increased load and rebounding when the load is removed. Such loads include the weight of the atmosphere, the gravitational pull of the Moon, and the weights of various bodies of water, of sediment, of volcanic material, and of glacial ice.
4. Much of the geologic record and many of the Earth's modern surface features have come about because of vertical movements of the lithosphere that take place without crumpling the strata. Downward movement is called *subsidence;* vertical uplift in which the strata are not deformed is called *epeirogeny.*
5. The condition of dynamic balance between the lithosphere and the asthenosphere is *isostasy,* which implies that the Earth's gravity causes segments of the lithosphere to shift upward or downward according to their densities.
6. In contrast to the vertical displacements of the lithosphere are the *horizontal* displacements that are implied by fold belts and great overthrusts.
7. The dynamic history of an orogenic belt involves both vertical displacements and horizontal displacements in six stages: *geosynclinical* stage of accumulation of strata; *terminal-orogeny* stage; *initial-elevation* stage, usually including a post-orogenic stage of block faulting and vertical movements; *long-continued-erosion* stage; *peneplanation* stage; and *recycling* stage.
8. Alfred Wegener and others became interested in the idea of moving continents on a global scale, based on the notion that Africa and South America had once been joined. Wegener challenged the established concept of fixed continents. He was not able to develop an acceptable new Earth model, and he could never establish a theoretical basis in support of his view of *continental drift.*
9. Studies of *remanent magnetism* and indications of *polar wanderings* revived interest in Wegener's theory of continental drift. Sea-floor research also suggested continental separation.

10. The concept of *plate tectonics* was born by the union of two hypotheses: (a) the idea, proposed in 1962 by Harry H. Hess, that the sea floor could move, and (b) the idea, by F. J. Vine and Drummond Matthews, that if the sea floor did move, it would lock in a record of the history of the Earth's magnetic field.
11. According to Hess, the continents did not plow through the ocean floor, as Wegener and others had supposed, but instead, movement was taking place at a deeper level, in an arrangement that carried not only the deep-sea sediments, but also the underlying oceanic crust and the continents with it. Hess also rejuvenated Arthur Holmes' earlier rejected idea that convection currents move the continents.
12. The name *sea-floor spreading* was proposed by Robert S. Dietz to describe the outward movement of new sea-floor crust away from a mid-oceanic ridge.
13. In the 1960s researchers not only confirmed the information that reversals of the Earth's magnetic polarity had taken place, but also were able to establish time scales based on magnetic reversals.
14. Lithosphere plates can be defined on the basis of three kinds of boundaries and motions: (a) *divergent* boundaries, where plates spread apart and new oceanic crust forms; (b) *convergent* boundaries, where plates come together and oceanic crust is presumed to be destroyed; and (c) *transcurrent* boundaries, where plates slip past each other and crust adjacent to the boundary is neither created nor destroyed.
15. The movement of oceanic crust downward into the mantle is *subduction,* and the places where it is supposed to happen, beneath the deep-ocean trenches, are *subduction zones.*
16. According to the theory of plate tectonics, continental drift is a normal part of the Earth's behavior; the great plates that compose the outermost shell of the Earth are always moving. The plate-tectonic concept reinforces Wegener's idea that a supercontinent, *Pangaea,* started breaking apart about 180 million years ago.
17. The *Vine-Matthews hypothesis* states that magnetic anomalies on the deep-sea floor can be used to predict the age of the crust underlying the sea floor at any point.
18. The *Wilson cycle* of disappearing and new oceans involves three stages: (1) a new ocean forms by spreading away from a mid-oceanic ridge; (2) a subduction zone appears at the ocean's western boundary, a new spreading center becomes active somewhere else, and when the now-inert ocean basin is consumed, the formerly separated continents are joined together; (3) spreading takes up again where it started, and a new ocean forms as in Stage 1.
19. According to the concepts of plate tectonics, orogenic belts are born at continental margins by the operation of Wilson cycles.
20. The concepts of plate tectonics can be used to explain the major sites of igneous activity, and these concepts may also be related to the systematic locations of ore deposits.

## QUESTIONS

1. What is meant by the term *tectonics?*
2. List four physical properties of the lithosphere and underlying parts of the mantle that affect tectonic activities. Using these properties, what large-scale groupings of lithosphere have been recognized?
3. List six kinds of natural loads on the lithosphere and describe the lithosphere's response to each.
4. Define *subsidence.* Compare and contrast subsidence with *epeirogeny.*
5. List and discuss three geologic proofs of subsidence and three of uplift. Give two examples of places where subsidence has been followed by uplift.
6. Define *isostasy.* Compare and contrast isostasy based on the *sial* and *sima* with the modern concept of isostasy based on the lithosphere and the asthenosphere.
7. List and explain five factors that can affect the lithosphere and the asthenosphere as related to isostasy and the vertical position of the surface of the lithosphere.
8. What kind of tectonic movements are implied by the folds and overthrusts of orogenic belts? Describe the typical result of overthrusts involving marine strata at the margin of a continent.
9. List and briefly describe the six stages in the dynamic history of an orogenic belt. Compare these stages with the stages of human lives.
10. Define *continental drift.* Who was responsible for bringing this concept to the attention of the world's geologists? Explain the arguments he used and describe how geologists at the time responded to these arguments.
11. Explain what is meant by *Pangaea.*
12. Define *paleomagnetism* and *remanent magnetism.* What was the impact of the first studies of paleomagnetism on the concept of continental drift?
13. What kind of evidence from the deep-sea floor provided arguments against the pre-1960 versions of continental drift?
14. Explain how the discovery of the world-encircling mid-oceanic ridge system was first related to the idea of continental drift.

15. What two ideas were joined together to form the basis of *plate tectonics?*

16. Explain what is meant by *sea-floor spreading* and the "magnetic tape recorder."

17. Discuss the three kinds of plate boundaries and related plate motions.

18. Define *mantle plume* and give an example in the modern world that is thought to be the effects of a mantle plume.

19. Explain how deep-sea trenches, Benioff zones, and *convergent plate boundaries* are thought to be related. Define *subduction.*

20. What is the plate-tectonic explanation of (a) continental drift, (b) age of the ocean-floor crust, (c) rows of volcanic islands, and (d) orogenic belts?

21. What is a *Wilson cycle?* How is a Wilson cycle thought to explain the origin of a fold-mountain chain?

22. Discuss the inferred relationships between plate tectonics and (a) igneous activity, and (b) natural resources.

## RECOMMENDED READING


Ahrens, T. J., 1980, Dynamic Compression of Earth Materials: *Science,* v. 207, p. 1035–1041.

Burgess, R. F., 1977, Submerged Forests: *Oceans,* v. 10, no. 5, p. 46–49.

Carr, M. J., 1977, Volcanic Activity and Great Earthquakes at Convergent Plate Margins: *Science,* v. 197, p. 655–657.

Christensen, M. N., 1966, Late Cenozoic Crustal Movements in the Sierra Nevada of California: *Geological Society of America, Bulletin,* v. 77, no. 1, p. 163–182.

Cook F. A., Brown, L. D., and Oliver, J. E., 1980, The Southern Appalachians and the growth of continents: *Scientific American,* v. 243, no. 4, p. 156–168.

Hamilton, Warren, 1979, Tectonics of the Indonesian Region: *U.S. Geological Survey, Professional Paper 1078,* 343 p.

Heezen, B. C., 1960, The Rift in the Ocean Floor: *Scientific American,* v. 203, no. 4, p. 98–110.

Matthews, R. K., 1974, *Dynamic Stratigraphy:* Englewood Cliffs, New Jersey, Prentice-Hall, Inc., 370 p. Isostasy and the Stratigraphic Record, p. 49–60.

McKenzie, D. P., 1972, Plate Tectonics and Sea-floor Spreading: *American Scientist,* v. 60, no. 4, p. 425–435.

Tanner, W. F., 1973, Deep-sea Trenches and the Compression Assumption: *American Association of Petroleum Geologists, Bulletin,* v. 57, no. 11, p. 2195–2206.

Tazieff, Haroun, 1970, The Afar Triangle: *Scientific American,* v. 222, no. 2, p. 32–40.

Vine, F. J., and Metthews, D. H., 1963, Magnetic Anomalies over Oceanic Ridges: *Nature,* v. 199, p. 947–949.

EXPLORATION OBJECTIVES

SURFACE CLUES ABOUT WHAT LIES BELOW THE LAND SURFACE

SUBSURFACE EXPLORATION METHODS USED ON LAND

SUBSURFACE EXPLORATION METHODS USED BENEATH WATER

# CHAPTER 19 EXPLORATION GEOLOGY

PLATE 19

Geologists examining coves brought up by drilling rig (at left) during evaluation of copper-zinc deposit near Silver City, New Mexico. (Shelly Katz, courtesy Exxon Corporation)

ONE OF THE IMPORTANT THEMES WE HAVE TRIED TO STRESS throughout this book is that the future of western civilization as we now know it , requires that new sources of minerals and energy be found, and soon. The task of finding these sources belongs to teams of skilled scientists including geologists, geophysicists, geochemists, and geobotanists. Every informed citizen should be aware of the importance of these searches and should understand some of the principles and techniques involved.

In this chapter, we summarize some of the kinds of geologic settings in which new sources are likely to be found and then review some of the methods available for carrying out the searches. In general, the objectives of the searches are to determine the composition and structure of subsurface materials. The techniques are indirect methods that enable one to infer certain things about what lies below or direct methods, namely, drilling holes or digging pits or shafts to obtain samples. The scientific challenges related to these searches are great and specific. The ultimate quantitative test of one's knowledge of geology is to be able to specify exactly *where* a given deposit may be found.

## EXPLORATION OBJECTIVES

From the material presented in Chapters 6, 7, 8, 9, and 16, it would be possible to compile a long list of particular kinds of geologic conditions favorable for the localization of economically valuable deposits, thus objectives for exploration. Our purposes in this chapter can be achieved by discussing only four major kinds: (1) various geologic structures, (2) buried surfaces of erosion, (3) plutons, and (4) buried valleys. The principles discussed in these examples can be applied as well to other kinds of objectives.

### Economic Deposits Related to Geologic Structures

The chief kinds of geologic structures that have localized important economic deposits include anticlines, salt diapirs, and faults.

**19.1 Fluid (petroleum) trapped in the crest of an anticline,** schematic profile of folded strata that include a sandstone between shales.

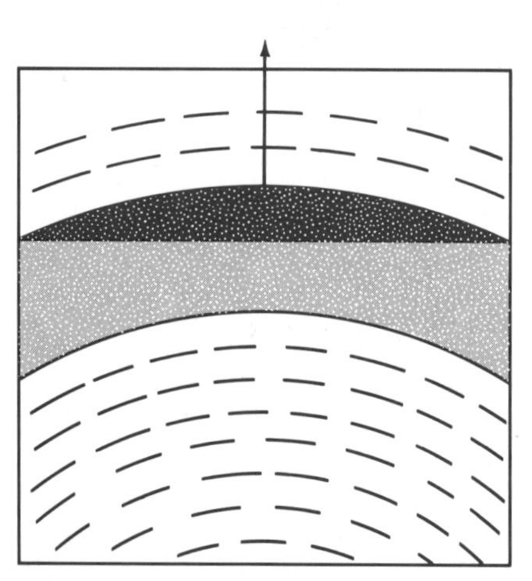

**Economic deposits related to anticlines** The important characteristics of anticlines that localize economic deposits are the reversal of dip at the crest and the seal provided by a shale or other impermeable layer overlying a permeable layer such as a sand. What is important is that fluids can migrate up the dip of the inclined limbs, but cannot escape from the crest. This means that the fluids, such as petroleum, which are valuable in themselves or which contain valuable materials dissolved in solution, can accumulate or deposit minerals in the crestal areas.

Figure 19.1 shows only a cross section to illustrate the concept of change of dip at the crest and the effect of shale in preventing the escape of fluids from the sandstone. Actually, the three-dimensional arrangement is all important. Thus, the best setting for holding in the fluids is provided by a simple circular dome (see Figure 16.1g) or by an elongate, doubly plunging anticline, in which the structure is closed in all directions (Figure 19.2). In a simple plunging anticline, which is not closed in all directions, fluids can migrate up the direction of plunge and thus can escape.

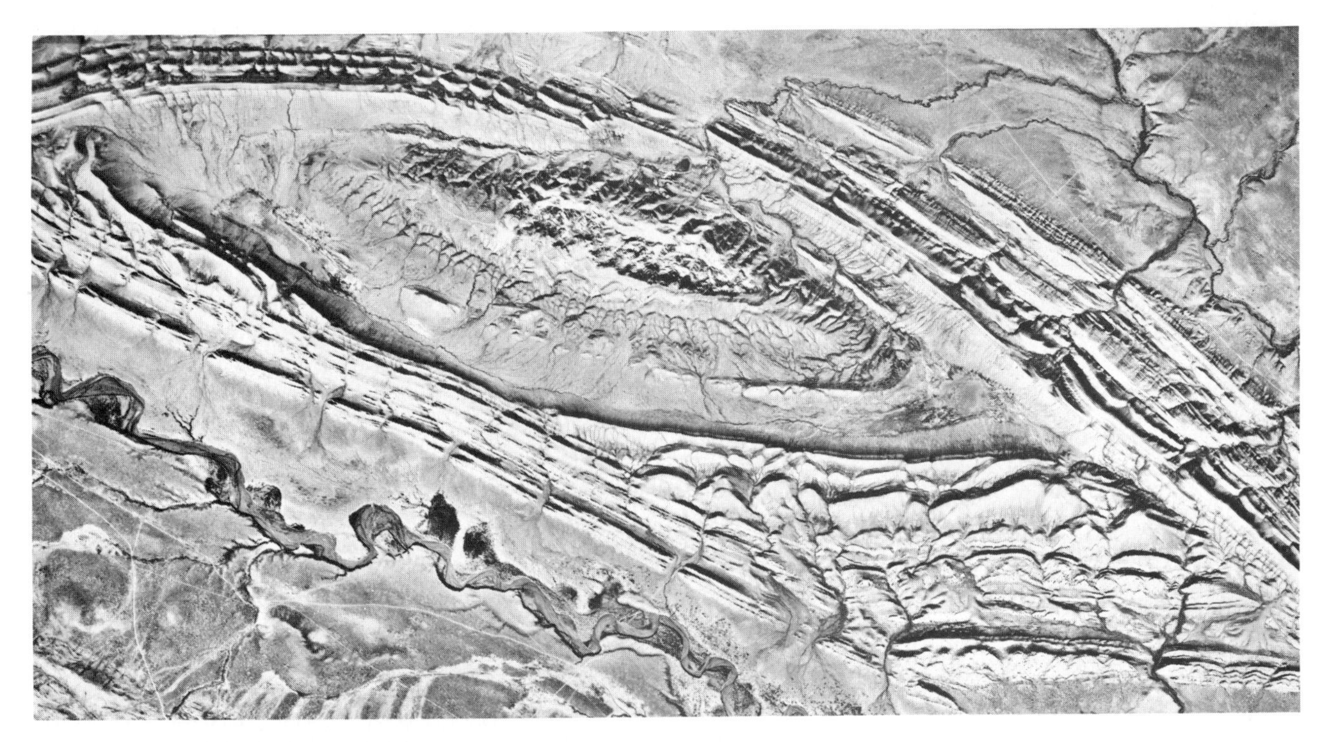

**19.2 Elongate doubly plunging anticline,** viewed from vertically above. Little Maverick dome, 58 kilometers NW of Riverton, Fremont County, Wyoming, in T5N, R1W (River Meridian), latitude 43°30′N, longitude 108°50′W, between Maverick Springs Creek on the SW and Fivemile Creek on the NE. North is at top. (U. S. Geological Survey)

We have assumed that in these anticlinal structures the folding took place entirely after uniformly thick layers of sedimentary strata had been deposited. If the history of folding began during sedimentation, that is, if the folds were growth structures (see Chapter 16), then the folds may have affected the distribution and continuity of the sands.

The origin of domes and of doubly plunging anticlines is not well understood. Various mechanisms are possible, including the interference pattern formed by the intersection of anticlines from two sets of non-plunging folds of different ages (Figure 19.3).

Not all closed domes in what appear to be favorable geologic settings contain petroleum. An example of a recently explored closed dome that proved to be nothing more than an expensive and spectacular failure is the Destin dome off the northwest coast of Florida (Figure 19.4). The explanation for such failures will not be known until the subject of the origin of petroleum is totally understood. An important point that needs to be emphasized and that is valid even at our present level of understanding of the origin of petroleum is the timing of petroleum migration and time of origin of the structure. Petroleum escapes if it migrates before the dome has formed.

Closed anticlinal structures can be exposed at the land surface or can be hidden from view beneath an unconformable cover that has not been folded. In some examples that have been drilled, several episodes of folding and erosion have taken place so that the structure at great depth is more pronounced than at shallow depths (Figure 19.5).

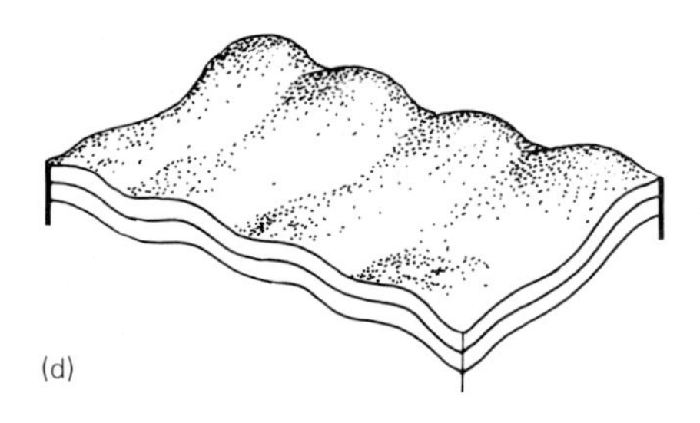

(d)

**19.3 Closed domes formed by "interference" of two sets of folds.** (a) First set of folds, parallel and upright; no plunge. (b) First set of folds is buried by younger strata. (c) Second set of folds, parallel and upright, but with axes at right angles to those of the first set. The effect on the first set of folds by the second episode of folding is to create closed domes where the anticlines cross. (d) Pattern of structural features in the older strata; upper group of strata arbitrarily removed to show shapes of all domes. (Suggested by subsurface relationships in Pennsylvanian strata of northeastern Oklahoma, based on unpublished studies by J. E. Sanders)

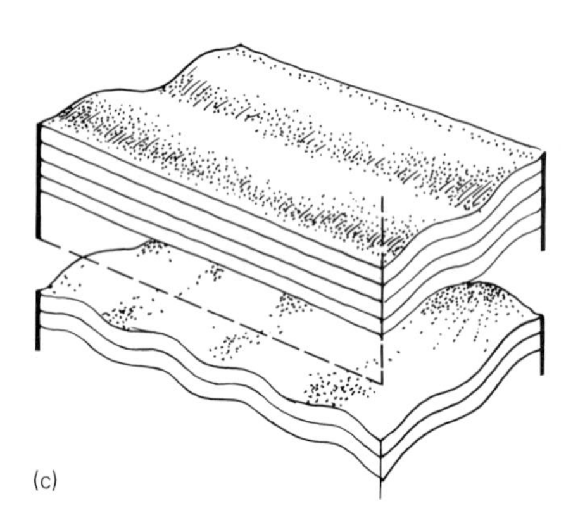

(c)

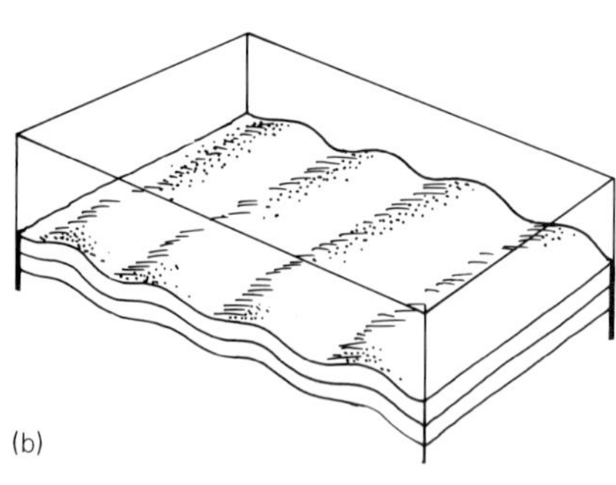

(b)

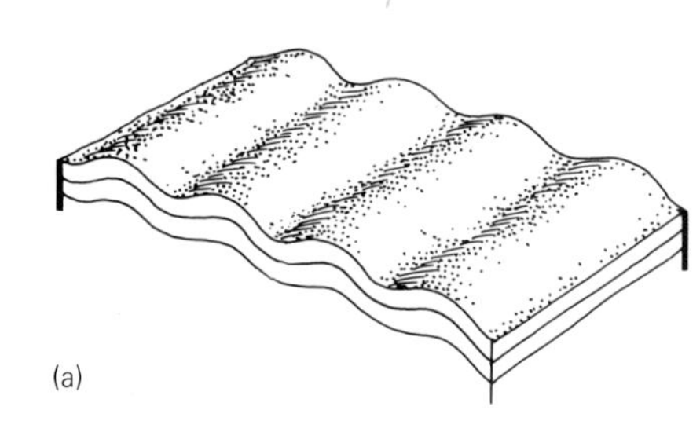

(a)

Separation of the layers being tightly folded creates curved, open spaces in which quartz and gold can be deposited, as in some of the gold deposits of Australia. Such curved mineral deposits have been named *saddle reefs* (Figure 19.6).

**Economic deposits related to salt diapirs** Salt diapirs are of economic interest for their materials, such as halite and sulfur, and for their effects on surrounding strata, which have caused petroleum to accumulate. (Figure 19.7). The growth of the diapirs during sedimentation may have affected the distribution of sands on the sea floor, and the doming of overlying strata plus upturning of strata penetrated have formed petroleum traps (see Figure 8.22).

The modern era of petroleum abundance began at the beginning of the twentieth century when fabulous amounts of petroleum were discovered at shallow depths east of Beaumont, Texas, above and in the caprock of a salt diapir named Spindletop. (Because of all the sordid deals that were made over control of this vast oil deposit, the place was nicknamed by many as "Swindletop.")

As with closed domes, not every salt diapir has been the site of petroleum accumulation.

**19.4 Destin dome,** off the west coast of Florida, generalized map. The high hopes for finding large petroleum deposits in this dome have been shattered by the drilling of more than 20 dry holes. The cost of the predrilling surveys, purchase of the leases, and drilling of the dry holes has been about one billion dollars.

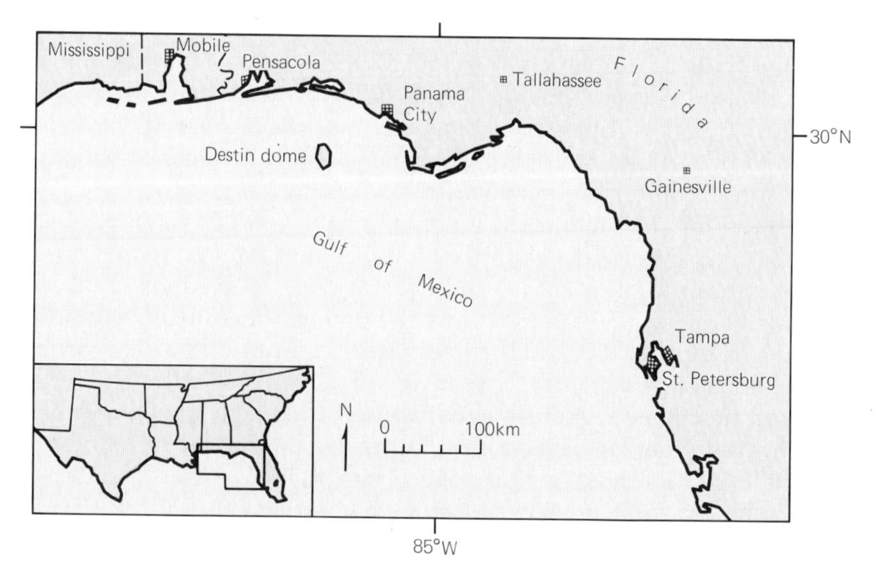

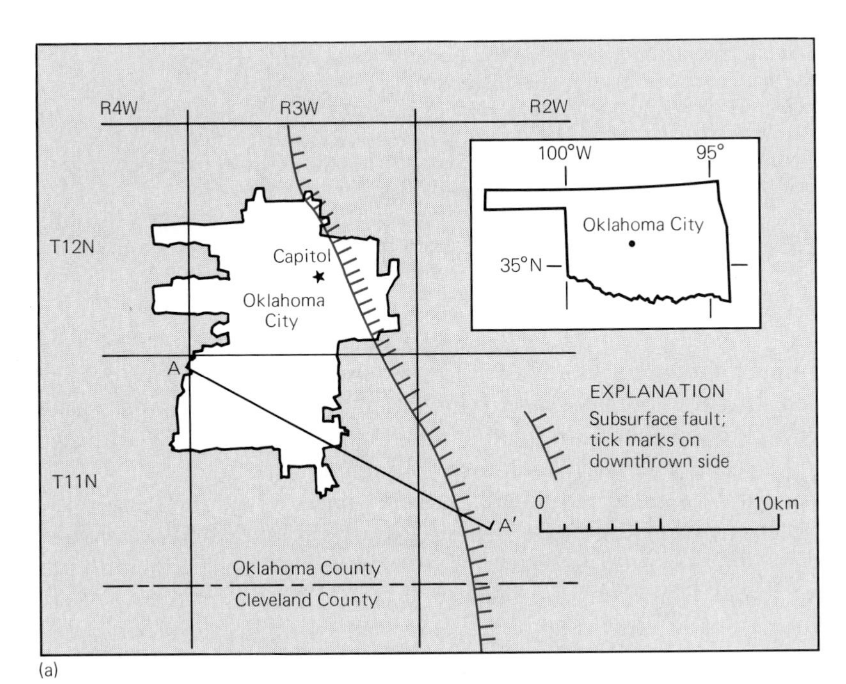


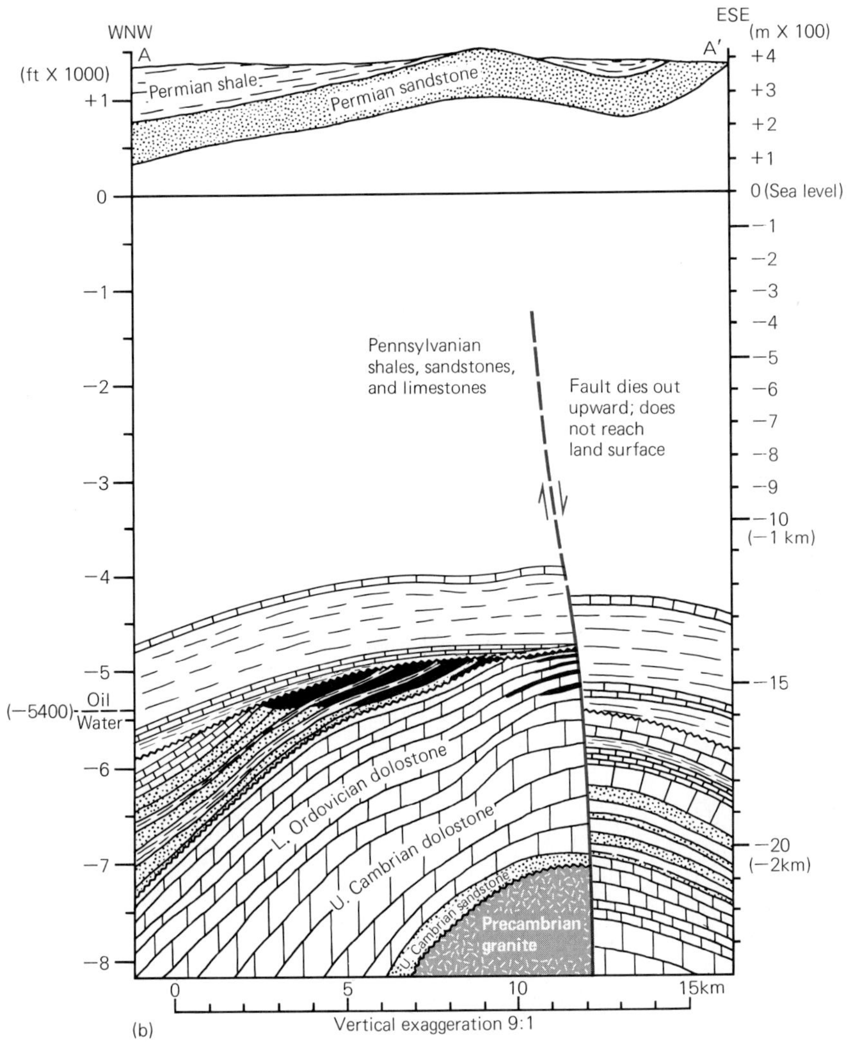


**19.5 Oklahoma City oil field.** (a) Location map. (b) Profile and section based both on surface mapping and on results of drilling. Many of the Ordovician producing sands were at the Earth's surface and being eroded during the post-Mississippian, pre-Pennsylvanian episode of uplift and deformation. In a few places, tar at the former land surface indicates that oil was naturally seeping out at the exposed undip edge. Elsewhere no tar is present and it is probable that most of the oil migrated after the Pennsylvanian shale had been deposited across the beveled edges of the older rocks. (L. E. Gatewood, 1970, Oklahoma City field—anatomy of a giant, p. 223–254 *in* Halbouty, M. T., *ed.,* Geology of giant petroleum fields. A symposium: Tulsa, Oklahoma, American Association of petroleum Geologists, Memoir 14, 575 p, Figures 7, 10, and 11, p. 233 and 235)

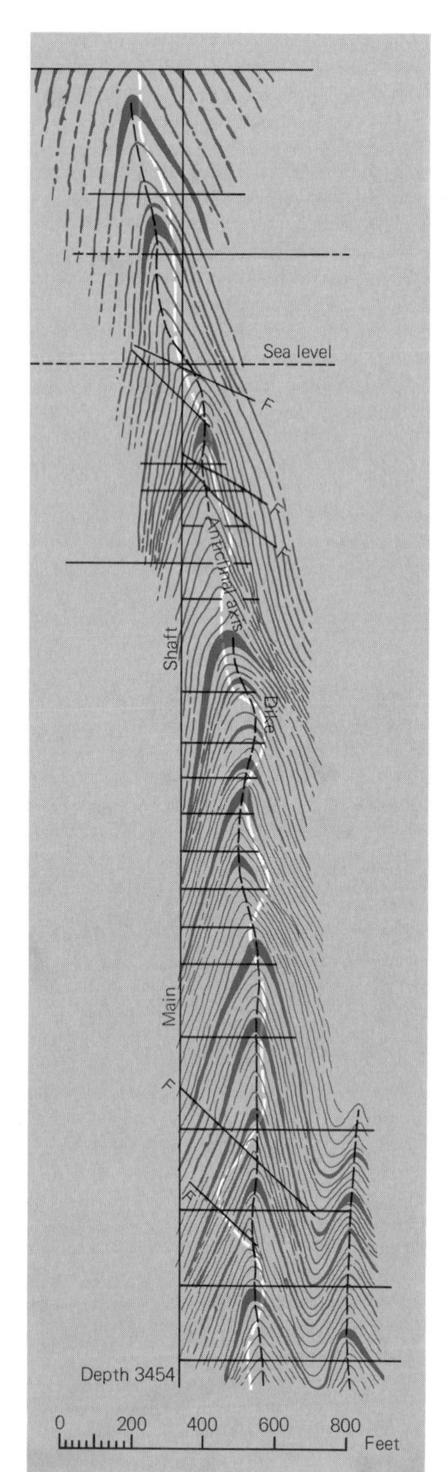


**19.6 Saddle "reefs," arcuate deposits of gold-bearing quartz,** Bendigo, Australia, main shaft of Great Extended Hustlers Mine. The name "saddle" comes from the similarity of appearance of these deposits with the sectional view of an equestrian's saddle. The word "reef" is used by miners for certain hard deposits of quartz; it is not related to the kinds of wave-resistant structures built by tropical corals and other kinds of organisms. (After Alan M. Bateman, 1950, Economic mineral deposits, 2nd ed.: New York and London, John Wiley and Sons, 916 p., Figure 12-14, p. 450, after Baragwanath, original not seen)

**Economic deposits related to faults** Economic deposits related to faults may have accumulated because of folds that formed during fault movement or because the fault created openings in fault breccia and also served as a channelway for migration of elements in solution.

Along the Mexia-Talco and Balcones fault zones in southeastern Texas and adjoining states, many elongate, narrow petroleum fields have been found in which the traps clearly have something to do with the fault (Figure 19.8a). These traps are classified as fault traps, but many of the structure maps of individual oil fields indicate that the petroleum accumulated in elongate domes which may have formed during faulting (Figure 19.8b).

**19.7** Spindletop dome and salt diapir, S of Beaumont, Jefferson County, Texas. (Donald C. Barton and Roland B. Paxson, 1925, The Spindletop salt dome and oil field, p. 478–496 *in* Moore, R. C., *ed.*, Geology of salt dome (*sic*) oil fields: Tulsa, Oklahoma, American Association of Petroleum Geologists, 797 p., Figure 3, p. 484)

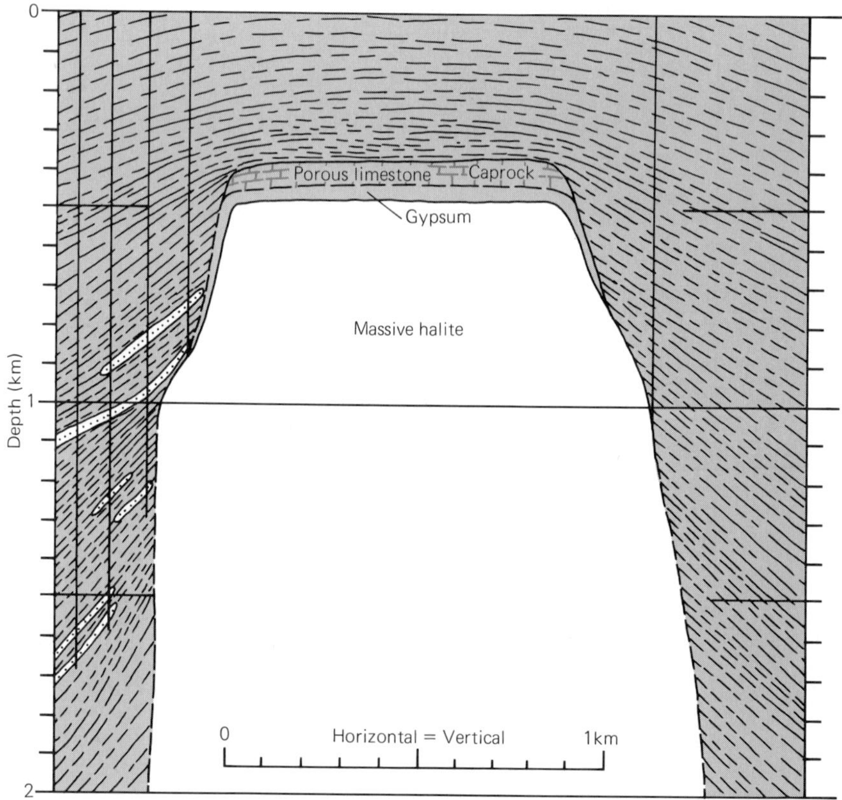

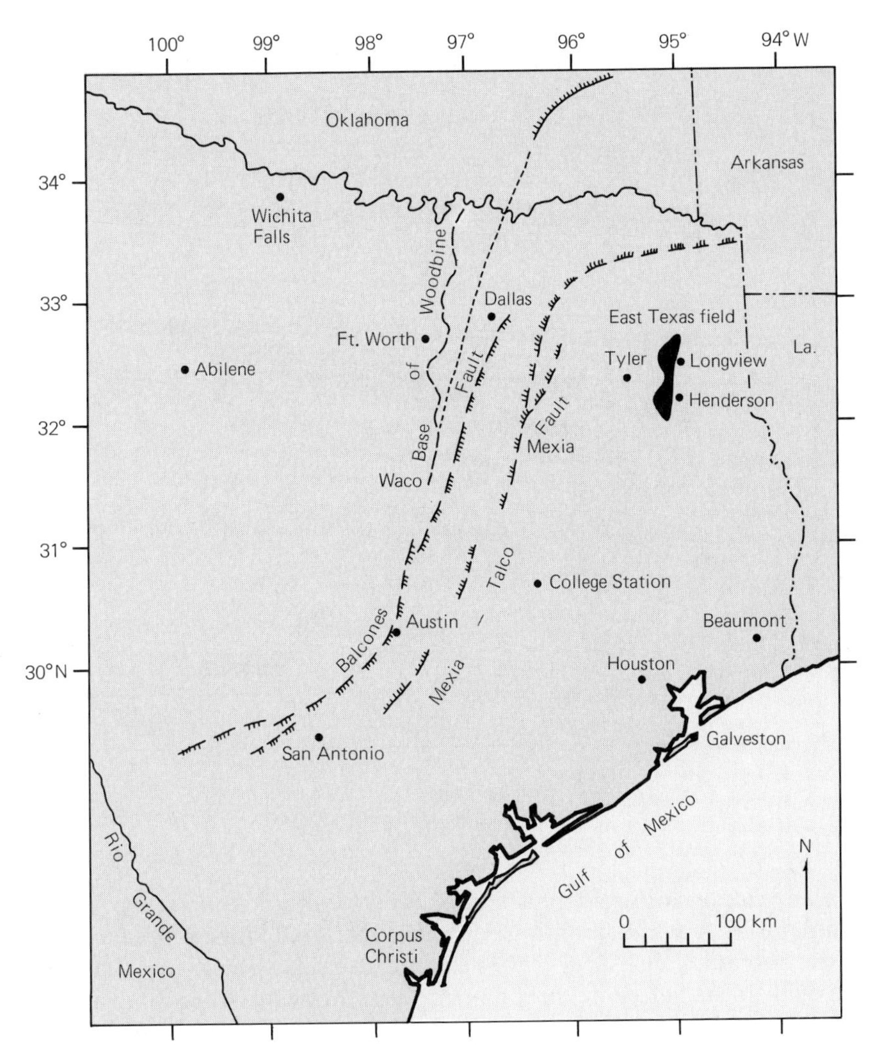


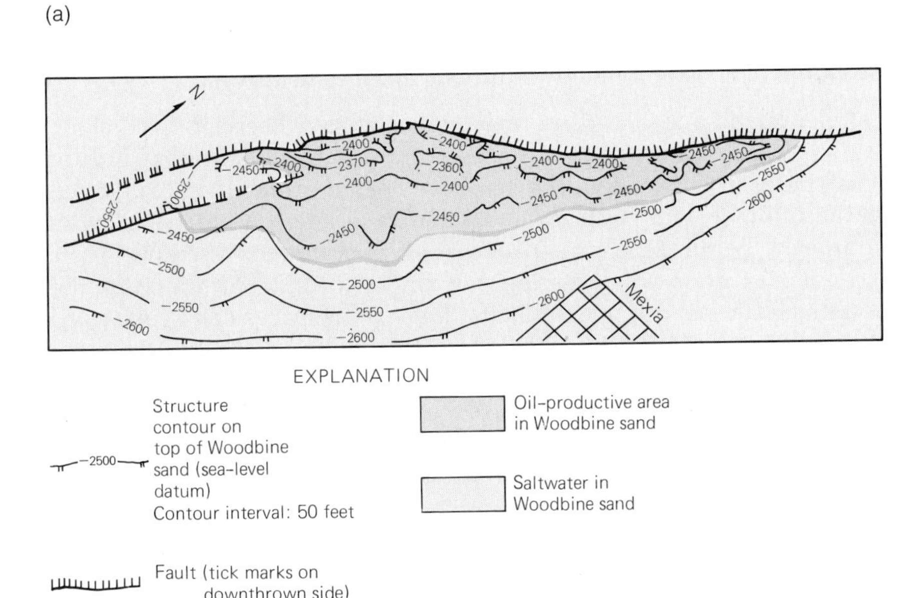


**19.8 Mexia oil field, along Mexia-Talco fault zone,** Limestone County, Texas. (a) Location map for this and some other figures. (b) Structure-contour map on top of Woodbine Sand shows an oil pool at the crest of an elongate dome on the upthrown side of the Mexia-Talco fault. (After Frederick H. Lahee, 1929, Mexia and Tehuacana fault zones, p. 304–388 *in* Powers, Sydney, *chm.,* Structure of typical American oil fields. A symposium: Tulsa, Oklahoma, American Association of Petroleum Geologists, v. 1, 510 p., Index map, Figure 1, p. 306; structure map, Figure 15, p. 336)

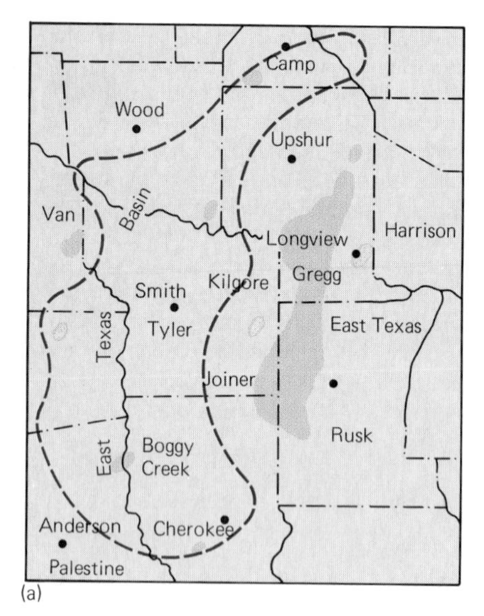


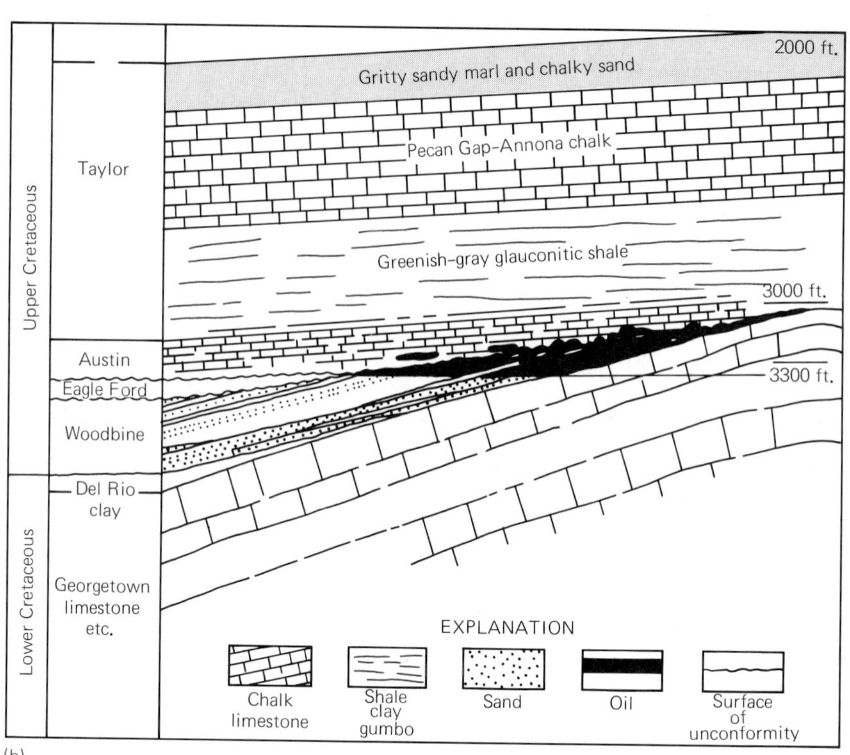


**19.9 East Texas oil field,** chiefly in Gregg and Rusk counties, Texas. (a) Location map. (b) Profile and section. Oil accumulated where the formerly eroded updip edge of the Woodbine Sand (lower Upper Cretaceous) on the E limb of the Tyler syncline (also named the East Texas basin, dashed blue line) has been sealed by the unconformably overlying Austin Chalk (upper cretaceous). On the W limb of this syncline, the Woodbine Sand comes to the present land surface (see Figure 19.8a). The length of the E limb from the center of the syncline to the truncated edge is about 110 kilometers. (A.M. Bateman, 1950, Economic mineral deposits, 2nd ed.: New York. John Wiley & Sons, Inc., 916 p., Figure 16–22, p. 680)

Many important ore deposits have been localized along faults. A few examples include the famous Comstock silver-gold lode in Nevada, gold in the Porcupine district of Ontario on the Canadian Shield, the lead-zinc deposits in the Paleozoic carbonate rocks in the Mississippi valley and in the Great Valley of the Appalachians, and the fluorite deposits of western Kentucky.

### Economic Deposits Related to Buried Surfaces of Erosion

Petroleum and a few kinds of metal deposits have been localized along buried surfaces of erosion. As explained in Chapter 7, various kinds of pore spaces in carbonate rocks that have been weathered and subjected to the influences of fresh groundwater have later become filled with petroleum. Because the pore space that later became the reservoir clearly did not exist prior to weathering, we can be absolutely certain that the petroleum did not migrate until after these pore spaces had been sealed and the porous strata had subsided in a second cycle of deposition (see Figure 7.17).

Similarly, many sands that have become reservoirs for some giant petroleum fields were at one time exposed at the land surface and weathered. At the time of exposure some of them clearly did not contain petroleum, yet others did. For example, early in their geologic histories, the future reservoir sands in the two largest oil fields in North America, the East Texas field (Figure 19.9) and Prudhoe Bay, Alaska (Figure 19.10), were weathered at the land surface. At the time of weathering, neither sand seems to have contained petroleum. By contrast, some of

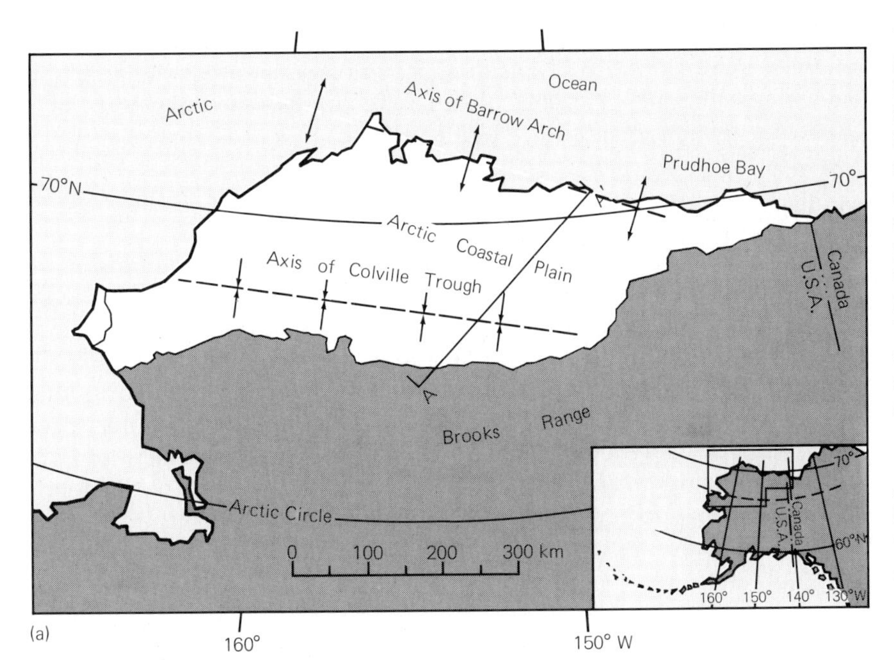


(a)

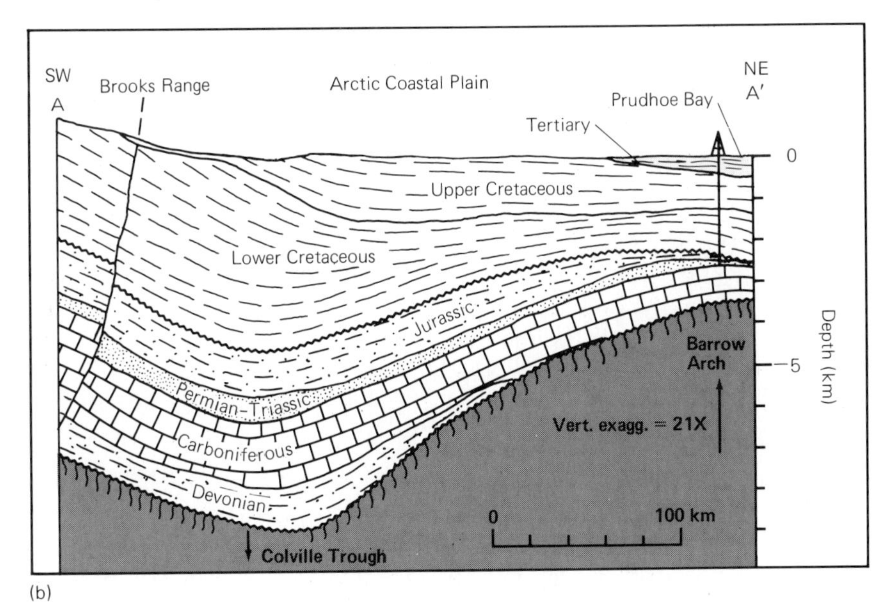


(b)

**19.10 Prudhoe Bay oil field, Alaska.** (a) Map showing major geologic features; inset map of Alaska shows regional location. (b) Profile and geologic section from Prudhoe Bay to Brooks Range. Repeated uplift of Barrow Arch, starting in post-Mississippian time, has resulted in thinner, truncated strata under Prudhoe Bay as compared with areas farther south in the Colville Trough. The length of the S-dipping common limb between the Colville Trough and the Barrow Arch, from the axis of the trough to Prudhoe Bay is about 200 km. (After H. P. Jones and R. G. Speers, 1976, Permo-Triassic reservoirs of Prudhoe Bay Field, North Slope, Alaska, p. 23–50 *in* Braunstein, Jules, *ed*., North American oil and gas fields; Tulsa, Oklahoma, American Association of Petroleum Geologists, Memoir 24, 360 p., Figure 2, p. 26)

the Ordovician sands that were exposed and weathered during Early Pennsylvanian time in the giant Oklahoma City field (see Figure 19.5) were leaking oil that formed a tar seal in some localities. In all three of these fields, burial by thick shales preceded migration of petroleum up the dip of the tilted sands to the trap located at the old erosion surface.

As was recommended repeatedly by A. I. Levorsen (1894–1965), a distinguished American petroleum geologist, these porous layers that have been truncated and sealed along buried surfaces of erosion should be systematically prospected for petroleum. As far as we are aware, no such systematic search has been made, possibly because petroleum fields of this kind are not always easily explained by existing ideas about the origin of petroleum.

Other kinds of economically valuable deposits that are related to buried surfaces of erosion include residual accumulations of bauxite and iron oxides (see Figure 7.15) and placers. We discuss buried placers in a following section devoted to buried valleys.

### Economic Deposits Related to Plutons

Many of the world's premier ore deposits are contained in plutons or are found in rocks which were altered by hydrothermal solutions that were associated with the emplacement of a pluton. In Chapter 6, we presented the relationships of the plutons at Bingham, Utah, and in Cornwall, England. Numerous other examples could be cited of ore deposits that have been localized in or near plutons of felsic rocks. In Chapter 9, we mentioned the magnetite deposits associated with the mafic pluton at Cornwall, Pennsylvania. Those who search for the sources of diamonds in the bedrock concentrate their attention on cylindrical mafic plutons known as diamond pipes (Figure 19.11).

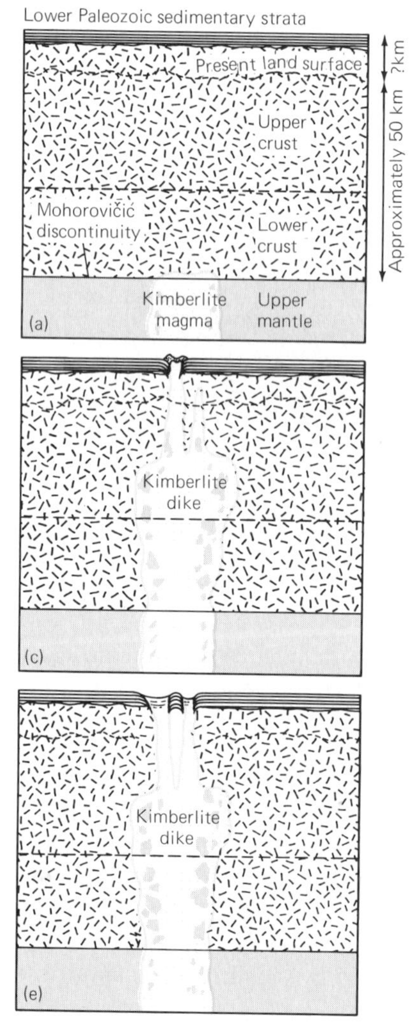


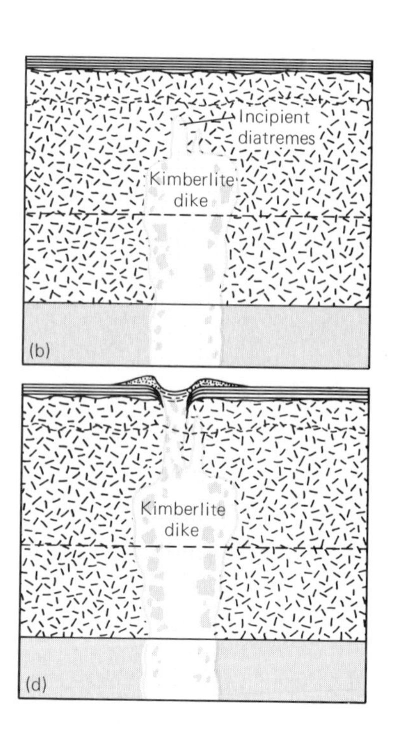


**19.11 Kimberlite pipes.** (a) Kimberlite magma (or fluidized gas-solid mixture), derived from depths of approximately 200 kilometers, moves upward in a dike. (b) Kimberlite penetrates through crust. Spalled-off pieces of crust and mantle become incorporated into the kimberlite. Cylindrical pipes form at top. (c) Of two adjacent pipes, only one breaks through to the surface. (d) Material from surface falls into the kimberlite, enabling xenoliths of the highest material penetrated to be mixed with xenoliths brought up from great depth. (After M. E. McCallum, 1976, An emplacement model to explain contrasting mineral assemblages in adjacent kimberlite pipes: Journal of Geology, v. 84, no. 6, p. 673–684, Figure 3, p. 682)

## Economic Deposits Related to Buried Valleys

Buried valleys contain filling sediments that may serve as valuable aquifers for supplying groundwater or that may include buried placers. In addition, the thicknesses and characteristics of the valley-filling sediments affect the designs and locations of engineering structures.

**Valley-fill sediments as potential aquifers** In many regions that were glaciated during the Pleistocene Epoch or that lie along stretches of seacoast crossed by major rivers, buried valleys are common. Examples include the buried preglacial Teays valley in central Ohio and central Indiana (Figure 19.12), various valleys in the New York metropolitan area that have not been given names and a large valley beneath New Haven harbor, Connecticut (Figure 19.13). Porous layers of sand and gravel lie near the bottoms of the valley-filling sediments. These layers can be valuable aquifers having capacities great enough to supply water to cities or large industrial installations.

**Placers in valley-fill sediments** Depending on the kinds of minerals in the drainage area, the valley-filling sediments may contain placers valuable enough to exploit. Along the western slope of the Sierra Nevada, California, the valley-filling gravels of Tertiary age contained large quantities of gold, most of which has now been mined. These gravels and the parts of the valleys not filled by the gravels were themselves buried by lava flows. After the valley-filling lava flows had cooled, the former valleys were the sites of resistant sheets of extrusive igneous rock. After

**19.12 Buried Teays valley, midwestern states,** contains filling sediments that are valuable aquifers. (After Raymond E. Janssen, 1952, The history of a river: Scientific American, v. 186, no. 6, p. 74–78; 80, map on p. 76–77; and W. D. Thornbury, 1965, Regional geomorphology of the United States: New York-London-Sydney, John Wiley and Sons, 609 p., Figure 11.16, p. 207)

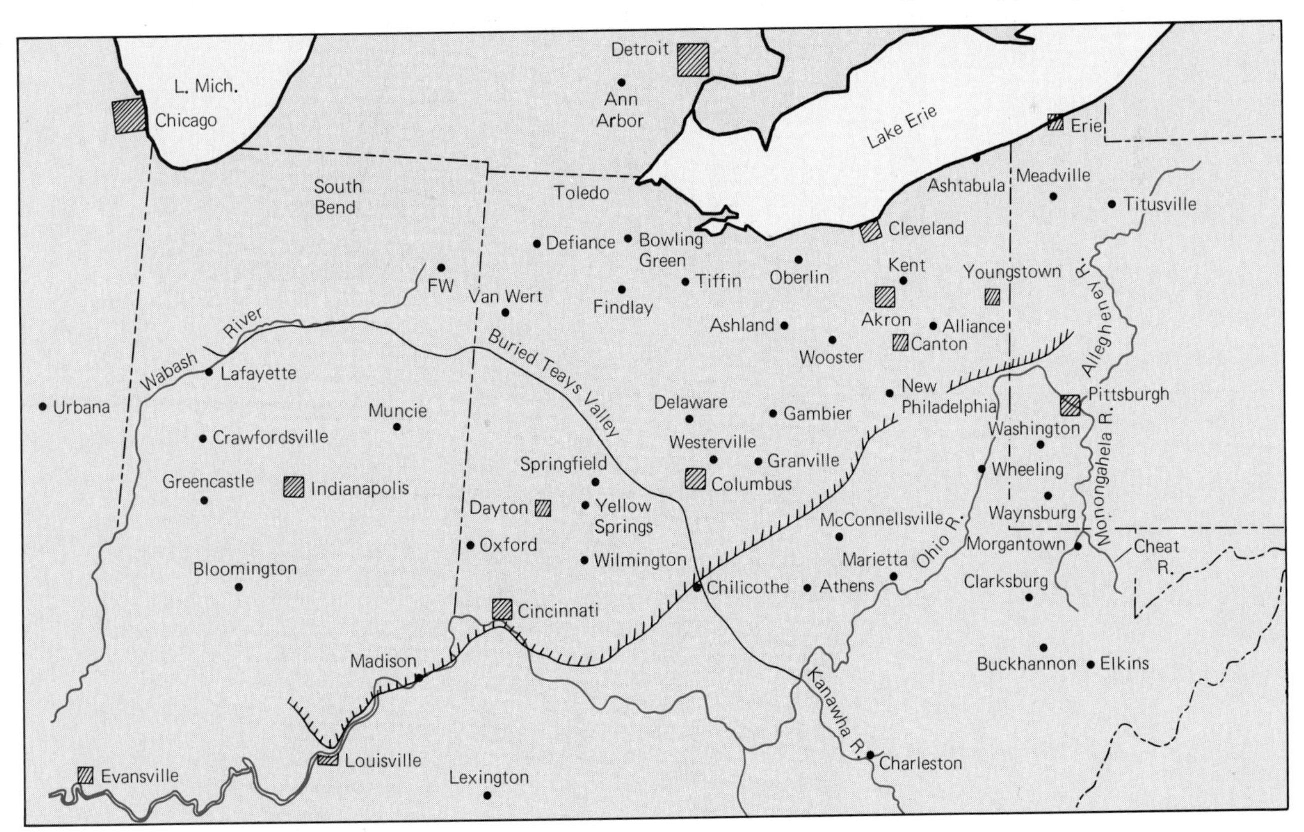

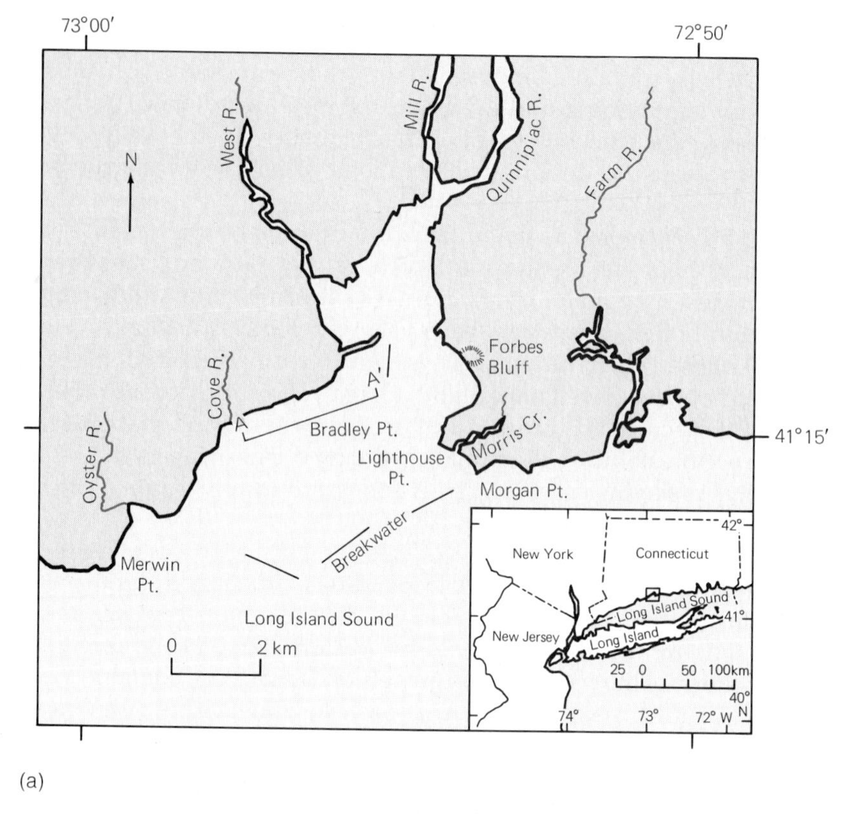


(a)

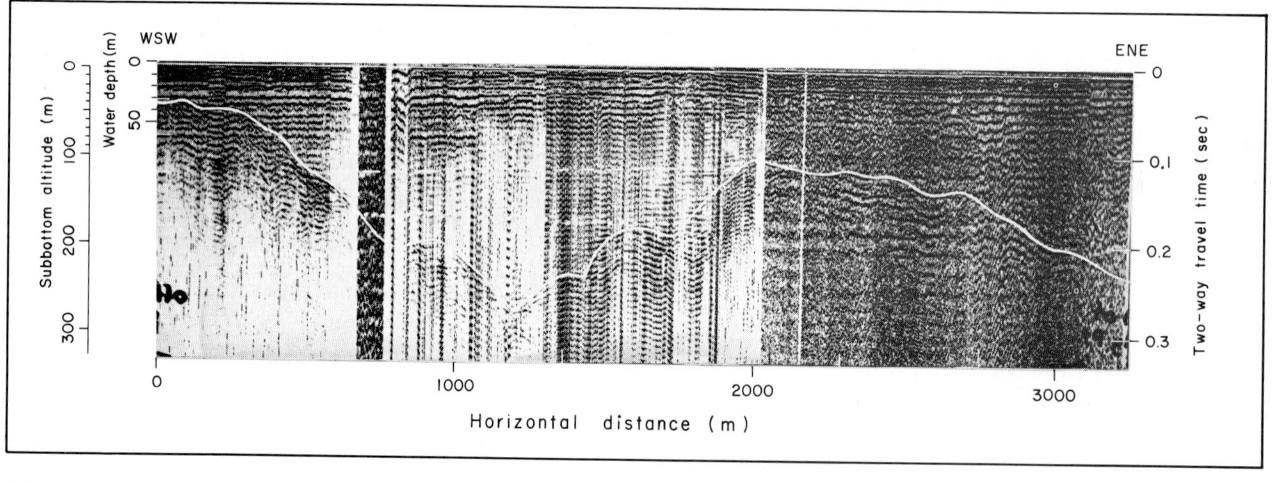


(b)

**19.13 Buried valley, New Haven, Connecticut.** (a) Location map. (b) Continuous seismic-reflection profile along AA′ of location map. (c) Model based on numerous seismic profiles. (top) View of New Haven harbor showing sea and land; contours of land based on U. S. Geological Survey quadrangle maps, with each step representing 50 feet of altitude, and the first 50 feet taken at the shoreline. (middle) Same, with water depths indicated by bathymetric contours in feet. (bottom) Same, with depth to bedrock in buried valley shown by 100-foot steps; surface morphology exaggerated four times. Valley that stops in the inner harbor is an artifact of the seismic survey; this valley continues well beyond the model. (J. E. Sanders; see also Fred D. Haeni and John E. Sanders, 1974, Contour map of the bedrock surface, New Haven-Woodmont quadrangles, Connecticut: U. S. Geological Survey, Map MF-557A)

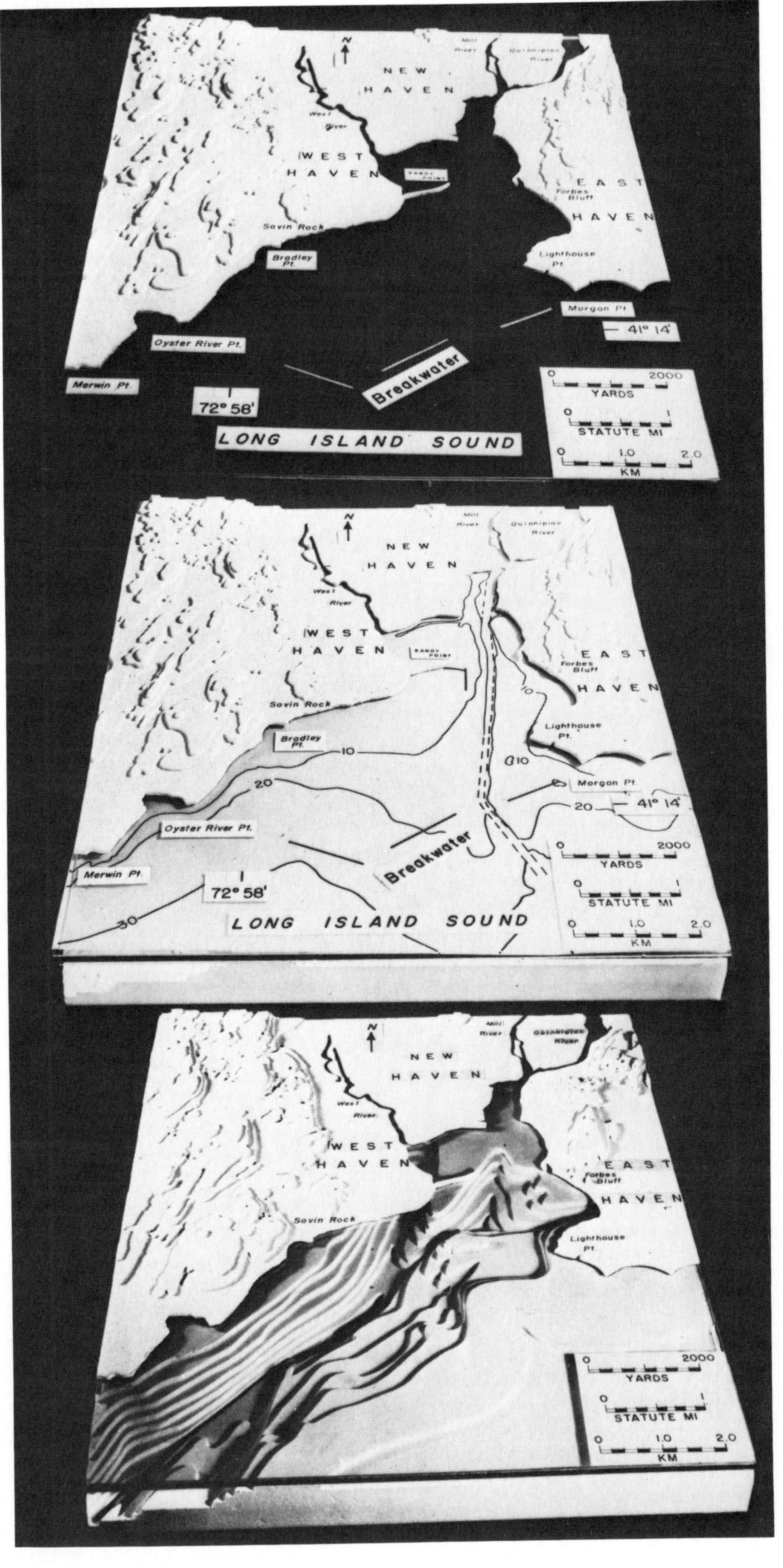
N
Mill River
Quinnipiac River
NEW HAVEN
West River
WEST HAVEN
Sandy Point
EAST HAVEN
Forbes Bluff
Savin Rock
Bradley Pt.
Lighthouse Pt.
Morgan Pt.
41° 14'
Oyster River Pt.
Merwin Pt.
72° 58'
Breakwater
LONG ISLAND SOUND
0 2000
YARDS
0 1
STATUTE MI
0 1.0 2.0
KM
10
20
30

(c)

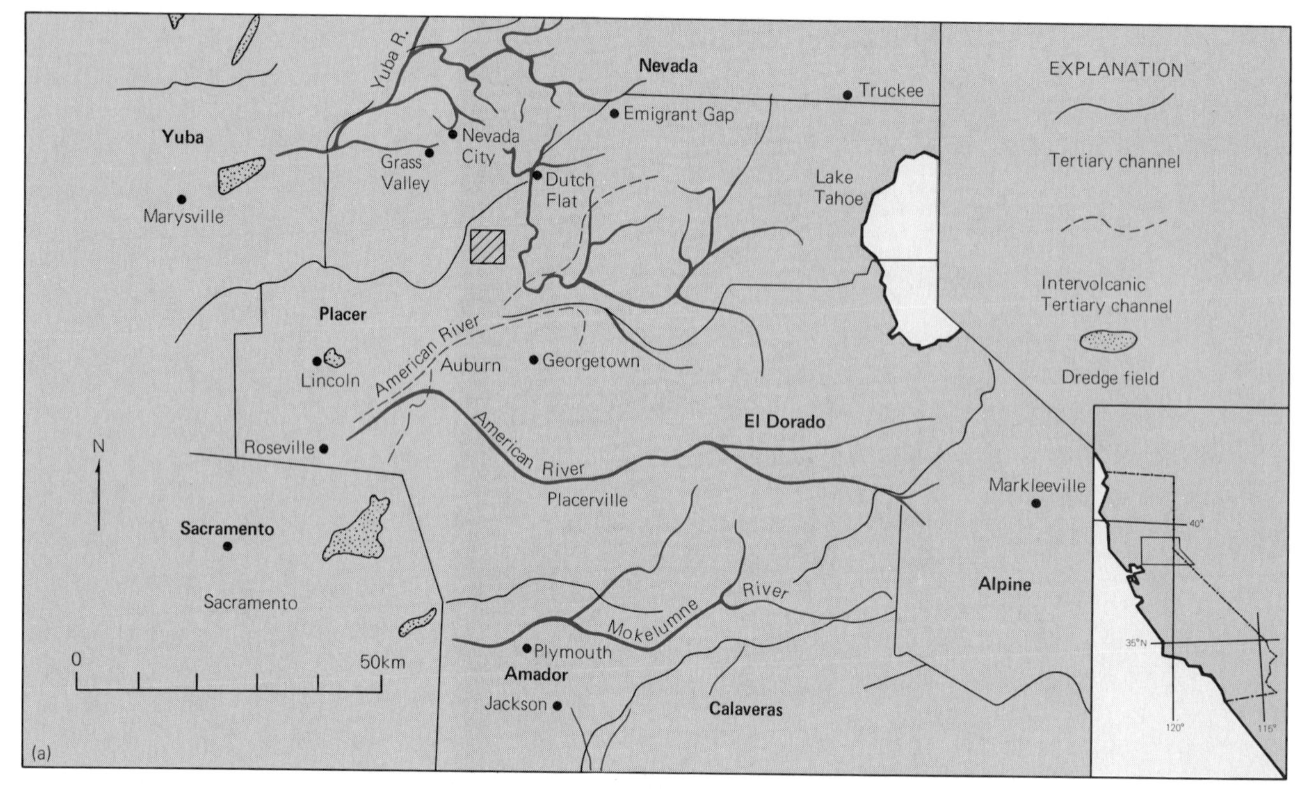


**19.14 Gold-bearing Tertiary gravels, western slope of Sierra Nevada,** California, many now exposed on sides of modern valleys. (a) General location map of area northeast of Sacramento; inset shows regional location. (b) Iowa Hill district, Placer County, showing locations of Tertiary channels and old gold workings. Line *AA'* shows location of profile and section. (c) Profile and section, showing successive steps in the geologic history of the area, which resulted in an inversion of the landscape. (After William B. Clark, 1970, Gold districts of California, *California Division of Mines and Geology, Bulletin 193;* a, Figure 5, p. 18; b, Figure 11, p. 68; profile by J. E. Sanders using U. S. Geological Survey Colfax 1/62,500 quadrangle, 1950)

later uplift and great erosion, these resistant volcanic rocks became the drainage divides between the modern valleys (Figure 19.14). As a result of the deep dissection that inverted the relief, the valley-filling gravels and their associated placer gold deposits became exposed along the sides of the modern valleys. At the times when the gravels were being vigorously dispersed by flood discharges, the particles of gold settled to the very bottom of the deposit. Many pieces of gold even became lodged in crevices in the underlying bedrock floor.

This inversion of the relief in California is exactly analogous to the inversion of relief in the Auvergne district of France mentioned in Chapter 5 (see Figure 5.2).

**Engineering aspects of valley-fill sediments** In the New York metropolitan area, many of the kinds of engineering projects have had to consider valley-filling sediments. For example, Manhattan Island is crossed by two major buried valleys, at 125th Street and 34th Street. The depth to bedrock beneath the valley at 125th Street is so great that the Broadway subway comes out from underground and crosses this valley on an elevated bridge. The valley at 34th Street is reflected in the heights of buildings. Numerous skyscrapers (including the Empire State Building) have been built where the solid bedrock (Manhattan Schist, of Cambro-Ordovician age) lies close at or below the surface. Two such clusters of tall buildings are located in midtown and at the southern tip of the island, in the Wall Street-Battery area, where the twin towers of the World Trade Center are located. In between, on the valley fill, no skyscrapers have been built.

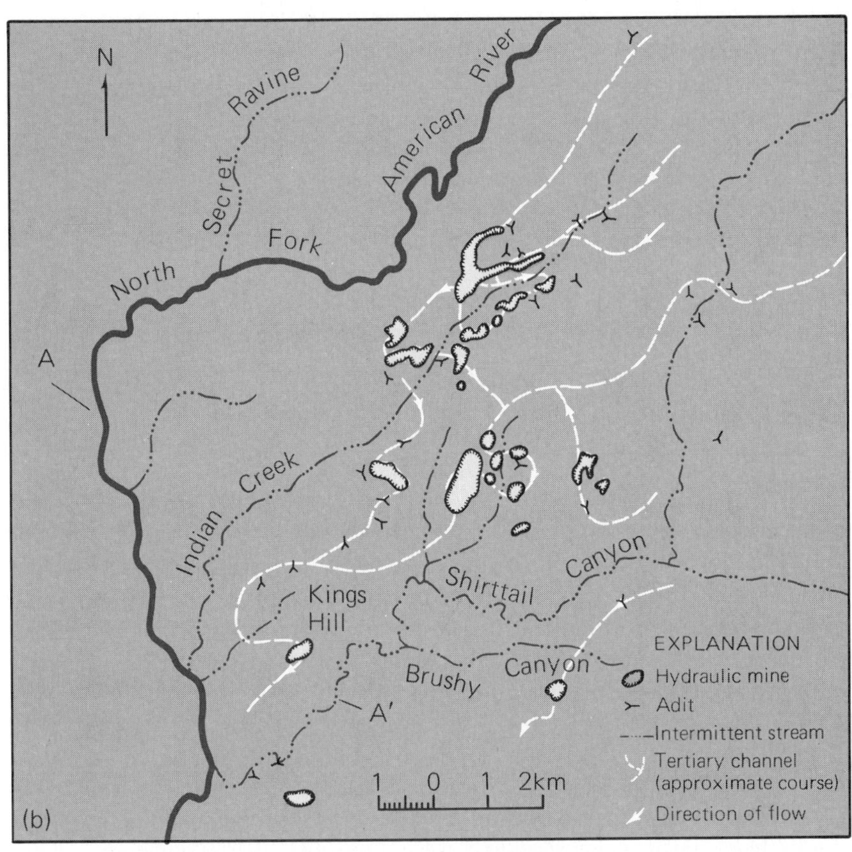
N
Ravine
Secret
American
River
North
Fork
A
Indian
Creek
Kings
Hill
Shirttail
Canyon
Brushy
Canyon
A′
1 0 1 2km
(b)
EXPLANATION
Hydraulic mine
Adit
Intermittent stream
Tertiary channel
(approximate course)
Direction of flow

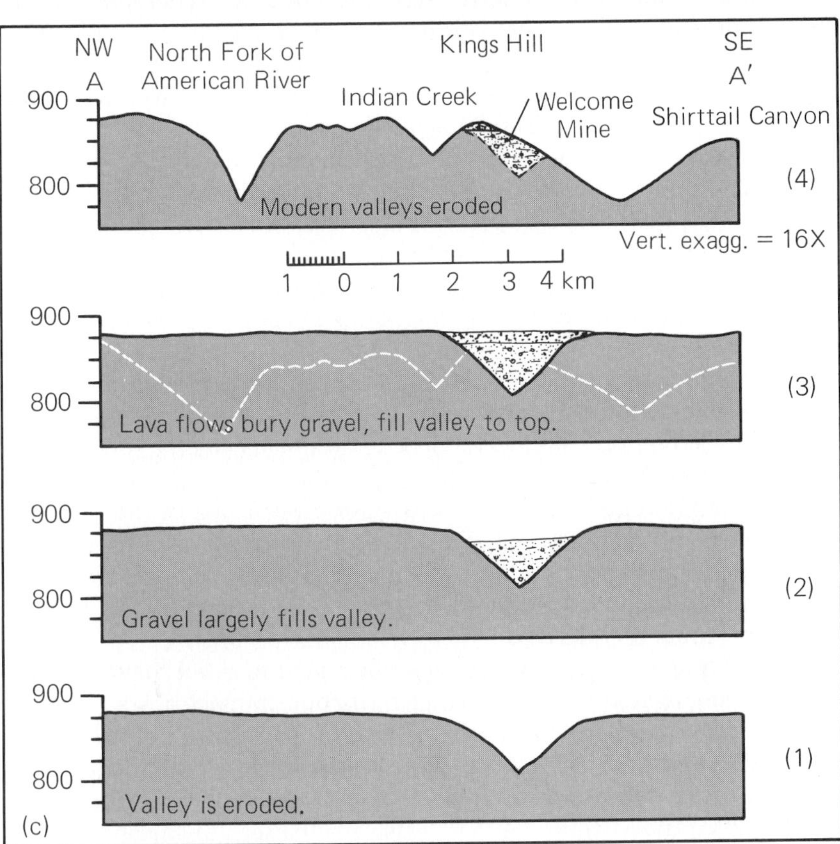
NW
A
North Fork of
American River
Kings Hill
SE
A′
Indian Creek
Welcome
Mine
Shirttail Canyon
900
800
(4)
Modern valleys eroded
Vert. exagg. = 16X
1 0 1 2 3 4 km
(3)
Lava flows bury gravel, fill valley to top.
(2)
Gravel largely fills valley.
(1)
Valley is eroded.
(c)

Several of the tunnels beneath the Hudson River are tubes that are literally floated in the watery tidal silts which form the thick upper part of the valley-filling sediments. These tubes move up and down slightly with changes in the tide and even in the atmospheric pressure.

## SURFACE CLUES ABOUT WHAT LIES BELOW THE LAND SURFACE

Various kinds of direct data and indirect clues about what lies below the land surface are available to guide exploration and to support inferences about possible locations of buried deposits. These range from radioactivity to the compositions of the leaves of plants.

### Radioactivity

No knowledge of geology is required for making successful searches for valuable deposits that are radioactive. One simply takes readings of radioactivity on some kind of detector that is activated by the emissions. This can be done on the ground or from an airplane. It is the best method for locating buried deposits, but its general use is limited in that not many kinds of valuable deposits are radioactive.

### Composition of the Regolith: Provenance Studies

Whether the regolith is residual or transported, it consists of samples of the bedrock or of minerals derived from the bedrock. Therefore, useful clues about the composition of buried bedrock can be obtained from studying the exposed regolith. The study of regolith from the point of view of determining the locations of the parent deposits from which the particles were derived is known as *provenance.*

The gold seekers of California's great nineteenth-century gold rush (the "forty-niners") wasted no time in looking upstream for the bedrock sources of the gold in the placers found in stream gravels. In short order they discovered the so-called mother lode, a zone of gold-bearing quartz veins in the western foothills of the Sierra Nevada. Copper mines have been found in Finland by tracing fragments of copper in the Pleistocene till. Similarly, pieces of native copper from the Keweenaw Peninsula, upper Michigan, have been found in localities far to the south.

The chemical composition of the regolith can be determined in the field by simple colorometric methods. Geochemical tests of stream sediments can provide useful data for guiding exploration for metal deposits.

If recycled kerogen proves to be a significant factor in the origin of petroleum deposits in detrital sands, then the provenance of sands will become an important part of petroleum exploration. The object of these studies will be to locate kerogen-bearing rocks in the drainage basin from which the sand was derived.

For example, studies of heavy minerals have indicated that the sands now entering the Gulf of Mexico come from four drainage areas: (1) rivers heading in the southern Appalachians and supplying sands to the eastern Gulf (=eastern-Gulf sands); (2) the Mississippi River drainage basin; (3) the rivers crossing the Texas coastal plain (=western-Gulf sands); and (4) the Rio Grande (Figure 19.15a). This same pattern of supply has pre-

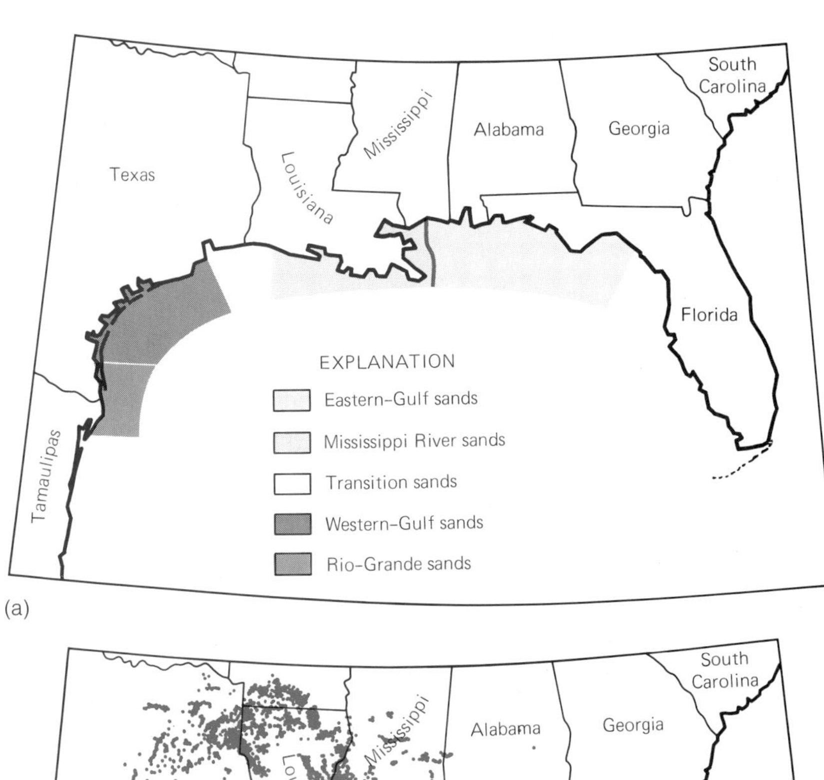


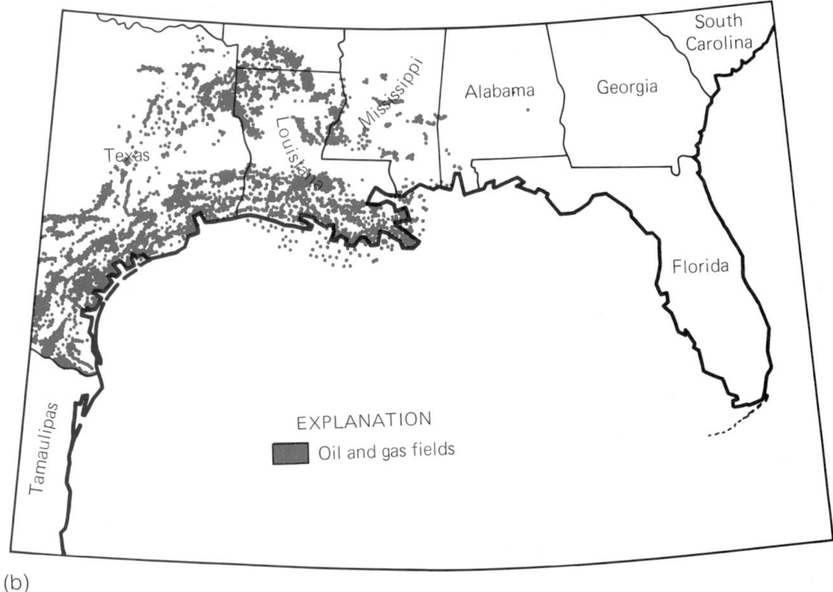


**19.15 Kinds of sands and oil fields, Gulf of Mexico.** (a) Distribution of the four chief varieties of modern sands. (August Goldstein, Jr., 1942, Sedimentary petrologic provinces of the northern Gulf of Mexico: Journal of Sedimentary Petrology, v. 12, no. 2, p. 77–84, Figure 1, p. 78) (b) Locations of oil fields in Cretaceous and younger terrigenous sands. Further explanation in text. (U. S. Geological Survey map 1967 in Ira H. Cram, 1971, Summary, p. 1–55 *in* Cram, I. H., *ed.*, Future petroleum provinces of the United States—their geology and potential: Tulsa, Oklahoma, American Association of Petroleum Geologists, Memoir 15, 2 vols., v. 1, 803 p., Figure 4, p. 9)

vailed since the Cretaceous Period, about 100 million years ago.

The eastern-Gulf sands, which come largely from the metamorphic rocks of the southern Appalachians, do not appear to drain an area that could supply much recycled kerogen. The Mississippi River drainage basin is dominated by Paleozoic shales and coal-bearing strata in its eastern part, and by Cretaceous shales and associated coals in its western part. Recycled kerogen could be abundant from these shales, and naturally recycled coal (in addition to coal now in the river from mining operations) is to be expected. Similarly, the dominant rock in the drainage basins crossed by the coastal-plain rivers of Texas are shales containing kerogen. Geochemists have positively evaluated the Cretaceous shales of the western interior of the United States as being excellent petroleum "source beds." Is it more than coincidence that the petroleum deposits in detrital sands of Cretaceous and younger ages in the Gulf of Mexico are situated in the western part of the Gulf (Figure 19.15b), exactly in the domain of the rivers draining kerogen-bearing shales? Notice that the Destin dome (see Figure 19.4) lies within the domain of the eastern-Gulf sands, those not likely to contain much recycled kerogen.

### Plants: Patterns of Species and Chemical Composition

Because the composition of residual regolith is related to that of the underlying and possibly buried bedrock, and because plants derive elements from the soil in which they grow, it should not be a surprise to learn that the study of plants is useful in exploration. *Geobotanical prospecting* is the search for minerals using the distribution and chemical compositions of plants.

Certain species of plants grow only where the soil contains threshold levels of particular elements. The distribution of these plants, known as *indicator plants,* serves as a clue to the presence of the elements. Other plants incorporate metallic elements in their stems or leaves in quantities that are related to the amounts of these elements in the soil. Thus, detailed chemical analyses of plants can be used to map areas of high metal content, and such areas have been shown to be related to hidden ore deposits.

### Chemical Composition of Natural Waters

As with regolith and plants, the chemical composition of natural waters is affected by the presence of hidden ore deposits. Groundwater near ore deposits acquires higher-than-normal concentrations of metals. Where such groundwater enters streams, either by diffuse seepage or at springs, the higher concentrations are incorporated into the surface water. In many cases, the temperature of groundwater that has encountered a hidden ore deposit is higher than normal.

### Oil Seeps and Gas Seeps

Many of the world's major petroleum-producing areas have been discovered as a result of drilling holes near oil seeps or gas seeps. No particular geologic expertise is required to recognize the locations: what is leaking out at the surface obviously is a clue to what lies beneath the surface.

Considerable effort has been devoted to developing methods for detecting methane gas that may be seeping upward to the surface from a buried petroleum deposit in amounts too small to form an obvious seep. The principle in looking for such methane is the same as in the visible seeps; the difference is that use is made of various kinds of sensitive chemical tests or geophysical measurements to identify the methane.

A major complication with such instrumental approaches is that the geologic distribution of small quantities of methane is not altogether controlled by buried petroleum deposits. Some bacteria generate methane, whereas other bacteria destroy it. Obviously, the possible effects of the bacteria have to be considered in exploration efforts that are based on sensitive gas detectors.

### Remote-Sensing Techniques Used From Aircraft or Satellites

The term *remote sensing* refers to any technique that can be used to find out about an object or an area without having to touch it or collect a sample. As commonly understood, remote sensing is taken to mean the study of the surface of the Earth or some other planet using instruments that are sensitive to some form of electromagnetic radiation. This means that the methods depend on the transmission or reflection of energy via waves that travel at the speed of light. Examples include light itself, radar, and infrared waves.

**Visual images: aerial photographs** The first remote-sensing technique to be used, and one of the most valuable, is that of making a picture of the land surface from overhead using a film-type camera or a television camera. The camera records the way the surface reflects light, either as shades of black and white or actual colors. Many such pictures are nothing more than pictorial records of what the human eye could see. However, as explained in Appendix F, overlapping aerial photographs taken from vertically above the land surface can be viewed in stereo by instruments that exaggerate the vertical scale, and thus provide views that cannot be seen by simply looking down from above.

The chief limitations on photographic methods involve distortion introduced by the camera lens, lack of light, and lack of visibility as a result of clouds or smog.

**Images presented by scanners** Other kinds of remote-sensing devices present their results as visual images, yet do not depend on light or on cameras. These devices are based on instruments known as scanners that sweep back and forth across the area being studied and build up visual images a line at a time. The scanner may record energy being emitted from the area directly or energy that is being backscattered upward, either from solar energy or from energy beamed downward by the instrument. Examples are side-looking radar, multispectral scanners, and infrared detectors.

In a *side-looking radar* survey, the instrument being flown over an area sends out radar waves (wavelengths of about 10 cm; see Figure D.1) whose reflections are recorded. The result is a visual image in black and white that records the relief features of the area (see Figure 12.25). Side-looking radar images are especially useful for displaying facture patterns (Figure 19.16) and folds. Radar surveys can be made under any conditions of light and cloud cover.

The images recorded by *multispectral scanners* (MSS) are line-by-line scans made by an instrument that is sensitive to a specific frequency of electromagnetic waves. Such scanners usually record the ways solar energy is backscattered. Images from multispectral scanners are particularly useful for showing contrasts between land and water and for displaying suspended sediment within the water (see Figures 3.5g, h).

The principle used in *infrared surveys* is that an object radiates infrared waves according to its temperature; the hotter the object, the more infrared waves it emits. Infrared detectors can record temperature dif-

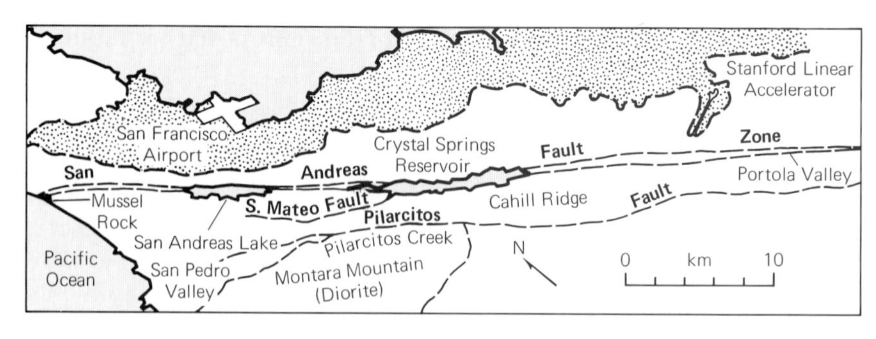


(a)

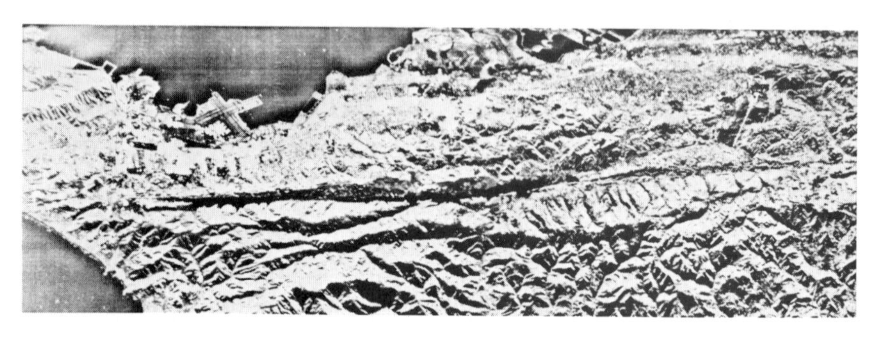

(b)

**19.16 Side-looking radar image of San Andreas fault** southeast of San Francisco, California. Sketch map above identifies prominent features. (U. S. Geological Survey)

ferences of fractions of a degree on the Celsius scale. The images presented by the infrared scanners are built up line-by-line with the tones of black or white varying as functions of the temperature. Infrared surveys have been used to record flow patterns in waters (cold springs of fresh water entering the warm seawater off Hawaii) and changes of temperature at volcanoes (Figure 19.17). In theory, infrared surveys could locate

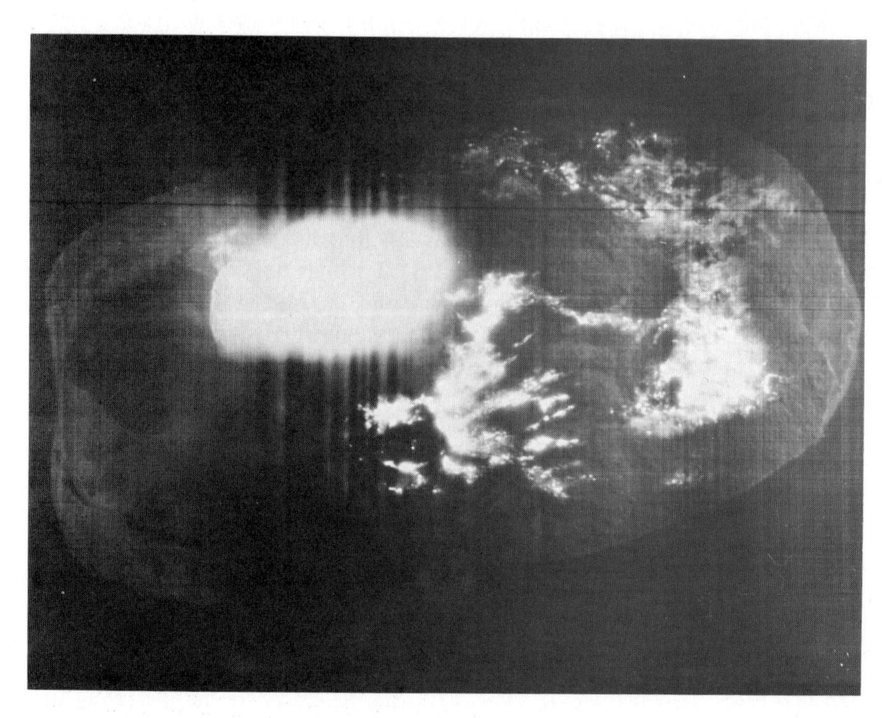

**19.17 Image formed by flying instrument sensitive to infrared energy waves** over Surtur I, Iceland, during 1965 eruption. Hot areas are displayed as white; cool areas, as gray. (U. S. Air Force Cambridge Research Laboratory, courtesy U. S. Geological Survey)

anomalously warm regolith where pyrite and other sulfide minerals of an ore deposit are being oxidized. However, the range of temperatures in regolith and bedrock as an effect of daytime solar heating overlaps with that of oxidizing sulfides. Thus, it is difficult to find the oxidizing-sulfide "signal" amidst the backscattered solar-heat "noise."

## SUBSURFACE EXPLORATION METHODS USED ON LAND

We divide subsurface exploration into two categories, methods used on land and those used beneath water, although some of the methods are used in both areas. We begin with drill holes and then summarize various geophysical methods based on measurements of seismic waves, gravity, magnetic fields, and electrical conductivity.

### Drilling Techniques

**Drill-hole samples** The ultimate method of subsurface exploraion is to drill a hole to obtain samples of the materials, to test the condition of the subsurface materials, or to recover fluids, such as water or petroleum. Various methods of drilling are available, depending on the depth and size of the hole, on the kinds of information needed, and, of course, on the expense.

The most widely used technique of exploratory boring for petroleum is called the rotary method. A special bit (Figure 19.18) is attached to the lower end of the drill pipe, which can be threaded together in sections to make whatever length is needed. The pipe and the bit are rotated from above and the bit cuts a hole that is slightly larger than the outside diameter of the pipe. A dense fluid, named drilling "mud," is pumped down the drill pipe under pressure. The mud passes through openings in the bit and returns to the surface in the annular space between the outside of the pipe and the wall of the hole. The circulation of the mud cleans out the hole and brings to the surface small chips of the rock that has been drilled. The weight of the mud keeps fluids from entering the hole and thus from escaping when the string of drill pipe has to be taken out of the hole to change the bit. Although the mud prevents blowouts of gas and oil, it also tends to obscure any petroleum that may be present in layers penetrated by the hole. As a result, other methods must be used to determine the fluid contents of the subsurface layers, as explained in a following section.

In borings intended to test the composition of ore minerals, cylindrical samples of the rock, known as *cores*, are recovered. Cores are useful, but very expensive; they are necessary for evaluating the quantities of ore minerals that are present or the characteristics of the subsurface pore spaces. This information is needed to make decisions about the value of the prospect. Decisions about opening mines are not usually made until after many cores have been recovered and studied.

**19.18 Rotary drill being welded** near a drilling site, near San Juan, Argentina. (United Nations, June 1969)

**Instrumental logs of bore holes: subsurface maps** The heavy equipment required to drill deep holes in petroleum exploration is very expensive; the costs can be as high as many tens of thousands of dollars per day. As a result, the chief objective is to drill the hole as rapidly as possible; recovery of samples is a secondary consideration. Drilling to the target depth is the first part of what may be a two-part project; the second part is known as setting production casing pipe. The large, expensive rig is required to handle the heavy pipe. Therefore, a decision about whether or not to set the pipe must be made while the rig is still set up on the hole.

The key information needed to make an intelligent decision is the fluid content of any porous layers penetrated by the hole. Because the drilling mud in the hole penetrates outward from the hole to prevent gas blowouts, special equipment is required to collect fluid samples. This equipment fits on to the lower end of the drilling pipe; its use requires many hours of rig time and is therefore expensive. The string of drilling pipe must be pulled out of the hole, the special test device placed on the drill pipe, the pipe sections put back into the hole, the test made, and the pipe sections retrieved from the hole. Direct fluid sampling is known as a *drill-stem test.*

A *drilling-time log* is recorded during drilling; it requires no additional rig time or services. From such a log, the geologist can find the depths to contacts between easily drilled layers, such as shale, salt, or anhydrite, and tough layers, such as chert, well-cemented sandstones, and certain massive carbonate rocks.

Several kinds of instrumental devices are available for recording conditions within the drilled hole. These devices are known as *logging tools;* they are brought to the hole by a special logging truck, which contains the logging tools, the special cable and winch for raising and lowering the tools, and an enclosed laboratory that houses the instruments and recorders. Before the logs are recorded, the hole is conditioned by circulating mud without further drilling, and the string of drill pipe is lifted out of the hole. The first step in logging is to rig the equipment and lower the logging tool to the bottom of the hole. The log is recorded as the tool is raised up the hole at a constant speed. Logging tools measure some property of the strata as they pass by. Electrical measurements yield the best data about the fluids; electric logs enable one to distinguish between salt water and petroleum in sands (Figure 19.19). Part of the graphical display of an electric log includes information about the porosity of the formations penetrated. Where only sands and shales are present, their thicknesses are shown on the log.

Radioactivity logs display the natural radioactivity of the strata penetrated by recording gamma rays and show how the strata respond to being bombarded by neutrons. The result of the neutron bombardment enables one to estimate the porosity (Figure 19.20).

Although the immediate use of logs from petroleum-exploration borings is to decide whether to complete the hole or not, many other kinds of studies can be made from the information presented on the logs. Some of these studies include geologic structure and various kinds of stratigraphic aspects, such as thicknesses of strata, thicknesses and distribution of sands, and porosity of petroleum-bearing formations.

**19.19 Induction-electrical log,** part of hole drilled into Cenozoic strata, Gulf coast, United States, depths 3450 to 3630 feet. Compare resistivity curves of two sands, tops at 3572 and 3608. Upper sand, 30 feet thick, appears to contain petroleum in its upper 8 feet, and saltwater below that. Lower sand, 14 feet thick, contains saltwater throughout. (Schlumberger-Doll Research Center, courtesy C. Clavier and V. Hepp, *in* Friedman, G. M., and Sanders, J. E., 1978, Principles of Sedimentology: New York-Santa Barbara-Chichester-Brisbane-Toronto, John Wiley and Sons, 792 p., Figure 13–15, p. 419)

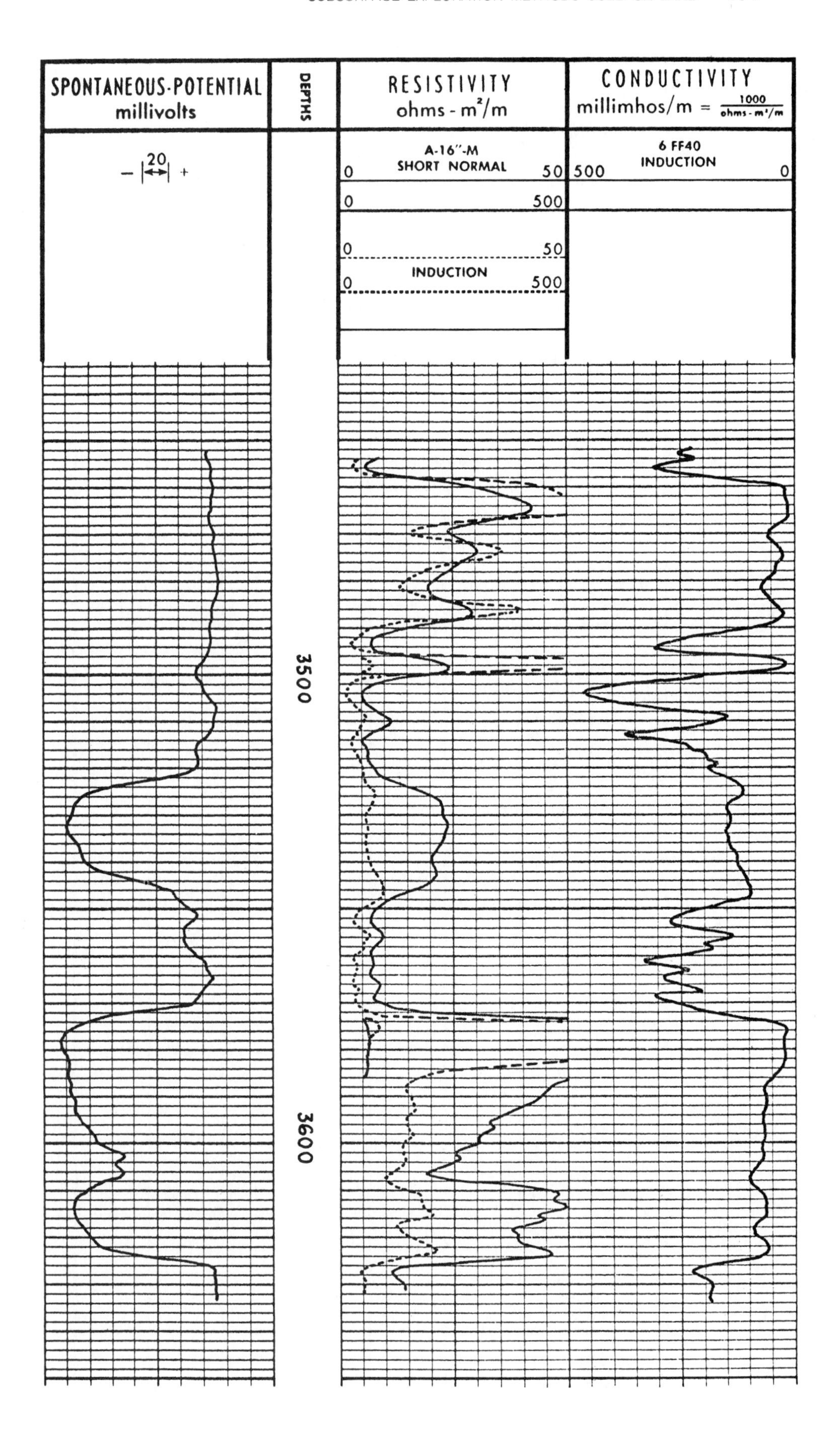
SPONTANEOUS-POTENTIAL
millivolts
DEPTHS
RESISTIVITY
ohms - m²/m
CONDUCTIVITY
millimhos/m = 1000/ohms-m²/m
20
A-16"-M
SHORT NORMAL
0
50
0
500
0
50
INDUCTION
0
500
6 FF40
INDUCTION
500
0
3500
3600

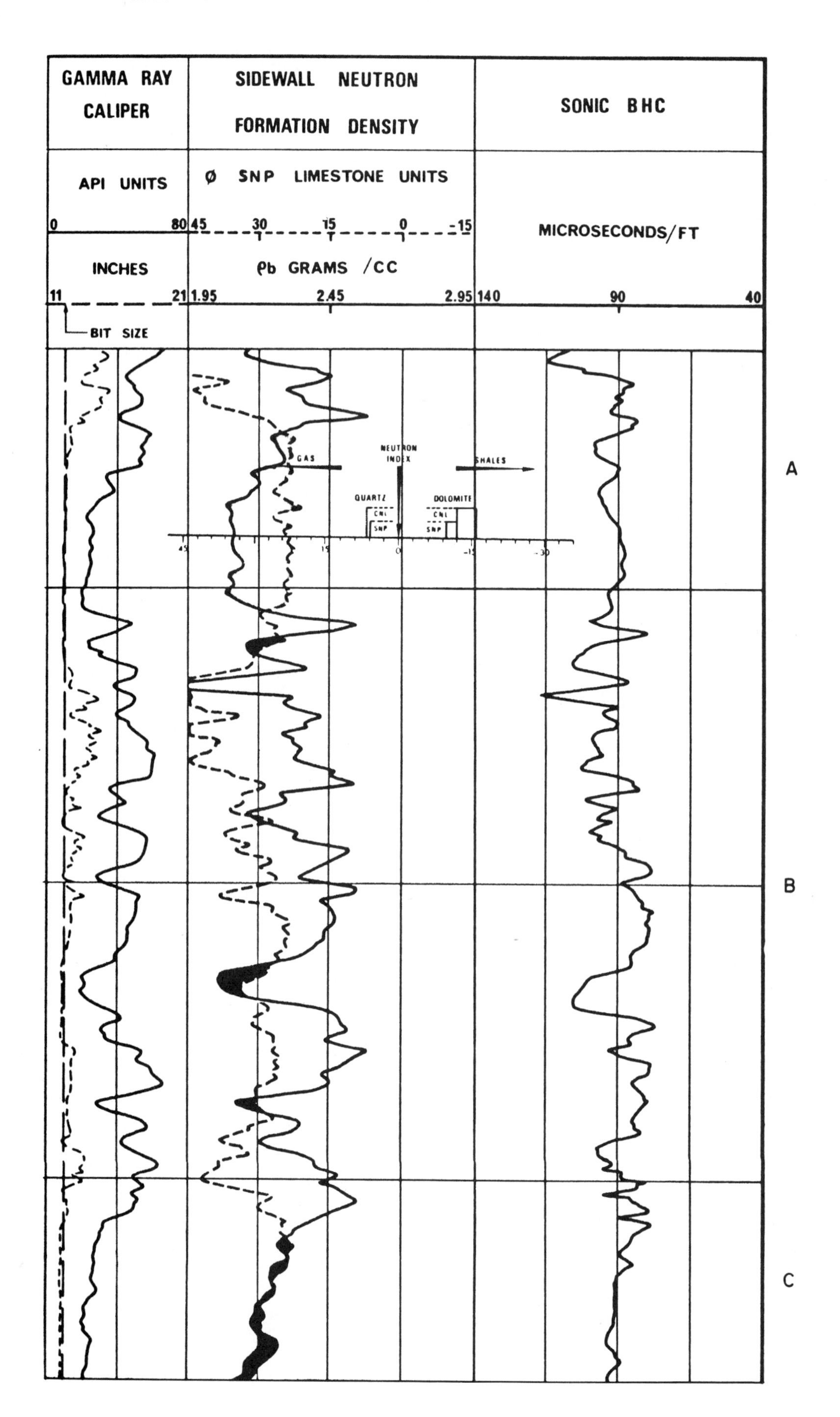
GAMMA RAY
CALIPER
SIDEWALL NEUTRON
FORMATION DENSITY
SONIC BHC
API UNITS
Ø SNP LIMESTONE UNITS
0
80
45
30
15
0
-15
MICROSECONDS/FT
INCHES
ρb GRAMS /CC
11
21
1.95
2.45
2.95
140
90
40
BIT SIZE
GAS
NEUTRON INDEX
SHALES
QUARTZ
DOLOMITE
CNL
SNP
A
B
C

**19.20 Radioactivity log, caliper log, and sonic log of part of hole penetrating Cretaceous of northern Saudi Arabia** (depths not available). Caliper log (dashed line at extreme left, scale in inches at top) registers diameter of hole. Radioactivity log consists of three traces: (1) a gamma-ray curve (solid line next to caliper trace), (2) a sidewall-neutron curve (dashed line in center), and (3) a gamma-ray density curve (solid line in center). Curve 1 (standardized API units devised by the American Petroleum Institute) shows natural gamma-ray activity of strata. Shales, which contain potassium in clay minerals and fine-grained feldspars, usually register high gamma-ray activity; sands and limestones typically are low. Curve 2 is generated by bombarding the strata with neutrons and by recording their reactions. The letters SNP refer to sidewall neutron porosity, a scale calibrated by limestone with pores filled with pure water. Curve 3 results from bombarding the strata with gamma rays; its units are density in grams per cubic centimeter. Three zones of strata are indicated by letters at right. In the lower part of zone A is a sand containing gas (dashed curve 2 to right of solid curve 3 by more than three porosity units). Zone B consists mostly of shale. In zone C, at bottom, is an oil-bearing sand (black area where dashed curve 2 lies to right of solid curve 3). Sonic log displays time for sound to travel one foot, with speed of sound increasing toward the right. The letters BHC at top, standing for bore-hole compensated, mean that the trace of the sonic log has been compensated for variation in diameter of hole as measured on the caliper log. (Schlumberger-Doll Research Center, courtesy C. Clavier and V. Hepp, *in* Friedman, G. M., and Sanders, J. E., 1978, reference cited in caption of Figure 19.23, Figure 13-16, **A,** p. 420)

Subsurface structure can be displayed on *structure-contour maps* (Figure 19.21). A structure-contour map resembles a topographic-contour map in that it displays information based on a given contour interval referenced to a datum surface, usually sea level (see Appendix F). A structure-contour map shows the changing elevations on the top of a given geologic unit that was horizontal initially. Because of tectonic move-

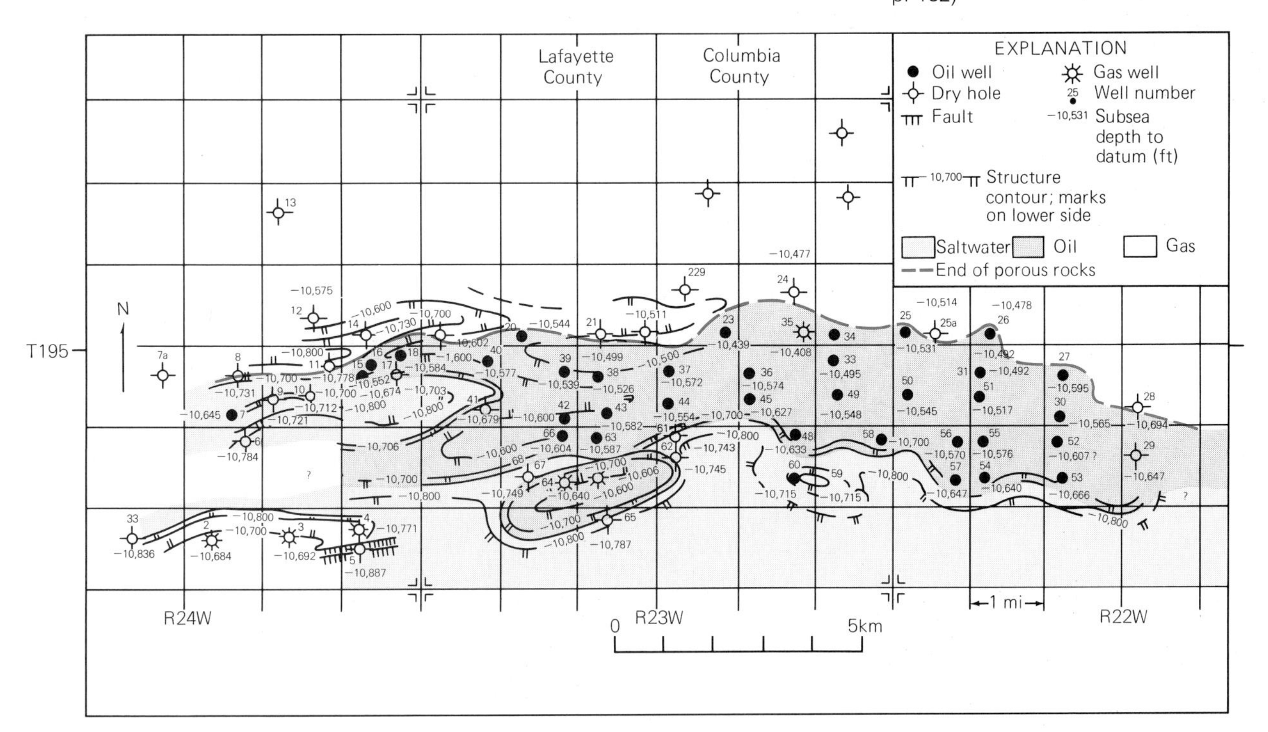


**19.21 Structure-contour map showing configuration of the top of the Smackover Formation (Jurassic)** in Walker Creek field area, near Arkansas-Louisiana border. Datum is mean sea level; contour interval, 100 feet (30 m). (Calvin A. Chimene, 1976, Upper Smackover reservoirs, Walker Creek field area, Lafayette and Columbia counties, Arkansas, p. 177–204 *in* Braunstein, Jules, *ed.,* North American oil and gas fields: Tulsa, Oklahoma, American Association of Petroleum Geologists, Memoir 24, 360 p., Figure 4, p. 182)

ments, emplacement of diapirs, removal of salt by solution, or large-scale subaqueous slumping, the top of a formation may be raised or lowered. After the depths to the mapping surface have been determined from drilling results, the elevations are reduced to a common basis by adjusting for the altitude of the drilling site, which is determined by a special survey. The numbers are entered on the map, and contours are drawn.

Maps of the thickness of a given stratigraphic interval are used to locate places that subsided rapidly during sedimentation and thus created *sedimentary basins*. Thickness maps are named *isopach maps* (Figure 19.22). If only thickness is desired, isopach maps can be made without reference to the elevation data.

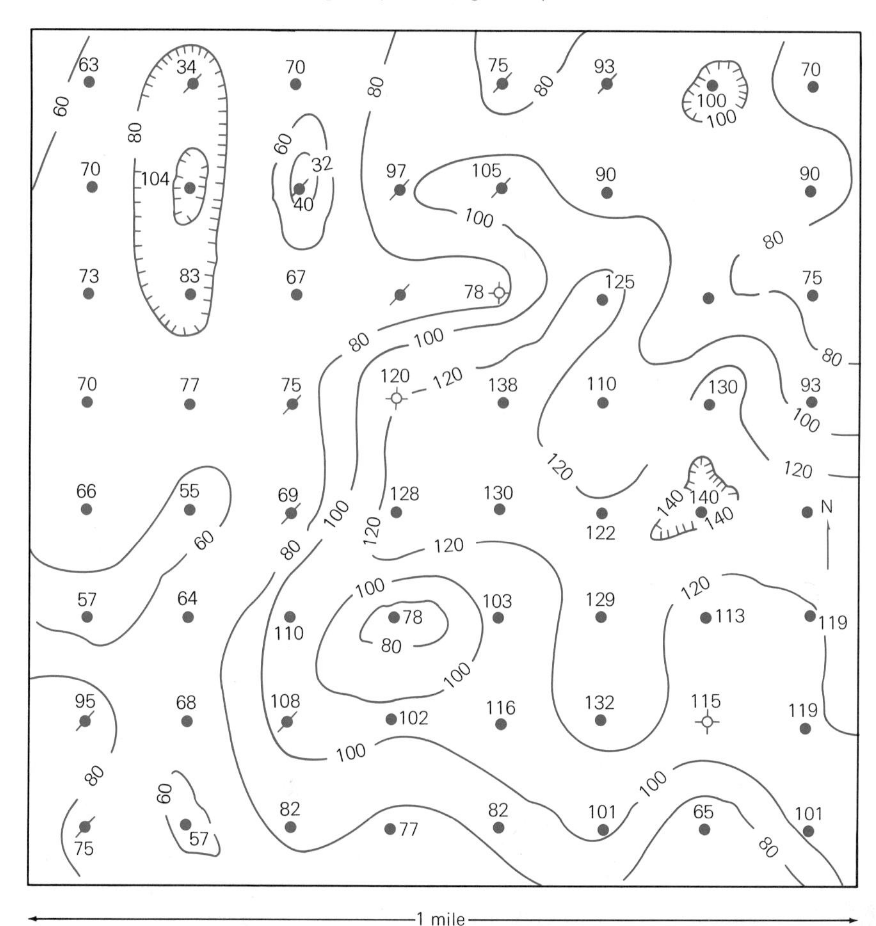


**19.22 Isopach map of Sylvan Shale (Ordovician)** in Bowlegs oil field, Section 14, T 8 N, R 6 E, Seminole County, Oklahoma. The variations in thickness represent the filling in of an irregular karst morphology formed by erosion of the top of the underlying Viola Limestone. (After Ira H. Cram, 1929, *in* Structure of typical American oil fields. A symposium: Tulsa, Oklahoma, American Association of Petroleum Geologists, p. 356, Figure 20)

Explanation

Oil well

Plugged oil well

Dry hole

—100— Isopach (feet)

The subsurface distribution of sands is usually shown by displaying the thickness of the sands. Where sands are confined to channels, as in a river system or network of tidal channels, the *sand-distribution map* resembles a photograph of channels (Figure 19.23).

### Geophysical Methods

**Seismic surveys** An important and widely used method for finding out about subsurface relationships without actually drilling deep holes is a *seismic survey*. The basic principle of seismic surveying is simple: an

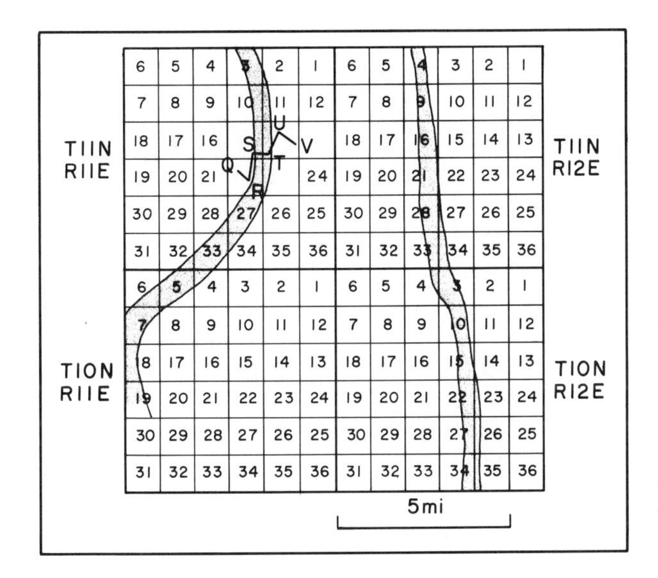


(a)

**19.23 Map of distribution of sand in part of the McAlester Formation (Lower Pennsylvanian),** central-eastern Oklahoma, based on electric logs of exploratory borings. (a) Map of four townships in Okfuskee and Okmulgee counties; shaded areas indicate sand in Booch interval is more than 20 feet thick. The pattern suggests former stream channels. (b) Cross section based on matching electric logs, with stipple added to show variation in thickness of Booch sandstone. (After G. M. Friedman and J. E. Sanders, 1978, reference cited in caption of Figure 19-23, Figure 10-25 C and D, p. 299)

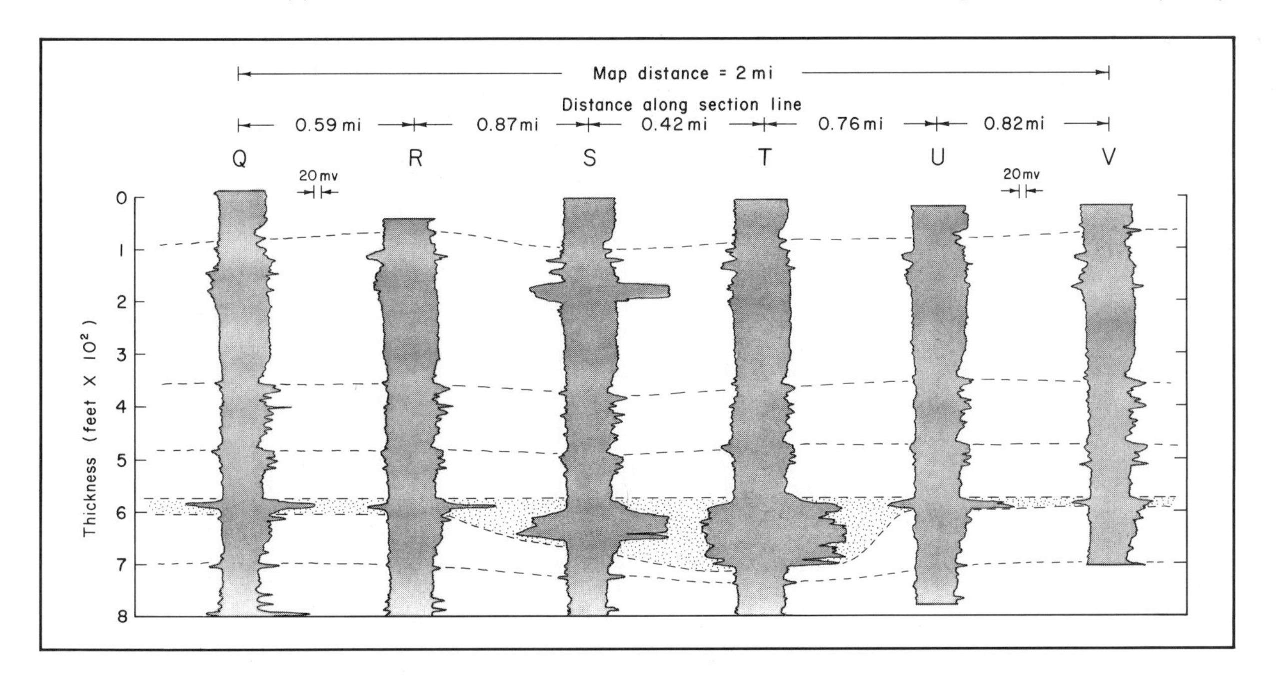


(b)

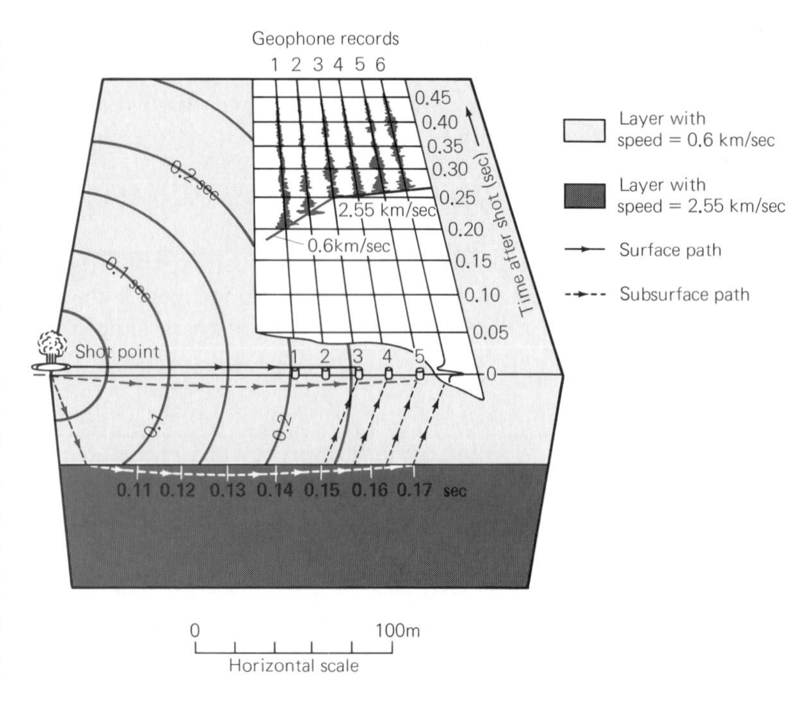


**19.24 Seismic survey to determine thicknesses of subsurface layers and speeds of travel of the seismic waves within them.** After six recording geophones have been emplaced at equal intervals along a line and all have been wired together, the shot is exploded. Waves travel along surface and downward. After 0.2 second, the surface waves reach geophone 1. However, after only 0.1 second, the body waves have reached the base of the upper layer, indicating that its thickness is 60 meters. Seismic body waves entering the lower layer in which they can travel faster, are speeded up, and 0.15 second after the shot, have reached a point beneath geophone 3. In another 0.1 second, they will arrive at geophone 3. Thereafter, the first arrivals at the higher-numbered geophones come from the fast-traveling body waves that passed through the lower layer. (Compare Figure 17.13) (Modified from F. E. Romberg, 1961, Exploration geophysics: a review: Geological Society of America, Bulletin, v. 72, no. 6, Figure 3, p. 889, based on C. A. Heiland, 1940, Geophysical exploration: Englewood Cliffs, New Jersey, Prentice-Hall, Inc., 1013 p.; reprinted 1963, New York, Hafner Publishing Co., Figure 2-10, p. 22)

energetic, low-frequency noise is made and a record is kept of the time required for the sound to penetrate into the ground and return to the surface again (Figure 19.24). In effect, what is done is to create a miniature earthquake — one that is so small that only special instruments can detect the seismic waves. Many seismic surveys use explosive charges; others generate mechanical vibrations from a special truck. Recording is done at different distances from the shot point by sensitive seismic listening devices known as *geophones*.

Where only the thickness of the regolith is sought, much smaller sounds can be generated and the results recorded on small, portable instruments. In some methods, the sound is made by striking a metal plate with a sledgehammer.

**Gravity surveys** The thicknesses and characteristics of the subsurface strata affect the value of the Earth's gravitational acceleration within local areas. As a result, careful measurement and analysis of the value of the Earth's $g$ can be used in exploration. Where the density of the subsurface materials remains the same, the value of $g$ measured at the surface will not change from one locality to another. However, if the densities of the subsurface materials change from place to place, then the value of $g$ will also change and the pattern of measured gravity determinations will indicate the density changes (Figure 19.25).

Gravity surveys have been particularly effective in locating salt diapirs in the Gulf coast and elsewhere and also in indicating where thick valley-fill sediments or buried ridges are to be found. Buried ridges are indicated by measurements of $g$ that are greater than background. These are known as *positive gravity anomalies;* they are caused by the relief on the surface of the basement rocks, along which dense basement rocks stand higher than less-dense strata. Other positive gravity anomalies are associated with kimberlite pipes, in which the ultramafic kimberlite is denser

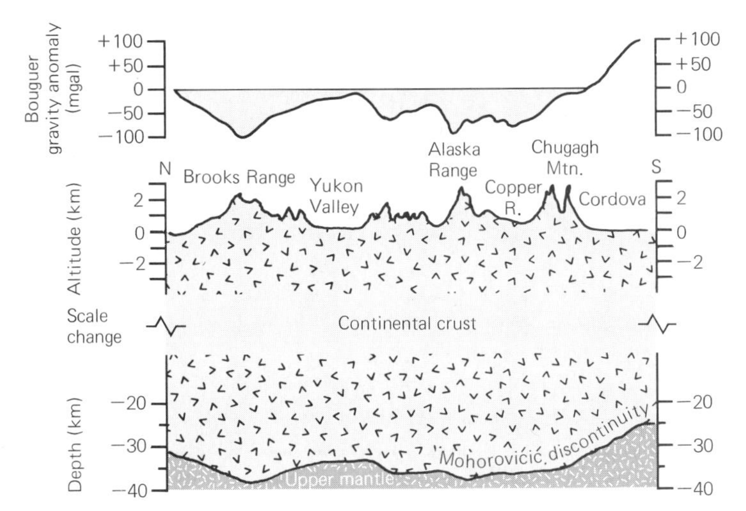


**19.25 Results of a regional gravity-profile survey,** crossing three mountain ranges and intervening lowlands, Alaska. The major adjustment in interpreting the subsurface relationships has been made at the level of the Mohorovičić discontinuity. (Modified from F. E. Romberg, reference cited in Figure 19.28, Figure 19, p. 912, based on G. P. Woollard, N. A. Ostenso, E. Thiel, and W. E. Bonini, 1960, Gravity anomalies, crustal structure, and geology in Alaska: Journal of Geophysical Research, v. 65, no. 3, p. 1021–1037, Figure 3, p. 1024)

than the surrounding basement rocks or sedimentary strata. Salt diapirs and thick valley-fill sediments are indicated by *negative gravity anomalies;* such materials are less dense than their surroundings.

**Magnetic surveys** As with gravity surveys, it is possible to make closely spaced, precise measurements of the Earth's magnetic field and thus to determine if the local rocks are more or less magnetic than a calculated generalized value for a given area. The chief magnetic variables are the amount of magnetite in the rocks and its remanent polarity. Some magnetite is present in most igneous rocks, and a great deal may be present in mafic rocks. As a result, magnetic surveys can be used to locate magnetite-bearing rocks. A body of rock containing abundant magnetite may display readings that are higher than normal (positive magnetic anomalies) if the remanent polarity is the same as the polarity of the Earth's magnetic field in that place; it may display lower-than-normal readings (negative magnetic anomalies) if the remanent polarity is opposite to the polarity of the Earth's magnetic field at that locality.

Magnetic surveys have been used to find buried dikes (Figure 19.26) and other rock bodies containing magnetite. In addition, surveys made from airplanes have proved to be useful in analyzing the geologic relationships of basement rocks, in showing thicknesses of strata, and the trends of buried basement ridges.

**Electrical surveys** Electrical surveys are responsive to factors that influence the electrical conductivity, or its inverse, electrical resistivity, of bedrock or regolith. In general, nonmetallic rock-making minerals are poor conductors of electricity, that is, their resistivities are high. Metallic copper, many metallic sulfide minerals, and salt water are good conductors, their resistivities are low. Thus, an electrical survey typically outlines areas within which conductivity is higher than normal (low resistivities), and such areas may coincide, for example, with buried sulfide-mineral ore deposits.

An electrical survey is made using a central point where electrical energy is put into the ground, a point that is analogous to the shot point of a seismic survey. Electrodes are placed in a line going away from the

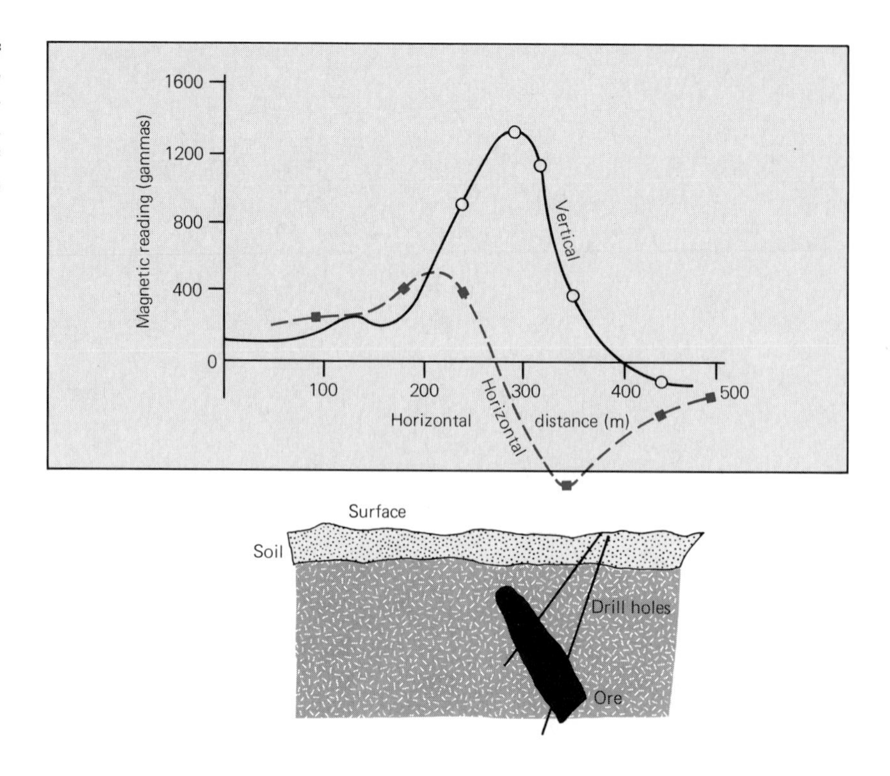


**19.26 Results of a magnetic survey of a buried ore body later confirmed by drill holes, northern Quebec.** Survey by A. M. Bateman, 1941. (After Alan M. Bateman, 1950, reference cited in caption of Figure 19.8, Figure 10-1, p. 379)

central point, and they receive some of the current. The pattern of readings on the various receiving electrodes is used to interpret the electrical properties of the materials between the central point and the receiving electrodes (Figure 19.27).

## SUBSURFACE EXPLORATION METHODS USED BENEATH WATER

Exploration carried out beneath a water-covered area presents certain advantages and disadvantages as compared with exploring on land. The most obvious complications in exploring beneath the water surface are the use of ships, which are expensive to operate; exposure to all kinds of weather; and the need for precise navigational techniques to identify the ship's position.

In some situations, exploration can be based on light waves, as in direct observation, photography, or television. Most underwater exploration, however, is carried out using sound waves. Continuous seismic profiling yields records that are not usually made on land. Samples of the bottom materials can be collected and also holes can be drilled to test the subbottom materials.

### Methods Using Light Waves

Our vision, which is so invaluable in exploring what is exposed on land, is strictly limited to short distances underwater. Water does not transmit light very far; thus one's range of vision underwater is never more than a few tens of meters at most.

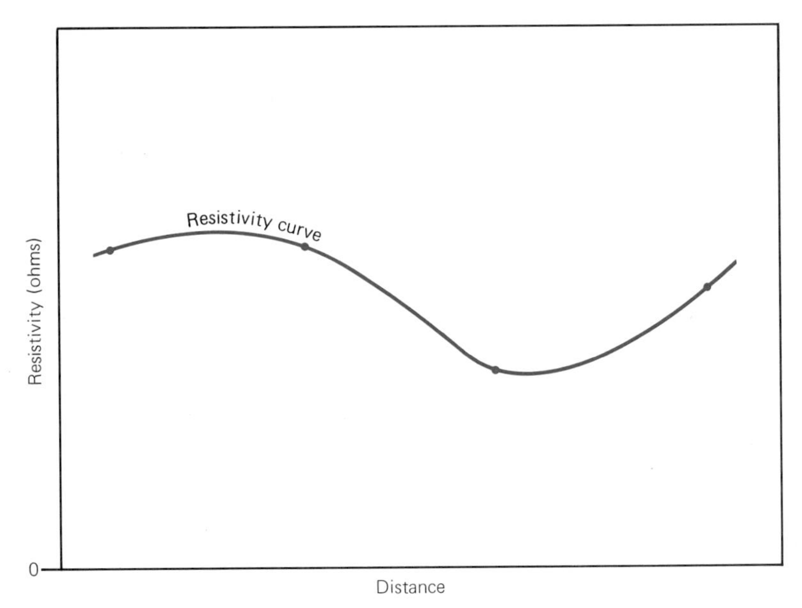


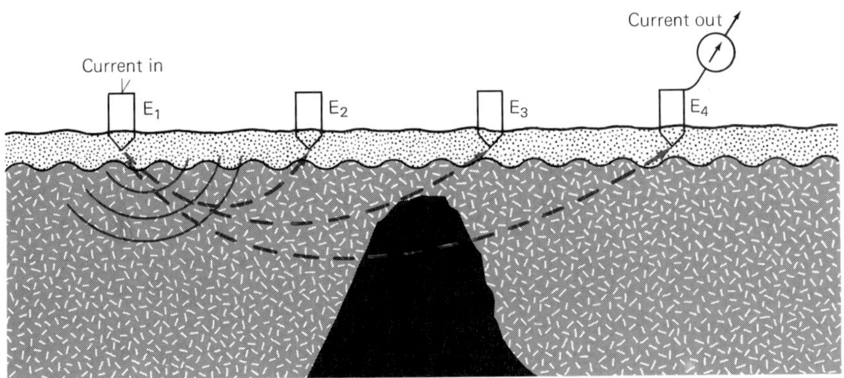


**19.27 Resistivity survey carried out by placing electrodes in the ground at equal intervals along a line.** An electrical current is sent into the ground at one of the end electrodes. Because the current moves away from the electrode along spherical fronts (solid blue lines), the paths of the current (blue dashed lines) curve downward and then return to the surface. The spacing of the electrodes determines the depth to which the current passes and thus the resistivity is measured. If a high-conductive (low resistivity) ore body is traversed by the current, the resistivity indicated by some electrodes will drop. (Modified from A. M. Bateman, 1950, reference cited in caption of Figure 19.8, Figure 10-4, p. 382)

Direct visual observations can be made by scientist-divers and from research submarines. Only small areas can be covered in a day's work. Nevertheless, some kinds of exploration, such as measuring the strike and dip of strata exposed on the sea floor, can be done reliably only by geologist-divers.

Photographs of the bottom (Figure 19.28) can be made by lowering cameras; videotapes can be made by lowering television cameras. Both kinds of cameras usually require special underwater lights. Many parts of the continental shelves have been recorded by movie and television cameras mounted on special surveying sleds that were dragged across the bottom.

**19.28** Photograph of sea floor at depth of 40 m on continental shelf SE of Wachapreague, Virginia, showing bottom littered with dead skeletal debris of marine invertebrates (white areas) and a cluster of living sand dollars (right). Foot ruler has been attached to blade of current-registering compass that served to trip the camera when the compass touched bottom. (David Owen, Woods Hole Oceanographic Institution)

### Methods Using Sound Waves

Sound waves are to underwater exploration what light waves are to exploration of the land. Sound waves are transmitted very well underwater and thus can be employed for many purposes. We examine three: (1) echo sounding, (2) side-scanning sonar, and (3) continuous seismic profiling.

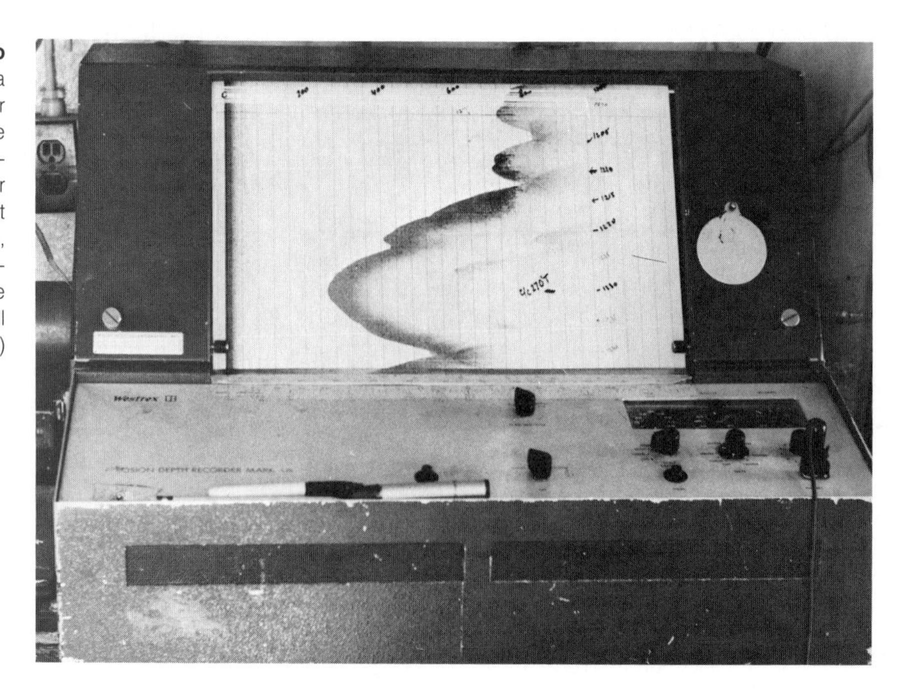

**19.29 Precision recording echo sounder,** at work off N coast of Jamaica in an area of steep bottom relief. Water surface is indicated by the trace of the outgoing signal (dark line at left). Numbers across top indicate depths of water (fathoms, about 1.8 meters). Marks at right are 5-minute time intervals. At 1230, R/V *Gosnold* changed course and established a heading of due W, (270° true bearing using a 0-to-360 basis for a full circle), parallel to shore. (J. E. Sanders)

**Echo sounding** The use of sound to determine depth began as an outgrowth of the sinking of the *Titanic* in the North Atlantic in 1912. Several inventors suggested that sound could be used to scan the water for icebergs in time for a ship to avoid a collision. When field trials were carried out in 1914, however, few icebergs were found. Instead, the sound waves which returned to the ship proved to be those that had bounced off the bottom.

Now, piezoelectric crystals (Chapter 4) are used in *echo sounding,* which is the determination of water depth using sound waves. Since 1953, precision recording echo sounders (Figure 19.29) have made it possible to record profiles of the bottom that are accurate to 1 meter in 3000 meters.

Recording echo sounders have been used in detailed mapping of the bottom. Such projects require exact navigation; the ship's position must be known with respect to the bottom every few minutes. A few sophisticated navigation tools are capable of plotting the position of the ship continuously. Where such equipment is available and can be kept in working order, it greatly simplifies the problem of underwater mapping.

Most of the deep-sea floor has been depicted on physiographic diagrams that were drawn from data recorded on precision echo sounders. Areas of the continental shelves are less well displayed. The shelves have not been mapped by echo sounders partly because the roughness of the bottom to be mapped usually is less than the roughness of the water surface as determined by the water-surface waves. A ship that is being elevated and depressed by several meters cannot be expected to record echo profiles from features whose relief is less than the heights of the water waves. Therefore, another technique, side-scanning sonar, is required to display shelf morphology.

**Side-scanning sonar** The underwater survey technique known as side-scanning sonar uses equipment that is similar in many respects to precision echo sounding, but with one major difference. In echo sounding, a narrow beam of sound is aimed directly at the bottom. In side-scanning sonar, a fan-shaped beam of sound is pointed partly at the bottom and partly across the bottom in a vertical plane at right angles to the direction the ship is moving (Figure 19.30). The sound waves move out in lines that form arcs to the fan; the sound grazes along the bottom and is reflected back toward the ship from changes of slope, from objects on the bottom, such as wrecked ships or cables and pipes that are approximately parallel to the ship's track, and from areas where bottom roughness changes. The display from a side-scanning sonar survey is analogous to a sonic map of the bottom (Figure 19.31).

**Continuous seismic profiling** The behavior of sound that strikes the bottom depends on the frequency of the sound. In the measuring scheme now generally adopted by physicists, frequency of waves is expressed in units called *hertz.* A hertz (hz) equals 1 cycle per second; a kilohertz (khz) equals 1000 cycles per second. High-frequency sound waves, having frequencies larger than about 5 kilohertz, tend to bounce upward from the bottom. Because the speed of all sound waves in water is the same (about 1500 meters per second), the wavelengths are inversely related to frequency, that is, high frequencies mean small wavelengths. For example, the wavelength of a sound wave having a frequency of 10 kilohertz is equal to 1500 meters per second divided by 10,000 cycles per second, or 0.15 meter (15 cm or about 6 in.). Thus, any object larger than 15 centimeters will reflect the sound. Sound waves having low frequencies, for example, 100 hertz, penetrate into the bottom sediments and are reflected by subbottom layers having physical properties that differ from those of the overlying layers. The wavelength of a sound wave having a frequency of 100 hertz is 1500 meters per second divided by 100 cycles per second, or 15 meters.

Continuous seismic profiling resembles echo sounding, but with the major difference that large, low-frequency sounds are used, such as those made by an explosion in the water, by the sudden release of compressed air from a small cylinder, by the arcing of a high-voltage electrical spark, or by the mechanical movement of a pair of metal plates (Figure 19.32). The graphic display of the survey is a continuous profile not only of the shape of the bottom itself, but also of the thicknesses and arrangements of subbottom reflecting layers. If the reflections are coming from bedding surfaces, then the profile shows geologic structure (Figure 19.33).

The excellent display of the subsurface structure that can be made by continuous seismic profiling is an important reason for the emphasis now being given to petroleum exploration in water-covered areas. The structural information from these profiles is far superior to that obtained from most seismic surveys conducted on land.

### Equipment for Sampling Bottom Materials

All kinds of samples can be collected from the bottom of water-covered areas. These range from small grab samples of sediment to continuous cores collected with special drilling ships.

**Grab samplers** For many purposes, only a small quantity of surface sediment is needed. In such cases, use can be made of several varieties of self-closing devices, named *grab samplers,* which cut into the bottom

**19.30 Use of vertical sound as contrasted with lateral sound in underwater surveying.** (a) In echo sounding, piezoelectric crystals are used to generate and detect underwater sound. When a pulsating electrical current enters the crystal of the sound transmitter-receiver, the crystal vibrates, forming sound waves. After the current has been switched off (after being on for a tiny fraction of a second), the crystal becomes quiet and serves as a detecting device. When sound reflected from the bottom reaches the crystal, the sound waves (a form of mechanical motion) make the crystal vibrate, and this causes the crystal to generate a small, pulsating electric current that is proportional to the amount and frequency of sound received. The total travel distance of the sound waves is twice the depth, once from the ship to the bottom, and again from the bottom to the ship. (b) In side-scanning sonar, some sound is used in a vertical path to determine depth, as in echo sounding. If the sonar equipment is towed on a stable device that is not attached to the ship, then the depth is the sum of the down-scanning lobe (P to B and back, 135 m on sketch 1) plus the up-scanning lobe (P to S and back, 75 m on the profile of sketch 1). As the sound from the main beam (white with blue arcs) of circles for fronts of sound waves) scans across the bottom, it may be reflected back to the source, or sent off in a different direction. The variation in darkness of the record (sketch 2, below) indicates amount of sound returned, which is related to the characteristics and the slope of the bottom: black, for a slope toward the sonar; and white, for areas where the sound did not strike the bottom ("shadow areas"). On the record, Y is the direction the ship traveled over the bottom. (J. E. Sanders and C. S. Clay, 1968, Investigating the ocean bottom with side-scanning sonar, p. 529–547 *in* Parker, D. C., *ed.*, Symposium on remote sensing of environment, 5th, Proceedings: Ann Arbor, Michigan, Infrared Physics Laboratory, Willow Run Laboratories, Institute of Science and Technology, University of Michigan, 946 p., Figure 7, p. 543)

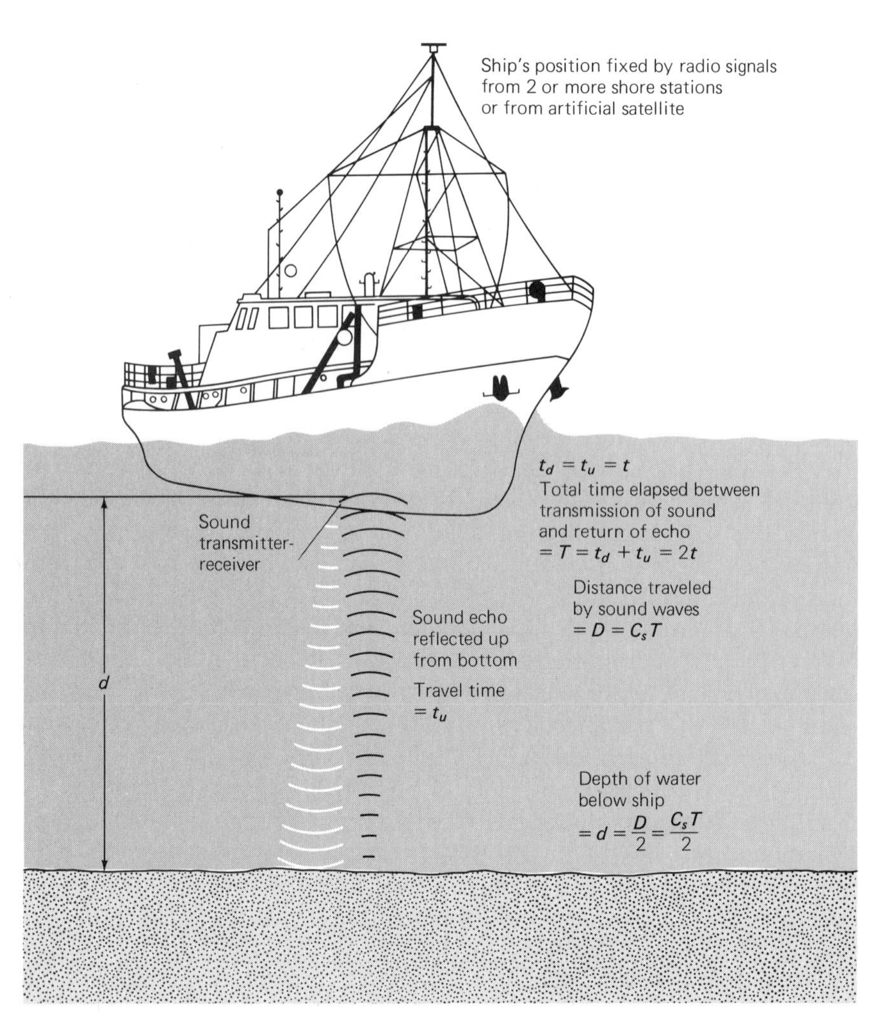


(a)

(b)

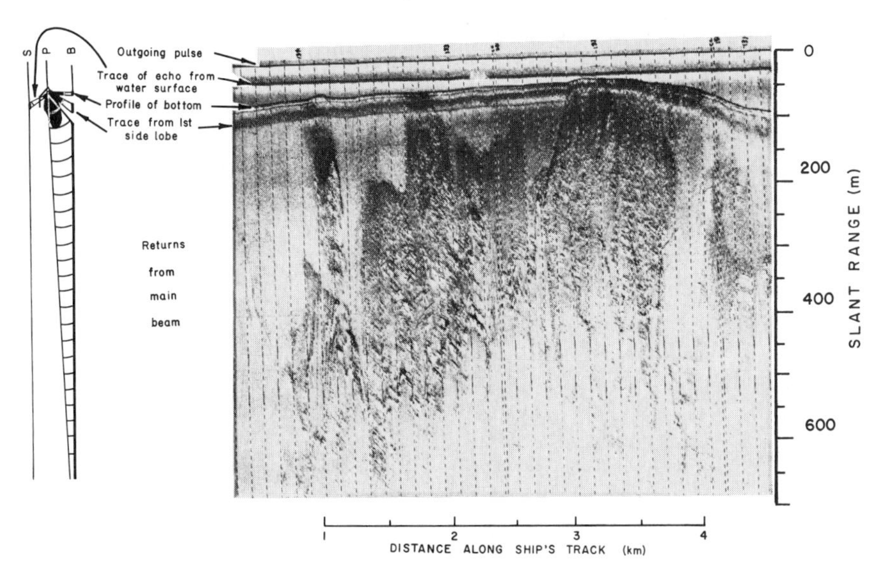


**19.31 Photograph of side-scanning sonar record of bottom near Cashes Ledge, Gulf of Maine.** Crossing pattern of dark-and-light line segments is inferred to indicate rippled sand. Distinct changes in darkness of the record are thought to mark boundaries between sand (high areas) and mud (lower areas with few reflecting features). (J. E. Sanders, K. O. Emery, and Elazar Uchupi, 1969, Microtopography of five small areas of the continental shelf by side-scanning sonar: Geological Society of America, Bulletin, v. 80, no. 4, p. 561–572, Plate 5, following p. 571)

sediment and enclose a small sample. The simplest of these does not even close; it is nothing more than a thick-walled, open-ended, bucket-like cylinder attached to a short section of chain.

**Rock dredges** Where the bottom consists of rock, it is sometimes possible to break loose a sample by dragging a special rock dredge along the bottom. The ship drifts or is maneuvered slowly across the area, with the dredge strung out behind. When the dredge catches on a corner of an exposed ledge, a great scientific tug-of-war takes place. Sometimes the

**19.32 Continuous seismic profiler,** the ultimate device for determining thicknesses and structure of subbottom sediments and rock units. The principle is the same as in echo sounding, but the sound energy is much greater and the interval between noises is longer than in most echo sounding. In addition, the returned signals are subjected to many manipulations in sophisticated computers. (U. S. Geological Survey)

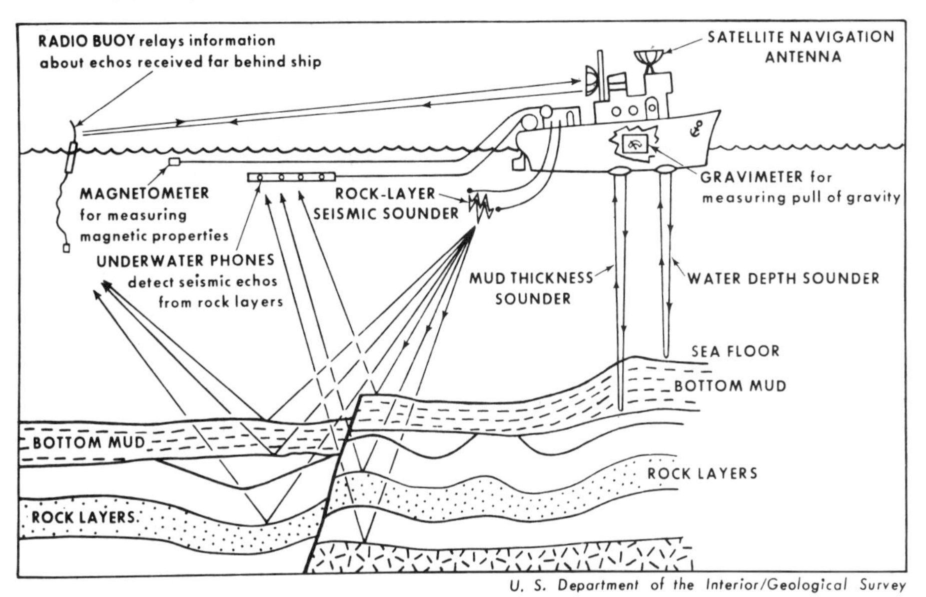

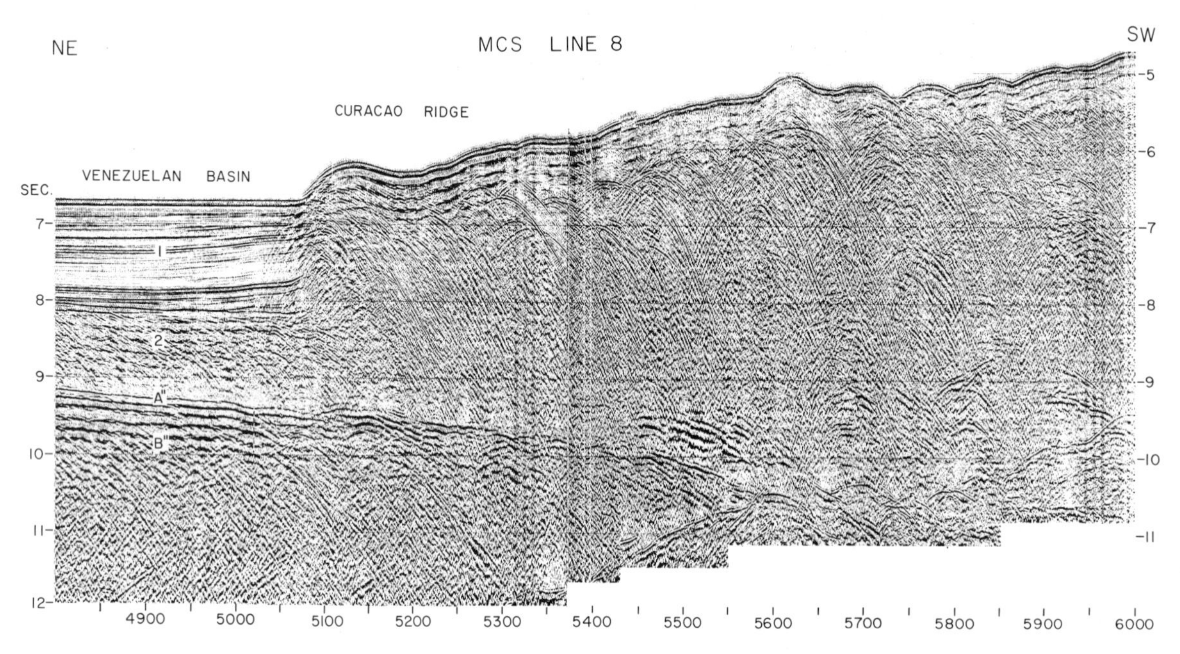


**19.33 Example of a record made by a continuous seismic profiler.** Thin, nondeformed strata underlying the Venezuelan Basin (left) give way abruptly along a vertical contact to thicker, folded strata beneath the Curacao Ridge. Horizon "A," the top of a layer of Eocene chert, extends continuously beneath both groups of younger sediments. "B" is thought to be the top of igneous basement. Water depth above the abyssal plain of the Venezuelan Basin at left (Lat. 13°30′ N., Long. 67°40′ W.) is 5100 meters. Depth above Curacao Ridge at right-hand end of section (Lat. 13°00′ N., Long. 67°40′ W.) is 3600 meters. Length of section is approximately 50 kilometers. Vertical exaggeration is 4.5x. (Manik Talwani, Charles C. Windisch, Paul L. Stoffa, Peter Buhl, and Robert E. Houtz, 1977, Multichannel seismic study in the Venezuelan Basin and The Curacao Ridge, p.83–98 *in* Talwani, Manik, and Pitman, w.c., III, *eds*. Island arcs, deep sea (*sic*) trenches and back-arc basins: Washington, D.C., American Geophysical Union, Maurice Ewing series, v. 1, 470 p.; right-hand side of Figure 2, facing p. 84, courtesy Paul Stoffa, Lamont-Doherty Geological Observatory of Columbia University.)

scientists win and are rewarded with a sample of the bedrock. Just as often, however, the scientists lose, and another rock dredge and various lengths of trawl wire are added to the enlarging collection in Davy Jones' Locker.

**Corers** For many kinds of geologic studies, it is necessary to collect samples of the bottom sediment that display the thicknesses and characteristics of the subbottom sediments. This is done by means of a cylindrical sample named a core, which can be collected by forcing in an open-ended tube known as a corer. The now-standard kind of corer used to collect samples up to lengths of 20 or 30 meters is rigged so that it drops freely through the last five meters of descent to the bottom and thus strikes the bottom with considerable speed and force (Figure 19.34). Such a freely dropped corer readily penetrates soft silts and clays to depths as long as the core barrel. However, in firmly packed sands, penetration ranges from zero to one or two meters at most.

**Drilling platforms and drilling ships** In response to the needs of science and of the petroleum industry, regular oil-field rotary-drilling rigs can now be taken out to sea on special drilling platforms or drilling ships. Some drilling platforms are floated out to sea and rest on long legs placed on the bottom. Others, called semisubmersibles, rest on submarinelike floats that are stable because they are submerged below the level of wave motion (Figure 19.35). Their work platforms are high enough out of the water to be above the level of the surface waves.

Drilling ships, such as the *Glomar Challenger* (Figure 19.36), are built so that a large center well is available for lowering and raising the drilling pipe. A drilling ship keeps its position above the hole by propulsion devices located on each side near the bow and near the stern. Because of the flexibility of the steel drilling pipe, no problem is encountered as long

(a)

Wire lowered

Wire stopped momentarily

Wire raised

1

2

3

4

Releasing device

Slack wire

Lever arm

Stabilizing fins

Weights

Large coring tube

Piston

Large coring tube starts free fall

Both corers extracted for return to ship

Trigger-weight coring tube penetrates into sediment

Large coring tube penetrates into sediment, sliding around piston held fast by wire to winch

(b)

**19.34 Sediment corer based on the principle of a free fall just prior to striking bottom.** (a) Ewing piston corer all rigged and ready to be lowered to the sea floor. Out of sight in the water is the small coring tube (the trigger-weight corer) whose weight times the length of the trigger arm exceeds the moment (weight times distance) of the main corer. Several gadgets have been invented to forestall premature release of the main corer before the trigger weight reaches bottom. (b) Stages in collecting a core, schematic sketches showing the Ewing corer near the bottom.

1. As trawl wire is lowered, corer approaches bottom.
2. Trigger weight touches bottom, allowing lever-release arm to be raised, releasing the main corer. The instant of release of heavy main corer is transmitted up the cable as a momentary interval of slack.
3. Heavy corer strikes bottom (if all goes well, not hitting the trigger-weight corer). Trawl wire becomes taut again when piston inside main core barrel is at level of bottom. Corer slides around piston, driven by its free fall and, after penetration begins, by the enormous hydrostatic pressure of the deep water.
4. Both corers retrieved and being lifted back to the ship.

**19.35 Offshore drilling platform of the semisubmersible variety,** at work off Nova Scotia, Canada. On semisubmersible rigs, an elevated work platform, built well above the level of storm waves, rests on three thick vertical columns, which extend downward to large submerged floats located deeper in the water than the orbital motion of wave action (see Figure 15.4 for decrease in orbits with depth). On other kinds of drilling platforms, the supports extend to the sea floor. (Mobil Corporation)

**19.36 *Glomar Challenger's* drilling well, pipe racks, and drilling tower.** This remarkable deep-sea drilling ship has collected cores of sediment and rock from many parts of the deep-ocean basins, enabling a nearly complete history of the oceans to be reconstructed. (Deep-Sea Drilling Project, courtesy Lamont-Doherty Geological Observatory)

as the ship stays within a circle having a radius of about 400 meters and whose center is directly above the hole.

Using regular rotary drilling equipment, it is possible to drill through any kind of materials. Thousands of cores have been collected for scientific study by the *Glomar Challenger* from all parts of the deep-sea basins, and thousands of holes have been drilled in offshore areas in search of petroleum.

# CHAPTER REVIEW

1. The important task of finding new sources of minerals and energy belongs to teams of skilled scientists, including geologists, geophysicists, geochemists, and geobotanists. In general, the objectives of such geologic searches is to determine the composition and structure of subsurface material.

2. The chief kinds of geologic structures that have localized important economic deposits include anticlines, salt diapirs, and faults.

3. The important characteristics of *anticlines* that localize economic deposits are the reversal of dip at the crest and the seal provided by a shale or other impermeable layer overlying a permeable layer such as a sand.

4. *Salt diapirs* are of economic interest for their materials, such as halite and sulfur, and for their effects on surrounding strata, which have caused petroleum to accumulate.

5. Economic deposits related to faults may have accumulated because of folds that formed during fault movement or because the fault created openings in fault breccia and also served as a channelway for migration of elements in solution.

6. Petroleum and a few kinds of metal deposits have been localized along buried surfaces of erosion. Various kinds of pore spaces in carbonate rocks that have been weathered and subjected to the influence of fresh groundwater have later become filled with petroleum.

7. Many of the world's premier ore deposits are contained in plutons or are found in rocks which were altered by hydrothermal solutions that were associated with the emplacement of a pluton.

8. Buried valleys contain filling sediments that may serve as valuable aquifiers for supplying groundwater or that may include buried placers. In addition, the thicknesses and characteristics of the valley-filling sediments affect the designs and locations of engineering structures.

9. Various kinds of direct data and indirect clues about what lies below the land surface are available to guide exploration and to support inferences about possible locations of buried deposits. These include radioactivity, composition of the regolith, compositions of the leaves of plants, chemical composition of natural waters, oil seeps and gas seeps, and remote-sensing techniques used from aircraft or satellites.

10. Subsurface exploration may be divided into two categories: methods used on land (such as drilling techniques and geophysical methods), and those used beneath water (such as methods using light waves, methods using sound waves, and sampling bottom materials), although some methods are used in both areas.

## QUESTIONS

1. Explain how anticlines can localize economic deposits.
2. Do all closed domes contain petroleum? Explain your answer.
3. Are all closed anticlinal structures visible at the Earth's surface? Explain.
4. What is a *saddle reef?*
5. List the economic deposits associated with salt diapirs and explain this association.
6. Discuss how faults and economic deposits can be related.
7. Explain the sequence of events that can be inferred from a petroleum deposit in which the reservoir was formerly exposed at the land surface.
8. Are economic deposits associated with plutons? Explain.
9. Discuss the potential economic significance of buried valleys.
10. Define *provenance.* Discuss some of the geologic relationships that may prevail between composition of the regolith and economic deposits.
11. What is *geobotanical prospecting?*
12. Define *remote sensing* and briefly explain four kinds of remote-sensing methods that are useful in exploration.
13. Describe some commonly used methods for recording subsurface conditions encountered in a drilled hole.
14. List four geophysical survey techniques and explain briefly the principle(s) upon which each is based.
15. Compare the use of light waves and sound waves in underwater exploration.

## RECOMMENDED READING


Davidson, C. F., 1964, On Diamondiferous Diatremes: *Economic Geology,* v. 59, no. 7, p. 1368–1380.

Fischer, R. P., 1974, Exploration Guides to New Uranium Districts and Belts: *Economic Geology,* v. 69, no. 3, p. 362–376.

Gustafson, L. B., 1978, Some Major Factors of Porphyry Copper Genesis: *Economic Geology,* v. 73, no. 5, p. 600–607.

Janssen, R. E., 1952, The History of a River: *Scientific American,* v. 186, no. 6, p. 74–78, 80 (Offprint No. 826)

Meyerhoff, A. A., 1976, Economic Impact and Geopolitical Implications of Giant Petroleum Fields: *American Scientist,* v. 64, no. 5, p. 536–541.

Moore, D. G., 1969, Reflection Profiling (*sic*) Studies of the California Continental Borderland. Structure and Quaternary Turbidite Basins: *Geological Society of America, Special Paper 107,* 142 p.

Romberg, F. E., 1961, Exploration Geophysics: A Review: *Geological Society of America, Bulletin,* v. 72, no. 6, p. 883–932.

Rona, P. A., 1973, Plate Tectonics and Mineral Resources: *Scientific American,* v. 229, no. 1, p. 86–95. (Offprint No. 909)

Rowan, L. C., 1975, Application of Satellites to Geologic Exploration: *American Scientist,* v. 63, no. 4, p. 393–403.

Sawkins, F. J., 1972, Sulfide Ore Deposits in Relation to Plate Tectonics: *Journal of Geology,* v. 80, no. 4, p. 377–397.

# CHAPTER 20
# PLANETARY GEOLOGY

PLATE 20
**Saturn and it s remarkable rings,** which lie in the planet's equatorial plane as seen from a range of 8 million kilometers in a mosaic image transmitted back to earth by Voyager I on 6 November 1980. Shadows of the outer rings, visible on the surface of Saturn, obscure the inner rings (lower right). (National Aeronautics and Space Administration, Jet Propulsion Laboratory, Pasadena, California)

IN THIS CHAPTER WE EXTEND OUR KNOWLEDGE OF GEOLOGY OUTward to other parts of the Solar System. We shall address two objectives: (1) completion of our discussion, begun in Chapter 1, about the descriptive data from the planets and their major satellites; and (2) to see what lessons derived from other planets can be applied to the problem of understanding how the Earth may have behaved at early stages in its history. We shall be dealing with the Solar System more or less as it is today, and shall not bother about such fascinating, but geologically remote, topics such as the origin of the Solar System or origin of the Universe.

We begin with a discussion of how it is possible to know about other parts of the Solar System. After that, we take up some geologic aspects of the Moon, the planets, and their chief satellites. Our general order will be from bodies closest to the Earth to those located progressively farther away.

**20.1 Near side of a nearly full Moon,** as photographed through a large telescope on Earth. Dark areas are *mare*, they are underlain by basalt that flooded into various irregular basins created by the impact of giant asteroids. Light-colored cratered areas (as at bottom and at upper right) are *lunar highlands*. At least three impact craters lie at the centers of radiating *rays*, light-colored streaks thought to be formed of debris thrown out as a byproduct of enormous impacts. Craters of several ages are visible. North is at top; Sun, at left. (Lick Observatory, University of California, Santa Cruz, California)

## METHODS OF STUDYING PLANETS AND THEIR SATELLITES

Knowledge of other planets began with visual observations. Late in the sixteenth century, William Gilbert (1544–1603), physician-in-ordinary to Queen Elizabeth I, drew the first map of the surface of the Moon, and did so without using a telescope. The telescope was invented shortly afterward, and in 1609, Galileo Galilei (1564–1642) became the first person to study the Moon using a telescope. Cameras attached to large telescopes make it possible to photograph the planets as well as the surface of the Moon (Figure 20.1).

Despite the obvious limitations associated with telescopic observations from the Earth, data from telescopes showed that Saturn is surrounded by a series of equatorial rings (Figure 20.2). More-distant planets, such as Neptune, show only as circular light spots (Figure 20.3).

Earth-based observations enabled the sizes of planets to be measured and their orbital characteristics to be determined. From orbital data, it has

**20.2 Saturn and its rings,** a composite made from 16 images photographed using the Catalina Observatory 155-centimeter reflector by S. M. Larson on 11 March 1974. Plane of rings is inclined 26.9° to line of sight. (National Aeronautics and Space Administration, photo 78-H-102)

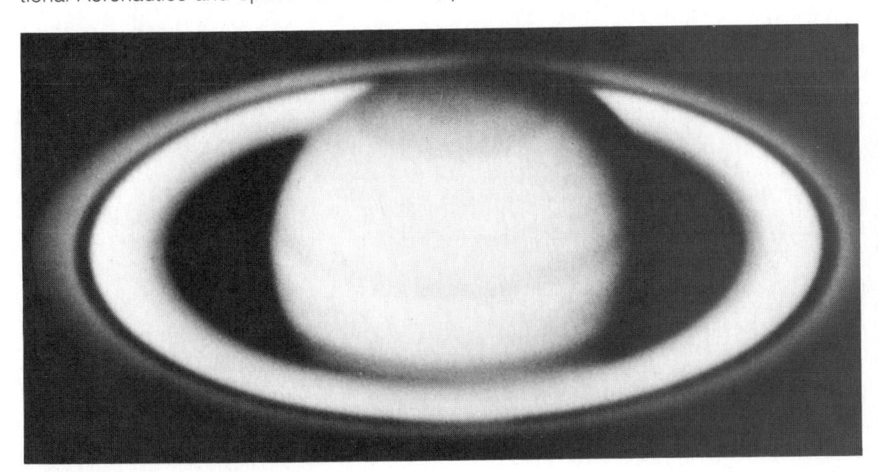

**20.3 Uranus, photographed in blue light** using 155-centimeter reflecting telescope of Catalina Observatory. (National Aeronautics and Space Administration photo 70-H-1654, by the Lunar and Planetary Laboratory, University of Arizona)

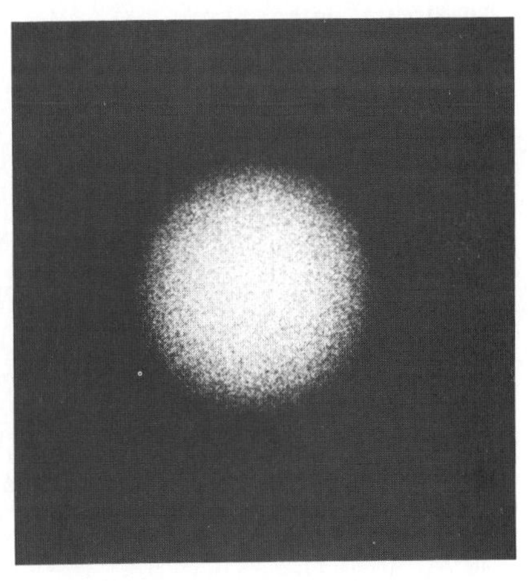

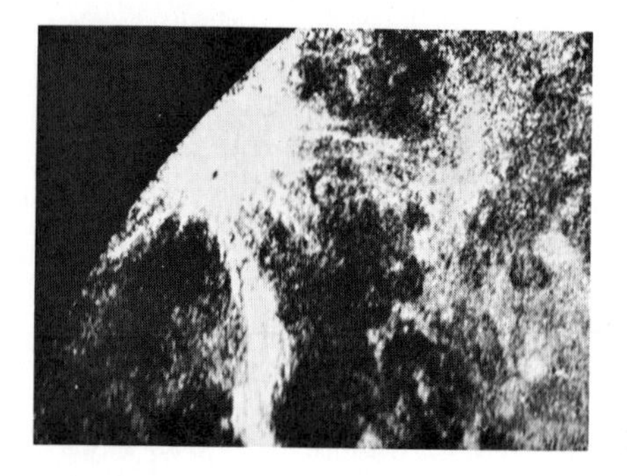

**20.4 Surface of Venus,** image using 12-centimeter wavelength radar from Arecibo Observatory, Puerto Rico. Radar waves penetrate the thick atmosphere that block out light rays. Viewed here are parts of the area between 9° and 31° N latitude and between 269° and 311° longitude; the average resolution distance is 11 kilometers. (National Aeronautics and Space Administration, photo 79-H-205)

been possible to calculate the masses of planets. Knowing a planet's mass and diameter, it is possible to compute its mean density, a characteristic that indicates much about the planet's composition. Based on computed mean densities, astronomers have divided the planets into two groups: (1) the inner group of Earthlike (or "rocky") terrestrial planets (Mercury, Venus, Earth and its Moon, and Mars) having densities ranging from 3.97 (Mars) to 5.52 (Earth); and (2) the outer group of low-density giant planets (Jupiter, Saturn, Uranus, Neptune, and Pluto), having densities ranging from 0.717 (Saturn) to 1.655 (Neptune) (Table 20.1).

Other kinds of Earth-based observations include spectroscopic analysis of the light reflected from the surfaces of planets and data obtained from radio telescopes and from radar transmitters-receivers (Figure 20.4).

Since the Space Age began, the amount of information about the Solar System has increased enormously. Powerful rockets are now available for launching instrument packages that can transmit images back to Earth from close approaches, or even from direct landings on distant planets (Figure 20.5). In addition, the United States succeeded in placing men and vehicles on the Moon (Figure 20.6) for direct observations and collection of samples.

| | Mercury | Venus | Earth | Mars | Jupiter | Saturn | Uranus | Neptune | Pluto | Sun | Moon |
|---|---|---|---|---|---|---|---|---|---|---|---|
| Mean distance from Sun (millions of km) | 57.8 | 108.2 | 149.6 | 227.9 | 778.3 | 1427 | 2870 | 4497 | 5900 | About 150 from Earth | About 382,171 km from Earth |
| Period of one revolution around Sun, or length of planet's year (Earth time) | 88 days | 224.7 days | 365.3 days | 687 days | 11.86 years | 29.46 years | 84 years | 164.8 years | 248 years | | 365.3 days |
| Rotation period, or length of day (Earth time) | 59 days | −243 days (retrograde) | 23.9 hours | 24.6 hours | 9.9 hours | 10.4 hours | −10.8 hours (retrograde) | 16 hours | 6.4 days | 27 days | 27.3 days |
| Equatorial diameter (km) | 4,847 | 12,118 | 12,756 | 6,761 | 142,870 | 119,399 | 51,790 | 49,494 | 1,280 | 1,390,000 | 3,460 |
| Mass (Earth = 1) | 0.055 | 0.815 | 1 | 0.108 | 317.9 | 95.2 | 14.6 | 17.2 | 0.0016 | 332,830 | 0.012 |
| Mean density (Water = 1) | 5.4 | 5.2 | 5.5 | 3.9 | 1.3 | 0.7 | 1.6 | 1.655 | 1.048 | 1.434 | 3.36 |
| Maximum surface temperature* (°C) | 315 | 315 | 60 | 24 | −145 | −168 | −183 | −195 | −217 | 5540 | 100 |
| Surface gravity (Earth = 1) | 0.37 | 0.88 | 1 | 0.38 | 2.64 | 1.15 | 1.15 | 1.12 | 0.04 | 27.9 | 0.17 |
| Weight of 70 kg person (kg) | 25.9 | 61.6 | 70 | 26.6 | 184.8 | 80.5 | 80.5 | 78.4 | 2.8 | 1953 | 11.9 |
| Number of known satellites | 0 | 0 | 1 | 2 | 15 | 12 | 5 | 2 | 1 | | 0 |
| Travel time from Earth | 146 days (Mariner 10) | 94 days (Mariner 10) | — | 304 days (Viking 1) | 607 days (Pioneer II) | 4 years (Pioneer II) | 9 years | 12 years | More than 12 years | | 3 days (Apollo) |

*Surface temperatures vary greatly. The maximum surface temperatures shown here represent the temperatures at one given time and one given place only. Daytime and nighttime temperatures vary considerably, as do polar and equatorial temperatures. No single system can possibly satisfy all conditions.

**20.5 Rock-strewn, desertlike surface of Mars,** photographed from Viking 1 Lander on 3 August 1976, view eastward at 0730 local Mars time. Large rock at left is about 8 meters from the spacecraft and measures about 1 by 3 meters. Sand dunes (light-colored, smooth areas) indicate wind transported sand from upper left to lower right. (National Aeronautics and Space Administration, photo 6-H-620)

## THE MOON

One of the most dramatic moments in the history of science occurred on 20 July 1969, when the American astronaut Neil A. Armstrong (1930– ) became the first person to set foot on the Moon (Figure 20.7). Although it is technically a satellite of the Earth, the Moon is sometimes considered as the Earth's sister planet. The Moon is closer in size compared to the Earth than any other satellite is to its primary. In addition, the Moon has provided us with more information about extraterrestrial conditions than any other planet, and so it will be treated as a planet here. So much has been learned about the Moon in such a short time that it would be impossible to present all of it here, but we will attempt to discuss the most important features of the Moon, from the outside in.

### The Atmosphere of the Moon

The density of the Moon's atmosphere is about one ten-thousandth of a billion of the density of the Earth's atmosphere; essentially, the Moon

**20.6 U. S. Astronaut James Irwin and lunar Rover vehicle, on surface of the Moon,** view to northeast with Mount Hadley in background; 1 August 1971, Apollo 15 mission. (National Aeronautics and Space Administration, photo 71-H-1413)

**20.7 Astronaut's footprint in lunar regolith,** 20 July 1969, Apollo 11 Mission during which Neil A. Armstrong and Edwin E. Aldrin, Jr., spent one day on the Moon. (National Aeronautics and Space Administration, photo 69-H-1258)

**20.8 Cratered far side of Moon,** oblique view southwest from Apollo 11 spacecraft in lunar orbit, July 1969. Large crater at center is International Astronomical Union (IAU) No. 308, diameter about 80 kilometers, located at 179° E longitude and 5.5° S latitude. At bottom center is an old crater, much modified by erosion and impact by smaller objects, which contrasts with much fresher crater of about the same diameter located along the near rim of IAU No. 308. (National Aeronautics and Space Administration, photo 69-H-1270)

**20.9 Cratered far side of Moon,** photographed by Orbiter IV on 32 May 1967 from an altitude of 3328 kilometers. Smooth, dark area at top is part of Mare Imbrium. (National Aeronautics and Space Administration, photo 67-H-1431)

does not have an atmosphere. Because the Moon lacks any significant atmosphere, it also lacks surface water, rain, glaciers, and winds; hence, little ordinary Earth-like erosion takes place. (It has been estimated that footprints left on the Moon by astronauts will remain undisturbed for at least 10 million years.) Whatever erosion does exist is caused mostly by thermal heating and cooling, and by downslope movement of regolith.

The Moon probably had an atmosphere early in its history, but because of its low gravitational attraction and the high volatility of the original atmospheric gases, the atmosphere has long since escaped into space. Without a protective atmosphere or a magnetic field such as the Earth's, the Moon is subject to the full force of the solar wind, high-energy X-rays and ultraviolet rays in solar radiation, low-energy cosmic rays, and *micrometeorites,* tiny specks of dust, which hit the Moon's surface at speeds up to 112,000 kilometers per hour (70,000 mph). The sky as seen from the Earth is blue because of the effects of the atmosphere, but with no lunar atmosphere, the Moon's sky always appears black, even in daytime.

## The Surface of the Moon

Galileo's telescopic observations of the Moon showed two distinct kinds of surfaces, one light and one darker (see Figure 20.1). The dark areas reminded him of the oceans of the Earth, so he called them *maria* (singular, *mare*), the Latin word for seas. The light-colored "mountainous" areas are known as the lunar *highlands.* Neither area of the lunar surface actually resembles the Earth's oceans or mountains, and the cratered surface of the Moon, not visible to Galileo, is also different from the Earth's relatively craterless surface.

Rimmed circular depressions, the so-called craters, dominate the surface of the Moon (Figure 20.8). Great debates have arisen as to whether these features were formed by impact from objects striking the surface of the Moon or from volcanic activity. As defined in Chapter 5, crater refers to a circular volcanic depression whose diameter is 1.5 kilometer or less. Circular volcanic features formed by collapse and having diameters substantially larger than 1.5 kilometer are calderas. The term astrobleme has been proposed for circular depressions formed by impact of meteorites (see Figure 9.5), but this term has not become widely adopted. Instead, the term crater has been applied in a general sense. Not many people would prefer to substitute "astroblemed surface" for "cratered surface" when describing the Moon (Figure 20.9) and some other planets.

Lunar impact craters come in all sizes, ranging in diameter from a few tens of meters to several hundred kilometers. Similar-looking craters are present on both the near side and the far side of the Moon (compare Figures 20.1 and 20.9). Impact craters having diameters of 10 kilometers or less usually display concave-up floors, simple steep walls, and sharp-crested rims (as in the pair of craters just below the center of Figure 20.8). Impact craters in the "large" class usually display terraced walls and central conical peaks or flat floors that may or may not have been cratered further (see large crater in center of Figure 20.8). A few exceptionally large craters lie at the center of a radiating pattern of linear features known as *rays* (see Figure 20.1).

Variations in crater morphology have been explained as functions of the size and timing of the impacting body and on the thickness of the

**20.10** Domes, rilles, and a linear ridge near impact crater Marius (upper right), which is located at 50°40′ W Long. and 11°55′ N Lat., on near side of the Moon. Marius impact crater is about 40 kilometers in diameter and 1.5 kilometers deep. The domes are 3 to 15 kilometers in diameter and 300 to 500 meters high. The area in view is about equal to the areas of Massachusetts, Connecticut, and Rhode Island. Photograph from Lunar Orbiter II on 25 November 1966. (National Aeronautics and Space Administration, photo 66-H-1634)

lithosphere of the planet struck. The terraced walls evidently indicate slumping of regolith, much of which may have formed by the impact-related effects. Similarly, the conical peak may be a remnant of upthrown debris after the shock of impact. Lava filled the floors of some craters. The lava may have formed from local melting related to the heat generated by impact in some cases or it may represent outflow of subcrustal magma, which had not yet cooled completely from an early time of general melting and which was released because the impact fractured the lunar crust. The rays have been interpreted as trails of debris sprayed outward from particularly violent impacts.

Other morphologic features on the Moon include domes, sinuous *rilles,* and linear ridges (Figure 20.10). The domes indicate upbulging of the surface accompanying volcanic activity. The rilles suggest flow courses of a fluid. Because water can be ruled out as ever having existed on the surface of the Moon, the most likely explanation is that the rilles were formed by flowing lava, possibly inside lava tubes whose tops later collapsed. The linear ridges suggest some kind of vertical crustal movement.

### Moon Rocks and Regolith

All the rocks returned to Earth by Apollo astronauts have been igneous rocks, which may be placed in three categories: (1) *basalt.* (2) *KREEP norite* (KREEP is an acronym for potassium, K; rare-earth elements, REE; and phosphorus, P; norite is a variety of gabbro discussed in Appendix C), and (3) *anorthosites* (Figure 20.10). The anorthositic group has been the most abundant type by far. The maria are made up of basalt, and the highlands are composed of KREEP norite and anorthosites. The mare basalt contains large amounts of the dark iron-titanium oxide mineral, ilmenite, which accounts for its relatively dark color and high content of titanium. The basalt and KREEP norite were produced by partial melting in the interior of the Moon. Anorthositic rocks were probably produced by *crystal fractionation,* which occurs when a magma begins to crystallize and a layered rock is formed of dense crystals that sank to the bottom or of light crystals that floated to the top.

If the crust of the Moon were indeed formed by crystal fractionation, it is possible that the entire surface of the Moon was once covered with a layer of lava. As the lava cooled, the plagioclase minerals that were to form the crust rose to the surface, whereas the other minerals, such as those rich in iron and magnesium, settled below.

The KREEP norite portion of the Moon was probably formed by the partial melting of the interior and the resultant volcanic activity. The KREEP norite, containing relatively high amounts of potassium, phosphorus, barium, the rare earths, uranium, and thorium, is found in the highlands.

The Moon also possesses a *regolith,* debris fragments that resulted from the bombardment of the Moon's surface by cosmic rays, solar wind, micrometeorites, and meteorites. (The lunar regolith is not a true soil comparable to the soil created on Earth, which accumulated as a result of the chemical and physical weathering of rocks.) One other kind of lunar surface material is *breccia,* composed of broken bits of igneous rock and regolith cemented together by the enormous heat and pressure produced by the impact of meteorites.

## The Interior of the Moon

The outer zones of the Moon are radioactive and are hotter than most scientists expected them to be. In fact, the heat flow is three to four times what was predicted and is about half the heat flow of the Earth. Although the outer areas may be heating up even further, temperatures in the interior are still not hot enough to melt ultramafic rocks. A weak magnetic field has been detected, but no lunar magnetic field is evident. Remanent magnetism in lunar rocks suggests that stronger magnetic fields may have existed about 3.2 to 3.9 billion years ago.

The probable internal structure of the Moon is shown in Figure 20.11. The Moon's crust and outer mantle appear to be rigid. Most moonquakes, too weak to be felt by an astronaut standing directly over the epicenter, occur between the lunar lithosphere and asthenosphere, at depths of 800 to 1000 kilometers, at far-greater depths than earthquake epicenters. Moonquakes, which liberate about one-millionth the seismic energy of earthquakes, seem to occur often when the Earth and Moon are closest together. This suggests that moonquakes are related to gravitational stresses, as are tides on Earth, for example. If such a tidal influence on moonquakes is real, it may prove that the occurrence of earthquakes is related to extra gravitational forces exerted on Earth by the periodic closeness of the Moon.

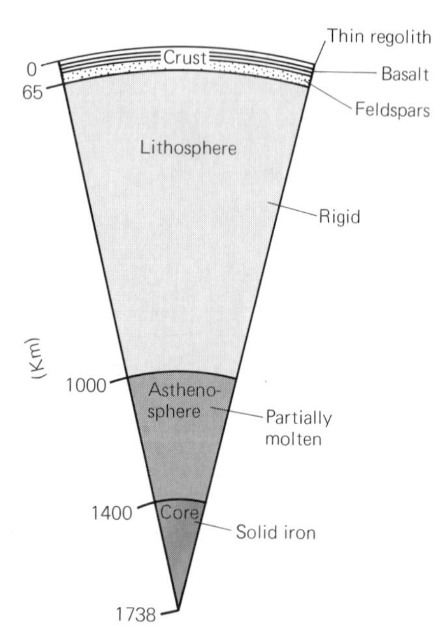


**20.11 Probable internal composition of the Moon,** schematic pie diagram. The lunar crust, estimated to vary in thickness between 25 and 70 kilometers, is probably not uniform.

Below the rigid rock of the lunar lithosphere, the asthenosphere appears to be partially molten. Although a solid iron core seems likely, the Moon's core may actually be partly fluid. Because the Moon is so inactive seismically and practically dead tectonically, it is difficult to obtain instructive data about the interior substance from permanent seismographs left on the Moon by Apollo missions.

## Evolution of the Moon

The oldest rocks from the Moon have been dated at 4.5 billion years, and no rocks younger than 3.1 billion years have been found. From these extremes of ages, we infer that all the developmental history of the Moon took place during the first 1.5 billion years or so of its existence and that not much has happened in the way of forming new rocks since then.

The date of origin of the Moon is considered to be the same as that of the Solar System, 4.6 to 4.7 billion years ago. The history of lunar rocks and morphologic features can be divided into six stages. After the Moon became a distinct body, its first stage of evolution included heating so that the entire body became molten. In stage 2, differentiation resulted in a rocky, silicate outer part and a central iron core. The lunar crust is thought to have formed by direct igneous activity during which crystal cumulates formed. During stage 3, the KREEP norite formed by partial melting of a solid deep-lying part of the crust.

During stage 4, bombardment by planetesimals formed its craters and pockmarked surface. Isotopic evidence in the lunar rocks sets this catastrophic period at about four billion years ago. We do not know for certain whether similar bombardments took place earlier than four billion years ago; the extreme heating caused by such major impacts would have erased the prior isotopic rock record. Certainly, the planetesimal bombardment had decreased sufficiently about four billion years ago so that a record of the decreased bombardment was preserved.

Stage 5 began after the intense bombardment had slowed down. A new period of volcanic activity began that lasted about a billion years. Lava flowed onto the Moon's surface and settled into the basins created by the planetesimal collisions. As these lava flows solidified, they created the overlapping mare basalt surfaces of the Moon.

Stage 6 was a time of general quiescence. The maria remained relatively smooth because of the decrease of planetesimal bombardment. However, the presence of some small craters indicates that some bombardment by comets and meteorites took place occasionally.

## THE TERRESTRIAL ("ROCKY") PLANETS

As mentioned previously, the Solar System consists of two groups of bodies having distinctly contrasting densities. The dense, inner planets have been named terrestrial planets because of their Earth-like densities. In addition to the Earth and its Moon, the terrestrial planets include Mercury, Venus, and Mars.

### Mercury

Until the *Mariner 10* flight in 1974, very little was known about Mercury. Its surface appeared dusky through telescopes, and because Mercury is the closest planet to the Sun, the solar glare made observation difficult. Mercury's proximity to the Sun also makes Mercury a hot planet; daytime temperatures reach as high as 425°C. Until only recently, scientists thought that Mercury always kept the same side facing the Sun, leaving a permanent hot, bright side and a permanent cold, dark side. Now, we know that Mercury rotates once every 59 Earth days instead of once every 88 Earth days. (The latter rotation would be required to match Mercury's orbital rate around the Sun, thereby keeping the same side of Mercury always facing the Sun.)

Only a thin atmosphere is present; it is composed mostly of helium, and its density is only about one-hundred billionth that of the Earth. No water or life occurs on Mercury.

**20.12 Cratered surface of Mercury,** a mosaic of 18 photographs taken at 42-second intervals by Mariner 10 spacecraft at a range of 210,000 kilometers after flyby on 29 March 1974. Diameter of the large circular basin partly in view at extreme left is about 1300 kilometers. Several prominent ray-type craters are present near the North Pole (at top). (National Aeronautics and Space Administration, photo 74-H-253)

High-resolution photographs from *Mariner 10* have shown that Mercury's surface is spattered with craters, as is the Moon's (Figure 20.12). Basic differences do exist, however, between the surfaces of Mercury and the Moon. The craters on Mercury are not as closely spaced as are those on the Moon, and there are many more relatively smooth areas on Mercury. Also, Mercury does not have craters as large as some of the craters on the Moon, and ejecta from planetesimal collisions on Mercury have not been thrown as far from the point of impact as on the Moon, probably because the force of gravity on Mercury is twice as strong. Mercury's surface does display at least one surface feature not evident on the Moon — long, shallow cliffs called *lobate scarps,* probably caused by early tectonic activity when Mercury's core was cooling and contracting.

Despite Mercury's surface resemblance to the Moon, Mercury's interior is thought to be similar to that of the Earth. The current evidence indicates that Mercury possesses a relatively thin silicate mantle and crust, and an iron core with a diameter equal to three-fourths of the planet's total diameter (Figure 20.13). This means that Mercury's core is about the size of the Moon. Like Earth, Mercury displays a distinct magnetic field aligned with its axis of rotation. The presence of a dipolar magnetic field, about one-hundredth as strong as the Earth's magnetic field, has presented several questions. Mercury's relatively slow rate of rotation would not seem to be forceful enough to produce a magnetic field by moving a molten core. At the same time, if Mercury's core is indeed molten, signs of some tectonic activity should be evident on the planet; no sign of such activity has been observed. And if a slowly spinning molten core can set up a magnetic field, why doesn't Venus, whose core is thought to be larger and hotter than Mercury's, show a magnetic field?

**20.13 Probable internal structure of Mercury,** schematic pie diagram. The iron core may be molten; its size (diameter 3600 kilometers) is such that it accounts for three-quarters of Mercury's total volume.

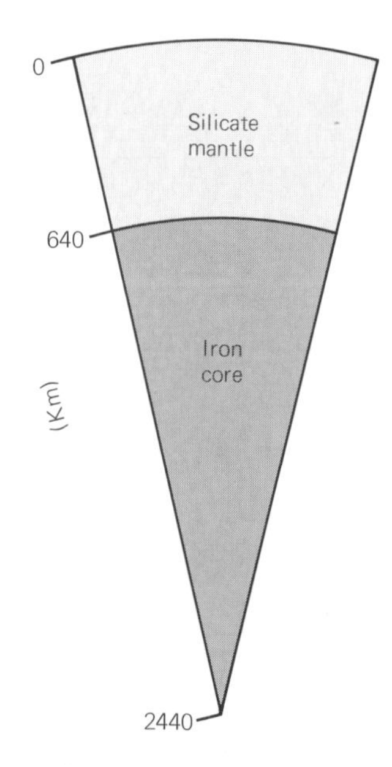


One of the biggest paradoxes created by data from *Mariner 10* is that despite their internal differences, the evolutionary histories of the Moon and Mercury apparently have been similar. The surface history of Mercury may be separated into five periods: (1) chemical fractionation into a large iron core surrounded by a silicate mantle; (2) early volcanism and interplanetary bombardment; (3) intense bombardment, creating craters, basins, and mountains; (4) widespread volcanism, producing broad plains similar to lunar maria; and (5) relative quiescence. No tectonic activity on Mercury appears to have taken place for at least three billion years.

### Venus

The atmosphere of Venus is about 100 times as dense as the Earth's and consists mostly of carbon dioxide (Figure 20.14). The rapid movement of the upper atmosphere and its content of acids formed the basis for the idea that the surface of Venus should show effects of great chemical corrosion. But when the Soviet spacecraft *Venera 9* penetrated the thick yellow veil of Venus in 1975, it sent back surface photographs that showed instead flat, angular rock fragments up to 50 centimeters long. The lack of chemical corrosion displayed by these rocks probably indicated that they had formed fairly recently, especially because later photographs from *Venera 10* showed an area of Venus where rounded rocks did confirm the predicted corrosion.

**20.14 Venus, full-disc view of dense atmosphere,** viewed from Pioneer-Venus Orbiter at a range of 65,000 kilometers on 19 February 1979. Dense atmospheric clouds move faster at the equator than at the poles. Because of the retrograde rotation of Venus (clockwise when viewed from above its North Pole), the clouds at the equator travel from west to east (left to right). (National Aeronautics and Space Administration, photo 79-H-142)

In December 1978, six United States space probes from the *Pioneer Venus* fleet penetrated the dense atmosphere of Venus, and four of these probes actually reached the Venusian surface. As with earlier Soviet probes, all of the *Pioneer Venus* spacecraft encountered surface temperatures high enough to melt such metals as lead or tin, but more information than ever before was transmitted to Earth before the probes succumbed to the extreme temperature and an atmospheric pressure that is about 100 times that of Earth. (*Venera 9* and *Venera 10* stopped transmitting photographs after about one hour on the Venusian surface.) Obviously, no life exists in the hostile environment of Venus.

Besides verifying the high content of carbon dioxide and sulfuric acid in the Venusian atmosphere, the new evidence indicates that larger-than-expected amounts of argon are present in the atmosphere. Although the atmospheres of Earth and Venus may contain similar amounts of argon, it appears that the Venusian atmosphere contains larger amounts of the primordial argon isotopes that may provide clues to the origin of Venus and the materials present at that time.

Although Venus is 42 million kilometers closer to the Sun than the Earth is, much of the solar heat is blocked by atmospheric clouds. Thus, at first glance it may seem surprising that the temperatures of the two planets are so different. The true cause of the extreme Venusian heat is not its proximity to the Sun, but rather the greenhouse effect. As we mentioned in Chapter 1, the greenhouse effect occurs when solar heat penetrates the atmosphere and reaches the surface, but cannot escape because it is trapped by such insulating substances as water vapor, ozone, or carbon dioxide (see Figure 1.18). The atmospheric carbon dioxide is so dense on Venus that it permits almost no heat to be reflected back into space. The carbon dioxide holds in heat the way an insulated attic holds in the heat of a house.

Obviously, Venus and Earth differ greatly. And yet, they have been called the twins of the Solar System. Their sizes, masses, gravitational attractions, and densities are remarkably similar, and they are probably both composed of a similar rocky exterior and partially molten iron interior. The Pioneer Venus Orbiter has been transmitting radar images from the surface of Venus. Resolution distance of this radar system is 30 to 60 km. The compiled results indicate that much of Venus consists of a rolling terrain having relief of one kilometer or less. Two higher-standing areas, referred to by some as proto-continents, display relief of 3 to 5 kilometers above the lowland. Several large volcanic cones and various impact craters are also present.

Venus probably possesses a molten iron core, but it is not surrounded by a noticeable magnetic field. Perhaps its slow rate of rotation, once every 243 days and in the opposite direction of the Earth's rotation (that is, retrograde), does not enable the fluid part of Venus' core to generate a magnetic field. Or, perhaps a former magnetic field has been lost. It can be speculated that Venus might be in the midst of a reversal in its magnetic field cycle, when the force of the field goes to zero. Many questions about Venus still remain to be answered.

## Mars

The United States unmanned spacecraft *Viking 1* landed on Mars on 26 July 1976 — exactly seven years to the day after Neil A. Armstrong set foot on the Moon. Since then, other spacecraft have landed on Mars, and new information about the planet continues to be received. So far, we have confirmed that the atmosphere of Mars contains water vapor, generates some cloud cover, and shows daily and seasonal wind patterns. At its closest approach to the Sun, near the end of the spring in the southern hemisphere, Mars experiences gigantic dust storms. Wind speeds in excess of 250 kilometers per hour (150 mph) raise dust into the atmosphere (Figure 20.15). The Martian atmosphere is about one-hundredth as dense as that of Earth and is composed of about 95.5 percent carbon dioxide, 2.5 percent nitrogen, 1.5 percent argon, 0.1 percent oxygen, and traces of water, krypton, xenon, neon, and helium.

**20.15 Mars, before and during a dust storm,** as viewed from Viking 1 and 2 Orbiters. In left frame, taken on 31 July 1976, the modified walls of a tectonic valley and a few impact craters are visible. In the right frame, taken on 25 March 1977, the entire surface is obscured by a cloud of dust about the size of the state of Colorado. (National Aeronautics and Space Administration, photo 77-H-205)

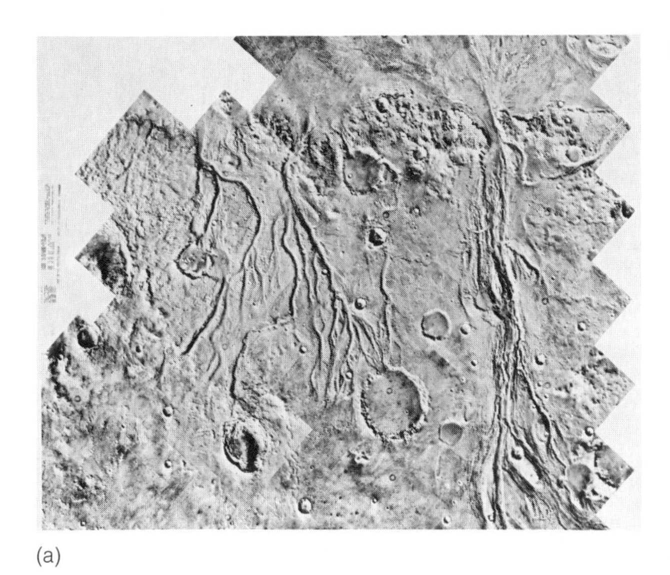
(a)

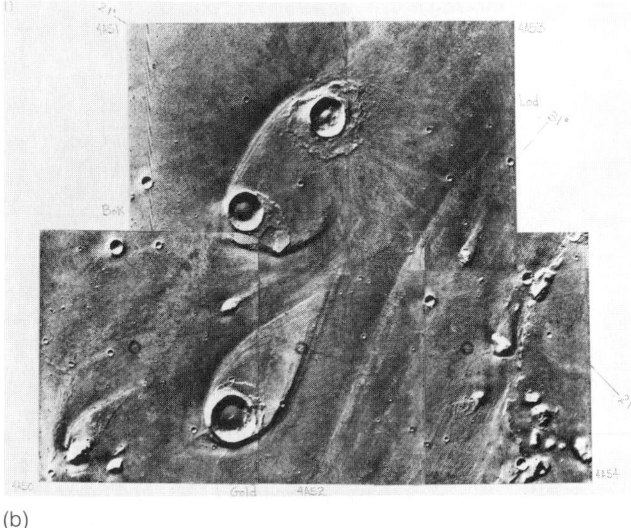
(b)

**20.16 Features on Mars eroded by flowing water.** (a) Network of channels formed by flow through a change of altitude of about 3 kilometers (from W, bottom, to E, top). (NASA, photo 76-H-697, mosaic from Viking Orbiter 1, 4 to 9 August 1976, centered at Lat. 17° N, Long. 55° W) (b) Impact craters on floor of channel, Chryse area, near equator, flow from lower L to upper R. (NASA photo 76-H-480, mosaic from range of about 1600 kilometers on 23 June 1976)

The reddish color of Mars is a result of atmospheric weathering and oxidizing of primitive crystalline rocks. The surface materials are subjected to wind erosion and various sedimentation processes, but are not now subjected to weathering by liquid water as on the Earth. The only forms of water on Mars now are water vapor and ice, but the photographs of the surface give every indication that running water was once a major erosional factor. Braided channels and teardrop islands (Figure 20.16) are probably the products of erosion and deposition by running water and catastrophic flooding which took place when the Martian climate was warmer and wetter.

Although Mars displays no signs of modern tectonic activity and seismographs placed on the surface of Mars have not detected any seismic activity, massive faulting that produced Martian gorges larger than the Grand Canyon must have taken place previously (Figure 20.17).

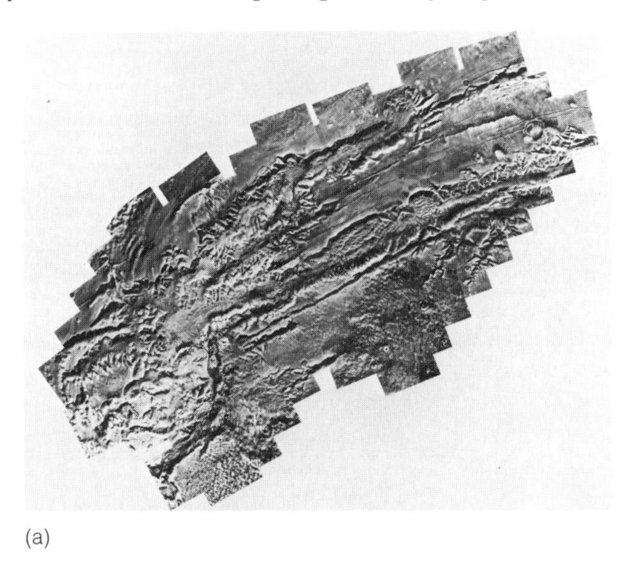
(a)

(b)

**20.17 Tectonic valley and modified side slopes,** Valles Marineris, near equator of Mars. (a) Mosaic showing full width of feature whose length is 5000 kilometers and depth, 2 kilometers. (b) Closer view of S wall; part of massive slump in center, remains and part moved toward camera along the parallel linear marks. (NASA; a, photo 76-H-726, 23 to 26 August 1976; b, photo 76-H-517, 3 July 1976, range 2000 kilometers)

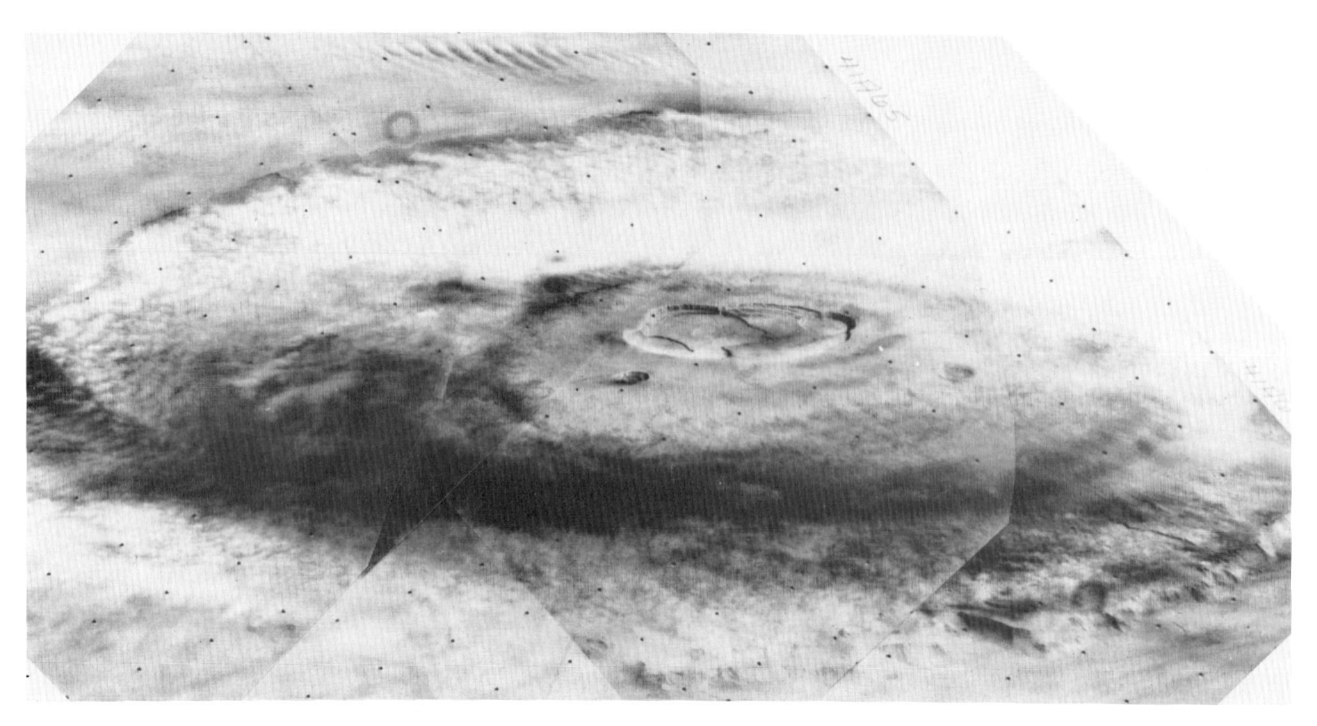

**20.18 The summit of Olympus Mons, the Solar System's largest-known volcano,** projecting through the Martian clouds, as photographed from Viking Orbiter 1 on 31 July 1976 at a range of 8000 kilometers. The caldera at the summit of this gigantic volcanic shield measures about 80 kilometers across. The diameter of the base of Olympus Mons is about 600 kilometers; its height is 24 kilometers. (National Aeronautics and Space Administration, photo 76-H-628)

Such tectonic activity probably ended about 2.5 billion years ago. Just as Valles Marineris is probably the deepest canyon in the Solar System, so the giant shield volcano Olympus Mons is probably its largest volcano (Figure 20.18).

Besides its familiar reddish color, one of the most distinctive features of Mars is the prominence of its polar ice caps. The caps are obviously affected by Martian climate changes; they are smaller in the summer than in the winter. Both poles are covered with layers of ice and dust, which are inferred to have formed when the Martian climate cooled down after the previously warm atmosphere had accumulated substantial amounts of water vapor. The cooling probably resulted when ammonia and methane were removed from the atmosphere, thereby eliminating the greenhouse effect that formerly trapped infrared radiation and produced high temperatures. Frozen water and carbon dioxide also settled into the cracks and loose regolith on the crust. If the water stored in the ice caps, crust, and regolith were to be melted, enough water would be released to flood the entire Martian surface.

The two hemispheres of Mars are morphologically distinct. The southern hemisphere is densely cratered and channeled, whereas the northern hemisphere is relatively smooth, has few craters, and is peppered with extinct volcanoes. The features of the southern hemisphere are thought to have formed during an early period of interplanetary bombardment, which probably started about 4.5 billion years ago and stopped about 4 billion years ago. The relative smoothness of the northern hemisphere was probably caused by lava floods that began after the most intense period of bombardment had ceased.

Martian rocks vary from basalt to breccia, and all appear to be volcanic in origin. *Viking 1* landed in a region called Chryse Planitia, which resembled the surface of a lunar mare. The landing site of *Viking 2*, Utopia Planitia, shows that rocks and dusty sediment have been moved by the wind (see Figure 20.5). The abundance of regolith derived from mafic igneous rocks rich in magnesium, calcium, and iron indicates that the Martian mantle has partially melted. Although the small core of Mars is probably molten, no magnetic field has been detected.

Despite great hopes by some scientists, no signs of Martian life have been found.

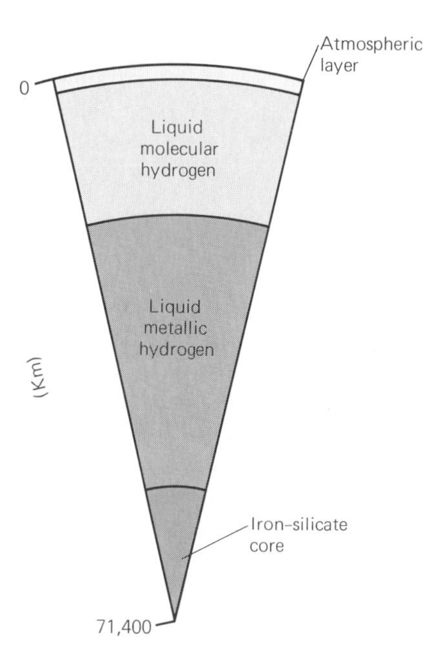


**20.19 Inferred internal composition of Jupiter,** schematic pie diagram.

## THE GIANT ("LOW-DENSITY") PLANETS

Of the five outer planets of the Solar System, four (Jupiter, Saturn, Uranus, and Neptune) are so significantly larger than the others that they qualify as giants. Pluto is assigned here on the basis of its low density. All these outer ("low-density") planets are orbited by satellites, three (Saturn, Jupiter, and Uranus) possess orbiting rings, and two (Jupiter and Saturn) could well be considered as tiny systems resembling the Solar System.

Spacecraft have recently penetrated out as far as Jupiter and are en route to Saturn. As a result, a flood of exciting new information is becoming available about the other planets.

### Jupiter

Jupiter is by far the largest planet in the Solar System. It contains almost two and a half times the mass of all the other planets put together. Because of its enormous mass, Jupiter probably has retained its original atmosphere. This is in contrast to the smaller planets, where because of the low gravitational attractions, the lighter gases in the original atmospheres escaped. In fact, Jupiter's density is relatively low compared with its mass; this situation exists precisely because Jupiter is composed mainly of the lighter elements, especially hydrogen and helium (Figure 20.19). The area of liquid metallic hydrogen is probably the source of Jupiter's strong magnetic field, and a radiation of internal heat is evident; such radiation may indicate nuclear decay, but could also result from phase changes or stored heat. Jupiter gives off twice as much heat as it receives from the Sun.

Views of Jupiter display only by atmospheric bands — the actual surface of the planet cannot be seen. Jupiter rotates very slowly, and the gaseous bands circle the planet at different speeds and altitudes. The alternating light and dark bands are called zones and belts, respectively, and range from white and yellow to several shades of brown. The light zones are made up of clouds of ammonia crystals suspended in gaseous hydrogen. The prominent Great Red Spot, larger in diameter than the Earth, (Figure 20.20) is probably a giant storm similar to a hurricane; it has persisted for 300 years. A much smaller spot with a life of about two years has been observed recently on Jupiter, supporting the theory that the Great Red Spot could be a gigantic storm that may not last forever.

Voyagers 1 and 2 passed close enough to Jupiter and five of its largest satellites to raise the number of known satellites from 12 to 15 and to

**20.20 Jupiter's Great Red Spot,** viewed from Pioneer 11 at a range of 1,100,000 kilometers, data received on 2 December 1976. View shows other oval spots that are smaller and less prominent than the Great Red Spot and are situated at higher latitudes both to the N and S. (National Aeronautics and Space Administration, photo A74-9200)

**20.21 Full-disk image of Io,** made from several frames by Voyager 1 on 4 March 1979 at a range of about 862,000 kilometers. The circular feature at center is an active volcano; other similar features can be seen. Both explosive and quiet eruptions have been confirmed. (National Aeronautics and Space Administration)

transmit clear visual images of the surfaces of these naked satellites. And such images as these are! Outward from Jupiter, the five satellites from which visual images have been newly transmitted back to Earth are Almathea, Io, Europa, Ganymede, and Callisto. Because these satellites were discovered by Galileo, astronomers commonly refer to them collectively as the Galilean satellites of Jupiter.

Io is remarkable for its molten-sulfur volcanic activity (Figure 20.21). Io thus joins Earth as being one of the two bodies in the Solar System known to be undergoing contemporary active volcanism. Europa's surface is crisscrossed by linear dark features, 20 to 300 kilometers wide and extending for more than 1000 kilometers (Figure 20.22). Ganymede features complex patterns consisting of parallel sets, 50 to 200 kilometers wide, of wavy, shallow grooves, each a few kilometers wide. In between these sets of curved grooves the surface is cratered (Figure 20.23). Callisto is remarkable for the completness of its cratered terrain (Figure 20.24). One large impact crater covers most of one hemisphere of this satellite whose equatorial diameter is 4840 kilometers.

In addition, Jupiter appears to be surrounded by a thin ring about 30 kilometers thick. The composition and origin of the ring are still not known, but it is probably made up of boulder-size material.

### Saturn

Saturn has 15 known satellites, now equal to Jupiter's 15, is second in size behind Jupiter, and displays colored atmospheric bands similar to Jupiter's. Saturn's most distinctive feature is its system of four rings (see Plate 20). Saturn's rings are thought to be made up of particles of snowlike ice, having diameters of 4 to 30 centimeters. One possible explanation for Saturn's thin rings is that they were derived from particles that were too close to Saturn's powerful gravitational attraction to form a satellite. Another possibility is that the concentric rings are the remains of a satellite that broke up when it came too close to Saturn's disruptive field of gravity.

To account for Saturn's low density — the lowest of all the planets, despite its huge size — scientists speculate that Saturn is composed mostly of hydrogen and helium, but does possess a small, hot core of metal. Saturn's density is so low, about the same as maple wood, that if the planet could be placed in a tub of water, it would float.

Saturn's largest satellite, Titan, is enveloped by an atmosphere containing hydrogen. Titan is the Solar System's only satellite possessing an atmosphere.

**20.22 Surface of Europa,** image made by Voyager 2 on 9 July 1979 from a range of 241,000 kilometers. The density of Europa suggests that it contains much ice. The complex pattern of fractures suggests that the icy surface has been fractured and the cracks were later filled with the dark material. The virtual absence of impact craters suggests that the surface has been modified since an early period when Europa, as the other bodies of the Solar System, were intensely bombarded. (National Aeronautics and Space Administration, photo 79-H-397)

### Uranus, Neptune, and Pluto

Little is known about Uranus, Neptune, and Pluto, mainly because they are so far away, reflect so little light from the Sun, and have not yet been visited by spacecraft.

The Uranian year is 84 Earth years long. During each Uranian year, the north and south poles face toward the Sun for 42 continuous years. The poles face the Sun because Uranus, unlike any other planet, rotates on an axis that is nearly parallel to the Mean Solar-System Plane rather than on an axis that is inclined at a large angle to the Mean Solar-System Plane.

But recently, major changes in the atmosphere of Uranus have been discovered, and perhaps an even more dramatic discovery has been made: Uranus, like Saturn, possesses a system of orbital rings. The substance of these rings is not yet known. Besides its newly discovered rings Uranus also has faint bands similar to those of Jupiter and Saturn.

During the past 10 years, radio emissions from within the atmosphere of Uranus have become 30 percent stronger. One possible cause is that the Uranian atmosphere is simply getting warmer, although a 30 percent increase in temperature in just 10 years does not seem likely. Another possibility is that the radio beams from Earth are detecting warmer areas of Uranus as the planet begins to turn its north pole toward the Sun after 42 years of darkness on that region of Uranus. If the polar-orbital hypothesis is correct, then the increased radio emissions may be caused by newly exposed warm areas as the north pole turns toward the Sun.

Neptune may have a deep atmosphere, and the variations in Pluto's brightness may be caused by the nature of its rotation and uneven surface features. Uranus and Neptune are both giant planets, and both are orbited by satellites; Pluto is smaller than the Earth and has a recently discovered satellite. Frozen methane has been detected on Pluto, and the presence of methane has been observed in the atmospheres of Uranus and Neptune; it is presumed that hydrogen and helium are present in the atmosphere of all the outer planets. Internally, Uranus and Neptune probably have rocky cores surrounded by ice and gaseous hydrogen.

Although Pluto is usually thought of as the planet most distant from the Sun, Pluto's elliptical orbit is moving that planet closer to the Sun temporarily. In fact, until about 1999, Pluto will be traveling closer to the Sun than Neptune's obtial path (see Figure 1.1c).

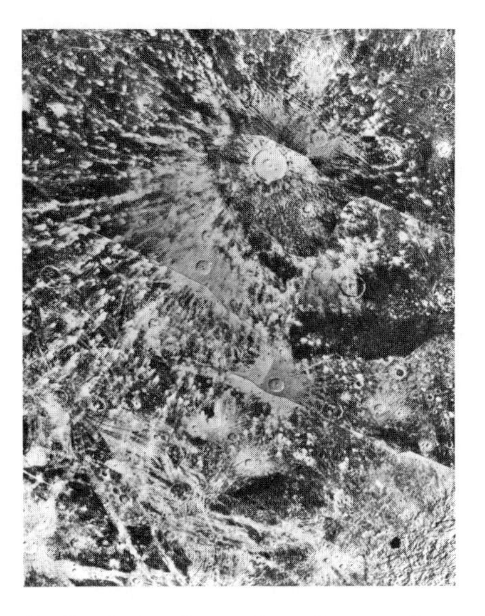

**20.23 Cratered, rough, and grooved surface of Ganymede,** a mosaic made on 9 July 1979 from Voyager 2 at a range of about 100,000 kilometers. The rough, mountainous terrain at lower right is the outer part of a large, fresh impact basin that postdates most of the other surface features. Within the densely cratered area in upper part of the view are black patches that are thought to be ancient icy material formed before the grooves were made. The diameter of the large crater at the center of the ray pattern in upper center is about 150 kilometers. (National Aeronautics and Space Administration, photo no. 79-H-502)

## SIGNIFICANCE FOR EARLY STAGES OF THE EARTH

The Earth is thought to have begun as a large accumulation of smaller particles. As more-and-more particles hit the Earth, they converted their energy of motion into heat, and the proto-Earth heated up. Further, the compression of the interior of the Earth by gravity produced more heat. By far the most important source of heat, however, is thought to have been from the radioactive decay of elements within the Earth, which is still in progress. Within a billion years after it had formed out of the cloud, the temperature of our planet reached the melting point of iron, a very common element which forms about one-third of the Earth. Throughout the globe, iron was melted into enormous "blobs." Because iron is denser than other common elements, it tended to differentiate, or sink toward the center of the Earth, displacing such lighter elements as silicon and aluminum. This melting and sinking of iron is thought to have been a catastrophic event. The sinking iron released large amounts of energy which heated the Earth to a temperature of perhaps 2000°C. This "iron catastrophe" was probably responsible for melting much of the Earth.

As the melting occurred, the various elements forming the Earth began to shift. As iron sank to the core, the less-dense substances rose to create the mantle and crust. A few elements, such as uranium, which are very dense in their pure forms, were also carried upward because they tend to form light compounds. The easily melted substances, especially feldspars, rose to the surface. Substances having slightly higher melting

**20.24 Densely cratered icy surface of Callisto,** photomosaic made from Voyager 2 at a range of 330,000 kilometers on 7 July 1979. Callisto is the most intensely cratered object yet seen in the Solar System. A few of the craters lie at the center of radial patterns of rays. At left is a probable impact structure consisting of a central bright spot and 15 concentric rings. (National Aeronautics and Space Administration, photo No. 79-H-378)

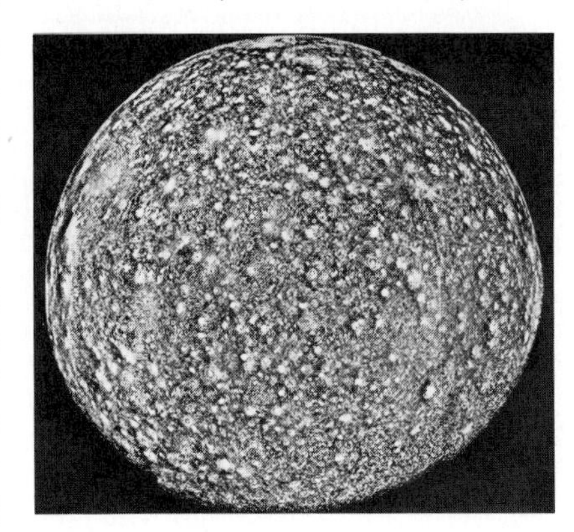

points, including the mafic minerals, became the Earth's mantle (see Figure 1.12).

The melting and differentiation is thought to have been completed by about 4 billion years ago. By then, the continents as we know them began to form. The oldest known rocks are about 3.8 billion years old. If we were transported back in time to that distant age, many of the geologic features would be familiar to us. Living organisms, of course, still would not have begun their arduous and fascinating journey through time to the present.

The Earth's surface was probably once as pitted as the Moon's, but after the period of intense bombardment of Earth from space had ceased, the erosive effects of air, water, and volcanic activity were able to erase permanently the traces of the craters formed when meteorites crashed into the Earth. Some fairly recent meteorite craters do exist, however, most notably the Arizona meteorite crater (Figure 20.25), which probably resulted from an impact about 30,000 years ago. Figure 9.7 shows the areas of North America where colliding meteorites are inferred to have formed craters (astroblemes) of varying sizes.

If comparisons are valid among the terrestrial planets, then it is possible to establish a kind of evolutionary sequence, which begins with melting and internal differentiation to segregate a core. After that comes a period of intense bombardment and formation of impact craters on the rocky surface. If no atmosphere was created or retained, action largely stopped with the craters. On Mars and Earth, at least, and possibly also on Venus, other processes have operated to keep the surface in a constant state of flux. If the Earth were ever as cratered as Mercury, for example, then all traces of this early condition have been obliterated by the operation of the geologic cycle. Thus, our sophisticated space-age research fully supports James Hutton's remarkable statement, as written by John Playfair: "I see no mark of a beginning." Such Earthly "beginning marks," sought in vain by James Hutton, may have indeed existed as impact craters, but erosion, sedimentation, and the operation of the geologic cycle and plate tectonics have erased all traces of them.

**20.25 Meteorite crater, Arizona,** photographed from an airplane looking west-southwest. The diameter of this crater is 1200 meters; its depth, about 180 meters; and the height of the rim above the surrounding plain is 60 meters. The white spots on the crater floor are from drill holes made by prospectors seeking a bonanza of nickel and iron from what they thought would be a buried metallic meteorite. The project was abandoned when the conclusion was reached that the meteorite had dissipated as a result of impact shock and the resulting explosion. (American Meteorite Laboratory, Denver, Colorado)

# CHAPTER REVIEW

1. The Solar System consists of two sets of bodies having contrasting densities: the inner *terrestrial* or Earth-like ("rocky") planets including Mercury, Venus, the Earth and its Moon, and Mars; and the outer, or giant planets having low densities (the "icy" planets), including Jupiter, Saturn, Uranus, Neptune, and Pluto.
2. Because the Moon has no significant atmosphere, it also lacks surface water, rain, glaciers, and winds; hence, lunar erosion is not very prominent.
3. The three main surface features of the Moon are the dark *maria,* the lighter *highlands,* and rugged *craters.*
4. All the rocks returned to Earth by Apollo astronauts have been igneous rocks, which may be placed in three categories: *basalt, KREEP norite,* and *anorthosites.* The maria are made up of basalt, and the highlands are composed of KREEP norite and anorthosites.
5. The outer zones of the Moon are radioactive. The interior of the Moon probably consists of a rigid crust and outer mantle. Most moonquakes occur between the lithosphere and asthenosphere, and may respond to tidal influences. The core of the Moon is probably solid iron, but it may be partly fluid.
6. The Moon probably underwent six evolutionary stages: (1) surface melting, (2) crustal differentiation through *crystal fractionation,* (3) origin of KREEP norite when deep crust melted partially, (4) intense bombardment, forming craters, (5) origin of mare basalts when upper mantle melted partially, and (6) quiescence.
7. *Mercury* probably possesses a relatively thin silicate mantle and crust and an iron core equal to three-fourths its total diameter. No water or life occurs on Mercury. Mercury's surface is spattered with craters. The interior of Mercury is thought to be similar to the Earth's, and there is a dipolar magnetic field.
8. The atmosphere of *Venus* is extremely dense and consists mostly of carbon dioxide. The surface of Venus shows effects of great chemical corrosion, and the surface pressure and temperature are very high. Venus probably possesses a rocky exterior and partially molten iron interior, but it has no noticeable magnetic field.
9. The atmosphere of *Mars* contains water vapor, generates some cloud cover, and shows daily and seasonal wind patterns; seasonal dust storms are prominent. Weathering on Mars is dependent mostly on wind erosion and sedimentation. The only forms of water on Mars now are water vapor and ice, but there is evidence that running water was once a major erosional factor; most of the Martian ice is stored in its polar ice caps. Martian rocks vary from basalt to breccia, and all appear to be volcanic in origin. No life exists on Mars.
10. *Jupiter* is the largest planet in the Solar System, but its density is relatively low. Views of Jupiter show atmospheric bands. The Great Red Spot on Jupiter's surface may be a giant storm.
11. *Saturn's* most distinctive feature is its system of four rings. Saturn has the lowest density of all the planets, and scientists speculate that it is composed mostly of hydrogen and helium, with a small, hot core of metal.
12. Little is known about *Uranus, Neptune,* and *Pluto,* mainly because they are so far away, reflect so little light from the Sun, and have not yet been visited by spacecraft. Internally, Uranus and Neptune probably have rocky cores surrounded by ice and gaseous hydrogen.

## QUESTIONS

1. List and comment on some of the methods available for studying the planetary bodies of the Solar System.
2. Summarize the distribution of mean density among the planets of the Solar System.
3. Does the Moon possess an atmosphere? How does this affect the Moon's surface?
4. Describe the surface of the Moon. What processes have shaped the Moon's surface?
5. What kinds of rocks have been found on the Moon? What is their age range?
6. Compare and contrast lunar regolith with the soil on Earth.
7. What is the interior of the Moon thought to be like? How does the measured lunar heat flow compare with that predicted? With heat flow on the Earth?
8. Summarize the six stages in the geologic history of the Moon.
9. Compare and contrast Mercury with the Moon and with the Earth.
10. What is known about the atmosphere and the surface of Venus?
11. Summarize the atmosphere and surface of Mars. Compare Mars with the Earth, the Moon, Mercury, and Venus.
12. Compare and contrast the generalized relationships of size, mean density, satellites, and orbiting rings among the five outer planets of the Solar System.
13. List the Galilean satellites of Jupiter and describe the distinctive features displayed on visual images recorded from them.

**14.** Summarize how knowledge of the terrestrial planets can be applied toward understanding the early history of the Earth. Do you think James Hutton would have been upset had he known what we do about the Solar System as it effects the Earth?

## RECOMMENDED READING


Allen, J. P., 1972, Apollo 15: Scientific Journey to Hadley-Apennine: *American Scientist,* v. 60, no. 2, p. 162–174.

Arvidson, R. E., Binder, A. B., and Jones, K. L., 1978, The Surface of Mars: *Scientific American,* v. 238, no. 3, p. 76–89.

Baker, David, 1978, *The Larousse Guide to Astronomy:* New York, Larousse and Co., Inc., 288 p. (paperback)

Gore, Rick, 1977, Sifting for Life in the Sands of Mars: *National Geographic Magazine,* v. 151, no. 1, p. 9–31.

Hartmann, W. K., 1980, Moons of the Outer Solar System Become Real, Although Weird Places: *Smithsonian,* v. 10, no. 10, p. 36–47.

Head, J. W., Wood, C. A., and Mutch, T. A., 1977, Geologic Evolution of the Terrestrial Planets: *American Scientist,* v. 65, no. 1, p. 21–29.

King, E. A., 1977, The Origin of Tektites: A Brief Review: *American Scientist,* v. 65, no. 2, p. 212–219.

Mutch, T. A., Arvidson, R. E., Head, J. W. III, Jones, K. L., and Saunders, R. S., 1977, *The Geology of Mars:* Princeton, New Jersey, Princeton University Press, 400 p.

Pollack, J. B., 1975, Mars: *Scientific American,* v. 233, no. 3, p. 107–112.

Runcorn, S. K., 1978, The Ancient Lunar Core Dynamo: *Science,* v. 199, p. 771–773.

Soderblom, L. A., 1980, The Galilean Moons of Jupiter: *Scientific American,* v. 242, no. 1, p. 88–100.

# APPENDIX A
# THE CHEMISTRY OF GEOLOGY

## MATTER

A general definition of *matter* is anything that possesses mass and occupies volume. Matter can be studied on many levels. In some cases, the important property of matter is simply whether it is a solid or not. The condition of matter, that is, whether matter is a solid, a liquid, or a gas, is referred to as the *state of aggregation* of the matter. Many important lessons about geology depend on the state of aggregation of matter. As one desires to be more and more specific about matter, one can try to understand how matter behaves in terms of its finer and finer structure. In the following sections, we review some aspects of matter that are especially significant for understanding geology. We include structure of matter, states of aggregation, bonding, and radioactivity.

### Structure of Matter

Many aspects of matter and energy can be understood only by means of an analysis of the smaller-and-smaller building blocks of which matter is composed. As long ago as 400 B.C., the Greek philosopher Democritus (470 or 460 B.C. to 380 or 390 B.C.) proposed that all matter was composed of tiny indivisible particles called atoms. Later, scientists found that individual atoms are so small that if one atom were enlarged 100 million times ($10^8$ times), it would be only the size of a pea, and its subatomic particles would not be visible, even with a microscope. In the twentieth century, scientists began to realize that atoms display characteristic structures.

**Structure of atoms** The structure of atoms has been described in terms of concentric spheres. At the center, most of the mass is concentrated in the *nucleus*. The nucleus is so dense that it holds 99.95 percent or more of the atom. Within an atomic nucleus are one or more protons and possibly also neutrons. A proton is so tiny that to make a mass weighing 0.5 kilogram would take 270 trillion trillion protons ($270 \times 10^{24}$).

Surrounding the nucleus are one or more electrons arranged in one or more *energy-level shells*. These shells are defined by the orbital paths of the electrons. Considering that the orbiting electrons travel at the speed of light ($297.6 \times 10^3$ kilometers per second; 186,000 miles per second), it is a wonder that they stay in their energy-level shells. They travel so fast within such a tiny orbit that if one could tag their trails in some way, the result would be a hollow sphere.

Atoms are described by two numbers that show the kinds and numbers of particles in the nucleus. The *mass number* gives the number of neutrons plus protons. The *atomic number* designates the number of protons (Figure A.1). The simplest atom is of the gas hydrogen; an atom of hydrogen consists of one proton and one electron (Figure A.2). Its mass number is 1.

The nucleus of a hydrogen atom is unbelievably small. For example, if we were to build a scale model of a hydrogen atom with the radius of curvature of the electron's energy-level shell equal to that of the roof of the Houston Astrodome, then the nucleus would be represented by a marble placed on second base. Our scaled-up model illustrates another important aspect of matter: despite the tiny sizes of atoms, they consist mostly of empty space.

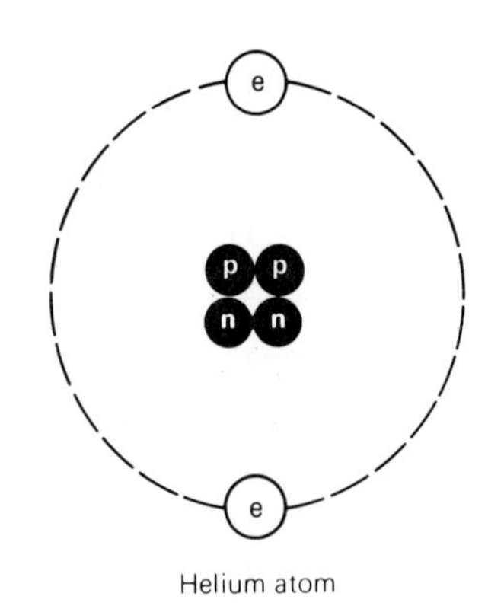


**A.1 Helium atom,** schematic sketch according to the Bohr model with electrons shown in simple concentric circles surrounding the nucleus. Protons in blue, neutrons, gray. Sizes of electrons greatly enlarged.

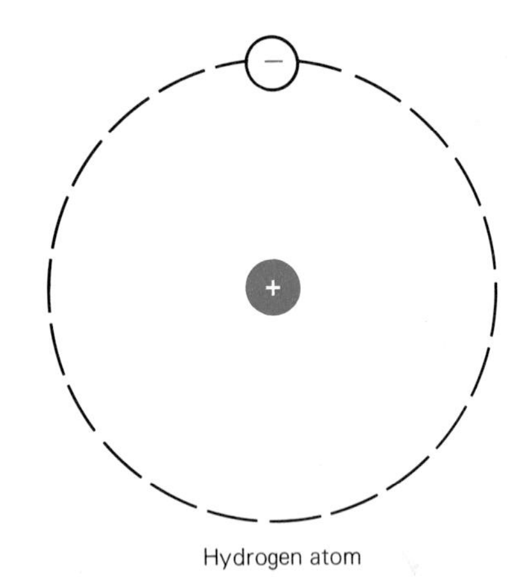


A.2 **Hydrogen atom,** schematic sketch according to the Bohr model.

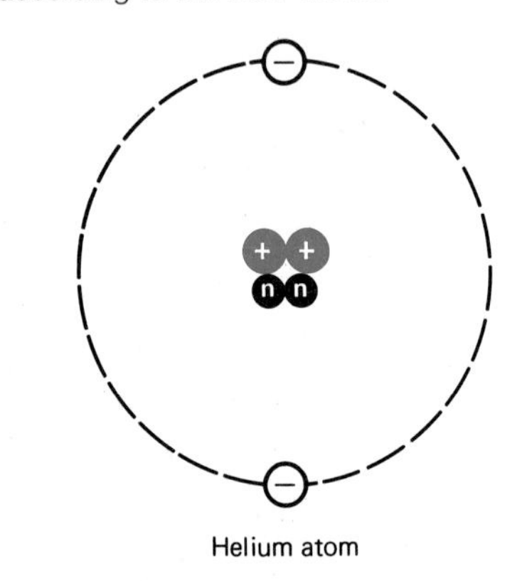


(a)

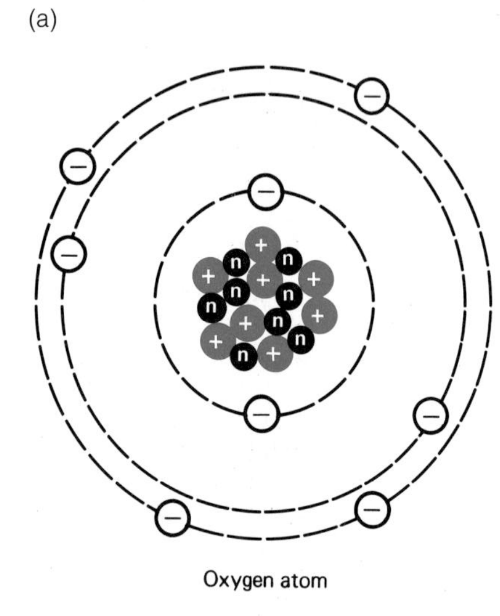


(b)

**A.3 Helium and oxygen atoms,** schematic sketches according to the Bohr model, showing equal numbers of protons in the nucleus and electrons in the orbits. (a) Helium atom, atomic number 2, mass number 4, 2 electrons. (b) Oxygen atom, atomic number 8, mass number 16, 8 electrons.

The next-larger atom after hydrogen is helium (Figure A.3a). A helium nucleus contains two protons and two neutrons; thus, its mass number is 4 and its atomic number is 2. The nucleus of oxygen contains 8 protons and 8 neutrons. Accordingly, the mass number of oxygen is 16 (8 + 8), and the atomic number is 8 (Figure A.3b).

**Elements and isotopes** All atoms having the same numbers of protons are considered to be the same kind of matter; they are named *elements.* Notice that the number of neutrons does not enter into the definition of an element. A variable number of neutrons can be associated with a particular number of protons. This gives rise to varieties of the element; each variety is characterized by its own distinctive mass number (which varies as the number of neutrons changes). Atomic nuclei having the same number of protons but a different number of neutrons are termed *isotopes* of a given element. For example, three isotopes of the element oxygen are known. The kind of oxygen we have mentioned previously is named oxygen-16 and abbreviated $^{16}O$. The isotope oxygen-17 ($^{17}O$) contains 8 protons and 9 neutrons; its mass number is 17 (8 for the protons + 9 for the neutrons). The isotope oxygen-18 ($^{18}O$) contains 8 protons and 10 neutrons; its mass number is 18 (Figure A.4).

**Electrical charges and ions** Electrical charge is a property somewhat akin to magnetism in that two opposite conditions are known: the likes repel and the opposites attract (Figure A.5). The intensity of the electrical charge may vary. Among subatomic particles we have been discussing, protons bear a unit-positive electrical charge, and electrons, a unit-negative electrical charge. Thus, the combination of one proton with one electron establishes a condition of electrical neutrality—the one unit positive charge of the proton is balanced by the one unit negative charge of the electron. Notice that these electrical charges are independent of mass. The single (+) charge of a dense proton is exactly neutralized by the single (−) charge of a low-mass electron. Neutrons, as their name suggests, lack electrical charge. Under certain conditions, a neutron within an atomic nucleus may break down to form a proton and an electron.

In an atom, the number of protons of the nucleus is exactly balanced by the number of electrons in the orbital shells. This balance establishes electrical neutrality and the electrical neutrality tends to inhibit atoms from taking part in chemical reactions.

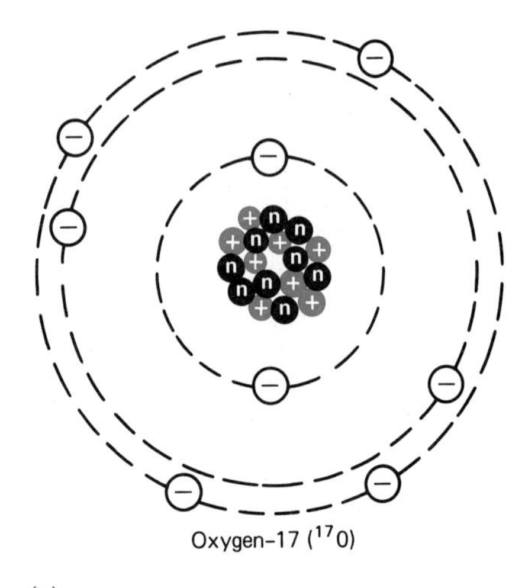


(a)

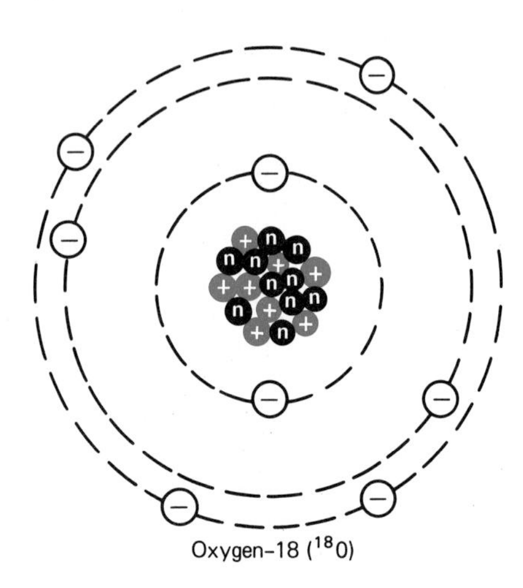


(b)

**A.4 Oxygen isotopes,** schematic sketches according to Bohr model. (a) Oxygen-17; nucleus contains 8 protons and 9 neutrons. (b) Oxygen-18; nucleus contains 8 protons and 10 neutrons. Placement of protons and neutrons in the nucleus and of electrons in circular orbits is schematic.

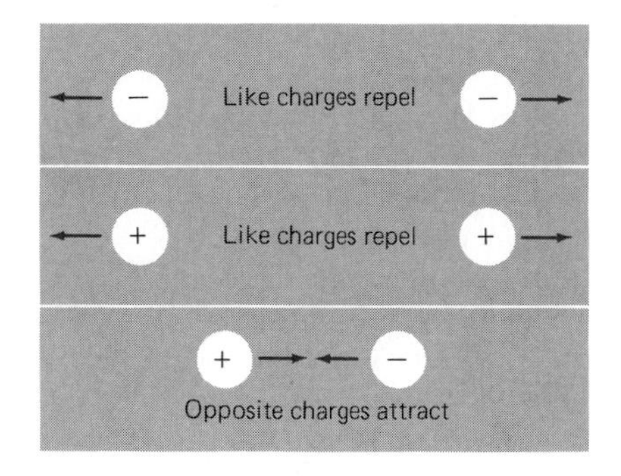


**A.5 Behavior of electrical charges.** Further explanation in text.

Under certain conditions, the number of orbiting electrons does not equal the number of protons in the nucleus. When this happens, the result is a charged particle known as an *ion*. If the number of electrons in the orbital shells exceeds the number of protons in the nucleus, the sign of the ion's charge will be negative, and the amount will be equal to the number of excess electrons. Thus, two excess electrons give a charge of $2^-$; three excess electrons, a charge of $3^-$; and so forth. If the number of electrons in the orbital shells is less than the number of protons in the nucleus, the sign of the ion's charge will be positive, and the amount will be equal to the number of deficient electrons. Thus, two deficient electrons give a charge of $2^+$; three deficient electrons, a charge of $3^+$; and so forth.

For designating ions, chemists have adopted a standard notation that we will use. An ion is indicated by the letter abbreviations for the one or more chemical elements involved. Following the letter abbreviations are numbers and plus signs or minus signs, indicating the electrical charges of the ions. These are written as superscripts. For example, the ion of silicon contains four positive charges; it is written: $Si^{4+}$. Ions of oxygen, containing two negative charges, are written: $O^{2-}$. If the ion bears only a single charge, then it is written with only a plus sign or a minus sign as the superscript. For example, the notation for hydrogen ions, which contain one positive charge, is $H^+$ and that of chlorine ions, which possess one negative charge, is $Cl^-$.

Many elements form only a single kind of ion. That is, all ions of that element bear the same electrical charges. For example, the only ions of sodium are $Na^+$. Because of the variable situations with their electron shells, however, some elements can form more than one kind of ion. For example, two ions of iron are $Fe^{2+}$ and $Fe^{3+}$; these are known as ferrous iron and ferric iron, respectively.

Ion complexes are composed of more than one element. Examples include the hydroxyl-ion complex $(OH)^-$, sulfate-ion complex $(SO_4)^{2-}$, and carbonate-ion complex $(CO_3)^{2-}$.

**Energy-level shells** The basic concentric structure of atoms and ions is displayed by both the protons and neutrons in the nucleus and by the orbital electrons.

In the nucleus, the numbers of protons and neutrons required to fill complete shells are 2, 8, 20, 28, 50, 28, and 126. Helium is an example of a filled 2-shell. $^{16}$Oxygen is an example of a filled 8-shell, and $^{40}$calcium, of a filled 20-shell.

Electrons are organized into energy-level shells surrounding the nucleus. The first shell outside the nucleus is the 2-shell; in it, two electron combinations are possible (Figure A.6). The next-outward shell is the 8-shell; in it, 8 combinations are possible, with subshells of 2 and 6. The next two shells are the third, or 18-shell (with subshells 2, 6, and 10) and the fourth, or 32-shell (with subshells of 2, 6, 10, and 14). The fourth shell is the outermost one in which known elements exist with full shells (Figure A.7). Shells 5, 6, and 7 are known, but

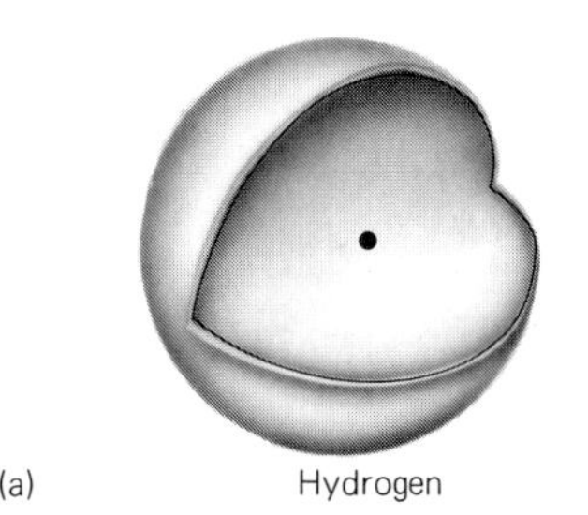


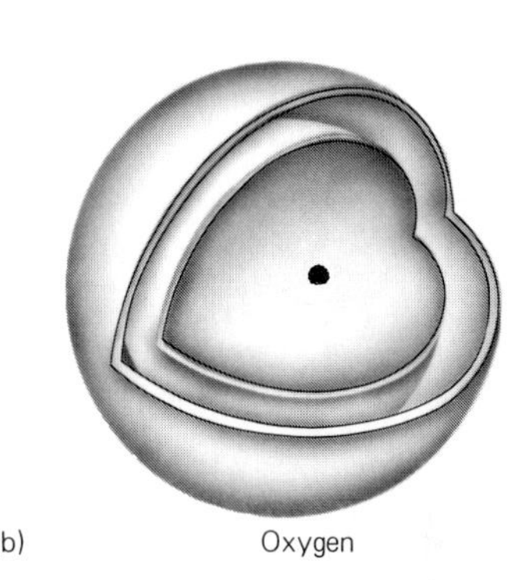


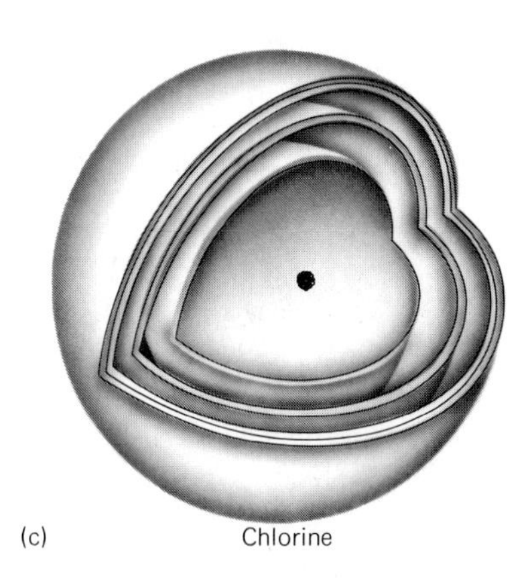


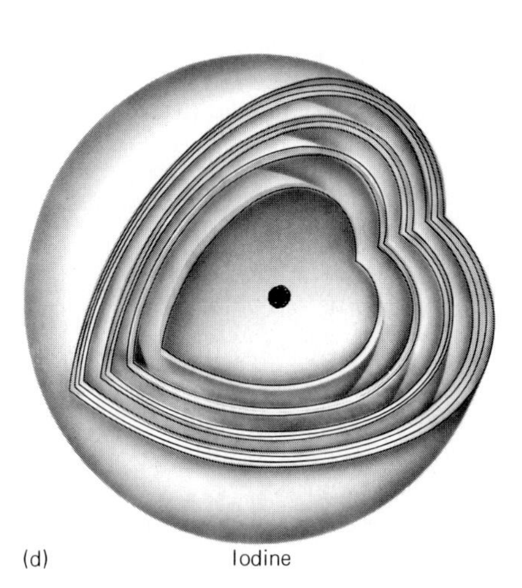


**A.6 Atoms; orbits of electrons shown as energy-level shells.** (a) Hydrogen atom with first energy-level shell containing only one electron. (b) Oxygen ion, with two energy-level shells, the second one containing two subshells. (c) Chlorine atom, with three energy-level shells, the third one containing three subshells. (d) Iodine atom, with four energy-level shells, the fourth one containing four subshells.

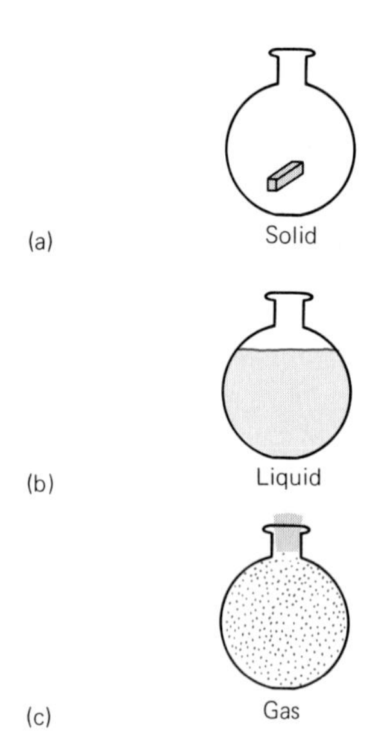


**A.7 A solid, a liquid, and a gas in a container.** Further explanation in text.

not with complete fillings (see Figure A.6). In the atomic "shell game," certain numbers of particles fill a shell and this tends to promote stability. Any ion whose energy-level shells are not filled tends to be more reactive than an ion having filled energy-level shells.

**Molecules and compounds** Many properties of matter can be explained by reference to clusters of two or more ions of one or more elements. For example, in oxygen gas, the fundamental units are pairs of oxygen ions that are joined together so that they behave as a unit. Such bound-together ions that function as units are *molecules*. Molecules of gas are the smallest units having the properties of that gas. Ions of different elements may combine in a liquid to make molecules of *compounds*. When an ion of sodium joins an ion of chlorine, the result is a molecule of the compound sodium chloride. This compound, known to most people as table salt and to geologists as *halite*, is not at all like either of the elements composing the molecules, the metal sodium and the gas chlorine. The molecules of some natural compounds contain dozens of ions. In summary, then, molecules are the smallest units having the properties of materials named compounds.

The coming together of ions to form molecules involves only the electron configuration in the energy-level shells; the nuclei of the ions are not affected. Changes in the energy-level shells do not involve large amounts of energy; therefore, such changes are quite common.

## States of Aggregation

For many purposes, the state of aggregation of matter can be defined by simple properties that are readily observable. Two states can be defined just by finding out whether or not matter can retain its own shape. Using this test, it is easy at first glance, at least, to distinguish solids from fluids. Solids retain their shapes, whereas fluids do not, but rather take on the shape of the container that holds them (Figure A.7a). Fluids, in turn, can be subdivided into liquids and gases. Liquids display a free surface within their containers, whereas gases do not and tend to occupy the entire container (Figure A.7b). A closer look will show the theoretical basis for these distinctions and indicate that situations exist in which these simple definitions require considerable explanation.

**Solids** The realm in geology dominated by solids is the lithosphere. We can consider that solids consist of rigid networks of ions. The ions of solids are packed close together and tend to remain in place. The properties of a solid depend on what kinds of ions are present and how they are packed.

A few simple tests of solids include mean density, elastic properties, and speeds of passage of various kinds of waves (Appendix D). *Mean density* is found by dividing the mass of a body of matter by its volume. Two contrasting density groups important to geology are most metals in a dense group and most nonmetals in a less-dense group. We discuss these and other properties of solids in Appendix D.

**Fluids: liquids and gases** The molecules of fluids do not form rigid networks, as do the ions in solids. Because fluids lack rigidity, they tend to flow. In geology, the flow of fluids is a fundamentally important process. Flowing fluids, such as the waters of rivers, erode valleys, characteristic features of the Earth's surface.

In a body of matter, the arrangement of the ions is an expression of the level of energy to which the matter is subject. Commonly, we can express this level of energy in the form of temperature. For example, consider water. At what most of us would consider comfortable temperatures, water is in liquid form. In liquid water, the ions and molecules are close together, but do not form a rigid network. However, at cold temperatures, defined as those below the freezing point, the ions form a rigid network; the water solidifies and becomes ice. At high temperatures, defined as those above the boiling point, all water liquid disappears and the water becomes a gas. In the gaseous state, the water's molecules are not so close together as in liquids and solids and the molecules move in random paths. The importance of changes of state in geology will become progressively more apparent as our study continues.

# BONDING

Many important properties of a solid, especially a mineral, are governed by the extent to which its ions cling together. Various kinds of bonds have been identified; their characteristics are related to the behavior of the electrons of the ions or to various electrostatic effects. Although the individual kinds of bonds display diagnostic characteristics, in a given mineral, more than one kind of bond may be present, and in most minerals, transitional kinds of bonds have been found. The chief kinds of atomic bonds are: covalent, metallic, ionic, hydrogen, and van der Waals'.

## Covalent Bonds

Covalent bonds in three directions yield strong, hard solids having high melting points and low electrical conductivity. Covalent bonds can be visualized only by considering that the orbitals of electrons can overlap. The more the orbitals overlap, the stronger is the bond. The ions most likely to form covalent bonds are those in which the outer electron shells are only half filled, and the electrons to fill the shells are shared. In other words, the shared electrons orbit around the nuclei of two ions. An example of a mineral having strong, covalent bonds is diamond.

### Metallic Bonds

Metallic bonds are so named because these bonds give metals their distinctive properties, such as ability to deform plastically, ductility, malleability, and high electrical conductivity. The basis for metallic bonding is the equal spacing among the nuclei of the ions and the availability of electrons that are able to move through the structure formed by the nuclei. In metallic bonding, the electrons are not confined to orbits around one or two nuclei, but instead are mobile and belong to the whole structure. Such electrons are said to be *delocalized*. Metals, therefore, can be described as a series of positive ions surrounded by numerous delocalized electrons.

### Ionic Bonds

Ionic bonds are formed by the forces of attraction between oppositely charged ions (see Figure A.5, lowest panel). Each ion tends to surround itself with as many oppositely charged ions as it can, the limiting number being determined by the sizes of the ions and the valence of the ions. Thus, the valence number of an ion indicates only the value of its positive or negative charges and does not refer to the number of oppositely charged ions to which it can be attached. Crystals displaying ionic bonding are poor conductors of electricity and of heat; their valence electrons are localized around individual ions.

### Hydrogen Bonds

The bonds we have been discussing previously involve electrons and their configurations in ions. Because the electrical charges are not evenly distributed over the surfaces of ions, other bonds are possible where charges become concentrated. An example is a molecule of water (see Figure 7.6), in which the uneven distribution of surface charges enables the molecules to behave as dipole magnets. The bonding among water molecules is a result of the concentration of charges by the tiny hydrogen ions. As a result, the bonding is named *hydrogen bonding*. Hydrogen bonds give water its distinctive properties of being able to dissolve solids (see Figure 7.8), its high heat capacity, and its ability to expand on freezing. Hydrogen bonds are also common in organic compounds.

### Van der Waals' Bonds

Even though neutral atoms and molecules lack valence electrons for forming covalent bonds, metallic bonds, or ionic bonds, nevertheless the electrical charges are not distributed uniformly but tend to shift as the nuclei and electrons move toward one or the other side of the atom. Displacement of the nucleus toward one side creates a fractional and temporary "excess" positive attraction, whereas shifting of the electrons toward one side creates a temporary excess negative attraction. As a result, an atom can function as a tiny dipole magnet and can be attached, though weakly and temporarily to other dipole atoms. These transitory "dipole-atom" bonds are van der Waals' bonds. They are not strong bonds and are usually only minor factors in the bonding of crystals.

## RADIOACTIVITY

The phenomenon named *radioactivity* is defined as the spontaneous disintegration of the nuclei of certain unstable isotopes, which converts these nuclei into the nuclei of other isotopes. The kinds of isotopes that disintegrate are known as *parent isotopes,* and the isotopes which thereby form are the *daughter isotopes*. Many daughter isotopes are stable; they do not disintegrate further. Other daughter isotopes, however, are themselves radioactive. They thus become parent isotopes and disintegrate into still-other daughter isotopes until a stable end product is formed. The reactions that lead from a radioactive parent isotope to a stable daughter isotope are known as a *radioactive decay series*.

If, for the moment, we skip over the details of the radioactive decay series involving daughter isotopes that become parents, then we can discuss what happens in general terms by considering only the radioactive parent isotopes and the stable daughter isotopes. Clearly, with time, the quantity of parent isotopes diminishes, and correspondingly, the amount of daughter isotopes increases. This relationship is what makes radiometric dating possible. Radioactive dating strategies vary according to the kinds of radioactive reactions that are available.

### Radioactive Reactions

Radioactivity is a process that affects the nuclei of isotopes. This means that radioactivity involves protons and neutrons, the particles that form atomic nuclei. Only a few kinds of reactions affect electrons. According to the kinds of participating nuclear particles, three contrasting kinds of radioactive reactions can be recognized: (1) nuclear fission, in which the nucleus of a heavy isotope splits and in so doing, forms the nuclei of two new isotopes; (2) emission of alpha particles; and (3) the emission or capture of beta particles (a name given to particles that turned out to be ordinary electrons).

**Nuclear fission** In nuclear fission, the splitting of the nucleus of the parent radioactive isotope gives rise to two stable daughter isotopes whose atomic numbers and mass numbers do not bear any systematic relationship to those of the parent isotope (Figure A.8). Nuclear fission has not yet formed the basis for dating mineral specimens in the ways the other two

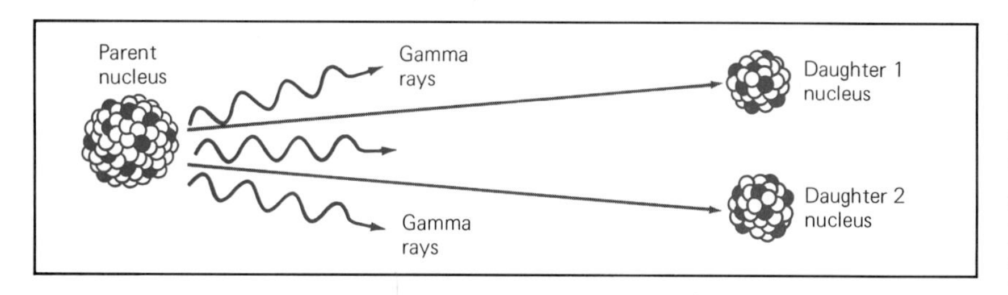


**A.8 Radioactivity involving nuclear fission.** Radioactive parent nucleus at left splits into two daughter nuclei, both having mass numbers exceeding 4; at the same time, the parent nucleus emits gamma rays.

kinds of radioactive reactions have been applied.

**Emission of alpha particles** This causes a parent isotope to become a daughter isotope having an atomic number smaller than the parent by the value of 2, and a mass number smaller than the parent by the amount of 4. These numerical changes result because an alpha particle is the nucleus of an atom of helium; it contains 2 protons and 2 neutrons (Figure A.9).

**Emission or capture of electrons** When the nucleus of a radioactive parent isotope emits or captures a beta particle (electron), the number of protons and neutrons in the nucleus changes by 1, but the mass numbers do not change. That is, the mass number of the parent isotope equals the mass number of the daughter isotope, even though these isotopes are of different elements (Figure A.10).

In the radioactive series involving rubidium, with atomic number 37 and a mass number of 87 ($^{87}Rb$), and strontium, atomic number 38 and a mass number of 87 ($^{87}Sr$), no change of mass takes place. One electron is captured by the nucleus of the parent isotope (37 protons and 50 neutrons) and this converts it into strontium (38 protons and 49 neutrons; Figure A.11). The decay of rubidium to strontium does not involve a gas, and this eliminates one of the possible problem areas in dating. However, this potential advantage of the rubidium-strontium series is offset by the fact that the half life of this reaction is 47 billion years and rubidium is not very abundant or widely distributed. Its chief occurrence is with potassium, thus it appears in lattices of muscovite, biotite, and orthoclase.

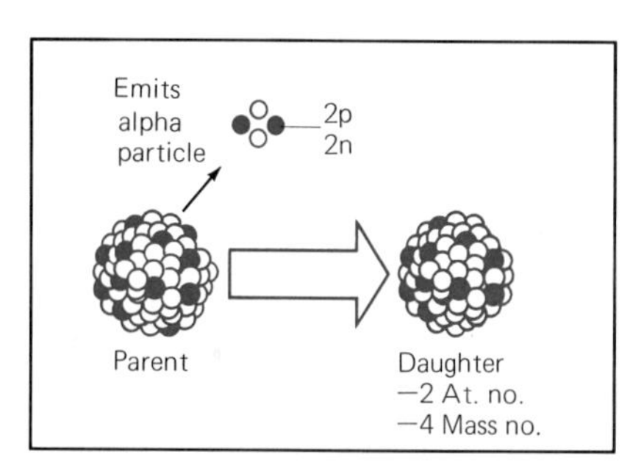


**A.9 Radioactivity involving emission of an alpha particle.** Radioactive parent nucleus at left emits an alpha particle (nucleus of helium atom) and thus transforms itself into a daughter nucleus whose atomic number is two less than the parent and whose mass number is 4 less than the parent.

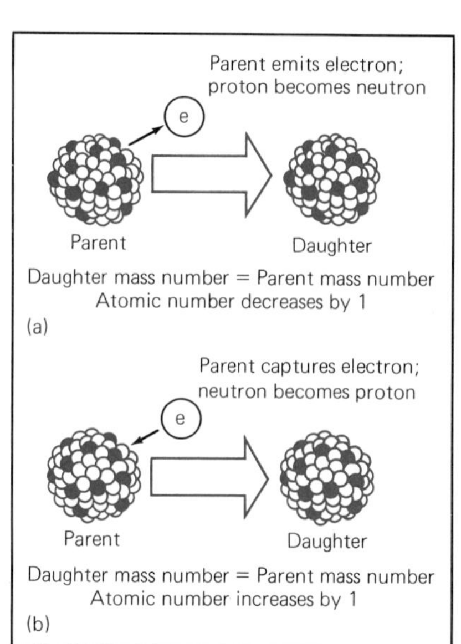


**A.10 Radioactivity by emission or capture of an electron.** (a) Radioactive parent nucleus at left emits an electron, thus becoming a daughter nucleus having the same mass number as the parent, but with an atomic number that is one less than that of the parent, because a proton became a neutron when the electron left the parent nucleus. (b) Radioactive parent nucleus at left captures an electron, becoming a daughter nucleus having the same mass number of the parent but whose atomic number is one larger than the parent's, because a neutron became a proton when the electron was captured.

In the radioactive series that starts with potassium-40 (atomic number 19, mass number 40, thus abbreviated $^{40}K$), two paths are possible (Figure A.12). About 11 per cent of the nuclei of $^{40}K$ give off electrons and thus become $^{40}Ar$. In this reaction, the 19 protons and 21 neutrons of the parent nucleus are converted into 18 protons and 22 neutrons of the daughter nucleus. The remaining 89 per cent of the nuclei of the parent isotope capture electrons. In this reaction, the 19 protons and 21 neutrons of the nucleus of the parent isotope become the 20 protons and 20 neutrons of the daughter isotope, calcium, having a mass number of 40 (abbreviated $^{40}Ca$). The stable end product of the first branch is the inert gas, argon. Geochemists have concluded that the only $^{40}Ar$ present on Earth has resulted from the radioactive decay of $^{40}K$. Because potassium is present in the lattices of many minerals, and because laboratory instruments can measure even the slightest amount of $^{40}Ar$, the method of dating by potassium-argon is very valuable.

The final radioactive series used for dating is based on the decay of radioactive carbon with a mass number of 14 ($^{14}C$) to nitrogen with a mass num-

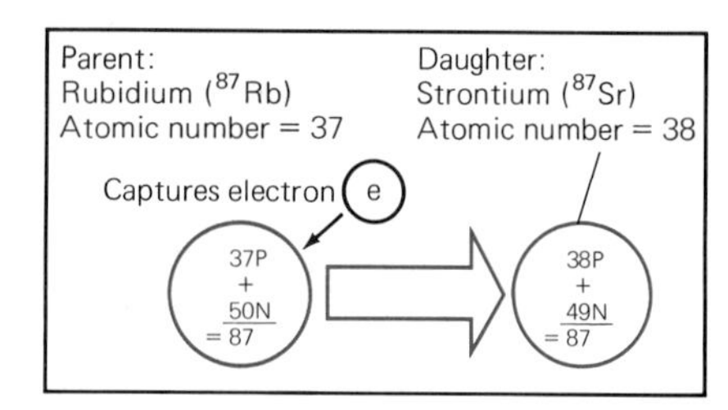


**A.11 By capturing an electron,** a nucleus of rubidium-87 becomes the nucleus of strontium-87.

# APPENDIX B IDENTIFICATION OF MINERALS

This appendix is devoted to minerals. Its main objective is to describe how to identify minerals in small specimens using visual observations and simple tests. Included are tables summarizing the properties of the common minerals. The final section deals with the crystal chemistry of the rock-forming silicate minerals.

## MEGASCOPIC PROPERTIES OF MINERALS

The term *megascopic* refers to features that can be seen plainly by the eye alone or by the aid of a pocket lens that magnifies 10 times or so. Megascopic contrasts with *microscopic,* which refers to features that can be seen only using a microscope.

In the following discussion, we describe some megascopic features of minerals that enable one to identify them, using only visual study and the application of a few simple tests. We begin with the visual features and progress to the simple tests.

### Visual Features of Minerals

The visual features of minerals include their external form, their color, and their luster. In the following paragraphs, we discuss these features and show how they can be made an important part of the systematic study of minerals. The properties of common minerals are summarized in Table B.1.

**External form: crystal faces, breakage features, and striae** One of the first features to notice about a mineral is its external form. Several possibilities exist. First of all, the surface of the mineral may consist of *crystal faces,* regular plane surfaces that have grown under the control of the crystal lattice. We have already mentioned some aspects of crystals. An important point to emphasize here is that crystal faces are the result of *growth* and that the geometric figures they form are the same whether the specimen is too small to be considered megascopic (Figure B.1), is of ordinary size (Figure B.2), or is very large (Figure B.3).

Some distinctive mineral crystal shapes include *cubes* (exemplified by halite, Figure B.4; pyrite, see Figure 4.10a; galena, see Figure 4.10b; and fluorite, Figure B.5), *rhombs* (exemplified by calcite, Figure B.6), *hexagons* (exemplified by quartz, see Figure B.2), and *12-sided figures* (exemplified by garnet, Figure B.7).

**B.1 Tiny crystals of quartz** growing on a fibrous manganese-oxide mineral. Specimens viewed through scanning-electron microscope. (U. S. Geological Survey)

The best-developed and most-spectacular crystals grow where they could enlarge without interference, as along open fractures or cavities in rocks where growth could take place into an empty space. In the Ural Mountains in Russia, a whole mineral quarry was opened on a single crystal of feldspar that was a rectangular solid 10 meters square and of unknown length (depth).

In some mineral specimens, the external form of the specimen is not a simple, direct expression of the crystal lattice, as is the case with crystals. In these other specimens, the regular internal arrangement typical of the lattice of that mineral is present, as it always is, but some factor other than simple unimpeded growth has controlled the external form. In many cases, for example, organisms secrete calcite mineral crystals. The internal arrangement of the calcite crystals is the normal lattice, but the organism is capable of controlling the external form, so that the outcome is an organic shape, not a crystal face. Such crystals having regular lattices internally, but organic shapes on the outside, are known as *biocrystals* (Figure B.8).

**B.2 Quartz of ordinary size and many varieties.** Specimens from Barnard College collection. (Frank Dunand)

1 Hexagonal crystal, smoky quartz.
2 Polished face of chalcedony, filling of a geode.
3 Massive rose quartz; well-developed fracture surface.
4 Hexagonal crystal, clear quartz.
5 Hexagonal crystal, polished slice from end of transparent hexagonal crystal.
6 Massive chalcedony, conchoidal fracture.
7 Small, glassy crystals growing into a cavity.
8 Chalcedony, conchoidal fracture chip.
9 Intergrowth of clear quartz and smoky quartz.
10 Quartz and potassium feldspar crystals.

**B.3 Large quartz crystal.** Specimen from Columbia University collection. (J. E. Sanders)

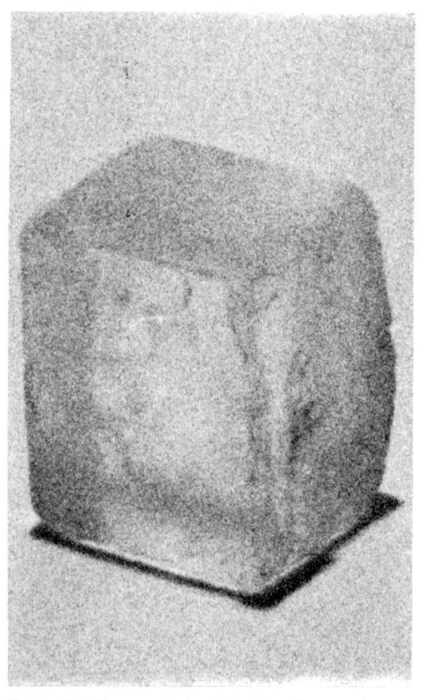

**B.4 Cubic crystal of halite.** Barnard College collection. (Frank Dunand)

**B.5 Cubic fluorite crystals.** Columbia University collection. (J. E. Sanders)

**B.6 Calcite forms rhombs when broken;** each pair of rhomb faces indicates a single cleavage direction. Cleavages in calcite are not at right angles. (Ward's Natural Science Establishment)

**B.7 Garnet, a 12-sided crystal,** each face of which is a parallelogram in a perfect crystal (two dipping faces beneath the paper clip). Barnard College collection. (J. E. Sanders)

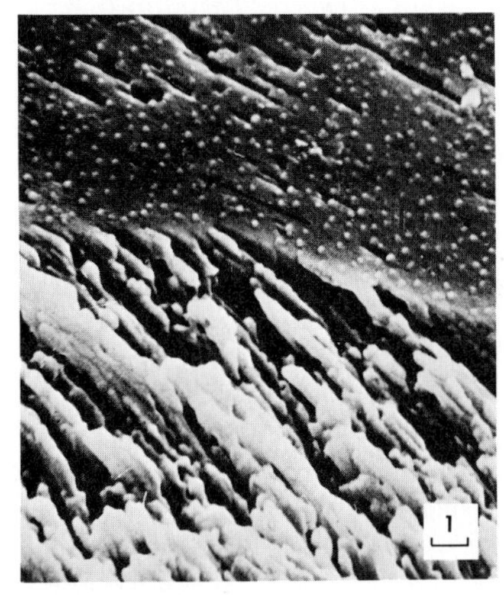


**B.8 Biocrystal;** scanning-electron micrograph reveals external shape built by the organism using building blocks composed of tiny crystals. (Lamont-Doherty Geological Observatory of Columbia University, Courtesy A.W.H.Bé)

**B.9 Botryoidal hematite,** where mineral grew as smooth, overlapping segments of spheres. Barnard College collection. (J. E. Sanders)

Some specimens form smooth, curved shapes resembling a closely packed bunch of grapes. This arrangement is named botryoidal (Figure B.9).

The external form of many specimens has resulted from breakage. The way a mineral breaks is controlled by the bonding among the ions in the lattice (see Appendix A). The bonding may be uniform in all directions or may be stronger along some directions and weaker along others. The breaking of a specimen may yield a splintery surface; smooth, curved surfaces named *conchoidal fractures* (Figure B.10); or plane surfaces that are parallel to lattice planes. Smooth planes formed by breaking that are parallel to planes of weakness in the lattice are named *cleavage planes;* the property of breaking into such smooth planes is known as *cleavage.*

At this point, you may well wonder, "What is the difference between a crystal face and a cleavage plane, and how is one supposed to be able to tell them apart?" This difference is a significant one, and the distinction, though difficult to observe, can be made and you must learn how to do it. The first point to remember is that

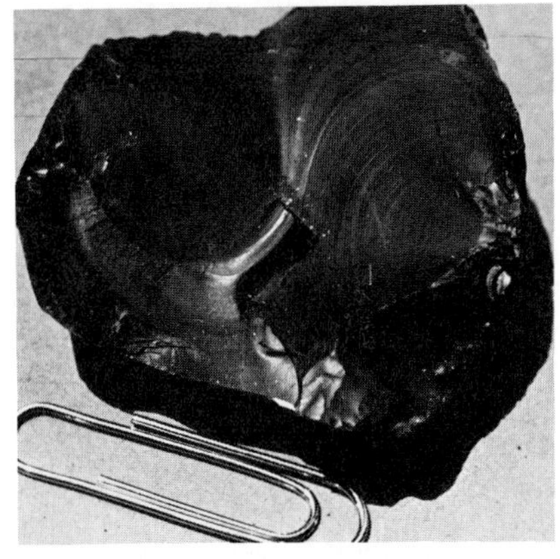

**B.10 Conchoidal fracture,** small specimen of chalcedony. Barnard College collection. (J. E. Sanders)

both crystal faces and cleavage planes are controlled by the lattice. This means that they are related to lattice directions. It also means that the two may coincide in some cases and not coincide in others. Consider the example shown in Figure B.11. In it, a direction of perfect cleavage is parallel to the plane on which the specimen rests. The cleavage of muscovite is so perfect in one direction that the specimen can be split into a series of

**TABLE B.1**
MINERALS AND PROPERTIES

| Mineral | Color | Luster | Form and Internal Features | Breakage | | Streak | Hardness | Magnetism | Chemical Tests | Remarks |
|---|---|---|---|---|---|---|---|---|---|---|
| | | | | Fracture | Cleavage | | | | | |
| **Amphibole** | ● Dark green to black | ● Shiny | | | ● 2, not at right angles | | ● 5 to 6 | | | Distinguished from pyroxene by two cleavages not at right angles. |
| **Biotite** | ● Black | | | | ● 1 perfect; cleavage flakes are flexible, elastic, and tough | | | | | Color and flexibility plus elasticity of cleavage flakes separate biotite from chlorite. |
| **Calcite** | | | | | ● Three, not at right angles | | ● 3 | | ● Effervesces with cold dilute acid | Three cleavages not at right angles, hardness of 3, and effervescence with cold dilute hydrochloric acid are diagnostic |
| **Chlorite** | ● Green | ● Silky | | | ● 1 perfect; cleavage flakes are flexible, soft, but not elastic | | | | | Green mica with cleavage flakes that are flexible but not elastic. |
| **Fluorite** | | ● Vitreous | | | ● Four directions | | ● 4 | | | Resembles calcite; distinguished from calcite by hardness of 4 and lack of reaction with acid. |
| **Galena** | ● Dark gray | ● Metallic; shiny | ● Cubes | | ● 3, at right angles | ● Black | | | | Specific gravity is 7.6. |
| **Garnet** | ● Mostly dark burgundy red | ● Vitreous | | ● Irregular | | | ● 6.5 to 7.5 | | | Red color, great hardness, and lack of cleavage, and color typify garnet |
| **Gypsum** | | ● If silky, then fibrous | | | ● 1 prominent cleavage | | ● 2 | | | Hardness of 2 is diagnostic; no salty taste. |
| **Halite** | | ● Vitreous to greasy | ● Cubes | | ● 3, at right angles | | ● 2.5 | | ● Salty taste | Salty taste is diagnostic |
| **Hematite** | | | | | | ● Reddish brown | | | | Many varieties; streak test is diagnostic for all. |
| **Limonite** | | | | | | ● Yellow-brown | | | | Streak test is diagnostic. |
| **Magnetite** | ● Black | ● Metallic; dull | | | | ● Black | | ● Only mineral that is strongly magnetic | | Use a magnet. |
| **Muscovite** | ● Silvery | ● Vitreous | | | ● 1 perfect; cleavage flakes are tough, elastic | | | | | Colorless ("white") mica. |
| **Olivine** | ● Green | ● Vitreous | | ● Conchoidal fracture | | | ● 6.5 to 7 | | | Green color, conchoidal fracture, vitreous luster, and hardness characterize olivine. |
| **Orthoclase feldspar (including microline)** | ● Creamy, pink, flesh-colored, green | ● Pearly | | | ● 2 excellent cleavages at right angles | | ● 5 | | | Separated from plagioclase by color and lack of striae. |
| **Plagioclase feldspar** | ● Dead white, gray, blue, or transparent | ● Pearly, iridescent blue, or glassy | ● Striae on cleavage faces | | ● 2 excellent cleavages at right angles | | ● 5 | | | Feldspars are generally recognized by pearly luster (exception: vitreous luster in some plagioclase), and two good cleavages at right angles. Striae distinguish plagioclase. |
| **Pyrite** | ● Brass yellow | ● Metallic | ● Chiefly cubes | | ● 3, at right angles | ● Black | ● 6 to 6.5 | | | Brassy yellow metallic luster and black streak are characteristics. |

**TABLE B.1**
MINERALS AND PROPERTIES *(Continued)*

| Mineral | Color | Luster | Form and Internal Features | Breakage: Fracture | Breakage: Cleavage | Streak | Hardness | Magnetism | Chemical Tests | Remarks |
|---|---|---|---|---|---|---|---|---|---|---|
| **Pyroxene** | | | | | ● 2, at right angles | | ● 5 to 6 | | | Distinguished from amphibole by angles of cleavages; from feldspars by dark color. |
| **Quartz** | | ● Vitreous | | ● Irregular and conchoidal | | | ● 7 | | | Identified by hardness, luster, and fracture. |
| **Talc** | | ● Greasy | | | | | ● 1 | | | Softness and greasy feel identify talc. |

**Note:** Boxes contain dots only if the property represented by that box is important in identifying the mineral. For example, the box for streak, which is not used to identify silicate minerals, is empty in the rows for those minerals.

**TABLE B.2**
PROPERTIES AND MINERALS

**Minerals Arranged by Method of Breaking**

| Fracture (No cleavage) | Cleavage (Number of cleavage directions and angles made by cleavages): Only one direction | Two directions: *At right angles* | Two directions: *Not at right angles* | More than two directions: *At right angles* | More than two directions: *Not at right angles* |
|---|---|---|---|---|---|
| Quartz<br>Garnet<br>Olivine<br>Bauxite<br>Serpentine | Muscovite<br>Biotite<br>Chlorite<br>Graphite<br>Talc | Feldspars<br>Pyroxene | Amphibole<br>Gypsum | Galena<br>Halite<br>Pyrite | Calcite<br>Dolomite<br>Fluorite<br>Sphalerite |

**Minerals Arranged by Luster**

| Metallic | Nonmetallic: Vitreous | Nonmetallic: Vitreous | Nonmetallic: Pearly | Nonmetallic: Earthy | Nonmetallic: Shiny |
|---|---|---|---|---|---|
| Galena<br>Graphite<br>Hematite (some varieties)<br>Magnetite<br>Pyrite | Quartz<br>Some plagioclase<br>Garnet<br>Calcite (some varieties)<br>Gypsum (some varieties)<br>Olivine | Halite<br>Malachite (crystals)<br>Sphalerite (some varieties)<br>Fluorite<br>Pyroxene (some varieties) | Feldspars (most)<br>Calcite (some varieties)<br>Dolomite | Hematite (some varieties)<br>Limonite (some varieties)<br>Bauxite<br>Malachite (some varieties) | Limonite (some varieties)<br>Sphalerite<br>Biotite<br>Amphibole<br>Malachite (some varieties) |

**Minerals Arranged by Streak**
**Note:** *Always* use streak test for minerals having metallic, shiny, or earthy luster.

| Black | Yellowish-brown | Reddish-brown | Green | Yellowish |
|---|---|---|---|---|
| Galena<br>Magnetite<br>Pyrite | Limonite | Hematite | Malachite | Sphalerite (when streak is fresh it gives off odor of sulfur) |

**Minerals Arranged by Hardness**

| Less than or equal to 2.5 | | More than 2.5, but less than 5.5 | | More than 5.5 | |
|---|---|---|---|---|---|
| Talc<br>Gypsum<br>Biotite<br>Chlorite<br>Galena | Graphite<br>Halite<br>Bauxite (to 3.5)<br>Muscovite (to 3) | Calcite<br>Fluorite<br>Limonite<br>Dolomite | Malachite<br>Serpentine (4 to 6)<br>Sphalerite | Amphibole<br>Pyroxene<br>Garnet<br>Hematite (5 to 6)<br>Feldspars | Quartz<br>Olivine<br>Magnetite<br>Pyrite |

**Minerals Arranged by Special Features**

| Striae | Specific Gravity | Magnetism | Taste | Dilute acid |
|---|---|---|---|---|
| Plagioclase<br>Quartz (on some crystal faces)<br>Pyrite (on some crystal faces)<br>Hornblende (on some crystal faces) | Minerals that "feel heavy"<br>Metallic luster: Galena, Pyrite, Magnetite<br>Luster other than metallic: Sphalerite, Fluorite, Barite | Only magnetite | Halite tastes salty | Calcite (effervesces readily)<br>Dolomite (effervesces only weakly and when assisted by powdering the mineral or by heating the acid) |

**B.11 Perfect cleavage in one plane** parallel to surface on which specimen rests allows several flakes of muscovite to split off; vertical sides are crystal faces, in this specimen not as smooth as the cleavage planes. (Ward's Natural Science Establishment)

paper-thin flakes. Notice the difference here between a cleavage direction and the plane surfaces of the flakes. Each flake has a top and bottom plane surface. Moreover, each of these top and bottom planes is parallel. Therefore, in the technical language of cleavage, there is only one *direction* of cleavage, the direction to which all the flakes are parallel.

Cleavage makes sense only when you fully realize that cleavages are defined by *directions of planes* in the lattice. Accordingly, the first and most important aspect of cleavage in a mineral is *how many directions?* It may be one, as in muscovite, or more than one, as in many other minerals. In addition to number of directions, one needs to know how each direction is oriented with respect to other cleavage directions. Are they at right angles or not at right angles? Finally, cleavage is defined by the smoothness of its surfaces. If the cleavage surface is a single plane, as in muscovite, the cleavage is described as perfect. If the cleavage surfaces do not form single planes, but a series of parallel plane segments, then such cleavage is referred to as imperfect.

Examples of minerals having two directions of cleavage at right angles are feldspars and pyroxenes (Figure B.12a). Examples of minerals having two cleavage directions not at right angles are amphiboles (Figure B.12b).

If a mineral breaks along three cleavage directions that are at right angles, the cleavage is said to be cubic.

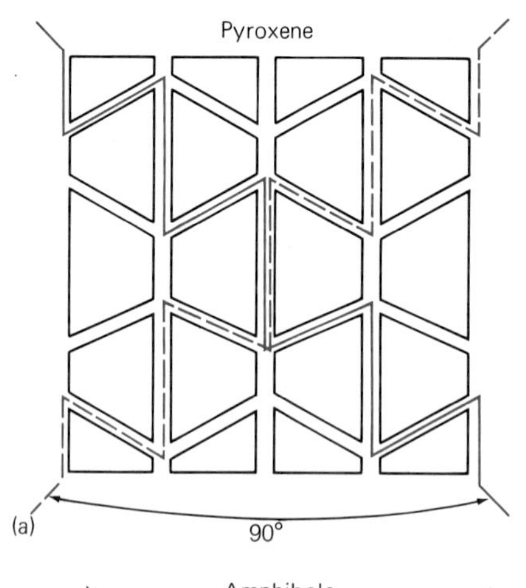


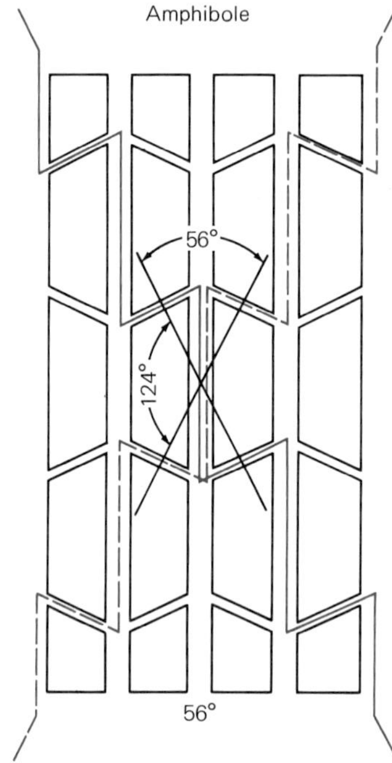


**B.12 Relationship between cleavage and internal arrangement of silica tetrahedra in pyroxene and amphibole.** The trapezoidal outlines are end views of the chains of tetrahedra, drawn to correct proportions (see Figure 4.9, p. 98).
(a) Breakage between single chains forms two cleavages that intersect at 90° in pyroxene.
(b) Breakage between double chains forms two cleavages that intersect at 56° and 124° in amphibole. (W. S. Fyfe, 1964, Geochemistry of solids. An introduction: New York, McGraw-Hill Book Company, 199p.; Figure 9–15, p. 124)

Two minerals having perfect cubic cleavage are halite (Figure B.13a) and galena (Figure B.13b).

The mineral calcite is characterized by its three directions of nearly perfect cleavage that are not at right angles. When calcite breaks, tiny rhombs are formed (see Figure B.6). A rhomb possesses six sides; the opposite pairs of sides are cleavage planes formed by breakage parallel to a single cleavage direction.

(a)

(b)

**B.13 Perfect cube cleavage.**
(a) Halite; light-colored areas inside cube are cleavages parallel to top and bottom surface of the cube and extending into it from each side. (E. J. Dwornik, U. S. Geological Survey)
(b) Galena. Specimen from Columbia University collection. (J. E. Sanders)

Fluorite is a mineral whose crystals are cubes but whose cleavage directions are not parallel to the faces of the cubes (Figure B.14).

The form of some minerals is fibrous (Figure B.15), whereas that of other specimens is massive. In some mineral specimens, the individuals have all

**B.14 Cleavage in fluorite cubes.** The crystal faces define the cubes; the noncubic cleavages extend into the crystals. Specimen from Columbia University collection. (J. E. Sanders)

grown together to form a mineral aggregate.

In some minerals, such as pyrite and plagioclase feldspars, the intersection of a series of parallel internal planes controlled by the lattice and the external planes of the specimen (either crystal faces or cleavage planes) define a series of closely spaced parallel lines. Such lines are named *striae* (from the Latin word for "furrow"); they are important in distinguishing among the major groups of feldspars. The potassium feldspars lack striae, whereas the plagioclase feldspars display well-developed striae (Figure B.16). Other minerals displaying striae include quartz and hornblende.

**B.15 Fibrous mineral, asbestos,** an extremely stable magnesium silicate that resists heat and chemical change. Fibers of asbestos are widespread in modern urban environments; tiny fibers in people's lungs and intestines have been found to trigger cancers. Barnard College collection. (Frank Dunand)

**Color** One of the most-obvious and striking properties of a mineral is its color. In some minerals, the color is a diagnostic property, but in many minerals, at least as far as identification is concerned, color is a snare and a delusion. Color is useful in distinguishing the major groups of feldspars, plagioclase and potassium feldspars. Plagioclase feldspars are white, colorless, or blue. By contrast, potassium feldspars are cream colored, pink, reddish, or light green. Among the copper minerals, malachite's green always distinguishes it from azurite's blue.

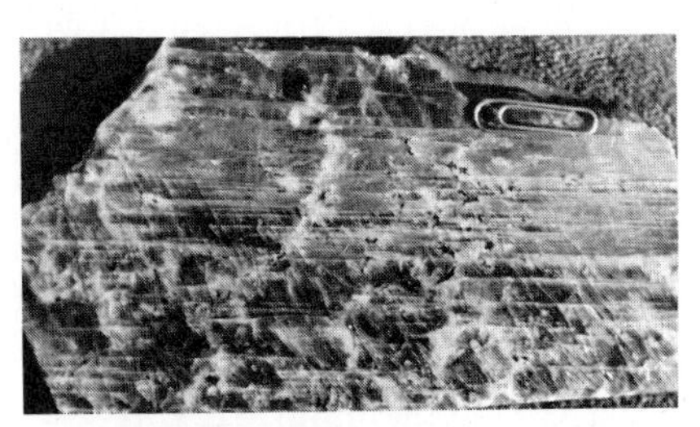

**B.16 Striae** (*not* striations) are parallel to the paper clip on this large cleavage fragment of plagioclase. The parallel features crossing the striae at an angle are cleavages within the specimen at right angles to the surface of the specimen. Columbia University collection. (J. E. Sanders)

Color is of no value in identifying quartz, calcite, and fluorite. Quartz displays many color varieties, most of them related to small amounts of various ions that can be considered as impurities within the diagnostic quartz lattice. Much quartz is colorless, but some is gray (smoky quartz), purple (amethyst), pink (rose quartz), white (milky quartz), or bluish (no special name). Calcite can be colorless, white, gray, bluish, pink, green, or black. Typical fluorite is purple, but other colors include light green, gray, or colorless.

The packing of the ions affects the appearance of carbon minerals. Diamonds are clear and brilliant, whereas graphite is black and greasy.

**Luster** The *luster* of a mineral is defined as the way in which the mineral reflects light (Figure B.17). Two major groupings are metallic and nonmetallic. Minerals having a metallic luster reflect most of the light that strikes them; the result is a shiny, metal-like appearance. By contrast, minerals having a nonmetallic luster absorb much of the light that strikes them. A common luster of nonmetallic minerals is clear and glassy, a characteristic named *vitreous*. The brilliant luster of a diamond is named *adamantine*. Other descriptive names for luster include such self-explanatory words as *silky, dull, earthy, shiny, pearly,* and *greasy*. Two other varieties are *porcellanous* (appearance comparable to that of glazed porcelain) and *resinous* (appearing like resin).

## Results of Simple Tests

In this category, we include a few simple tests that give results useful for distinguishing between minerals that may otherwise look very similar. Our list includes streak, hardness, specific

**B.17 Luster.**
1 Shiny luster (black, limonite).
2 Metallic luster (steel gray, galena).
3 Metallic luster (also shiny, specular hematite).
4 Dull luster (magnetite; notice tiny iron filings adhering to specimen in lower right corner).
5 Earthy luster (oölitic hematite).
6 Silky luster (fibrous gypsum).
7 Vitreous luster (glassy quartz crystal).
8 Pearly luster (cleavage fragment of feldspar).
9 Vitreous luster (massive quartz).
10 Vitreous luster (tiny halite cube).
11 Metallic luster (pyrite).

gravity, magnetic susceptibility, taste, feel, and reaction with dilute acid. Table B.2 summarizes these properties and the common minerals displaying each.

**Streak** A special test for the color of finely powdered minerals is named streak. The *streak* of a mineral is determined by rubbing the specimen across a flat surface of nonglazed porcelain. Streak is especially valuable in identifying metallic minerals. For example, all varieties of hematite display reddish-brown streak, whereas limonite streaks yellowish brown. Magnetite, galena, and pyrite yield black streaks. The streak of sphalerite is yellowish and freshly streaked sphalerite smells of sulfur.

**Hardness** The property known as *hardness* refers to the mineral's resistance to scratching. This property is determined by the ways the ions are packed. Hardness is a test that most geologists make as a matter of course. It is a quick and reliable way for distinguishing quartz from calcite, for example. In 1822, the German mineralogist Friedrich Mohs (1773–1839), who eventually became a professor in Vienna, arranged 10 minerals in a standard scale of relative hardness, starting with talc at the soft end (hardness number of 1) and progressing to diamond at the hard end (hardness number of 10). Simple groupings on the Mohs scale can be made by one's fingernail (hardness about 2.2), a copper penny (hardness of 3), a pocket knife, or nail, or a piece of glass (hardness of 5.5 to 6). The Mohs scale of hardness is shown graphically in Figure B.18.

**Density and specific gravity** The "heaviness" of a mineral is a function of its *density* (defined as the mass per unit volume), which is numerically equal to the *specific gravity*, an expression of the weight of a mineral compared with the weight of an equal volume of water at 4° C. For example, the density of quartz is 2.65 grams per cubic centimeter, and the specific gravity of quartz is 2.65. This means that quartz is 2.65 times as heavy as the same volume of water. The specific gravity of a mineral specimen is determined by weighing a mineral sample in air, by weighing the specimen sus-

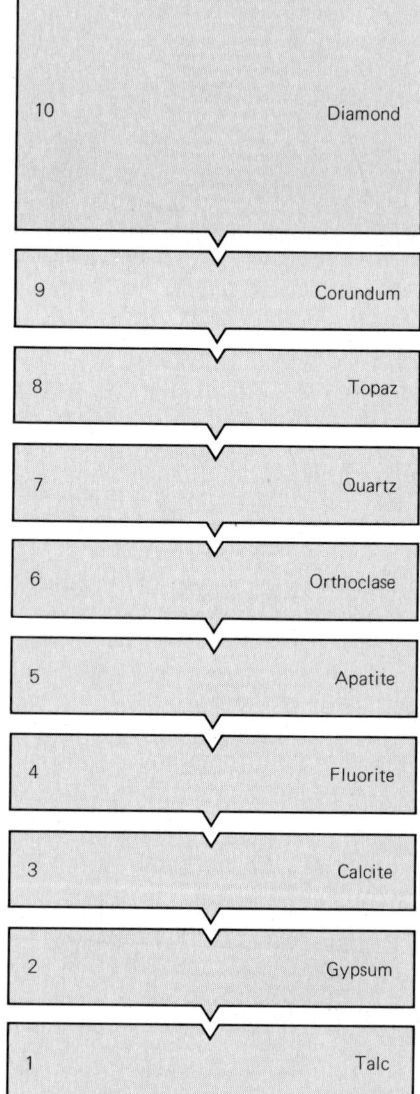


**B.18 Mohs scale of relative hardness.** Minerals having high numbers will scratch all minerals having lower numbers. The difference in hardness between diamond (10) and corundum (9) is greater than the difference between corundum (9) and talc (1).

pended in water, and then by solving the equation: specific gravity = weight in air weight in water − weight in air. The density can be determined by weighing the specimen in air and then by dividing this weight by the volume of the specimen, which can be measured by its displacement of water or of air, as determined by measuring the volume of water displaced by the specimen or by air displacement in a special apparatus.

Table B.3 lists common minerals in order of their densities. The density of a mineral affects how it will be transported by currents of water or of air (Chapters 12 and 14) and also controls the densities of rocks. The density of a rock composed chiefly of potassium feldspar and quartz (such as granite), is about 2.6. By contrast, the density of a rock composed chiefly of plagioclase feldspar and pyroxene (such as basalt) is about 3.0.

**TABLE B.3**
DENSITIES (ATMOSPHERIC PRESSURE), SELECTED METALS AND MINERALS

| | | Minerals | | | |
|---|---|---|---|---|---|
| Metals | Density (grams/cm³) | Metallic Luster | Density (grams/cm³) | Nonmetallic Luster | Density (grams/cm³) |
| Gold | 19.3 | | | | |
| Mercury | 13.6 | | | | |
| Copper | 8.9 | | | | |
| Nickel | 8.6 | | | | |
| Iron | 7.9 | Galena | 7.57 | | |
| Tin | 7.3 | | | | |
| Zinc | 7.2 | | | | |
| | | Hematite | 5.26 | | |
| | | Magnetite | 5.2 | | |
| | | Pyrite | 5.01 | | |
| | | | | Barite | 4.3 to 4.6 |
| | | | | Corundum | 4.0 to 4.1 |
| | | | | Sphalerite | 3.9 to 4.1 |
| | | | | Garnet | 3.58 to 4.31 |
| | | | | Olivine | 3.2 to 4.3 |
| | | | | Staurolite | 3.7 to 3.8 |
| | | | | Kyanite | 3.53 to 3.65 |
| | | | | Topaz | 3.49 to 3.57 |
| | | | | Augite | 3.2 to 3.4 |
| | | | | Fluorite | 3.18 |
| | | | | Biotite | 2.7 to 3.3 |
| | | | | Hornblende | 2.9 to 3.4 |
| | | | | Aragonite | 2.94 |
| | | | | Muscovite | 2.77 to 2.88 |
| Aluminum | 2.7 | | | Dolomite | 2.85 |
| | | | | Wollastonite | 2.8 to 2.9 |
| | | | | Calcite | 2.71 |
| | | | | Talc | 2.58 to 2.83 |
| | | | | Chlorite | 2.6 to 2.9 |
| | | | | Feldspars | 2.56 to 2.76 |
| | | | | Plagioclase | 2.62 to 2.76 |
| | | | | Albite | 2.62 |
| | | | | Orthoclase | 2.56 to 2.59 |
| | | | | Quartz | 2.65 |
| | | | | Kaolinite | 2.6 |
| | | | | Chalcedony | 2.57 to 2.64 |
| | | | | Gypsum | 2.3 |
| | | | | Serpentine | 2.2 to 2.6 |
| | | | | Graphite | 2.02 to 2.23 |
| | | | | Bauxite | 2.0 to 2.55 |
| | | | | Opal | 2.0 to 2.2 |
| | | | | Halite | 2.16 |

[a]Metals from Handbook of Chemistry and Physics; minerals from Vanders and Kerr, 1967.

**Magnetic susceptibility** This property refers to the magnetic characteristics of minerals. The most-magnetic mineral is magnetite, which is readily attracted to a small hand magnet (Figure B.19). Magnetite is the only common mineral that is significantly magnetic; it is easily identified because of its attraction to a hand magnet.

**Taste** The mineral for which the taste test is applied is halite. Everyone can identify halite by its salty taste.

**Feel** This term refers to the "feel" of a mineral when one touches a specimen. The category distinguished by touch is slipperiness or greasiness. For example, talc and graphite feel greasy, whereas most other minerals do not.

**Reaction with dilute acid** A drop of dilute hydrochloric acid (HCl) gives a positive test for the calcium-carbonate minerals, calcite and aragonite. When the acid touches the specimen, many tiny bubbles froth upward, a phenomenon named *effervescence.* The magnesium-calcium carbonate, dolomite,

**B.19 High magnetic susceptibility of massive magnetite** shown by its ability to keep an adhering hand magnet from falling. The demonstration did not work the other way around—the magnetite specimen was too heavy. Specimen from Barnard College collection. (Nat Messick)

will effervesce only slowly and only after the mineral has been powdered or the acid heated. Other minerals are not readily affected by dilute hydrochloric acid, hence do not effervesce when acid is applied to them.

## MAJOR GROUPS OF ROCK-FORMING SILICATE MINERALS

The commonest rock-forming silicate minerals can be arranged into various groups that are based either on chemical composition or on the structure of the tetrahedra. The major subdivision based on chemical composition includes two categories: (1) ferromagnesian silicates and (2) nonferromagnesian silicates. The ferromagnesian silicates are defined as those minerals containing iron and magnesium, to which may be added calcium and various other ions. The nonferromagnesian silicates lack iron and magnesium and contain sodium, calcium, and potassium. Examples of ferromagnesian silicate minerals are olivine, pyroxene, amphiboles, and biotite. These minerals are all dark colored, and their specific gravities range from 2.7 to 4.3 (Table B.3). Examples of nonferromagnesian minerals include muscovite, feldspars, and quartz. These minerals are light colored, and their specific gravities range from 2.56 to 2.88.

Subdivision of the rock-making silicates according to the structure of the tetrahedra results in five categories: (1) isolated tetrahedra, (2) single chains, (3) double chains, (4) sheets, and (5) complex three-dimensional networks. The ferromagnesian silicates form isolated tetrahedra, single chains, double chains, and sheets. The nonferromagnesian minerals form sheets and complex three-dimensional networks.

In the mineral olivine, $Mg^{2+}$ and/or $Fe^{2+}$ ions are joined to isolated individual tetrahedra. What we shall refer to simply as olivine is actually a group of minerals in which the proportions of these two ions varies from zero to 100 per cent. One end member of the olivine group consists entirely of magnesium ions joining isolated silica tetrahedra; its formula is $Mg_2(SiO_4)$. The other end member consists entirely of ions of ferrous iron and isolated silica tetrahedra; its formula is $Fe_2(SiO_4)$. The intermediate varieties consist of some mixtures of $Mg^{2+}$ and $Fe^{2+}$. The chemical formula for the varieties is $Mg,Fe(SiO_4)$. Olivine is an example of a mineral in which within a given framework of tetrahedra, continuous substitution between $Mg^{2+}$ and $Fe^{2+}$ can take place. As we shall see, such substitution involving these and other ions is possible in other groups of silicate minerals.

Single-chain silicates are illustrated by *pyroxenes.* In single-chain silicates, each tetrahedron shares two oxygen ions (see Figure 4.9e). The ion formula of the linked tetrahedra in single chains is $(SiO_3)^{2-}$. In single-chain silicates, two kinds of bonds are present.

Double-chain silicates are represented by the *amphiboles.* In double-chain silicates, alternate pairs of tetrahedra share two oxygen ions and the intervening pairs of tetrahedra share three oxygen ions, the third point of sharing being with the other member of the pair to join the chains together (see Figure 4.9f). The ion formula for the linked tetrahedra in double-chain silicates is $(Si_4O_{11})^{5-}$.

The minerals having chain structures cleave in two directions that are along the directions of the chains. In single-chain minerals, such as pyroxenes, the two cleavages form a right angle. In double-chain minerals, such as amphiboles, the two cleavages meet at oblique angles (60° or 120°) (see Figure B.12b).

In pyroxenes, the metallic ions joined to the single chains of tetrahedra are $Mg^{2+}$, $Fe^{2+}$, and $Ca^{2+}$. In amphiboles, the double-chain structure can participate in complex substitutions, involving not only the metallic ions as well as hydroxyl ions, $(OH)^-$, that join the chains together, but also the $Al^{3+}$ and $Si^{4+}$ ions within the tetrahedra. When one $Al^{3+}$ substitutes for one $Si^{4+}$ within a tetrahedron, the overall electrical charge for the tetrahedron is decreased by one unit. This change is compensated for by the substitution among the ions outside the tetrahedra so that one positive charge is gained. For example, $Ca^{2+}$ might take the place of $Na^+$. A substitution within the tetrahedra that goes hand in hand with a charge-matching switch outside the tetrahedra is known as a *coupled substitution.*

Sheet-structure silicate minerals (see Figure 4.9g) are typified by the micas, including the ferromagnesian mineral, biotite, and the nonferromagnesian mineral, muscovite. In the sheet-structure silicates, each tetrahedron shares three oxygen ions and, in addition, some coupled substitution takes place because within a few tetrahedra $Al^{3+}$ ions replace $Si^{4-}$. In biotite, the ions added outside the tetrahedra include $Fe^{2+}$, $Mg^{2+}$, $K^+$, $(OH)^-$, and $F^-$. In muscovite, the ions added outside the tetrahedra include $K^+$, $Al^{3+}$, and $(OH)^-$.

The fifth major group of rock-forming minerals is known as complex three-dimensional networks. In such networks, each tetrahedron shares all four of its oxygen ions with four other tetrahedra. The formula for the linked tetrahedra is either $(Si_4O_8)$ or simply $(SiO_2)$. Notice that the electrical charge of either formula is balanced. Using the presence or absence of coupled substitutions, we can make two major subdivisions: (1) feldspars and (2) quartz.

Within the tetrahedra of feldspars, some $Al^{3+}$ substitutes for $Si^{4+}$; this change requires a coupled charge-balancing substitution outside the tetrahedra. In potassium feldspars, the balance is made by simple addition of one $K^+$ ion for every $Al^{3+}$ that enters a tetrahedron. The resulting chemical formula is $K(AlSi_3)O_8$. In plagioclase feldspars, the composition ranges between the sodium end member, albite, and the calcium end member, anorthite. The formula for albite is $Na(AlSi_3)O_8$. The formula for anorthite is $Ca(Al_2Si_2)O_8$.

In quartz, no substitution takes place within the tetrahedra. As a result, no ions are required outside the tetrahedra to balance the charge. The resulting formula is simply $SiO_2$.

# APPENDIX C
# ROCKS

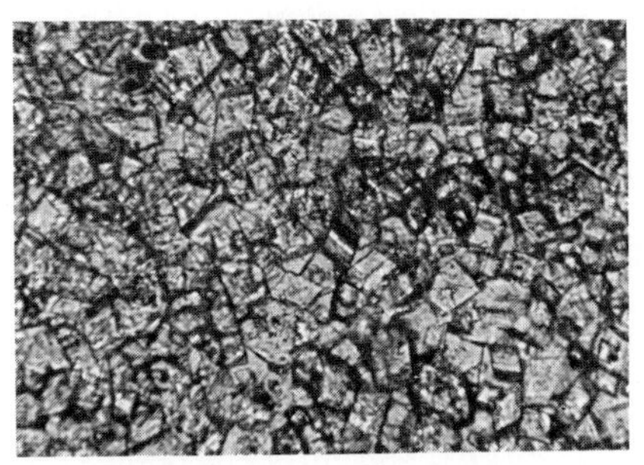

**C.1 Crystalline texture in a soft-mineral sedimentary rock** (dolostone), paper-thin slice of Madison Formation (Mississippian) from Wyoming as seen with a polarizing microscope. Longest dimensions of rhombs are about 0.1 mm (Gerald M. Friedman, Rensselaer Polytechnic Institute)

**C.2 Crystalline texture in a soft-mineral metamorphic rock,** a marble specimen from Columbia University collections. (J. E. Sanders)

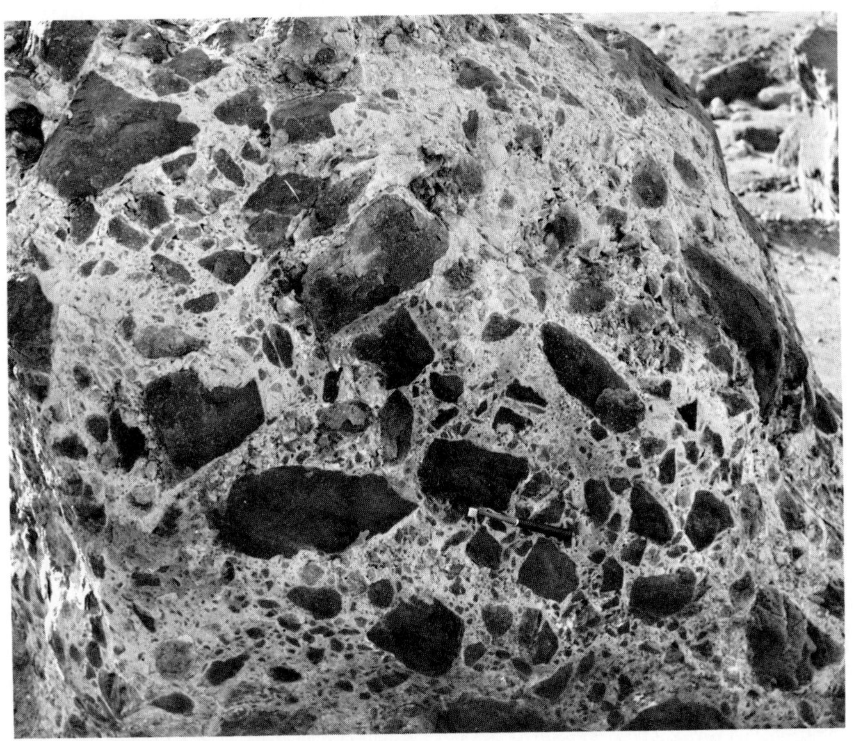

**C.3 Volcanic breccia;** dark-colored broken pieces in a light-colored matrix. Augustine Volcano, Alaska. (Hans-Ulrich Schmincke, Ruhr University, Bochum, W. Germany)

**C.4 Bioclastic texture** in Pleistocene coquina, Florida. Broken skeletal debris mostly of clams. (Smithsonian Institution)

**C.5 Cataclastic texture in a metamorphic rock,** view of paper-thin slice using polarizing microscope. Specimen from pre-Triassic, Branford, Connecticut, collected by J. E. Sanders. (B. M. Shaub)

(a)

(b)

**C.6 Organic textures in limestones.**
(a) Parts of coral colonies have been broken away, showing the internal structure that resembles an apartment house with one wall removed. (Gerald M. Friedman, Rensselaer Polytechnic Institute)
(b) Remains of clams, left as imprints after the skeletal material had been dissolved. Pleistocene limestone, NW of Miami, Florida. (J. E. Sanders)

**C.7 Oriented fabric** in streaked-out feldspars (white) and other minerals in a metamorphic rock contrasts with fabric of pegmatite. Precambrian N of Parry Sound, Ontario, Canada. (J. E. Sanders)

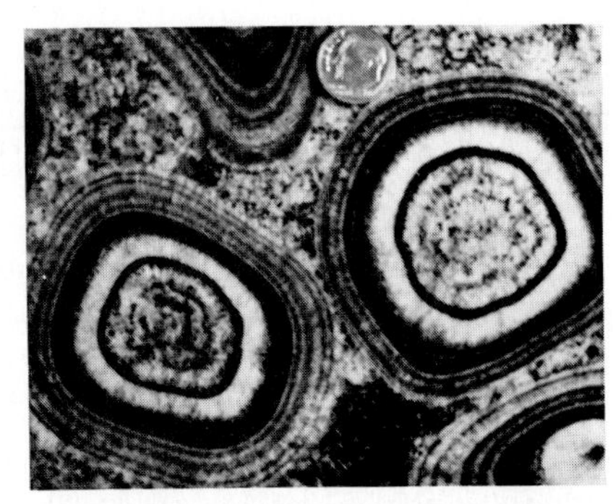

**C.8 Orbicular granite,** polished section of specimen in University of Groningen, The Netherlands, courtesy Professor L. M. J. U. van Straaten. (J. E. Sanders)

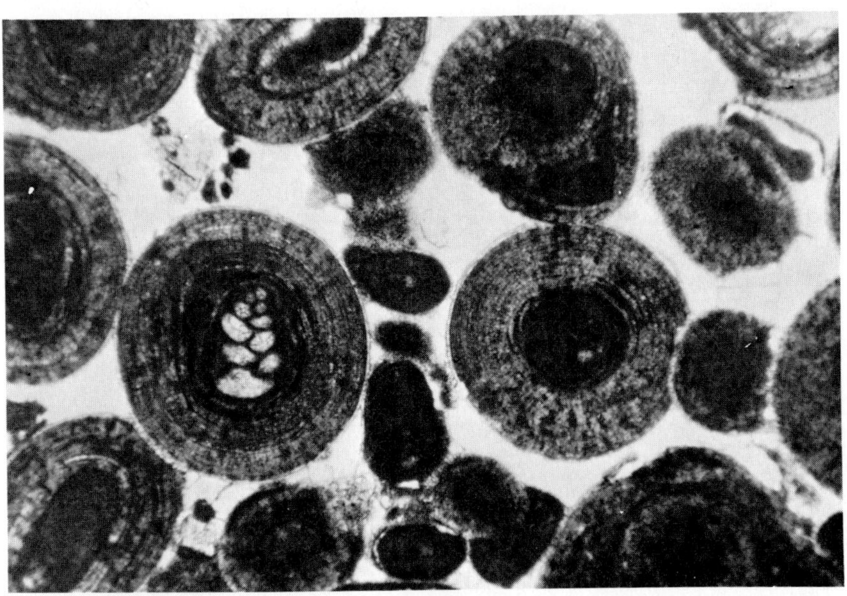

**C.9 Oölite,** from the reservoir rock of one of the world's largest oil fields, Arab-D Formation (Jurassic), Gwahar field, Saudi Arabia. (R. W. Powers)

**C.10 Graphic granite;** narrow strips of quartz and wider sections of albite. Specimen from Barnard College collection. (Nat Messick)

**C.11 Conchoidal fracture,** well developed in this specimen of obsidian, Barnard College collection. (Nat Messick)

**C.12 Thin partings parallel to vertical strata,** a fissile shale in the Big Cottonwood Formation, Tintic Junction quadrangle, Juab County, Utah. (H. T. Morris, U. S. Geological Survey, ca. 1952)

C.13 **Slaty cleavage,** closely spaced, nearly horizontal partings cutting across steeply dipping strata at a high angle. Moe River, Quebec. (K. C. Bell, Canada Geological Survey)

C.14 **Organic shapes of skeletal debris,** surface of ancient limestone displaying numerous bryozoa (rodlike features with many small pits) and brachiopods. Silurian, Lockport, New York. (Smithsonian Institution)

C.15 **Pegmatite** composed largely of potassium feldspar; longest crystal is 22 centimeters (film cartridge is 5 centimeters long). Pelham Bay Park, Bronx, New York. (J. E. Sanders)

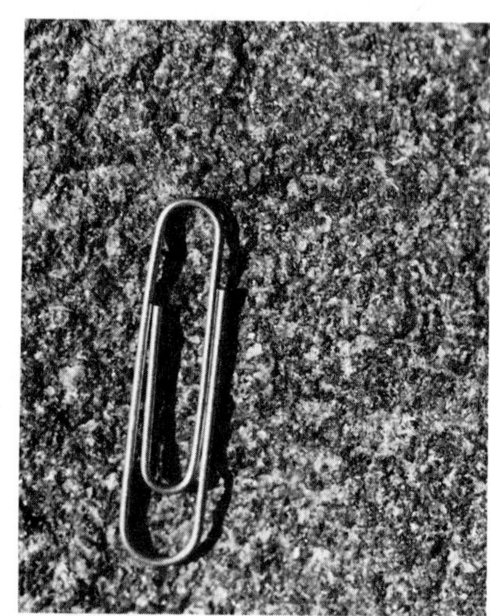

C.16 **Gabbro,** uniform crystalline texture with plagioclase (white) and pyroxene (gray) individuals 2 to 3 millimeters long. Palisades sheet, Edgewater, New Jersey. (J. E. Sanders)

**C.17 Dolerite,** uniform crystalline texture with individuals about 1 millimeter in size. Palisades sheet, Edgewater, New Jersey. (J. E. Sanders)

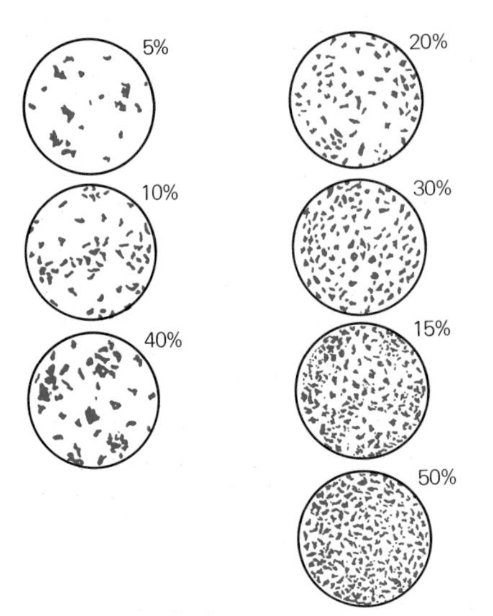


**C.18 Area-estimator charts** showing proportion in percent of area occupied by black fragments.

**C.19 Rounded fragments in a gravel containing broken shell debris** (white). Modern beach, Prince's Bay, Staten Island, New York. (J. E. Sanders)

**C.20 Conglomerate containing rounded boulders.** Roadcut on Route 128, S of Boston, Massachusetts. (J. E. Sanders)

**C.21 Sedimentary breccia,** with sharply angular, light-colored fragments of feldspar and carbonate rocks. Shuttle Meadow Formation (Lower Jurassic of Newark Group), N side of U. S. 1, East Haven, Connecticut. (J. E. Sanders)

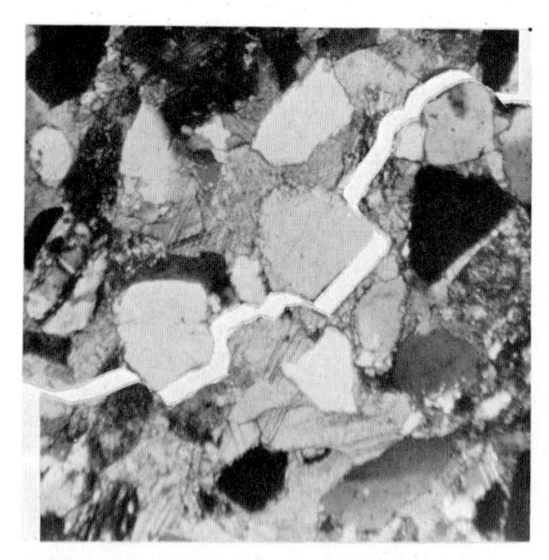

**C.22 Breaking of sandstone around particles** illustrated by cutting apart a photograph made of view of paper-thin slice of sandstone using a polarizing microscope. (J. E. Sanders)

**C.23 Calcarenite,** a rock composed of sand-size pieces of calcium carbonate, in this example, skeletal debris. Mississippian limestone, southern Indiana. (J. E. Sanders)

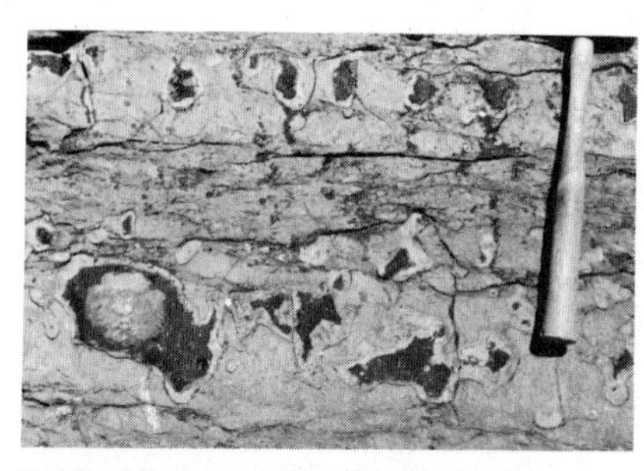

**C.24 Chert nodules (black) in limestone (light gray),** Kalkberg limestone (Devonian), Austin's Glen, NW of Catskill, New York. (J. E. Sanders)

**C.25 Bituminous coal.** Estimated thickness of this view is about 7 meters. Locality not known. (U. S. Bureau of Mines)

**C.26 Slate,** diagnosed by well-developed slaty cleavage (parallel to top of specimen on which dime is resting) that is not parallel to the stratification (dipping steeply to left, as indicated by the dark gray layer). Ordovician, Northumberland County, Pennsylvania. Specimen from Columbia University collection. (J. E. Sanders)

**C.27 Schist** composed of mica (darker gray), plagioclase (white), and porphyroblasts of garnet (somewhat circular gray areas within the plagioclase). Garnet porphyroblasts have been cut across by a continental glacier. Manhattan Schist (Ordovician), W side of Riverside Drive at 168th Street, New York City. (J. E. Sanders)

**C.28 Gneiss displaying prominent foliation based on contrasting minerals.** Erratic of Precambrian from Adirondacks, specimen in outdoor geological museum, Cornell University, Ithaca, New York. (J. E. Sanders)

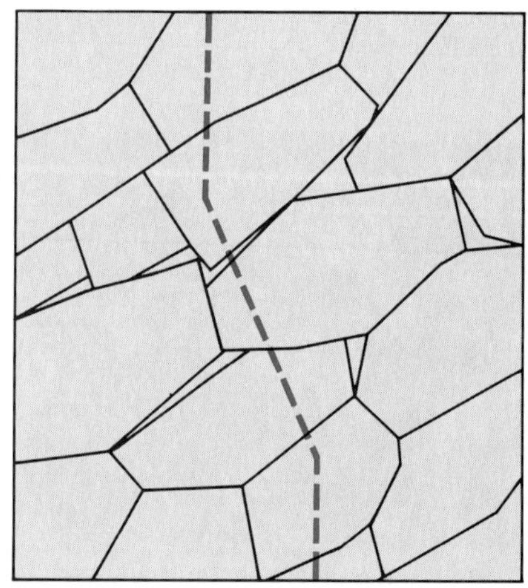

**C.29 Quartzite breaks across the particles;** surface of breakage is the dashed blue line. Schematic. Compare with Figure C.22.

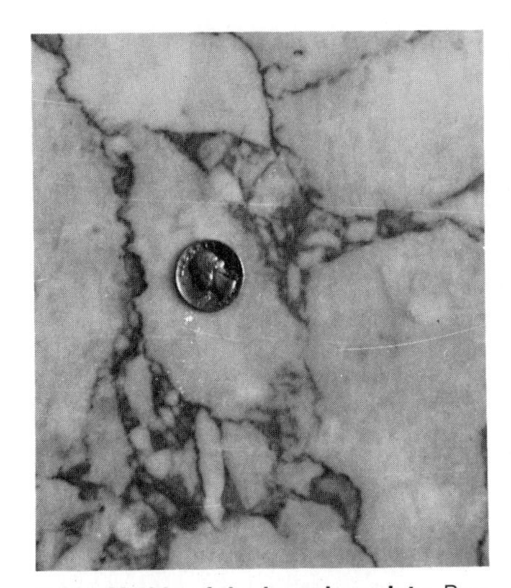

**C.30 Marble of the breccia variety.** Pressure between adjacent fragments has formed stylolites. Interior building stone of Uris Hall, School of Business, Columbia University; locality of stone not known. (J. E. Sanders)

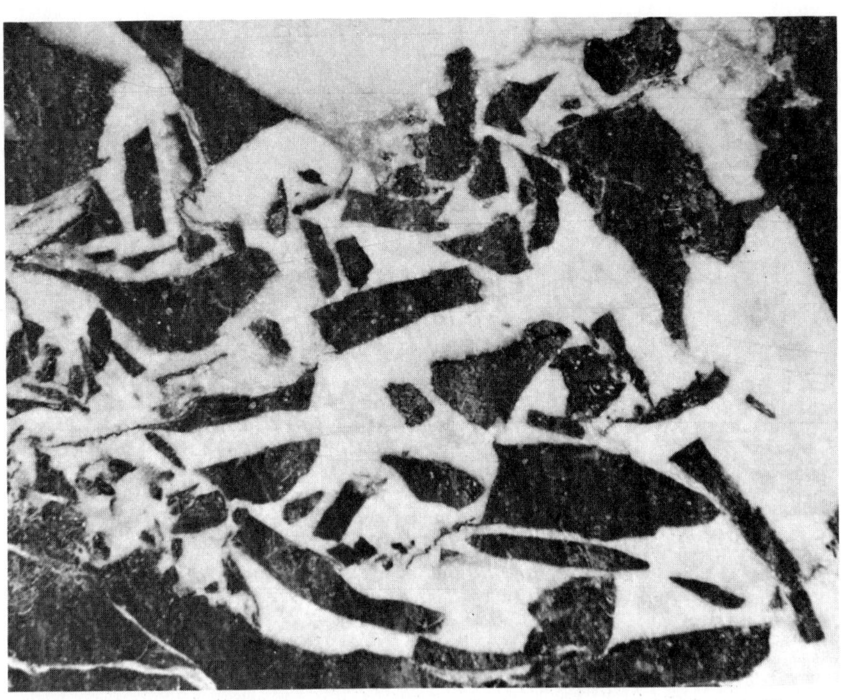

**C.31 Cataclastic breccia.** Exterior building stone, Hotel Roosevelt, Manhattan, New York City; locality of stone not known. (J. E. Sanders)

**Table C.1** Some characteristics of rocks arranged by their occurrence in the three great groups of rocks. Particularly distinctive features for identifying a given group of rocks are shown by blue color.

| | CHARACTERISTICS | | | | | |
|---|---|---|---|---|---|---|
| | Texture | | | | Fabric | |
| Kind of rock | Crystalline | Clastic | Pyroclastic | Organic | Random | Oriented |
| IGNEOUS | Typical (See figure 5a) | Present in some breccias (Figure C.3) | Explosive volcanic; typified by a combination of features formed from breaking of cooling lava | Generally absent | Fabric usually is random | Phenocrysts of porphyries may be aligned (See figure 6.3b) |
| SEDIMENTARY | Present in some varieties composed of soft (hardness ≤ 3.5) minerals (Figure C.1) | Typical, especially if particles have been abraded; broken skeletal debris is named **bioclastic** (Figure C.4) | | Typical; all varieties (Figure C-6) | Random fabrics common | Common; platy minerals aligned parallel to stratification; elongate particles may be oriented by currents |
| METAMORPHIC | Typical; includes rocks composed of both hard and soft minerals (Figure C.2) | Present in some cataclastic varieties (Figure C.5) | May survive in some kinds of metamorphism | May be present where parent rocks were sedimentary | Random fabrics are present in many marbles, quartzites, and hornfelses | Oriented fabrics are typical of schists, queisses, amphibolites, granulites, and mylonites; also present in other metamorphic rocks (Figure C.7) |

| | BULK CHARACTERISTICS | | | |
|---|---|---|---|---|
| | Fabric (con't) | | Special features | Partings |
| Kind of rock | Minerals in concentric or radial patterns | Complex intergrowths | Vesicles | Conchoidal fracture |
| IGNEOUS | When present in granites is named **orbicular granite** (Figure C.8); concentric patterns and rosettes are common in fillings of vesicles and other cavities | Cuneiform intergrowths of feldspar and quartz form **graphic granite** (Figure C.10) | Characteristic of many volcanic rocks; named **amygdales** when filled with minerals | Present in aphanitic rocks and glasses (obsidian and basalt glass) (Figure C.11) |
| SEDIMENTARY | Common in oolitic (Figure C.9) and pisolitic carbonates; concentric patterns in geodes and concretions; many rosettes | No cuneiform patterns | Rare; only in sediments buried by lava | Present in **cherts, lithographic limestones, argillites,** some fine-grained sandstones, and siltstones |
| METAMORPHIC | May be present, as when garnets grew while they were being deformed and rolled | No cuneiform patterns | Not present | Present in many quartzites |

| Kind of rock | BULK CHARACTERISTICS | | CONSTITUENTS |
|---|---|---|---|
| | Layers | Partings (con't.) Direction of splitting | Shapes |
| IGNEOUS | Not present in usual plutons; but, typical of crystal-cumulate rocks (see Figure 6.5) and volcanic rocks | May split along oriented platy minerals such as micas | Usually irregular because of compromise boundaries (see Figure 4.14); crystal shapes in most phenocrysts. |
| SEDIMENTARY | Typical | Typically splits parallel to stratification; spacing of partings may be from millimeters (Figure C.12) to meters | Rounded or angular (see Figure 8.10); organic shapes in skeletal debris (Figure C.14); some crystal shapes in cavity fillings and in some authigenic minerals |
| METAMORPHIC | Foliation is typical; may be inherited stratification or not related to former stratification but created during metamorphism | Splits parallel to oriented platy minerals, usually not parallel to stratification; diagnostic when rock splits along slaty cleavage that is not parallel to stratification (Figure C.13) | Crystal shapes in some minerals such as garnet; mostly irregular because of compromise boundaries or cataclastic deformation |

| Kind of rock | CONSTITUENTS |
|---|---|
| | Minerals |
| IGNEOUS | Silicates, especially feldspars and ferromagnesian silicates; quartz not dominant nearly all minerals harder than 6 |
| SEDIMENTARY | Silicates, carbonates, oxides, sulfates, halides; quartz dominant; many minerals softer than 3.5 |
| METAMORPHIC | Silicates predominate, but carbonates are common (in marbles); micas, green minerals and garnets typical; feldspars abundant; hardness spans range of hardness of igneous rocks and sedimentary rocks |

**Table C.2** Plot of sizes of constituents (at top, size increasing upward) against mineral composition (shown in rectangle below) forms basis for classifying equigranular igneous rocks into three families and for naming individual rocks within each family. Formal definitions of each rock type are in glossary.

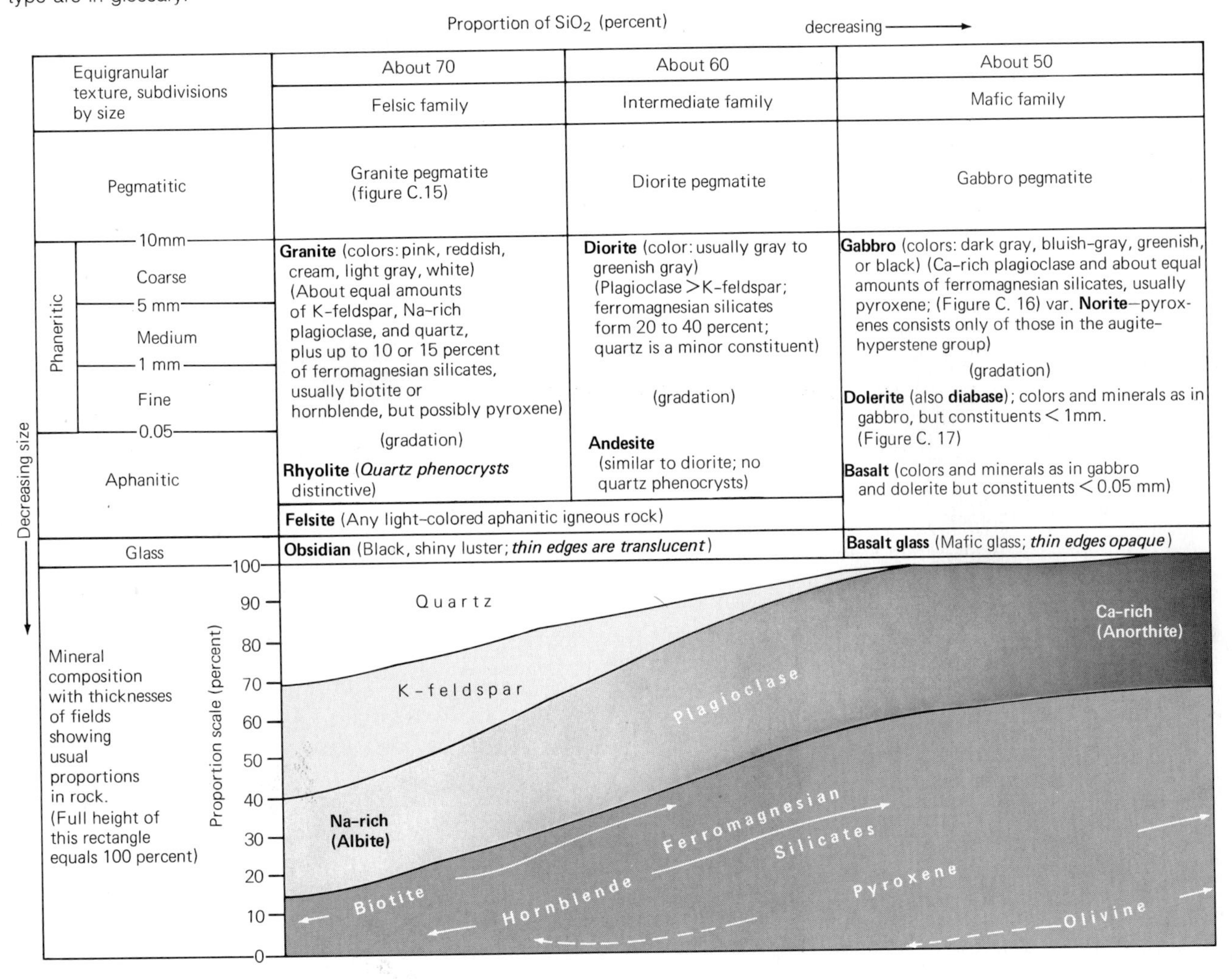

Proportion of $SiO_2$ (percent) decreasing →

| Equigranular texture, subdivisions by size | About 70<br>Felsic family | About 60<br>Intermediate family | About 50<br>Mafic family |
|---|---|---|---|
| Pegmatitic | Granite pegmatite (figure C.15) | Diorite pegmatite | Gabbro pegmatite |
| Phaneritic: 10mm — Coarse — 5 mm — Medium — 1 mm — Fine — 0.05 | **Granite** (colors: pink, reddish, cream, light gray, white) (About equal amounts of K-feldspar, Na-rich plagioclase, and quartz, plus up to 10 or 15 percent of ferromagnesian silicates, usually biotite or hornblende, but possibly pyroxene)<br>(gradation) | **Diorite** (color: usually gray to greenish gray) (Plagioclase > K-feldspar; ferromagnesian silicates form 20 to 40 percent; quartz is a minor constituent)<br>(gradation) | **Gabbro** (colors: dark gray, bluish-gray, greenish, or black) (Ca-rich plagioclase and about equal amounts of ferromagnesian silicates, usually pyroxene; (Figure C. 16) var. **Norite**—pyroxenes consists only of those in the augite-hyperstene group)<br>(gradation)<br>**Dolerite** (also **diabase**); colors and minerals as in gabbro, but constituents < 1mm. (Figure C. 17) |
| Aphanitic | **Rhyolite** (*Quartz phenocrysts* distinctive)<br>**Felsite** (Any light-colored aphanitic igneous rock) | **Andesite** (similar to diorite; no quartz phenocrysts) | **Basalt** (colors and minerals as in gabbro and dolerite but constituents < 0.05 mm) |
| Glass | **Obsidian** (Black, shiny luster; *thin edges are translucent*) | | **Basalt glass** (Mafic glass; *thin edges opaque*) |

**Table C.3** Crystal-cumulate rocks, showing distribution of constituent minerals. Notice that crystal-cumulate rocks form only phaneritic varieties.

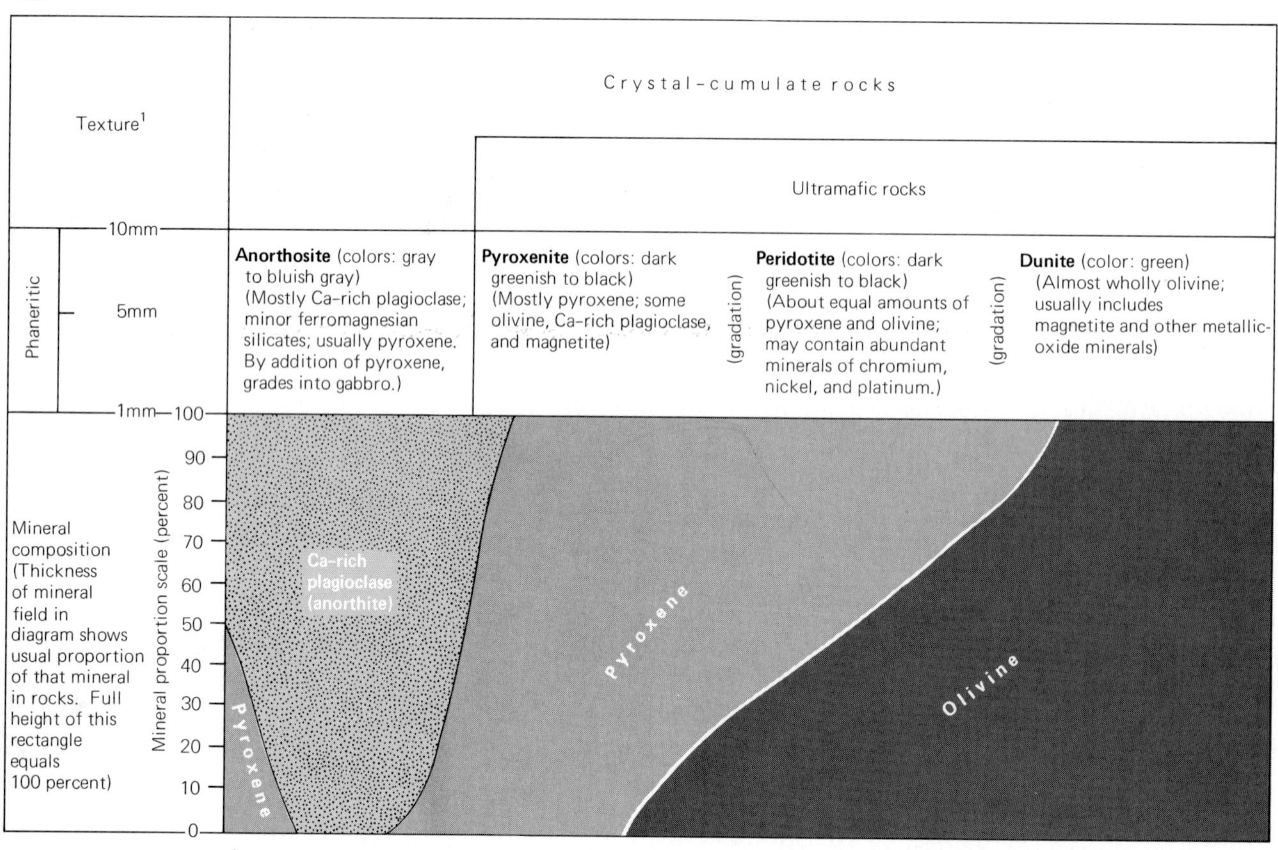

| Texture[1] | Crystal-cumulate rocks | | | |
|---|---|---|---|---|
| | | Ultramafic rocks | | |
| Phaneritic (10mm – 5mm – 1mm) | **Anorthosite** (colors: gray to bluish gray) (Mostly Ca-rich plagioclase; minor ferromagnesian silicates; usually pyroxene. By addition of pyroxene, grades into gabbro.) | **Pyroxenite** (colors: dark greenish to black) (Mostly pyroxene; some olivine, Ca-rich plagioclase, and magnetite) (gradation) | **Peridotite** (colors: dark greenish to black) (About equal amounts of pyroxene and olivine; may contain abundant minerals of chromium, nickel, and platinum.) (gradation) | **Dunite** (color: green) (Almost wholly olivine; usually includes magnetite and other metallic-oxide minerals) |
| Mineral composition (Thickness of mineral field in diagram shows usual proportion of that mineral in rocks. Full height of this rectangle equals 100 percent) | | | | |



[1] No pegmatites, no aphanitic varieties, no glasses.

**Table C.4** Bases for classifying sedimentary rocks and pyroclastic rocks. Left-hand part is for varieties where the name is based on sizes (and in some cases, on shapes) of particles. Right-hand part is for rocks named on the basis of composition in which size is not a factor in giving the name.

| Name of pyroclastic rock having constituents chiefly in size class shown | Names of particle-size groups | | | Names of sedimentary rocks formed by lithification of physically transported sediments | | |
|---|---|---|---|---|---|---|
| | Tephra | Sediments | | Rocks based on size only | Composed of detritus (chiefly silicates) | Composed of carbonate minerals |
| Agglomerate | Bombs (molten when ejected) | Boulders | Gravel (Fragments angular or rounded; Figure C.19) | Rudite | Conglomerate (>50 percent rounded particles) (Figure C.20)<br>Sedimentary breccia (>50 percent angular particles) (Figure C.21) | Calcirudite |
| | | 256 mm | | | | |
| | Blocks (solid when ejected) | (Volleyball) Cobbles | | | | |
| | | 64 mm | | | | |
| | | (Tennis ball) | | | | |
| | 32 mm | | | | | |
| Lapilli tuff | Lapilli | Pebbles | | | | |
| | 4 mm | 4 mm | | | | |
| Tuff | Ash | (Granules) | | | | |
| | | 2 mm | | | | |
| | | Very coarse | Sand | Arenite | Sandstone (Figure C.22) (quartz predominant)<br>Feldspathic sandstone (feldspar to 25 percent)<br>Arkose (feldspar >25 percent)<br>Graywacke (rock firmly cemented by chlorite and micas) | Limestone (calcite or aragonite predominate)<br>Dolostone (>50 percent dolomite) (See Figure C.1)<br>Calcarenite; (Figure C.23)<br>Oölite (See Figure C.9) |
| | | 1 mm | | | | |
| | | Coarse | | | | |
| | | 0.5 mm | | | | |
| | | Medium | | | | |
| | | 0.25 mm | | | | |
| | | Fine | | | | |
| | | 0.125 mm | | | | |
| | | Very fine | | | | |
| | 0.063 mm | | | | | |
| | Dust | Silt | "Mud" | Lutite | Siltstone | Calcilutite; chalk; lithographic limestone;<br>Lime mudstone; marlstone |
| | | Clay | | | Claystone / Argillite (firmly cemented hard rock) / Shale (possesses fissility) (See Fig. C.12) | |

| Sedimentary rocks based on composition (size not involved in name) | |
|---|---|
| Composed of hematite and other iron minerals | Ironstone formation (Used for Precambrian rocks) |
| Composed of plant detritus | Peat<br>Lignite<br>Bituminous coal (various ranks) (Figure C.25) |
| Composed of silica | Chert (Figure C.24)<br>Jasper<br>Flint<br>Chalcedony rock |
| Composed of evaporite minerals[1] | $CaSO_4$: Gypsum rock, Anhydrite rock<br>NaCl: Halite rock ("salt") |
| Composed of carbonate minerals | Limestone<br>Travertine<br>Dolostone |

[1] Recent studies have shown that many evaporite rocks display features and textures indicative of physical transport of the crystals. Such features include laminae, cross strata, and graded beds.

# APPENDIX D THE PHYSICS OF GEOLOGY

This appendix includes the following: Some properties of ellipses, energy, solar energy, elastic properties of matter and elastic waves, gravity, magnetic lines of force, and the age of Earth expressed in seconds using powers of 10.

## SOME PROPERTIES OF ELLIPSES

An ellipse is a closed circular figure having two mutually perpendicular axes whose lengths are not equal, as they are in a circle. The longer axis is the *major axis* of the ellipse and the shorter axis, the *minor axis*. In mathematical analyses, the distance from the point of intersection of the axes (0 in Figure D.1) to the ellipse is particularly important. These two distances are exactly half of the total length of the axes; they are referred to as the *semimajor axis* (symbolized by the letter $a$) and the *semiminor axis* ($b$).

If a circle having a radius equal in length to $a$ is centered at B, it will intersect the major axis ($AA'$) at two points, $F$ and $F'$. These two points are the *foci* of the ellipse.

The ratio of the distance $OF'$ to $OA'$ ($OA'$ equals the semimajor axis $a$) is the *eccentricity* (or *ellipticity*) of the ellipse (symbolized by $e$). Eccentricity may be expressed as a decimal or as a percent (for example, 0.17 or 17 percent). One finds the value of the ellipticity, by first computing the distance $OF'$. This is done by using the Pythagorean theorem for analyzing the right triangle having $a$ as hypotenuse, $b$ as one of the legs, and $OF'$ as the other leg:

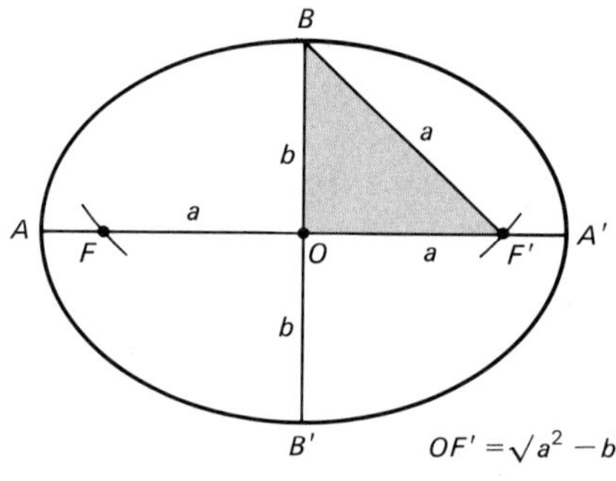


$OA = OA' = a$ = semimajor axis
$OB = OB' = b$ = semiminor axis
$F$ and $F'$ = foci

**D.1 Ellipse,** with ellipticity of about 70 percent, showing definitions of fundamental parts. Further explanation in text.

$$a^2 = b^2 + \overline{OF}'^2$$

$$\overline{OF}' = \sqrt{a^2 - b^2}$$

$$\text{Eccentricity } (e) = \frac{\overline{OF}'}{a}$$

To draw an ellipse, place a pin or a tack at each focus, and tie a piece of thread or string around each tack with the length just passing point $B$ ($FBF'$). Place a pencil or pen inside the loop and move the pencil or pen slowly around from $A$ to $A'$ to draw half the ellipse. To draw the other half, shift the string loop to the opposite side of the two pins or tacks and repeat.

## ENERGY

In a general physical sense, *energy* is defined as the capacity to do work. By "work" is usually meant some effect that is mechanical, electrical, or thermal (heat). Table D.1 gives the units and definitions of these three kinds of energy. Box D.1 summarizes various kinds of energy.

An entirely different insight into energy and matter was made by Albert Einstein (1879–1955) in his now-famous equation:

$$E = m \cdot c^2$$

where $E$ = energy (expressed in ergs)
$m$ = mass (in grams)
$c$ = speed of light (or other electromagnetic waves; expressed in cm/sec)

From measurements of the speed of light, $2.998 \times 10^{10}$ cm/sec, we can calculate the amount of energy in a one-gram mass of matter:

$$E = 1 \times (2.998 \times 10^{10})^2$$
$$= 9 \times 10^{20} \text{ ergs}$$

With the Einstein relationship in

**TABLE D.1**
DEFINITIONS OF ENERGY UNITS

| Kind of Energy | Name of Unit | Definition |
|---|---|---|
| Mechanical | Erg | The mechanical energy required for a force of 1 dyne to act through 1 centimeter. A graphic example of one erg's worth of work is a housefly doing a single push-up. |
| | Dyne | The force required to accelerate a mass of 1 gram by 1 centimeter per second per second. |
| | Joule | $10^7$ ergs. |
| Heat | Calorie | The amount of heat energy required to raise the temperature of 1 gram of water by 1°C at 15°C. One calorie equals 4.18 joules. |
| Electrical | Electron volt | The energy possessed by 1 electron that has fallen through a potential difference of 1 volt. One electron volt equals $1.6 \times 10^{-6}$ ergs. |

mind, we can define energy in some forms as the equivalent of mass.

### Energy and Electromagnetic Waves

As modern physicists have examined matter and energy in greater detail, they have found that it is difficult, if not impossible, to make a clear-cut distinction between tiny particles and *electromagnetic waves,* the general name for all waves that travel at the same speed as light waves. Tiny energy particles are photons.

The fundamental equation for

BOX D.1

## Kinds of Energy

The various kinds of energy are related to different scales of sizes of the matter involved. Energy that is particularly important to large objects includes potential energy and kinetic energy. Other kinds of energy are related to molecules, ions, or subatomic particles. Such energy includes heat energy, change-of-state energy, bonding energy, electrical energy, nuclear energy, and radiant energy.

In mechanics, two important forms of energy are (1) potential and (2) kinetic. **Potential energy** is non-realized energy or stored energy. One form relates to position. A boulder perched on top of a cliff has potential energy equal to its mass times the height of the cliff. Other potential energy is elastic; that is energy that is stored in a body being deformed, as in a rubber band being stretched. Sudden release of the *elastic energy* stored in deformed parts of the lithosphere is responsible for earthquakes.

**Kinetic energy** is the energy of a moving object. It may be thought of as actualized former potential energy. A boulder falling down a hill or water flowing in a stream possesses kinetic energy.

**Heat energy** affects the temperature of an object; what we call heat is the motion of ions (or molecules in a gas). The greater the heat energy, the greater the motion of the ions and the higher the temperature.

**Change-of-state energy** is also known as *latent heat.* It is the energy (usually measured in calories per gram) added or released when matter changes its state of aggregation (such as liquid water freezing to solid ice) without a change in temperature of the matter involved. This qualification is necessary because released latent heat can raise the temperature of nearby substances. The following definitions apply:

| States of aggregation | Kind of latent heat | Value for water (calories/gram) |
|---|---|---|
| Liquid-gas | **Latent heat of vaporization** | 539.0 |
| Liquid-solid | **Latent heat of fusion** | 79.7 |
| Solid-gas | **Latent heat of sublimation** | 618.7 |

For example, when a crystal forms from a magma, the latent heat of fusion is released. By contrast, to melt a crystal, one must first heat it to its melting temperature and then add additional energy equal to the latent heat of fusion before the crystal will melt.

**Bonding energy** is the energy associated with the making and breaking of bonds among ions, ion complexes, and molecules.

**Electrical energy** is the flow of electrons in a conducting medium, such as along a copper wire. A common method for generating electrical energy is to spin a copper coil inside a magnetic field. As the wire cuts the magnetic lines of force, an electrical current flows along its length. What is more, the flow of electrons along the wire creates its own surrounding magnetic field.

**Nuclear energy** is the energy related to changes in the nuclei of atoms (thus it is also known as *atomic energy*). The spontaneous splitting of an atomic nucleus, the emission of alpha particles from a nucleus, or the expulsion or capture of electrons are known as radioactivity (see Appendix A). Enormous amounts of heat and electromagnetic waves are released during nuclear reactions. Electromagnetic waves are commonly referred to as **radiant energy** (see section on Solar Energy).

waves of any kind states that the frequency of the waves is the product of the periods of the waves and the wavelengths. The period of a wave is the time required for a complete wave form to pass a given point. The wavelength is the distance between successive wave peaks (or between wave troughs; see Figure 15.2). In mathematical notation, the fundamental wave equation becomes:

$$\mu = n\lambda$$

where $\mu$ = frequency
$n$ = wave period
$\lambda$ = wavelength

In 1901, the German physicist Max Planck (1858–1947) proposed that radiant energy is emitted or absorbed in discrete entities (named photons, or *quanta*) and that the energy of the photons is proportional to the frequency of the radiation. By introducing what became known as Planck's constant, this relationship may be stated mathematically:

$$E = h\mu$$

where $E$ = Energy of photons (ergs)
$h$ = Planck's constant ($6.624 \pm 0.002 \times 10^{27}$ erg·sec)
$\mu$ = frequency (cycles·sec$^{-1}$)

A remarkable characteristic of electromagnetic waves is that they all travel at the same speeds, but their frequencies are controlled by the vibrations of the generating object; the larger the vibrating object, the longer the wavelengths of the electromagnetic waves (Figure D.2). The long-wavelength group is generated by the mechanical vibration of objects larger than molecules. Such objects include crystals, metal plates, and so forth; they give rise to what are known as

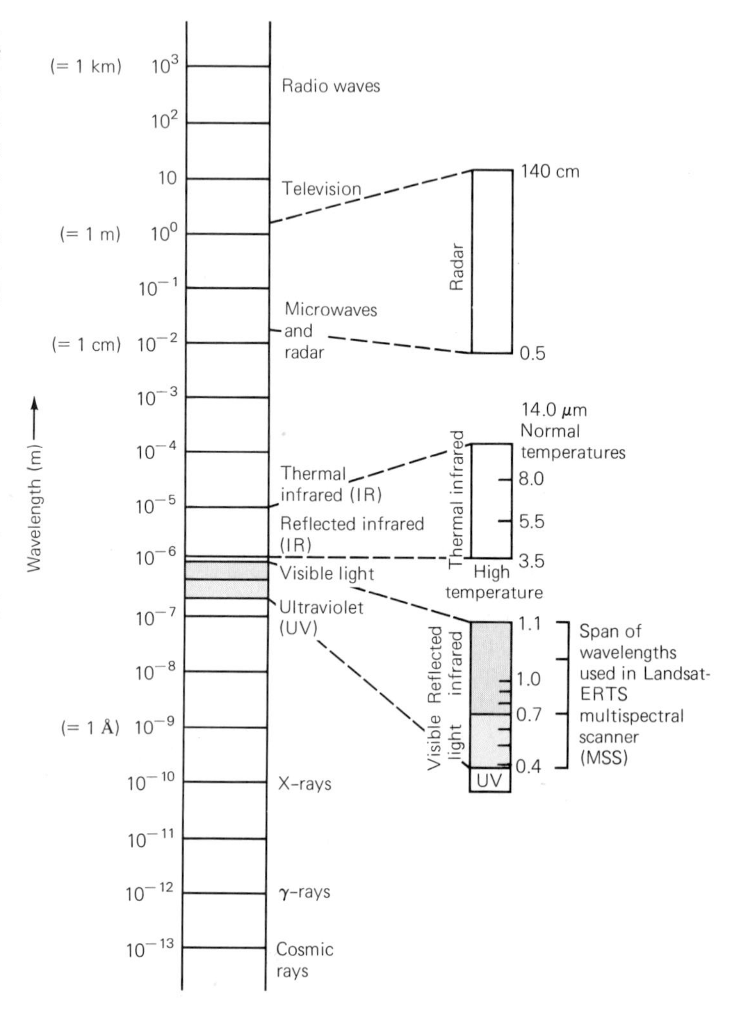


**D.2 Electromagnetic spectrum,** showing enlarged regions spanned by radar, thermal infrared, and visible light with adjoining waves.

*Hertzian waves* (named after the German physicist Heinrich Rudolf Hertz, 1857–1894). In Hertzian waves, the atomic structure within the generating object does not exert the major control on the wavelengths of the waves.

Vibrations of molecules, and to a certain extent, of the ions forming the molecules, create *infrared waves.* Within the range of infrared waves are irregular, discrete changes of frequency that are related to the kinds of waves that are generated by energy from vibrating electron shells of ions. Such discrete changes of frequency are named quantum jumps.

If the vibrating energy source consists of the electron shells of ions, the resulting electromagnetic waves lie in the range of visible light and ultraviolet light (see Appendix Figure D.2). If the inner shells of ions are the sources, the result is *X-rays*. When the atomic nuclei are the sources, the resulting waves are *gamma rays*.

The shortest-wavelength waves of all are the *cosmic rays*. The source of these waves is not known.

### Electromagnetic Waves as Analytical Tools

The science of astrophysics and many branches of analytical chemistry have been built on the principle that predictable and quantitative relationships exist between the kinds of electromagnetic waves and chemical molecules, ions, or nuclei. This principle likewise forms the basis for identifying ions by means of a spectroscope, which is an instrument for making detailed measurements of electromagnetic waves generated by ions being tested.

Since the field of solid-state physics has developed in the 1960s, the study of electromagnetic waves has expanded enormously. It is now possible to make detectors that respond to any given kind of electromagnetic waves and that express this response as an electric signal. The electric signal can be measured, displayed, stored on tape, passed through computers, or manipulated in a nearly endless variety of ways.

As a result of the existence of these remarkable detectors, a whole new field of science important to geology and to space studies has grown up. It is known as *remote sensing* and consists of finding out important environmental information about the surface of the Earth or about any other body in the universe by analyzing the electromagnetic waves that the body may be giving off on its own or generating in response to some other kind of radiation. This other kind of radiation may be solar energy as reflected by the planets within the Solar System, or some kind of energy beamed outward from the Earth or a satellite, such as radio waves or radar waves.

For example, matter radiates infrared waves in proportion to its temperature. Infrared mapping of bodies of water is capable of indicating temperature changes as small as 0.01°C. Such mapping has resulted in the discovery of freshwater springs in Hawaii, where the temperature of the spring water is much less than that of the surrounding tropical ocean water. Infrared detectors have been used to study volcanic eruptions (see Figure 19.17), and to search for places where sulfide minerals associated with deposits of valuable metals may be oxidizing, thus generating heat.

By passing the light emitted by distant stars through a spectroscope, it is possible to infer the chemical composition of the stars. Astrophysicists have used electromagnetic waves not only for inferring the chemical compositions of stars, but also for drawing important conclusions about the nature of the universe. Their conclusions are based on a phenomenon that has been named the *red shift*. This "shift" refers to the spectral lines of hydrogen, which form a characteristic pattern and normally occupy a fixed position in the electromagnetic spectrum. Analysis of the light from distant stars has shown that the pattern of hydrogen spectral lines does not appear in its usual wavelength position. Instead, the distinctive hydrogen pattern has been found at higher-than-normal frequencies, within the range of infrared waves. This is the basis for the name red shift; the spectral lines have shifted into the red range.

The red shift has been explained by a phenomenon affecting waves that is known as the *Doppler shift*. The Doppler shift refers to a change in the characteristics of waves that comes about if the source and receiver are moving. If source and receiver are not moving, then frequency and wavelength remain fixed. However, if the source is moving toward the receiver, then the frequency of the waves increases (hence in electromagnetic waves, the wavelength decreases). By contrast, if source and receiver are moving away from each other, then the frequency of the waves decreases (and, in electromagnetic waves, the wavelength *increases*). The shift of the spectral lines of hydrogen from their "standard" wavelengths to the greater wavelengths of the infrared range has been taken to mean that the light from stars exhibiting the red shift is coming from a star that is moving *away* from the Earth. Another factor in the Doppler shift is speed: the faster the source is moving with respect to the detector, the greater is the shift of frequency. Stars at greatest distances from the Earth display the greatest amounts of red shift. This finding has been interpreted as meaning that the farther from the Earth the star is located, the faster the star is traveling away from the Earth. The interpretations related to the red shift form the basis for the concept of an expanding universe.

## SOLAR ENERGY

Because solar energy is one of the things that makes life possible on Earth, it is important for us to become acquainted with some of the principles related to this energy. We summarize the characteristics of solar energy and its decrease in intensity outward from the Sun.

### Characteristics of Solar Energy

Solar energy can be described by its spectrum and intensity. The spectrum of solar energy features a significant peak in the range of visible light, but also includes infrared radiation on the side of the spectrum where wavelengths are shorter than visible light (Figure D.3).

From our previous discussion about energy and electromagnetic waves, we can conclude that on the Sun, the energy is being radiated outward from a series of vibrating molecules, electron shells, and atomic nuclei, and, in addition, from whatever gives rise to the somewhat-mysterious cosmic rays.

The element helium was first discovered in the spectroscopic analysis of solar radiation. Only afterward was helium found on Earth. The presently preferred interpretation of the origin of solar energy is that hydrogen is transformed by fusion into helium. As a result, $4.7 \times 10^6$ tons of the Sun's hydrogen vanish each second, and

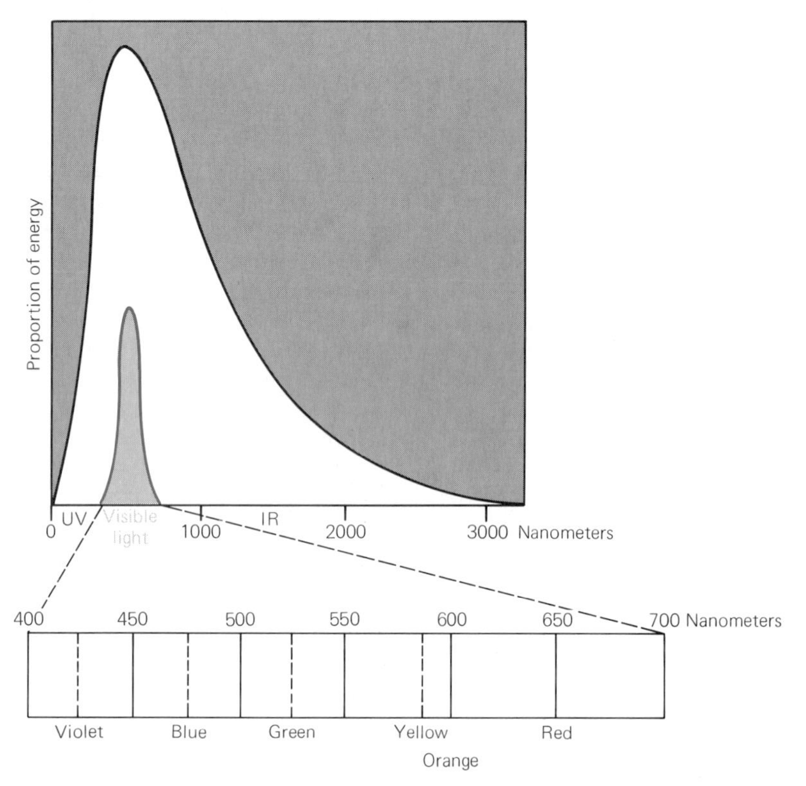


**D.3 Spectrum of sunlight** (black curve), showing range of sensitivity of the human eye (blue curve), and expanded scale at bottom to indicate wavelengths of colors.

enormous quantities of energy are created and radiated outward.

### Decrease in Intensity of Energy Outward From the Sun

The energy from any concentrated source, such as the Sun, typically radiates outward in all directions uniformly. This radially outward distribution of the energy involves an important physical principle known as Snell's law (named for its discoverer, the Dutch astronomer and mathematician Willebrord Snell, more commonly known as Snellius, 1591–1626).

We can visualize how Snell's law works by reference to concentric "energy spheres" that we can imagine surround a fictitious Sun that is a single point at the center of the concentric spheres. The same amount of energy spreads across the surface of each "energy sphere." However, because the surface area of each "energy sphere" becomes larger outward, the intensity of the energy on a unit area of each sphere diminishes.

For example, suppose the radius of one "energy sphere" is 3 million kilometers and that of the second "energy sphere" is 6 million kilometers, which is twice as large as the first. The formula relating the surface areas of a sphere to its radius is:

$$\text{Surface area} = 4 \cdot \pi \cdot (\text{radius})^2$$

Using this formula, we can calculate the surface areas of the two "energy spheres" and compare them. The surface area of the first "energy sphere" is:

$$4 \cdot \pi \cdot (3 \times 10^6)(3 \times 10^6) = 4 \cdot \pi \cdot 9 \times 10^{12}\ \text{km}^2$$

The surface area of the second "energy sphere" is:

$$4 \cdot \pi \cdot (6 \times 10^6)(6 \times 10^6) = 4 \cdot \pi \cdot 36 \times 10^{12}\ \text{km}^2$$

In doubling the radius of the "energy sphere," we have increased the surface area by four times (from 9 to 36 times $4\pi \times 10^{12}$ km$^2$). We can visualize 4 as being the square of 2. Expressed as a general principle, therefore, *Snell's law* states that the intensity of radiating energy decreases with the square of the distance from the source. In physics, this is also known as the *inverse-square law.* Because of their varying distances from the Sun, the intensity of solar energy received on one square meter at the surface of Mercury is about 3 times that on Venus, 6.5 times that on Earth, and 15.5 times that on Mars.

Moreover, planetary orbits are ellipses not circles (see Figure 1.1b and c). Therefore, the distance between any planet and the Sun, hence the intensity of solar energy, varies during each orbit. For example, the distance between the Earth and the Sun ranges between about 149.6 million kilometers and 152.0 million kilometers (see Table 1.1, p. 20). The intensity of solar radiation on Earth at a distance of 149.6 million kilometers is 6.7 percent greater than the intensity at 152.0 million kilometers.

## ELASTIC PROPERTIES OF MATTER AND ELASTIC WAVES

The *elastic properties* of matter refer to the circumstances involving changes in shape or in volume. Elastic properties are very important in the use of solids for construction purposes. Because of this, the elastic properties have been studied in great detail by engineers, and the fundamental mathematical relationships have been established. In the language of the engineers, a numerical coefficient expressing some property is commonly designated as a *modulus.* As far as elastic properties are concerned, two such numbers are the *modulus of rigidity* and *bulk modulus.* The modulus of rigidity is an expression of the elasticity of shape of a solid; it is defined as the pressure required to cause a unit change of shape. In this definition, pressure is defined as force per unit area. The bulk modulus is an expression of the elasticity of volume; it applies to both solids and fluids. The bulk modulus is defined as the pressure required to cause a unit change of volume.

*Elastic waves* are so named because their speeds depend on the elastic properties of the materials through which they pass. In other words, the same waves may travel fast through some materials and slow through other materials. Box D.2 shows the mathematical relationships among elastic properties, densities, and speeds of two kinds of elastic waves.

## GRAVITY: *G* AND *g*

The two *G*'s involved in analyses of gravity are a potential source of confusion. Here they are defined and explained.

*G* is the Universal gravitational constant; it is the attractive force between two masses of 1 gram each (which can be considered as points) that are exactly 1 centimeter apart

---

**BOX D.2**

## Elastic properties, densities, and speeds of elastic waves

In many geologic situations, it is not possible to examine something directly. For example, one cannot collect a specimen of the lithosphere from deeper than about 10 kilometers, the depth of the deepest boring. Even though a body of material may not be readily accessible, one can learn important things about it by measuring the speeds of passage of certain kinds of waves. We can illustrate with two kinds of waves: (1) longitudinal waves, and (2) shear waves.

The speed with which longitudinal waves pass through a homogeneous solid is a function of three variables: (1) modulus of rigidity, (2) bulk modulus, and (3) density. The equation is as follows:

$$C_1 = \sqrt{\frac{(\frac{4}{3})\mu + \kappa}{d}}$$

where: $C_1$ = celerity of longitudinal waves
$\mu$ = modulus of rigidity
$\kappa$ = bulk modulus
$d$ = density

The speed with which shear waves ($C_s$) pass through a body is governed only by two variables: (1) modulus of rigidity and (2) density:

$$C_s = \sqrt{\frac{\mu}{d}}$$

Expressed mathematically, the inability of fluids to maintain their own shapes means that the modulus of rigidity of a fluid is zero. If the value of $\mu$ in the equation for the speed of shear waves is set equal to zero, then the value of $C_s$ becomes zero. This is the mathematical basis of the point that fluids do not transmit shear waves. Thus, shear waves can be used as a conclusive means of distinguishing solids from fluids.

Notice that the expression $\mu/d$ appears in both of the equations for wave speed. If we square both sides of each equation, multiply both sides of the second equation by $\frac{4}{3}$, and rearrange terms, we have:

$$(C_1)^2 = \frac{4}{3}\frac{\mu}{d} + \frac{\kappa}{d}$$

$$\frac{4}{3}(C_s)^2 = \frac{4}{3}\frac{\mu}{d}$$

$$\frac{4}{3}\frac{\mu}{d} = (C_l)^2 - \frac{\kappa}{d} = \frac{4}{3}(C_s)^2$$

$$\frac{\kappa}{d} = (C_l)^2 - \frac{4}{3}(C_s)^2$$

What this last equation enables us to do is to measure the speeds with which longitudinal waves and shear waves pass through a body, and, from these speeds, to calculate the ratio of the bulk modulus and the density. Knowing this ratio, we can narrow down the field considerably in an attempt to understand what the material may be. This approach forms the basis for understanding much about the interior of the Earth (Chapter 17) and of other bodies in the Solar System where one can measure these waves.

(this does not apply to concentric hollow spheres).

$g$ is the value of the acceleration caused by the mass of the Earth on other bodies. It is the inward-acting force that tends to pull all objects on the surface of the Earth toward the center of the Earth (see Figure 11.2).

Newton's famous law of gravitational attraction included $G$ but not $g$:

$$F = \frac{GM_1M_2}{d^2}$$

where $F$ = force of attraction
$M_1$ = mass of one body (whose mass can be considered as concentrated at a single point, the body's center of mass)
$M_2$ = Mass of a second body (as in $M_1$)
$d$ = distance between the centers of mass of the bodies

In order to find $g$, one makes use of Newton's equation with substitutions for examples on the Earth. The mass of the Earth, $M_E$, replaces $M_1$; the radius of the Earth (considered as a sphere), $r$, replaces $d$; and the mass of an object on the surface of the earth, $M$, replaces $M_2$. We can now write:

$$F = M \cdot G \cdot \frac{M_E}{r^2}$$

Newton's Second Law of Motion states that the force acting on a body, its mass, and its acceleration are related as follows:

$$F = M \cdot a$$

or, the acceleration equals the force divided by the mass:

$$\frac{F}{M} = a$$

The gravitational force of the Earth's attraction on a unit mass resting on its surface is found by:

$$\frac{F}{M} = G \cdot \frac{M_E}{r^2}$$

Because $F/M = a$ it follows that $a = G \cdot (M_E/r^2)$. We defined $g$ as the acceleration caused by the attraction of the mass of the Earth on other bodies; therefore:

$$a = G \cdot \frac{M_E}{r^2} = g \quad \text{or} \quad G = \frac{g \cdot r^2}{M_E}$$

The mean value of $g$ is 980 dynes per gram, but this value varies according to latitude, altitude, and the densities of near-surface rock bodies.

## MAGNETIC LINES OF FORCE

Magnetic lines of force are diversely oriented. In order to analyze any magnetic field, it is necessary to establish some standard angles of reference. The orientation of a magnetic force vector of a magnetic field is shown in Figure D.4. The two standard angles are:

**Declination** (symbol **D**): a compass bearing from true North (clockwise angle from true North measured in the horizontal plane)

**Inclination** (symbol **I**): an angle from the horizontal measured in the vertical plane passing through the magnetic force vector

In the Earth's present magnetic field, the inclination is +90° at the North Magnetic Pole; 0° at the Magnetic Equator, and −90° at the South Magnetic Pole (see Figure 1.15b).

In analyzing the past magnetic fields of the Earth, geophysicists do two things. First, they make detailed measurements of the inclination and declination of the remanent magnetism in specimens. Second, they reconstruct ancient positions of the Earth's magnetic poles by making plots of inclination *versus* declination.

## AGE OF THE EARTH IN AN ORDER-OF-MAGNITUDE SCALE

Using radioactive methods, geologists are now convinced that the age of the Earth is about 4.6 billion years. This is such an enormous length of time that it is not easily comprehended. However, we can try to give some idea by referring to an important kind of scientific scale based on powers of 10—an order-of-magnitude scale. Let us use one second of time as the basic unit. Because there are 60 seconds in a minute, 60 minutes in an hour, and 24 hours in a day, the length of a day is 86,400 seconds. This lies between $10^4$ and $10^5$ on the order-of-magnitude scale (Figure D.5). The length of a year (365¼ days) in seconds is 31,557,600, which lies between $10^7$ and $10^8$ on the order-of-magnitude scale. A human lifetime of 70 years is a bit more than 2 billion seconds, or just about $10^9$ on the order-of-magnitude scale. Thus, the amount of time between 1 second and 70 years spans 9 orders of magnitude.

To compare how a single year relates to 4.6 billion years, notice that 4.6 billion can be expressed as $4.6 \times 10^9$, which means 9 orders of magnitude. Expressed on the order-of-magnitude scale in seconds, the age of the Earth lies between $10^{16}$ and $10^{17}$. From the scale we can conclude that one second compares to two long human lifetimes (158 years) in the same way that one year compares to the age of the Earth. Using this ratio of one second to 158 years, it is possible to express the events in the history of the Earth within the time frame of an ordinary year, beginning 1 January and ending 31 December.

The oldest rocks (about 3.8 billion years old) would date from mid-March. The first living creatures would appear in May, and land plants and animals in late November. The Rocky Mountains would not arise until 26 December, and the first humanlike animals (geologic age about five million years) would appear at approximately 3 P.M. on 31 December. The last glaciers of North America would begin their retreat from the Great Lakes about one minute before midnight. The seeds of modern geology would be sown just one second before the end of the year.

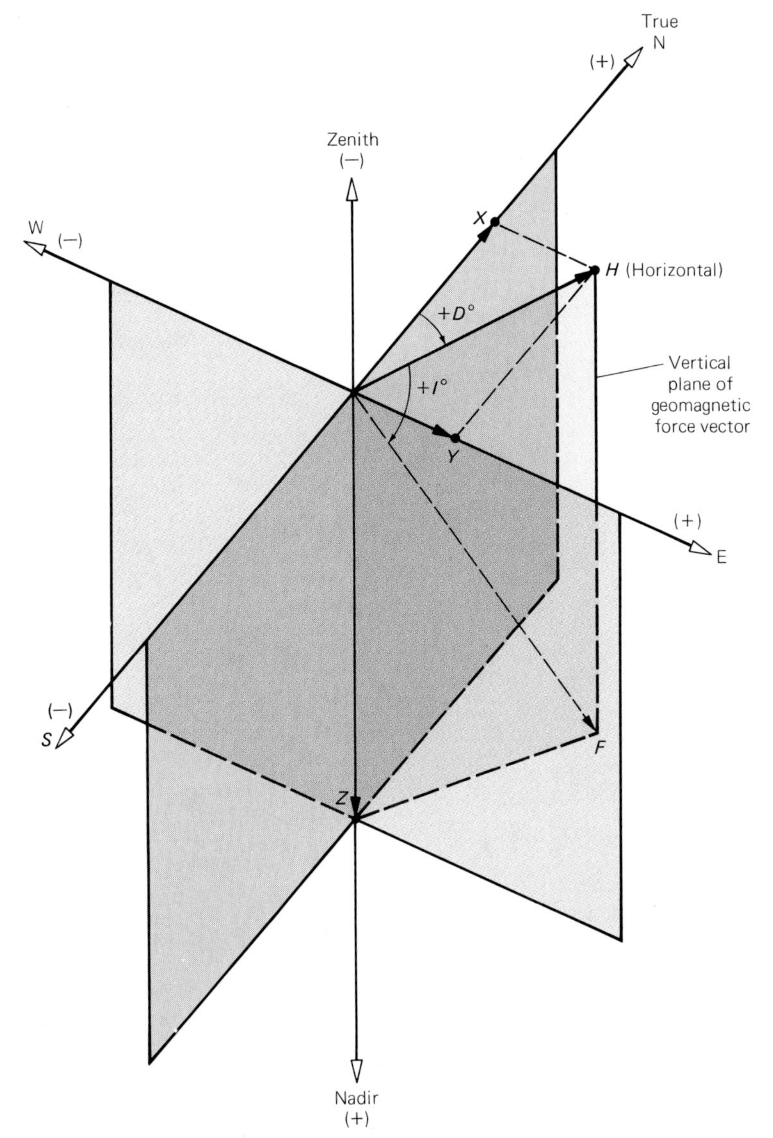


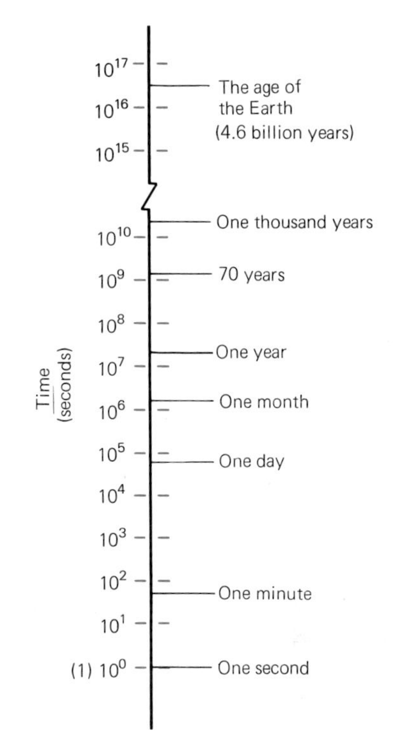


**D.5** The age of the Earth in seconds shown on an order-of-magnitude graphic scale.

**D.4 Geomagnetic line-of-force** vector (*OF*) shown with standard references axes and standard angles and components. The *X*-axis is true North-South, with North positive (+) and South, negative (−). The *Y*-axis is East-West, with East positive (+) and West, negative (−). In the *Z*-axis, nadir (down) is positive (+), and zenith (up) is negative (−). The geomagnetic line-of-force vector makes an *angle of inclination* (of *I* degrees) with the horizontal plane, as measured in the vertical plane of the vector, using the convention of positive (+) if downward, as shown here, but negative (−) if upward from the horizontal. A geomagnetic line-of-force vector can be resolved into a *horizontal component* (*OH*) and a vertical component (*OZ*). The vertical plane containing the line-of-force vector makes an *angle of declination* (*D* degrees) with true North that is positive if clockwise, as shown here, but negative if counterclockwise. The horizontal component (*OH*) can be resolved into a North-South component (*OX*) and an East-West component (*OY*).

# APPENDIX E METRIC CONVERSION FACTORS AND POWERS OF TEN

| ENGLISH TO METRIC | METRIC TO ENGLISH |
|---|---|
| *Weight (Mass)* | |
| Ounces × 28.3495 = grams | Grams × .03527 = ounces |
| Pounds × .4536 = kilograms | Kilograms × 2.2046 = pounds |
| Tons × 1.1023 = tonnes (1000 kg) | Tonnes × .9072 = tons |
| *Length* | |
| Inches × 25.4 = millimeters | Millimeters × .0394 = inches |
| Inches × 2.54 = centimeters | Centimeters × .3937 = inches |
| Feet × .3048 = meters | Meters × 3.2809 = feet |
| Yards × .9144 = meters | Meters × 1.0936 = yards |
| Miles × 1.6093 = kilometers | Kilometers × .621377 = miles |
| *Area* | |
| Square inches × 6.4515 = square centimeters | Square centimeters × .155 = square inches |
| Square feet × .0929 = square meters | Square meters × 10.7641 = square feet |
| Square yards × .8361 = square meters | Square meters × 1.196 = square yards |
| Square miles × 2.59 = square kilometers | Square kilometers × .3861 = square miles |
| Acres × .405 = hectares | Hectares × 2.471 = acres |
| Square miles × 259.07 = hectares | Hectares × .00386 = square miles |
| *Measure* | |
| Fluid ounces × 29.47 = milliliters | Milliliters × .0339 = fluid ounces |
| Quarts × .9433 = liters | Liters × 1.06 = quarts |
| Gallons × 3.774 = liters | Liters × .265 = gallons |
| Cubic inches × 16.4 = cubic centimeters | Cubic centimeters × 0.06 = cubic inches |
| Cubic feet × .02832 = cubic meters | Cubic meters × 35.3156 = cubic feet |
| Cubic yards × .7645 = cubic meters | Cubic meters × 1.308 = cubic yards |
| Cubic miles × 4.19 = cubic kilometers | Cubic kilometers × 0.24 = cubic miles |
| *Pressure* | |
| $lb/in.^2 \times .0703 = kg/cm^2$ | $kg/cm^2 \times 14.2231 = lb/in.^2$ |

*Temperature*

*When you know the Fahrenheit temperature:*

$$°C = \frac{(°F - 32)}{1.8}$$

*When you know the Celsius temperature:*

$$°F = (1.8 \times °C) + 32$$

| °C | | °F |
|---|---|---|
| 100 | Water boils | 212 |
| 90 | | 194 |
| 80 | | 176 |
| 70 | | 158 |
| 60 | | 140 |
| 50 | | 122 |
| 40 | | 104 |
| 37 | Normal body temperature | 98.6 |
| 30 | | 86 |
| 20 | | 68 |
| 10 | | 50 |
| 0 | Water freezes | 32 |
| −5 | | 22 |
| −10 | | 14 |
| −15 | | 4 |
| −20 | | −4 |
| −25 | | −13 |
| −30 | | −22 |
| −35 | | −31 |
| −40 | | −40 |
| −45 | | −49 |
| −50 | | −58 |

MULTIPLES (POWERS OF 10) AND PREFIXES

| Multiples and Submultiples | Prefix | Symbol |
|---|---|---|
| $1{,}000{,}000{,}000{,}000 = 10^{12}$ | tera | T |
| $1{,}000{,}000{,}000 = 10^{9}$ | giga | G |
| $1{,}000{,}000 = 10^{6}$ | mega | M |
| $1{,}000 = 10^{3}$ | kilo | k |
| $100 = 10^{2}$ | hecto | h |
| $10 = 10^{1}$ | deka | da |
| Base Unit $1 = 10^{0}$ | | |
| $0.1 = 10^{-1}$ | deci | d |
| $0.01 = 10^{-2}$ | centi | c |
| $0.001 = 10^{-3}$ | milli | m |
| $0.000001 = 10^{-6}$ | micro | μ |
| $0.000000001 = 10^{-9}$ | nano | n |
| $0.000000000001 = 10^{-12}$ | pico | p |
| $0.000000000000001 = 10^{-15}$ | femto | f |
| $0.000000000000000001 = 10^{-18}$ | atto | a |

# APPENDIX F
# UNDERSTANDING TOPOGRAPHIC MAPS

## SOME GENERAL BACKGROUND ABOUT MAPS

A *map* is a scaled two-dimensional (*i.e.*, flat) representation of some part of the Earth's curved surface. Because some distortion is inevitable when one tries to convert the curved surface of the Earth into a single flat plane of a map, various schemes of map projection have been devised for shifting the information from the surface of the Earth to a flat map in such a way as to reduce the distortion for parts of the map. (The earliest map makers were not troubled by projections or distortion—they thought that the Earth is flat.)

No matter what features it displays, any map involves certain common information that all users must understand. To start with, you should ask four important questions about any map.

1. What part of the Earth's surface is represented on the map?
2. What kind of projection was used to construct the map?
3. What kind(s) of grid system(s) is (are) shown?
4. What is the scale of the map and how is the map oriented?

The following paragraphs answer these four questions. After we have discussed the answers, we summarize map symbols and explain what various lines mean.

### Location of Points on the Surface of the Earth: Longitude and Latitude

Points on the surface of the Earth can be located by using various systems of spherical coordinates. The scheme most widely used today involves the longitude-latitude coordinate system. This system was invented by the ancient Greeks, who were skilled in geometry.

**Meridians of longitude** One part of this coordinate system consists exclusively of great circles that pass through the pole of rotation and that cross the Equator at right angles. The halves of each of these great circles that extend from one pole to the other pole are *meridians* (Figure F.1). *Longitude* is defined as the angle, in the equatorial plane, between the meridian passing through the Greenwich Observatory, outside London, England (known as the Zero or Prime Meridian), and the meridian passing through a given point. The value of longitude ranges from 0° to 180° E or from 0° to 180° W of Greenwich. The 180-degree meridian is the International Date Line. The Prime Meridian and the International Date Line are the boundaries between the Eastern Hemisphere and the Western Hemisphere (Figure F.2). If slices were made to the center of the Earth along successive meridians, the result would be a series of curved wedge-shapes pieces, as when a cantaloupe is cut.

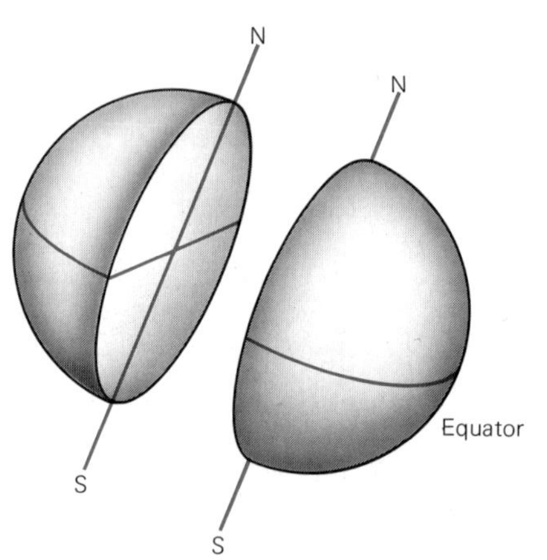


**F.1 The Earth divided into Eastern and Western Hemispheres** by a great circle passing through the pole.

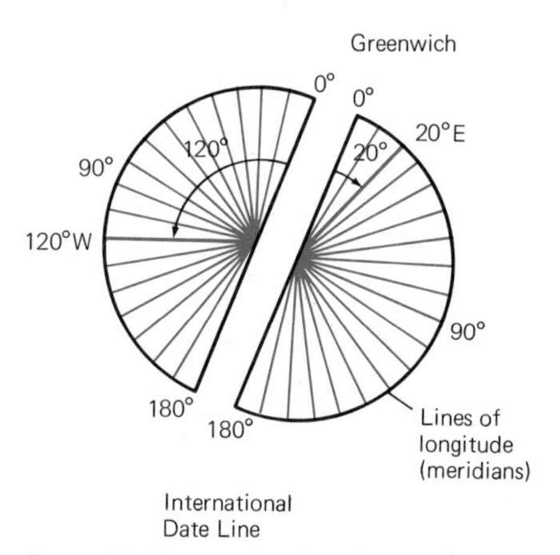


**F.2 Eastern and Western Hemispheres sliced along the Equatorial plane** to show measurement of longitude.

**Parallels of latitude** The reference circles of the second part of the Greeks' spherical-coordinate system contain only one great circle, the Equator itself. A slice through the Earth at the Equator divides the world into Northern Hemisphere and the Southern Hemisphere (Figure F.3). The other reference circles are small circles parallel to the Equator. These reference circles are named *parallels*. Within each of the two hemispheres bounded by the Equator, parallels are defined by angles from the center of the Earth, measured from the Equator to a given point. This angle is known as *latitude*. The value of latitude ranges from 0° at the Equator to 90° N at the North Pole and 90° S at the South Pole (Figure F.4). To show how these two angles define a point, Figure F.5 illustrates the longitude and latitude of New York City (Figure F.5).

It must be obvious to every reader that this system may be mathematically elegant, but that clearly one cannot go to the center of the Earth to measure the angles. Instead, one measures angles by various means at the surface, such as sighting on the Sun at a particular moment, measuring the angle from the horizon of a given star, and so forth.

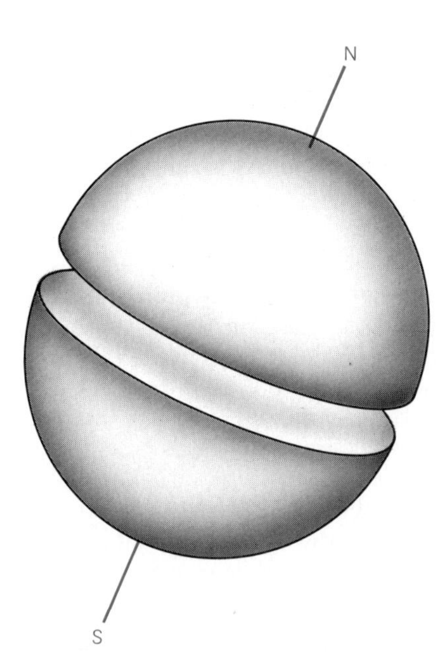


**F.3 The Earth divided into Northern and Southern Hemispheres** by slicing along the equatorial plane.

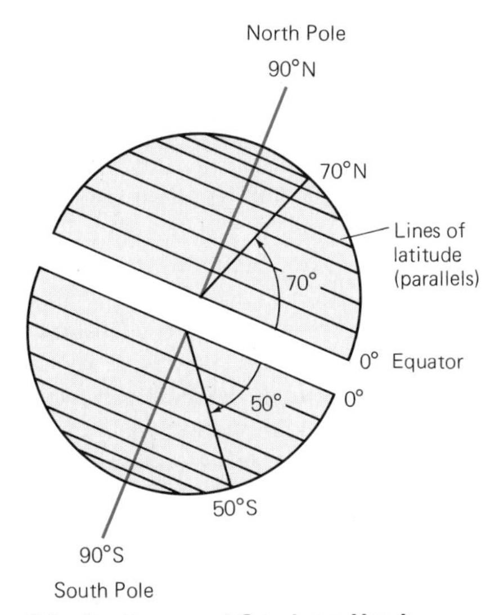


**F.4 Northern and Southern Hemispheres sliced along a plane perpendicular to the equatorial plane** and passing through the pole to show measurement of latitude.

## Map Projections

By bringing in the subject of map projections it is our intention to give only a general orientation, concentrating on the kinds of projections that are employed for drawing detailed topographic maps. We do not consider special map projections that try to show the whole world.

In drawing detailed maps, four kinds of projections are widely used: (1) *Lambert conformal conic* and a variety known as *polyconic*, (2) *equatorial Mercator*, (3) *transverse Mercator*, and (4) *polar stereographic*. We describe each of these briefly.

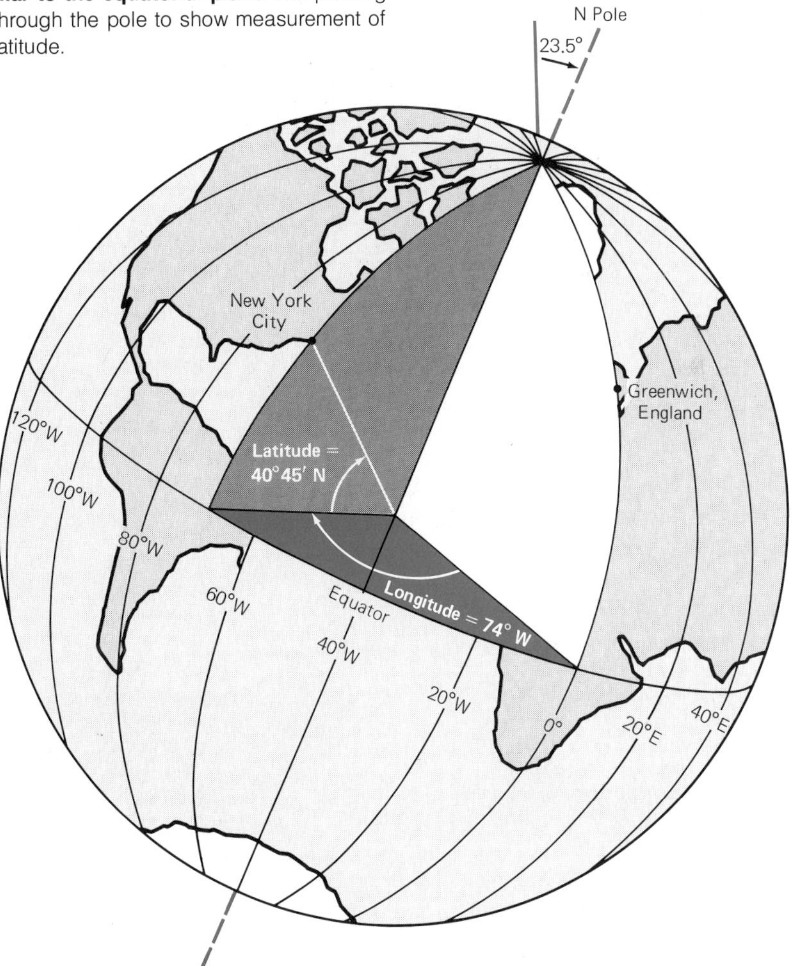


**F.5 Latitude and longitude of New York City** shown on a sketch of the Earth with a segment removed by slicing along the equatorial plane, along the Greenwich meridian and along the meridian passing through New York City. Internal structure of the Earth not shown.

**Lambert conformal conic projection and polyconic projections.** The principle of a conic projection is that points on a sphere can be projected upward onto the surface of a cone whose apex has been placed over the pole (Figure F.6a). Points are transferred to the cone by drawing lines from the center of the sphere through the points on its surface and then by extending the lines until they intersect the cone. To prepare a map from a conic projection, one slices the cone and spreads it out so that it is flat (Figure F.6b).

On a map drawn using a conic projection, the meridians of longitude appear as straight lines that converge toward the apex of the cone. The parallels of latitude are arcs of concentric circles (Figure F.6b).

The cone may be tangent to the surface of the Earth or may cut through the surface of the Earth along two circles, forming a *secant cone* (Figure F.7a). The Lambert conformal conic projection is a secant cone that intersects the surface of the Earth along two standard parallels, 45° N and 35° N (Figure F.7b). Between these standard parallels, distortion is minimal. For example, large-scale maps of the conterminous United States (the "lower 48") based on the Lambert conformal conic projection show only small distortion. In addition, straight lines drawn on such maps closely approximate great circles, the preferred routes followed by airplanes.

A polyconic projection is one having more than one size of secant cone, each having two standard parallels. One selects the cones so as to keep the area of interest between the two standard parallels, where distortion is least. Polyconic projections are used for most topographic maps published by the United States Geological Survey.

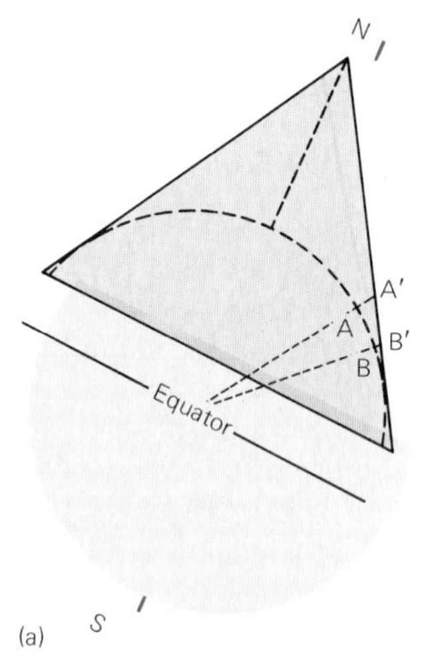


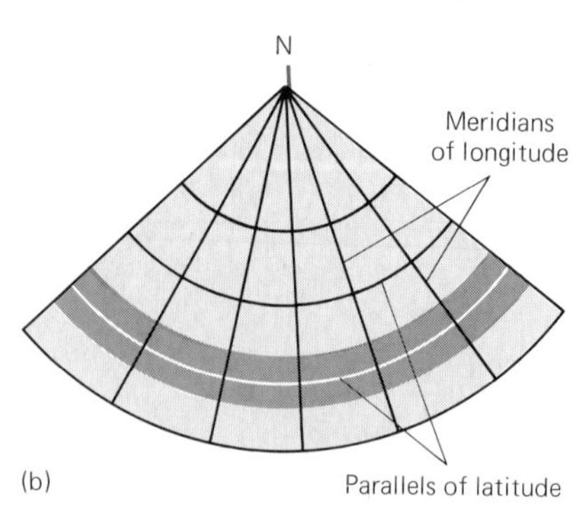


**F.6 Basis of a conical projection.**
(a) Cone tangent at a single latitude. Points A and B on the surface of the Earth appear at A′ and B′, respectively, on the cone. (b) Cone unrolled and laid flat as a map. Gray area is where distortion is least.

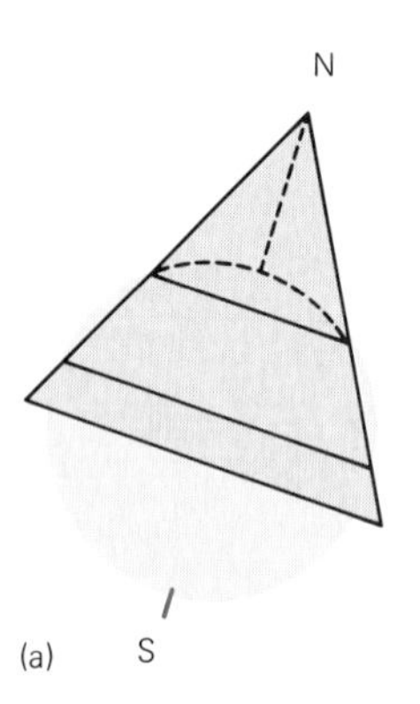


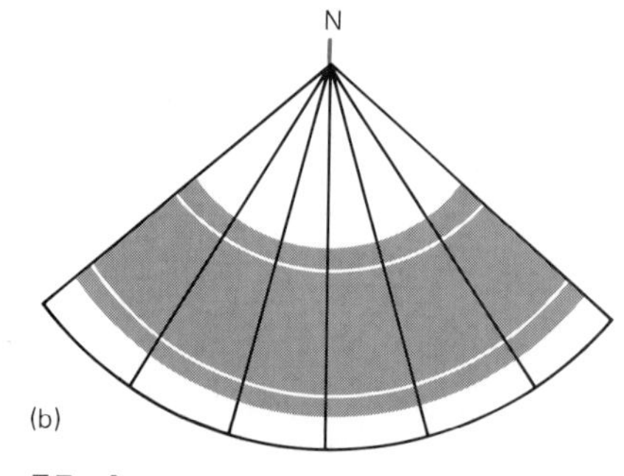


**F.7 A secant cone.**
(a) The cone cuts sphere along two widely separated latitudes. For many map purposes, a secant cone is designed to cut the Earth along 35° N and 45° N.
(b) A secant cone unrolled and laid flat as a map. Gray area is where distortion is least; white arcs indicate the parallels where the cone cut the sphere.

**Equatorial Mercator projection** A Mercator projection uses a cylinder. Two varieties are recognized, based on where the cylinder touches or cuts the spherical Earth. In an *equatorial Mercator projection,* the cylinder is tangent to (i.e., "wraps around") the Earth at the Equator, and the long axis of the cylinder is parallel to the Earth's pole of rotation (Figure F.8a). On an equatorial Mercator projection, the N-S lines (longitude) and E-W lines (latitude) form a rectangular grid system (Figure F.8b).

The unique and valuable feature of an equatorial Mercator projection is that one can draw a line between two points on the map, measure the angle that this line makes with the north-south lines of longitude, that is the compass bearing, and then if one starts out at point A and travels in the direction of the measured compass bearing, one will arrive at point B. Accordingly, the equatorial Mercator projection has been and still continues to be widely used for navigation. The chief drawback of an equatorial Mercator projection is the great distortion, which increases progressively as distance from the Equator increases.

**Transverse Mercator projection** In a transverse Mercator projection, the cylinder is not tangent to the sphere at the Equator, but along a great circle of longitude (Figure F.9a). The transverse Mercator projection used for many maps is based on a secant cylinder that intersects the Earth's oblate surface along two small circles that closely parallel two meridians at the Equator (Figure F.9b). On such projections, the distortion is small for belts of longitude about 6° wide, and extending from latitude 80° S to latitude 80° N. Worldwide U.S. military maps are

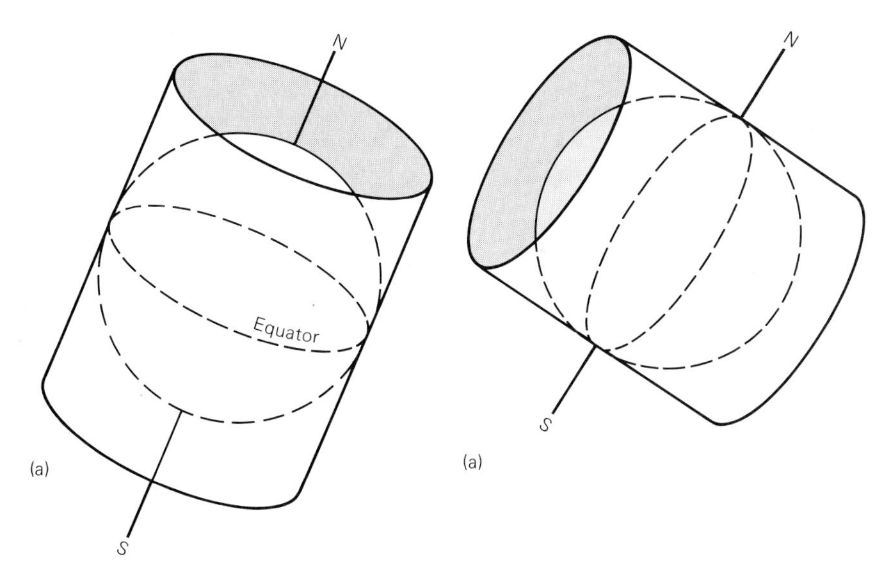


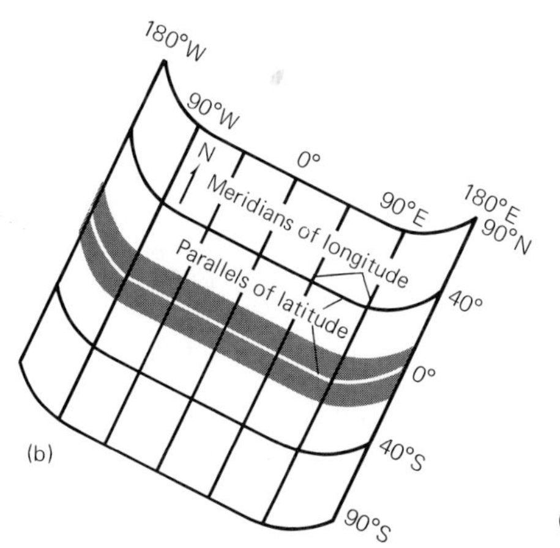


**F.8 Equatorial Mercator projection.** (a) Cylinder fitted over sphere, touching it at the Equator; axis of cylinder parallels pole of rotation. (b) Cylinder unrolled in part to show flat area as map. Gray area is region of least distortion; white line is Equator.

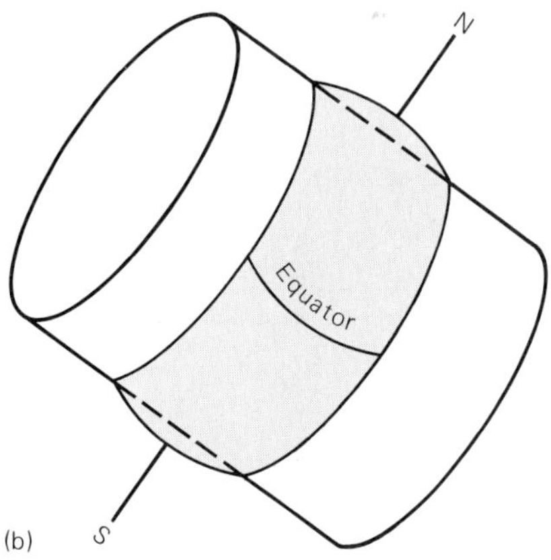


**F.9 Transverse Mercator projection.** (a) Cylinder fitted over sphere, tangent to sphere along a great circle of longitude; axis of cylinder at right angles to pole of rotation. (b) Secant cylinder cutting sphere in two places that are parallel to great circles of longitude but a bit smaller in diameter.

based on two projections, one of which is known as the Universal Transverse Mercator, or UTM projection. The other is a stereoscopic projection explained in the next section.) The UTM projection is very useful in the age of computers and is the basis for the metric grid system that we explain in a following section.

**Polar stereographic projection** A stereographic projection is one in which points on the surface of a sphere are projected onto a plane surface by means of lines drawn from the opposite side of the sphere (Figure F.10a). On a polar stereographic projection, meridians appear as a series of radiating straight lines that pass through the pole, as the spokes of a wheel, and parallels of latitude, as concentric circles whose spacing increases outward (Figure F.10b). The U.S. military map scheme for regions of the world lying north or south of latitude 80° is based on a Universal Polar Stereoscopic projection (UPS system).

## Grid Systems

On the detailed topographic maps geologists use, various rectangular coordinate systems, or *grid systems* are employed for locations. In the United States, four kinds are common: (a) map quadrangles; (b) the United States Land Survey; (c) the 10,000-foot state grid; and the (d) the universal (metric) grid system.

**Map-quadrangle grid** The map-quadrangle grids are based on longitude and latitude measured in minutes or in degrees (60 minutes equal 1 degree). In the United States, the standard map-quadrangle sizes are: 7.5 minutes; 15 minutes; 30 minutes; and 1 degree.

**United States Land Survey grid system** Much of the land surveying in the central and western states is based on a scheme established by Congress in 1785 when the Northwest Territory was settled. The fundamental grid unit is a square, 6 miles on a side, that has become known as the "Congressional township." These townships are parts of a larger system of surveyed lines. The N-S lines of the survey are *principal meridians,* and the E-W lines, *base lines.* The first principal meridian is the Ohio-Pennsylvania border. Others west of this line are numbered consecutively from 1 through 6. After 6, the principal meridians are named after geographic features. The base lines are just E-W lines along various latitudes.

The center of each Congressional township is defined by the intersection of a N-S line, named a *range line,* and an E-W line, named a *township line.* (This usage of township line is a trifle confusing, because it is the 36-square-mile townships that are being defined by range lines and township lines.) Township designation is made by simple counting of township lines N or S of the base line and range lines

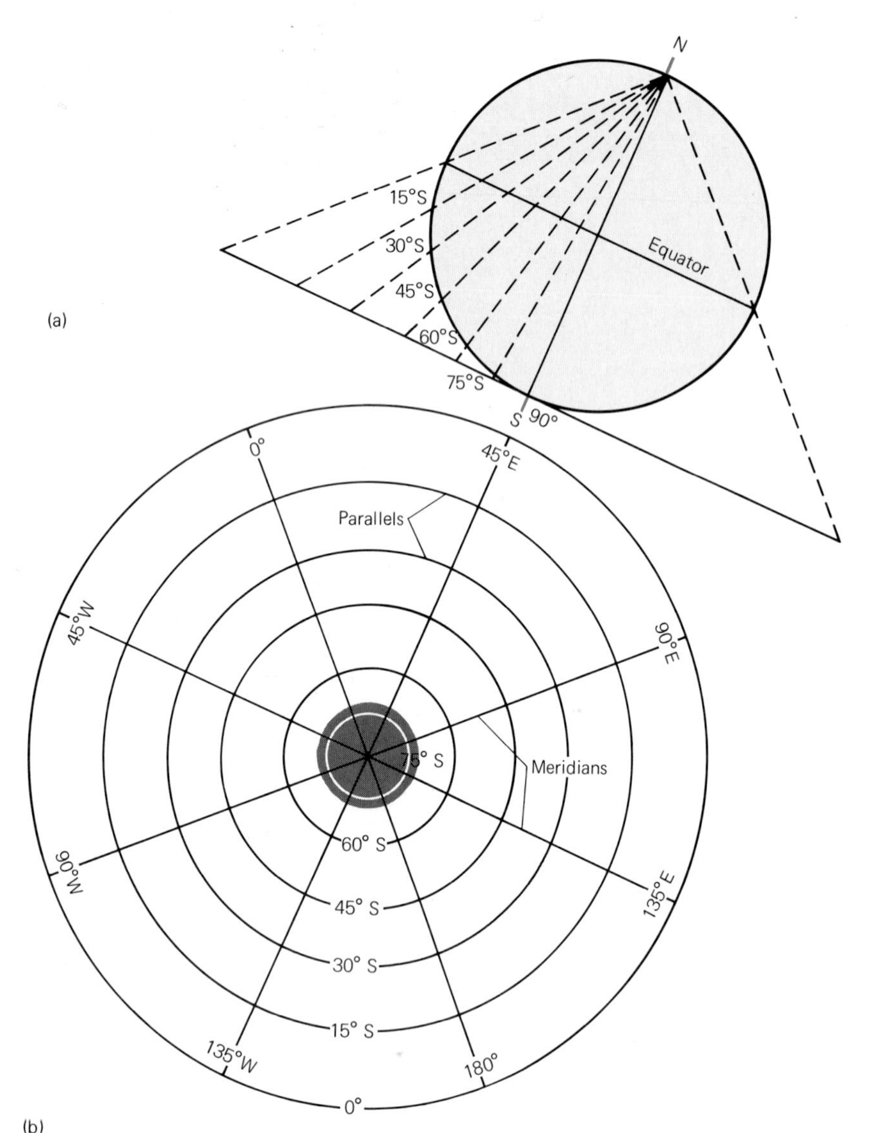


**F.10 Polar stereographic projection,** centered over South Pole. (a) Principles of a stereographic projection; lines from opposite pole are drawn through points on Earth's surface and continued until they reach the plane of the stereographic projection. (b) Map view showing parallels of latitude as concentric circles and meridians of latitude as radiating lines passing through the South Pole.

(or simply "ranges") E or W of the principal meridians. Some idea of the sizes of these territories can be gained from the situation in the largest system of numbers, based on the fifth principal meridian. The townships go as high as T163N and the ranges to R104W.

Every four tiers of townships are bounded by what are named "standard parallels." Along these lines, surveying problems are resolved by offsetting the N-S lines. Many a straight N-S road, built along the boundaries of square-mile sections, enounters two 90° curves and a short E-W stretch along standard parallels.

Within the 36-square-mile townships, subdivisions are made into square-mile sections, each having an area of 640 acres. (Notice that the metric system has not yet invaded U.S. land surveying.) At present, the sections are numbered in a standard way, starting in the NE corner with section 1, and increasing westward across the top row to section 6. Then one drops S a row for section 7 and continues the count in this second row eastward to section 12. After section 12, one moves another row S for section 13, then west to 18, and so forth, zig-zagging back and forth to section 36, in the SE corner of the township (Figure F.11). The intersection of the defining township line and range line always forms the common boundary of sections 15, 16, 21, and 22. (In the first version of section-numbering schemes, still found on many maps in southeastern Ohio, for example, the numbering is up the columns, starting with section 1 in the SE corner, moving one row W and starting up again with section 7 at the S border of the township just W of section 1, and ending with section 36 in the NW corner of the township.)

The U.S. Land Survey grid system can be employed for locating points or for indicating areas within a section by

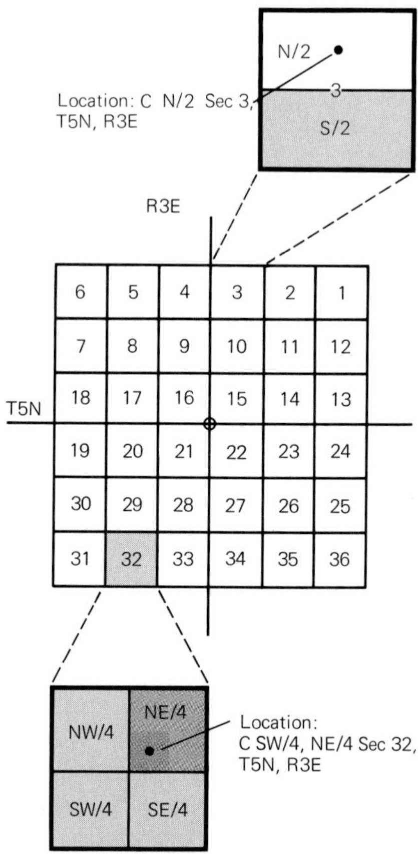


**F.11 A congressional township and its numbered square-mile sections.**

means of fractions of a section, usually one-half or one-quarter. For example, the area of a half section is 320 acres. One could specify the E half, N half, S half, or W half of section 3, for example (see Figure F.11, top). A specific point might be the exact center of one of these half sections.

The area of a quarter section is 640 divided by 4, or 160 acres. The quarter sections are indicated by directions, as NE 1/4, NW 1/4, SE 1/4, or SW 1/4. Each quarter section can be subdivided into smaller halves or quarters. For example, one might locate a point as the center of the SW/4, NE/4 sec. 32, T 5 N, R 3 E (see Figure F.11, bottom). In using such multiple quarter-section locations, it is convenient to read them backwards. They are listed in order of *increasing* area, but in trying to find the point or area, one proceeds to *decreasing* areas (i.e., starting with the full quarter section, then going to the subquarter sections). Thus, in the example given above, one would find section 32, then go to the NE/4, and then the SW/4.

The U.S. Land Survey grid system is widely used in the petroleum industry for locating and for spacing wells.

**The 10,000-foot state grid system** On many of the 7½-minute quadrangle maps, a 10,000-foot state grid system is indicated. To use this grid, one reads right (to the E) and then up (to the N). The 10,000-foot grid system was used for a few purposes before the universal metric grid system had been widely circulated. At present, the trend is to abandon the 10,000-foot grid system in favor of the universal metric grid system next to be described.

**The universal metric grid system** As mentioned, the universal grid system is based on two projections: (1) the UTM projection between 80° S and 80° N, and (2) the UPS projection between 80° S and the South Pole and between 80° N and the North Pole.

Where the UTM projection applies, 60 zones are used, each spanning 6° of longitude, and each using its own transverse Mercator projection. The grid zones are numbered consecutively from 1 to 60, starting with longitude 180° W and proceeding counterclockwise as viewed from above the Equator. This is in keeping with the general grid rule: "read right (i.e., E) up" (i.e., N). From S to N, the latitudes are divided into 20 rows, each spanning 8°. (This gives 160° or twice 80°, the span of the UTM projection.) The rows are lettered, starting at 80° S latitude and proceeding northward. Only 20 of the 26 letters of the alphabet are used. The letters I and O are usually omitted because they can be confused with numerals. The letters A and B are reserved for designating major zones near the South Pole, and Y and Z, for major zones near the North Pole.

For example, the 6° by 8° quadrilateral that includes New York City (see Figure F.5) is defined by 78° to 72° W longitude and 40° to 48° N latitude. In the universal grid system, New York City lies in quadrilateral 18T.

Within each quadrilateral, further designation is made starting with a series of squares measuring 100 kilometers on a side that are referred to by two letters. Each 6° by 8° quadrilateral contains 54 of the 100-kilometer squares plus 18 narrow marginal rectangles that measure 100 kilometers in a N-S direction, but much less than 100 km in an E-W direction. The N-S columns of squares are designated by capital letters, starting with A at longitude 180° W and proceeding alphabetically eastward, but excluding I and O and *including* A, B, Y, and Z. This sequence is repeated every three grid zones (i.e., every 18° of longitude). The E-W rows are also lettered, but in contrasting pairs of sequences. The odd-numbered zones start with the letter A and proceed northward alphabetically, repeating every 20 rows (i.e., A through V, but excluding, as usual, I and O). The even-numbered zones start at the Equator with the letter F and repeat after reaching V, with I and O omitted.

Within each 100-kilometer square, further identification is numerical, based on a metric grid. The southwest corner of the C row of grid quadrilaterals (i.e., bounded on the S by 80° S longitude) is the origin for the Southern Hemisphere, and that of the N row (i.e., bounded on the S by the Equator), for the Northern Hemisphere. The Equator is assigned a value of 10,000,000 meters N in the Southern Hemisphere, and zero in the Northern Hemisphere. The central meridian of each grid zone is assigned a value of 500,000 meters E. By this convention, the smaller values lie in the SW corner of each grid quadrilateral and the numbers increase both eastward and northward. Grid squares are identified by the coordinates of their SW corners. For many purposes, only two digits are used. For example, a location of 867,000 meters E and 3,800,000 meters N would be expressed as: 6780. Further refinement can be added by using more digits, but always including the identical number of digits for each direction so that the resulting string of numbers can always be divided evenly in half. Therefore, no matter how many digits are used, the first half refers to distances E and the second half, to distances N.

## Scale and Measurement of Distance

A **map scale** is the ratio between a distance shown on the map to the actual corresponding distance on the ground. To be a valid ratio, both distances must be expressed in the same linear units. For example, if 1 cm on the map represents 5000 cm in the field, the ratio is 1:5000.

For practical map use, the scale can be expressed in other ways, all derived from the ratio. In the simplest cases, map scales may be shown as 1 cm = 1 km, for example, thereby varying the units. Most maps show a scale directly in what is known as a **graphic scale.** A line is marked off with any convenient lengths, such as meters, kilometers, or feet, yards, or miles. A graphic scale can be used as a special ruler to measure directly the distance between points on the map. The advantage of a graphic scale is that it is still valid even if the map is photographed and then enlarged or reduced in size.

## Orientation and the Expression of Direction

Orienting a map is the process of placing it horizontally so that map lines are parallel to Earth lines radiating from the point of observation. Two general methods of orientation are:

1. **Alignment by compass.** The compass is placed in a horizontal position and rotated until the north-seeking end of the compass needle points to the proper angle of magnetic declination on the compass dial. A meridian on the map is then aligned parallel to the N-S direction on the compass dial (true N-S direction), making certain that the north ends of both lines coincide.
2. **Visual alignment.** A well-defined linear feature, such as a known stretch of road on the map, is aligned with the actual road in the field. This method is more practical near buildings, power lines, and railroads, where the compass may be affected by local magnetic disturbance resulting in a deviation of the compass needle from the magnetic N-S line.

When the map has been oriented, geographic features appear on it in proper angular relationship to the corresponding features in the field.

**Expression of direction** A direction is an angle, measured in a horizontal plane, which any given line makes with a standard reference line. If both ends of the standard reference line are used, the directions are called **bearings.** A given line is referred either to the north or south end of the reference line, whichever makes an acute angle with the given line. The acute angle may lie either east or west of the reference line. All bearings are expressed by three parts in the order indicated: (1) North or South, (2) the acute angle, (3) East or West. Figure F.12 shows a line bearing North 40° West (N40°W).

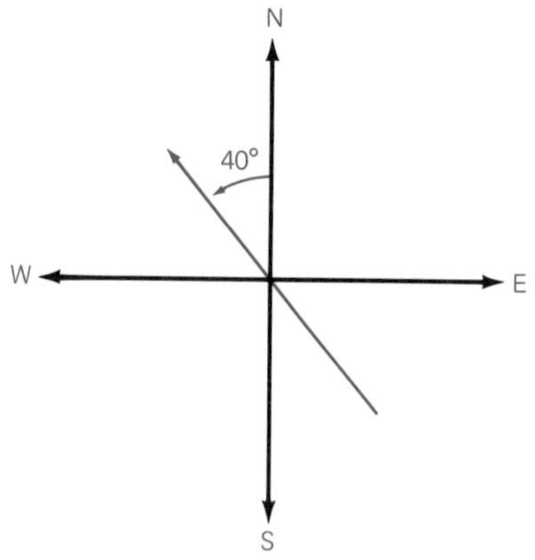

For many hundreds of years people have been drawing maps, scaled representations on flat surfaces, (e.g., on sheets of paper), of one or more attributes of the Earth's curved surface that could be seen by eye. Maps have been drawn showing boundaries of lands and seas, locations of rivers, hills, and valleys and human-built cities, roads, monuments, fences, and boundaries of nations or of other political units. Some maps show locations of hidden things, such as buried treasure.



**F.12 Compass bearing.**

## Modern Mapmaking

With the invention in the twentieth century of the airplane and perfection of aerial surveying cameras and of special photographic emulsions, map-making capabilities took a quantum jump forward. Aerial photographs can be used in two ways. (1) When a print from only one frame from an aerial camera is viewed the scene differs little from what the human eye could see by looking vertically downward out the window of an airplane (e.g., see Figure 13.20). (2) However, when adjacent pairs of aerial photographs having appropriate overlapping coverage of the ground are viewed stereoscopially, one sees an image of the Earth's surface that has been "stretched" upward, or, in technical terms, an image that shows vertical exaggeration. These unique, vertically exaggerated, stereoscopic views of the ground are scenes that the unaided human eye cannot possibly see. The reason for this is that between successive openings of the shutter the airplane moves the camera lens perhaps hundreds of meters along the flight line. Thus, the "interpupillary" distance of the aerial survey is many times greater than that of humans. Instead of "seeing" the ground through eyes spaced only a few centimeters apart as humans do, the aerial camera "looks," as it were, through the "eyes" of a super giant. If this "giant" were a scaled-up human being using the interpupillary distances for the sizing ratio, and the individual aerial photographs were snapped every 210 meters along the flight line, then the giant would be 3000 times the size of a human being having an interpupillary distance of 7 centimeters.

With precision laboratory instruments that have been invented it is possible to trace level lines around on the vertically exaggerated stereoscopic images of the ground. And, as the movable leveling device is moved on the stereoscopic image, another part of the gadget traces a line on a map to a predetermined scale. In this way it is possible to draw precise maps showing the details of the Earth's morphology. Such maps are *topographic contour maps;* we shall discuss them at greater length presently.

In recent years, photographic emulsions sensitive to colors have been perfected. Some color films record all colors. That is, they respond to a wide band of frequencies of light waves (Light and other waves can be specified by using either their frequencies or their wavelengths.) Other films respond to light having only a limited range of wavelengths. (For example, Figure 3.5 is a black-and-white print of a view of the Earth made using a wavelength of five millimeters.

The main disadvantage of photographic methods is that they can be used only when clouds do not obscure the camera's view of the ground.

One of the greatest outgrowths of space-age developments in solid-state physics and electronics has been the perfection of sensors that have two properties. (1) They respond to some parts of the entire range of the spectrum of electromagnetic energy waves, all of which travel at the "speed of light." (2) They convert their responses into electric signals, and electric signals can be made into visual images of the various electromagnetic energy waves. Most of the electromagnetic energy waves lie outside the range of human vision. Therefore, we humans are not aware of the movements of many kinds of energy waves. Appropriate instruments are now in use that display visually all kinds of energy waves. Of particular value in geology are images created by sending and receiving radar waves in oblique paths from an airplane to the Earth's surface. Such images are known as side-looking aerial radar (SLAR). A side-looking radar images of the Earth's surface in Kentucky is shown in Figure 12.25. Side-looking radar images can be recorded whether the sky is clear or cloudy.

Making maps is one of the fundamental quantitative procedures in geology and related Earth sciences. Maps have been drawn showing the distribution of various surface characteristics, such as soils, vegetation, deserts, the weather (as expressed in temperature, rainfall or snowfall, barometric pressure, wind speed and direction, and cloud cover), snow cover, and so forth. Ordinary dimensions, such as length, breadth, and thickness may not seem very exotic and, for many map purposes, do not have to be determined to the nearest nanometer. Nevertheless, a map showing these ordinary dimensions presents quantitative geologic data of lasting value.

The following sections describe (1) two-dimensional planimetric maps and (2) three-dimensional maps of the Earth's surface features (topographic contour maps). The three-dimensional maps showing both the Earth's surface relief, and in addition, the subsurface arrangement of bodies of rock and surface distribution of regolith (geologic maps) are discussed in Appendix G.

## Planimetric Maps

A planimetric map is one that attempts to show only the two-dimensional aspect of the Earth's surface, such as locations and areas. The problems of constructing planimetric maps are common to all kinds of maps; hence, a discussion of these problems on these simplest, two-dimensional maps forms a basis for understanding the more complicated three-dimensional maps that follow.

**Map symbols** Many map symbols are in use; it would be difficult to reproduce them here. However, an excellent pamphlet, "Topographic Maps," is available without charge from the Map Information Office, U.S. Geological Survey, Washington, D.C. 20242. This pamphlet contains a comprehensive color representation of map symbols.

As many as five colors may be used with map symbols, and the four most important are listed below:

Red and black: Human-made features.

Blue: Water forms.

Brown: Contours, other relief symbols. (Contours indicating depth of water or elevation on the surface of glaciers are shown in blue.)

## Topographic Contour Maps

**Topographic maps** are scale models of portions of the Earth's surface that show three dimensions of space in two dimensions. A map view of the ground is from vertically above. Such maps show the configuration of the Earth's surface and aid in making geologic interpretations without actually visiting the region shown. With some basic geologic knowledge and the ability to read such maps, it is often possible to determine general or specific rock structures, the agents which have modified them at the surface, and the extent of the modification.

In addition to their use for geologic interpretation, topographic maps are invaluable for getting around in the field, and serve as bases for plotting the results of geologic observations. The following information is intended to help the student read and understand topographic maps.

To determine a quantitative picture of heights, slopes, and the shapes of land features from a map, a device for showing relative and absolute elevations of all points on the map must be used. The device most commonly employed is the **contour line.** A contour line traces the shape that would be formed by the intersection of a horizontal plane with the land surface at a stated altitude above a reference surface. Successive lines are separated by a uniform vertical distance, called the **contour interval.**

**Some conventions about contour lines** In drawing and reading contour maps the following conventions are applied:

1. All points on a given contour line lie at the same elevation.
2. All contours are closed lines but may run off any particular map.
3. Two contours of different elevations cannot cross each other, except at overhangs, which are rarely mapped.
4. Contour lines of different elevations may meet or merge at a cliff.
5. The spacing of contours indicates the degree and kind of slope as follows:
   (a) The closer the spacing, the steeper the slope.
   (b) Even spacing indicates a uniform slope.
   (c) Uneven spacing indicates an irregular slope which may be either concave or convex.
6. When contours cross a stream, they always form a V whose apex points **up the valley.** The apex of a V-shaped contour crossing a ridge line points **down the ridge.**

**Constructing topographic profiles** For many purposes, it is necessary to construct topographic profiles from a contour map.* The points at the ends

*Indexes showing topographic maps published for each state, Puerto Rico, the Virgin Islands, Guam, and American Samoa are available free on request from the U.S. Geological Survey, Washington, D.C. 20242, or Federal Center, Denver, Colorado 80225.

of the profile are selected, a line drawn between them, and the edge of a piece of paper is laid along the line. The intersections of each contour line with the profile line are marked on the edge of the paper. The paper is now moved to a piece of graph paper, on which the vertical scale has been selected. Each point where a contour line crosses the profile line is now drawn at the appropriate height on the graph paper. When the points are connected in a smooth line the profile is finished. In many profiles, the vertical and horizontal scales are not the same. Therefore, it is necessary to label both scales on every profile. In Figure F.13, the length of 1 kilometer on the horizontal scale is equal to only 40 meters on the vertical scale. Therefore, the ratio between the two scales is 1000/40 = 25. This ratio is known as the vertical exaggeration.

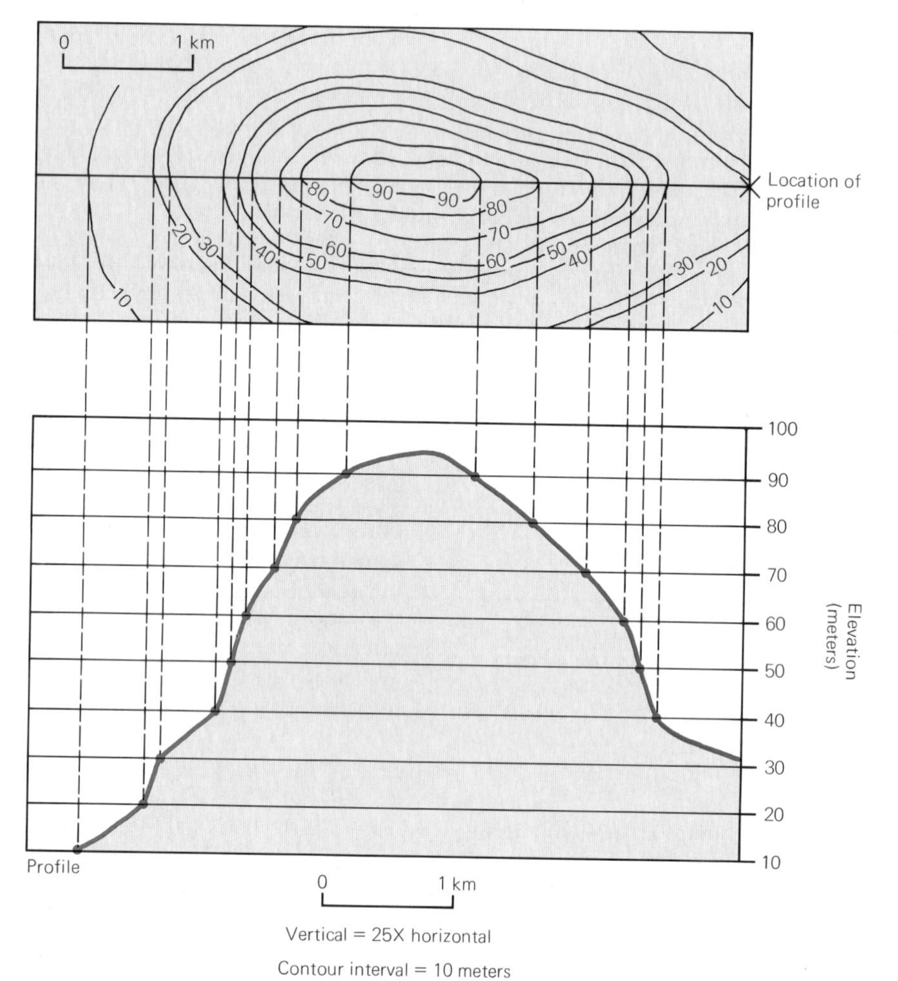


F.13

# APPENDIX G GEOLOGIC MAPS

A geologic map displays the distribution of rock types; the boundaries between rock units where these units come up to the surface of the Earth, or "crop out," to use the geologists' term; and the configuration and distribution of geologic structures. The distinctive feature of a geologic map is that the lines on it, which represent boundaries or *contacts* between rock bodies, enable a geologist to infer the arrangement of rocks underground. A geologic map includes in addition to the patterns of outcrop areas and boundaries between rock units a formal explanation on which all of the units are listed in their inferred order of age, the oldest at the bottom of the column and the youngest at the top.

The most useful geologic maps are those that display both the topographic contours and the geologic contacts. Geologic contacts can be plane surfaces within a sequence of layered rocks; curved, plane, or irregular boundaries of plutons along which variable metamorphic effects are present; plane, curved, or irregular faults; or plane, curved, or irregular surfaces of burial.

The following discussion concerns plane surfaces that are contacts between units forming a layered sequence. The starting point in most geologic mapping projects is a full, detailed knowledge of the succession of the strata present in an area. A geologist determines the characteristics, thicknesses (distance perpendicular to the layer boundaries), order, and geologic age. All this comes under the heading of "knowing the section." One must be able to recognize the layers when they are horizontal, inclined, or vertical. Some useful clues for determining the correct facing directions of strata are presented in Box G.1.

In addition to the contact lines, a geologic map displays the attitudes of the strata or contacts by means of standard symbols. These are shown in Box 16.1, where the concept of the *attitude* of a surface is explained and where strike and dip are defined.

The following paragraphs introduce the principles of the lines that represent geologic contacts. The emphasis is on plane surfaces separating sedimentary formations. Contacts that are plane surfaces are formation boundaries within a conformable sequence or the boundaries of a sill or a sheet of extrusive igneous rock. The three cases are considered: (1) horizontal; (2) vertical; and (3) inclined.

A *horizontal geologic contact* appears on a geologic map with the same aspect as a contour line (Appendix F). This follows from the definition of a contour line as a line connecting points of equal altitude. A horizontal geologic contact can be considered as a special kind of contour—one whose altitude is determined by the rock strata rather than by a standard altitude of the map contour interval. Horizontal contacts follow the contours; indeed, the contours can be employed to determine the thickness of horizontal geologic units (top at 103 meters, base at 98 meters, thus thickness is 5 meters in Figure G.1a).

A *vertical geologic contact* of a plane surface is a straight line on a geologic map. If this line represents a strata-bounding plane surface, then its direction is the strike of the strata (Figure G.1b). Notice that a vertical contact crosses contours without any deviation. The thickness of a vertical unit can be determined by scaling off the map distance between the contacts (800 meters in Figure G.1b).

An *inclined geologic contact* of a plane surface crosses the contours. Where an inclined contact line crosses a valley it forms a V that *points down the dip*. In some cases, this V will point upstream (dip in the upstream direction; Figure G.1c). In other cases, this V will point downstream (dip in the downstream direction; Figure G.1d).

The map information enables one to determine the strike and dip. One finds the strike line by finding two points where the contact crosses the same contour (line *AB* connecting the contact's crossing of the 105-meter contours in Figure G.1c). When these two points have been connected by a line, they show the strike line; measure the angle E or W of true N to get the strike direction (N 40° E in Figure G.1c). The dip can be calculated by

**BOX G.1**

**Some useful features (tops to left):**

Tree trunk in growth position

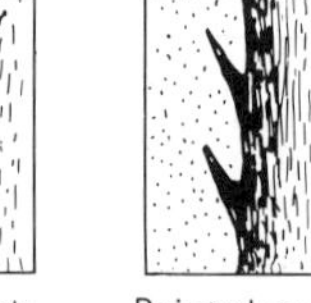

Pointed mud wisps at base of sandstone

Graded layer

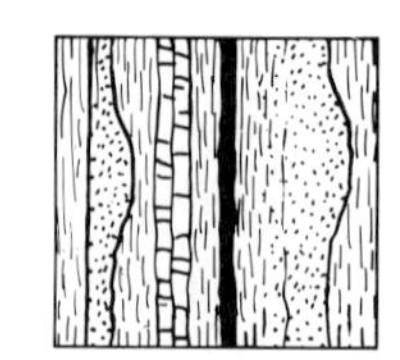

Cyclothem (or other patterned succession)

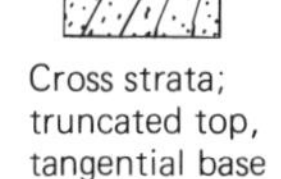

Cross strata; truncated top, tangential base

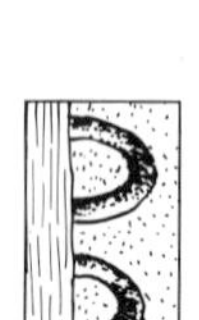

Curled-up mud layer

Worm burrows

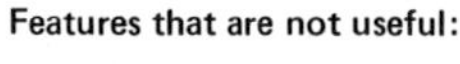

**Features that are not useful:**

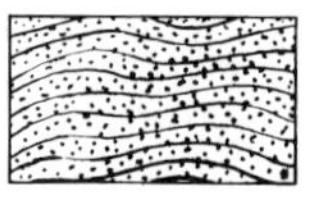

Planar cross strata; truncated top and bottom

Incomplete ripples

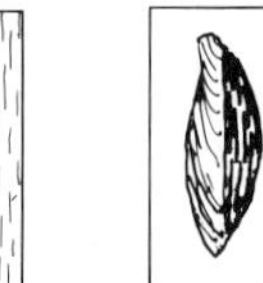

Footprint in mud, covered by sand

Shell partly filled with mud (mud on bottom half)

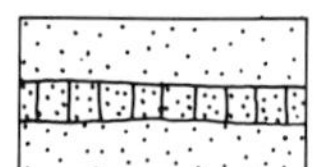

Ripple-drift cross laminae; not truncated

Symmetrical ripples, pointed crests

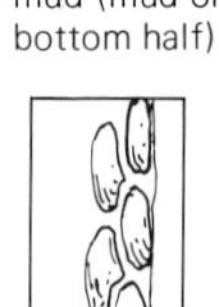

Channel cut into shale

Pillows; points on bases

finding a third point where the contact crosses some other contour (point C, 95 meters, in Figure G.1c). Use the map scale and find the distance perpendicular to the strike line from this third point to the strike (line *CD*, 800 meters, in Figure G.1c) line. Read the contours to find the vertical difference in altitude between this third point and the two lines used to find the strike (105 − 95 = 10 meters, in Figure G.1c). Draw a profile at right angles to the strike line using the same scale for horizontal and vertical and measure the dip angle directly from the profile. Alternatively, use trigonometric relationships to calculate the dip angle. The vertical difference in altitude (10 meters, in Figure G.1c) divided by the perpendicular map distance from the thid point to the strike line (800 meters in Figure G.1c) defines the tangent of the dip angle (10/800 = 0.0125; dip angle is about 1°).

Accompanying geologic maps are an EXPLANATION or columnar section. Examples of symbols used on geologic maps and cross sections are shown in Figure G.2.

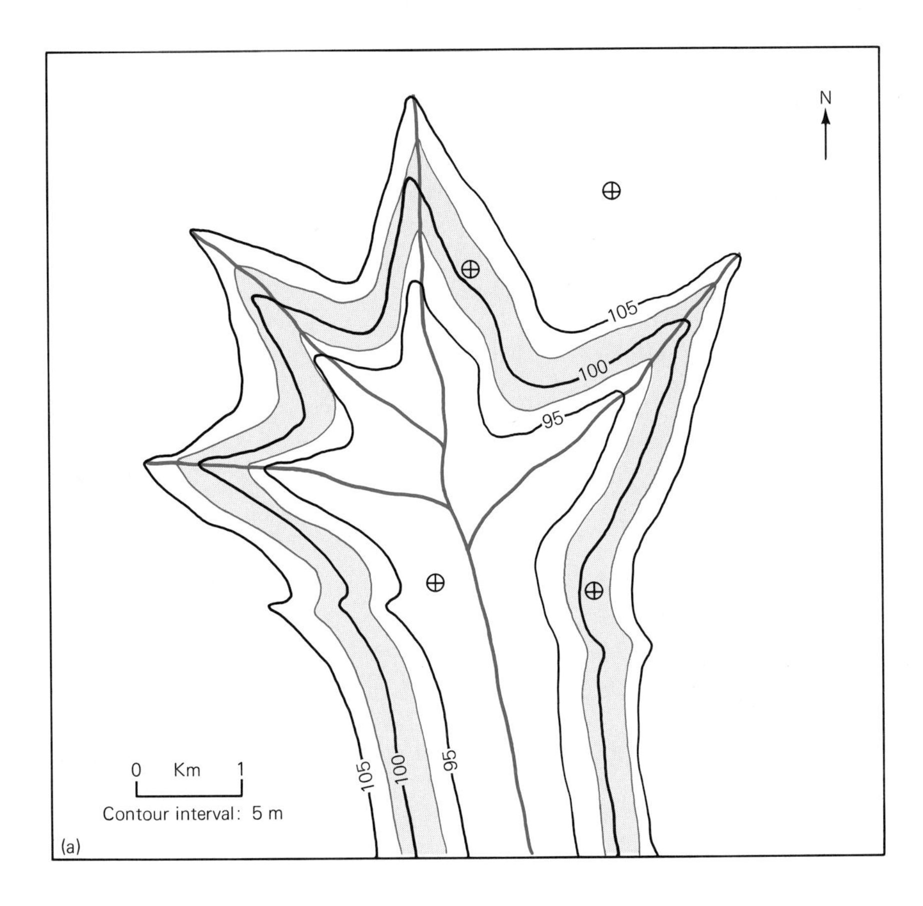
N
105
100
95
105
100
95
0 Km 1
Contour interval: 5 m
(a)

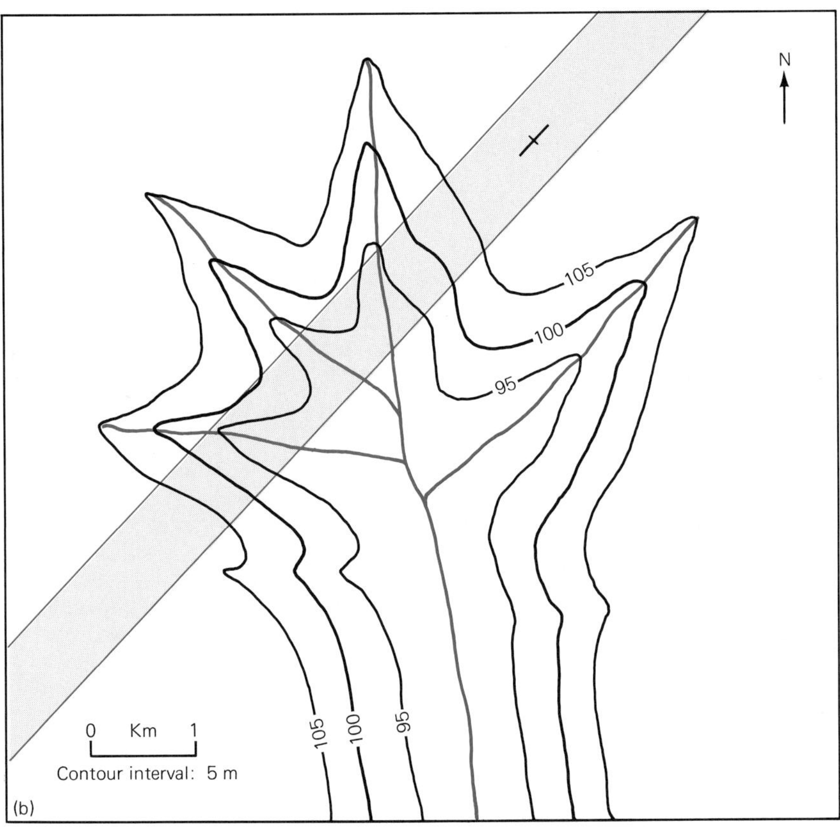
N
105
100
95
105
100
95
0 Km 1
Contour interval: 5 m
(b)

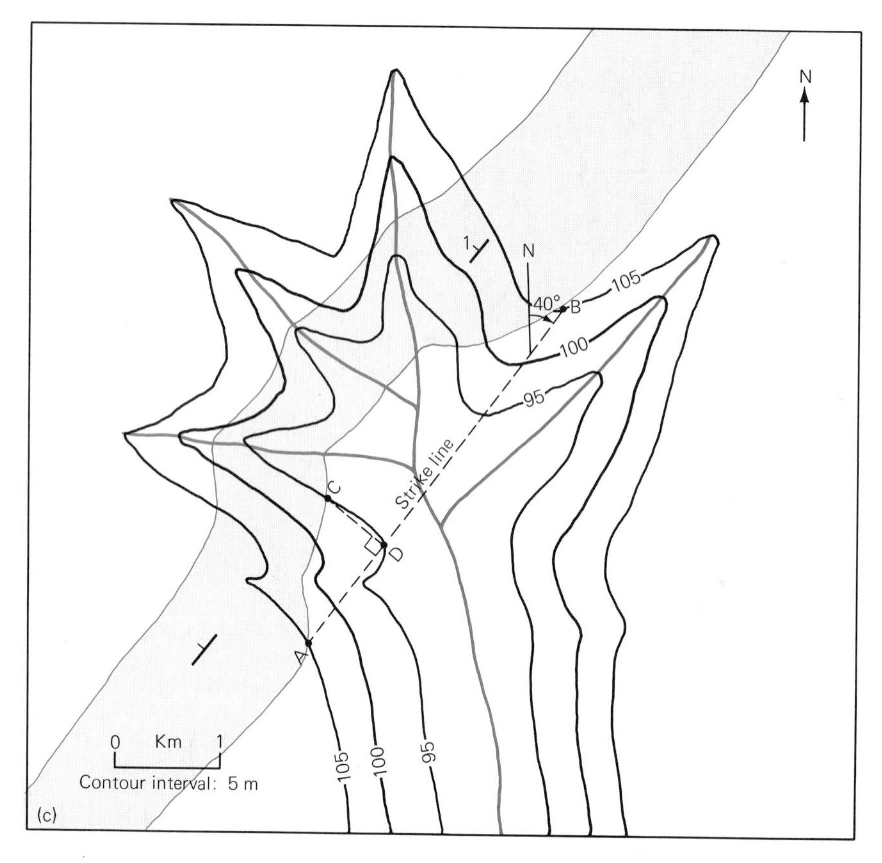

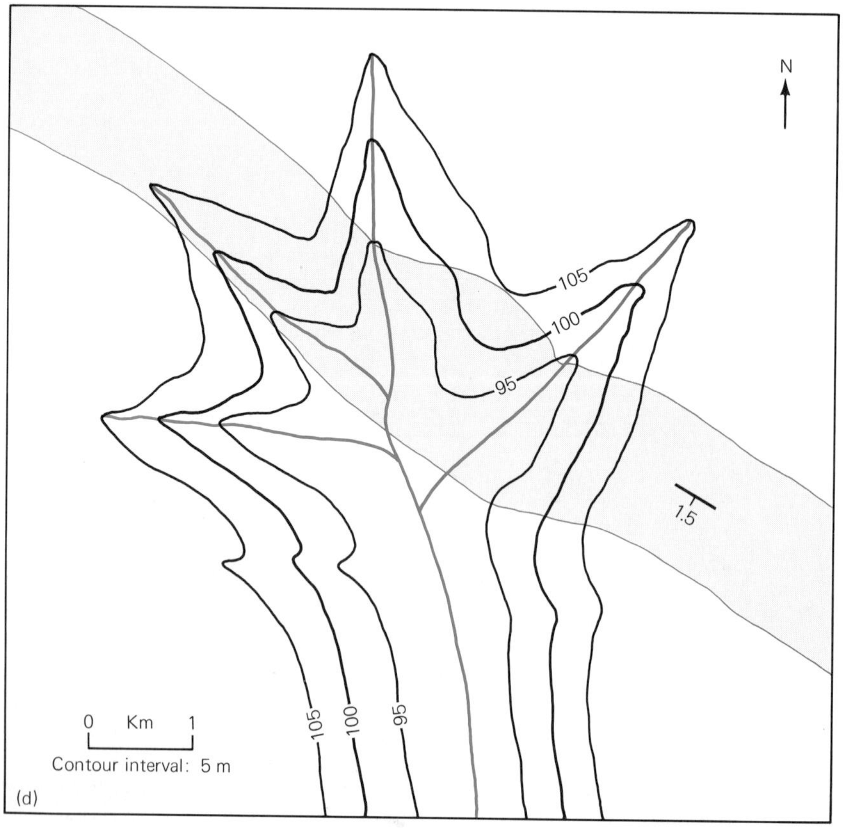

| (a) Symbol | EXPLANATION |
|---|---|
| 18 | Strike and dip of strata |
| 90 | Strike of vertical strata; tops of strata are on side marked with angle of dip |
| ⊕ | Structure of horizontal strata; no strike, dip = 0 |
| 43 | Strike and dip of foliation in metamorphic rocks |
| | Strike of vertical foliation |
| | Anticline; arrows show directions of dip away from axis |
| | Syncline; arrows show directions of dip toward axis |
| 21 | Anticline, showing direction and angle of plunge |
| 15 | Syncline, showing direction and angle of plunge |
| | Normal fault; hachures on downthrown side |
| | Reverse fault; arrow shows direction of dip, hachures on downthrown side |
| 50 U D | Dip of fault surface; D, downthrown side; U, upthrown side; arrow shows dip |
| | Directions of relative horizontal movement on strike-slip fault |
| T | Low-angle thrust fault (**T** on upper block) |

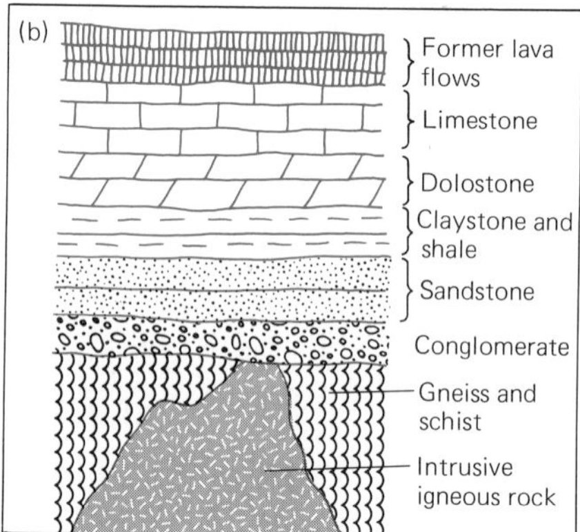


**G.2** Symbols and patterns commonly used on geologic maps and cross sections. (a) Symbols used to show structures. (b) Patterns in common used to show kinds of rock on cross sections.

**G.1 Contours and geologic contacts,** schematic geologic maps. (a) Horizontal unit (blue) is 5 meters thick, with base at 98 meters altitude and top at 103 meters altitude. (b) Vertical unit, 800 meters thick, strikes N 45° E. (c) Inclined unit dips upstream; strike N 40° E, dip 1° NW. Further explanation in text. (d) Inclined unit dips downstream; strike N 60° W, dip 1.5° SW.

# GLOSSARY

**Aa (ah-ah)** The jagged, uneven surface on viscous mafic lava. Also applied to extrusive igneous rocks that were formed by the solidification of such lava.

**Abyssal fan** A basin-floor fan formed in water deeper than 2000 meters.

**Abyssal plain** A flat (gradient less that 1:1000) surface on the deep-sea floor (depths greater than 2000 meters) that is underlain by gravity-transported sediment.

**Active volcano** A volcano that is erupting or one that has erupted within the past 50 years.

**Aeration, zone of** See **Zone of aeration.**

**Aftershock** A group of seismic waves that are felt after the main shock of an earthquake and that decrease with time.

**Agglomerate** A coarse-grained fragmental volcanic rock composed of pieces larger than 32 millimeters in diameter.

**Aggradation** The process of upward growth of a body of alluvium by deposition from a stream channel.

**A horizon** See **Zone A.**

**Algal mat** A sticky mucilaginous layer of algal cells that spread over moist surfaces of many kinds of water-laid sediments. Algal mats trap and encapsulate extraneous objects that settle onto or wash across the layer of cells. Algal mats are one of the important sources of high-hydrogen kerogen, the kind capable of generating crude oil.

**Algal stromatolite** A layered deposit formed by the upgrowth of successive algal mats.

**Alluvial fan** A redundancy because all subaerial fans consist of alluvium. See **fan.**

**Alluvium** Sediment deposited by a stream.

**Alpha decay** A radioactive reaction in which the parent nucleus expels an alpha particle, thus decreasing the mass of the parent nucleus by 4 and its atomic number by 2.

**Alpha particle** A particle released during certain kinds of radioactive transformations that consists of two protons and two neutrons, thus it is identical to the nucleus of a helium atom. Compare **beta particle.**

**Alpine glacier** See **valley glacier.**

**Amorphous** Without form; applied to noncrystalline solids whose ions do not form symmetrical patterns.

**Amphibolite** A coarse-grained mafic metamorphic foliate containing abundant hornblende or other amphiboles and plagioclase.

**Amphibolite facies** (of metamorphic rocks) A group of metamorphic rocks having the bulk chemical composition of basalt and characterized by amphiboles; formed in the middle range of temperatures of metamorphism.

**Amplitude** The vertical distance between the highest part of a wave crest and the lowest part of a wave trough. See also **tidal amplitude.**

**Amygdale** A mineral-filled former vesicle in an extrusive igneous rock.

**Amygdaloidal** An adjective describing a volcanic rock containing amygdales.

**Andesite** An aphanitic igneous rock of intermediate composition; the fine-grained equivalent of diorite. Porphyritic andesite contains feldspar phenocrysts but no quartz phenocrysts.

**Andesite line** A narrow belt around the margins of the Pacific Ocean within which volcanoes discharge andesitic lava.

**Angle of repose** The steepest natural slope angle assumed by a body of cohesionless particles.

**Ångstrom** A unit of length equal to $10^{-10}$ meters and abbreviated by Å. Used for measuring extremely small entities: 10 Ångstroms equal 1 nanometer.

**Angular momentum** The product of the mass times the speed times the angular distance that is traveled by one body orbiting about a point; an important property of the individual members of the Solar System.

**Angular unconformity** The relationship of the unconformity along a contact of burial in which the older strata are not parallel to the younger strata.

**Annular pattern** A drainage pattern in which master streams flow in curving belts of easily eroded strata of a dome or circular basin.

**Anorthosite** An igneous cumulate rock composed almost completely of plagioclase.

**Antecedent stream** A stream, established in its course, that was able to cut downward as fast as a tectonic structure athwart its course grew upward. As a result, the stream cut a gap through a ridge that otherwise it would have flowed around. Compare **superposed stream.**

**Anthracite coal** A lustrous hard coal that has conchoidal fracture, consists of more than 90 percent fixed carbon, and burns with almost no smoke. The highest rank coal.

**Anticline** A wavelike convex-up upfold of rock strata in which the oldest layers occupy the center and dip outward beneath younger layers. Compare **syncline.**

**Apatite** A calcium-phosphate mineral that also contains fluorine, chlorine, and hydroxyl ions; a scale-of-hardness mineral with hardness 5 on the Mohs' scale. The only common phosphate mineral.

**Aphanitic** A texture of igneous rocks in which the individual constituents of the rock are smaller than 1 millimeter.

**Aquiclude** An impermeable formation that creates a barrier to the flow of groundwater.

**Aquifer** A porous, permeable, and saturated subsurface formation through which groundwater can flow readily and from which groundwater can be extracted by drilling a well.

**Arc-trench system** An elongate zone along an ocean-basin margin that includes a deep-sea trench, adjacent rows of both nonvolcanic and volcanic islands, and related basins. See also **island arc.**

**Arenite** A general name for any sedimentary rock consisting of sand-size particles of any composition.

**Arête** (ah-RET) A thin, sharp-crested bedrock ridge, eroded by glaciers in adjacent cirques.

**Argillite** A tough, firmly cemented, fine-grained sedimentary rock, lacking fissility, and more indurated than mudstone and claystone.

**Arkose** A sandstone that contains more than 25 percent feldspar. A sandstone that contains some feldspar (less than 25 percent) is a **feldspathic sandstone,** not an arkosic sandstone.

**Arroyo** A Spanish term for the dry channel of an intermittent stream (synonyms: **box canyon** and **wadi**).

**Artesian water** The groundwater derived from a well that penetrates an aquiclude to an aquifer whose contents rise higher than the aquiclude.

**Artesian well** A well penetrating a confined aquifer.

**Ash** Tephra smaller than 4 millimeters.

**Ash flow** A body of volcanic solids and gases that is gravity transported along the ground, for example, an avalanche.

**Asthenosphere** A concentric shell of the Earth, within the upper mantle and underlying the lithosphere, that consists of solid material that has such little strength it flows easily and, thus, makes isostasy and horizontal motions of lithosphere plates possible.

**Astrobleme** A circular feature, such as an impact crater, formed on the Earth by the impact of a meteorite.

**Astronomic theories** Theories that climate changes are the result of variations in the Sun's output of energy or in the factors which affect how this energy is received on Earth.

**Asymmetric fold** A fold on which the maximum inclination of the strata on one limb exceeds that of the strata on the other limb.

**Atmosphere** A gaseous envelope surrounding a planet. As a pressure unit, 1 atmosphere equals about 1 kilogram per square centimeter (14.7 pounds per square inch).

**Atoll** A group of low, circular islands surrounded by coral reefs and enclosing a circular body of water known as a lagoon. Atolls form because of the upward growth of corals on a subsiding circular base, usually a wave-truncated volcanic cone.

**Atomic energy** See **nuclear energy.**

**Atomic mass** See **mass number.**

**Atomic number** The number of protons in the nucleus of an atom, an isotope, or an ion; symbolized by the letter Z. Compare **mass number.**

**Attitude** The orientation of layers, of fractures, or of boundaries between bodies of rock (known as **contacts**).

**Aureole** See **contact-metamorphic aureole.**

**Authigenic mineral** A mineral that grew within a sediment at some time after the body of the sediment had been deposited.

**Authigenic sediment** A sedimentary deposit formed in situ, not from physically transported material but rather from minerals that crystallized out of the water of the depositional basin or the interstitial water within the sediment.

**Axial plane** (of a fold) The plane passing through points where the strata of a fold bend around, including the fold axis.

**Axis** (of a fold) See **fold axis.**

**Backwash** The return flow from the spent swash of a sheet of water down a sloping part of the shore.

**Badlands** A stream-dissected landscape, generally lacking vegetative cover and consist- of closely spaced V-shaped valleys.

**Bajada** (ba-HA-da) A gentle slope inclined toward the center of a desert basin and underlain by the sediments of coalesced fans.

**Bar** (1) Sediment: a low, rounded ridge of sediment whose top is nearly always submerged. (2) Physics: a unit of pressure equal to $10^6$ dynes per square centimeter or about 1 atmosphere.

**Barchan dune** A crescent-shaped dune that is concave in the downwind direction. Barchans are the typical kinds of dunes in deserts with steady moderate winds and limited supplies of sand.

**Barrier island** (or simply **barrier**) A narrow, elongate island composed of cohesionless sediment, the seaward side of which closely parallels the general trend of a coast.

**Barycenter** The center of mass of two or more orbiting bodies. See also **Earth-Moon barycenter** and **Solar-System barycenter.**

**Basal slip** Glacial motion by physical translation of the overlying body of the glacier along a discrete zone of displacement near the base of the glacier.

**Basalt** A dark-colored, aphanitic mafic igneous rock consisting of plagioclase and pyroxene. See also **dolerite** and **gabbro.**

**Basalt glass** Opaque mafic glass formed by the rapid chilling of mafic lava or mafic magma; distinguished from obsidian by its opaque thin edges.

**Base flow** The component of stream discharge, derived from groundwater, that keeps a stream flowing between rains.

**Base level** The level of water at the mouth of a stream, a level that controls the depth to which a stream can erode. See also **local base level** and **ultimate base level.**

**Basement rocks** (1) Ancient metamorphic rocks and igneous rocks, created in former orogenic belts, that underlie the sedimentary strata of continents (synonym: basement complex). (2) Any bedrock beneath a body of sediment.

**Base surge** An outward-moving dense mixture of hot gases and/or solid particles that is triggered by nuclear explosions or volcanic blasts.

**Basin** (1) Morphologic: an enclosed low area surrounded on all sides by higher ground. (2)Bathymetric: a morphologic basin on the floor of the sea or a lake. (3) Structural: a dishlike configuration of strata in which strata dip radially inward beneath a central point. A structural basin compares with a dome as a syncline to an anticline. (4) Stratigraphic: a region within which a stratigraphic unit is significantly thicker than in adjacent 0NRIAAAA ceeds 100 square kilometers.

**Bauxite** Hydrated aluminum oxide, formed by the intense leaching of other metallic ions from residual regolith.

**Bay barrier** A beach built completely across the entrance to a bay.

**Beach** A body of cohesionless sediment particles, sand-size or coarser, subject to the effects of breaking waves along the shore of a lake or the sea and extending from the outer line of breakers to the upper limit of wave action or to places within this zone where cohesionless sediment gives way to other material.

**Beach drifting** The longshore transport of sediment operating on a beachface which is caused by the oblique approach of waves that create a parabolic flow pattern of the swash and backwash. In no way related to **continental drift** or **glacial drift.** Compare **longshore drifting.**

**Beachface** A gently sloping surface that extends from the crest of a berm to the parts of a beach that are submerged (at least at high tide).

**Beach plain** A low plain built along a coast by the outward progradation of a beach or a barrier.

**Beach scarp** A vertical slope 1 meter to greater than 10 meters high, undercut by the water during an episode of beach erosion.

**Bed** A sedimentary layer thicker than 1 centimeter (synonym: **stratum**).

**Bed form** A feature formed by a current flowing over sand or other cohesionless sediment; usually part of a group of features that are regularly spaced.

**Bed load** The sediment being physically transported along the base of a fluid current (such as a stream or the wind) that moves by sliding, rolling, or bouncing and impacts with other bed-load particles.

**Bedding-surface parting** (or **bedding-plane parting**) A parting surface along the contact between adjacent strata.

**Bedding thrust** An overthrust along which great bodies of initially flat-lying strata have been shifted over other initially flat-lying strata.

**Bedrock** Continuous solid rock that everywhere underlies the regolith and in some places is exposed and, thus, forms the Earth's surface.

**Benioff zone** An inclined zone extending through the crust and upper mantle along a continental margin; defined by earthquake foci that become systematically deeper beneath the continental mass.

**Berm** The upper part of a beach, usually dipping gently landward and covered by water only during spring tides or by the uprush from waves that are larger than normal.

**Beta decay** A radioactive reaction in which the parent nucleus expels a beta particle (electron), thus, not changing the mass number of the parent nucleus but decreasing its atomic number by 1.

**Beta particle** A subatomic particle (subsequently discovered to be the same as an electron) expelled energetically from the nucleus of a parent radioactive isotope that disintegrates by converting a neutron into a proton.

**B horizon** (of a soil) See **Zone B.**

**Bicarbonate-ion complex** An ion complex having the formula $HCO_3$ and a charge of −1.

**Biocrystal** Part of the mineral skeleton of an organism: the exterior shaped by the organism but the internal structure conforming to that of crystal lattices.

**Bioerosion** The erosion of coastal bedrock by organisms.

**Biosphere** The part of the Earth in which organisms can live; consists of regions providing sufficient liquid water and solar energy.

**Bituminous coal** A coal of middle rank that contains 50 to 90 percent fixed carbon and that burns with a smoky flame.

**Block lava** Viscous lava that forms a solid crust by rapid cooling and that breaks into angular blocks by continued forward movement.

**Blocks** Angular particles of extrusive igneous rock, larger than 32 millimeters, that were exploded from a volcano as solids.

**Blowout** See **deflation basin.**

**Body waves** Seismic waves that travel within the Earth.

**Bog soil** Saturated soil consisting of partly decomposed plant matter that becomes peat by bacterial action.

**Boiling mud pit** A hot spring in which the hot water has been mixed with fine-grained sediment to make mud.

**Bolson** A desert basin surrounded by mountains.

**Bombs** (volcanic) Smoothly curved pieces of extrusive igneous rock, larger than 32 millimeters, that were ejected from a volcano while molten and that were solidified in flight.

**Botryoidal** An adjective that refers to minerals displaying smooth, rounded forms that resemble grapes closely bunched.

**Boulder** A sedimentary particle (usually rounded) whose diameter is larger than 256 millimeters (about the size of a volleyball).

**Boulder field** An accumulation of boulders; commonly applied to collections of angular blocks broken from the bedrock of mountain tops by repeated cycles of freezing and thawing.

**Bowen's reaction principle** A statement of the relationships between magma and the minerals crystallizing from it during the formation of an igneous rock. See also **continuous reaction series** and **discontinuous reaction series.**

**Box canyon** See **arroyo** and **wadi.**

**Braided stream** A wide, shallow stream that lacks substantial base flow, that flows on abundant cohesionless sediment, and whose floor is interspersed with islands and bars around which the channels divide and reunite in a braidlike pattern.

**Breaker** A collapsing wave.

**Breaker zone** The zone where waves collapse.

**Breccia** (BREH-cha) Any rock consisting of angular coarse fragments. See **collapse breccia, fault breccia, impact breccia, intrusion breccia, sedimentary breccia,** and **volcanic breccia.**

**Brown clay** A pelagic sediment on the deep-sea floor that contains less than 30 percent of skeletal remains of planktonic microorganisms.

**Bulk modulus** The stress required to cause a unit change in the volume of a solid.

**Calcareous** An adjective describing material composed of calcium carbonate in the form of calcite or aragonite.

**Calcium-carbonate compensation depth** The oceanic depth below which the water is so corrosive to calcium carbonate that little if any calcareous skeletal material accumulates on the sea floor.

**Caldera** (cahl-DAY-ruh) A large, circular, closed depression that has a diameter of more than 1.5 kilometers and is created by volcanic explosion or collapse following an explosive eruption. Compare **crater** (volcanic).

**Caliche** (kuh-LEE-chee) A layer of nodules composed of calcite within or calcite crusts upon the surface of a soil formed in a semi-arid climate.

**Capacity** The total load-carrying capability of a fluid current, such as a stream of water. Compare **competence.**

**Capillary cohesion** The cohesion among the particles of a body of moist sand caused by the surface tension of water.

**Capillary fringe** A zone of capillary water located in the lower part of the zone of aeration.

**Capillary water** The water held within regolith or sediment by the surface tension of the water in narrow pore spaces whose widths are 0.0025 to 0.25 centimeter.

**Carbonate-ion complex** An ion complex having the formula $CO_3$ and a charge of −2.

**Carbonate mineral** A mineral built of various combinations of other ions with carbonate ions, which are clusters of three oxygen ions surrounding a tiny carbon ion. Common examples of carbonate minerals are calcite, aragonite, dolomite, and siderite.

**Carbonate rock** A sedimentary rock composed predominantly of carbonate minerals (calcite, aragonite, or dolomite). The commonest carbonate rocks are limestone and dolostone.

**Carbonation** A chemical reaction between carbonic acid and minerals, usually those forming carbonate rocks.

**Carbonic acid** A weak acid, $H_2CO_3$, formed by dissolving carbon dioxide gas ($CO_2$) in water ($H_2O$).

**Cataclastic** An adjective describing metamorphic rocks whose constituents were broken by internal shearing during metamorphism.

**Catastrophism** See **doctrine of catastrophism.**

**Catastrophist** One who believes in the doctrine of catastrophism.

**Cave** A natural underground chamber dissolved out of soluble rocks, such as limestones, that can be reached from the surface and is large enough for a person to enter. See also **cavern.**

**Cavern** A large cave; a series of interconnected underground caverns within a cave system.

**Celerity** (wave) The speed of advance of a wave, such as a wave on the surface of a body of water or a seismic wave. Celerity is the speed component of velocity.

**Cement** The minerals precipitated from low-temperature solutions that bind together the framework particles of a sedimentary rock.

**Cementation** A process of lithification of sediments that involves crystallization of minerals, from ions carried by interstitial fluids, within the pore spaces among the framework particles.

**Chalk** A soft earthy variety of limestone composed of calcite skeletal debris from pelagic microorganisms that is only weakly cemented.

**Chemical energy** Energy released or absorbed during reactions involving changes in the electron configuration of atoms or of ions. Compare **nuclear energy.**

**Chemical weathering** Chemical changes that affect rock materials subject to abundant oxygen, moisture, and organisms within the environment of weathering (synonym: **decomposition**).

**Chert** A nondetrital sedimentary rock composed of silica, mostly in the form of microcrystalline quartz. Chert typically occurs as nodules within carbonate rocks, but it also forms widespread layers that may or may not be associated with carbonate rocks.

**C horizon** (of a soil). See **Zone C.**

**Cinder cone** See **tephra cone.**

**Cinders** Tephra having diameters ranging between 0.06 and 4 millimeters.

**Circle of illumination** The boundary between the half of a planet that is illuminated by the Sun's rays and the dark part of the planet (synonym: terminator).

**Cirque** (seerk) A steep-walled, bowl-shaped niche in the bedrock that is above the present or a former snowline and that is formed by erosion at the head of a valley glacier.

**Clastic** An adjective describing sediment particles that have been broken. See also **clastic texture.**

**Clastic texture** A sedimentary texture characterized by broken particles.

**Clay** (1) Size: sedimentary particles smaller than 1/512 millimeter. (2) Mineral: a hydrated layer-lattice silicate that has the silicate sheets more closely spaced than they are in micas.

**Claystone** A detrital sedimentary rock that consists chiefly of clay-size particles.

**Cleavage** (1) Minerals: the tendency of a mineral to break along smooth planes parallel to zones of weakness among the ions in the crystal lattice. (2) Rocks: the tendency of a rock to break along smooth planes parallel to zones of weakness related to the direction(s) of preferred orientation(s) of micaceous minerals.

**Clockwise rule (Coriolis effect)** The deflection toward the right (as viewed in direction of motion) of a moving body or a current traveling in a frame of reference that is rotating clockwise. In the Southern Hemisphere of the Earth, the sense of rotation is clockwise; therefore, in the Southern Hemisphere, currents are deflected toward their right.

**Coal** A combustible organic sedimentary rock composed of altered plant material and containing fixed carbon in excess of 50 percent.

**Coast** The general area where land and a body of water meet.

**Coastal zone** A strip of land of varying width that is affected by the marginal parts of a large lake or the sea.

**Cobble** A sedimentary particle greater than 64 millimeters (size of a tennis ball) and less than 256 millimeters (size of a volleyball).

**Coesite** A high-pressure form of $SiO_2$ that has a density of 2.92 grams per cubic centimeter and that is formed during impact metamorphism.

**Cohesion** The ability of a body of particles to stick together. See **capillary cohesion** and **electrostatic cohesion.**

**Cohesionless particles** A body of sediment, of sand size or coarser, within which cohesive forces among the particles are negligible.

**Col** (coll) A gap along the crest line of an arête caused by headward erosion of cirques.

**Collapse breccia** (BREH-cha) A sedimentary breccia whose angular blocks, usually composed of one or more kinds of carbonate rocks, were formed when groundwater dissolved underlying soluble formations.

**Collapse crater** A crater formed by collapse around a volcanic vent.

**Column** (1) Cave deposit: a cylindrical feature formed by the joining together of a stalactite and a stalagmite. (2) Stratigraphy: all the stratigraphic units present in a particular area. (3) Igneous rocks: a geometrically regular prism of rock bounded by columnar joints.

**Columnar joint** A joint, part of a geometrically regular patterned set, in a body of igneous rock that is formed as a result of cooling. The intersection of the columnar joints divides the rock into geometrically regular prisms or columns.

**Colluvium** Sediment that has been transported by gravity down a subaerial slope.

**Compaction** The reduction in volume by loss of pore space in a body of sediment subjected to the weight of overlying strata or to tectonic forces.

**Competence** (1) Stream: an expression of the force of a stream exemplified by the size of the largest particle a stream can transport. Compare **capacity.** (2) Tectonic: the ability of a geologic formation to withstand deformation. A competent formation displays great strength.

**Complex mountains** Mountains resulting from the elevation and differential erosion of orogenic belts having complex internal structures, such as folds, great overthrusts, batholiths, and belts of regionally metamorphosed rocks.

**Complex volcanic cone** A volcanic cone composed of inclined layers of tephra and reinforcing layers consisting of tabular bodies of intrusive igneous rock or of extrusive igneous rock; may stand alone or be the upper part of a composite cone.

**Complex weathering** Weathering processes that involve more than one kind of activity, such as the crystallization of minerals, the combined effects of physical weathering and chemical weathering, and the effects of organisms.

**Composite cone** A two-part volcanic cone consisting of: (1) an upper, steep-sided complex cone built of alternating layers of tephra and igneous rock and (2) an underlying gently sloping volcanic shield.

**Composite pluton** A pluton containing the products of more than one pulse of intrusion.

**Compressional wave** See **P wave.**

**Compressive stress** Stress that tends to push material closer together.

**Compromise boundary** (minerals) An irregular boundary of a mineral formed when enlarging individual crystal lattices interfere with one another so that no individual displays its characteristic crystal faces.

**Conchoidal fracture** Breakage along a smooth curved surface, as on the inside of a shell (conch). Compare **cleavage.**

**Concordant contact** A geologic contact that is parallel to the layers within the adjacent rocks.

**Concordant pluton** A pluton whose boundaries are parallel to the layers of the country rock.

**Concretion** A solid body of mineral matter, in sedimentary strata, that is chemically precipitated from circulating groundwater in concentric layers outward from a nucleus. It is commonly a whole fossil or a broken fossil. Compare **geode.**

**Conduction** The transfer of heat by the mechanism of ion-to-ion or molecule-to-molecule impacts. In this way, hotter, faster moving ions or molecules cause slower-moving ions or molecules to move faster. Thus, the temperature of the former decreases, whereas that of the latter increases.

**Conductivity, electrical** See **electrical conductivity.**

**Conductivity, thermal** See **conduction.**

**Cone of depression** A conical depression in the groundwater table that is formed by the effects of drawdown from a well.

**Cone sheet** A curved dike that is inclined radially inward to form a funnel-shaped structure.

**Cone, volcanic** See **volcanic cone.**

**Confined aquifer** An aquifer in which the groundwater is held in by an overlying stratum that has low permeability.

**Conformable** A stratigraphic succession deposited without significant interruptions.

**Conglomerate** A clastic sedimentary rock containing more than 50 percent of rounded gravel-size particles. Compare **sedimentary breccia.**

**Connate water** Water, formerly from an ocean, that is buried with sediments.

**Contact** (geologic) A boundary between two bodies of rock that are different.

**Contact-metamorphic aureole** A zone surrounding a pluton within which the country rocks display the effects of metamorphic changes.

**Contact-metamorphic ore** An ore deposit formed by contact metamorphism during the emplacement of a pluton.

**Contact metamorphism** The metamorphism related to the immediate vicinity of a pluton.

**Continent** A major land mass that stands above sea level. See also **continental mass.**

**Continental drift** The horizontal displacement of continents. Not related to **beach drifting, glacial drift,** or **longshore drift.**

**Continental glacier** A massive, thick ice sheet covering more than 50,000 square kilometers.

**Continental interior desert** A desert, formed in regions far from the sea, that is rarely if ever affected by precipitation from clouds that derive their moisture from the sea.

**Continental mass** A major high-standing part of the lithosphere underlain by thick felsic crust **(sial).**

**Continental rise** The gentle slope (gradient between 1:100 and 1:700) on the top of a wedge-shaped body of sediment that lies seaward of the continental slope.

**Continental shelf** The submerged outer part of a continental mass, beginning at the shoreline and extending to the first prominent change in the bottom slope at a depth of 200 meters or less.

**Continental shield** See **shield, continental.**

**Continental slope** The relatively steep (3° to 6°) slope that lies seaward of the continental shelf.

**Continuous reaction series** In Bowen's reaction principle, the condition in which ions are exchanged continuously between growing lattices of one kind of mineral and a magma.

**Contour current** A subsurface current consisting of a thin tongue of dense water flowing along the margins of the ocean basins and following submarine contours.

**Contour interval** The constant vertical interval between adjacent contour lines on a contour map.

**Contour line** A line on a map that traces the shape of the landscape as defined by the intersection of a given horizontal plane at a stated altitude above a reference datum level and the land surface.

**Contour map** A map showing the Earth's morphology by means of contour lines.

**Convection** The motion of a fluid caused by the uneven density related to heat; the hotter, less-dense fluid rises, whereas the cooler, denser fluid sinks.

**Convergent** (plate boundary) The boundary between two lithosphere plates that are coming together.

**Coquina** A clastic limestone consisting of broken shells.

**Cordillera** One of the great mountain belts of the Earth.

**Core** (1) Earth's: the dense (presumably) metallic sphere at the center of the Earth whose outer part, about 2200 kilometers thick, is liquid and whose inner part, about 1200 kilometers in diameter, is solid. (2) Subsurface material: a cylindrical sample of subsurface sediment or subsurface rock collected for scientific study.

**Coriolis effect** The systematic deflection in the travel paths of bodies or currents that are moving across the surface of a rotating object. See **clockwise rule** and **counterclockwise rule.**

**Corrosion** A reaction in which a fluid dissolves a solid. Examples include minerals in the zone of weathering, in contact-metamorphic aureoles, and in bedrock stream beds.

**Counterclockwise rule (Coriolis effect)** The deflection toward the left (as viewed in direction of motion) of a moving body or a current traveling in a frame of reference that is rotating counterclockwise. In the Northern Hemisphere of the Earth, the sense of rotation is counterclockwise; therefore, in the Northern Hemisphere, currents are deflected toward their left.

**Country rock** A general term for rock that has been penetrated by a pluton.

**Covalent bond** A kind of bonding among ions whose shared electrons orbit more than one nucleus.

**Crater** (1) Volcanic: a circular volcanic orifice having a diameter of less than 1.5 kilometers. Compare **caldera.** (2) Impact: a circular steep-walled closed depression, large or small, that has an elevated rim composed of fragmented rock and a flat or concave-up floor that may display a central mound. Impact craters are accompanied by evidence of shock metamorphism, which may include coesite, a high-pressure form of $SiO_2$. The walls of large impact craters may be terraced.

**Creep** (1) Gravity transport: the movement of regolith downslope that takes place so slowly as to escape notice. (2) Tectonic: the slow, steady displacement along a fault.

**Crevasse** (1) Glacier: a deep fissure in the upper zone of a glacier. (2) Natural levee: a break in a natural levee that enables flood waters to spread away from the channel.

**Cross strata** Inclined layers of cohesionless sediment that are deposited by a fluid current at the forward side of a migrating bed form or an advancing embankment and that are enclosed between layers that were originally horizontal or nearly horizontal. Cross strata are common both in sediments that were physically transported and in sedimentary rocks formed by the lithification of such sediments.

**Crust** (Earth's) The outermost part of the solid Earth that lies above the Mohorovičić discontinuity; the thickness of the crust varies from about 5 kilometers in ocean basins to more than 50 kilometers under some parts of the continents.

**Crystal** A mineral bounded by natural plane surfaces whose orientation and symmetry reflect a regular internal ionic structure.

**Crystal chemistry** The scientific study of the relationships between the properties of minerals and the kinds of ions present and the ways the ions are arranged.

**Crystal cumulate** An igneous rock in which crystals of early formed minerals, such as olivine, pyroxene, magnetite, chromite, and plagioclase, become physically separated from the rest of the magma by sinking, floating, or the action of density currents to form loosely packed aggregates that make layers comparable to sedimentary strata.

**Crystal form** The natural distinctive shape of a mineral that reflects the regular, symmetrical internal structure of its ions.

**Cubic packing** An arrangement of spheres (such as the ions in a mineral) placed in planes in such a way that lines connecting the centers of the spheres meet at right angles.

**Cumulate** See **crystal cumulate.**

**Curie point** The temperature above which a magnetic substance loses its magnetism and below which the substance is magnetic. At atmospheric pressure the Curie point for iron is 760°C, for magnetite 578°C.

**Cyclothem** A patterned succession of strata, typically including marine strata, nonmarine strata, and peat or coal. Many cyclothems were formed by the forward growth of delta lobes and the later submergence by the sea and burial by marine strata of the inactive deltas.

**Darcy's law** The relationship that states: the rate of flow of groundwater through porous material is proportional to the pressure driving the water and inversely proportional to the length of the flow path. See **hydraulic gradient.**

**Datum plane** An established horizontal reference surface used as the zero point in surveys. For example, a common datum plane for contour maps is mean sea level.

**Daughter isotope** The isotope formed by radioactive decay of a parent isotope. Some daughter isotopes in turn become parent isotopes. At the end of each radioactive series is a stable daughter isotope that does not undergo further radioactive decay.

**Debris flow** The rapid downslope movement of a viscous mass of water-saturated regolith.

**Debris slide** The rapid downslope movement of a body of sand that broke loose, initially as a cohesive block, by slump along a steep surface and immediately lost its cohesion.

**Decay constant** A number expressing the rate of a radioactive reaction that is based on the proportion of the nuclei of the parent isotope that decay in a unit of time.

**Declination** The angle, measured in the horizontal plane, between true North (determined by a meridian of longitude) and the local lines of force of the Earth's magnetic field.

**Decomposition** See **chemical weathering.**

**Deep-water wave** A water-surface wave traveling in water deeper than $L/2$ that is not influenced by the bottom ($L$ = wavelength).

**Deflation** The erosion of dry regolith by the wind.

**Deflation armor** A lag concentrate of pebbles and larger stones formed by the selective deflation of finer particles that serves as a protective layer preventing further deflation.

**Deflation basin** A closed depression eroded out of regolith by wind action (synonym: **blowout**).

**Deformation** A change in shape, in volume, in both shape and volume, or in the level of the surface of a body of rock as a result of tectonic forces acting within the Earth.

**Degradation** (the opposite of **aggradation)** The deepening of a stream channel by erosion.

**Deleveling** A long-term change of the relative position of a water surface against the land.

**Delta** A body of sediment deposited at the mouth of a river that tends to build upward to a flat surface controlled by the water level of the lake or the sea into which the river flows.

**Dendritic pattern** A drainage pattern formed in material that erodes equally in all directions and in which streams branch repeatedly to form a network resembling the veins of a leaf.

**Density** Mass per unit volume; a useful attribute related to chemical composition and packing of ions within a substance.

**Density current** A current that flows because of unequal density of parts of the body of a fluid. The difference in density may result from temperature differences (convection current), from salinity differences (salinity current), from differences in content of suspended sediment **(turbidity current),** or from some combination of all of these.

**Desert** A region characterized by a great excess of evaporation over precipitation and where the yearly rainfall is usually less than 25 centimeters.

**Desert pavement** A lag concentrate of closely packed and tightly fitting stones formed by the selective removal by wind or by water of formerly present finer particles. See also **deflation armor.**

**Detrital** (deh-TRY-tal) **sedimentary rock** A sedimentary rock formed by the lithification of detritus.

**Detritus** (deh-TRY-tus) A collective term for particles—derived from the breaking up of bedrock—that were transported and physically deposited as solids of varying sizes.

**Diagenesis** A general term designating all the changes that deposited sediments undergo, short of the kinds of reactions encompassed under metamorphism.

**Diapir** A body of solid or semisolid material that has punched its way upward through overlying materials. For example, a body of halite that has moved upward through the strata covering a thick layer of salt. A diapir differs from a pluton in that the material to make a pluton (the magma) (1) moved upward as a liquid, not as a solid and (2) from a diatreme owing to the absence of explosive activity associated with its emplacement. Some bodies of igneous rock are thought to have been emplaced as diapirs.

**Diatom** A group of single-celled aquatic plants that secrete siliceous skeletons. Some diatoms live in the sea; others live in lakes or streams.

**Diatreme** A cylindrical pluton, formerly connected to a volcano, that contains explosion breccia.

**Differential erosion** The uneven removal of rock material by agents of erosion.

**Differential weathering** The uneven wearing away of rock material. See also **differential erosion.**

**Dike** A discordant tabular pluton that fills a fracture in a body of older material, such as bedrock or regolith.

**Dip** The angle, in the vertical plane at right angles to the strike, between the horizontal (0°) and an inclined surface. The dip of a vertical surface is 90°.

**Dip-slip fault** A fault on which the direction of displacement parallels the direction of dip of the fault.

**Discharge** (1) The volume of water flowing through the cross section of a stream channel during a given time. (2) Discharge of groundwater (see **groundwater discharge).**

**Disconformity** (1) The unconformable relationship between parallel groups of marine strata that are separated by an erosion surface. (2) A surface marking a gap in the record within a sequence of parallel strata.

**Discontinuous reaction series** In Bowen's reaction principle, the relationship between growing crystals and a melt in which the crystals of one mineral phase disappear and their places are taken by crystals of a new mineral phase as the temperature decreases. Compare **continuous reaction series.**

**Discordant contact** A term describing a geologic contact, such as the margin of a pluton, that is not parallel to the layers of the adjacent rock (country rock in the case of a pluton). Compare **concordant contact.**

**Divergent plate boundary** The boundary between two lithosphere plates that are moving away from each other.

**Doctrine of catastrophism** The view, widely held in the early nineteenth century, that the Earth was formed as a result of numerous violent catastrophes.

**Doctrine of uniformitarianism** The view, advocated by James Hutton, Charles Lyell, and others, that the Earth's past history can be understood by making comparisons with modern processes and their products and by establishing fundamental relationships between processes and materials.

**Dolerite** A medium-grained mafic igneous rock consisting chiefly of pyroxene and plagioclase. Compare **basalt** and **gabbro.**

**Dolostone** A sedimentary carbonate rock composed of more than 50 percent dolomite.

**Dome** A geologic structure that has the oldest rocks at the center and away from which younger strata dip radially outward in all directions. Compare **basin,** (3) structural.

**Dormant volcano** A volcano known to have been active but that has not erupted during the past 50 years.

**Drainage basin** The area that supplies water to a given stream network.

**Drawdown** The lowering of the water table in the vicinity of a well from which water has been removed. See also **cone of depression.**

**Drift, glacial** A general term for any sediment related to glaciers, such as till, outwash, and varved lake deposits. In no way related to **beach drifting, continental drift,** or **longshore drifting.**

**Dripstone** A speleothem formed by deposition from dripping water. Compare **flowstone.**

**Drumlin** An elongated, streamlined hill shaped by a glacier; consists of till, bedrock, or both.

**Ductile** A material capable of undergoing deformational strain of at least 10 percent without breaking.

**Dune** A hill or ridge of sand built of windblown sand.

**Dynamic metamorphism** Metamorphism that results from intense shearing deformation.

**Earth flow** A slow downslope movement in which saturated regolith sags downward in a series of irregular terraces.

**Earth-Moon barycenter** The center of mass of the Earth-Moon pair, located inside the Earth within a depth zone 611 kilometers thick and ranging between 1436.5 and 2047.5 kilometers beneath the Earth's surface. The mean depth of the Earth-Moon barycenter is 1707 kilometers.

**Earthquake** An intense motion of the ground resulting from the passage of seismic waves that have been released suddenly inside the Earth.

**Ebb tide** A falling tide.

**Echo sounder** A device that generates pulses of sound to determine the depth of water; based on the time required for the sound pulses to travel from the instrument to the bottom of the water and back to the instrument again.

**Ecliptic, Plane of the** See **Plane of the Ecliptic.**

**Ecliptic Pole** A line perpendicular to the Plane of the Ecliptic.

**Eclogite** A garnet-pyroxene rock formed under high pressures and elevated temperatures, such as those within the upper part of the Earth's mantle.

**Economic geology** The branch of geology that deals with the origin, location, evaluation, extraction, and economic impacts of mineral deposits.

**Economic mineral deposit** A deposit from which the extraction and processing of the material can be carried on at a profit.

**Elastic energy** The potential energy stored within a solid that is being deformed at levels below its elastic limit (synonym: strain energy).

**Elastic limit** The upper limit of strength of a material; the maximum stress a body can withstand without undergoing permanent deformation.

**Elastic rebound** The sudden return to a predeformation configuration of a body that is being deformed to the point of rupture.

**Elastic waves** So named because the properties of the waves are related to the elastic properties of the matter through which they pass (synonym: seismic waves).

**Electrical conductivity** The property of a substance that enables it to conduct an electrical current.

**Electrical energy** Energy transmitted by the flow of electrons along a conductor, such as a copper wire.

**Electrical resistivity** The property of a substance that resists the flow of an electrical current; the inverse of electrical conductivity.

**Electron** A fundamental subatomic particle of matter containing a unit negative electrical charge and negligible mass ($0.55 \times 10^{-3}$ atomic mass units). Compare **proton.**

**Electron-capture decay** A kind of radioactive transformation in which the atomic nucleus captures an orbital electron, thus enabling a proton to become a neutron. The mass number of the resulting daughter isotope is the same as that of the parent isotope, but the daughter's atomic number is 1 less than the parent's (reflecting the loss of a proton).

**Electrostatic cohesion** The cohesion among particles of silt and clay sizes (smaller than about 1/16 of a millimeter) that results from the lack of uniformity in the distribution of ionic charges.

**Emergence** A relative drop of water level against the land.

**End moraine** A ridge of till deposited at the outer margin of a glacier (synonym: **recessional moraine**). The end moraine most distant from the source of the glacier is a **terminal moraine.**

**Energy** (1) The capacity for doing work. (2) The equivalent of mass according to the Einstein equation, $E = mc^2$, where $E$ is the energy in ergs, $m$ is the mass in grams, and $c$ is the speed of light in centimeters per second. See also **chemical energy, elastic energy, electrical energy, geothermal energy, kinetic energy, nuclear energy, potential energy, radiant energy,** and **solar energy.**

**Environment** The conditions (physical, chemical, and biological) that prevail within a given location.

**Environment, subaerial** See **subaerial environment.**

**Environment, subsurface** See **subsurface environment.**

**Environment of weathering** The subaerial environment within which weathering reactions take place.

**Eolian** An adjective describing a feature caused by wind action.

**Eolian sediment** A sediment that is transported and deposited by the wind.

**Eon** The largest subdivision of geologic time, includes two or more eras.

**Epeirogeny** Vertical subsidence or elevation of the Earth's crust, which takes place without crumpling the strata.

**Epicenter** The place on the surface of the Earth vertically above the focus of an earthquake.

**Epidote-amphibolite facies** (of metamorphic rocks) A group of metamorphic rocks that have the bulk chemical composition of basalt and that are characterized by epidote and amphiboles; formed in the lower range to middle range of temperatures of metamorphism.

**Epoch** A subdivision of a geologic period.

**Equigranular** A textural term that describes igneous rocks in which the minerals are all about the same size.

**Era** A geologic time interval that includes several periods and that is based on the major developments in the history of life. Examples of eras are Paleozoic, Mesozoic, and Cenozoic.

**Erg** A tiny energy unit equal to the force of 1 dyne acting through a distance of 1 centimeter. An erg is about the amount of energy expended by a housefly in doing a single push-up.

**Erosion** A collective term for any process of physical disintegration, chemical dissolution, or movement of affected rock materials from one place to another on the surface of the Earth.

**Erosion cycle** The cycle of destruction of a landmass by the effects of streams and the downslope gravity-transport processes that are thought to begin with the uplift of a landmass, ranging from an island to a mountain range to an entire continent. The cycle is presumed to end when the uplifted area has been eroded away and all that is left is an almost flat plain (known as a **peneplain**), lying near base level.

**Erratic** A transported rock fragment that is unlike the bedrock underlying it.

**Esker** A long continuous, but sinuous, ridge of stratified drift created by a stream of meltwater flowing beneath a body of stagnant ice that is formed by the wasting of a former glacier.

**Estuary** An arm of the sea formed by the submergence of a river valley; commonly V-shaped and pointing landward.

**Eustatic change of sea level** A worldwide change of sea level caused by changes in the capacity of the ocean basins or in the total amount of sea water.

**Euxinic sediment** Black mud deposited in basins where the bottom waters became stagnant and hydrogen sulfide killed all organisms except the bacteria that could live without dissolved oxygen.

**Evaporite** A group of sedimentary rocks composed of minerals that were precipitated out of supersaturated water solutions. See also **evaporite mineral.**

**Evaporite mineral** A collective name for any mineral that was precipitated out of a supersaturated water solution.

**Exchange reaction** A chemical reaction in which positive ions change partners with negative ions and thus form new compounds.

**Exfoliation** The spalling off of bodies of rock along sheeting joints.

**Extinct volcano** A volcano in which all activity has forever ceased.

**Extrusive igneous rock** An igneous rock that formed at the Earth's surface by the cooling and solidification of lava.

**Facies** A term that designates a rock having a distinctive aspect or appearance as a result of its being formed under particular conditions. Usually applied to sedimentary materials (see **sedimentary facies**) or to metamorphic rocks (see **metamorphic facies**).

**Fan** A low, fan-shaped cone of alluvium, ranging from a few meters to many kilometers in extent. See also **abyssal fan** and **basin-floor fan.**

**Fault** A fracture along which the opposite sides have been displaced in a direction parallel to the fracture. Compare **joint.**

**Fault-block mountain** A part of the Earth's surface displaying mountainous relief (more than 400 meters), resulting from vertical elevation along faults and possibly also from differential erosion of an area elevated along one or more faults.

**Fault breccia** (BREH-cha) A rock consisting of angular pieces broken by movement along a fault. Compare **fault gouge** and **mylonite.**

**Fault creep** The slow, steady displacement along a fault.

**Fault gouge** Various ground-up rock material created by fault motion. Compare **fault breccia** and **mylonite.**

**Fault plane** The surface along which the fault blocks have been displaced.

**Feldspathic sandstone** See **sandstone, feldspathic**

**Felsic** An adjective designating light-colored igneous rocks that contain abundant feldspars (of the kinds rich in sodium or potassium), quartz, and minor amounts of ferromagnesian-silicate minerals. Compare **intermediate rocks, mafic rocks,** and **ultramafic rocks.**

**Felsic magma** A magma that results from melting continental-type crust in places where tectonic forces caused such crust to be depressed to great depths.

**Felsite** Any aphanitic, light-colored felsic igneous rock.

**Fetch** The length of the stretch of open water across which the wind from a particular direction can flow.

**Firn** A body of granular particles derived from the melting of snow and subsequent refreezing of the water.

**Fission** The splitting of an atomic nucleus and the resulting formation of two elements out of one.

**Fission-fragment track** A small linear feature within a crystal which marks an area that has been damaged by the passage of an atomic nucleus released during radioactive decay by nuclear fission.

**Fissure eruption** A volcanic eruption in which the lava emerges at several locations along a crack (the fissure, commonly a fault).

**Fixed carbon** The carbon content of a coal that is in the elemental state, as contrasted with combined carbon, such as that in methane gas (a compound containing carbon and hydrogen, $CH_4$).

**Fjord** A glaciated U-shaped valley that has been flooded by the sea after the glacier which eroded the valley has melted.

**Flank eruption** An eruption of lava from the side of a volcanic cone, as contrasted with the discharge of lava from the apex of the cone or from beneath its base.

**Flood bulge** A long kinematic wave of water that travels downstream at its own speed, independently of and added to the speed of the base flow.

**Flood plain** A low flat area adjacent to a stream channel, which is occasionally inundated with water that overflows the banks of the channel.

**Flood-plain river** A river flowing in a lowland and displaying channels that form great sweeping curves **(meanders).**

**Flood tide** A rising tide.

**Flowstone** A speleothem formed by deposition from flowing as opposed to dripping water. Compare **dripstone.**

**Focus** The place within the Earth where the energy of an earthquake is first released (synonym: hypocenter).

**Fold** A wavelike undulation of strata, usually the result of tectonic movements. Most folds can be classified as anticlines or synclines. See also **asymmetric fold, isoclinal fold, overturned fold, plunging fold, recumbent fold, symmetrical fold,** and **upright fold.**

**Fold axis** The line around which the strata of a fold curve.

**Folding** The bending of rock strata.

**Fold mountains** Mountain belts whose major relief is related to the eroded parts of extensive groups of folds.

**Foliation** The layered arrangement of many metamorphic rocks in which the layers consist of different minerals or parallel arrangements of the same mineral.

**Footwall block** A body of rock beneath a nonvertical fault.

**Foraminifera** A class of one-celled marine animals that move by extruding long filaments of protoplasm named pseudopods. Many groups of Foraminifera secrete calcite skeletons.

**Foraminiferal ooze** A fine-grained pelagic sediment composed of more than 30 percent of the skeletal remains of Foraminifera.

**Foreset bed** An inclined layer formed by the forward growth of the front of a delta or an advancing dune.

**Foreshocks** Seismic waves, beginning before the main shock of an earthquake, starting with small tremors that become increasingly large.

**Formation** The basic unit for geologic mapping; a formation consists of rocks sharing one or more common characteristics.

**Formation water** The saline water found in the same rock formations that contain oil or gas.

**Form ratio** The ratio of a stream channel's depth divided by its width.

**Fossil** The naturally preserved remains or traces of an ancient organism.

**Fossil fuel** A fuel formed from the altered remains of organisms; two examples are coal and petroleum.

**Fracture** (1) Minerals: the irregular breaking of a mineral. Compare **cleavage.** See also **conchoidal fracture.** (2) Tectonics: a surface along which a brittle rock has been broken by failure in response to tensile stresses.

**Fracture zone** A great linear break or series of breaks in the Earth's crust.

**Framework particles** Sand-size or larger particles that form the bulk of a body of coarse-grained regolith.

**Frost heaving** The lifting of a rock or the surface of a body of regolith by the growth of subsurface ice crystals.

**Frost wedging** The prying apart of bedrock by the pressure created when water expands upon freezing.

**Fumarole** A volcano that discharges only gases.

**Fusion** The joining of atomic nuclei.

**Gabbro** A coarse-grained mafic igneous rock composed about equally of pyroxene and plagioclase. See also **norite.**

**Gangue** The materials associated with an ore deposit that are not of immediate concern to the mining operation. Compare **ore.**

**Geode** A cavity in a sedimentary rock that is fully or partially filled by crystals that have grown inward from the wall. Compare **concretion.**

**Geologic cycle** All of the interactions between the Earth's materials and its sources of energy. See also **hydrologic cycle** and **rock cycle.**

**Geologic record** The systematic arrangement of rocks, regolith, and the Earth's morphology that forms a natural archive from which the history of the Earth may be inferred.

**Geologic structure** A feature delineated by the three-dimensional arrangement of strata.

**Geology** The scientific study of the Earth. See **historical geology** and **physical geology.**

**Geomorphology** The scientific study of landforms and their development.

**Geosyncline** A major linear or curvilinear trough in the Earth's crust (usually longer than 1000 kilometers) that subsides (or has subsided) deeply and within which sediments have accumulated (or may accumulate) to great thickness (about 10 kilometers or so). The strata deposited in ancient geosynclines have been deformed into the world's major mountain belts.

**Geothermal energy** The heat energy within the Earth.

**Geothermal gradient** The regular increase of temperature downward within the Earth. The usual value of the geothermal gradient away from hot spots is 30°C per kilometer.

**Geyser** A hot spring that periodically gushes a fountain of steam-charged water.

**Geyserite** A deposit of minerals around a geyser that were precipitated from the hot water discharged by the activities of the geyser.

**Glacial drift** See **drift, glacial.**

**Glacial ice** The ice of a glacier.

**Glacial grooves** Linear grooves eroded in bedrock by a glacier. The long axes of the grooves parallel the direction of glacial flow.

**Glacial-marine sediment** A body of sediment deposited on the sea floor and composed chiefly of debris dropped from melting icebergs.

**Glacial striae** Parallel sets of tiny scratches made on bedrock or on a boulder by a glacier. Bedrock striae parallel the direction of glacier flow. Incorrectly referred to as striations.

**Glacial till** A needless redundancy for **till.** The adjective, glacial, is superfluous; by definition *all* till is of glacial origin—no other kind exists.

**Glaciation** The advance and retreat of a glacier across an area.

**Glacier** A flowing mass of ice formed by the recrystallization of snow that has attained a density of at least 0.84 grams per cubic centimeter, that has become impermeable to air, that is powered by gravity, and that has moved outward and downward beyond the snowline.

**Glacier surge** A short-lived rapid advance of a glacier by one or more mechanisms that are not yet fully understood.

**Glass** A noncrystalline igneous rock formed by the extremely rapid chilling of lava or magma.

**Gneiss** A coarse-grained felsic metamorphic foliate characterized by alternating layers of light-colored and dark-colored minerals.

**Gondwanaland** The name of a large continental mass of Paleozoic age that is thought to have consisted of the joined-together areas now forming South America, Africa, Australia, India, and Antarctica and that began to break apart during the Mesozoic Era.

**Graben** A downdropped block between two more-or-less parallel normal faults that dip beneath the downdropped block. Compare **horst** and **rift valley.**

**Graded layer** A layer within which the sizes of the particles vary systematically. In normal grading, coarse particles at the base of a layer grade into fine particles at the top.

**Graded stream** A stream in a condition of balance established by the interaction of factors, such as energy relationships, discharge, speed, slope, load, and the shape of the channel's cross section.

**Gradient stream** The local slope on which a stream flows.

**Granite** A coarse-grained felsic igneous rock whose chief constituents (about 75 to 90 percent) include about equal parts of potassium feldspar, sodic plagioclase (albite), and quartz. Its lesser constituents may include ferromagnesian minerals, usually biotite or hornblende.

**Granitization** The formation, by processes not including melting, of a metamorphic rock containing the minerals of and having the general appearance of a granite.

**Granular disintegration** A process of physical weathering in which the individual minerals of a rock separate from one another.

**Graphic granite** A patterned intergrowth of quartz and potassium feldspar in which the long axes of the darker-colored quartz crystals are parallel to the axes of the lighter colored feldspar crystals. The resulting appearance resembles the inscriptions on ancient cuneiform tablets.

**Gravel** Sediment particles coarser than 2 millimeters that include the size ranges of both cobbles and boulders.

**Gravity anomaly** A measured value of the Earth's gravitational acceleration that is not the same as the predicted value. Measured values larger than predicted are positive anomalies; measured values less than predicted are negative anomalies.

**Gravity-displaced sediments** Basin-floor sediments that have been displaced down an underwater slope by any one of numerous gravity-transport processes, including turbidity currents, liquified cohesionless-particle flows, slumps, and debris flows.

**Gravity** (Earth's) The inward-acting force that causes the Earth to attract all objects to its center.

**Graywacke** A kind of sandstone, usually dark colored, in which the sand-size particles have been tightly bound together by the growth of layer-lattice silicate minerals, such as chlorite.

**Greenhouse effect** The atmospheric entrapment of heat caused by carbon dioxide, which allows the wavelengths of incoming solar energy to pass downward but does not permit the upward escape of the longer wavelengths of the heat reflected by a planet's surface. In this way, carbon dioxide functions as the glass in a greenhouse, hence the name.

**Greenschist facies** (of metamorphic rocks) A group of metamorphic rocks that has the bulk chemical composition of basalt and that is characterized by chlorite; formed in the lowest range of temperatures of metamorphism.

**Groundwater** Water below the surface of the Earth.

**Groundwater discharge** The flow of groundwater to the Earth's surface or into a well.

**Growth structure** A tectonic structure, such as a fold or a fault, that was active where sediments were being deposited and, thus, affected the thickness and characteristics of the sediments.

**Gully** A deeply eroded stream channel in regolith.

**Guyot** (ZHEE-oh) A flat-topped, truncated volcanic cone on the deep-sea floor whose top has been submerged.

**Gyre** (JHY-er) A major circulation cell—around a vertical axis—of the surface waters of each ocean basin that is caused by uneven heating of the Earth's surface, the prevailing winds, and the Coriolis effect.

**Hackly fracture** The breaking of a mineral along an irregular jagged surface that has sharp edges.

**Half life** The time required for one half the atomic nuclei of a radioactive parent isotope to disintegrate into those of a daughter isotope.

**Halide** A member of a mineral group characterized by ions of a metal, such as sodium, potassium, or magnesium, joined together with chlorine.

**Hanging valley** A tributary valley whose floor lies higher than the floor of the main valley so that the tributary stream enters the main valley over a waterfall.

**Hanging-wall block** A body of rock beneath which a nonvertical fault dips.

**Hardness** A mineral's resistance to being scratched. See **Mohs' scale of hardness.**

**Heat energy** The energy related to the temperature of a body, which is an expression of the kinetic energy of its ions or molecules. Heat energy can be transferred from one place to another by **conduction, convection,** or radiation.

**Hexagonal packing** The spheres in a single plane are arranged in parallel rows, but with the spheres of every other row offset from those in the adjacent row, so that the centers of the spheres attain their closest possible spacing.

**High-altitude suspension** A body of wind-blown sediment suspended by the fast-moving eddies within a jet stream flowing in the upper atmosphere. Compare **low-altitude suspension.**

**Historical geology** The scientific study of the history of the Earth, including the Earth's origin, the development of its geologic record, and the history of life.

**Hooke's law** The statement that strain within the elastic limit is proportional to the stress.

**Horn** A mountain pinnacle formed by the wearing away of several arêtes and overlooking three or more cirques.

**Hornfels** A metamorphic nonfoliate that originates in contact-metamorphic aureoles.

**Horst** An elevated upthrown block bounded by two more-or-less parallel normal faults that dip away from the upthrown block. Compare **graben.**

**Hot spot** A localized area within the lithosphere that is being heated from below by a chimneylike mantle plume.

**Hot spring** A spring that emits water that is hotter than the mean annual air temperature of the surrounding region.

**Humus** Dark-colored, partially decomposed plant material formed when dead plant matter is oxidized and slowly altered but not destroyed by bacteria.

**Hydraulic conductivity** The ability of a porous, permeable body to transmit groundwater.

**Hydraulic geometry** (of a stream) An expression for the important variables in a stream system and how they interact. The independent variable is discharge; the dependent variables are shape and depth of the channel, speed of the flow, slope of the water surface, and gradient (slope of the stream bed).

**Hydraulic gradient** The slope of the groundwater table.

**Hydrocarbon** A compound created by organisms and consisting mainly of hydrogen and carbon; the altered fossil form, petroleum, is a widely used fuel.

**Hydrologic cycle** The collective designation for all the motions of water and the sequence of changes that water undergoes as it is evaporated from the ocean, is precipitated onto the land as rain or snow, and returns to the ocean by flowing overland or by moving below the ground.

**Hydrology** The scientific study of water.

**Hydrolysis** A reaction in which a mineral combines with water; an important process of chemical weathering.

**Hydrosphere** A collective designation for all the Earth's water, including the oceans, lakes, rivers, groundwater, snowfields, glaciers, icebergs, and the water vapor in the atmosphere.

**Hydrothermal deposit** A mineral deposit formed by precipitation out of a hot water solution (see **hydrothermal solution**).

**Hydrothermal solution** A hot water solution; one that usually deposits minerals.

**Ice sheet** A glacier of wide extent that has spread out uniformly over a land surface.

**Igneous** Literally translated from the Latin as fire-formed; refers to all rocks formed by the solidification of molten material. See **extrusive igneous rock, intrusive igneous rock,** and **volcanic rock.**

**Igneous rock** Rock formed by the solidification of molten material.

**Impact erosion** A kind of abrasion in which wind-driven sand and tiny pebbles chip shallow irregularities into bedrock.

**Inactive volcano** A volcano that has not erupted within historic time (about 10,000 years).

**Inclination** (magnetic) The angle, measured in a vertical plane, between a magnetic line of force and the horizontal. Compare **declination.**

**Inclusion** An early formed small crystal that becomes engulfed within the rapidly growing lattice of a larger crystal.

**Infiltration** The seepage of rainwater and other surface water into the pores of the regolith and the bedrock.

**Interfacial angle** The constant angle between adjacent faces of a crystal. See **law of constancy of interfacial angles.**

**Interior drainage** A network of streams that flow into an enclosed desert basin and that commonly disappear by evaporation and infiltration into the basin sediments and, thus, do not drain into the ocean.

**Intermediate rocks** Igneous rocks having silica contents intermediate between felsic rocks and mafic rocks.

**Intermittent stream** A stream that flows only occasionally.

**Interstitial space fillers** Minerals that finish filling the remaining available space after the framework minerals have formed.

**Interstitial water** The water contained within the pore spaces of regolith, sediment, or sedimentary rock.

**Intrusive igneous rock** An igneous rock that solidified within the Earth.

**Ion** A charged atomic particle in which the numbers of protons and electrons are not equal.

**Ion complex** A group of ions that stick together and behave as a unit (synonym: **radical**). Two common examples are $(SO_4)^{2-}$ and $(CO_3)^{2-}$.

**Ionic bond** A bond in crystals formed by the electrostatic attraction between ions or ion complexes that have opposite charges.

**Island arc** A linear or curving chain of islands, some volcanic and others nonvolcanic, that are parallel to and on the continental side of a deep-sea trench.

**Isoclinal fold** A tight fold in which the strata on the limbs are parallel to each other.

**Isograd** A line on a map that marks the geographic limit of a particular mineral that formed during metamorphism.

**Isoseismal** A line on a map that connects points that experienced equal intensities of seismic shaking during an earthquake.

**Isostasy** (eye-SAHSS-tuh-see) The condition of flotational balance maintained by the Earth's gravity among segments of the lithosphere.

**Isotope** A variety of an element that differs from other varieties by the number of neutrons in the nucleus and, thus, in mass number.

**Joint** A crack in rock along which the opposite sides have not shifted relative to each other.

**Joint set** A group of joints that are more or less parallel.

**Joint system** Several intersecting joint sets.

**Jøkulhlaup** An Icelandic term for a flood that is the result of the catastrophic release of water formed by the melting of a glacier from a subglacial volcanic eruption.

**Juvenile water** Water originating from the mantle that joins the hydrologic cycle for the first time.

**Kame** A short, steep-sided mound of outwash that was deposited in contact with a block of stagnant ice left by a wasting glacier.

**Kame terrace** A terracelike body of outwash built against the margin of a former glacier.

**Karst morphology** A type of land surface that is found in regions underlain by soluble rocks, such as limestones in a humid climate, and that is characterized by sinkholes, underground streams, and caves.

**Kerogen** A complex, natural solid hydrocarbon, not soluble in organic solvents, that yields petroleum on heating. Kerogen is the most abundant and widespread form of natural organic matter.

**Kettle** A depression in glacial drift formed after the burial or melting of a large block of ice.

**Kilobar** A unit of pressure equal to 1000 bars.

**Kimberlite** An ultramafic igneous rock derived from the upper mantle and composed of pyroxene, olivine, and garnet; kimberlite commonly contains diamonds.

**Kinematic wave** A wave within which the water particles are moving forward along with the wave form. Compare **wave of oscillation** and **wave of translation.**

**Kinetic energy** The energy possessed by a moving object; computed by multiplying one half the object's mass times its celerity squared ($\frac{1}{2}mc^2$).

**Laccolith** A concordant lenticular pluton having a flat floor and an arched roof.

**Lamina** (singular; plural laminae, not laminations) A layer of sediment thinner than 1 centimeter.

**Laminar flow** A type of flow in which parallel layers of water shear past one another as smooth planes or smooth curves.

**Landslide** A popular term for any slope failure.

**Lapilli** (lah-PILL-ee) Tephra ranging in size from more than 2 to less than 64 millimeters.

**Large dome exposing basement rocks** A kind of mountain resulting from the vertical uplift of a cylindrical segment of the lithosphere and from later differential erosion.

**Latent heat of crystallization** The amount of energy per unit mass that is given off without change of temperature when a crystal forms from a melt. Compare **latent heat of fusion.**

**Latent heat of fusion** The amount of energy per unit mass that is required for converting a solid at its melting point into a liquid without increasing the temperature. Latent heat is usually expressed as calories per gram. Compare **latent heat of crystallization.**

**Latent heat of sublimation** Heat given off when minerals form directly from vapors.

**Lateral moraine** A ridgelike body of sediment that is deposited along the nonleading edges of a valley glacier.

**Laterite** A tropical regolith that is rich in iron oxides. See **latosol.**

**Latosol** A reddish soil of the humid tropics, typically lacking humus and rich in iron oxides and aluminum oxides.

**Lattice** The regular arrangement of ions in a crystal.

**Lava** Molten silicate material that has been extruded onto the surface of the Earth.

**Lava tube** A hollow, tunnellike space formed when the exterior of a lava flow has solidified and the liquid interior part has drained away.

**Law of constancy of interfacial angles** The law that states: the angles between corresponding faces on different crystals of one substance are constant.

**Law of faunal succession** The geologic principle that states: the order of appearance of the various groups of fossil organisms in the geologic record is not haphazard but follows a well-defined progression.

**Law of original continuity** The geologic principle, first formulated by Nicolaus Steno, that states: strata were initially continuous throughout their basin of deposition.

**Law of original horizontality** The geologic principle, first formulated by Nicolaus Steno, that states: newly deposited strata are horizontal (exception: cross strata).

**Law of superposition** The geologic principle, first formulated by Nicolaus Steno, that states: strata are deposited in a systematic sequence, starting with the oldest and lowest at the base and the progressively younger added at the top.

**Leachate** Heated groundwater that is contaminated with chemical materials dissolved from garbage and other wastes.

**Leaching** The removal from rock of materials in water solution.

**Left lateral** A sense of relative horizontal movement of a strike-slip fault or a shearing couple, which is determined as motion of the far block toward one's left when one is facing across the fault or the centerline of the couple. Compare **right lateral.**

**Levee** See **natural levee.**

**Lignite** A brownish combustible solid composed of altered plant debris that is intermediate between peat and bituminous coal; it consists of about 40 percent fixed carbon, 20 percent combustible gases, and 40 percent moisture.

**Limb** (of a fold) The side of a fold where the curvature of the strata is at a minimum.

**Limestone** A sedimentary carbonate rock consisting of more than 50 percent calcite or aragonite.

**Liquefaction** A general name for the change of a body of stable solidlike cohesionless particles, in which all particles are in continuous contact with other particles, into a liquidlike condition, in which the particles have moved out of continuous contact with one another.

**Lithification** The conversion of loose sediments into sedimentary rock by cementation, crystallization, or replacement.

**Lithosphere** The solid outermost part of the Earth, consisting of the crust and the uppermost part of the mantle that overlies the asthenosphere.

**Lithosphere plates** The large and presumably movable segments of the lithosphere.

**Local base level** The level of a lake or other body of water that is situated above sea level.

**Loess** (luhss) Nonstratified silt, commonly mixed with scattered particles of sand and clay, that has been deposited from suspension by the wind.

**Longitudinal dune** A long, ridgelike sand dune aligned parallel to the general direction of the prevailing wind.

**Longitudinal wave** A wave whose passage imparts a back-and-forth motion in the direction of wave travel. See also **P wave.**

**Longshore current** A current flowing parallel to shore, caused by the oblique approach of waves.

**Longshore drifting** A general term for the movement of sediment parallel to the shore as a result of the activities of shoaling waves and breaking waves. In no way related to **continental drift** or **glacial drift.** Compare **beach drifting.**

**Low-altitude suspension** A body of wind-transported particles kept aloft by the turbulent eddies within the lowest 10 kilometers of the atmosphere and not involving the high-altitude jet streams. Compare **high-altitude suspension.**

**Luster** The way a mineral reflects light. Common varieties include metallic, shiny, vitreous (glassy), earthy, and pearly.

***L* wave** (long wave) A seismic surface wave that has a wavelength much greater than seismic body waves. Two common kinds of *L* waves are Rayleigh waves and Love waves.

**Mafic** (MAY-fik) **magma** A magma containing less than 50 percent $SiO_2$; the product of partial or complete melting of mantle material.

**Mafic** (MAY-fik) **rocks** Igneous rocks rich in minerals, composed of iron and magnesium and containing less than 50 percent $SiO_2$.

**Magmatic differentiation** Any process by which one kind of magma gives rise to more than a single kind of igneous rock.

**Magmatic water** Water dissolved in a magma.

**Magnetic anomaly** A measured value of the local magnetic field that differs from the predicted value.

**Magnetic-polarity reversal** A change in the Earth's magnetic field that involves its temporary disappearance and subsequent reappearance, but with switch of the polarity, that is, the polarity of the positive pole becomes negative and that of the corresponding negative pole becomes positive. At the present time, the Earth's North Pole is defined as positive.

**Magnetism** A force of attraction or repulsion exerted through space by certain substances, such as iron.

**Magnitude** (earthquake) See **Modified Mercalli scale** and **Richter magnitude scale.**

**Manganese nodule** A subaqueous concertionary deposit of manganese oxides that may also include small amounts of iron, copper, and nickel.

**Mantle** The layer, forming the main bulk of the solid Earth, between the core and the crust. The top of the mantle is placed at the Mohorovičić discontinuity.

**Mantle-core boundary** The seismic interface at the depth of 2900 kilometers.

**Mantle plume** A cylindrical part of the mantle that has high heat flow.

**Marble** A granular metamorphic nonfoliate composed of calcite, dolomite, or both.

**Mass** An expression of the quantity of matter; a numerical value for the mass of a body on Earth is found by dividing its weight by the Earth's gravitational acceleration.

**Mass number** The number of protons plus neutrons in an atomic nucleus. Compare **atomic number.**

**Matrix** Fine particles that are distributed among the coarser framework particles of a sedimentary rock.

**Maturation** The conversion of kerogen to petroleum.

**Maturity** The middle stage in the cycle of erosion, characterized by well-developed networks of streams flowing in V-shaped valleys. Compare **old age** and **youth.**

**Meander** A great sweeping curve in the course of a stream.

**Mean Solar-System Plane** The mean plane of the orbiting planets within the Solar System.

**Mechanical weathering** The physical breaking of bedrock into smaller pieces with resulting increase in surface area (synonym: disintegration).

**Medial moraine** A moraine that is the product of converging lateral moraines, which flow together where two valley glaciers meet, and that occurs atop and within the glacier itself.

**Metallic bonding** The distinctive bonding of metals in which the electrons are not tightly bound to individual atomic nuclei but rather are free to move around among all the nuclei.

**Metamorphic facies** Metamorphic rocks that have characteristic mineral assemblages as a result of experiencing particular conditions of temperature, pressure, and volatiles. Compare **sedimentary facies.**

**Metamorphic rock** A rock created as a result of extreme modification of already formed rock by heat, pressure, or chemical action.

**Metamorphism** The changes in the mineral composition, texture, and fabric of rocks that result from the effects of temperature and pressure within the Earth or as a result of meteorite impact at the Earth's surface.

**Meteor** A piece of extraterrestrial matter from outer space that in falling toward the Earth's surface becomes incandescent from friction with the Earth's atmosphere.

**Meteoric water** All water that is precipitated from the atmosphere as rain or snow.

**Meteorite** A meteor that strikes the Earth's surface. Meteorites are classified by composition—metallic, stony, or mixed metallic and stony (stony-iron)—and by circumstances of discovery—falls and finds.

**Migmatite** A body of rock, partly igneous and partly metamorphic, that consists of intermixed granitic rock and metasedimentary rock, such as schist.

**Milanković curve** A plot of the variation of the intensity of solar energy received in the Northern Hemisphere as a result of astronomic factors affecting the Earth's orbit and axial inclination; expressed as the location of the intensity of solar energy usually received at latitude 65° N.

**Mineral** A naturally occurring solid that has a definite chemical composition or a composition that varies within a fixed range and an ordered crystalline structure, resulting in a set of relatively uniform chemical and physical properties.

**Mineralogy** The scientific study of minerals.

**Modified Mercalli scale** A scale that uses the Roman numerals I through XII to express earthquake intensity based on phenomena experienced by people.

**Mohorovičić discontinuity** The seismic interface between the Earth's crust and mantle.

**Mohs' scale of hardness** A relative numerical scale of the hardness of minerals that begins with 1 (talc, the softest) and ends with 10 (diamond, the hardest). A mineral can scratch any other mineral having a smaller hardness number and, in turn, can be scratched by any mineral having a larger hardness number.

**Molecule** The smallest unit displaying the properties of a compound.

**Monadnock** An isolated peak projecting above the level of a peneplain.

**Moraine** A large body of drift that has been shaped into a rounded ridge.

**Mountain** Any landmass that stands higher than its surroundings by 400 meters or more.

**Mountain chain** An elongated morphologic unit that includes many ranges or groups of ranges and implies no conditions about similarity of form, equivalence in age, or uniformity of conditions of origin.

**Mountain range** A large complex elongated ridge or a series of related ridges that form an essentially continuous and compact morphologic unit.

**Mountain system** A group of mountain ranges whose forms, orientations, and geologic structures are closely similar, presumably because they are products of the same general causes that operated within a narrow span of geologic time.

**Mudflow** A debris flow containing an abundance of fine particles.

**Mud-lump island** An island formed by the upward growth of a diapir composed of gas-charged mud; common where a delta advances over its thick fine-grained foreset beds.

**Mud volcano** The point of discharge at the Earth's surface of gas-charged mud from a mud diapir.

**Mylonite** A cataclastic metamorphic rock composed of rock material ground into small bits by fault motion. Compare **fault breccia** and **fault gouge.**

**Native element** A mineral that consists of only a single element that is not combined with any other kind of ion(s).

**Natural gas** Combustible hydrocarbon gases, chiefly methane, that constitute the non-liquid fraction of petroleum.

**Natural levee** A broad, low ridge of fine sand or silt that has been deposited along the banks of a channel by repeated overflow.

**Neap tide** A tide whose amplitude is less than normal.

**Neptunists** Geologists who adhered to the theory that the only way any rock having a crystalline texture could form was by chemical precipitation of ions held in solution, as in the crystallization of sugar out of a saturated sugar-water solution.

**Neutron** A subatomic particle within an atomic nucleus that has no electrical charge and a mass number of 1.

**Nonconfined aquifer** An aquifer in which the water table forms the upper surface of the zone of saturation.

**Nonconformity** A relationship of unconformity in which sedimentary strata overlie basement rocks.

**Nontectonic structure** A structure formed by various processes other than large-scale movements of the lithosphere.

**Norite** A variety of gabbro in which the pyroxene is not augite but rather one of the orthorhombic pyroxenes, such as hypersthene.

**Normal fault** A nonvertical dip-slip fault that dips beneath the hanging-wall block and along which the hanging-wall block has been relatively downdropped; the opposite of a reverse fault.

**Nuclear energy** The energy released by changes in the nucleus of an atom.

**Nucleus** The central part of an atom composed of protons and neutrons.

**Nuée ardente** (noo-AY ar-DAHNT) A glowing avalanche of white-hot fragments ejected from a volcano.

**Oblate spheroid** A spherical object in which the length of the polar axis is less than that of the equatorial diameter. The Earth's shape is thought to approximate an oblate spheroid. Compare **prolate spheroid.**

**Obliquity of the ecliptic** The angle between the Mean Solar-System Plane and the Plane of the Ecliptic.

**Obsidian** An igneous rock composed of felsic glass and breaking with conchoidal fracture.

**Oceanic rises and ridges** The higher standing rocky parts of the floors of the deep-ocean basins.

**Oil** Liquid hydrocarbons constituting the chief fraction of petroleum liquids.

**Oil shale** A sedimentary rock containing abundant kerogen. Oil shale is a misnomer because oil shale contains no oil but only the raw material for making oil, which can be obtained by heating the kerogen to about 500°C.

**Old age** The final stage in the cycle of erosion characterized by a surface of low relief across which streams meander; a few monadnocks may be present. Compare **maturity** and **youth.**

**Oöid** (OH-oh-id) A small (less than 2 millimeters) spherical particle composed of calcium carbonate that has a concentric or radial internal structure.

**Ooze** A pelagic fine-grained sediment on the deep-sea floor that consists of more than 30 percent of the skeletal remains of planktonic microorganisms.

**Open-water evaporites** Evaporite minerals precipitated from the open waters of a basin having restricted circulation. Compare **sabkha evaporite.**

**Ophiolite suite** A succession of mafic rocks and ultramafic rocks, up to 10 kilometers thick, now exposed on land but formerly part of the oceanic crust and upper mantle beneath a deep-sea basin. A full ophiolite succession includes a fossil Mohorovičić discontinuity and contains, from top downward, pillowed basalt and associated dikes, a sheeted-dike complex intruding dolerite, gabbro, and ultramafic rocks (crystal cumulates). The ferromagnesian minerals may display varying degrees of hydration so that the rocks appear more metamorphic than igneous.

**Ore** The valuable part of an ore deposit. See also **gangue.**

**Ore deposit** A natural deposit from which a metallic element can be extracted at a profit.

**Orogenic belt** A mountain belt.

**Orogeny** (or-AH-juh-nee) A general term for the deformation of orogenic belts and the origin of belts by folding, faulting, regional metamorphism, and emplacement of granitic batholiths.

**Outcrop** An exposure of bedrock at the Earth's surface.

**Outer ridge** A long ridge of sediment that extends upward 200 to 2000 meters above the adjacent sea floor.

**Outwash** The sediments deposited by streams that issue from a glacier.

**Overthrust** A reverse fault that has a low angle of inclination.

**Overturned fold** A fold in which the strata of one limb have been rotated by more than 90° so that their original top sides face downward.

**Ox-bow lake** A crescent-shaped lake formed when a meander has been cut off.

**Oxidation** A chemical process in which oxygen combines with another element.

**Oxide mineral** A mineral in which ions, usually of a metal, are combined with oxygen and do not form tetrahedra, as in silicates.

**Ozone** A gas composed of three oxygen atoms in the molecules instead of the usual two.

**Pahoehoe** (pah-hoey-hoey) The smooth, curved, or ropy surface of fluid mafic lava.

**Paleosol** An ancient soil buried within the geologic record.

**Pangaea** A single landmass composed of all the modern continents that were united until about 200 million years ago.

**Parabolic dune** A sand dune resembling an elongated crescent with the horns extending upwind from the main body. Also known as a U-shaped dune. Compare **barchan dune.**

**Peat** Partially decomposed plant matter that can be burned as fuel.

**Pebble** A sedimentary particle larger than 2 millimeters and smaller than 64 millimeters (size of a tennis ball).

**Pedalfer** (peh-DAL-fur) Soil of the humid-temperate regions, rich in aluminum and iron. It is divided into three varieties: conifer soils, leafy-tree soils, and prairie or grassland soils. Compare **pedocal.**

**Pedestal** A smooth, rounded column, abraded by saltating windblown sand, that supports a larger body of rock that has not been thus abraded.

**Pediment** A gently sloping eroded surface between the foot of a mountain and a bajada.

**Pedocal** (PEH-doh-cal) Soil of arid or semiarid regions in which calcium has been precipitated as calcite in the A zone. It is divided into grassland soils, forest soils, and desert soils. Compare **pedalfer.**

**Pegmatite** An extremely coarse igneous rock in which the constituent crystals are larger than 10 millimeters. Many pegmatites are granitic, but other compositions are possible.

**Pelagic sediment** An open-sea deposit that contains predominantly skeletal remains of microorganisms and clays or products derived from clays.

**Peléan eruption** A violently explosive volcanic eruption accompanied by the discharge of **nuées ardentes.** Named for Mont Pelée on the Caribbean island of Martinique.

**Peneplain** (PEE-nuh-plain) A land area eroded to an almost flat surface at or near sea level.

**Peridotite** An igneous cumulate composed chiefly of olivine and pyroxene.

**Period** (1) Geologic time: a subdivision of an era defined on the basis of distinctive fossil assemblages, for example, Ordovician Period, Devonian Period. (2) Wave: the time required for one wavelength to pass a given point. (3) Orbital: the time required for one full orbit.

**Permafrost** Permanently frozen regolith.

**Permeability** The capacity of a porous material to transmit a fluid. Compare **hydraulic conductivity.**

**Petrified** Turned to stone by a process in which organic molecules are replaced by inorganic molecules.

**Petroleum** A combustible natural mixture of hydrocarbons, including both liquids (chiefly crude oil) and natural gas.

**Phenocryst** One of the large crystals of a porphyry.

**Phosphates** A group of minerals containing phosphorus.

**Photosynthesis** A natural photochemical reaction in which chlorophyll-bearing plants combine water, carbon dioxide, and solar energy to make various organic compounds, including oxygen as a byproduct.

**Phyllite** A lustrous fine-grained metamorphic foliate intermediate between slate and schist and consisting of micaceous minerals large enough to impart a sheen to the rock but not large enough to be distinctly visible.

**Physical geology** A branch of geology that deals with the dynamics of the Earth's materials, processes, and environments as related to the development of the Earth's morphology and its tectonic structures.

**Physiographic province** A large part of the Earth's surface within which the surface morphology and kinds of underlying rocks form consistent associations.

**Piedmont glacier** A glacier of semicircular ground plan formed by the joining together of two or more valley glaciers that have flowed beyond the ends of their confining valleys.

**Piezoelectricity** The property that refers to the connection between the mechanical movement of the deformation of a crystal lattice and the charge of electricity within the lattice.

**Pillow** A saclike blob of lava discharged into water. Pillows accumulate in piles. See **pillow structure.**

**Pillow structure** A volcanic structure that consists of clusters of ellipsoidal pillowlike forms; created when lava erupts underwater.

**Placer** A detrital sedimentary deposit in which chemically resistant minerals having high density (more than 3 grams per cubic centimeter) have been concentrated by physical processes. Examples of the minerals found in placers include chromite, diamond, gold, ilmenite, magnetite, and monazite.

**Plane of the Ecliptic** The plane of the Earth's orbit.

**Plateau** A deeply dissected area in which the stratified rocks, either sedimentary or volcanic, are essentially horizontal.

**Plate tectonics** A series of concepts based on the idea that the large-scale dynamics of the lithosphere can be understood in terms of the behavior of great plates that move with respect to the underlying asthenosphere.

**Playa** (PLY-ya) The flat floor of a desert basin.

**Playa lake** An ephermeral lake in a desert basin.

**Playfair's law** The principle that streams do not simply occupy valleys but create them.

**Plinian eruption** A volcanic eruption that consists of the explosive emission of large amounts of lava and that causes the upper part of the cone to collapse and form a caldera.

**Plucking** See **quarrying.**

**Plunge, angle of** The angle between the horizontal and the axis of a plunging fold.

**Plunging fold** A fold whose axis is not horizontal.

**Pluton** Any body of intrusive igneous rock. Plutons are classified as concordant, discordant, tabular, lenticular, or massive. The word, intrusion, is widely, but incorrectly, used as a noun form and synonym for pluton. Correctly used, intrusion is a verb designating the process of emplacing plutons.

**Plutonists** Followers of Hutton's theory that igneous rocks formed because, under one set of conditions, the Earth's internal heat had caused some rock material to melt and, then, under other conditions, the molten material had cooled and become solid.

**Pluvial episode** A part of the Quaternary Period during which regions that are now deserts experienced abundant rainfall.

**Pluvial lake** A desert lake whose former large size is evidence of past abundant rainfall.

**Point bar** A body of sandy alluvium deposited on the inside of a meander bend on a slip-off slope.

**Polarity reversal** A reversal of the Earth's magnetic field that occurs every 500,000 years or so.

**Polar wandering** The movement of the magnetic North Pole from place to place through time.

**Polymerization** The process of linking carbon and hydrogen ions to form chains, as in kerogen, or the linking of silicon and oxygen ions to form various structural patterns, as in rock-forming silicate minerals.

**Poorly sorted** Sediment in which the silt and clay have not been separated from the coarser material.

**Pore pressure** The pressure of interstitial water.

**Porosity** The proportion, in percent, of the total volume of rock or sediment not occupied by solid particles.

**Porphyrin** A complex organic compound formed only in the chlorophyll of green plants or in the hemoglobin of blood. Not related in any way to and should not be confused with **porphyry** or **porphyritic.**

**Porphyritic** A texture of igneous rocks in which crystals of two distinctly different sizes are present. As an adjective in a rock name, porphyritic designates rocks in which the volume of the phenocrysts is less than 25 percent of the total volume.

**Porphyry** An igneous rock having crystals of two contrasting sizes and in which the volume of the phenocrysts exceeds 25 percent of the total.

**Potential energy** Stored energy related to elevation or to elastic deformation.

**Potential resource** A deposit, rich or lean, that may be inferred to exist but that has not yet been discovered or a deposit that is known to exist but cannot be extracted profitably under existing conditions.

**Pothole** A rounded depression caused by the erosion of bedrock by falling water.

**Precession** The toplike motion of the Earth's axis in which the North Pole describes a circle by moving clockwise around the Ecliptic Pole (as seen from above the Plane of the Ecliptic). The time required for the completion of each such circle is 25,920 years.

**Preferred orientation** A fabric displaying a systematic parallelism of constituents having long axes or flat planes (as in flake-shaped crystals).

**Primary migration** The name given to a hypothetical process by which oil is presumed to migrate in liquid form underground out of a supposed source-bed shale and to move into the nearest sand.

**Primary wave** See **P wave.**

**Proglacial lake** A lake that receives meltwater from a glacier.

**Prolate spheroid.** An elongated spherical object in which the length of the polar axis exceeds that of the equatorial diameter. The ball used in the game of rugby is an example of a prolate spheroid. Compare **oblate spheroid.**

**Proton** A subatomic particle within an atomic nucleus that has a unit electrical charge of +1 and a mass number of 1. Compare **electron** and **neutron.**

**Pseudomorph** A product of the replacement of one mineral by another in which the replacing mineral takes the crystal form of the crystal it has replaced.

**Pulling force** See **slope-parallel component** (of the Earth's gravity).

**Pumice** A porous volcanic rock formed by the solidification of frothy felsic lava.

**P wave** A seismic body wave that alternately pushes and pulls particles of the material through which it travels in a direction parallel to the direction of wave travel (synonym: **compressional wave**).

**Pyroclastic** Designating particles broken by volcanic activity.

**Pyroxene-hornfels facies** (of metamorphic rocks) A group of metamorphic rocks having the bulk chemical composition of basalt and characterized by pyroxene; formed in the highest range of temperatures of metamorphism.

**Quarrying** The removal of large blocks of bedrock by an advancing glacier; also called **plucking.**

**Quartzite** A metamorphic nonfoliate consisting predominantly of quartz. Quartzite is distinguished from sandstone by the breaking across rather than around the quartz individuals.

**Radial pattern** A drainage pattern in which the main streams flow away from a central point.

**Radiant energy** Heat energy that travels by means of infrared waves (of the electromagnetic spectrum).

**Radiolaria** A class of tiny, spherical one-celled marine planktonic animals in which the protoplasm is segregated into two distinct parts separated by a perforated capsule. Most Radiolaria secrete skeletons of silica, but one group secretes skeletons of strontium sulfate.

**Radiolarian ooze** A fine-grained pelagic sediment containing more than 30 percent of skeletal debris of Radiolaria.

**Rain shadow** A region on the leeward side of a mountain range where little rain falls because in rising upward to cross the mountain crest, moisture-bearing winds blowing off an ocean dropped their water on the windward side.

**Rain-shadow desert** A desert located in a rain shadow.

**Reaction series** See **Bowen's reaction principle.**

**Recessional moraine** A ridge, deposited during retreat, along the margins of a glacier at places where the rate of retreat is temporarily slowed.

**Recharge** The replenishment of groundwater from surface sources.

**Recrystallization** A reaction within the solid state involving simple reorganization of the material already present.

**Rectangular pattern** A trellislike drainage pattern in which the master stream makes many right-angle bends.

**Recumbent fold** An isoclinal fold in which the axial plane is horizontal.

**Reef** A wave-resistant ridge of rock in the ocean, especially one made of coral.

**Refraction** See **wave refraction.**

**Regolith** Any body of loose, noncemented particles overlying and usually covering the bedrock. Compare **sediment.**

**Relief** The vertical distance between the highest and lowest points in an area.

**Remanent magnetism** Evidence of ancient magnetic fields preserved in rock.

**Remote sensing** The determination of the characteristics of the Earth or other bodies in the Solar System by indirect means, without making direct contact, by using sensors responsive to energy waves within the electromagnetic spectrum.

**Replacement** The exchange of one solid element for another, as when one material is dissolved and another is deposited in its place.

**Reserves** Known, identified deposits from which some material can be extracted profitably with existing technology and under present economic conditions.

**Reservoir rock** A porous rock containing petroleum.

**Residual regolith** Regolith formed by the weathering of underlying bedrock.

**Resisting force** See **slope-normal component** (of the Earth's gravity).

**Resistivity, electrical** See **electrical resistivity.**

**Resources** Reserves plus other mineral deposits that may become available in the future as well as known deposits that cannot be mined profitably with existing technology or under present economic conditions but that might become profitable with new technology or under different economic conditions.

**Retreat** (of a glacier) The condition in which the outermost edge of a glacier is receding faster than the ice is flowing outward.

**Reverse fault** A nonvertical dip-slip fault that dips beneath the footwall block and along which the footwall block has been relatively elevated; the opposite of a **normal fault.**

**Rhyolite** An aphanitic felsic igneous rock that has the mineral composition of granite; distinguished from andesite by possessing quartz phenocrysts.

**Richter magnitude scale** (of earthquakes) A numerical scale based on logarithms to base 10 that ranks the amount of energy released during an earthquake as determined by the extent of ground motion.

**Rift valley** A graben having surface relief resulting from fault displacement.

**Right lateral** A sense of relative horizontal movement of a strike-slip fault or a shearing couple, which is determined as motion of the far block toward one's right when one is facing across the fault or the center line of the couple. Compare **left lateral.**

**Rill** A tiny channel in soil carved by sheet flow resulting from heavy rainfall.

**Ring dike** A cylindrical dike.

**Rip current** A current flowing away from the land by means of one of a group of regularly spaced gaps through the breaker zone and powered by mounds of water piled up within the surf zone.

**Roche moutonné** (ROASH moo-tahn-AY) An asymmetrical streamlined hillock created by glacial abrasion on one side and by quarrying on the other. The slope of a roche moutonné is smooth and gentle on the side from which the glacier flowed and steep and jagged on the side toward which the glacier flowed.

**Rock avalanche** Catastrophic downslope movement involving large amounts of rock debris traveling at high speed.

**Rock cycle** A continuous cycle of change that includes the creation, weathering, transporting, deposition, and alteration of natural materials in the creation of regolith, soils, sediments, and rocks.

**Rock-dilatancy theory** A theory explaining the relationship between the dilation of rocks and the changes in the speeds of P waves before an earthquake.

**Rock fall** Rapid downslope movement involving the relatively free falling of detached blocks of bedrock.

**Rock flour** Fine chips and powder that are embedded in the basal and side layers of a glacier; created by the abrasion of rocks against one another and against solid bedrock.

**Rock fragments** Particles that are large enough to retain recognizable minerals and textural characteristics of the original bedrock.

**Rock glacier** A tongue-shaped body of large rocks undergoing slow downslope movement that is found in some alpine and arctic mountain regions. Interstitial ice may be an important factor in the motion of a rock glacier.

**Rock slide** Rapid downslope slippage of many fragments newly detached from the bedrock. Compare **rock fall.**

**Roughness** The microrelief characteristics of the surface over which a fluid flows, usually expressed by the sizes of the sediment particles or bed forms present.

**Runoff** The flow of water, from rain or melting snow, over the surface of the Earth.

**Sabkha** A flat-topped body of permeable sediment along the margin of a sea or lake in an arid climate and from which saline interstitial water evaporates and deposits evaporite minerals.

**Sabkha evaporite** An evaporite deposited within the pores of the sediment underlying a sabkha that results from evaporation of the saline interstitial water.

**Saltation** A process by which bed-load particles being driven by water or by wind move downcurrent in a series of short excursions or jumps.

**Salt diapir** A cylindrical body of rock salt that has forced its way upward into the sedimentary strata that initially overlay the salt layer (synonym: salt plug).

**Salt dome** A dome formed above a salt diapir.

**Saltwater encroachment** The movement of saltwater into an aquifer that formerly contained fresh groundwater.

**Saltwater wedge** An extension of water from the sea, which tapers landward beneath the outflowing fresh water at the mouth of a river.

**Sand** Sediment particles more than 1/16 and less than 2 millimeters in diameter.

**Sandstone** A detrital sedimentary rock formed by the lithification of sand-size constituents. The chief mineral of sandstone is quartz. Varieties of sandstone include **arkose** and **graywacke.**

**Sandstone, feldspathic** A sandstone that contains less than 25 percent feldspar; the term "arkosic sandstone" for such rocks should not be applied. See **arkose.**

**Schist** A coarse metamorphic foliate that contains large amounts of mica (muscovite, chlorite, or biotite) and that also includes quartz, feldspar, and other minerals, such as garnet or kyanite.

**Scoria** Porous mafic volcanic rock formed by the solidification of vesicular mafic lava.

**Sea arch** A natural arch excavated by wave action, one side of which is a wave-cut cliff.

**Sea-floor spreading** The outward movement of newly formed crust away from a mid-oceanic ridge.

**Sea-floor trench** An elongated, narrow, steep-sided depression, generally deeper than the adjacent sea floor by 2000 meters or more, that extends parallel to the margin of an ocean basin.

**Seamount** See **guyot.**

**Sea waves** Steep, irregular water-surface waves that are actively blown by the wind.

**Secondary enrichment** Concentration of ions by downward migration during weathering.

**Sediment** Regolith that has been transported at the surface of the Earth and deposited as strata, usually in low places.

**Sedimentary breccia** (BREH-cha) A clastic sedimentary rock containing more than 50 percent of angular gravel-size particles. Compare **conglomerate.**

**Sedimentary facies** Distinctive-appearing sedimentary materials that were deposited under particular environmental conditions. Compare **metamorphic facies.**

**Sedimentary rock** A rock formed by the accumulation and solidification of layers of sediment.

**Sedimentary structure** A structural feature formed in sediments during deposition.

**Sediment yield** The amount of sediment derived from a unit area of land in a given time; usually expressed in terms of tonnes per square kilometer per year or tons per square mile per year.

**Seepage pressure** The force exerted on particles of regolith by outward-flowing interstitial water.

**Seif dune** See **longitudinal dune.**

**Seismic waves** Elastic waves associated with the kinetic energy that travels within the body of the Earth and along the Earth's surface when accumulated strain within the lithosphere or asthenosphere is suddenly released.

**Seismogram** The record of seismic waves made by a seismograph.

**Seismograph** An instrument for detecting and recording seismic waves.

**Seismology** The scientific study of seismic waves, earthquakes, and the interior of the Earth.

**Shale** A fine-grained, fissile detrital sedimentary rock composed predominantly of silt-size and clay-size particles.

**Shallow-water wave** A wave traveling in water where the depth ranges between $L/8$ and $L/20$ ($L$ equals wavelength).

**Shearing stress** Stress that tends to make materials slip sideways along a series of parallel planes.

**Sheet flow** A flow of surface water in a rather uniform, thin layer.

**Sheeting joint** A joint parallel to the ground surface, formed by the upward expansion of bedrock.

**Sheetwash** A broad surface runoff laden with suspended sediments.

**Shield, continental** The low-lying nucleus of a continent, composed of a worndown basement complex of the Precambrian age.

**Shield volcano** See **volcanic shield.**

**Shoaling transformations** A series of systematic changes that start when deep-water waves begin to drag bottom.

**Sial** (SEE-al) A felsic layer composing and underlying the continents, made primarily of rocks rich in silicon and aluminum.

**Silicate** A mineral composed primarily of silicon and oxygen. Silicates are either ferromagnesians or nonferromagnesians.

**Siliceous** Composed of silica, $SiO_2$.

**Sill** A concordant tabular pluton formed when magma has spread laterally between layers of weak rock.

**Silt** Sediment particles, intermediate in size, between clay and sand that have diameters more than 1/512 and less than 1/16 millimeter.

**Sima** (SEE-ma) A world-encircling layer, composed mainly of mafic rocks, that forms the crust of the deep-sea basins and underlies the sial of continents.

**Similar fold** A fold that resulted from flowage of materials along a series of parallel planes, as in a glacier.

**Sinkhole** A closed depression in the land surface, formed by the collapse of a cave roof.

**Slate** A fine-grained metamorphic foliate characterized by closely spaced, smooth cleavage surfaces that cut across the bedding.

**Slaty cleavage** The breaking of slates along smooth surfaces that are parallel to oriented micas.

**Slickensides** Sets of parallel striae formed where solid material rubs against other solid material, as next to a fault.

**Slip face** The steep lee-side, angle-of-repose slope of a dune along which particles blown over the crest cascade downward.

**Slope angle** The angle, between 0° (horizontal) and 90° (vertical), measured in a vertical plane at right angles to the trend of the slope, between the horizontal and the surface of the slope.

**Slope-normal component** (of the Earth's gravity) The resisting force of gravity that pushes material toward a slope and, thus, tends to keep it from moving.

**Slope-parallel component** (of the Earth's gravity) The pulling force of gravity that acts along and down a slope.

**Slump** The downward and outward movement of a mass of bedrock or regolith along a distinct surface of failure.

**Snowfield** A large area of snow that lasts from one winter to the next.

**Snowline** The altitude above which snow lasts throughout the year.

**Soil** The upper part of the regolith that is capable of supporting the growth of rooted plants.

**Solar energy** The energy radiated outward by the Sun; the chief source of energy at the surface of the Earth.

**Solar System** The Sun and the nonluminous bodies held within the Sun's gravitational field.

**Solar-System barycenter** The center of mass of the Solar System, the point around which the planets and the Sun orbit.

**Solifluction** A type of creep that involves water-saturated soil, occurring in cold regions where the ground freezes deeply.

**Sorting** The degree of uniformity of sizes of aggregates of sedimentary particles. Sediments are said to be well sorted if the fine particles have been separated from the coarser material, poorly sorted if they have not.

**Spatter cone** A small volcanic cone produced by the sticking together of small clots of liquid lava.

**Specific gravity** A numerical comparison between the density of a substance and the density of water at 4°C.

**Specific heat** The amount of heat required to raise the temperature of a given mass of a substance by a stated amount. If the specific heat is expressed in calories, the mass is 1 gram and the temperature rise is 1°C.

**Speleothem** A mineral deposit found in caves, usually having the shape of an icicle, a slab, or a mound.

**Spheroidal joint** A crack in rock occurring along concentrically curved shells so as to make a cube of rock more and more spherical.

**Spit** A small sediment peninsula connected to a land or large island at one end and terminating in open water at the other end.

**Spontaneous liquefaction** The abrupt attainment of a condition of liquefaction as a result of rapid jolts or a sudden increase in pore pressure.

**Spreading boundary** See **divergent plate boundary.**

**Spring** A surface stream of flowing water that emerges from the ground.

**Spring tide** A tide whose amplitude is larger than normal.

**Stack** A small, steep-sided bedrock island seaward of a wave-cut cliff; a remnant of a former narrow promontory, most of which has been eroded by wave action.

**Stalactite** A speleothem that has grown vertically downward from the ceiling of a cave.

**Stalagmite** A speleothem that has grown upward from a cave floor toward a drip source on the ceiling.

**Steppe** A semiarid region surrounding a desert.

**Stock** An irregularly shaped or massive discordant pluton whose exposed area is less than 100 square kilometers.

**Strain** The changes in shape or in volume that result from deformation.

**Strata** (singular: **stratum**) Layers of sediment, sedimentary rock, or volcanic rock that were spread out, one at a time, with the oldest at the base.

**Stratified drift** A term for all glacially derived sediments deposited by streams.

**Stratigraphic trap** (petroleum) A subsurface location where petroleum can accumulate because an inclined permeable stratum within which petroleum has migrated ends as a result of the original depositional pattern of units or by being cut off at a contact of burial and covered by an impermeable layer.

**Stratigraphy** The scientific study of strata.

**Streak** The color of the powder of a mineral that is determined by dragging a specimen across a piece of nonglazed porcelain.

**Stream discharges** The water, sediment, and dissolved ions carried by a stream. When used by itself, the term discharge refers to the volume of water transported by a stream.

**Stream flow** Surface water that has become organized into a flow within a definite channel.

**Stream terrace** A flat, benchlike area, part of a former flat valley floor left behind after a stream has cut downward.

**Strength** (of materials) The ability of a material to resist deformation.

**Stress** The internal response within a body subjected to external deforming forces. See **compressive stress, shearing stress,** and **tensile stress.**

**Striae** (singular: **stria**) Tiny linear grooves scratched in a rock by a glacier parallel to the direction of glacier flow. Also the parallel lines on the surface of a mineral caused by the intersection with the surface of closely spaced parallel internal planes of the lattice. Incorrectly referred to as striations.

**Strike** The compass bearing of the horizontal line formed by the intersection with the horizontal plane of the nonhorizontal plane being measured.

**Strike angle** The angle between the strike line and true North; located in the horizontal plane.

**Strike line** A horizontal line formed by the intersection between a horizontal plane and the dipping plane being measured.

**Strike-slip fault** A fault in which the adjacent blocks shift laterally with respect to each other.

**Strike valley** An elongate valley following the strike of an easily eroded part of a sequence of dipping strata.

**Strombolian eruption** A term applied to relatively regular volcanic eruptions occurring as moderate explosions that throw out incandescent cinders, lapilli, and bombs.

**Structure** (tectonic) A feature formed when strata are changed from their original positions as a result of tectonic deformation. Examples include various folds and faults.

**Subaerial** Existing or taking place on the land surface, that is, in contact with the atmosphere.

**Subaerial environment** A collective term for all environments exposed to the atmosphere.

**Subduction** A plate-tectonic process in which one lithosphere plate consisting of oceanic crust is thought to be driven beneath an adjacent lithosphere plate at a deep-sea trench. An active subduction zone is presumed to coincide with a **Benioff zone.**

**Subduction zone** In plate-tectonic theory, the convergent boundary between two plates, at which point crustal material is thought to be destroyed.

**Sublimation** The change of a solid directly into a vapor, without involving the liquid state.

**Submarine canyon** A sinuous, V-shaped valley on the sea floor that cuts the continental shelf and continental slope and ends on the continental rise.

**Submergence** A relative rise of water level against the land.

**Subsidence** Downward movement of the surface of the lithosphere.

**Subsolar point** The point on a planet where the Sun's rays strike the surface perpendicularly and, thus, where solar energy is most intense.

**Subsurface environment** The environmental conditions prevailing beneath the surface of the Earth.

**Subsurface geology** A subspecialty, of particular importance to the petroleum industry, that is concerned with plotting the characteristics, thicknesses, fluid contents, and depths of layers that lie beneath the Earth's surface.

**Subtropical desert** A desert created primarily by the cool, dry descending air just outside the tropics.

**Sulfide mineral** A mineral in which one or more sulfur ions are combined with a metallic ion.

**Sunspot** A dark area on the Sun coinciding with a solar prominence and appearing dark because its temperature is less than that of its surroundings. The number of sunspots varies cyclically within a period of about 22 years for a full cycle and about 11 years for a half cycle.

**Superposed stream** A stream that has eroded downward into a complex block from an unconformably overlying covering layer.

**Surf** Waves of translation, landward of the breakers and seaward of the backwash.

**Surface sealing** The clogging of soil pores during a rainstorm.

**Surface wave** A wave that travels along the surface of a body of water or of the Earth.

**Surface of unconformity** A contact of burial between two groups of unconformable strata.

**Surf zone** A nearshore zone lying landward of the breaker zone and seaward of the zone of swash and backwash.

**Surge, glacial** See **glacial surge.**

**Suspended load** Particles supported by upward-flowing components of the eddies within a turbulent current of water or of air.

**Superposition** (1) Principle of stratigraphy: the principle, first stated by Nicolaus Steno, that strata are deposited in a definite order with the oldest and originally lowest layers at the base and progressively younger strata in overlying positions. (2) Geomorphology: the process involved in the origin of a superposed stream by which a stream cutting down from an upper stratigraphic unit across a contact of burial occupies an anomalous position with respect to the underlying resistant rocks.

**S wave** A seismic shear wave, the second group of body waves to arrive at a station after an earthquake. S waves travel only through solids and cannot pass through fluid.

**Swash** The mass of water from a collapsed wave that surges toward the land and flows up the gently sloping shore. Some swash infiltrates, some may flow to the top of the berm, and the rest flows down the shore as backwash.

**Swell** A long, low, regular water-surface wave that has traveled away from a storm center and, thus, is no longer being actively blown by winds.

**Symmetrical fold** A fold having a vertical axial plane and limbs on which the dip of the strata is the same.

**Syncline** A wavelike, concave-up downfold of rock strata in which the youngest strata occupy the center and older layers dip inward beneath them. Compare **anticline.**

**System, stratigraphic** All the strata deposited during a geologic period.

**Talus** (TAY-luss) A body of large, angular blocks that have fallen from above and have accumulated at the base of a cliff.

**Tectonics** The scientific study of deformation of the lithosphere.

**Tensile stress** Stress that tends to pull materials apart.

**Tephra** (TEFF-ruh) A general term for various pyroclastic materials, including cinder, lapilli, ash, pumice, and bombs.

**Tephra** (TEFF-ruh) **cone** A volcanic cone consisting of fine-grained, usually uniformly sized, tephra that were ejected from a volcanic vent and then fell back nearby.

**Terminal moraine** A ridge formed at the outer margin of a glacier that has reached its maximum extent, when the ice pushes up debris into a ridge whose trend follows the edge of the ice.

**Terminal orogeny** The deformation, metamorphism, and plutonism of orogenic belts.

**Terrestrial planet** Designates the Earth and other Earthlike bodies of the inner part of the Solar System having mean densities greater than 3.97 grams per cubic centimeter.

**Terrestrial theories of climate** Theories that state: climate changes are the result of factors on Earth.

**Terrigenous sediment** Sediment derived from erosion of the lands.

**Tetrahedron** The arrangement of spheres (ions) in a mineral in which three spheres form a triangle and are set in a single basal plane, with a fourth sphere occupying a central position among these three but in the next-higher plane.

**Texture** The appearance of a rock, based on the sizes, uniformity of the sizes, and shapes of the constituents.

**Thrust fault** (or simply **thrust**) A low-angled reverse fault in which older strata have been emplaced over younger strata.

**Tidal amplitude** The vertical difference in altitude of the water surface between the level of high water and that of the preceding or following low water.

**Tidal current** A current created by differences in water level related to tidal action.

**Tide** A rhythmic rise and fall of water level resulting from astronomical causes and from the ways in which extremely long waves move into and out of basins.

**Till** Sediment deposited directly by a glacier.

**Tillite** Sedimentary rock created by the lithification of till.

**Tombolo** A beach connecting an island to a mainland or to another island.

**Topographic map** See **contour map.**

**Topography** The scientific description of the Earth's morphologic features; commonly and loosely used as a synonym for these morphologic features themselves.

**Traction** A collective name for all processes of bed-load transport.

**Transcurrent plate boundary** A boundary between two lithosphere plates in which the two plates shift laterally, one past the other.

**Transform fault** A fault that ends where expansion takes place and new material is added, as on the fractured crust of a lava flow or between parts of the lithosphere where new material is being added from the mantle, as at the centers of mid-oceanic ridges.

**Transpiration** The movement of water into the atmosphere through the pores of plants.

**Transported regolith** Regolith that has come from elsewhere and may be altogether different from the underlying bedrock.

**Transverse dune** A sand dune that is linear, usually short, and aligned at right angles to the prevailing wind.

**Travel-time curve** A plot of travel time versus distance to the epicenter of seismic waves.

**Travertine** A nonmarine carbonate deposit precipitated out of carbonate-bearing water around springs or on stream beds.

**Trellis pattern** A drainage pattern in which the master streams are long and straight, following belts of weak rock, and the tributaries enter at right angles.

**Trench** A long narrow trough on the sea floor where water depths usually exceed 6 kilometers.

**Tropical desert** A desert lying between 5° and 30° N or S latitude, caused by the persistent downdrafts of air that were heated, rose, and dropped their moisture at the Equator.

**Tropic of Cancer** The northernmost latitude reached by the subsolar point in the Northern Hemisphere. In 1979, the latitude of the Tropic of Cancer was 23°26′31″; through time, this latitude varies within a range bounded by 20°53′36″ and 25°26′30″. It changes about half a second of arc per year and at present is decreasing.

**Tropic of Capricorn** The southernmost latitude reached by the subsolar point in the Southern Hemisphere; the values and changes match those of the **Tropic of Cancer.**

**Tsunami** (tsoo-NAH-mee) A huge sea wave, not related to tide or wind, that is set off because of the displacement of the sea floor by a volcanic eruption, earthquake, or undersea avalanche. The plural form is spelled the same as the singular (as in sheep).

**Tufa** (TOO-fuh) A siliceous product deposited by a hot spring around a vent.

**Tuff** A fine-grained pyroclastic rock having particles smaller than 2 millimeters.

**Tundra soil** Soil consisting of sandy clay and raw humus, characteristic of arctic regions.

**Turbidite** Any deposit from a turbidity current.

**Turbidity current** A density current in which the excess density results from turbulenty suspended sediment.

**Turbulent flow** Fluid flow characterized by a variety of vortices and eddies that are continually forming and disappearing.

**Ultimate base level** Sea level and its inland projection.

**Ultramafic rock** An igneous cumulate composed predominantly of ferromagnesian silicate minerals, such as olivine or pyroxene.

**Unconformity** A stratigraphic relationship involving a gap in the geologic record. See **angular unconformity** and **surface of unconformity.**

**Uniformitarians** Followers of the ideas of Hutton and Lyell and the doctrine of uniformitarianism.

**Unit cell** A regular arrangement of ions that forms the basic building block of a mineral.

**Upright fold** A fold in which the axial plane is vertical.

**U-shaped valley** A valley, modified by a valley glacier, that has nearly vertical walls and a flat floor.

**Vadose water** The water occupying the zone of aeration.

**Valley glacier** A glacier occupying a valley; usually found in mountain regions (synonym: **alpine glacier**).

**Van der Waals bond** A weak bond among ions, based on the attraction of opposite fractional electrostatic charges that are not evenly distributed over the surfaces of the ions.

**Varve** A layer of sediment deposited during one year.

**Vein** A tabular, fracture-filling mineral deposit in which the individual minerals grew outward from the walls toward the center. Compare **dike.**

**Ventifact** A pebble or larger rock with smooth facets that have been cut and polished by the abrasive action of windblown sand.

**Vertical exaggeration** The ratio between the horizontal scale and the vertical scale of a topographic profile.

**Very-shallow-water waves** Water-surface waves traveling in depths of less than *L*/20 (*L* equals wavelength).

**Vesicle** A small cavity made in volcanic rock by expanding gas. A rock containing many vesicles is described as vesicular.

**Vine-Matthews hypothesis** A hypothesis explaining the relationships among sea-floor spreading, magnetic anomalies, and magnetic-polarity reversals.

**Viscosity** A property of matter, in a body subject to flow, that tends to act in such a direction as to oppose the flow.

**Vitreous luster** Designates a mineral having a glassy appearance.

**Volcanic breccia** (BREH-cha) Consists of angular fragments of volcanic rock.

**Volcanic cone** A conical feature formed by material that accumulated around a volcanic vent as a result of volcanic activity. See also **complex volcanic cone, composite cone, spatter cone, tephra cone,** and **volcanic shield.**

**Volcanic dust** Tephra smaller than 0.063 millimeters.

**Volcanic mudflow** A debris flow consisting of fine-grained tephra, which may have been hot when transported.

**Volcanic neck** A narrow cylindrical discordant body of intrusive igneous rock that solidified in the throat of an ancient volcano.

**Volcanic plain** A low-lying area underlain by horizontal sheets of extrusive igneous rocks.

**Volcanic plateau** A high-standing area usually underlain by a thick sequence of horizontal layers of extrusive igneous rocks.

**Volcanic shield** A broad, gently sloping conical mound of volcanic rock created by repeated effusive eruptions of liquid mafic lava.

**Volcano** A vent through which hot molten material or gas passes from the Earth's interior onto its surface.

**V-shaped valley** A stream-carved valley having sides of nearly uniform slope.

**Vug** An irregular cavity in a rock, commonly lined with one or more minerals whose composition differs from that of the surrounding rock.

**Wadi** Steep-sided, flat-bottomed valley (synonyms: **arroyo** and **box canyon**).

**Water cycle** See **hydrologic cycle.**

**Water table** The upper boundary of the zone of saturation.

**Wave base** The depth, *L*/2 (*L* equals wavelength), at which a deep-water wave first begins to interact with the bottom.

**Wave-built terrace** The nearly flat upper surface of a body of sediment deposited seaward of a wave-cut bench.

**Wave celerity** *(c)* The speed of advance of a wave form.

**Wave-cut bench** A nearly horizontal surface cut across bedrock by wave action.

**Wave-cut cliff** A coastal cliff whose base has been undercut by wave action. See also **bioerosion.**

**Wave height** *(H)* The vertical distance between the top of a wave crest and the bottom of a wave trough.

**Wavelength** *(L)* The horizontal distance between two adjacent wave crests or between two adjacent wave troughs.

**Wave of oscillation** A wave whose form travels along the water surface without causing any appreciable net displacement of the water masses in the direction of wave travel; such a wave causes the water to oscillate in circular orbits. Compare **wave of translation.**

**Wave period** *(T)* The time required for a complete wave form to pass a given point.

**Wave refraction** The bending of a train of waves as a result of a change of wave celerity. Seismic body waves are refracted because they encounter a zone in which the elastic properties change. Light waves are refracted when they pass into a medium having greater or lesser ability to transmit light. Water-surface waves are refracted when they undergo shoaling transformations or encounter a current. Refraction related to shoaling transformations causes a series of waves that move in shallow water at an angle to the shoreline and change direction to become more nearly parallel to the shore.

**Wave steepness.** *H*/*L*, the ratio of wave height to wavelength.

**Wave of translation** A wave that displaces the water masses within the moving crest (synonym: **kinematic wave**). Compare **wave of oscillation.**

**Weathering** A general term for all the changes in rock material that take place as a result of the rock's exposure to the atmosphere.

**Welded tuff** A tough, usually fine-grained, rock formed by the melting together of the tephra from a **nuée ardente.**

**Well sorted** Sediments in which the silt and clay have been separated from coarser materials.

**Wilson cycle** In plate-tectonic theory, the cycle of the creation, destruction, and recreation of oceans and marginal mountain belts.

**Wind-formed arch** A natural arch excavated by the abrasive effects of windblown sand.

**Xenolith** A block of country rock that has been broken loose and engulfed in magma so that it is now surrounded on all sides by igneous rock.

**Yield point** The maximum stress a body being deformed can withstand before undergoing permanent deformation (such as creep, plastic flow, or rupture).

**Youth** The initial stage in the cycle of erosion when streams are eroding headward actively; valley walls are steep, even vertical; waterfalls and rapids as well as lakes and swamps may be numerous; and large areas in between streams have not yet been dissected. Compare **maturity** and **old age.**

**Zone A** The top layer of a soil profile.

**Zone B** The next-to-the-top layer of a soil profile, below zone A.

**Zone C** The lowest zone of a soil profile; the subsoil or parent material from which the soil is formed.

**Zone O** An upper part of zone A of some soil profiles, rich in decaying organic matter and, thus, dark colored.

**Zone of aeration** The shallow subsurface zone in which the pore spaces are occupied partially by water and partially by air.

**Zone of cementation** The subsurface zone in which groundwater reacts with rock material to precipitate minerals that become the cements of sedimentary rocks.

**Zone of flow** A deep subsurface zone in which the lithostatic pressure is so great that rocks undergo solid flow and do not maintain open fractures. A zone of flow also exists in the lower parts of a glacier. Compare **zone of fracture.**

**Zone of fracture** The Earth's surface and shallow subsurface zone where intersecting cracks and partings along bedding surfaces have cut the rock into innumerable blocks of various sizes. Also the upper part of a glacier characterized by crevasses. Compare **zone of flow.**

**Zone of saturation** The part of the ground in which all the subsurface pore spaces are filled with groundwater under hydrostatic pressure.

**Zone of soil moisture** The top of the zone of aeration.

**Zone of swash and backwash** The zone of a beach where swash and backwash are active. See also **beachface.**

**Zone R** The bedrock underlying any soil profile.

# INDEX



Page numbers with asterisks(*) indicate definitions. Page numbers in **boldface** indicate illustrations. The appendices are not indexed.

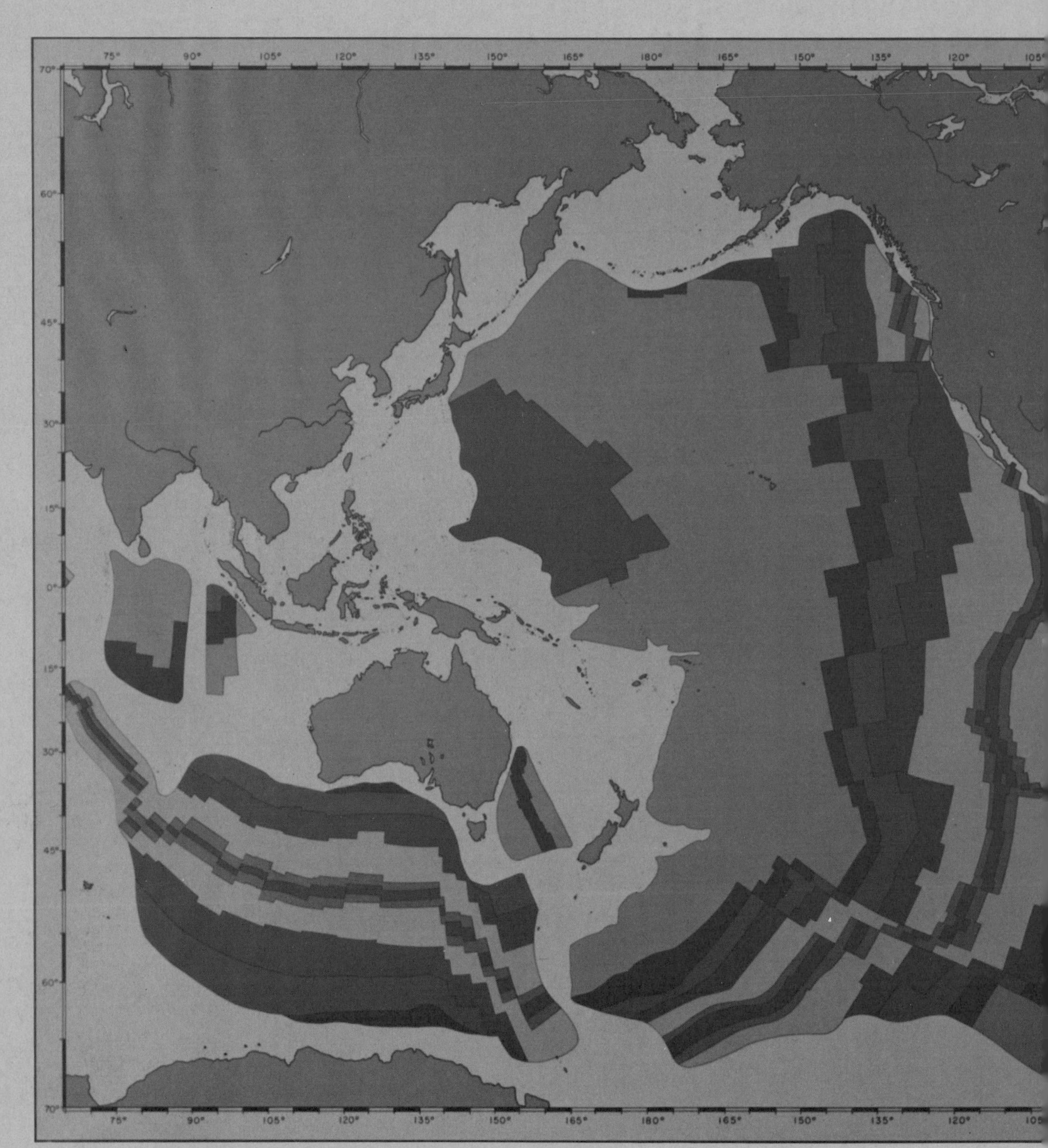
75° 90° 105° 120° 135° 150° 165° 180° 165° 150° 135° 120° 105°
70° 60° 45° 30° 15° 0° 15° 30° 45° 60° 70°